University of Brighton

Aldrich Library
Cockcroft Building
Moulsecoomb
Brighton BN2 4GJ
Telephone 01273 642770

Online renewal http://library.brighton.ac.uk

This item must be returned on or before the last date stamped
A fine may be charged if items are returned late

LR.04.08.ISR.vL1

Physical Chemistry from a Different Angle

Georg Job • Regina Rüffler

Physical Chemistry from a Different Angle

Introducing Chemical Equilibrium, Kinetics and Electrochemistry by Numerous Experiments

 Springer

Georg Job
Job Foundation
Hamburg
Germany

Regina Rüffler
Job Foundation
Hamburg
Germany

Translated by Robin Fuchs, GETS, Winterthur, Switzerland
Hans U. Fuchs, Zurich University of Applied Sciences at Winterthur, Switzerland
Regina Rüffler, Job Foundation, Hamburg, Germany

Based on German edition "Physikalische Chemie", ISBN 978-3-8351-0040-4, published by Springer Vieweg, 2011.

Exercises are made available on the publisher's web site:
http://extras.springer.com/2015/978-3-319-15665-1

By courtesy of the Eduard-Job-Foundation for Thermo- and Matterdynamics

ISBN 978-3-319-15665-1   ISBN 978-3-319-15666-8   (eBook)
DOI 10.1007/978-3-319-15666-8

Library of Congress Control Number: 2015959701

Springer Cham Heidelberg New York Dordrecht London

Springer International Publishing AG Switzerland is part of Springer Science+Business Media (www.springer.com)

Preface

Experience has shown that two fundamental thermodynamic quantities are especially difficult to grasp: entropy and chemical potential—entropy S as quantity associated with temperature T and chemical potential μ as quantity associated with the amount of substance n. The pair S and T is responsible for all kinds of heat effects, whereas the pair μ and n controls all the processes involving substances such as chemical reactions, phase transitions, or spreading in space. It turns out that S and μ are compatible with a layperson's conception.

In this book, a simpler approach to these central quantities—in addition to energy—is proposed for the first-year students. The quantities are characterized by their typical and easily observable properties, i.e., by creating a kind of "wanted poster" for them. This phenomenological description is supported by a direct measuring procedure, a method which has been common practice for the quantification of basic concepts such as length, time, or mass for a long time.

The proposed approach leads directly to practical results such as the prediction—based upon the chemical potential—of whether or not a reaction runs spontaneously. Moreover, the chemical potential is key in dealing with physicochemical problems. Based upon this central concept, it is possible to explore many other fields. The dependence of the chemical potential upon temperature, pressure, and concentration is the "gateway" to the deduction of the mass action law, the calculation of equilibrium constants, solubilities, and many other data, the construction of phase diagrams, and so on. It is simple to expand the concept to colligative phenomena, diffusion processes, surface effects, electrochemical processes, etc. Furthermore, the same tools allow us to solve problems even at the atomic and molecular level, which are usually treated by quantum statistical methods. This approach allows us to eliminate many thermodynamic quantities that are traditionally used such as enthalpy H, Gibbs energy G, activity a, etc. The usage of these quantities is not excluded but superfluous in most cases. An optimized calculus results in short calculations, which are intuitively predictable and can be easily verified.

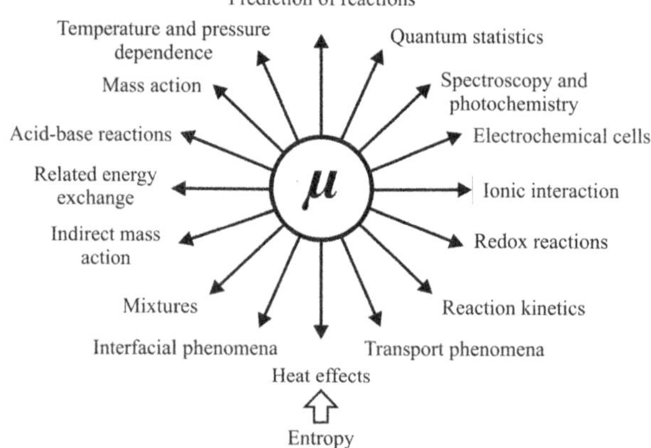

Because we choose in this book an approach to matter dynamics directly by using the chemical potential, application of the concept of entropy is limited to the description of heat effects. Still, entropy retains its fundamental importance for this subject and is correspondingly discussed in detail.

The book discusses the principles of matter dynamics in three parts,

- Basic concepts and chemical equilibria (statics),
- Progression of transformations of substances in time (kinetics),
- Interaction of chemical phenomena and electric fields (electrochemistry)

and gives at the same time an overview of important areas of physical chemistry. Because students often regard physical chemistry as very abstract and not useful for everyday life, theoretical considerations are linked to everyday experience and numerous demonstration experiments.

We address this book to undergraduate students in courses where physical chemistry is required in support but also to beginners in mainstream courses. We have aimed to keep the needs of this audience always in mind with regard to both the selection and the representation of the materials. Only elementary mathematical knowledge is necessary for understanding the basic ideas. If more sophisticated mathematical tools are needed, detailed explanations are incorporated as background information (characterized by a smaller font size and indentation). The book also presents all the material required for introductory laboratory courses in physical chemistry.

Exercises are made available on the publisher's web site. A student manual with commented solutions is in preparation. Detailed descriptions of a selection of demonstration experiments (partly with corresponding videos clips) can be found on our web site (www.job-foundation.org; see teaching materials); the collection will be continuously extended. Further information to the topics of quantum statistics and the statistical approach to entropy, which would go beyond the scope of this book, can also be called up on the foundation's home page.

We would particularly like to thank Eduard J. Job[†], the founder of the Job Foundation, who always supported the goals of the foundation and the writing of the current book, with great personal commitment. Because efficient application of thermodynamics played an important role in his work as an internationally successful entrepreneur in the field of fire prevention and protection, he was particularly interested in a simplified approach to thermodynamics allowing for faster and more successful learning.

We gratefully acknowledge the constant support and patience of the board of the Job foundation. Additionally, we would like to thank the translators of the book, Robin Fuchs and Prof. Hans U. Fuchs, for their excellent collaboration, and Dr. Steffen Pauly and Beate Siek at Springer for their advice and assistance. Finally, we would like to express our gratitude to colleagues who gave their advice on the German edition and reviewed draft chapters of the English edition: Prof. Friedrich Herrmann, Prof. Günter Jakob Lauth, Prof. Friedhelm Radandt, and Dr. Uzodinma Okoroanyanwu.

We would be very grateful for any contributions or suggestion for corrections by the readers.

<table>
<tr><td>Hamburg, Germany</td><td>Georg Job</td></tr>
<tr><td>November 2014</td><td>Regina Rüffler</td></tr>
</table>

List of Used Symbols

In the following, the more important of the used symbols are listed. The number added in parentheses refers to the page where the quantity or term if necessary is described in detail. Special characters as prefix ($|$, Δ, Δ_R, $\Delta_{s \to l}$, ...) were omitted when ordering the symbols alphabetically.

Greek letters in alphabetical order:

$A\alpha$ $B\beta$ $\Gamma\gamma$ $\Delta\delta$ $E\varepsilon$ $Z\zeta$ $H\eta$ $\Theta\theta\vartheta$ $I\iota$ $K\kappa$ $\Lambda\lambda$ $M\mu$ $N\nu$ $\Xi\xi$ Oo $\Pi\pi$ $P\rho$ $\Sigma\sigma\varsigma$ $T\tau$ $Y\upsilon$ $\Phi\varphi$ $X\chi$ $\Psi\psi$ $\Omega\omega$.

Roman

A, B, C, ...	Substance A, B, C, ...
$\|$A, $\|$B, ...	Dissolved in A, in B, ... (240)
Ad	Acid (188)
a, $\|$a	Amorphous (19) (also subscripted or superscripted)
Bs	Base (188)
C	Catalyst (462)
c, $\|$c	Crystalline (19) (also subscripted or superscripted)
d, $\|$d	Dissolved (19) (also subscripted or superscripted)
E	Enzyme (466)
e, e^-	Electron(s) (7, 553) (also subscripted)
e	Eutectic (367) (also subscripted or superscripted)
F	Foreign substance (320)
g, $\|$g	Gaseous (19) (also subscripted or superscripted)
J	Ion, unspecific (533)
l, $\|$l	Liquid (19) (also subscripted or superscripted)
M	Mixture (homogeneous) (346)
$\mathcal{M}$	Mixture (heterogeneous) (348)
Me	Metal, unspecific (533)
m, $\|$m	Metallic (conducting electrons) (553) (also subscripted or superscripted)
Ox	Oxidizing agent (537)

P	Products, unspecific (462)
p	Proton(s) (187) (also subscripted)
Rd	Reducing agent (537)
S	Solvent (97), solution phase (535)
S	Substrate (466)
s, $|$s	Solid (19) (also subscripted or superscripted)
w, $|$w	Dissolved in water (20) (also subscripted or superscripted)
$|\alpha, |\beta, |\gamma, \ldots$	Different modifications of a substance (20)
$\square$, $\boxed{\text{B}}$	Adsorption site ("chemical") empty, occupied (394)
$\lbrack\text{\textvisiblespace}\rbrack$, $\lbrack\text{B}\rbrack$	Adsorption site ("physical") empty, occupied (394)
‡	Transition complex (450) (also subscripted or superscripted)

Italic

A	Area, cross section
A	Helmholtz (free) energy (only used exceptionally) (595)
$\mathcal{A}$	(Chemical) drive, affinity (108)
$\mathcal{A}^{\ominus}$	Standard value of the chemical drive (109)
$\overset{\circ}{\mathcal{A}}$	Basic value of the chemical drive (159)
$\overset{\times}{\mathcal{A}}$	Mass action term of the chemical drive (159)
a	Acceleration (32)
a	Length of box (281)
a	(First) van der Waals constant (299)
a	Temperature conductivity (491)
a, a_{B}	Activity (of a substance B) (only used exceptionally) (604)
B	Matter capacity (182)
B_{p}	Buffer capacity (201)
$\mathcal{B}, \mathcal{B}_i$	Substance in general (with subscript i) (25)
b, b_{B}	Molality (of a substance B) (18)
b	(Second) van der Waals constant (321)
$\mathnormal{b}$	Matter capacity density (182)
$\mathnormal{b}_{\mathrm{p}}$	Buffer capacity density (212)
C, C_p	Heat capacity (global, isobaric) (254, 591)
C_{m}	Heat capacity, molar (isobaric) (254)
C_V	Heat capacity (global, isochoric) (254, 587)
$\mathcal{C}, \mathcal{C}_p$	Entropy capacity (global, isobaric) (75)
$\mathcal{C}_{\mathrm{m}}$	Entropy capacity, molar (isobaric) (75)
$\mathcal{C}_V$	Entropy capacity (global, isochoriv) (77)
c	Speed of light (13)
c, c_{B}	Molar concentration (of a substance B) (17)
c, c_{s}	Heat capacity, specific (isobaric) (254, 491)
c_{r}	Relative concentration $c/c^{\ominus}$ (156)
c_{ξ}	Density of conversion (163)

$c^{\ominus}$	Standard concentration (1 kmol m^{-3}) (103, 156)
$c^{\dagger}$	Arbitrary reference concentration (416)
$\mathscr{c}$	Entropy capacity, specific (isobaric) (76, 491)
D	Spring stiffness (39)
D, D_{B}	Diffusion coefficient (of a substance B) (480)
d	Thickness, diameter
$E, \vec{E}$	Electric field (strength) (500)
E	Electrode potential, redox potential (558)
ΔE	Reversible cell voltage ("zero-current cell voltage") (568)
e_0	Elementary charge, charge quantum (16)
F	Force, momentum current (31, 45, 486)
$\mathcal{F}$	Faraday constant (504)
f, f_{B}	Fugacity (of a substance B) (only used exceptionally) (606)
G	Weight (according to everyday language) (9)
G, G_Q	(Electric) conductance (494, 508)
G	Gibbs (free) energy (only used exceptionally) (596)
$\mathcal{G}$	Arbitrary quantized quantity (15)
g	Gravitational acceleration (46)
g_i	Content number of the ith basic substance (6)
g	Quantum number (15)
H	Enthalpy (only used exceptionally) (589)
h	Height
h	Planck's constant (451)
I	(Electric) current (494)
J	Current (of a substance-like quantity) (493)
J_{B}	Matter flux, current of amount of a substance B (479)
J_{S}	Entropy flux, entropy current (490)
j	Current density (of a substance-like quantity) (493)
j_{B}	Flux density, current density (of matter) (478)
j_S	Entropy flux (or entropy current) density (490)
$\overset{\circ}{K}$	Conventional equilibrium constant (167, 176)
$\overset{\circ}{\mathcal{K}}$	Numerical equilibrium constant, equilibrium number (166, 176)
K_{M}	Michaelis constant (466)
k	Rate coefficient (417)
$k_{+1}, k_{-1}, \ldots$	Rate coefficient for forward or backward reaction (No. 1, etc.) (430)
k_{B}	Boltzmann constant (280)
k_{∞}	Frequency factor (444)
l	Length
M	Molar mass (16)
m	Mass
N	Number of particles (15)
N_{A}	Avogadro constant (15)
n	Amount of substance (15)

n_p	Amount of protons (in a reservoir for protons) (203)
P	Power
p	Pressure (41)
p	Probability (291, 307)
p	Steric factor (449)
p_{int}	Internal pressure (298)
p_r	Relative pressure $p/p^\ominus$ (171)
p_σ	Capillary pressure (387)
$p^\ominus$	Standard pressure (100 kPa) (72, 103)
$\wp$	Momentum (44)
Q	(Electric) charge (16)
Q	Heat (only used exceptionally) (80)
q	Fraction of collisions of particles having minimum energy w_{min} (448)
R	General gas constant (148, 277)
R, R_Q	(Electric) resistance (494)
$\mathcal{R}, \mathcal{R}', \mathcal{R}''$	Arbitrary reaction (28)
$r, r_{AB}, \ldots$	Radius, distance from center, distance between two particles A and B
r	Rate density (419)
$r_{+1}, r_{-1}, \ldots$	Rate density for forward or backward reaction (No. 1, etc.) (430)
r_{ads}, r_{des}	Rate (density) of adsorption or desorption (395)
S	Entropy (49)
$\Delta_{fus}S$	Molar entropy of fusion (75, 312)
$\Delta_R S$	Molar reaction entropy (232)
$\Delta_{vap}S$	Molar entropy of vaporization (75, 309)
$\Delta_\rightarrow S$	(Molar) transformation entropy (234)
S_c	Convectively (together with matter) exchanged entropy (65)
S_e	Exchanged entropy (convectively and/or conductively) (65)
S_g	Generated entropy (65)
ΔS_ℓ	Latent entropy (84)
S_m	Entropy demand, molar entropy (71, 229)
S_t	Transferred entropy (85)
S_λ	Conductively (by conduction) exchanged entropy (65)
s	Length of distance traveled
T	(Thermodynamic, absolute) temperature (68)
$T^\ominus$	Standard temperature (298.15 K) (71, 103)
$\mathcal{T}, \mathcal{T}_O$	Duration of conversion, observation period (404)
$t, \Delta t$	Time, duration
$t_{1/2}$	Half-life (420)
t, t_i, t_+, t_-	Transport number (of particles of type i, of cations, of anions) (517)
$U, U_{1\rightarrow 2}$	(Electric) voltage (from position 1 to position 2) (502)
U	Internal energy (only used exceptionally) (582)
U_{Diff}	Diffusion (Galvani) voltage (548)
u, u_i	Electric mobility (of particles of type i) (503)
V	Volume

$\Delta_R V$	Molar reaction volume (228)
$\Delta_{\rightarrow} V$	(Molar) transformation volume (228)
V_m	Volume demand, molar volume (220)
V_W	Co-volume (van der Waals volume) (298)
$v, \vec{v}$	Velocity (magnitude, vector)
v_x, v_y, v_z	Velocity, components in x, y, z direction (281)
W	Energy (36)
W	Work (only used exceptionally) (581)
W_A	Molar (Arrhenius) activation energy (581)
$W_A, W_{\rightarrow A}$	Energy expended for a change of surface or interface (385)
$W_B, W_i, \ldots$	Abbreviation for $W_{\rightarrow n_B}, W_{\rightarrow n_i}, \ldots$ (346)
W_b	Burnt energy (78)
W_e	Energy transferred together with exchanged entropy (79)
W_f	Free energy (only used exceptionally) (592)
W_{kin}	Kinetic energy (43)
$W_n, W_{\rightarrow n}$	Energy expended for a change of amount of substance (124)
W_{pot}	Potential energy (46)
W_t	Energy expended for transfer (of an amount of entropy, of matter …) (85, 235)
$W_S, W_{\rightarrow S}$	Energy expended for a change of entropy ("added + generated heat") (81)
$W_V, W_{\rightarrow V}$	Energy expended for a change of volume ("pressure–volume work") (81)
$W_\xi, W_{\rightarrow \xi}$	Energy expended for a change of conversion (236)
w, w_B	Mass fraction (of a substance B) (17)
w	Energy of a particle (278, 287)
x, x_B	Mole fraction (of a substance B) (17)
x, y, z	Spatial coordinates
Z_{AB}	Collision frequency between particles A and B (446)
z, z_i, z_+, z_-	Charge number (of a type i of particles, cations, anions) (16, 535)
α, α_B	Temperature coefficient of the chemical potential (of a substance B) (131)
α, α_ξ	Degree of dissociation, degree of conversion (513, 163)
α	Temperature coefficient of the drive (of a transformation of substance) (131)
β, β_B	Pressure coefficient of the chemical potential (of a substance B) (140)
β, β_B	Mass concentration (of a substance B) (17)
β_r	Relative pressure coefficient (271)
β	Pressure coefficient of the drive (of a transformation of substance) (140)
γ	Concentration coefficient of the chemical potential (154)
γ	Cubic expansion coefficient (256)
γ	Activity coefficient (only used exceptionally) (604)
η	Efficiency (85)
η	(Dynamic) viscosity (486)

Θ	Degree of filling (degree of protonation, etc.), fractional coverage (201, 396)
θ	Contact angle (387)
ϑ	Celsius temperature (70)
κ	Dimension factor (167, 173)
ϑ_F	Faraday temperature
Λ, Λ_i	Molar conductivity, (molar) ionic conductivity of ions of type i (519)
λ	Thermal conductivity (490)
$\lambda, \lambda_1, \lambda_2, \ldots$	Wave length, wave lengths of fundamental and harmonics (483)
λ, λ_B	Chemical activity (of a substance B) (only used exceptionally) (605)
μ, μ_B	Chemical potential (of a substance B) (98)
μ_d	Decapotential (abbreviation for $RT \ln 10$) (157)
$\mu_e, \mu_e(\text{Rd}/\text{Ox})$	Electron potential, of a redox pair Rd/Ox (529, 537)
$\mu_p, \mu_p(\text{Ad}/\text{Bs})$	Proton potential, of an acid–base pair Ad/Bs (191)
$\mu^{\ominus}$	Standard value of the chemical potential (103, 157)
$\overset{\circ}{\mu}$	Basic value of the chemical potential of a dissolved substance (156)
$\Delta_{\ddagger} \overset{\circ}{\mu}$	Activation threshold (451)
$\overset{\circ}{\mu}_c, \overset{\circ}{\mu}_p, \overset{\circ}{\mu}_x, \ldots$	Basic value of the chemical potential in the $c, p, x, \ldots$ scale (340)
$\overset{\bullet}{\mu}$	Chemical potential of a substance in its pure state (345)
$\overset{\times}{\mu}$	Mass action term of the chemical potential (157)
$\overset{+}{\mu}$	Extra potential (extra term of the chemical potential) (345)
$\tilde{\mu}, \tilde{\mu}_i$	Electrochemical potential (of a substance i) (528)
$v, v_B, v_i, \ldots$	Conversion number, stoichiometric coefficient (of a substance B or $i \ldots$) (26)
v	Kinematic viscosity (486)
ξ	Extent of reaction (26)
ρ, ρ_B, ρ_i	(Mass) density (of a substance B or i) (9)
ρ, ρ_Q	(Electric) resistivity (494, 509)
$\sigma, \sigma_{g,l}, \ldots$	Surface tension, interfacial tension (383, 387)
σ, σ_Q	(Electric) conductivity (493, 509)
σ_B	"Matter conductivity" (for a substance B) (527)
σ_S	Entropy conductivity (490)
τ	Elementary amount (of substance), quantum of amount (of substance) (15, 16)
$\tau_1, \tau_2, \ldots$	Decay time of fundamental and harmonic waves, respectively (483)
$\tau_{\ddagger}$	Lifetime of the transition complex (450)
ϕ	Fugacity coefficient (only exceptionally used) (612)
φ	Electric potential (90, 500)
φ	Fluidity (494)
χ	Compressibility (268)
ψ	"Gravitational potential" (90)
ω, ω_B	Mechanical mobility (of a substance B) (476)
ω	Conversion rate (407)

                                                                                    xv

Subscript

ads	Concerning adsorption (396)
c	Critical (304)
d→d, dd	Transition of a dissolved substance from one phase to another (181)
des	Concerning desorption (395)
eq.	In equilibrium (166)
g→d, gd	Transition from gaseous to dissolved state (180)
l→g, lg	Transition from liquid to gaseous state (boiling) (75, 228)
ℓ	Latent (84, 243)
m	Molar
mix	Mixing process (351)
osm	Osmotic (325)
R	Concerning a reaction (228)
r	Relative (156)
s→d, sd	Transition from solid to dissolved state (176, 228)
s→g, sg	Transition from solid to gaseous state (sublimation) (137)
s→l, sl	Transition from solid to liquid state (melting) (75, 228)
s→s, ss	Transition in the solid state from one structural modification to another (change of modification) (228)
use	Useful (87, 240)
→	Concerning a transformation (228)
□	Concerning an adsorption process (396)
0	Value interpolated to vanishingly low concentration (477) (also superscript)
+, −	Concerning cations, anions (also superscript) (517)

Superscript

⊖	Standard value (71, 103)
●	Value for a substance in its pure state (329, 333)
▲	Characterizes a homogeneous or heterogeneous mixture of intermediate composition, the "support point" by the application of the "lever rule" (348)
*, **, . . .	Characterizes different substances, phases, areas [e.g., the surroundings (239)]
*	Characterizes "transfer quantities" (492)
′, ″, ‴, . . .	Characterizes different substances, phases, areas

Character Above a Symbol

$\rightarrow$ Vector
$-$ Mean value
$\cdot$ Derivative with respect to time
$\circ$ Basic term, basic value (156)
$\bullet$ Basic value of a quantity for a substance in its pure state (320)

$\times$ Quantity caused by mass action (154,157)
$+$ Extra term, extra value (345)
$*$ Residual term, residual value (residual without basic term)

General Standard Values (Selection)

$b^{\ominus} = 1\,\mathrm{mol\,kg^{-1}}$ Standard value of molality
$c^{\ominus} = 1,000\,\mathrm{mol\,m^{-3}}$ Standard value of concentration
$p^{\ominus} = 100,000\,\mathrm{Pa}$ Standard value of pressure
$T^{\ominus} = 298.15\,\mathrm{K}$ Standard value of temperature
$w^{\ominus} = 1$ Standard value of mass fraction
$x^{\ominus} = 1$ Standard value of mole fraction

Physical Constants (Selection)

$c = 2.998 \times 10^{8}\ \mathrm{m\ s^{-1}}$ Speed of light in vacuum
$e_0 = 1.6022 \times 10^{-19}\ \mathrm{C}$ Elementary charge, charge quantum
$\mathcal{F} = 96,485\,\mathrm{C\,mol^{-1}}$ Faraday constant
$g_n = 9.806\ \mathrm{m\ s^{-2}}$ Conventional standard value of gravitational acceleration
$h = 6.626 \times 10^{-34}\ \mathrm{J\ s}$ Planck constant
$k_B = 1.3807 \times 10^{-23}\ \mathrm{J\ K^{-1}}$ Boltzmann constant
$N_A = 6.022 \times 10^{23}\ \mathrm{mol^{-1}}$ Avogadro constant
$R = 8.314\ \mathrm{G\ K^{-1}}$ General gas constant
$T_0 = 273.15\ \mathrm{K}$ Zero point of the Celsius scale
$\tau = 1.6605 \times 10^{-24}\ \mathrm{mol}$ Elementary amount (of substance), quantum of amount

Contents

Chapter 1
Introduction and First Basic Concepts

In this first chapter, we will be introduced briefly to the field of *matter dynamics*. This field is concerned in the most general sense with the transformations of substances and the physical principles underlying the changes of matter. As a consequence, we have to review some important basic concepts necessary for describing such processes like substance, content formula and amount of substance, as well as homogeneous and heterogeneous mixture and the corresponding measures of composition. But in this context, the *physical state* of a sample is also of great importance. Therefore, we will learn how we can characterize it qualitatively by the different states of aggregation as well as quantitatively by state variables. In the last section, a classification of transformations of substances into chemical reactions, phase transitions, and redistribution processes as well as their description with the help of conversion formulas is given. The temporal course of such a transformation can be expressed by the extent of conversion ξ. Additionally, we will take a short look at the basic problem of measuring quantities and metricizing concepts in this chapter.

1.1 Matter Dynamics

The term *dynamics* is derived from the word "dynamis," the Greek word for "force." In physics, dynamics is the study of forces and the changes caused by them. The field of mechanics uses this word in particular when dealing with the motion of bodies and the reasons why they move. This term is then expanded to other areas and is reflected in such expressions as *hydrodynamics*, *thermodynamics*, or *electrodynamics*. When we discuss the field of *matter dynamics* we will generally be talking about transformations of substances and the "forces" driving them. States of equilibrium (treated in the field of statics, also called "chemical thermodynamics") will be covered in addition to the temporal course of transformations (kinetics) or the effects of electrical fields (electrochemistry).

© Springer International Publishing Switzerland 2016
G. Job, R. Rüffler, *Physical Chemistry from a Different Angle*,
DOI 10.1007/978-3-319-15666-8_1

What makes this field so valuable to chemistry and physics as well as biology, geology, engineering, medicine, etc., are the numerous ways it can be applied. Matter dynamics allows us to predict in principle

- Whether or not it is possible for a given chemical reaction to take place spontaneously,
- Which yields can be expected from this,
- How temperature, pressure, and amounts of substances involved influence the course of a reaction,
- How strongly the reaction mixture heats up or cools down, as well as how much it expands or contracts,
- How much energy a chemical process needs to run or how much it releases, and much more.

This kind of knowledge is very important for developing and optimizing chemical processes, as well as preparing new materials and active ingredients by using energy carriers efficiently and avoiding pollution, etc. It plays an important role in many areas of chemistry, especially in chemical engineering, biotechnology, materials science, and environmental protection. Moreover, this knowledge can equally help us to understand how substances behave in our everyday lives at home, when we cook, wash, clean, etc.

Although we will mainly deal with chemical reactions, it does not mean that matter dynamics is limited to this. The concepts, quantities, and rules can, in principle, be applied to every process in which substances or different types of particles (ions, electrons, supramolecular assemblies, and lattice defects, to name a few) are exchanged, transported, or transformed. As long as the necessary data are also available, they help in dealing with and calculating various types of problems such as

- The amount of energy supplied by a water mill,
- Melting and boiling temperatures of a substance,
- Solubility of a substance in a solvent,
- The construction of phase diagrams,
- How often lattice defects occur in a crystal,
- The potential difference caused by the contact between different electric conductors

and much more. Matter dynamics can also be very useful in discussing diffusion and adsorption processes or questions about metabolism or transport of substances in living cells, as well as transformation of matter inside stars or in nuclear reactors. It is a very general and versatile theory whose conceptual structure reaches far beyond the field of chemistry.

Now we can ask for the causes and conditions that are necessary for the formation of certain substances and their transformations into one another. This can be done in different ways and on different levels:

1. *Phenomenologically*, by considering what happens macroscopically. This means directly observing processes taking place in a beaker, reaction flask, carius tube, or spectrometer when the substance in it is shaken, heated, other substances are added to it dropwise, poured off, filtered, or otherwise altered.
2. *According to molecular kinetics*, by considering the reacting substances to be more or less orderly assemblies of atoms where the atoms are small, mutually attracting particles moving randomly but always trying to regroup to attain a statistically more probable state.
3. *According to chemical bonding theory*, by emphasizing the rules and laws according to which different types of atoms come together to form assemblies of molecules, liquids, or crystals in more or less defined relationships of numbers, distances, and angles. The forces and energies that hold the atoms together in these associations can also be investigated.

All of these points of view are equally important in chemistry. They complement one another. In fact, each is inextricably interwoven with the others. To give an example, we operate at the third level when the structural formula of the substance to be produced is written down. On the second level, one might make use of plausible reaction mechanisms for planning a synthesis pathway. The first level is applied when, for instance, the substances to be transformed are put together in a laboratory. To work economically, it is important to be able to switch between these different points of view unhindered. Our goal is not so much a concise explication of the individual aspects mentioned above, as it is a unified representation in which the knowledge gained from these differing points of view merges into a harmonic overall picture. Conversely, the individual aspects can also be easily derived from this overall picture.

One might say that the phenomenological level forms the "outer shell" of the theory. It relates the mathematical structure to phenomena observed in nature. The first step toward expressing such relationships is to prepare the appropriate concepts, which helps the facts gained by experience to be formulated, put into order, and summarized. It follows that these expressions will appear in farther-reaching theories as well. The phenomenological level constitutes the natural first step into a chosen area of investigation.

In the next sections as well as in the next chapter, important fundamental terms and concepts will be discussed. Among these will be substance, amount of substance, measures of composition, and energy, all of which students are probably familiar with from high school. For this reason, it should be easy to start right in with Chap. 3 (Entropy) or even Chap. 4 (Chemical Potential). Chemical potential puts us right at the heart of matter dynamics. Using this as a starting point opens up a multitude of areas of application. Chapters 1 and 2 can then be considered reference work for fundamental terms and concepts.

1.2 Substances and Basic Substances

When we think about *substances*, we think about kinds of matter and their actual or imagined constituents. Simply stated, we think of substances being what the tangible things of our environment are made up of. They are that formless something that fills up space when we disregard the shape of things. There are a multitude of substances around us that we give names to such as iron, brass, clay, rubber, soap, milk, etc. We characterize these substances individually or as members of a category. We use the term *matter* when it is unimportant what kind of substance we are discussing.

Some things appear totally uniform materially, such as a glass or the water in it. If the macroscopic characteristics of a substance such as its density, index of refraction, etc., are the same overall, it is considered *homogeneous*. Wine, air, stainless steel, etc., are other examples of homogeneous substances. Aside from these, there are *heterogeneous* substances that are composed of dissimilar parts, i.e., they are made up of clearly different components. Examples are a wooden board or a concrete block. On the one hand, we tend to think of even these materials as substances on their own. On the other hand, we imagine them to be made up of several substances. We do this even when we consider the sweetened tea or diluted wine that look to be homogeneous. This ambivalence is a striking characteristic of our concept of substance that reflects a noteworthy aspect of the world of substances.

Imagine breaking down some matter into certain components. We find that these components can be broken down into their own components, as well. These subcomponents can also be called substances. The process can be repeated at different levels and in varying ways.

At the heart of the matter lies the following characteristic, one we will need later on: on every level, we can choose certain *basic substances* A, B, C, ... from which all the other substances on this level can be produced. Moreover, none of the basic substances can be made up of any other basic substance. In a way, the basic substances form the coordinate axes of a "material" reference system comparable to the more familiar spatial coordinate systems. In the same way a point in space can be described by three coordinate values in a spatial reference system, a substance can be characterized by its coordinates in a material reference system. The coordinate values of a substance are given by the amounts or the fractions of its individual components.

Therefore, on a given level, every substance can be assigned a *content formula*

$$A_\alpha B_\beta C_\gamma \ldots$$

which gives its composition in terms of the basic substances. The content numbers $g_i = \alpha, \beta, \gamma \ldots$ express the ratios of amounts,

Experiment 1.1 *Polished cross section of granite*: Magnification shows clearly different minerals: the dark mica, the brownish-red alkali feldspar, the sallow beige soda–lime feldspar, and the translucent quartz (the colors of the minerals can vary strongly depending upon tiny amounts of impurities).

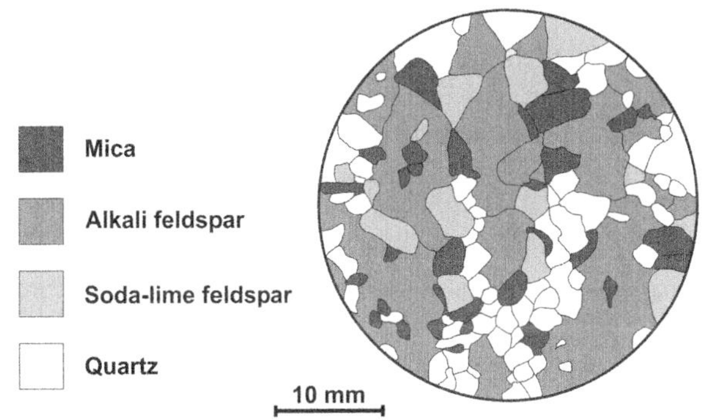

$$\alpha : \beta : \gamma : \ldots = n_A : n_B : n_C : \ldots,$$

with which every basic substance participates in the chemical structure. They correspond to the coordinate values in the chosen material reference system. At the moment we will leave open the question of how to determine the amount n of a substance. In principle, the content numbers can also be negative although we attempt to choose the basic substances so that this does not occur.

Let us consider a concrete example. If a geologist were asked what a paving stone is made up of, he might say granite, basalt, or some other rock. The substances of his world are *rocks*, and his basic substances are *minerals*. From these minerals a multitude of rocks can be formed, depending upon the types, proportions, and grain formation of the individual minerals involved. Let's take a look at a polished cross section of some typical granite (Experiment 1.1).

The granite pictured above may serve as an example for a "petrographical content formula":

$$[Q_{0.3}AlkF_{0.15}Plag_{0.4}Bi_{0.15}].$$

Here, the numbers indicate the fraction by volume of the "basic geological substances": Q = Quartz, AlkF = *Alk*ali *F*eldspar, Plag = *Plag*ioklas (soda–lime feldspar), Bi = *Bi*otite (magnesium mica).

Mineralogists, on the other hand, see these individual rock components (the basic substances for geologists) as themselves being made up of other components. A mineralogist will see that the soda–lime feldspar, one of the main components of basalts and granites, is a mixed crystal with changing fractions of both soda feldspar and lime feldspar. On the next lower level, these crystals can be considered to be unions of various oxides ("earths"), silicium oxide (siliceous earth), aluminum oxide, calcium oxide, and sodium oxide (chemically SiO_2, Al_2O_3, CaO, Na_2O).

What we have found out about minerals can be used in discussions about a myriad of substances, such as resins, oils, wine, schnaps, etc. These substances are also made up of simpler components that they can decompose into and they can be formed from again by the process of *mixing*. Chemists call the basic substances of such homogeneous mixtures "pure" substances or *chemical substances*. An example for a "content formula" of a mixture is that of schnaps: [$Ethanol_{0.2}Water_{0.8}$]. In this case, the relative amounts are not given as volume ratios, as is done in the liquor business, but as it is done in chemistry by stating the ratios of the physical quantity called *amount of substance*, which we will go into more deeply in Sect. 1.4.

On a higher level of complexity, we can produce heterogeneous mixtures—in analogy to rocks—from homogeneous mixtures (more about this in Sect. 1.5) by using these mixtures as basic substances, such as whitewash from chalk dust and sizing solution, or egg white foam from air and egg whites. In a similar fashion, we can, given the right means, decompose the chemical substances into lower level basic substances or we can form them from these basic substances. For chemists, the basic "building blocks" are made up of the roughly 100 chemical elements. Some of these are hydrogen H, helium He, . . ., carbon C, nitrogen N, oxygen O, etc. A special characteristic here is that the ratios of the amounts of elements in the content formulas of individual substances cannot vary continuously; rather, they are quantized in integer multiples. This is known as the "law of multiple proportions." If the measure of amount of substance is suitably chosen, the content numbers introduced above will themselves become integers. Examples are the formulas for water or lime,

$$H_2O = H_2O_1 \ \text{ or } \ CaCO_3 = Ca_1C_1O_3.$$

At the time it was made, this discovery was one of the most important reasons why matter was not considered continuous, but quantized. Indeed, matter was thought of in a simplified mechanistic way to be made of small, mobile geometric entities called atoms. These atoms can assemble into small groups called molecules, which then can merge into extensive networks and lattices creating the matter we know.

On this level, the content formula corresponds in the most simple and frequent case to the so-called *empirical* or *stoichiometric formula*. However, for a more unambiguous identification of the substance in question, it can be suitable to consider the actual number of atoms of each type in a molecule meaning the content formula can be a (integer) multiple of the empirical formula. For example, the content formulas for formaldehyde, acetic acid, and glucose can be given as CH_2O, $C_2H_4O_2\left(= (CH_2O)_2\right)$, and $C_6H_{12}O_6\left(= (CH_2O)_6\right)$.

Just giving the type and proportion of the constituents is often not sufficient to describe a substance completely. More characteristics are necessary. In addition, the spatial arrangement of the atoms of the basic substances is important. In chemical formulas this "structure" is often indicated by dashes, brackets, etc., or by a particular grouping of element symbols. The pair made of ammonium cyanate and urea (carbamide) (Fig. 1.1) is an example. Both of these substances have the same content formula, CH_4ON_2, but their structural formulas differ. This is called *structural isomerism*.

In general, we expect a substance to be something that can be produced in "pure form" and, maybe, filled into a bottle. However, there are substances that cannot be

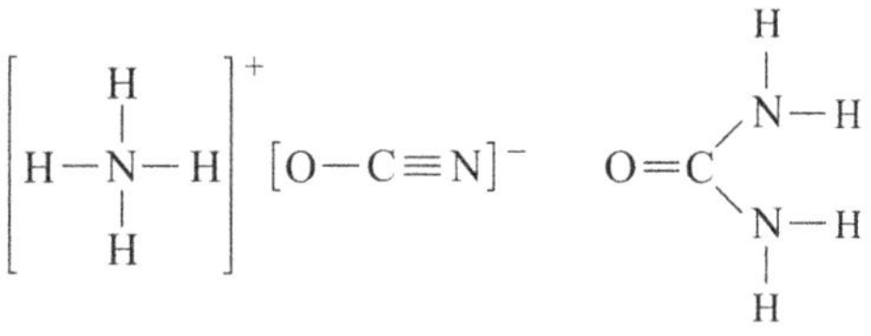

Fig. 1.1 Structural formulas of ammonium cyanate (*left*) and urea (*right*) as examples of two different substances with the same composition (*above*: detailed "valence dash formula," *below*: condensed formula).

$$NH_4OCN \qquad\qquad OC(NH_2)_2$$

understood this way even though they resemble what we normally call a substance in all other chemical and physical characteristics. This category contains the actual carbonic acid H_2CO_3 which forms in trace amounts in aqueous carbon dioxide solutions. The carbonic acid is stable enough to be detected within the thousandfold excess of CO_2, but it is too short-lived to be produced in its pure form.

We consider many substances to be produced from a type of lower level substances, the so-called *ions*. The symbols for table salt $NaCl$ or limestone $CaCO_3$ can be formulated to emphasize their ionic structures

$$[Na]^+[Cl]^- \text{ and } [Ca]^{2+}[CO_3]^{2-}.$$

The brackets are usually left off when dealing with simple ions. For clarity we leave them in because substances of differing rank appear next to each other. We can also include metals such as silver and zinc in this scheme,

$$[Ag]^+[e]^- \text{ and } [Zn]^{2+}[e]_2^-,$$

where the electrons e form the negative partner. In homogeneous mixtures such as crystalline phases, solutions, or plasmas, the individual types of ions, including the electrons, basically behave like independent substances. It is therefore advisable to treat them as such even though they can concentrate into pure form only temporarily and in imponderably small amounts. Their electric charge inevitably drives them apart. The electromagnetic interaction forces electrical neutrality of all parts of matter and allows only a trace of excess of positive relative to negative ions or vice versa. Apart from that, they have all the freedom that uncharged substances have.

There is a substance appearing in the formulas for metals whose composition cannot be expressed by the chemical elements: the *electrons* e. One must therefore introduce a new basic substance. The most obvious candidate would be the electrons themselves. Consequently, negative ions like chloride ions or phosphate ions would obtain the content formulas

$$[\text{Cl}]^- = \text{Cl}_1\text{e}_1 \quad \text{and} \quad [\text{PO}_4]^{3-} = \text{PO}_4\text{e}_3.$$

Positive ions such as sodium ions and uranyl ions, which lack electrons, correspondingly obtain the formulas

$$[\text{Na}]^+ = \text{Na}_1\text{e}_{-1} \quad \text{and} \quad [\text{UO}_2]^{2+} = \text{UO}_2\text{e}_{-2}.$$

Here we have negative content numbers.

The concept of basic substances and material coordinate systems is used for making order of the great multitude of substances. It is only possible to make quantitative descriptions of transformation processes by use of content formulas.

1.3 Measurement and Metricization

Before we turn to the first important quantity, amount of substance, we will take a short look at the basic problem of measuring quantities and metricizing concepts.

Measurement To measure a quantity means to determine its value. Very different methods are used when measuring the length of a table, the height of a mountain, the diameter of the Earth's orbit, or the distance between atoms in a crystal lattice. Length, width, thickness, and circumference are different names for quantities that we consider to be the same *kind of quantity* that we call length. Already in everyday language, length is used in the sense of a metric concept, meaning it quantifies an observable characteristic. *Values* are given as wholes or fractions of a suitably chosen unit.

In 1908, Wilhelm Ostwald already stated that "[It is] extremely easy to measure extensity factors (lengths, volumes, surface areas, amounts of electricity, amounts of substance, weights …). One arbitrarily chooses a piece of it to be the unit and connects so many units together until they equal the value to be measured. If the chosen unit is too rough a measure, correspondingly smaller ones can be created. The simplest way to do this would be 1/10, 1/100, 1/1000, etc. of the original unit." Ostwald's method is valid for *direct* measurement processes. But what does this mean? Let us return to the example of length. In the past, it was customary to *directly* measure the length of a path by counting how many steps were necessary to walk its distance (Fig. 1.2). The arbitrarily chosen unit in Ostwald's sense was one *step*. If one step corresponds with one meter, we get our results in SI units [SI stands for the international metric unit system (from the French Système Internationale d' Unités)].

But quantities are often determined *indirectly* as well, meaning they are found by calculation from other measured quantities. In the field of geodesy, the science of measuring and mapping the Earth's surface, lengths and altitudes have mostly been determined through calculations based on measured angles (Fig. 1.3). When he

Fig. 1.2 Length of a path measured directly by number of steps.

Fig. 1.3 Indirectly determining distance and altitude in impassable terrain by measuring angles.

used this method to measure the acreage of his sovereign, the German mathematician Carl Friedrich Gauss developed error analysis and non-Euclidean geometry.

It is generally necessary when working in industrial arts, engineering, and the natural sciences to have agreement about how quantities will be applied, what units will be used, and how the numbers involved will be assigned. The process of associating a quantity with a concept (that usually carries the same name)—which is the basis of constructing this quantity—is called *metricization*. Determination of values of this quantity is called *measurement*. Measurement can take place only after a corresponding metricization has been established.

Most physical quantities are established through *indirect metricization*, which means they are explained as *derived concepts*. We specify how they are gained from known, previously defined quantities. This is how the density (more exactly, mass density) ρ of a homogeneous body can be defined as the quotient of mass m and volume V, $\rho = m/V$, and the velocity v of a body moving uniformly in a straight line as the quotient of the distance traversed s and the time needed for it t, $v = s/t$.

A totally different method for defining quantities is the *direct metricization* of a concept or characteristic. A concept, initially only understood qualitatively, is then quantified by specifying an appropriate instruction for measurement. This is the usual procedure for quantities considered basic concepts (length, duration, mass, etc.), from which other quantities such as area, volume, velocity, etc., are derived. However, this procedure is not limited to just these quantities.

Direct Metricization of the Concept of Weight A simple example for the direct metricization of a concept would be the introduction of a measure for that what is called *weight* in everyday language. When we talk about a low or high (positive) weight G of an object, we are expressing how strongly the object tends to sink downward. (We use the letter G instead of the usual W in order to avoid confusion with other quantities such as energy W.) There are essentially three stipulations that must be met in order to determine a measure for weight:

(a) *Algebraic sign.* The weight of an object that, if let go, sinks downward is considered to be positive, $G > 0$. Consequently, a balloon flying upward will have a negative weight, $G < 0$. The same applies to a piece of wood floating upward toward the surface after being submerged in water. Something just floating has zero weight, $G = 0$.

(b) *Sum.* If we combine two objects with the weights G_1 and G_2 so that they can only rise or fall as a unit (for example, by putting them together onto the plate of a scale), then we assume that the weights add up: $G_{\text{total}} = G_1 + G_2$.

(c) *Unit.* In order to represent the unit of weight γ, we need something whose weight never changes (when appropriate precautionary measures are taken). For example, we might choose the International Prototype Kilogram in Paris. This is a block of a platinum–iridium alloy representing the unit of mass of 1 kilogram.

The weight G in the sense we are using it here is not an invariable characteristic of an object, but depends upon the milieu it is in. A striking example of this would be a block of wood (Wd), floating to the surface of water (W), $G(\text{Wd}|\text{W}) < 0$, while in air (A), it sinks downward, $G(\text{Wd}|\text{A}) > 0$. As a first step, we will consider the environment to be constant so that G is also constant. In the second step, we can investigate what changes when different influences are taken into account.

These few and roughly sketched specifications for

(a) Algebraic sign
(b) Sum
(c) Unit

are sufficient to *directly metricize* the concept of *weight* as we speak of it in everyday language. This means that we do not need to refer to other quantities in order to associate a measure with the concept. *Measuring* the weight G of an object means determining how many times heavier it is than the object representing the weight unit γ. *Direct measurement* means that the value is determined by direct comparison with that unit and not by calculations from other measured quantities. Figure 1.4 shows how this can be done even without using a scale. First, an object has to be chosen that represents the weight unit γ (Fig. 1.4a). Then, along with the object of unknown weight G, one looks for things that have a weight of $-G$, helium balloons for instance, that will hold the object just enough for it to float in the air

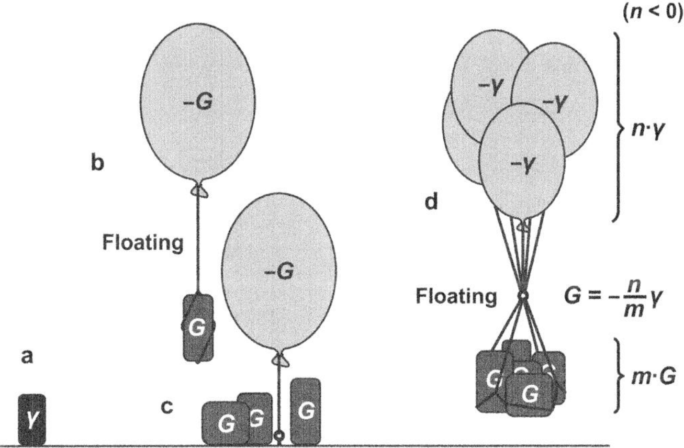

Fig. 1.4 Direct measurement of weights: (**a**) Object representing the weight unit γ, (**b**) "Bundle" consisting of an object with the unknown weight (+G) and a balloon (−G) which just floats in the air, (**c**) Searching further objects with weight +G by means of the balloon, (**d**) Combination of m objects having the same weight +G with n balloons representing the negative weight unit $-\gamma$ to a "bundle" just floating in the air.

(Fig. 1.4b). One of these balloons can be used to easily find further objects with a weight $+G$, meaning ones that the balloon can just lift (Fig. 1.4c). The weight units $+\gamma$ and $-\gamma$ can be multiplied correspondingly. Now in order to measure the weight of an object, for example, a sack, we need only as many things (balloons) representing the negative weight unit $-\gamma$, to bind to the sack until it floats. If n specimens are needed, then $G = n \cdot \gamma$. The number of objects with negative unit weight is expressed in terms of a negative n. If we now want to determine a weight G more accurately, say to the mth part of the unit, we only need to connect m objects having the same weight G with a corresponding number of (positive or negative) unit weights (Fig. 1.4d). If the entire "bundle" floats, it has a total weight equal to 0 according to the specification above:

$$G_{\text{total}} = m \cdot G + n \cdot \gamma = 0 \ \text{ or } \ G = (-m/n) \cdot \gamma.$$

Because any real number can be approximated arbitrarily closely by the quotients of two integers, this method can be used for measuring weights to any desired degree of precision without the use of any special equipment. The measurement process can be simplified if a suitably graded set of weights is available. Negative weights are unnecessary if an equal arm balance can be used because an object can be placed on one side of the balance so that the other side automatically becomes a negative weight. These are, however, all technicalities that are important for practical applications, but are unimportant to basic understanding.

Indirect Metricization of the Concept of Weight Metricization can also be accomplished indirectly. For example, the weight of an object can be determined

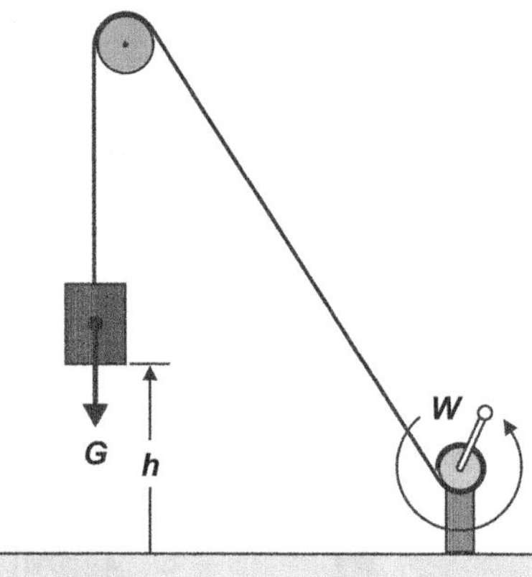

Fig. 1.5 Determining indirectly the weight G through the energy W and lifting height h.

via the energy W needed to lift the object a height h counter to its own weight (Fig. 1.5). (We will go more deeply into the concept of energy and its metricization in Chap. 2.) Both the amount of energy W expended at a winch to lift a block from the ground up to a height h and the height h itself are measurable quantities. The greater the weight, the more energy W is necessary to lift it, so it is possible to find the weight of the block by using W. Because W is proportional to h (as long as h remains small), it is not W itself that is suitable as a measure for the weight, but the quotient $G = W/h$. Using the unit Joule (J) for the energy and the unit meter (m) for the height, we obtain the unit of weight $J\,m^{-1}$. The object embodying the weight unit γ mentioned above can also be measured this way so that the old scale can be related to the new one.

When the lifting height h measured above ground level is too high, W and h are no longer proportional. At great heights, the tendency of the weight to fall decreases due to the lessening of the Earth's gravitational pull and the increase of centrifugal force caused by its spinning. If $G = \Delta W/\Delta h$ is used where ΔW means the additional energy needed to increase the lifting height by a small amount Δh, the definition of the quantity G can be expanded to include this case. Thereby, the symbol Δ indicates the difference of final value minus initial value of a quantity, for example, $\Delta W = W_2 - W_1$. In order to indicate that the differences ΔW and Δh are intended to be small, the symbol for difference, i.e., Δ, is replaced by the differential symbol d. One writes

$$G = \frac{\mathrm{d}W}{\mathrm{d}h} \quad \text{or more detailed,} \quad G(h) = \frac{\mathrm{d}W(h)}{\mathrm{d}h}.$$

For the sake of simplicity, although it is not completely mathematically sound, we will consider the differentials to be very small differences. This will suffice for all or nearly all of the applications we will present in this book. Above and beyond this, it gives us an effective (heuristic) method of finding a mathematical approach to a physical problem. Dealing with differentials is described in more detail in Sect. A.1.2 in the Appendix.

Note that in the expression on the left in the equation above, W and G take the roles of the variables y and y'. In the expression on the right, they appear in the roles of the function symbols f and f'. It is actually a common but not fully correct terminology to use the same letters for both cases, but if one is careful, it should not cause serious mistakes.

In order to lift something, we must set it in motion and this takes energy too. The greater the velocity v attained, the more energy is needed. Therefore, W does not only depend upon h, but upon v as well. This is expressed by $W(h, v)$. In order to introduce a measure for weight also in this case, we must expand the definition above:

$$G = \frac{\partial W(h, v)}{\partial h}.$$

Replacing the straight differential sign d by the curved ∂ means that when calculating a derivative, only the quantity in the denominator (in this case h) is to be treated as variable, while the others appearing as argument (in this case only v) are kept constant (so-called *partial derivative*, more about this in Sect. A.1.2 in the Appendix). A constant v, and therefore, $dv = 0$, means that the increase of energy dW has only to do with the shift in height dh and not with change of velocity.

There is another notation which is preferred in (physical) chemistry where the dependent variable (here it is W) is in the numerator while the independent variables (in this case h and v) appear in the denominator and the index, respectively. The variable to be kept constant is added to the expression of the derivative (placed inside parentheses) as index:

$$G = \left(\frac{\partial W}{\partial h}\right)_v.$$

We can go a step further and imagine the object in question to be like a cylindrical rubber plug with changeable length l and cross section A. These changes also consume energy and the total amount of energy needed now depends upon four variables h, v, l, A. In order to eliminate possible additional contributions due to changes to l and A, they (along with v) must be kept constant. This can be expressed as follows:

$$G = \left(\frac{\partial W}{\partial h}\right)_{v,l,A}.$$

We find that defining the weight G in terms of energy becomes increasingly more complicated, the more generally one attempts to comprehend the concept. This is why we will introduce important quantities such as energy (Chap. 2), entropy (Chap. 3), and chemical potential (Chap. 4) through direct metricization.

1.4 Amount of Substance

There are various measures used for amount of substance, so we need to consider what properties we expect to see in the quantity we are looking for.

It seems reasonable to claim that a certain amount of a substance within a specified volume can only change if parts of it are emitted, are added from outside, or are consumed or produced by chemical reaction. Just displacing, heating, compressing, segmenting, removing accompanying substances, etc., should not change the amount of substance in question. If we wish to adhere to these properties, we automatically eliminate certain measures of amount such as volume which is the one most often used in everyday life. Examples of this are a solid cubic meter of wood, a liter of water, a cubic meter of gas, etc. According to Einstein's relation ($W = m \cdot c^2$; W: energy, c: speed of light), the mass m of a substance that grows also when only energy is added to it, must be—strictly speaking—excluded as well. Because such changes are much smaller than what can be measured with any precision, mass is nevertheless widely used in science and commerce. It is, however, not fully satisfactory if one considers that when 1 cm^3 of water is heated by $1°$, its mass grows only by 5×10^{-14} g, but this change corresponds to about one billion water molecules.

It is plausible to assume that two amounts of the same substance are identical if they occupy the same space or have the same weight if conditions such as the form of the area, temperature, pressure, field strength, etc., are all identical. In order to measure a certain amount of substance, it is enough to fill the substance into equal sized containers or segment it into equal parts and then to count them (Fig. 1.6)—again under uniform conditions. This *direct* measurement of amounts of substance by dividing it into equal unit portions and then counting them has been used since ancient times and is still used today in the household, in trade, and in business. Unit portions have mostly been established by filling and emptying a defined "cavity" such as a bushel basket. Other types of measurements have also been created. Examples would be 1 pinch of salt, 2 teaspoons of sugar, 3 bunches of radishes,

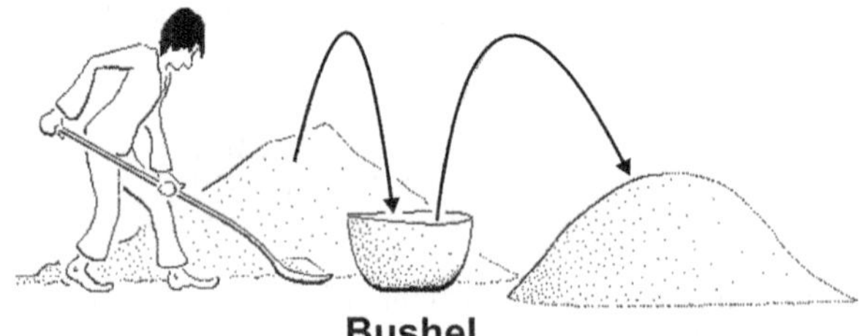

Fig. 1.6 Direct measurement of amounts of substance by dividing it into unit portions and counting them (for example, in the past determination of the amount of harvested grain by use of bushel baskets).

or 10 scoopfuls of sand. Alternatively, the unit portions can be established and counted automatically (such as in the water or gas meters in every household).

Due to the atomic structure of matter, there is a natural division into atoms, or rather, the constantly repeating groups of atoms described in the chemical formula. It is therefore obvious to have the unit be such an elementary entity, such a "particle." The amount of substance corresponds to a number of units (like 24 apples or 120 cars). However, in macroscopic systems, the numbers of particles are very high and this can be problematic. For instance, 10 g of water contains about 10^{23} particles. Therefore, a more suitable unit must be found that is comparable to the everyday dozen (12 units) or score (20 units). In chemistry, the measure of amount called *mole* (derived from the Latin word "moles" meaning "huge mass") has been determined as follows:

One mol is a portion of substance made up of 6.022×10^{23} particles (units).

or more exactly stated:

One mol is a portion of substance made up of as many elementary entities (particles) as there are atoms in exactly 12 g of the pure isotope carbon-12 (^{12}C).

$N_A = 6.022 \times 10^{23}$ mol^{-1} is also called the *Avogadro constant*. Because it is possible to directly or indirectly count the atoms or groups of atoms in a given sample, the so-defined *amount of substance n* is in principle a measurable quantity.

The fact above can be stated differently. Instead of saying that a substance is made up of countable particles, one might say that there is a smallest possible portion of substance, an *elementary amount (of substance) τ*. The following is valid for this elementary amount:

$$\tau = \frac{1}{N_A} = \frac{1}{6.022 \times 10^{23} \ \text{mol}^{-1}} = 1.6605 \times 10^{-24} \ \text{mol}. \tag{1.1}$$

The amount of substance therefore is given by

$$n = N \cdot \tau, \tag{1.2}$$

where N represents the number of particles in the given portion of substance. Quantities $\mathcal{G}$ with real but discrete and therefore countable definite values are called *quantized*. We introduce a *quantum number g* to number the values. If the values are not only discrete, but equidistant as well, the quantity is said to be integer quantized. In the simplest case, the values are integer multiples g of a universal quantum γ:

$$\mathcal{G} = g \cdot \gamma. \tag{1.3}$$

In the case of the variable $\mathcal{G}$, which represents various physical quantities, we use another font in order to avoid confusion (for example, with weight G). We do this as well for g (instead of g) and γ (instead of γ).

The amount of substance n is therefore integer quantized with N as the quantum number and τ as the corresponding quantum of amount of substance. This is comparable to the more familiar integer quantization of the charge Q of an ion,

$$Q = z \cdot e_0 \tag{1.4}$$

with the charge number z in the role of the quantum number and the elementary charge e_0 in that of the charge quantum ($e_0 = 1.6022 \times 10^{-19}$ C).

The relation between the amount of substance n and the mass m is determined by the *molar mass M*. This quantity corresponds to the quotient of the mass of a sample of the substance and the amount of substance in the sample:

$$M = \frac{m}{n} \; \left(\text{SI unit: kg mol}^{-1}\right). \tag{1.5}$$

With the help of the molar mass we can convert the more easily measured mass to amount of substance:

$$n = \frac{m}{M}. \tag{1.6}$$

1.5 Homogeneous and Heterogeneous Mixtures, and Measures of Composition

We will now look more closely at the term homogeneous mixture, mentioned in Sect. 1.2. We will also contrast it with the concept of heterogeneous mixture. There is, unfortunately, no unified way of wording this, so we will give a short explanation of how we will use the terminology. *Mixture* is used as superordinate term. A *homogeneous mixture* is homogeneous in the sense that it has a molecular dispersion with granularity of <1 nm, and all its constituents A, B, C, ... are considered equal. If there is an excess of one of the components in a homogeneous mixture, we will speak of a *solution*. The main component A is then called the *solvent* and the minor constituents B, C, ... the *dissolved substances* or *solutes*. Unlike a homogeneous mixture, a *heterogeneous mixture* is coarsely dispersed with granularity > 100 nm. Microheterogeneous *colloids* are a special case (granularity 1 ... 100 nm). However, not every kind of material system made up by two and more different substances fits into this scheme.

A homogeneous region, meaning a region that is uniform in all of its parts, is called a *phase*. One can differentiate between pure phases made up of one substance and *mixed phases* made up of more than one. Homogeneous mixtures are always single phases. Examples of single-phase systems are air, wine, glass, or stainless steel. Heterogeneous mixtures are, by contrast, multi-phase, wherein the equal homogeneous parts form together a phase. Fog, construction steel, soldering tin (Pb-Sn), slush, etc., are all examples of two-phase heterogeneous mixtures. A very esthetic example of a two-phase system is a so-called lava lamp with its wax-water

Experiment 1.2 *Lava lamp*: When the lamp is turned on, blobs of heated wax ascend slowly from the bottom to the top where they cool and then descend to the bottom again, causing a constant movement of both phases.

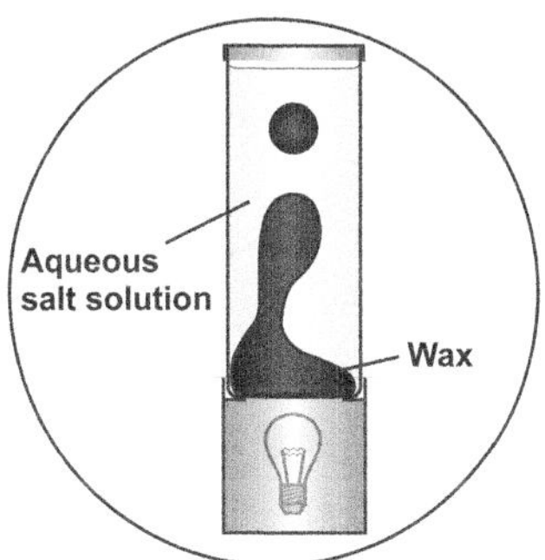

filling (Experiment 1.2). The granite shown in Experiment 1.1, however, is essentially composed of four phases.

As a rule, one does not specify the amounts of substance of all the components in order to characterize homogeneous mixtures, but the *content* of selected components. The superordinate concept "content" meaning the material fraction of a substance in the mixture can be quantified by various measures of composition. Several of these measures will be introduced in the following.

The *mole* (or *amount*) *fraction x* of a component B corresponds to the quotient of the amount of substance n_B and the total amount n_{total} of all the substances present in the mixture:

$$x_B = \frac{n_B}{n_{total}} \quad \left(\text{SI unit: 1 or } mol\,mol^{-1}\right). \tag{1.7}$$

The mole fraction is a relative quantity where $0 \leq x \leq 1$. The sum of all the mole fractions must always result in 1, so for a complete characterization of a binary mixture (a mixture of two components A and B), only *one* mole fraction is necessary. The second one will result according to $x_A = 1 - x_B$.

If the amount of substance is replaced by the mass, another measure of composition results, the *mass fraction w*:

$$w_B = \frac{m_B}{m_{total}} \quad \left(\text{SI unit: 1 or } kg\,kg^{-1}\right). \tag{1.8}$$

The composition of solutions is often expressed by concentration. The *molar* (or *amount*) *concentration c* (formerly called molarity) of a dissolved substance B results from the quotient of the amount of solute n_B and the volume of solution V:

Table 1.1 Conversion of the most common measures of composition for binary mixtures.

$x_B =$	x_B	$\dfrac{M_A c_B}{\rho - c_B(M_B - M_A)}$	$\dfrac{M_A b_B}{1 + M_A b_B}$
$c_B =$	$\dfrac{\rho x_B}{M_A + x_B(M_B - M_A)}$	c_B	$\dfrac{\rho b_B}{1 + b_B M_B}$
$b_B =$	$\dfrac{x_B}{M_A(1 - x_B)}$	$\dfrac{c_B}{\rho - M_B c_B}$	b_B

$$c_B = \frac{n_B}{V} \left(\text{SI unit: mol m}^{-3}\right). \tag{1.9}$$

The unit mol L^{-1} ($= \text{kmol m}^{-3}$) (abbreviation M) is often used in place of the SI unit. When referring to *the* concentration in chemistry one usually means the quantity c.

Sometimes the *mass concentration* β is used that can be calculated from the quotient of substance mass m_B and volume V of solution:

$$\beta_B = \frac{m_B}{V} \left(\text{SI unit: kg m}^{-3}\right). \tag{1.10}$$

The disadvantage of these two easily accessible concentrations is that they are both temperature and pressure dependent. This is due to corresponding changes of total volume and can be avoided if the mass of the solvent is used instead. The *molality b* corresponds to the quotient of amount of substance n_B of the solute B and the mass m_A of the solvent A:

$$b_B = \frac{n_B}{m_A} \left(\text{SI unit: mol kg}^{-1}\right). \tag{1.11}$$

In Table 1.1, the relations for converting the individual measures of composition have been compiled using molar mass M and density ρ.

1.6 Physical State

System and Surroundings We tend to consider (material) *systems* as strongly simplified, often idealized, parts of the natural world around us in which we have a special interest. For example, we can be interested in a rubber ball, a block of wood, a raindrop, the air in a room, a solution in a test tube, a soap bubble, a ray of light, or a protein molecule. We assume that systems can appear in various (physical) states, where the word *state* means a momentary specific condition of a sample of matter determined by macroscopic characteristics. States can differ qualitatively due to characteristics such as state of aggregation or crystal structure or quantitatively in the values of suitably chosen quantities such as pressure, temperature, and amount of substance.

Everything outside of the system in question we call the *surroundings*. If the system is completely *isolated* from its surroundings, we can ignore anything happening there (in the surroundings). However, this requirement is hardly ever fulfilled, so we must deal to a certain extent with the conditions in the surroundings. When considering the pressure or temperature of a system as prescribed, one usually thinks of some equipment in the surroundings that establishes these values. In a lab, this is generally a cylinder with a moveable piston that will allow us to set up the pressure and a "heat reservoir" with a defined temperature that is connected to the system through heat conducting walls.

Types of States The classical *states of aggregation*, *solid*, *liquid*, and *gaseous*, serve as a first rough characteristic for distinction. Seen macroscopically, the following is valid for a substance enclosed in a container:

- A *solid* has a fixed volume and withstands shear. This means that it retains its volume and shape regardless of the shape of the container it occupies.
- A *liquid* has a fixed volume and is able to flow. It retains its volume, but its shape is unstable and adapts to the walls of the container.
- A *gas* is able to flow and fills the whole space it is in. It assumes the volume and shape of the container.

Information about the state of aggregation of a substance can be added to its formula by using a vertical stroke and the abbreviation s for solid, l for liquid, and g for gaseous. Ice is then characterized by $H_2O|s$, liquid water by $H_2O|l$, and water vapor by $H_2O|g$. We prefer this way of writing to the usual form with parentheses because it prevents the confusing overabundance of parentheses occurring when formulas of substances appear in the argument of a quantity [for example, the mass density $\rho(H_2O|l)$ instead of $\rho(H_2O(l))$].

A deeper look into the nature of states of aggregation can be gained if one leaves the phenomenological level and moves to the molecular-kinetic level (Fig. 1.7). A particle model lets us create a relation between the macroscopic properties of matter and the behavior of the particles—atoms, ions, or molecules of which it is composed.

- From the atomistic point of view, the particles in *solids* are packed closely and well-ordered due to their strong reciprocal attraction. They have only very limited space to move in, meaning that they essentially stay on one site, but oscillate somewhat around that position.
- The particles of *liquids* are still rather closely packed, but not in an orderly way. Motion of the particles is so strong that the reciprocal attraction is not intense enough to hold them in one position. Although they stay in contact, they are able to slide by each other freely.
- The particles in *gases* are packed very loosely and disorderly. Their constant movement is quick and chaotic and they tend to be far apart from each other except for occasional collisions. The typical distance between the particles of the air in a room is about ten times the diameter of the particles themselves.

There is a somewhat different but comparable point of view which uses *crystalline, amorphous*, and *gaseous* for classification.

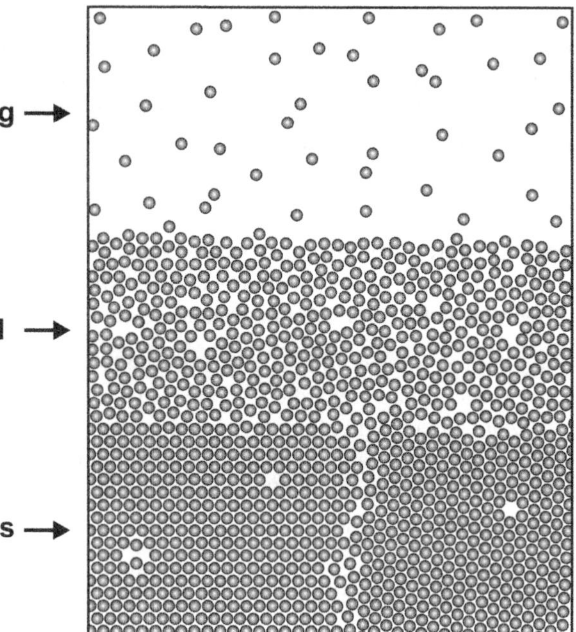

Fig. 1.7 Molecular-kinetic illustration of the three states of aggregation, solid (s), liquid (l), and gaseous (g). The strict order of a solid can be disturbed by defects or by a grain boundary (a fault zone where differently oriented regions that have otherwise identical crystal structure (grains) adjoin each other).

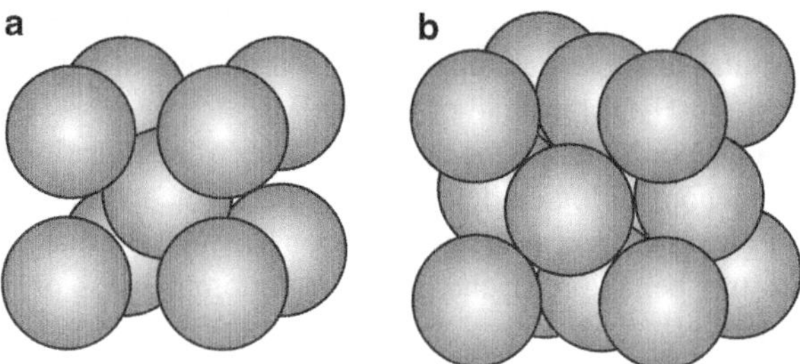

Fig. 1.8 (**a**) Body-centered cubic crystal lattice, (**b**) Face-centered cubic crystal lattice.

- A substance can be categorized as *crystalline* if it is inherently stable and if its components are packed in a regularly ordered, repeating pattern that continues in all three directions over long distances (long-range order). This crystalline state is generally characterized by |c. Different crystal structures appearing at the level of chemical bonding theory and which are created by different ways of packing their components are indicated by Greek letters or the appropriate mineral names. For example, iron can have a body-centered cubic structure (Fe|α) or a face-centered cubic structure (Fe|γ) (Fig. 1.8), and carbon can exist in the form of hexagonal graphite (C|Graphite) or in the cubic diamond structure (C|Diamond). These different forms of a substance are called *modifications*.

- The components in an *amorphous* substance only show some short-range order. Macroscopically, the substance can be either solid or liquid. The symbol used for amorphous is |a. A typical amorphous solid would be glass, but also spun sugar (candyfloss, cotton candy) can be counted in this category.

- The *gaseous* state is defined in the usual sense.

If a substance appears in dissolved form in a homogeneous mixture, it is characterized by |d. Water is by far the most common solvent. We therefore give substances dissolved in water their own symbol |w.

State Quantities Besides the qualitative description of a system discussed above its quantitative properties can be characterized by physical quantities. A quantity that describes the state of a system or is defined by the instantaneous state of a system is a *state variable*. Depending upon the conditions involved, the same type of quantity might or might not be a state variable. The time-dependent amount of air $n(t)$ in a ventilated room is a state variable, but the amount of incoming air n_{in} supplied by ventilation, or the amount of outgoing air n_{out}, escaping around windows or doors, is not:

$$\Delta n = n_{in} - n_{out} \text{ or in more detail } \Delta n(t) = n_{in}(t) - n_{out}(t).$$

Analogously, the volume of water V in a bathtub is a state variable, but the volume of water flowing in through the water tap (V_A) or shower head (V_B) and spilling over (V_O) is not (Fig. 1.9):

$$V = V_A + V_B - V_O. \tag{1.12}$$

The water level h in the bathtub is a more complex example of a state variable. h not only depends upon the water volume V but upon the volume V_d displaced by the person standing, sitting, or lying in it, $h = h(V, V_d)$. If the tub had straight walls, i.e., if the cross section A were constant, we could easily write:

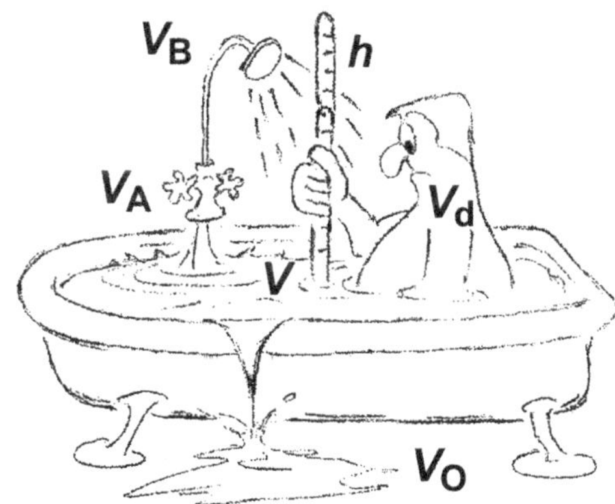

Fig. 1.9 Volume of water V and water level h in a bathtub as examples of state variables.

$$h = (V + V_d)/A \qquad \text{``Equation of state.''} \qquad (1.13)$$

In contrast to Eq. (1.12), Eq. (1.13) relates only state variables (the quantities whose values are determined only by the momentary state of the system, independent of the path by which that state was reached). The advantage of applying such relations is that one can reach conclusions without having to know the details that led to this state. This type of equation is called an *"equation of state"* and functions such as $h = h(V, V_d)$ are called *state functions* (or *state variables*). Even though each of these quantities can have a value that changes with time, the equations of state relating these quantities are timeless. However, the division of V into the three values V_A, V_B, and V_O, as shown in Eq. (1.12), depends upon the *process* by which the bathtub was filled. Such quantities can be called *"process quantities"* to distinguish them from the state variables.

The considerations that hold for the volume of water in a tub also hold for the energy of a system if several possibilities exist for inflow and outflow. The excess energy in hot cooking water or in a charged car battery is a state variable, but the energy we are charged for every year by the electricity company or that our kitchen stove consumes when we are preparing food, is not. We will deal with this in more detail in later chapters.

Different Forms of Notation State variables are easier to handle mathematically. Therefore, if possible, all calculations are carried out with these quantities and unknowns, results, and parameters are expressed by them. This is especially valid when dealing with an abstract quantity that we cannot imagine or only imagine insufficiently. The characteristic of being a state variable is an important and helpful orientation device. We will now use an example to illustrate the approach that is also graphic and understandable.

A small increase of water volume dV_A flowing out of a water tap and into the tub can be expressed as the increase dV of the water volume in the tub if inflow through the showerhead and outflow over the edge of the bathtub are not allowed (either in reality or just in our model). We will express this increase by the symbol $(dV)_{B,O}$. When we correspondingly deal with V_B and V_O, an equation results where the V in all the terms is the same quantity, the water volume in the bathtub:

$$dV = (dV)_{B,O} + (dV)_{A,O} - (dV)_{A,B}.$$

Sometimes instead of d, the symbol δ or đ is used in the case of "process quantities." One then writes δV_A or đV_A, respectively. However, in the following we will avoid doing this.

The changes of volume per time on the right describe the strengths of the water currents J into and out of the bathtub. These are the water flowing out of the water tap J_A and the shower head J_B, and the water spilling over the edge of the tub J_O:

$$\frac{\mathrm{d}V}{\mathrm{d}t} = \left(\frac{\mathrm{d}V}{\mathrm{d}t}\right)_{\mathrm{B,O}} + \left(\frac{\mathrm{d}V}{\mathrm{d}t}\right)_{\mathrm{A,O}} - \left(\frac{\mathrm{d}V}{\mathrm{d}t}\right)_{\mathrm{A,B}}.$$

When $\mathrm{d}V/\mathrm{d}t$ is abbreviated to $\dot{V}$, the resulting equation is the same but has a more compact form. It can also be more easily understood:

$$\dot{V} = J_{\mathrm{A}} + J_{\mathrm{B}} - J_{\mathrm{O}} \qquad \text{``Continuity equation.''}$$

Expressed in words: "The rate of increase of the amount of water in the tub equals the sum of the water currents flowing into it (and out of it)." This is a very simple example of applying an equation that appears in a myriad of ways in various areas of physics.

However, another aspect is more important for us here. In Sect. 1.3, we were introduced to similar expressions where instead of the straight d, the somewhat differently written curved ∂ appeared. Although we could actually do without the curved ∂ and always use d, without making a formula wrong, this does not work both ways. In an expression like $(\partial y/\partial u)_{v,w}$, it is always assumed that the quantity in the numerator can be written as a function of the quantities in the denominator and the index, i.e., $y = f(u, v, w)$. Thereby, the index $_{v,w}$ means that only u occurs as an independent variable, while v and w are treated as constant parameters. We will illustrate this by crossing out these quantities in the argument of the function: $y = f(u, \cancel{v}, \cancel{w})$, but just this one time, and not in general. The derivative can now be calculated as usual by applying the rules of school math (compare to Sect. A.1.2 in the Appendix). If we indicate the derivative with the usual $'$, i.e., $y' = f'(u, \cancel{v}, \cancel{w})$, we obtain:

$$\left(\frac{\mathrm{d}y}{\mathrm{d}u}\right)_{v,w} = \left(\frac{\partial y}{\partial u}\right)_{v,w} = f'(u, \cancel{v}, \cancel{w}).$$

The derivative with respect to the other variables is calculated accordingly:

$$\left(\frac{\mathrm{d}y}{\mathrm{d}v}\right)_{u,w} = \left(\frac{\partial y}{\partial v}\right)_{u,w} = f'(\cancel{u}, v, \cancel{w}) \quad \text{and so on.}$$

Unlike u, v, w, the indices A, B, and O in the expressions above do not denote quantities. However, that is not important at this point. What is important is that both cases express that the increase of the numerator is caused by changes of the quantity in the denominator, while all other influences are eliminated.

Coupled Changes The bathtub can help us once again to understand another aspect that we will need to deal with later on. In the above, we discussed the special features resulting when the same entity (measured by the same quantity: water volume) of a system can be exchanged with its surroundings simultaneously via various paths. Let us now replace the person in the tub with an expandable rubber

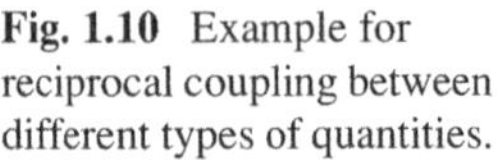

Fig. 1.10 Example for reciprocal coupling between different types of quantities.

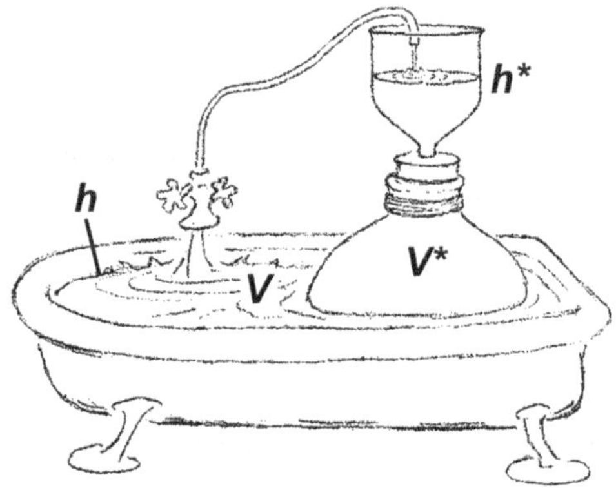

sack (Fig. 1.10). The water volume V in the tub is separated from that in the sack V^*, so that the water levels in the bathtub and the sack, h and h^*, can be different. In this case, all four quantities V, h, V^*, h^* are state variables. They are all "geometric" and therefore it appears to us that they are comparably simple.

Despite their separation, the two parts of the system influence each other so that an increase of water level in one causes an increase of the water level in the other and vice versa. This reciprocal coupling between different types of quantities, mechanical, thermal, chemical, electrical, etc., is central to thermodynamics and matter dynamics. We will go into this more deeply in Chap. 9 when we have acquired the necessary background.

Extensive, Intensive, Substance-Like The concept of substance is primary to the discussion of systems of matter dynamics. The simplest case would be a *homogeneous* domain, meaning one where all the parts are uniform and where the form and size are unimportant. Such a formless domain represents what we call substance, whether it is pure or a mixture of various components. Some quantities describing the state of a domain, such as mass, volume, concentration, energy, entropy, etc., add up when two uniform domains are merged into one. Other quantities such as mass density, pressure, temperature, concentration, refractive index, etc., do not change. The first group is made up of the so-called *extensive* parameters and the second, of *intensive* parameters. One group describes *global* characteristics for the entire domain in question. The second group describes *local* characteristics, attributed to one place. These concepts can be applied analogously and with good accuracy to material systems that are not homogeneous in their entirety but are approximately uniform in small sections—at close range.

Classification is not always clear. Let us consider the surface A of a liquid droplet in fog. When we put two sections of the fog together, the surfaces add up so that A appears to be an extensive quantity. However, if we unite two round droplets into one, A is clearly not additive.

The extensive quantities $\mathcal{G}$ which we can imagine as a "something" distributed in space are called *substance-like*. Mass m, amount n_B of a substance B, and electric charge Q are some of these, and so are energy, momentum, and entropy, which are often considered quite abstract concepts. We will get back to them later in more detail (Chaps. 2 and 3). Their distribution need not to be uniform and the density ρ_G of the "something" can change in space and time by being used up, produced, or redistributed. A quantity that distinctly exhibits this behavior is the amount n_B of a substance B that diffuses and reacts. If G is a so-called conserved quantity, like energy, for example, the "something" can be neither created nor can it decay. It can only migrate inside the system or be exchanged with its surroundings. When the "something" disappears in one location, it must reappear somewhere nearby. From there it can be transfered further and further. This process can be considered a kind of flow.

1.7 Transformation of Substances

In Sect. 1.2, we saw that the multitude of substances can be understood as combinations of relatively few basic substances, where the ratio is quantitatively expressed by the content formula. In chemistry, we have seen that chemical elements play the role of basic substances and when indicated, electrons e, if the totality of ions is considered a "charged" substance.

We use *transformation* here and in the following as the superordinate term for processes that are otherwise more differentiated, such as *reaction, transition* (change of state of aggregation, etc.), and the (spatial) *redistribution* of substances. This is done simply because these processes can all be described using the same paradigm. Whether a transformation of substances is chemical or physical, it can be expressed by a *conversion formula*, also called a *reaction equation* or *reaction formula*. Usually, the content formulas of the starting substances (also called reactants) are to the left of the reaction arrow, and the end products are to the right. The term "reaction equation" is not exactly apt because we do not have an equation in the usual sense here. The name comes from the fact that the amounts of the chemical elements—either free or bound—remain unchanged during the transformation from initial to final substances. The number of symbols for each element must, therefore, be the same on both the left and the right.

A simple example of a reaction is synthesizing ammonia from nitrogen and hydrogen. One usually finds different substances B_i ($i = 1, 2, 3 \ldots$) participating in such processes, each of which can be assigned a number. For example, nitrogen N_2 can be given the number 1, hydrogen H_2 the number 2, and ammonia NH_3 the number 3. This can be accomplished by simply setting the numbers above the substances in the conversion formula:

$$\overset{1}{N_2} + 3\,\overset{2}{H_2} \rightarrow 2\,\overset{3}{NH_3}\,,$$

but also by the order they are put into in the formula, a table, or some other type of list, for instance:

$$\mathcal{B}_1 = N_2, \mathcal{B}_2 = H_2, \mathcal{B}_3 = NH_3, \mathcal{B}_4 = CO_2 \ldots .$$

Note that the index for $\mathcal{B}$ is arbitrary, while the content numbers in the content formulas are well defined. In the case of the variable $\mathcal{B}$, we use another font in order to avoid confusion but also because we use B as well as B for other purposes.

It is a good idea to write all of the initial and final substances on one side in order to avoid having to distinguish between different cases, for example as follows:

$$0 \rightarrow -1\,N_2 - 3\,H_2 + 2\,NH_3$$

or, in general, for various substances A, B, C, ... of the set $\mathbb{S}$ of all possible substances

$$0 \rightarrow \nu_A A + \nu_B B + \nu_C C + \ldots \quad \{A, B, C \ldots\} \subset \mathbb{S}.$$

A symbolic "0" appears on the left that does not actually represent the number 0, but a substance represented by a content formula in which all content numbers disappear. If we consider the substances to be numbered as discussed above, the expression can be written more compactly using the summation operator $\sum$:

$$0 \rightarrow \sum_{i=1}^{n} \nu_i \mathcal{B}_i.$$

The chemical elements participating in a transformation are always conserved. This means that their total amount does not change, whether they are free or bound. Therefore, the *conversion numbers (stoichiometric coefficients)* ν_i in front of the content formulas should be chosen so that the number of element symbols is the same on both sides. This also holds for electrons when they appear, so to speak, as an additional basic substance in the conversion formula where ions participate. They appear openly with the symbol e or more hidden with the superscript charge numbers. The more unusual way of writing that uses "0" on the left side has the advantage that the conversion numbers appear with the correct algebraic signs as factors in front of the content formulas. ν is negative for starting substances and positive for final products, for example, $\nu_{N_2} = -1, \nu_{H_2} = -3, \nu_{NH_3} = +2, \nu_{CO_2} = 0$, ... If the index itself contains subscripts it is better to refer back to the numbering scheme $\nu_1 = -1, \nu_2 = -3$, etc., or to use the form with arguments $\nu(N_2) = -1$, $\nu(H_2) = -3$, etc. As the lower formulas above imply, we have to sum over all the basic substances, meaning the chemical elements, the electrons, as well as their combinations. However, for the overwhelming number of substances $\nu_i = 0$ and the corresponding substances can be ignored so that, in the example above, $\nu_{CO_2} = 0$, and the same would hold for ν_{Fe}, ν_{NaCl}, etc.

The amounts of the substances change in the course of a reaction and these changes can be used to measure the progression of the process. Substances are not all formed or consumed in the same ratio. A look at the conversion formula for ammonia synthesis shows hydrogen converting at three times the rate of nitrogen. Changes of amounts are proportional to the conversion numbers v. In order to attain a quantity that is independent upon the type of substance B, the observed changes Δn_B are divided by the corresponding conversion numbers v_B:

$$\xi = \frac{\Delta n_B}{v_B} = \frac{n_B - n_{B,0}}{v_B} \qquad \text{``basic stoichiometric equation.''} \qquad (1.14)$$

n_B is the instantaneous amount of substance and $n_{B,0}$ stands for the initial amount. Note that both Δn_B and v_B are negative for reactants, so the quotient is positive which is also true for the products. Of course, the reacting material system must be isolated from its surroundings, meaning that no exchange of substances or secondary reactions may be allowed to occur so that the amounts of substance converted during the process can be clearly identified.

The following is valid for different substances A, B, C, ...:

$$\xi = \frac{\Delta n_A}{v_A} = \frac{\Delta n_B}{v_B} = \frac{\Delta n_C}{v_C} \quad \text{or} \quad |\xi| = \frac{|\Delta n_A|}{|v_A|} = \frac{|\Delta n_B|}{|v_B|} = \frac{|\Delta n_C|}{|v_C|} = \dots \qquad (1.15)$$

Just one quantity, the time-dependent quantity ξ, is enough to describe the temporal course of the reaction. We will call it the *extent of reaction* or *extent of conversion*. In the case of ammonia synthesis, the value of ξ indicates what the momentary extent of the production of ammonia is after a given period. The same unit used for amount of substance, the mol, is normally used for this quantity. Take $\xi = 1$ mol. This means that since the process started, 1 mol of nitrogen and 3 mol of hydrogen have been consumed and 2 mol of ammonia have been produced. For the same value ξ (the same extent of reaction), the changes of amounts Δn can be very different (with regard to both the absolute value and the sign involved). It is important to remember that the ξ values only make sense in relation to a certain conversion formula. If the same reaction is described by another formula, such as

$$\tfrac{1}{2}\,N_2 + \tfrac{3}{2}H_2 \rightarrow NH_3,$$

the meaning of the ξ values changes. At any given moment, the extent of reaction ξ is only half as great for the same amounts of converted substances. The conversion formula must therefore always be specified.

The usual stoichiometric calculations can be carried out directly or indirectly using Eq. (1.15). Therefore, we call it and its source, Eq. (1.14), the *basic stoichiometric equations*. These equations allow us to calculate the change of amount of a substance C from the change of amount of a substance A. For example, knowing the consumption of an acid in titration Δn_A, we can calculate the original amount of a base Δn_B, or knowing the amount of a precipitate Δn_P, we can find the amount of the substance Δn_S which was precipitated out from the initial solution. Here, the conversion numbers are to be taken from the neutralizing or the precipitation reaction, respectively. Since in most cases only the absolute values of Δn matter, and not their algebraic signs, the simpler variant where no algebraic sign need be dealt with [Eq. (1.15) right] is the one most often used.

Often, it is not directly the amount of substance n that is interesting or known to us but rather a volume ΔV of a reagent solution that is either given or consumed, the concentration c of a sample or standard solution, or the increase of mass Δm of a filtering crucible where a precipitate was collected, etc. The Δn values are then expressed by the given, measured, or the sought quantities. For example, $\Delta n_A = c_A \cdot \Delta V$ for the consumed acid or $\Delta n_P = \Delta m/M_P$ for the weighted precipitate, where M_P stands for the molar mass of P.

Here is a short example showing this. For the titration of 25 mL of sulfuric acid, 20.35 mL of sodium hydroxide solution (0.1 mol L^{-1}) is consumed. We are looking for the concentration of the acid. The conversion formula for this is

$$H_2SO_4 + 2\,NaOH \rightarrow Na_2SO_4 + 2\,H_2O$$

and the basic equation (1.15), where A stands for the sulfuric acid and B for the sodium hydroxide solution, is

$$\frac{|c_A \cdot \Delta V_A|}{|\nu_A|} = \frac{|c_B \cdot \Delta V_B|}{|\nu_B|} \quad \text{or} \quad c_A = c_B \cdot \left|\frac{\Delta V_B \cdot \nu_A}{\Delta V_A \cdot \nu_B}\right|.$$

Using numerical values, we obtain

$$c_A = 0.1 \,\text{mol} \,\text{L}^{-1} \frac{20.35 \,\text{mL} \cdot 1}{25 \,\text{mL} \cdot 2} = 0.041 \,\text{mol} \,\text{L}^{-1}.$$

The basic stoichiometric equation can be transformed somewhat so that for any arbitrary substance i:

$$n_i = n_{i,0} + \nu_i \xi. \tag{1.16}$$

This equation can basically be used for any substance, even when it does not participate in the reaction taking place, because $\nu_i = 0$. In this respect, this equation is more general than our initial equation (1.14), for which $\nu_i \neq 0$ needs to be the case for the denominator.

We call a change in the extent of a reaction $\mathcal{R}$, $\Delta\xi$, the *conversion* of reaction $\mathcal{R}$ or the conversion *according to* reaction $\mathcal{R}$. Every conversion leads to changes in the amounts of the participating substances, which are proportional to their conversion numbers:

$$\Delta n_i = \nu_i \cdot \Delta\xi. \tag{1.17}$$

Conceptually, the *extent* and *conversion* of a reaction are related to each other in the same way that *location* and *displacement* of a mass point are.

Equation (1.17) can be expanded easily so that it remains useful when several reactions $\mathcal{R}$, $\mathcal{R}'$, $\mathcal{R}''$, ... run simultaneously where each one is described by its own quantity ξ, ξ', ξ'', etc.:

$$\Delta n_i \;=\; v_i \cdot \Delta\xi + v_i{}' \cdot \Delta\xi' + v_i'' \cdot \Delta\xi'' + \ldots \quad \text{for all substances } i. \qquad (1.18)$$

Not only can chemical reactions be described in this way, but also a simple exchange of a substance B with the surroundings can as well,

$$\text{B}|\text{outside} \rightarrow \text{B}|\text{inside} \quad \text{or} \quad 0 \rightarrow -1\ \text{B}|\text{outside} + 1\ \text{B}|\text{inside},$$

so that we can apply equations of the type (1.18) very generally for calculating changes in amounts of substance.

Chapter 2
Energy

Energy is a quantity that not only plays a dominant role in the most diverse areas of the sciences, technology, and economy, but is omnipresent in the everyday world around us. For example, we pay for it with every bill for electricity, gas, and heating oil that arrives at our homes. But we are also confronted more and more with questions about how we can save energy in order to cover our current and future demands. At the beginning of the chapter, the conventional indirect way of defining energy is briefly presented. A much simpler way to introduce this quantity is characterizing it by its typical and easily observable properties using everyday experiences. This phenomenological description may be supported by a direct measuring procedure, a method normally used for the quantification of basic concepts such as length, time, or mass. Subsequently, the *law of conservation of energy* and different manifestations of energy like that in a stretched spring, a body in motion, etc., are discussed. In this context, important quantities such as *pressure* and *momentum* are introduced via the concept of energy.

2.1 Introducing Energy Indirectly

Energy is a quantity that not only plays a dominant role in the most diverse areas of the sciences, technology, and economy, but is omnipresent in the everyday world around us. We buy it in large amounts and pay for it with every bill for electricity, gas, and heating oil that arrives at our homes. There is also information on every food package about the energy content of the food inside. We are confronted more and more with questions about how we can save energy in order to cover our current and future demands.

Ironically, this everyday quantity is defined and explained in a very complicated way. To start with, it is dealt with as a special concept within the subject of mechanics and then gradually expanded and generalized. The quantity called energy is almost always introduced *indirectly* through mechanical work. The relation

© Springer International Publishing Switzerland 2016
G. Job, R. Rüffler, *Physical Chemistry from a Different Angle*,
DOI 10.1007/978-3-319-15666-8_2

Fig. 2.1 Interaction of force and distance when doing work (here, as seen by the person involved).

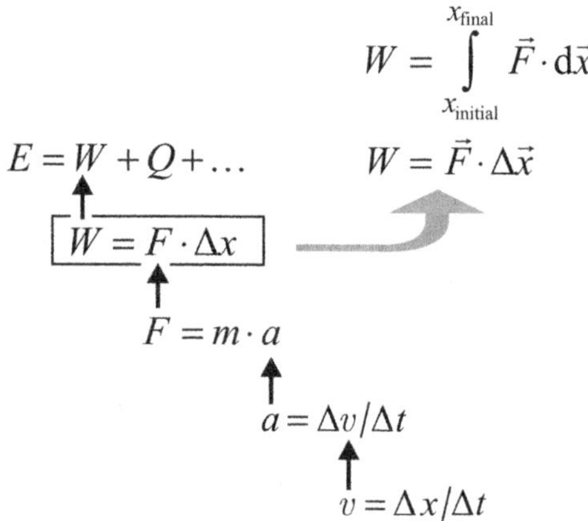

$$W = \int_{x_{\text{initial}}}^{x_{\text{final}}} \vec{F} \cdot d\vec{x}$$

$$E = W + Q + \dots \qquad W = \vec{F} \cdot \Delta\vec{x}$$

$$\boxed{W = F \cdot \Delta x}$$

$$F = m \cdot a$$

$$a = \Delta v / \Delta t$$

$$v = \Delta x / \Delta t$$

Fig. 2.2 The usual indirect way over many steps to energy (a simplified image here). The formulas above on the right clarify the framed equation at step 4. *a* acceleration, *E* energy, *F* force, *m* mass, *Q* heat, *t* time, *v* velocity, *W* work, *x* position.

"work = force times distance" is our access route. It tells us that a lot of work must be *done* if, for instance, one wishes to cover a distance using great force against a strong resistance. Examples of this might be riding a bicycle against a strong wind or towing a car (Fig. 2.1) across a sandy surface. The figure shows the interaction of force and distance from the perspective of the person performing the work. The force being applied by him in the x-direction is positive for the example on the left: $F_x = F$; on the right, it is negative: $F_x = -F$. Correspondingly, the work $W = F_x \cdot \Delta x$ done by the person in the figure is positive on the left and negative on the right. Seen from the car being pulled, all the forces and work have the opposite algebraic signs.

The unit for work is the *Joule* (J), named for the British beer brewer and private scholar James Prescott Joule, who lived in Manchester, England, in the nineteenth century. One Joule corresponds to the product of the units of force (Newton, N) and length, N m = kg m^2 s^{-2}.

The path to the concept of work leads through many steps (Fig. 2.2). The quantity called force is also defined *indirectly* (force = mass times acceleration). The same holds for acceleration (= change of velocity divided by time interval) as well as velocity (= distance covered divided by time needed to cover it). Mechanical work is only one form of energy input. There are other forms as well, the most important of which is *heat*. The name *energy* is an umbrella term for these various forms. This new quantity has its own symbol, mostly E, with the unit Joule. Furthermore, each of the numerous variants has its own name and symbol. Along with work W and heat Q, there are internal energy U, enthalpy H, Gibbs (free) energy G, exergy B, etc.

In order to avoid this convoluted derivation, we will introduce energy directly by metricization. As a first step, we will characterize the concept using typical and easily observable properties. We will see that we can do without the large number of energy terms because a single one is basically enough to do what is necessary to describe all processes we deal with in physical chemistry.

2.2 Direct Metricization of Energy

Basic Idea Almost everything that we do requires *effort* and *strenuousness*. We notice this especially when doing something that is so strenuous, it makes us sweat and gasp. So let us imagine all devices and things we use to be so large and heavy that we feel the consequences of dealing with them. We will take a look at a few *strenuous* activities that we accomplish *without* tools or *with* the help of levers, ropes, pulleys, winches, etc. (Fig. 2.3). All the activities in the figure belong to the subject of mechanics.

We can also include thermal processes (such as a "heat pump"), electric processes (an "electrostatic machine," for instance), or chemical processes (such as a water electrolysis apparatus) in this (Fig. 2.4). However, because these kinds of processes are less familiar, we will come back to them later on in more detail. For now, we will limit ourselves to mechanical processes.

It is noteworthy that the effort we have expended for these activities does not just disappear, but can be used to accomplish other strenuous activities. For example,

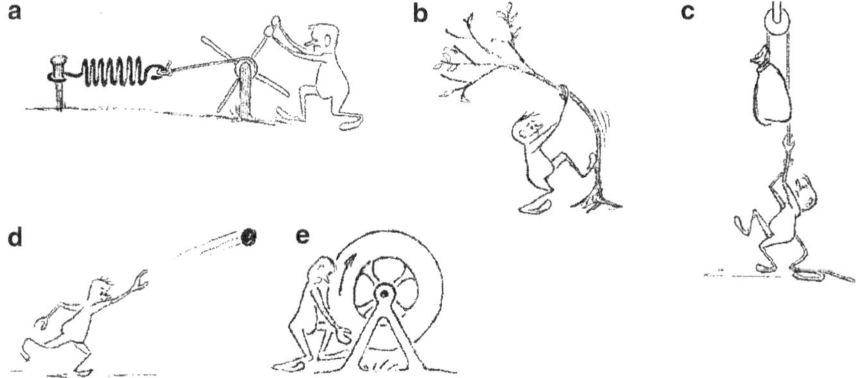

Fig. 2.3 Strenuous activities: (**a**) stretching, (**b**) bending, (**c**) lifting, (**d**) throwing, (**e**) starting a wheel turning.

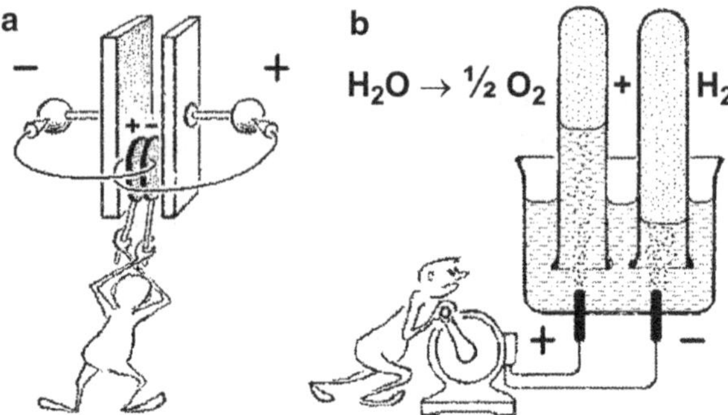

Fig. 2.4 (**a**) A primitive "electrostatic machine": progressive charging of a capacitor by separating charge in the already existing field and transporting the charge to the corresponding opposite plate, (**b**) A "water electrolysis apparatus": forcing a spontaneous reaction in the opposite direction is strenuous, here using the example of decomposing water into its elements.

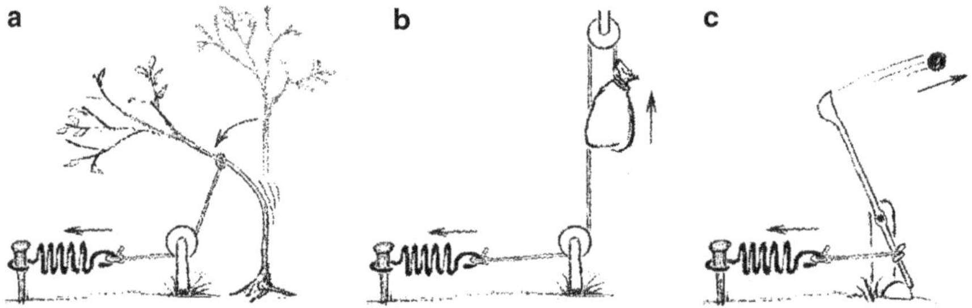

Fig. 2.5 Utilizing the effort needed for stretching the spring in order to (**a**) bend a tree, (**b**) lift a sack, (**c**) hurl a stone.

we can directly or indirectly use a stretched spring that is trying to contract, in order to bend a tree, to lift a sack, or to hurl a stone, etc. (Fig. 2.5).

Conversely, we can stretch a spring by using a bent tree, a lifted sack, or even a hurling stone (if it can be caught correctly). Any combination can be accomplished, if the right tools are put to use. What is important here is that the effort put into something can be used to perform other activities.

It appears that the effort expended to change something is *stored* within the changed objects. We can imagine it to be contained in the stretched spring, the bent tree, the lifted sack, the flying stone, etc. It can be taken out again by reversing the change and then reused to change other things.

Lost Effort All of us experience activities where all effort apparently seems to disappear (Fig. 2.6). We rub our hands together and they get warm, but we cannot lift a sack with this heat. It is hard work to pull a heavy cart across a sandy surface. Not only do we start to sweat, but the sand also becomes warm even if we do not notice it. Even when we do not accomplish anything, for example, when we try to hold back an attacking dog or try to pull a firmly rooted bush out of the ground, we get hot.

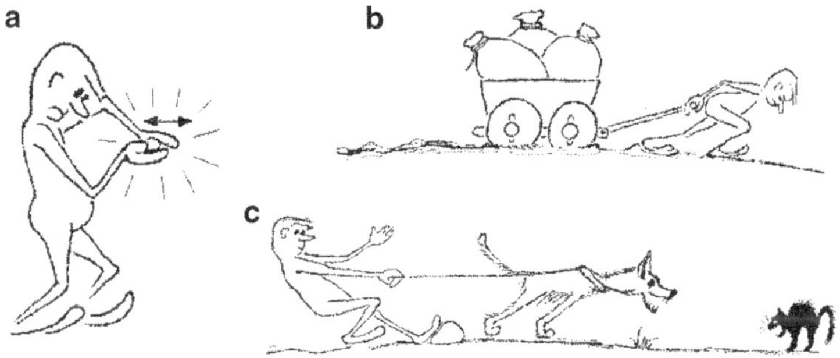

Fig. 2.6 Activities where all effort seems lost: (**a**) rubbing, (**b**) pulling, (**c**) restraining.

In such cases, it seems there is nothing left of the effort that was expended that can then be put to use. At least under the given circumstances, this appears to be the case.

Although the effort needed for this does not get reused, it does not simply disappear without a trace. *We* sweat, and not only do we get hot, but also the things around us: the rubbed hands, the sand under the wheels, the squeaking wheel bearings. The more effort wasted, the more pronounced the warming. This is a significant trace left by the lost effort. However, even this trace gradually fades so that eventually it seems that all the effort disappears into nothing.

We must take into account that with everything we do, a part of our effort does not accomplish anything but will get lost in some unintended secondary activity. Friction almost always hinders our activities. It takes a lot of effort to overcome friction and it is something we seek to avoid, if possible. To give another example, we ourselves are affected by the activity of our muscles where—just like with rubbing—heat is produced. This is an unwanted but unavoidable secondary activity.

Measuring Effort   The question now arises of whether it is possible (unencumbered by what we feel) to determine how much effort a certain activity entails— stretching a spring, lifting a sack, or charging a capacitor? We expect that an objective measure of the effort expended for the same activity will always have the same value, no matter who does the work, where it is done, and when. If something is done twenty times in exactly the same way, it should mean, all in all, twenty times the amount of effort.

The unit used (as for example, the unit of length) can in principle be chosen arbitrarily. One could for instance select a coil spring. The position of the end of the spring in its relaxed state receives a value of 0, and at an arbitrary but defined stretched state, a value of 1 (Fig. 2.7). In this manner, an "amount" of effort is defined that can serve as our private unit. Naturally, we can also choose the unit to be equivalent to the SI unit. Such springs can be made so that the initial and final states are easily recognizable, comparable to a spring balance whose scale is limited to values of 0 and 1 (compare Fig. 2.7a, lower spring).

We can set the flywheel in motion using the stretched "unit-spring." The spring then returns to its rest position and the flywheel rotates. The effort stored in the

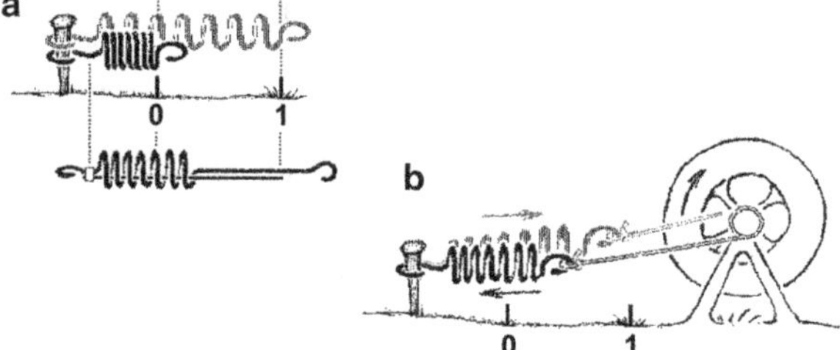

Fig. 2.7 (**a**) Showing the "unit of effort" by a spring stretched between the marks 0 and 1, (**b**) Transferring the effort to a flywheel (front, *black*) and back to the spring again (back, *gray*).

spring is now in the turning wheel and can be given back to the spring by using the spin of the wheel to stretch the spring (Fig. 2.7b). This interplay can in principle be repeated indefinitely and in various ways. Unfortunately, air and bearing friction gradually consume the stored effort, or in other words, friction continuously takes some of it for other purposes.

If we were able to prevent such unwanted losses, it would be easy to measure the effort necessary for our activity. We will therefore assume that by using appropriate measures, losses can be avoided. Ball bearings help against axle friction, a vacuum helps against air friction, thicker wires help against line resistance, and the friction of wheels on a surface can be compensated for by a harder surface, or better yet, air cushions. Later on we will see how we can deal with things when such compensating measures are unavailable or insufficient.

Another type of error arises when a part of the effort gets "caught" in the storage. This is the case where the tree is being bent and the sack is being lifted by springs (Fig. 2.5a, b). The process comes to a halt when the pull of the spring upon the rope reduces to the point where it can no longer overcome the counter-pull of the tree or the sack. The rope and deflection pulleys alone are not enough to make the best use of the effort stored in a spring during stretching. To do so, it would be necessary to make use of a somewhat more complicated construction which we will not do at this time. We will just assume that it is, in principle, possible to utilize the stored effort entirely.

Measuring effort simply means counting how many unit portions it can be divided into. One counts the number of unit-springs that can be stretched or one counts the number of already stretched springs needed to obtain a desired change (compare Fig. 2.8).

Energy We call the quantity introduced by the process described above, *energy*. Of course there are more accurate and more easily reproducible means to represent the energy unit (or an arbitrary multiple of it) than our spring. As an example, consider the energy of a photon emitted by a hydrogen atom when its electron drops from the 2p state to the 1s state. This sounds a little bit strange at the beginning, too. But, also the meter, the fundamental unit of length in the SI system, is meanwhile no longer defined by the international prototype meter bar composed of an alloy of platinum and iridium. In 1960, the meter was fixed as equal to 1,650,763.73 times the wavelength in a vacuum of the radiation corresponding to the transition between the levels $2p^{10}$ and $5d^5$ of the krypton-86 atom. In order to enable traceability

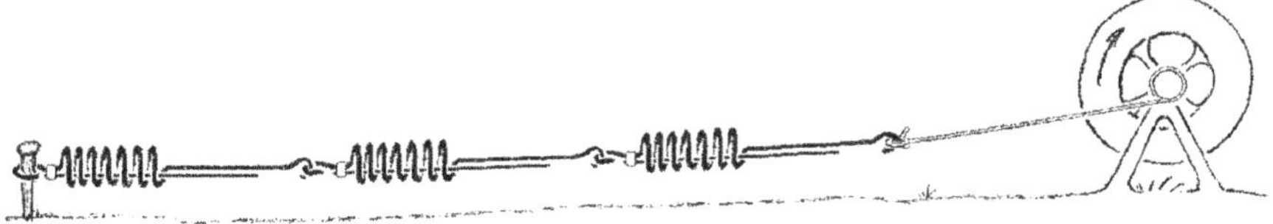

Fig. 2.8 Creating a multiple of the "unit of effort" by connecting unit-springs in series.

between this new definition and the old prototype, the numeric value (1,650,763.73) was chosen (according to the measurement accuracy of that time). To further reduce uncertainty, the length of the meter was currently fixed indirectly in terms of the second and the speed of light. However, precision of measurement is not what is actually important when first learning about a quantity, here the quantity called energy.

Depending upon the purpose, different symbols such as $E, W, U, Q, H, G, \ldots$ are in use for the quantity called energy. We will use only the symbol W because E will be used for electric field strength. Moreover, there is no good reason to give different symbols and names to stored and transported energy.

The quantity W has been introduced by *direct metricization* of the everyday concept of *effort*. Although we have relied on our senses, the quantity W is ultimately independent of subjective feelings. This fact is essential for dealing with it objectively because the same activity can, for example, seem more strenuous to one person than to another or more exhausting to a tired person than to a well-rested one. The quantity called energy defines and quantifies what we call *effort* in everyday life, a term relating to an activity. On the other hand, it also denotes what is *stored* in an object that is deformed, moved, lifted, charged ... and that can be retrieved when needed. That means energy also quantifies the ability to do something, what in everyday language could be circumscribed vaguely by "power to do something."

We should, though, be wary of taking this comparison of energy (in physics) and "power to do something" (in everyday life) too literally. In the economy, the more money you have, the more you can accomplish, but coins and notes do not have any intrinsic power. Almost everything we do involves some kind of turnover of energy. Energy might be considered the *price* paid for an activity or, conversely, what can be gained. If we know the price of an activity, we also know whether or not it is possible to pay for it.

It is a matter of preference whether one wishes to describe processes as *dynamic*, meaning as a result of forces working with and against each other or simply as a form of *accounting* by considering credits and debits on a balance sheet. The former takes into account everyday images and what we sense and feel. The latter makes use of our experiences dealing with cash and non-cash money values.

2.3 Energy Conservation

One of the most important insights of nineteenth century physics was that energy—or "force" or "power," as it was called then—never gets lost, i.e., cannot disappear into nothingness. A hundred years before that people knew that no matter how ingenious you are, energy cannot be created from nothing. In Sect. 2.2, we saw several examples of effort seeming to get lost. This loss was always accompanied by the evolution of heat. In the so-called "Joule's apparatus" (Fig. 2.9), a falling weight was used to spin a paddle wheel. The cold water in the insulated vessel (a kind of calorimeter) was warmed up by the rotating paddles and the corresponding temperature was measured.

An ice calorimeter (Fig. 3.23b and Experiment 3.5) can also be used to determine how much heat has evolved in a process by showing us how much ice it melts. The amount of melt water is proportional to the amount of energy expended, independent of where the energy comes from or how it gets into the ice. Naturally, this is assuming that nothing comes into it from other sources or drains off to some sink—through a leak in the heat insulation, for example.

It was therefore concluded already at that time that a certain amount of energy is needed to heat a body, whether or not this happens intentionally. If the amount of energy used for this is included, one finds that the entire "stock" of energy remains unchanged. Energy can be moved from one storage to another, but the amount remains the same. This finding is called the *"law of conservation of energy"* or short *energy principle*. One result of the law of conservation of energy is that any energy expended must be independent of the path or the tools used. Otherwise, in contradiction to this law, it would be possible to create energy from nothing (or let energy disappear into nothing) by delivering it by one path and releasing it via another.

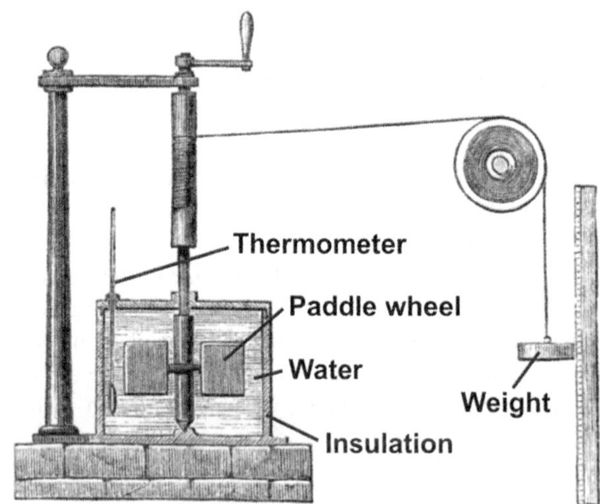

Fig. 2.9 Joule's apparatus for demonstrating the equivalency of energy and heat (from: Abbott (1869) The new theory of heat. Harper's New Monthly Magazine 39:322–329).

Until now, we have assumed that the energy consumed producing a change can be retrieved if the process involved is reversed. However, we run into a problem with processes associated with the evolution of heat: they cannot be reversed, or at least not entirely so. The energy has not actually disappeared, but is somehow not quite accessible any longer. This situation caused a lot of headaches in the nineteenth century and still does today. We will look more closely at this subject in Chap. 3.

Before we do that, though, we will discuss some simple cases showing how to find values for energy. Many quantities derived from energy are easier to measure than the energy itself, so we usually calculate energy indirectly through these quantities.

2.4   Energy of a Stretched Spring

A stretched spring has the tendency to contract, and the more strongly stretched it is, meaning the more the length l exceeds the value l_0 in its relaxed state, the more strongly it tends to contract. Depending upon how much the spring is already stretched, it becomes increasingly strenuous to continue stretching the spring by the small amount Δl (Fig. 2.10). In other words, the energy ΔW needed for this increases by l proportionally to $l - l_0$ (at least within certain limits) as long as the changes ΔW and Δl are small enough. This condition can be expressed by replacing the differences with differentials:

$$\frac{\Delta W}{\Delta l} = D \cdot (l - l_0) \ \text{ or } \text{ rather } \ \frac{dW}{dl} = D \cdot (l - l_0). \tag{2.1}$$

The graphic shows that, when greatly magnified, the curve around the point l seems to have become straight. We then calculate the slope of this extremely small section of the curve (see the magnified section in Fig. 2.10). There is a more detailed description of this in Sect. A.1.2 in the Appendix. When we now plot the values gained in this manner as a function of the corresponding l values, we obtain a linear function as we would expect from Eq. (2.1).

The proportionality factor D quantifies the characteristic called *spring stiffness*. When the factor D is great, a spring is *hard* or *stiff*; when it is small, the spring is *soft*. dW/dl is a measure of the "force" with which the spring resists stretching. We take W and l as measurable quantities so we can express this force—as usual symbolized by F—as follows:

$$F = \frac{dW}{dl} \ \text{ or in more detail } \ F(l) = \frac{dW(l)}{dl}. \tag{2.2}$$

The corresponding SI unit is $J\,m^{-1} = N$ (Joule/Meter $=$ Newton). If we insert F into Eq. (2.1), we obtain the usual form of a familiar law,

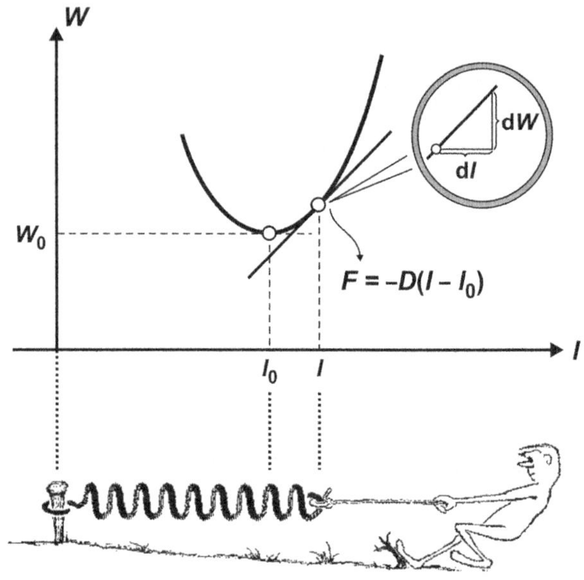

Fig. 2.10 The energy $W(l)$ of a spring as a function of its length l. The graph is close to a parabola for the range around the rest position l_0. The force by which the spring resists stretching at position l corresponds to the slope dW/dl of the graph at this point. This is illustrated by the inset "magnifying glass".

$$F(l) = D \cdot (l - l_0) \quad \text{Hooke's law.} \tag{2.3}$$

$F(l)$ then describes the slope of the graphical representation of the function $W(l)$ at point l. In order to find $W(l)$, we need only to find the antiderivative of $F(l)$. In this case, it is the function whose derivative with respect to l results in $D \cdot (l - l_0)$. Antiderivatives are gone into in more detail in Sect. A.1.3 in the Appendix. Now we see that:

$$W(l) = \tfrac{1}{2}D \cdot (l - l_0)^2 + W_0. \tag{2.4}$$

To start with we have assumed that D is constant and that $F(l)$ is a linear function, meaning that the graph (the *characteristic curve*) of the spring is a straight line. However, this does not need to be the case. In Eq. (2.1) a different derivative dW/dl would appear on the right, which can be measured like the one before. Now the spring stiffness D that still corresponds to the slope of graph $F(l)$ is itself dependent upon l. Although it can be mathematically more challenging to find the antiderivative of $F(l)$, the method remains the same.

Let us now imagine two different springs stretched and connected in series (Fig. 2.11). The one on the right can only contract as it causes the one on the left to stretch. The process can take place as long as the spring supplies more energy $-\Delta W'$ when shortened by the small length $-\Delta l'$ than the other spring consumes (ΔW) when expanding by the same length $\Delta l = -\Delta l'$. If the spring supplies less than this, the process reverses. The process will come to a standstill or to an *equilibrium of forces* when the energy supplied by a small shift of one spring is compensated by the energy consumption of the other:

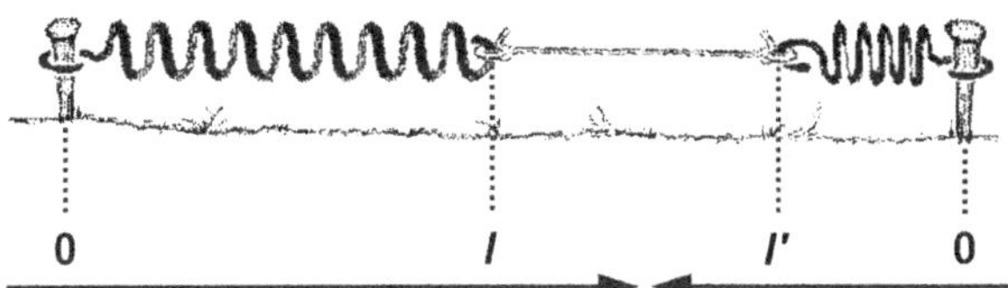

Fig. 2.11 Equilibrium between two stretched springs connected in series.

$$\frac{\Delta W}{\Delta l} = \frac{\Delta W'}{\Delta l'} \quad \text{or better} \quad \frac{dW}{dl} = \frac{dW'}{dl'} \quad \text{this means} \quad F = F'.$$

Springs can therefore be easily used as force meters. In order to find the force acting upon a spring by a rope or a rod, for example, it would be sufficient to use a scale that shows the stretching $l - l_0$. If the characteristic curve of the spring is linear, the scale will be equidistant and can, for the sake of simplicity, be labeled with the unit Newton (N). We are only mentioning this well-known method because other quantities can be measured in a very similar way, among them pressure, temperature, and even the chemical potential.

2.5 Pressure

The same paradigm used for force F can be used to introduce pressure p. Pressing water into a pressure vessel costs energy (Fig. 2.12). The container resists a change ΔV of the volume V of the water in it which manifests itself in some kind of counter pressure. The expenditure of energy ΔW related to the same increase of volume ΔV can be considered the measure for pressure p:

$$\frac{\Delta W}{\Delta V} = p \quad \text{or more exactly} \quad \frac{dW}{dV} = p. \tag{2.5}$$

As we know, when two such vessels with pressures p and p' and water volumes V and V' are connected by a hose, the pressures will equalize when the water flows from the vessel with higher pressure into the one at lower pressure. Energy considerations can be used to explain this. Energy is transported along with the water flowing in and out. This will take place in the mentioned direction as long as more energy is released on the side with water flowing out at higher pressure than is consumed on the other side; otherwise it will reverse. It will only come to a standstill when supply and consumption reach equilibrium, $dW + dW' = 0$. If we divide $dW = -dW'$ by $dV = -dV'$, this leads to

Fig. 2.12 Pumping water into a vessel, in this case a rubber bladder, against the pressure within it. Containers of this kind, mostly of steel with a rubber membrane inside, serve as equalizing vessels in heating systems.

$$\frac{dW}{dV} = \frac{dW'}{dV'}, \quad \text{i.e.} \ \ p = p',$$

which means that we have *pressure equilibrium*. The rather simple case discussed here belongs to the subject of hydraulics where water is considered *incompressible*. Volume takes the role of a substance-like quantity, as a substitute for the amount of water being discussed.

We encounter the quantity called pressure also in another more complex relationship where *compressibility* is concerned. It costs energy to compress an elastic body. The more an object is pressed from all sides, the more strongly volume V decreases. The effort dW needed to cause a small change of volume $-dV$ increases according to how compressed the body is to begin with. More precisely: the quotient $dW/(-dV)$ increases with a decrease of V, at first linearly (proportionally to $V - V_0$) and then more and more steeply. We might say that the body increasingly resists compression, which is expressed by the growing counter pressure p felt when compressing the body. Similar to the case of hydraulics, the quotient $p = dW/(-dV)$ lends itself well as a measure of this kind of pressure:

$$p = -\frac{dW}{dV}. \tag{2.6}$$

The expended energy dW can be retrieved in the course of expansion. One might consider the energy to be contained in the body and that it can be recalled from there when needed. However, the change of energy is not necessarily a measure of the part needed for changing the volume—the only part we are interested in at this point. In order to get this part, it would be necessary to block all the other pathways the energy might use to flow in or out. If an energy exchange similar to that resulting from changes of V is possible for other quantities $q, r, \ldots$, then $W(V, q, r, \ldots)$, and the latter must be kept constant:

$$p = -\left(\frac{\partial W}{\partial V}\right)_{q,r\ldots}. \tag{2.7}$$

We were introduced to this method with the indirect metricization of weight in Sect. 1.3. However, we are still missing a crucial quantity. This missing quantity, which we call entropy, will be the subject of the next chapter.

Although we have chosen a totally different way to find the pressure p, it is nevertheless identical to the quantity usually introduced as the force per area, $p = F/A$. The SI unit of the pressure is $\mathrm{J\,m^{-3}} = \mathrm{N\,m^{-2}} = \mathrm{Pa}$ (Joule/Meter^3 = Newton/Meter^2 = Pascal). A non-SI unit still widely used is the bar, where 1 bar $= 10^5$ Pa. The isotropic pressure p is a quantity that only describes one of the possible stress states of bodies. It describes a very simple but especially important one, which is just about the only one we will be dealing with.

2.6 Energy of a Body in Motion

Energy is needed to accelerate a body, for instance a car or a projectile, and the faster the body is already moving, the more energy (relative to Δv) will be necessary to accelerate it. The effort ΔW is proportional to the velocity v (= distance covered Δx/time needed Δt):

$$\frac{\Delta W}{\Delta v} = m \cdot v \quad \text{or better} \quad \frac{\mathrm{d}W}{\mathrm{d}v} = m \cdot v. \tag{2.8}$$

The proportionality factor m quantifies a characteristic called *inertia, inertial mass*, or simply the *mass* of the body. We can consider m to be invariable in all the cases that will interest us. It is easy to see how W depends upon v. We only have to find the antiderivative for the derivative $\mathrm{d}W/\mathrm{d}v$ in Eq. (2.8):

$$W(v) = \tfrac{1}{2}mv^2 + W_0. \tag{2.9}$$

W_0 is the energy contained by a body at rest, i.e., at $v = 0$. The energy in a moving body $W(v) - W_0$ is called *kinetic energy*. If the body moves uniformly (when v is constant), $W(v)$ also remains constant. When this is not the case, then v and therefore indirectly also W is dependent upon position x: $W(v(x))$. By applying the chain rule (compare Sect. A.1.2 in the Appendix) we easily find the force F with which a body resists change of position:

$$F = \frac{\mathrm{d}W(v(x))}{\mathrm{d}x} = \frac{\mathrm{d}W(v)}{\mathrm{d}v} \cdot \frac{\mathrm{d}v(x)}{\mathrm{d}x} = mv \cdot \frac{a}{v} \quad \text{or} \quad F = m \cdot a, \tag{2.10}$$

where $a = \mathrm{d}v/\mathrm{d}t$ represents acceleration. One can prove that $\mathrm{d}v(x)/\mathrm{d}x = a/v$ by taking the derivative of $v(x(t))$ with respect to t and then solving the equation obtained for $\mathrm{d}v(x)/\mathrm{d}x$:

$$a = \frac{\mathrm{d}v(x(t))}{\mathrm{d}t} = \frac{\mathrm{d}v(x)}{\mathrm{d}x} \cdot \frac{\mathrm{d}x(t)}{\mathrm{d}t} = \frac{\mathrm{d}v(x)}{\mathrm{d}x} \cdot v.$$

There is a shorter way to the same result by using the rules for differentials [expanding and inverting a derivative, Sect. 9.4 (transformation of differential quotients)]:

$$F = \frac{\mathrm{d}W}{\mathrm{d}x} = \frac{\mathrm{d}W}{\mathrm{d}v} \cdot \frac{\mathrm{d}v}{\mathrm{d}t} \cdot \frac{\mathrm{d}t}{\mathrm{d}x} = \frac{\mathrm{d}W}{\mathrm{d}v} \cdot \frac{\mathrm{d}v}{\mathrm{d}t} \bigg/ \frac{\mathrm{d}x}{\mathrm{d}t} = mv \cdot a/v = m \cdot a.$$

The equation $F = m \cdot a$ is usually used to define the force F, which is then used to introduce the concept of work and, as a generalization, energy (compare Sect. 2.1).

2.7   Momentum

Equation (2.8) can be read another way when *momentum* þ is introduced in place of velocity. Instead of the usual symbol p that is already being used for pressure, we will use the symbol þ (Thorn) which is similar to p and comes from the Icelandic language. Momentum plays a decisive role in modern physics, in quantum mechanics and in relativity, for example. Therefore, this would be a good moment to familiarize ourselves with this quantity. This concept is essential to describing interactions between moving bodies such as collisions in kinetic gas theory (Chap. 10) or in the kinetics of elementary chemical reactions.

Momentum is a substance-like quantity. The total momentum of an assembly of moving bodies or of parts of a body is the sum of the momenta of all the parts. It can be transferred from one moving body to another. The total momentum is conserved in such processes. If the momentum of a part has decreased, the momentum of another part must have increased irrespective of how the transfer took place. Knowing this can save us a lot of detailed work. In everyday life, one has to learn to recognize conservation of momentum. If we push a car and it is gaining momentum, or if it is losing momentum when coasting, it is not quite clear where the momentum comes from or goes to (Fig. 2.13). It turns out that the momentum comes from or goes into the Earth. Our planet is so big that we do not notice if it loses or gains a little momentum—just as we do not notice a change in the ocean if we take a bucket of water from it.

Momentum is a vector quantity, which makes it somewhat difficult to deal with. However, it is not more complicated than other vector quantities such as velocity, acceleration, force, etc. In fact, it is less difficult because of its substance-like character. Firstly, it is enough to simply observe motion in one direction, for

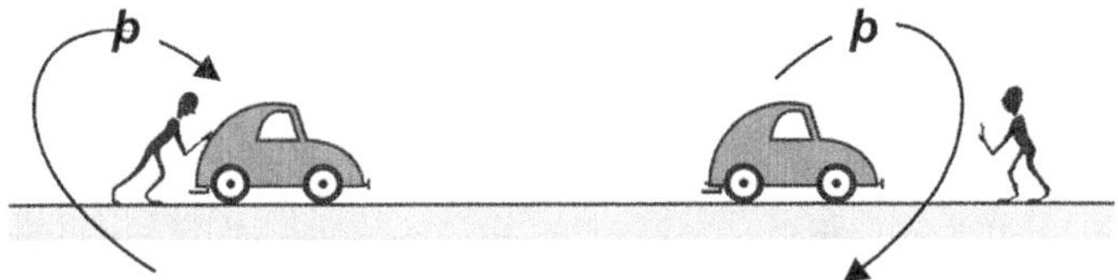

Fig. 2.13 Momentum p when a car is pushed, and when coasting.

example, along the x-axis. The momentum of a body moving in the direction of increasing x values is counted as positive. Motion in the opposite direction is considered negative. This way of seeing things must be learned, though, because we usually speak in absolute values of these quantities. Who would ever say that a vehicle moving down a street is doing so at negative speed?

The momentum contained in a moving body increases with its mass m and its velocity v. In fact, it is proportional to both, $p \sim mv$. The proportionality factor is set equal to 1:

$$p = mv \quad \text{SI unit: } \mathrm{kg\,m\,s^{-1}} = \mathrm{Ns}. \tag{2.11}$$

If $v = p/m$ is inserted in $W(v)$ and the derivative of the function $W(v(p))$ is taken with respect to p, one obtains

$$\frac{\mathrm{d}W(v(p))}{\mathrm{d}p} = \frac{\mathrm{d}W(v)}{\mathrm{d}v} \cdot \frac{\mathrm{d}v(p)}{\mathrm{d}p} = mv \cdot \frac{1}{m} \quad \text{or} \quad \frac{\mathrm{d}W}{\mathrm{d}p} = v. \tag{2.12}$$

This equation agrees with Eq. (2.8), except that the factor m in the denominator is moved from the right side to the left and combined with $\mathrm{d}v$ to make $\mathrm{d}p$. It can also be explained using a similar pattern to the one we used for discussing the force of a stretched spring or pressure in hydraulics: The faster a body is already moving, the more it opposes increasing its momentum. The energy $\mathrm{d}W$ needed for this, relative to the same amount $\mathrm{d}p$, grows proportionally to the velocity v. In this case, v appears in a role similar to that of force or pressure.

For more than a hundred years, quantities that appear in this role have been called *intensive factors*, *intensive quantities*, or simply *intensive*. Unfortunately, this description does not agree completely with the definition in Sect. 1.6. In order to avoid misunderstandings, we have no choice but to look for a new name. The German physicist and physician Hermann von Helmholtz came up with one that would be helpful to us. Using Joseph Louis De Lagrange's concept of "forces" in the field of mechanics, he generalized it. We refer to this and call the quantities "force-like."

Each one of these quantities has a counterpart. These are called *extensive factors*, *extensive quantities*, or simply, *extensive* and appear in the form of differentials. x belongs to F, V or $-V$ belongs to p (as the case may be), and p belongs to v, just to name the ones we have already discussed. Each pair describes a path over which energy can be exchanged:

$$dW = F dx, \quad dW = p dV, \quad dW = v dp, \quad \text{etc.}$$
$$\text{Spring} \qquad \text{Hydraulics} \quad \text{Motion}$$

This term does not agree with the earlier definition either, so a new name must be found here as well. Using the concept of position coordinates to give position and orientation of one or more bodies in a space, Helmholtz expanded this concept analogously to quantities outside of mechanics (electric, chemical, etc.). These are the quantities called "extensive factors" above. For a rough characterization of the role of these quantities, the term *position-like* would fit nicely to *force-like*, as a counterpart.

Let us return for a moment to the equation $F = m \cdot a$. If we take the derivative of the expression for momentum $p = mv$ with respect to time t, and because of $dv/dt = a$, we obtain a relation that the famous English physician and mathematician Sir Isaac Newton already used to start his description of classical mechanics in the seventeenth century:

$$\frac{dp}{dt} = ma = F.$$

Because p is a conserved quantity, its amount can only increase in the body when it decreases somewhere else. In other words, this quantity must flow in from there. Therefore, F describes the momentum flowing in from the surroundings. This idea may be a bit unusual, but it can be very useful.

2.8 Energy of a Raised Body

We will take another look at the body being hoisted upward by a rope and pulley (Fig. 1.5). In the following, we will ignore buoyancy by imagining the surroundings to be void of air. When we release the body, it falls, as we know, at constant acceleration $a = -g$ (g *gravitational acceleration*), independent of its size, weight, or composition. After a time t, it will have reached a velocity of $v = at = -gt$ and will have fallen the distance $h_0 - h = \frac{1}{2}gt^2$, where h is the height above ground at time t and h_0 is the initial height. The energy necessary for the body with a mass m to accelerate from $0 \rightarrow v$,

$$W = \frac{1}{2} mv^2 = \frac{1}{2} m(-gt)^2 = mg(h_0 - h), \tag{2.13}$$

comes from the gravitational field of the Earth. However, the energy W, which is released by falling and is used here to accelerate the body, is usually attributed to the raised body itself. The energy stored in a raised body (or in the gravitational field, respectively) is called *potential energy*, W_{pot}; the one in the moving body is called *kinetic energy*, W_{kin}, as mentioned. During free fall, energy is transferred from one storage to the other. According to the law of conservation of energy, the sum of both of these contributions remains constant as long as no energy is diverted (such as during impact or falling through air):

$$W_{\mathrm{kin}} + W_{\mathrm{pot}} = \tfrac{1}{2}mv^2 + mgh = \mathrm{const.} \tag{2.14}$$

The term *potential* energy is often transferred to similar cases. For example, the potential energy W_{pot} of a charged body increases by ΔW when it is moved in a static electric field against the field forces by expending the energy ΔW. The stored energy in a stretched spring at rest (Sect. 2.4) is called potential to distinguish it, if necessary, from the contributions from movements of the spring or other parts.

Chapter 3
Entropy and Temperature

In phenomenological description (comparable to a kind of "wanted poster"), the entropy appears as a kind of "stuff" which is distributed in space, can be stored or transferred, collected or distributed, soaked up or squeezed out, concentrated or dispersed. It is involved in all thermal effects and can be considered their actual cause. Without it, there would be no hot and no cold. It can be easily generated, if the required energy is available, but it cannot be destroyed. Actually, entropy can be easily recognized by these effects. This direct understanding of the quantity S is deepened by a simplified molecular kinetic interpretation.

In addition to the *first law* of thermodynamics, a version of the law of conversion of energy (Sect. 2.3), the *second law* will be formulated in the following without recourse to energy and temperature. On the contrary, the absolute temperature can be introduced via energy and entropy. The *third law* is also easily accessible, and heat engines and heat pumps are analyzed after this introduction, without discussing process cycles, gas laws, or energy conversion processes. In closing, the entropy generation as a consequence of entropy conduction will be discussed.

3.1   Introduction

Misjudged and Avoided The central concepts of thermodynamics are *entropy S* and *temperature T*. While everyone is familiar with temperature, entropy is considered as especially difficult, in a way the "black sheep" among physicochemical quantities. School books avoided it totally in the past, introductory physics books often only "mention" it, and even specialists in the field like to avoid it.

© Springer International Publishing Switzerland 2016

G. Job, R. Rüffler, *Physical Chemistry from a Different Angle*,

DOI 10.1007/978-3-319-15666-8_3

But why is the subject of entropy avoided when it is actually something rather simple? It is just what is considered "heat" in everyday life (Fig. 3.1)!

Unfortunately, the name "heat" was given to another quantity (compare Chap. 24) which robbed S of its natural meaning, making S an abstract concept that is difficult to understand and deal with. Therefore, entropy could only be introduced abstractly, i.e., *indirectly* by integrating a quotient formed from energy and temperature, making it difficult to deal with. Furthermore, it is customary to interpret entropy atomistically as a measure of the probability of a certain state of a system composed of numerous particles. In chemistry, we must be able to infer our actions in the laboratory from atomistic concepts. In other words, we must be able to transfer the insight gained on one level to a different one as *directly* as possible. In the following we will demonstrate how to accomplish this.

Macroscopic and Microscopic View To illustrate this, we will characterize entropy at first by use of some of its typical and easily observable properties— similarly as already the energy. In the same way, a wanted person would be described by a list of easily distinguishable ("phenomenological") characteristics like height, hair color, eye color, etc. This group of characteristics is basically what makes up the person and his or her name is just an identification code for this group of characteristics. A "wanted poster" is an example for such a group of characteristics in strongly abbreviated form. Our intent is to design such a "wanted poster" for entropy that allows it to be defined as a measurable physical quantity. After that has been done, we will substantiate it by reverting to ideas actually foreign to macroscopic thermodynamics: particle concepts (atomistic concepts) usually only

Fig. 3.1 Entropy in everyday life: Generally stated, it is that which hot coffee loses when it cools down in a cup and what is added to a pot of soup to heat the food. It is what is generated in a hot plate, a microwave oven, and an oil heater. Entropy is also what is transported in hot water and distributed by a radiator. It is what is conserved by the insulating walls of a room and by the wool clothing worn by the body.

construed as thoughts. The idea of "entropy $\approx$ everyday 'heat'" is always kept in mind as an additional aid to understanding. After the phenomenological characterization we will discuss how a measure for entropy can be introduced, and that directly, meaning without recourse to other quantities (direct metricization) (Sect. 3.7).

3.2 Macroscopic Properties of Entropy

Thermal Effect Let us begin with the characteristics that are important in our everyday experience. Entropy can be understood as a weightless entity that can flow and is contained in everything to one extent or another. In physical calculations it represents like mass, energy, momentum, electric charge, and amount of substance a *substance-like* quantity, meaning that it is like the other quantities a measure for the amount of something which can be sought as distributed in space. Thereby, it is not important whether this "something" is material or immaterial, stationary or flowing, unchangeable or changeable. It can be distributed in matter, it can accumulate, and it can be enclosed. Entropy can also be pumped, squeezed, or transferred out of one object and into another one. The entropy density is high if a lot of entropy is accumulated in a small area and low if it is widely distributed.

Entropy changes the state of an object noticeably. If matter, for example, a piece of wax or a stone, contains little entropy, it is felt to be cold. If, however, the same object contains more or a lot of entropy, it can feel warm or even hot. If the amount of entropy in it is continuously increased, it will begin to glow, firstly dark red, then bright white, subsequently melt, and finally vaporize like a block of iron would, or it may transform and decompose in another way, as a block of wood might. Entropy can also be removed from one object and put into another. When this is done, the first object becomes cooler and the second, warmer. To put it succinctly: Entropy plays a role in all thermal effects and can be considered their actual cause. Without entropy, there is no warm and cold and no temperature. The obvious effects of entropy allow us to observe its existence and behavior quite well even without measurement devices.

Spreading Entropy tends to *spread*. In a uniform body, entropy will distribute itself evenly throughout the entire volume of the body by flowing more or less rapidly from locations of higher entropy density (where the body is especially warm) to areas where the body is cooler and contains less entropy (Fig. 3.2).

If two differently warm bodies touch each other, entropy will flow from the warmer one to the cooler one (Fig. 3.3).

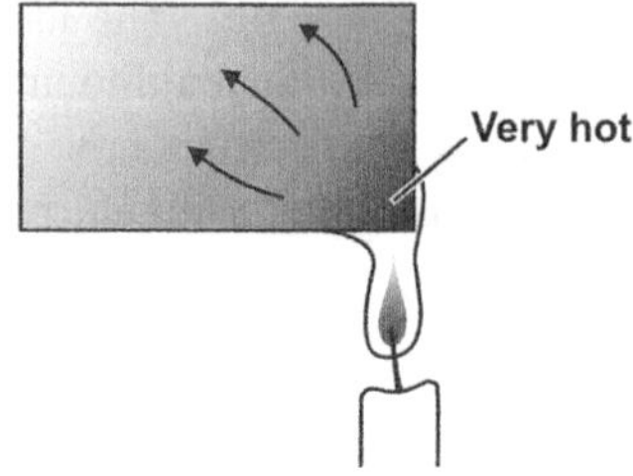

Fig. 3.2 Spreading of entropy within a uniform body.

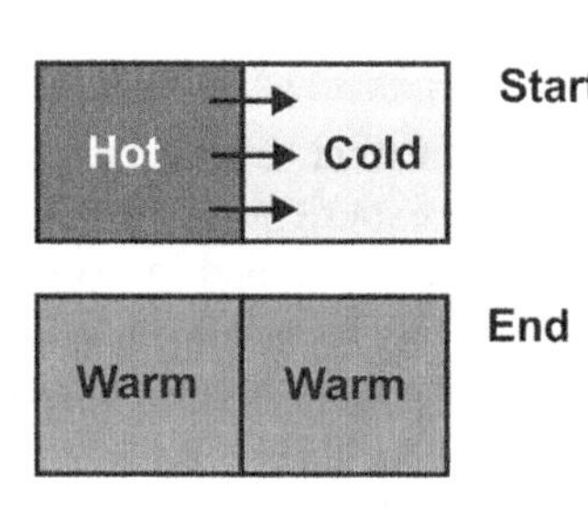

Fig. 3.3 Spreading of entropy from one body to another (entropy transfer).

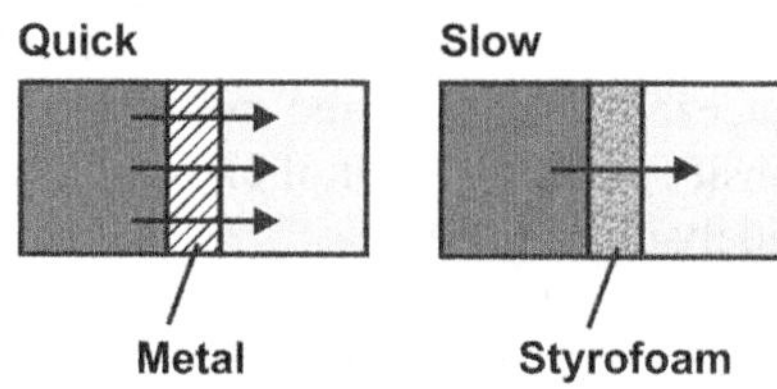

Fig. 3.4 Good and bad entropy conductors.

There are substances which conduct entropy very well, such as silver, copper, aluminum, and diamond, and others, such as wood, foamed plastic, or air, which only allow entropy to pass through them very slowly (Fig. 3.4).

Good entropy conductors are used to transfer entropy over a short distance. In order to overcome distances of decimeters and more—for example, for regulating the temperature of a room or an apartment or for cooling a motor—the conductivity is too small, the conductive transport of entropy—meaning the transport by conduction alone—is too slow. If one would like to transfer entropy from the furnace in the basement or the solar collector on the roof this has to be done convectively—meaning the entropy is transported by circulating water to the radiator and from there by circulating air into the room. To remove excess entropy out of a combustion engine, water is pressed through its cooling channels or air is blown over its cooling fins. If distances of meters and more are to be overcome like in industrial plants or even distances of kilometers like in the atmosphere or the oceans, convection is the dominant type of transport.

Bad conductors, however, are used to contain entropy. A vacuum acts like an especially good insulation. Entropy is also able to penetrate layers without matter by radiation, but this process takes place rather slowly at room temperature or below. This property is used in thermoses to keep hot beverages hot and cold

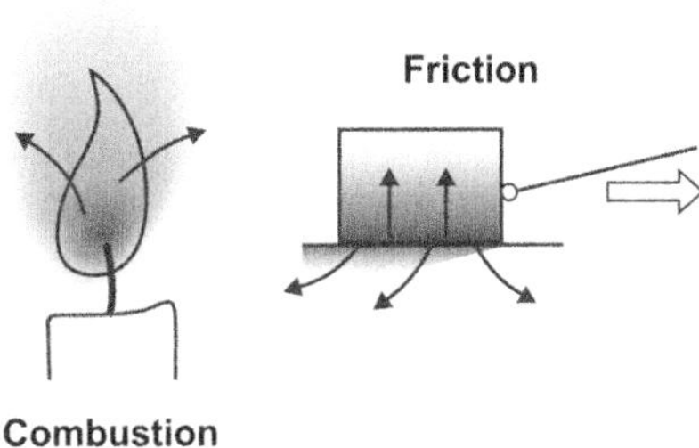

Fig. 3.5 Entropy generation: Locations where entropy was generated are generally noticeable by increased temperature.

beverages cold. Entropy transfer by radiation can be minimized by silvering the surfaces of the flask.

Generation and Conservation Entropy can be easily *generated*. For instance, great amounts of it are generated in the heating coils of a stove plate, in the flame of an oil burner, and on the surfaces rubbing together in a disc brake, but also by the absorption of light on a sunlit roof, in the muscles of a runner, and in the brain of a person thinking. In fact, entropy generation occurs almost every time something changes in nature (Fig. 3.5).

The most remarkable characteristic of entropy, however, is this: While it is generated to some extent in every process, there is no known means of destroying it. The cumulative supply of entropy can increase, but can *never decrease*! If entropy has been generated in a process, one cannot consequently reverse this process as one would rewind a film. The process is *irreversible* as one says. This does not mean, however, that the body in question cannot attain its initial state again. This may be possible by way of detours, but only if the entropy which was generated can flow out of it. If there is no such disposal available or accessible, because the system is enclosed by entropy-insulating (= heat-insulating or adiabatic) walls, the initial state is indeed inaccessible.

Laws of Thermodynamics Since it takes energy to generate entropy—which cannot disappear again—it seems as if energy is lost. This was the commonly held belief until the middle of the nineteenth century. Only in the second half of that century did the concept take hold that even under these circumstances, energy is conserved (compare to Sect. 2.3). Since then, this has been referred to as the *first law of thermodynamics* and is the basis of all teachings in the field of thermodynamics.

The statement that entropy can increase but can never decrease is the subject of the *second law of thermodynamics* which will be discussed in more detail in Sect. 3.4.

Let us conclude:

- Energy can neither be created nor destroyed (first law).
- Entropy can be generated but not destroyed (second law).

3.3   Molecular Kinetic Interpretation of Entropy

Atomic Disorder   So what is this entity that flows through matter and, depending upon how much is contained in it, causes it to seem warm or hot to the hand? For more than two hundred years, one has attempted to explain thermal phenomena by the movements of atoms. The image is as follows: The warmer a body is, the more intensely and randomly the atoms oscillate, spin, and swirl—so the idea, the greater the agitation and the worse the *atomic disorder*.

In the particle view, the quantity called entropy is a measure of

- The *amount* of atomic disorder in a body
- With regard to *type, orientation, and motion* of the atoms, or more exactly, with regard to any characteristic which differentiates one group of atoms from another.

Two questions arise here:

- What does disorder mean regarding type, orientation, and motion of atoms?
- What is meant by amount of disorder?

To clarify the first question, one might consider a park on a sunny summer Sunday. There are children playing, a soccer game taking place, and joggers, but also people just resting or even sleeping—a mass of running, sitting, lying people without order to their distribution or motion (Fig. 3.6). The opposite would be the dancers in a revue—or soldiers marching in lockstep. In this case, position, motion, and dress are strictly ordered. Disorder grows when the motion becomes random, but it also grows if the orientation in rank and file is lost or the type of people becomes nonuniform. All three: randomness of type, orientation, and motion of the individuals cause the total disorder.

The same holds for the world of atoms (Fig. 3.7). Not only disorder in the type and distribution of atoms, but also disorder in their motion, which can be expressed in how *agitated* they are, makes an important contribution to entropy. In this sense, the atoms in a hot gas are similar to children romping in the schoolyard. Motions are completely free and without order, and therefore the agitation, meaning the disorder concerning motion, is great. The atoms of a crystal, in contrast, can be compared to tired pupils in a school bus. Motion is more or less limited to fixed locations, so the disorder and agitation stay small.

Amount of Disorder   In order to get an impression of what is meant by *amount* of disorder, one might imagine a collection of, say, one hundred books at someone's home. A visitor comes, starts rummaging through the books, and makes a total jumble of them. Although the disorder appears great, the old order can be reinstated within a few hours. This means that even though the density of disorder is high, its amount is small. Compare this to just every hundredth book being falsely placed in a large university library. At first glance, there would appear to be almost no disorder. However, the amount of disorder, measured by the effort needed to

Fig. 3.6 Examples of groups of people becoming increasingly disordered in type, orientation, and motion.

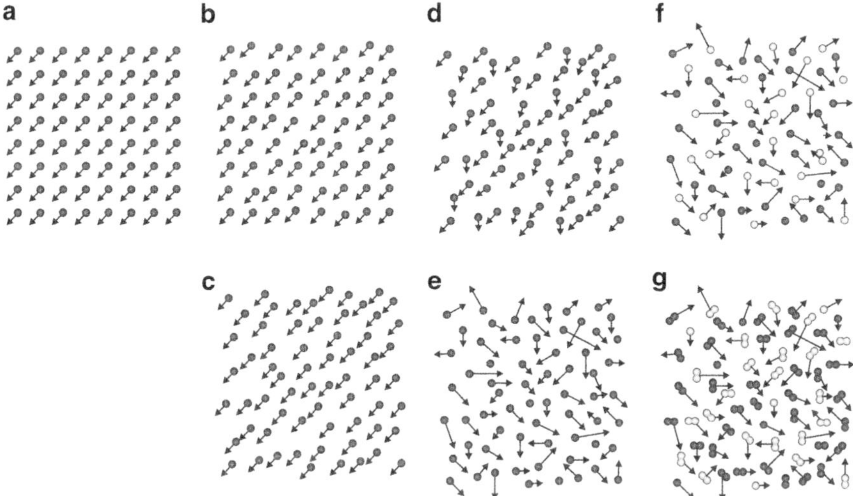

Fig. 3.7 An assembly of particles in states of increasing entropy: (**a**) Assembly is well ordered in every way, (**b**, **c**) Positions become increasingly perturbed, (**d**, **e**) Motion is increasingly disordered, (**f**, **g**) Particles become increasingly different (type, orientation, agitation, ...). The *arrows* show magnitude and direction of momentum (and not of velocity) (This differentiation is important when the entropy of particles of different mass shall be compared.).

place all the books back in their rightful places, is much greater. The density of disorder is small, but the total amount of it is very great.

3.4 Conservation and Generation of Entropy

The atomic disorder in a warm object and, therefore, the entropy in it have remarkable and well-defined characteristics, some of which have already been mentioned. They will be described in more detail in the following.

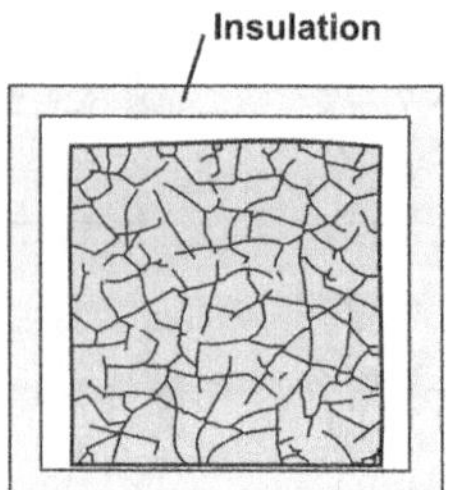

Fig. 3.8 Conserving entropy in a thermally insulated system. (Entropy is depicted by an irregular hatching in reference to the standard interpretation of entropy as atomic disorder. The amount of printing ink symbolizes the amount of entropy, the density of hachures, however, the entropy density. In objects made of the same material and in the same state of aggregation, a higher entropy density correlates with a higher temperature.)

Experiment 3.1 *Brownian motion*:
Brownian motion is a tremulous, random movement of tiny particles distributed in a liquid (e.g., drops of fat in milk) or particles stirred up in a gas (e.g., smoke particles in air). This kind of movement can be observed under a microscope for indefinite amounts of time without it letting up.

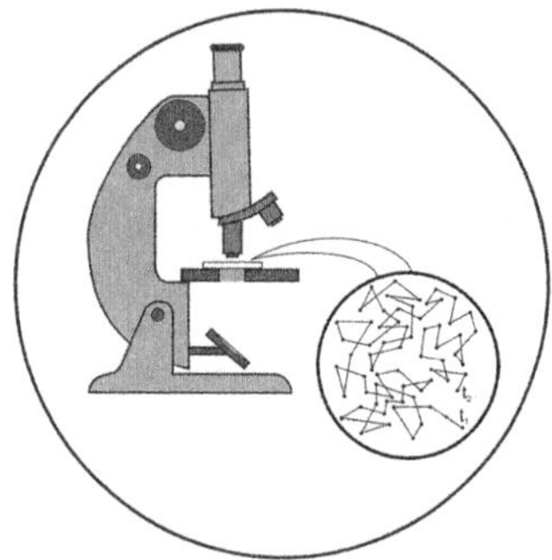

Conservation The atomic disorder and agitation in a thermally insulated body which is left to itself remain undiminished for an unlimited amount of time. An object *contains* entropy—we can say—whose amount S cannot decrease if it is in a thermally insulating (adiabatic) envelope, because entropy cannot penetrate thermally insulating walls (Fig. 3.8).

The agitation manifests itself among others by the microscopically visible Brownian motion (Experiment 3.1). Therefore, it can be regarded not only as theoretically constructed but as directly observable.

The amount of entropy an object contains depends upon its state. Identical objects in the same state contain identical amounts of entropy. The entropy contained in an object that is composed of pieces is the sum of the entropies of

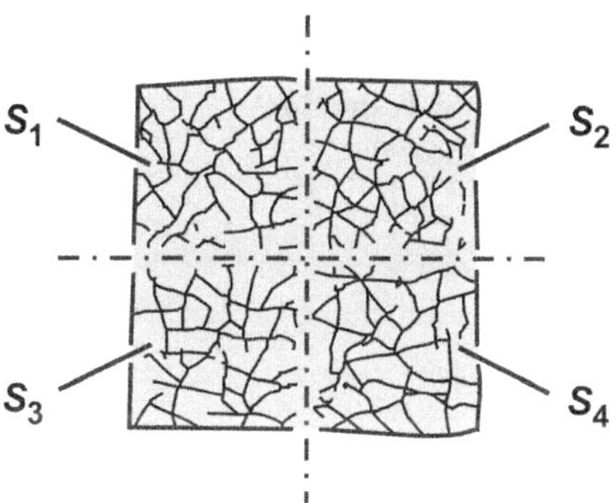

Fig. 3.9 Entropy as substance-like state variable ($S_1 \approx S_2 \approx S_3 \approx S_4$, and $S_{\text{total}} = S_1 + S_2 + S_3 + S_4$).

its parts. This is a direct result of the substance-like character of this quantity. In summary, it might be said: The entropy in an object is a *substance-like* (or *extensive*) quantity which—together with other quantities—determines its state (Fig. 3.9).

If a thermally insulated piece of matter, such as an iron block, is cautiously and slowly compressed with the help of a hydraulic press, or a gas in a cylinder with a piston (meaning the external pressure is only very slightly higher than the internal pressure of the confined gas), the interior agitation increases, and the motion of the particles becomes faster. This is easy to understand: An atom colliding with another particle moving toward it is hit like a tennis ball by a racquet and speeds backward. During compression, this process takes place simultaneously at innumerable interior locations so that agitation increases evenly overall. If the piece of matter is gradually relieved of pressure, the atoms quiet down, and it reaches its original state again. This is understandable as well, because the impact upon a receding particle lessens the rebound. No matter how often the process of compressing and subsequent releasing of tension is repeated, the original state of agitation is attained at the end—cautious action provided.

The atomic disorder in these types of *reversible* processes is conserved. Agitation is stronger in the compressed state—as mentioned, and motion, therefore, less ordered. At the same time, the range of motion for the atoms is decreased so that their positions are perforce more orderly than before. Therefore, it is plausible to assume that the extent of atomic disorder does not first increase and then decrease upon *cautious* compression and expansion, but remains constant (Fig. 3.10). This is an important fact that we should mention explicitly: Entropy is conserved in reversible processes.

Generation However, disorder in a thermally insulated body increases if the atomic structure is permanently disturbed. This can happen mechanically by simply hitting an object with a hammer, or more gently by rubbing two objects against each other. If an object can conduct electricity, an electric current can be sent through it. This means that electrons that have been accelerated by applying a voltage collide with the atoms. Another way would be the collision of fast particles

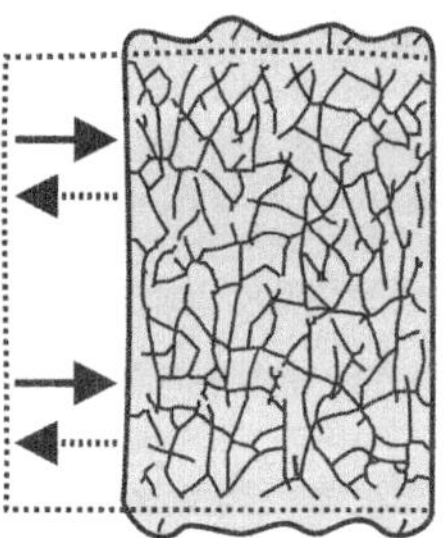

Fig. 3.10 Conservation of entropy during cautious compression and expansion (reversible process).

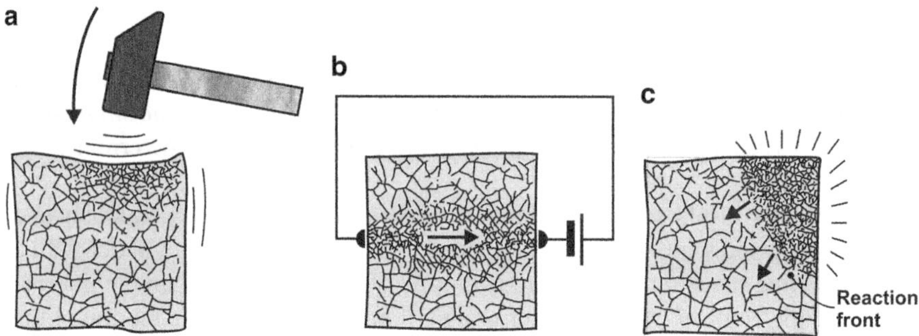

Fig. 3.11 Examples of entropy generation: (**a**) mechanically by hammering, (**b**) electrically by electron impact, (**c**) chemically by collisions of atoms shooting off in a reaction.

which have been formed by numerous chemical or nuclear transformations, irradiation by light, treatment with ultrasound, and many others (Fig. 3.11).

Entropy distributes more or less quickly over the entire body from the point where it is created. This process is also connected with the generation of entropy even if it is not directly obvious (see Sect. 3.14). All of these *entropy generating* processes are *irreversible*. If entropy was created in this way we will not get rid of it again, unless we could transfer it in the surroundings. But this is inhibited by the thermal insulation.

Entropy and Arrow of Time   To sum up: In a thermally insulated system, entropy can increase but never decrease; at best its amount remains constant. As mentioned before, this is what the *second law of thermodynamics* states. We can also formulate: For a thermally insulated system entropy always increases for irreversible processes. It remains, however, constant for reversible processes. We can write in abbreviated form

$$\Delta S = S(t_2) - S(t_1) \overset{\text{irrev.}}{\underset{\text{rev.}}{\geq}} 0 \quad \text{for } t_2 > t_1 \text{ in a thermally insulated system,} \quad (3.1)$$

where t represents time. At the more, this is valid for a so-called *isolated* system that does not interact with the surroundings, meaning that it can exchange neither entropy nor energy or matter.

The inequality (3.1) obviously interlinks an increase in entropy with the direction of time. If $S(t_2) > S(t_1)$, then $t_2 > t_1$ has to be valid, meaning that t_2 indicates a later point in time, t_1, however, an earlier one. It seems that the second law of thermodynamics determines what is future and what is past.

3.5   Effects of Increasing Entropy

If the entropy and thereby the atomic disorder inside a piece of matter is continuously increased, certain external effects soon become noticeable.

Main Effect   The main effect is that the matter becomes *warmer* (Fig. 3.12). To demonstrate this, entropy can be increased for example mechanically by strong hits with a hammer (Experiment 3.2).

Another way of formulating this effect: Of two otherwise identical objects, the one with more entropy is the warmer one. An object with no entropy is absolutely cold (Fig. 3.13).

As mentioned, entropy always moves spontaneously from warmer locations to colder ones (Fig. 3.14). When fast moving atoms collide with ones moving more slowly, they are themselves slowed while their collision partners speed up. As a result, the agitation and, therewith, the total disorder at the warmer locations gradually decrease while they continuously increase at the colder locations. In a homogeneous body, the process continues until the level of agitation is the same everywhere and the body is equally warm everywhere. This state is called *thermal equilibrium*.

Fig. 3.12 Warming as the main result of increase of entropy.

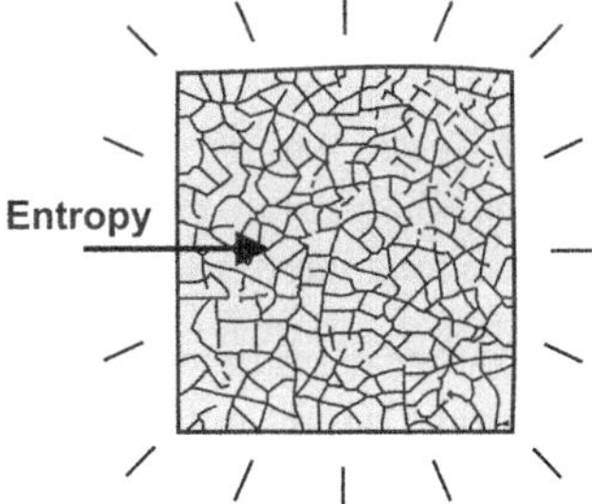

Experiment 3.2 *Heating of metal by forging*: A block of copper having a volume of a few cubic centimeters will become so hot after about 15–20 strong hits with a heavy hammer that it will hiss when put into water. A strong blacksmith can even forge a piece of iron of similar size in a few minutes to red heat.

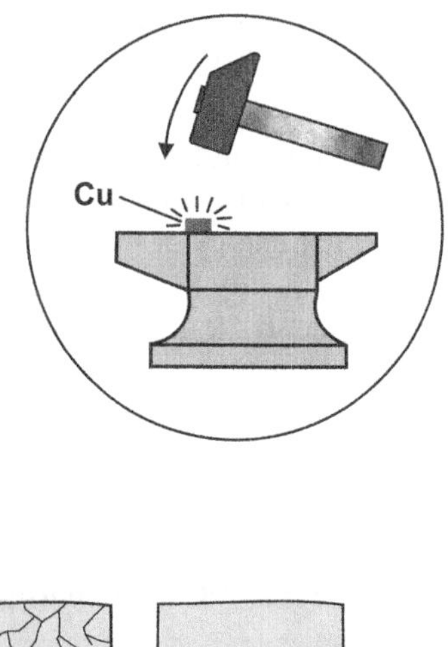

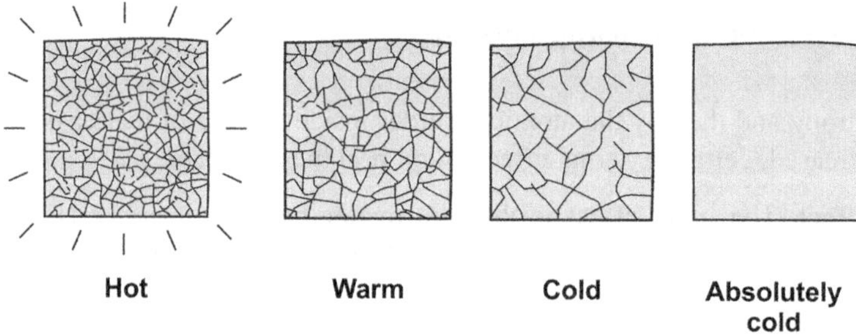

Fig. 3.13 Otherwise identical objects with different entropy content.

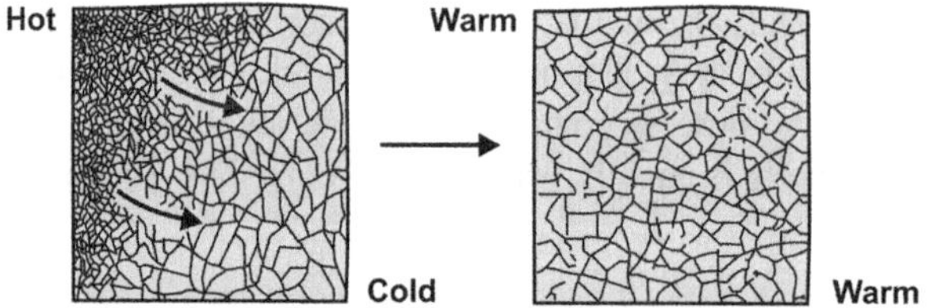

Fig. 3.14 Distribution of entropy in a homogeneous body.

Side Effects An increase of entropy can cause numerous side effects: Changes of volume, shape, state of aggregation, magnetism, etc., can result. Let us look at how a continuous increase of entropy affects a substance in general.

(a)   Matter continuously *expands* (Fig. 3.15). This seems logical because moving atoms would need more space depending upon how strong and random their motion is. This process is called *thermal expansion*.

   Experimentally, entropy can for example be increased by sending an electric current through the matter (Experiment 3.3).

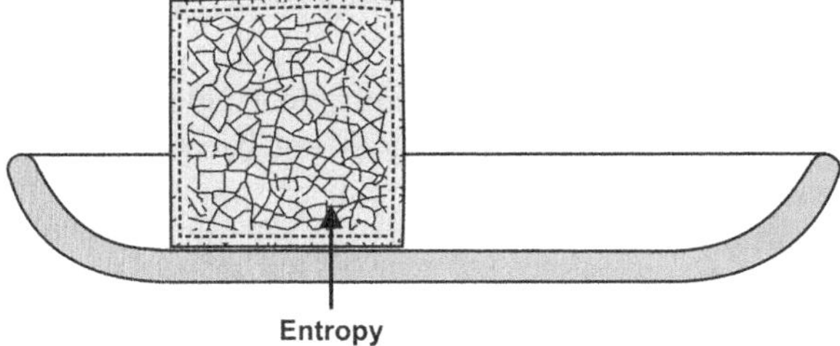

Fig. 3.15 Expansion due to the addition of entropy. The initial state is indicated by the *dashed line*.

Experiment 3.3 *Expansion of a wire caused by electric current*: A wire with a weight hanging from it lengthens noticeably when an electric current flows through it. The lowering of the weight can easily be observed. If the electric current is turned off, the entropy in the wire flows off into the air and the wire shrinks again.

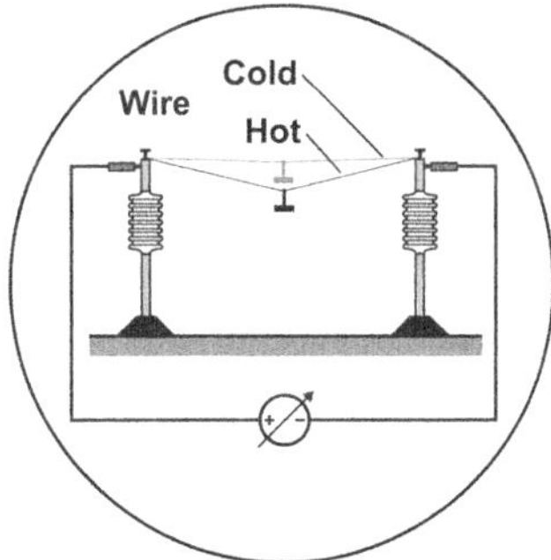

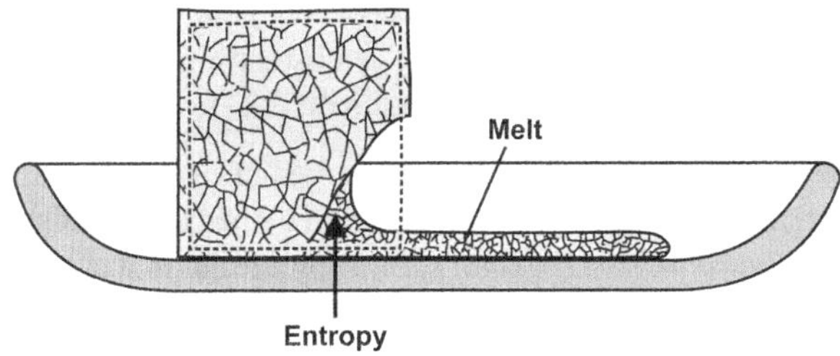

Fig. 3.16 Melting as an example of a change of state of aggregation with increasing entropy.

A substance that expands when entropy is added to it will, inversely, become warmer when compressed. This was mentioned in the previous section. Ice water is one of the few exceptions of volume decreasing with an increase of entropy. Therefore, it becomes colder ($< 0\,^\circ$C) when compressed.

(b) The substance will finally *melt, vaporize,* or *decompose* (Fig. 3.16). This begins when the disorder and the motion with it reach a level where the atoms can no longer be held together by the bonding force in a lattice or particle union, but try to break out of them. A melt that has been produced in this way from atoms or groups of atoms that still hold together but are easily shifted against each other is much less orderly than a crystal lattice in which the atoms generally remained fixed in their places. This melt contains more entropy than the identically warm solid substance. As long as part of the solid substance is available, the entropy flowing in will collect in the resulting liquid so that the melting substance itself does not become warmer. When this happens, the main effect of entropy remains unnoticeable. If a substance changes completely at its melting point from solid to liquid state, the entropy inside it increases by a given amount. As we will see, this characteristic can be made use of to determine a unit for amounts of entropy.

Analogously, the vapor formed at the boiling point absorbs the additional entropy, preventing the boiling liquid from becoming hotter.

3.6   Entropy Transfer

Entropy and with it atomic disorder can also be transferred from one object to another. If two objects with variously strong particle motion touch each other, the agitation in one of them will decrease because of a slowing down of the atoms, while in the other, the opposite occurs. Figuratively speaking, the disorder flows from one body into the other. This process, as well, only continues until the agitation has reached the same level everywhere and thermal equilibrium has been reached (Fig. 3.17).

The thinner the surrounding walls are, the greater their area is and the better the substance of which such a wall is composed conducts entropy, the easier the entropy runs through the walls (Fig. 3.18). The correlation is similar to that of the current of electric charge through a wire (see Sect. 20.4).

Zero-Point Entropy   All entropy capable of movement will escape an absolutely cold environment, meaning that any atomic motion comes to a standstill. This is the subject of the *third law of thermodynamics*. Entropy caught in a lattice defect is just about unmovable at low temperatures. It can therefore neither escape nor contribute

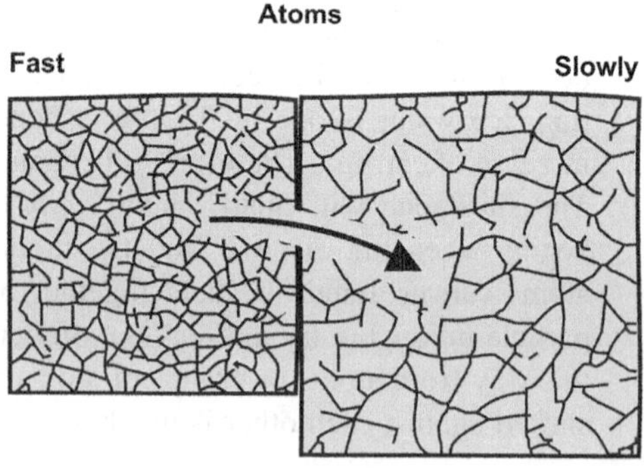

Fig. 3.17 Conduction of entropy from a warmer, entropy richer body where the atoms are moving fast to another cooler, entropy poorer one where the atomic motion is slow.

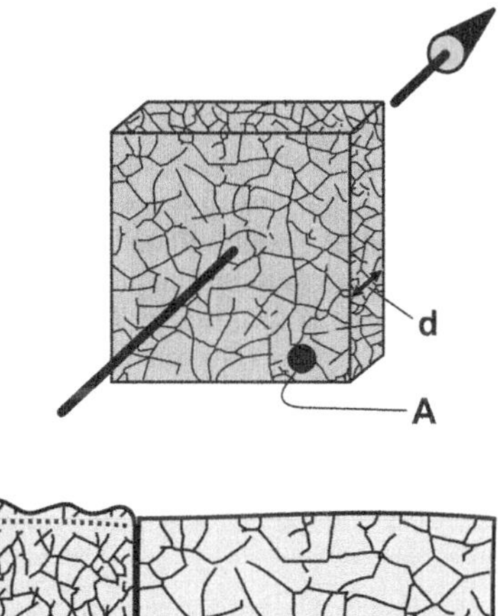

Fig. 3.18 Entropy current through a wall. The resistance which the wall imposes to the flux depends upon the thickness d, the area A through which the entropy flows, and the conductivity of the material.

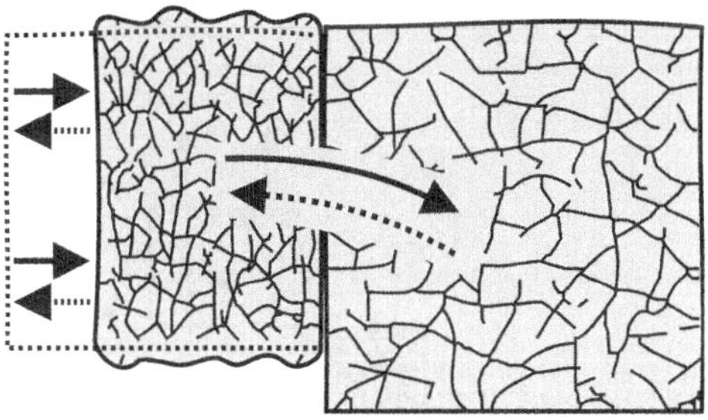

Fig. 3.19 Directed exchange of entropy between two bodies touching each other.

in any noticeable way to the warmth of an object. Whoever fails to leave a building or a park before closing is in danger of being locked in for the night. In this sense, the entropy stuck in the lattice defects can only escape as long as the particle motion is strong enough for the atoms to relocate. If the atomic motion in cold surroundings quiets down too quickly, the atoms do not have time to relocate into an ordered lattice structure, or to *crystallize*, as we say. The object then just solidifies into a more or less amorphous state. This unmovable entropy that does not flow off even in an absolutely cold environment is called "*zero-point entropy*." Therefore, we have to formulate the third law of thermodynamics as follows: The entropy of every pure (also isotope pure) *ideal* crystallized substance takes the value of zero at the absolute zero point. Only if the substance crystallizes ideally there is no spatial disorder and therefore also no residual "zero-point entropy."

Directed Entropy Transfer   Let us now return to entropy transfer. Even when the atomic motion is equalized everywhere in the manner described above, it is still possible for disorder to pass from one object to another. It is only necessary to compress one of the objects to raise the agitation of the atoms, and the desired flow process takes effect. The more the object is compressed, the more disorder "flows out" (just like pressing the water out of a sponge). If the body is slowly relaxed, the atoms gradually quiet down and the disorder begins to flow back in (the "entropy sponge sucks up entropy") (Fig. 3.19).

These elastic expansion and compression effects can be especially well observed in substances that can be easily compressed such as gases (Experiment 3.4).

Experiment 3.4 *Compression and expansion of air*: If air is compressed with a piston in a plexiglass cylinder having a thermocouple built in, the atoms become accelerated making the gas warmer (phase 1). After a while, the gas cools down to its original value because it is not insulated from the cylinder walls (phase 2). The piston's expansion leads to further cooling (phase 3). Then, entropy begins to flow back in and the gas begins to warm up (phase 4). The more slowly this is done, the more the difference between the compression and expansion disappears.

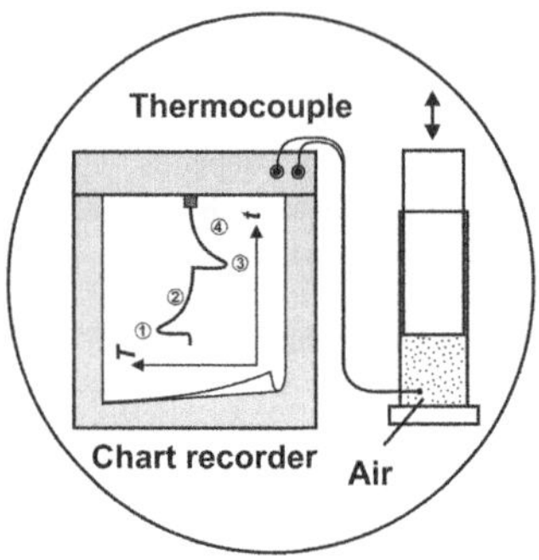

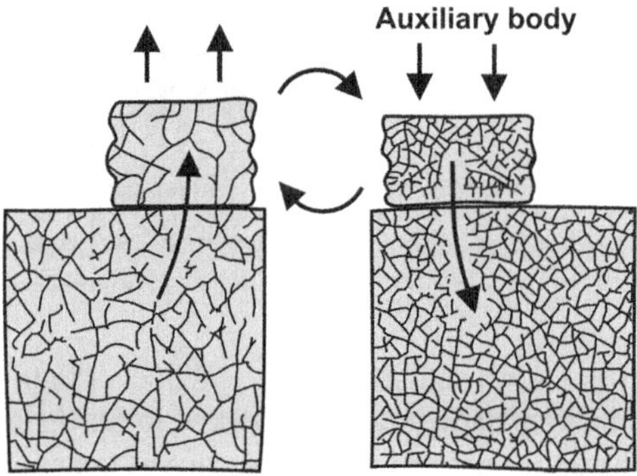

Fig. 3.20 Transfer of entropy with an auxiliary body. On the *left side*, the auxiliary body expands thereby absorbing entropy from the object. On the *right side*, it is compressed thereby adding entropy to another object.

As we have seen, entropy always flows spontaneously from an object with a higher level of agitation to one with a lower level. However, it is not difficult to make this happen in the opposite direction (Fig. 3.20). An *auxiliary body* is needed, a kind of "entropy sponge" which can easily be compressed and expanded. A gas contained in an expandable envelope is suitable for this purpose. When such a body touches an object and expands, it absorbs disorder from it. This absorbed disorder can now be transferred to any other object. The "sponge" is brought in contact with this second body and compressed. This process can be repeated at will, and as much entropy can be transferred as desired.

Ideal and Real Transfer   Every refrigerator uses this principle to pump entropy from its interior into the warmer air outside, while the low-boiling coolant (operating as the auxiliary body) circulates in a closed circuit (Fig. 3.21). The entropy

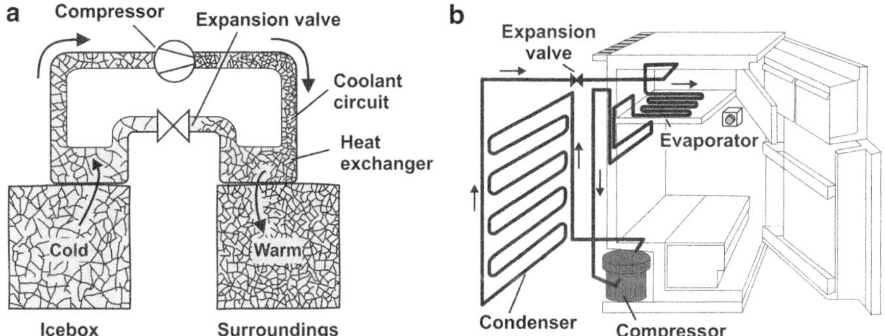

Fig. 3.21 (**a**) Principle of operation of a refrigerator, (**b**) Technical realization (according to: Leitner E, Finck U, Fritsche F, www.leifiphysik.de).

transfer takes place by the coiled pipe inside the refrigerator (heat exchanger) that is made of well-conducting material such as copper or aluminum. In older models, this coiled pipe is easy to see; in newer models, it is built into the back wall. The liquid vaporizes, taking up entropy in the process. The compressor sucks the gaseous coolant in and compresses it. The entropy is emitted into the air through the second coiled pipe that takes up most of the back of the refrigerator. This can be easily detected because the coil remains warm as long as the refrigerator is running. The coolant condenses, becoming a liquid again. Finally, the pressure of the liquid is brought back to the original value through an expansion valve, and the cycle is complete.

With skill and enough cautiousness during the compression and expansion processes, an (almost) reversible process can be attained where it is possible to keep the disorder during transfers from increasing noticeably. In this way, disorder is like a kind of substance that can be taken from one body and decanted into another. For instance, the entropy in a piece of chalk could be taken out of it and transferred to an ice cube. In the process, the chalk would cool down and the ice cube would begin to melt.

In summary, we have determined that the entropy content S of a body can basically increase in two ways: through the entropy generated inside it $S_{g(enerated)}$ (cp. Sect. 3.4) and, as described in this section, by the entropy exchanged with the surroundings $S_{e(xchanged)}$ (and that *conductively* by "conduction" in matter at rest, S_λ, or *convectively*, carried by a flow of matter, S_c):

$$\Delta S = S_g + S_e = S_g + S_\lambda + S_c. \tag{3.2}$$

3.7   Direct Metricization of Entropy

Selection of a Unit for Entropy The transferability of entropy opens up a possibility of measuring the amount of it in a body—at least theoretically. Measuring a quantity means determining how much more of it there is than its unit. Any amount

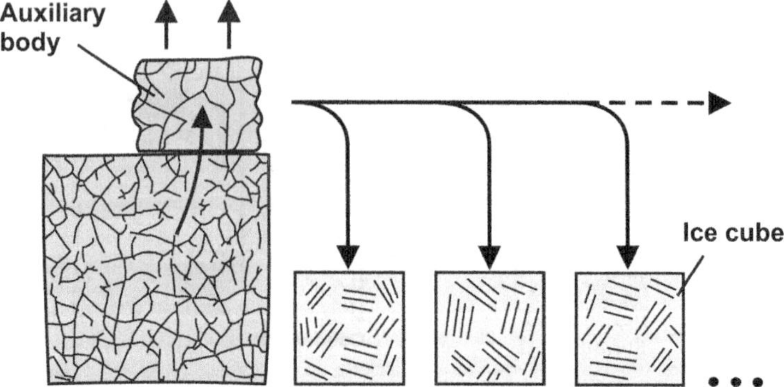

Fig. 3.22 Measuring entropy by counting ice cubes that melt when entropy is added to them.

of entropy can be used as the unit. For example, the amount needed to warm up a certain quantity of water by 1 °C (possibly from 14.5 to 15.5 °C), to evaporate a given volume of ether, or to melt an ice cube (Fig. 3.22). In order to accurately determine this unit, the size and state of the body in question must be exactly specified. For example, the ice cube would need to be 1 cm^3 in size, bubble-free, not undercooled, and the resulting water not be warmed up. However, instead of 1 cm^3, the somewhat smaller value of 0.893 cm^3 lends itself well because it yields exactly the amount of entropy that corresponds to the international unit. This unit has been fixed by a special method which we will come back to later. A certain amount of entropy contained in a body will be referred to as z units when z standard ice cubes can be melted with it. This procedure is comparable to the determination of the amount of harvested grain by using a bushel (Sect. 1.4) or that of an amount of water by scooping it out with a measuring cup.

Ice Calorimeter Instead of counting ice cubes, it is easier to use the amount of melt water produced as measure. A simple "entropy measurement device" can be built for this purpose. Melt water has a smaller volume than ice because of the anomalous behavior of the density of water and the decrease of volume can be measured. A bottle with a capillary on it and filled with a mixture of ice and water (ice-water bottle) (Fig. 3.23a) can then be used to show the change in volume. The lowering of the water level is simple to observe. Unintended entropy exchange can be avoided by using good insulation, and unintended entropy generation can be avoided by paying attention to reversibility.

This principle is also used by "Bunsen's ice calorimeter" (Fig. 3.23b). The glass container is filled with pure water and the U-shaped capillary with mercury. The central tube is cooled to below the freezing point of water, possibly by pouring in ether and sucking off the vapor, so that an ice mantle is formed on it. Then the sample to be measured is inserted into it. The amount of ice melted is noted by the volume decrease indicated by the mercury in the capillary. If no entropy escapes, is exchanged, or is generated during the measurement process, the height difference in

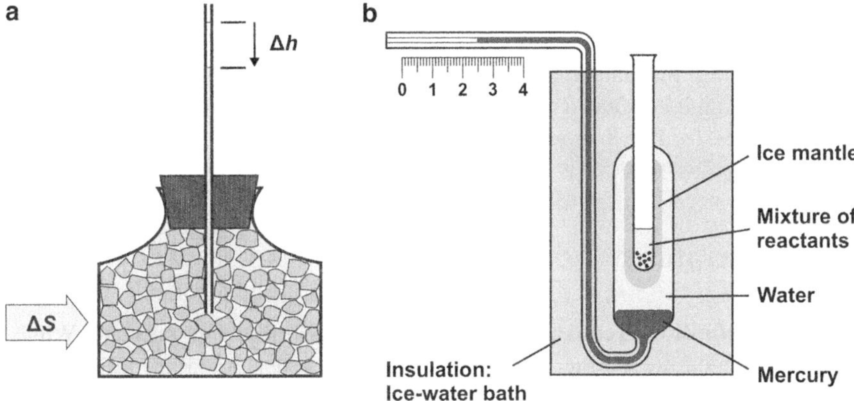

Fig. 3.23 (**a**) Principle of direct entropy measurement with the ice-water bottle, (**b**) Bunsen's ice calorimeter.

Experiment 3.5 *Measuring the entropy emitted during a reaction*: For example, the entropy emitted by the chemical reaction of iron and sulfur into iron sulfide can be measured by a simple ice calorimeter. A mixture of iron powder and sulfur powder is put into a test tube, and the test tube is subsequently placed in the calorimeter vessel filled with crushed ice. The reaction is initiated by a preheated glass rod or a sparkler. The melt water is collected in a graduated cylinder whereby 0.82 ml of melt water corresponds to the unit of entropy.

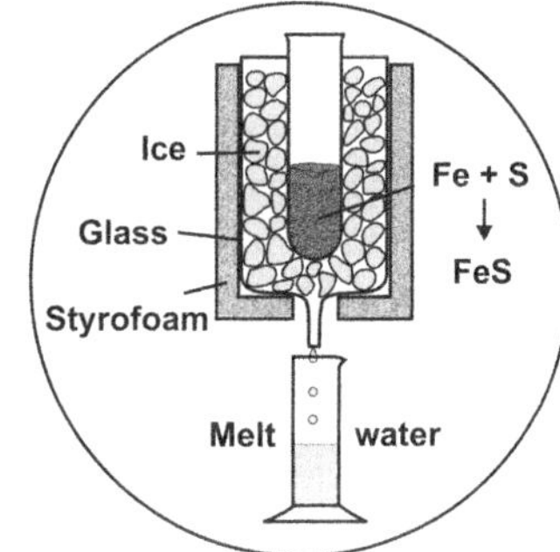

the capillary is proportional to the change of entropy in the sample or that of a reaction mixture, and the scale can be directly calibrated using entropy units.

Another way of determining the volume of the produced amount of water is to pour it into a graduated cylinder (Experiment 3.5).

Return to Macroscopic View The remarkable thing here is that this entire process has been developed using atomistic considerations, but the execution of it makes no use of them. Indeed, only macroscopic bodies are moved, brought into contact, separated, compressed, and expanded. Finally, ice cubes are counted. These are all manipulations that can be carried out when nothing is known about atoms. In order to have a well-directed approach, it is enough to remember the concept mentioned in Sect. 3.2 that all things contain a movable, producible, but indestructible something that generally makes the things warmer depending upon how much there is of it. What one actually imagines it is or what it would be called is unimportant when measuring or manipulating it. The German physicist Rudolf Clausius suggested calling it *entropy* in the middle of the nineteenth century, and the symbol S has been used for it ever since.

3.8 Temperature

Role and Definition Temperature and entropy are closely connected. While entropy is a measure of the amount of atomic disorder contained in a body, temperature describes how *strong* the atomic agitation, which means the intensity of random particle *motion*, is. Temperature is something like a level of agitation that is low when the atoms and molecules are gently oscillating and rotating. It is high when atomic motion becomes hectic and turbulent. The temperature in a body is therefore comparable to the strength of winds in the atmosphere with low values when the leaves rustle, but higher ones when the branches start swinging. Just as high winds can break branches or even whole trees, high temperatures can cause atoms to tear away from their bonds.

So how can temperature be defined? We will use the following statement as a basis: The more disorder is put into a body (meaning the more entropy there is), the higher the temperature will be in general. To generate entropy (or to increase the disorder in a body by the amount S_g), a certain amount of energy W must be expended. This is understandable considering that for example gas particles are accelerated, particle oscillations initiated, rotations increased, and bonds between atoms broken. The energy W needed will be greater depending on how many atoms are to be moved, and how many bonds torn. This means,

$$W \sim S_g.$$

Moreover, the warmer the body is, the more energy is needed. An example will show this. We imagine a body made up of some loosely and some tightly bound particles. Atomic disorder can be increased by breaking the particles and scattering the fragments. When the body is cold and the level of agitation low, the particles move slowly. Only the weakest connections break during collisions because very little energy is necessary to split them. Under such circumstances, it does not take much energy to increase the disorder by causing weak bonds to break by an increase in agitation. If agitation is already strong, the weakest connections will already have broken. If the disorder should be increased even more, the strong bonds left over need to be separated and this takes a lot of energy.

So now we know that increasing the entropy in a body takes more energy the higher the level of agitation is, meaning the warmer it appears to us. This fact can be used to make a general definition of temperature, a definition that remains *independent* of any thermometric substance (e.g., mercury or alcohol).

Fig. 3.24 Relation between the energy needed, the entropy generated, and the thermodynamic temperature.

This quantity is assumed to be proportional to the energy needed. It is called the *thermodynamic temperature* or *absolute temperature* and symbolized by the letter T:

$$W \sim T.$$

Because the more entropy that is generated the more effort is needed to generate it, the amount of energy used depends upon the amount of entropy created. Therefore, we define:

$$T = \frac{W}{S_g}. \tag{3.3}$$

The relation is clarified by Fig. 3.24.

The entropy generated generally changes the temperature in a body, so when applying this definition, only very small amounts of entropy may be generated in order to be able to ignore the perturbation. The exact temperature value is obtained in the limit of infinitesimally small contributions of entropy:

$$T = \frac{\mathrm{d}W}{\mathrm{d}S_g}. \tag{3.4}$$

By the way, energy conservation guarantees that the energy W needed does not depend upon which method we employ to increase the entropy. In each case, T has a well-defined value.

Because energy and entropy are both measurable quantities independent of any atomistic considerations, the temperature T is also measurable. The zero point of the temperature scale cannot be arbitrarily chosen, meaning temperature can be determined in an absolute sense. From experience we know that entropy is only generated when energy is expended. No entropy is generated when energy is gained. From $W > 0$ and $S_g > 0$ (third law of thermodynamics) follows $T > 0$. Therefore, negative temperatures do not exist. As a concrete example, let us discuss the determination of the melting temperature of ice (Experiment 3.6).

Experiment 3.6 *Determination of the absolute melting temperature of ice*: We start with a beaker filled with pieces of ice into which an immersion heater has been inserted. When the immersion heater is switched on, entropy is generated in the heating coil by the collisions of electrons and then emitted through the metal casing to the ice. The ice melts and the volume of the resulting melt water shows us how much entropy has flowed into the ice. The amount of energy needed for generating the entropy can be determined from the power P of the immersion heater and the measured period of time t according to $W = P \cdot t$. The ratio of measured values of energy and entropy yields the temperature.

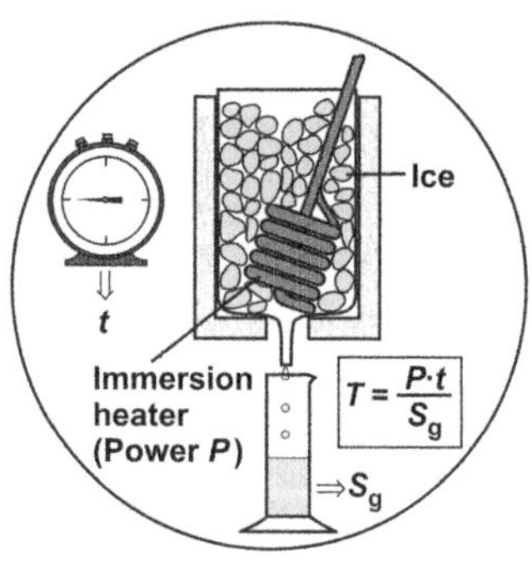

SI Unit The basic unit used in the SI system is not the unit of entropy, but the temperature unit called *Kelvin*, abbreviated to K. This was done by giving the melting temperature of pure airless water in a sealed container with pure water vapor (no air) above it a value, namely

$$T_0 = 273.1\underline{6} \ \text{K.} \tag{3.5}$$

This is based upon the so-called triple point of water, where all three states of aggregation (ice, water, water vapor) coexist and where pressure can be ignored. [When water is at the triple point, the pressure is fixed (see Sect. 11.5).] This odd numerical value is chosen so that the temperature difference between the normal freezing and boiling points of water is close to 100 units, as it is in the Celsius scale. For this reason, one Kelvin is one 273.16th of the thermodynamic temperature of the triple point of water. The zero point of the Kelvin scale lies at the absolute zero point which is indicated by an absence of entropy in the body. When one wishes to establish the relation between thermodynamic temperature T and Celsius temperature ϑ, it is important to be careful to set the zero point of the Celsius scale to the freezing point of water *at normal pressure*. This lies nearly exactly 0.01 K under the temperature of water's triple point, so that:

$$\frac{T}{\text{K}} = \frac{\vartheta}{^\circ\text{C}} + 273.1\underline{5}. \tag{3.6}$$

The Fahrenheit temperature scale used mostly in the USA can be converted into the absolute temperature scale in the following way:

$$\frac{T}{\text{K}} = \left(\frac{\vartheta_\text{F}}{^\circ\text{F}} + 459.67\right) \times \frac{5}{9}.$$

The unit of entropy is indirectly determined by the stipulation above [Eq. (3.6)] and our definition for T. The unit for energy is called Joule (J), and the temperature unit Kelvin (K), resulting in the entropy unit Joule/Kelvin (J K^{-1}). This is exactly the amount of entropy needed to melt 0.893 cm^3 of ice at the temperature T_0. The fact that entropy plays such a fundamental role in thermodynamics justifies giving it its own unit. Hugh Longbourne Callendar (Callendar HL (1911) The Caloric Theory of Heat and Carnot's Principle. Proc Phys Soc (London) 23:153–189) suggested naming it in honor of S. Carnot and calling it a *"Carnot,"* abbreviated to Ct $=$ J K^{-1}. Through his work with heat engines, the French engineer Nicolas Léonard Sadi Carnot (1796–1832) made important contributions to the development of thermodynamics.

3.9 Applying the Concept of Entropy

Molar Entropy We will look at some examples that give an impression of the values of entropy: A piece of blackboard chalk contains about 8 Ct of entropy. If it is broken in half, each half will contain about 4 Ct because entropy has a substance-like character. (Entropy is also generated in the breaking process, but this is so little that it can be ignored.)

A 1 cm^3 cube of iron also contains about 4 Ct, although it is much smaller. Therefore, the *entropy density* in iron has to be greater. If the amount of entropy in such a cube is doubled (by hammering, friction, or radiation, for example), it will begin to glow (Fig. 3.25). If the amount of entropy is tripled, the iron will begin to melt.

There is about 8 Ct of entropy in 1 L of ambient air. This is the same amount as in the piece of chalk. The reason that there is so little despite a volume more than 100 times as great lies in the fact that the air sample has far fewer atoms in it than the piece of chalk with its densely packed atoms. If the air is compressed to 1/10 of its original volume, it will become glowingly hot (Fig. 3.26).

This effect is utilized in pneumatic lighters to ignite a piece of tinder (flammable material) (Experiment 3.7), but also in diesel engines to ignite the fuel–air mixture. The compression must happen quickly because the entropy flows immediately from the hot gas into the cold cylinder walls and the gas cools down quickly.

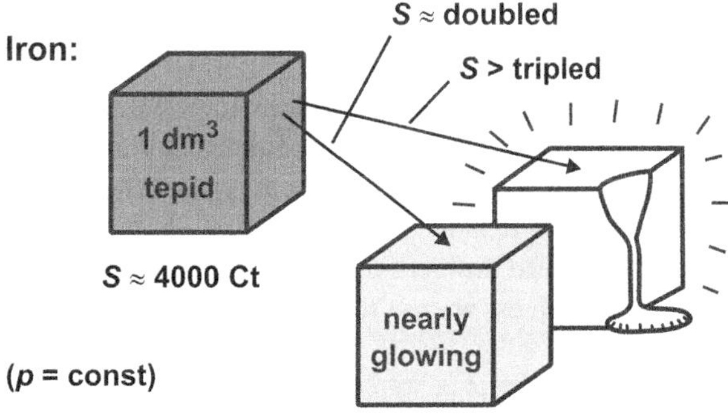

Fig. 3.25 The effects of raising the entropy content in a cube of iron with a volume of 1 dm^3.

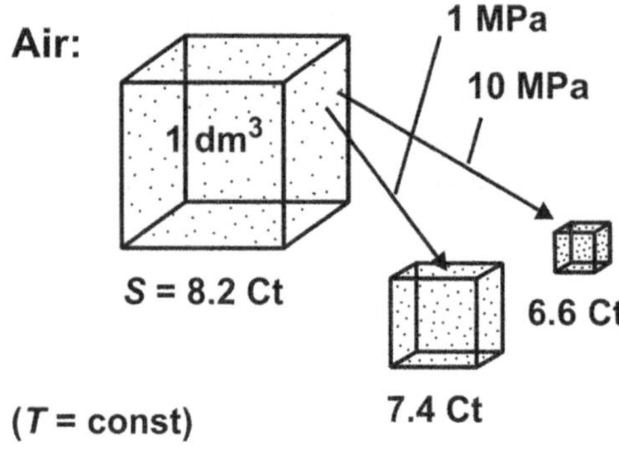

Fig. 3.26 Change of entropy content in air (1 dm³) with rising pressure (The gas molecules are represented by *points*.)

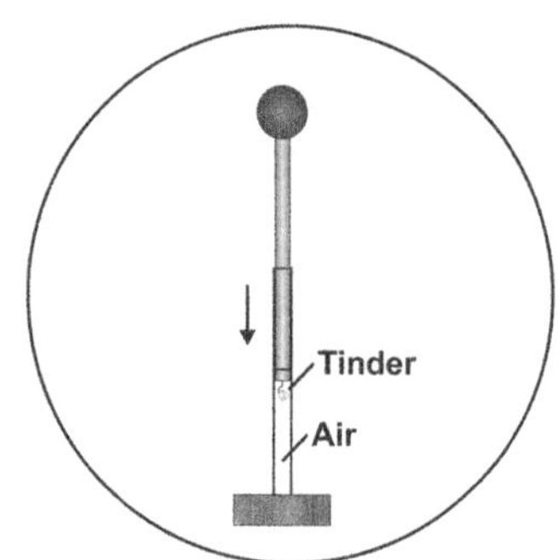

Experiment 3.7 *Pneumatic lighter*: If the piston is moved down quickly and powerfully, the tinder (for example, a piece of nitrocellulose foil or a piece of cotton wool impregnated with a highly flammable liquid) bursts into flame.

1 L of gas loses almost 1 unit of entropy if it is compressed to 1/10 of its original volume. If the gas is compressed to 1/100 of its volume, one more entropy unit can be squeezed out of it.

Chemists tend to relate entropies to the amount of a substance, i.e., how much entropy is contained in 1 mole of the substance in question. This quantity is called *molar entropy*:

$$S_\mathrm{m} \equiv \frac{S}{n} \qquad molar\ entropy\ of\ pure\ substances. \qquad (3.7)$$

S and n symbolize the entropy and amount of substance of the sample. The formula or name of the substance is usually enclosed in parentheses, for example, $S_\mathrm{m}(\mathrm{Fe}) = 27.3\ \mathrm{Ct\ mol}^{-1}$.

Molar entropy depends upon both temperature and pressure. For this reason, an additional stipulation is necessary if the values are to be tabulated. In chemistry, one generally refers to *standard conditions*, i.e., 298 K (more precisely 298.15 K) and 100 kPa [this corresponds to room temperature of 25 °C and normal air pressure, so-called standard ambient temperature and pressure (SATP)]. For characterizing the standard values, we use the symbol $^\ominus$, so for example,

$$S_\mathrm{m}^\ominus(\mathrm{Fe}) = 27.3\ \mathrm{Ct\,mol}^{-1} \qquad at\ 298\ K\ and\ 100\ kPa.$$

Table 3.1 Molar entropies of some pure substances at standard conditions (298 K, 100 kPa).

Substance	Formula	$S_\mathrm{m}^\ominus$ (Ct mol^{-1})
Graphite	C|graphite	5.7
Diamond	C|diamond	2.4
Iron	Fe|s	27.3
Lead	Pb|s	64.8
Ice	H_2O|s	44.8
Water	H_2O|l	70.0
Water vapor	H_2O|g	188.8

The value of ice was extrapolated from lower temperatures to 298 K

The values of some substances are listed in Table 3.1.

However, molar entropy not only depends upon the kind of substance in question, characterized by its content formula but also on the state of aggregation, as it is proved by the example of water. In order that the values are unambiguously given the aggregation state of the substance in question is added to the formula by a vertical stroke and the abbreviations s for solid, l for liquid and g for gaseous (cp. Sect. 1.6), for example, H_2O|l for liquid water. Because we do not want to overload the expressions, we stipulate that the most normal case is meant if there is no further information. Therefore, H_2O generally symbolizes the liquid and not vapor or ice. Entropy also depends upon the crystal structure. Modifications can be indicated for example by their names like graphite, diamond, etc.

A rule to bear in mind is that, at the same pressure, temperature and particle number, the entropy of a body will be greater, the *heavier* the atoms and the *weaker* the bonding forces. Diamond, which consists of atoms that are rather light and very firmly linked in four directions, has an unusually low entropy per mole. Lead, on the other hand with its heavy, loosely bound atoms, is rather rich in entropy. The characteristics of iron lie somewhere in between; it has a medium value of molar entropy. Using the example of water, the table shows how entropy increases by transition from a solid to a liquid state and even more by transition from a liquid to a gaseous state.

Determining of Absolute Entropy Values How are the values in Table 3.1 actually determined? It would be possible to find the entropy content of a sample by "decanting" the entropy from it into the ice-water bottle with the help of an auxiliary body. However, this would require that each step be configured reversibly, as discussed in Sect. 3.7, so that the entropy cannot increase during transfer, and this is very difficult to accomplish in practice. The goal is reached more easily by taking a detour. First, all the entropy contained in the sample must be removed. Favorable circumstances would allow simply immersing the sample in liquid helium (4.2 K). With the entropy having flowed off, the sample would be just about empty of entropy. For very accurate applications, the sample would have to be further cooled to reduce the remaining entropy. However, the entropy from the disorderly distribution of isotopic atoms cannot be gotten rid of in this way. This value can be easily determined by other means. Afterward, the sample is thermally insulated, and

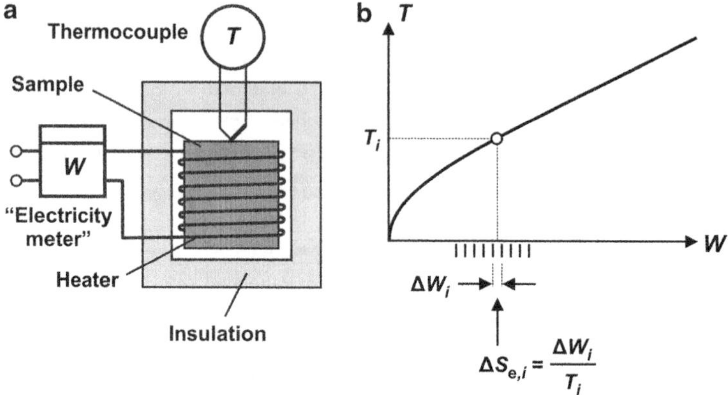

Fig. 3.27 Indirect measurement of entropy by heating up a sample previously cooled down to almost 0 K. (**a**) Measuring setup, (**b**) Corresponding experimental curve.

entropy is generated inside it in a controlled way. This might be done by electric heating (Fig. 3.27a). Energy consumption W and temperature T should be constantly measured (Fig. 3.27b) until the sample has attained the desired end temperature. The entropy generated during a small time span simply results from reversing the definition equation of temperature as the quotient of energy consumption and average temperature during this time period:

$$S_g = \frac{W}{T}. \tag{3.8}$$

The total amount of entropy contained in the sample at the end can be obtained by adding up all the amounts of entropy that have been generated over all time spans. To abbreviate, the symbol for summation $\sum$ will be used:

$$S_g = \sum_{i=1}^{n} \Delta S_{g,i} = \sum_{i=1}^{n} \frac{\Delta W_i}{T_i}. \tag{3.9}$$

The smaller the chosen time span, the more exact the result. If the time interval is allowed to approach zero, we have the definite *integral* (cp. Sect. A.1.3 in the Appendix):

$$S_g = \int_{\text{initial}}^{\text{final}} dS_g = \int_{\text{initial}}^{\text{final}} \frac{dW}{T} = \int_{0}^{t'} \frac{P(t)dt}{T}. \tag{3.10}$$

Because of the convention $S = 0$ at $T = 0$ for ideally crystallized solids (third law of thermodynamics), it is possible to determine not only differences but also absolute values of entropies and therefore absolute molar entropies as well. This determination is not only possible for substances in the state stable at 0 K but also for

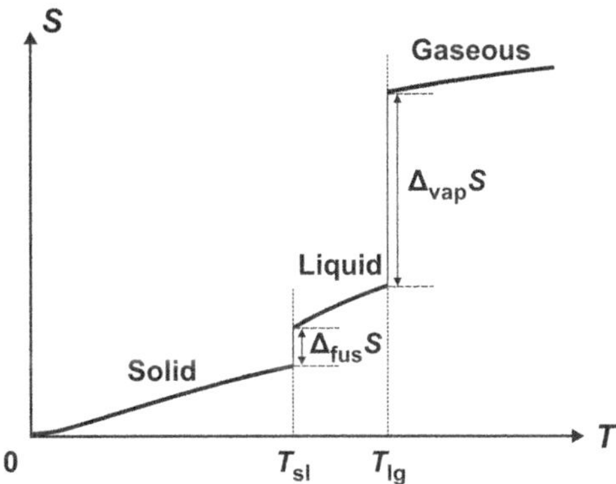

Fig. 3.28 Entropy of a pure substance as a function of temperature (without change of modification).

states which are formed during heating (other modifications, melt, vapor). The experimental curves $T = f(W)$ have horizontal parts when such phase transitions take place which means that energy is consumed, generating entropy without the temperature changing. If the resulting entropy content of a substance is plotted as a function of temperature (at constant pressure), we obtain the relationship shown in Fig. 3.28.

The entropy of a solid increases with an increase of temperature. It takes a jump at the *melting point* because the melting process causes the order of the solid to break and a noticeably higher disorder is produced in the liquid (see Sect. 3.5). Generally, we will symbolize the transition from the solid (s) to the liquid state (l) by the abbreviation s→l, the melting point therefore by $T_{s \to l}$ (the freezing point correspondingly by $T_{l \to s}$). Because melting and freezing points are identical for pure substances, we write in short T_{sl}. The change of entropy per mole of substance at the melting point is indicated as *(molar) entropy of fusion* $\Delta_{fus}S$. Subsequently, the entropy increases again up to the *boiling point* T_{lg}, at which point, another jump takes place [*(molar) entropy of vaporization* $\Delta_{vap}S$]. The entropy increases much more strongly during vaporization than during melting because disorder grows more strongly as a result of the transition from liquid to gas than from solid to liquid. We will deal in more detail with the entropy of fusion and of vaporization in Chap. 11.

Entropy Capacity Let us return again to the entropy content of a solid. As we have seen, it grows always with rising temperature. The curve is different for various substances, though. The increase of entropy per temperature increase is called the *entropy capacity* $\mathcal{C}$ in analogy to electric capacity $C = $ charge Q/voltage U (or if this is not constant, $C = \Delta Q / \Delta U$):

$$\mathcal{C} = \frac{\Delta S}{\Delta T} \quad \text{or for infinitesimally small changes} \quad \mathcal{C} = \frac{dS}{dT}. \tag{3.11}$$

The steeper a section of the curve, meaning the faster it rises at a given temperature, the greater the entropy capacity. The entropy content of a body is not generally proportional to temperature so its entropy capacity does not only depend upon the

Table 3.2 Molar entropy capacities of some pure substances at 298 K and 100 kPa.

Substance	Formula	$\mathcal{C}_{\mathrm{m}}$ (Ct mol^{-1} K^{-1})
Graphite	C|graphite	0.029
Diamond	C|diamond	0.020
Iron	Fe|s	0.084
Lead	Pb|s	0.089
Ice	H$_2$O|s	0.139
Water	H$_2$O|l	0.253
Water vapor	H$_2$O|g	0.113

The value of ice was extrapolated from lower temperatures to 298 K

Table 3.3 Specific entropy capacities of some construction materials at 298 K and 100 kPa.

Substance	ε (Ct kg^{-1} K^{-1})
Window glass	3.1
Concrete	3.7
Styrofoam	4.4
Wood (pine)	5.1
Particleboard	6.6
Wood (oak)	8.8

substance but, in general, also upon temperature. The pressure should be constant. This is important because a body can lose entropy during compression—like a sponge the absorbed water. Instead of $\mathcal{C} = \mathrm{d}S/\mathrm{d}T$, it is more correct to write

$$\mathcal{C} = \left(\frac{\partial S}{\partial T}\right)_p \qquad \text{or even more detailed} \qquad \mathcal{C} = \left(\frac{\partial S}{\partial T}\right)_{p,n}. \tag{3.12}$$

Because $\mathcal{C}$ is directly proportional to the amount of substance n, one divides it by n and obtains the *molar entropy capacity* $\mathcal{C}_{\mathrm{m}}$:

$$\mathcal{C}_{\mathrm{m}} = \frac{\mathcal{C}}{n} = \frac{1}{n}\left(\frac{\partial S}{\partial T}\right)_{p,n}. \tag{3.13}$$

The values of some substances are listed in Table 3.2. Usually, the corresponding *molar heat capacities* $C_{\mathrm{m}} = \mathcal{C}_{\mathrm{m}} \cdot T$ are given in tables instead of entropy capacities. We will discuss the reasons for this in more detail in Chap. 24.

In the field of engineering, entropy capacities are mostly related to mass. The result is the *specific entropy capacity* ε. The specific entropy capacities of some common construction materials are given in Table 3.3. (Because the composition of the particular material can vary significantly, the given values are averages.)

The specific entropy capacity plays an important role for the "heat storage capacity" of a material. Therefore, it has for example consequences for the behavior of construction materials during heating such as wood in case of fire.

Experiment 3.8 *Vaporization of liquid nitrogen by graphite and lead*: If samples of equal amounts of different substances (possibly 0.1 mole of graphite and 0.1 mole of lead) are put into small flasks filled with liquid nitrogen ($N_2|l$) which are cooled in a Dewar vessel, an amount of nitrogen corresponding to the entropy capacity will evaporate, and the balloons will be inflated differently. Additionally, a considerable amount of entropy is generated; therefore the volume of the balloons is bigger than expected from the entropy exchanged. But the result remains qualitatively correct.

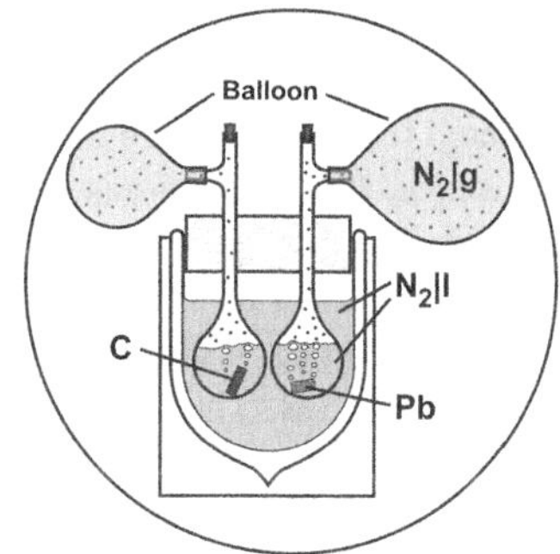

The effect of different entropy capacities of various substances can be illustrated by Experiment 3.8.

The entropy capacity depends not only—like the entropy—upon temperature and pressure, but also upon the conditions under which the substance is heated. A substance will absorb more entropy when it is allowed to expand freely than it would if its expansion were hindered. Depending on whether in most cases the pressure or more rarely the volume remains constant during the temperature increase, a different change of entropy content and therefore also a different entropy capacity can be observed. One characterizes the two different coefficients, if necessary, by indices: $\mathcal{C}_p$ and $\mathcal{C}_V$, respectively. If there is no index we always refer to $\mathcal{C}_p$.

3.10   Temperature as "Thermal Tension"

The atomistic image of entropy was given a short and qualitative description in this chapter. This was sufficient for introducing it. A formal version of the concept of entropy based upon this model would, however, be time-consuming. For the moment, referring to the particle image should just serve as an orientation. Phenomenologically or macroscopically, all the activities to be carried out in order to calculate quantities are well defined. The question arises here of whether or not these activities can be understood without recourse to the atomistic image. This has been hinted at in Sect. 3.7, and is, in fact, possible.

An image developed already in the eighteenth century seems especially simple. It imagines temperature as a kind of "pressure" or "tension" weighing upon entropy. However, at that time the word entropy was not used. One imagined a fluid like entity that warms a body, and considered it a kind of weightless substance comparable to electric charge. The temperature equalization of two bodies was described as a pressure equalization of this "heat substance" (caloric) in which this substance migrated from places of higher "pressure" to places of lower "pressure." If we accept this image, then it becomes obvious that energy is needed to generate entropy in a body against this "pressure" or "tension," or to force it into a body (comparable to filling a tire with air against an interior pressure p, or charging a body against its electric potential φ). The higher this "pressure" (the higher the temperature), the more energy is needed. The amount of energy also grows the more entropy is generated (S_g) or added (S_e). The following types of relations could be expected:

$$W = T \cdot S_g \qquad (3.14)$$

and

$$W = T \cdot S_e. \qquad (3.15)$$

We imagine the two entropies to be small, meaning

$$dW = T \cdot dS_g \qquad (3.16)$$

or

$$dW = T \cdot dS_e, \qquad (3.17)$$

so that the temperature will not change much as a consequence of the increase of entropy. Because we will discuss the energy exchange connected with the addition or removal of substances later on we suppose in this chapter that there is no convective exchange of entropy, $dS_c = 0$, meaning that the whole exchange takes place by "conduction," $dS_e = dS_\lambda$.

The first equation follows directly from the equation defining the absolute temperature if it is solved for dW. With help from the law of conservation of energy, the second equation follows easily from the first one. This law states that the same effect, no matter how it comes about, always requires the same energy. Whether a certain amount of entropy is generated in a body or added to it, the effect upon the body is identical. It expands, melts, vaporizes, or decomposes in the same manner. It must follow, then, that the energy needed for these processes must be the same.

3.11 Energy for Generation or Addition of Entropy

"Burnt" Energy Despite their similarity, the two equations above, $dW = TdS_g$ and $dW = TdS_e$, describe two rather different processes. Because entropy can increase but cannot be destroyed the process that generates it can only run in one direction and never in the other. As already mentioned, it is *irreversible*. The energy used cannot be retrieved (except indirectly). It is said that when entropy is generated and something is heated by it—noticeable such as in the heating coils of a stove plate or imperceptible when paddling in a lake—the energy needed is *devalued, wasted,* or "*burnt,*" or that it gets *lost.* "Burnt" energy is found again in a state of random molecular motion. Statistically, it is distributed in tiniest portions over the innumerable oscillating and rotating atoms or groups of atoms. In view of these circumstances, one can speak of *dissipation of energy* instead of energy loss, waste, devaluation, etc. There are, as we have seen, plenty of terms to choose from depending upon which aspect is being emphasized. We will call the wasted and, therefore, no longer retrievable amount of energy that appears in the first of the two equations "*burnt*" *energy* $W_{b(urnt)}$ (because of the close relation to the generation of entropy S_g):

$$dW_b = T dS_g \qquad (3.18)$$

or summed up

$$W_b = \int_{initial}^{final} T dS_g. \qquad (3.19)$$

The fact that energy is needed to generate entropy does not mean that special efforts or equipment are necessary. Quite the contrary:

- In every process, a certain amount of dissipation of energy is unavoidable.
- Entropy is readily generated all the time and everywhere.

Just consider friction. On the contrary, special caution and devices are necessary to avoid this—such as ball bearings, lubricants, etc., in the case of cars.

The energy expended for generating entropy can come from inside a region itself, i.e., as if from an inner source. A compressed gas is an energy source that can be tapped. When a gas cools as it expands, the tapped energy W can be used to generate entropy S_g, which is then conducted back into the gas (along with W), making it warm again. In the ideal case, it will become as warm as it was at the beginning (Fig. 3.29). Entropy then appears to have been created without expending any energy. The total amount of stored energy in the system is exactly the same as at the beginning. Nevertheless, we can assume that *whenever* entropy is generated, it occurs at the cost of energy which might have been used more intelligently in

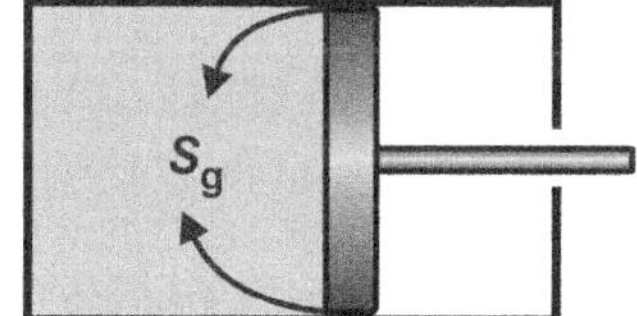

Fig. 3.29 A piston pushed out by gas enclosed in the cylinder. Entropy is generated by the friction between the cylinder and the wall. The entropy flows into the cooling gas, warming it up.

countless other ways. Energy that we can freely use is called *available energy* or *useful energy*. When referring to *energy production,* we actually mean available energy. This is also the case with so-called *lost* or *wasted energy.* The total amount of energy always remains the same, but it is of no use to us if we cannot draw it from its sources or if it disappears into sinks from where it is inaccessible.

Energy and Entropy Exchange In contrast to this, the second equation above, $dW = TdS_e$, describes a process that is fundamentally reversible. When entropy S_e from one body is transported into another at constant temperature T, the energy $W = T \cdot S_e$ is transferred with it. We will refer to this energy as W_e, if a differentiation from W_b seems necessary. The energy which was transferred returns to the original body again along with the entropy flowing back to it. This process is, therefore, *reversible.* The process described here corresponds to what is usually called *heat supply* and *heat removal.* Energy and entropy are exchanged together:

$$dW_e = TdS_e \tag{3.20}$$

or summed up

$$W_e = \int_{\text{initial}}^{\text{final}} TdS_e. \tag{3.21}$$

In order to understand the importance of this equation, we will take a short detour to look at the development of the concept of heat. In the early days of the "science of heat" (thermodynamics), there were very diverse ideas about its nature. In the eighteenth century, heat was conceived of as a weightless "something" that heats things and could be exchanged between bodies. It was considered a kind of "heat substance" called caloric. The first successful qualitative and quantitative descriptions of effects such as heating and cooling, melting and evaporation, condensation and freezing were created based upon the concept of caloric at that time. Following the spirit of the time, it was assumed that this something could be neither created nor destroyed, just like chemical elements. In the nineteenth century, it became increasingly evident that this "substance" could actually increase indefinitely. When, however, energy appeared as a quantity that corresponded to the ideal of an entity that could neither be created nor destroyed, the view of things changed. From then on, "heat" was considered to be the energy transferred by random collisions of molecules, which could even penetrate seemingly rigid walls. This is exactly the energy described by W_e above that, even today, is usually symbolized by Q. When Rudolf Clausius introduced entropy S in the middle of the nineteenth century (under another name), neither he nor his contemporaries appear to have realized that he was only reconstructing the old quantity but with the new characteristic of being producible while remaining indestructible. Only later on, in 1911, did Hugh Longbourne Callendar allude to this fact.

Clausius derived a relation for determining the change of entropy ΔS in a body—an iron block, for instance—while it heats from a temperature T_1 up to T_2. We can find this relation much more easily. We can assume that T as well as $Q = W_e$ can be measured. T can be determined using suitably calibrated thermometers and Q can be measured calorimetrically. The entropy within a body can increase by generation or addition: $dS = dS_g + dS_e$ [compare with Eq. (3.2)]. If the addition of energy Q should be reversible, indicated by the index $_{rev}$, then $dS_g = 0$ and therefore

$$dS = \frac{dQ_{rev}}{T} \tag{3.22}$$

or correspondingly added up

$$S = \int_0^T \frac{dQ_{rev}}{T}. \tag{3.23}$$

In this way, Clausius defined the quantity entropy for the first time a century and a half ago.

Entropy plays an important role in all thermal effects. Along with temperature, it is the quantity that characterizes this field of study. Energy also plays its role, not only here but on (nearly) all stages of physical chemistry or physics. It is important but nonspecific. Although the exchanges of entropy and energy are so closely linked that it is not easy to clearly distinguish between their separate roles, it would be disastrous to mix them up. For this reason, we will avoid the word *heat* for any kind of energy, especially for the quantity Q, which we will no longer use. This word is the cause of grave misunderstandings that have been difficult to dispel. It leads us to believe that these energy-related quantities are the measure of what we imagine heat to be based upon our everyday experience. This does not really work and at the same time hinders the quantity S from being related to everyday life.

To improve understanding, we will contrast an entropy conserving process with one that generates entropy in two simple experiments. In order for an undesired exchange of entropy with the surroundings not to falsify the results, the samples must be well insulated, or the experiments must be carried out very quickly. Let us begin with the *entropy conserving* process, the expansion of rubber (Experiment 3.9).

The experiment can be adapted in a simplified manner to everyday life: We touch a thick rubber band with the upper lip and after waiting a short while for equalization of temperature, it is stretched quickly and powerfully and immediately pressed again against the upper lip. The band feels noticeably warm. When the stretched band is allowed to contract to its original length and then quickly pressed against the upper lip, there is a noticeable cooling.

Bending an iron rod, however, is an example for an *entropy generating* process (Experiment 3.10).

Energy Exchange Along Different Paths   In the systems we will be investigating, the exchange of energy will generally occur simultaneously along several paths rather than a single path. The simplest and most important paths are changes of volume V and entropy S. We were introduced to the relation between energy and

Experiment 3.9 *Temperature as a function of time in expanding rubber*: If a rubber band is expanded and then relaxed, the temperature that rises during expansion sinks again no matter how often the experiment is repeated. The energy expended at the beginning is retrievable. The temperature change $T(t)$ resembles a square wave. The process is reversible. Entropy is scarcely generated because the band is as cool at the end as it was at the beginning.

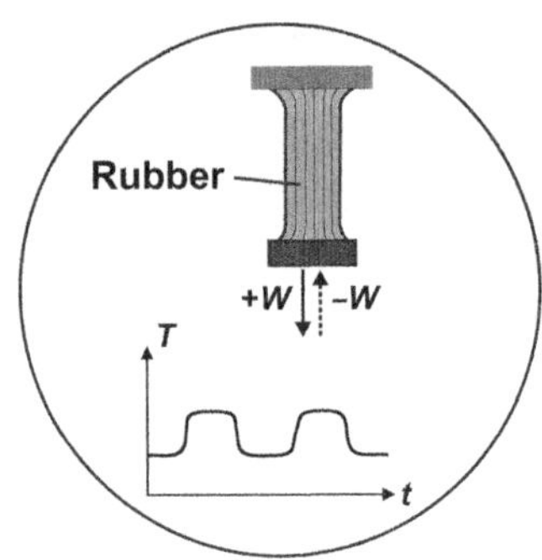

Experiment 3.10 *Temperature as a function of time in bending iron*: Bending an iron rod back to its original state (after previous bending) costs again energy, and therefore, the temperature rises in steps. This bending process is irreversible. Although the iron rod returns to its original position, it is now warmer. In this case, entropy is obviously being generated and the energy involved is used up. It is not retrievable.

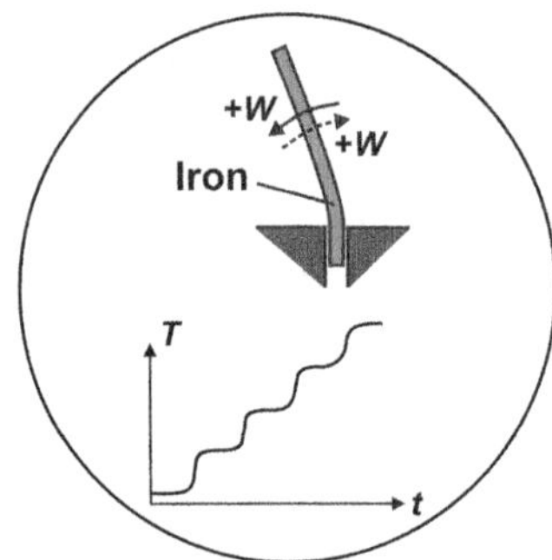

volume in Sect. 2.5. If only infinitesimal changes dV and dS are considered, the following is valid:

$$dW = \underbrace{-p\,dV}_{dW_{\rightarrow V}} + \underbrace{T\,dS}_{dW_{\rightarrow S}} \; . \tag{3.24}$$

These kinds of equations, which describe the energy paths of a system, will be gone into in more detail in Chap. 9. At this point it will suffice to say that the increase of energy dW in our example is composed of one part $dW_{\rightarrow V} = (dW)_S$ in the V direction if all other parameters are kept constant (in this case it is only S) and a second part $dW_{\rightarrow S} = (dW)_V$ in the S direction, meaning at a constant V. In a graph of the function $W(V, S)$, the negative pressure $-p$ appears as the slope in the V direction and the temperature T as the slope in the S direction (Fig. 3.30). To visualize the foregoing: the slope of a mountainside m in the direction of north equals the increase of altitude Δh in this direction divided by the corresponding horizontal distance Δs in the same northerly direction, $m = \Delta h/\Delta s$ or more precisely, $m = \mathrm{d}h/\mathrm{d}s$ (compare Sect. A.1.2 in the Appendix). The following is correspondingly valid:

$$-p = \frac{\mathrm{d}W_{\to V}}{\mathrm{d}V} = \left(\frac{\partial W}{\partial V}\right)_S \quad \text{and} \quad T = \frac{\mathrm{d}W_{\to S}}{\mathrm{d}S} = \left(\frac{\partial W}{\partial S}\right)_V . \tag{3.25}$$

The increase of energy ΔW over longer paths, from location $P_1 = (V_1, S_1)$ in the (V, S) plane to a second location, $P_2 = (V_2, S_2)$, for example, can be found by adding up all the tiny segments along path $\mathcal{W}$. Curved paths can be approximated by zigzag curves made up of paraxial segments (dotted lines in the (V, S) plane in Fig. 3.30). In the case of infinitesimally small curve segments, the sum becomes an integral. For the increase $\Delta W = W(V_2, S_2) - W(V_1, S_1)$, we obtain:

$$\Delta W = \underbrace{-\int_{\mathcal{W}} p\,\mathrm{d}V}_{W_{\to V}} + \underbrace{\int_{\mathcal{W}} T\,\mathrm{d}S}_{W_{\to S}} . \tag{3.26}$$

$W_{\to V}$ is the resulting sum from all the segments running from right to left on the zigzag course, and correspondingly, $W_{\to S}$ results when all the parts along the segments running from front to back are added up. The path could be expressed in parametric form by assigning the coordinates of all the points being traversed $(V(t), S(t))$ as functions of a parameter such as time t.

In the cases we will generally be dealing with, ΔW is independent of the path, but the individual parts such as the mechanical $W_{\to V}$ and the thermal $W_{\to S}$ are not. This can most easily be seen when the paths from P_1 to P_2 along the outer edge of the

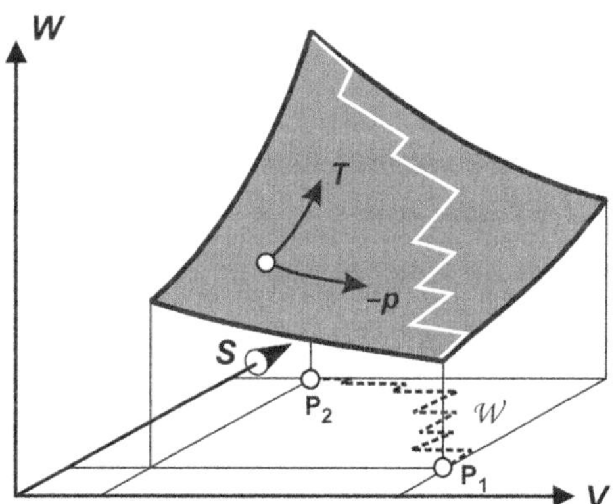

Fig. 3.30 Energy W as a function of volume V and entropy S.

gray surface are compared, first along the path on the left of the diagonal and then along the one on the right. The increase corresponding to $W_{\rightarrow V}$ on the left-hand path is small and the $W_{\rightarrow S}$ is correspondingly large, whereas it is just the opposite on the right-hand side.

If one knows that ΔW is independent of any path it takes, a lot of mathematical work can be saved by carefully choosing the path for determining ΔW. The arrows inserted for clarity's sake into the index above will generally be left out later on.

3.12 Determining Energy Calorimetrically

In Sect. 2.2 we discussed a method for measuring amounts of energy that resembles the one used since ancient times for quantifying lengths, time spans, and amounts of substance. This method involved dividing them into unit portions and then counting them. The unit portion we chose for energy was the amount necessary for stretching a so-called unit-spring. This method is easy to understand, but it is unfeasible because loss is unavoidable. The most common cause for energy loss is obstruction due to friction and the unwanted generation of entropy associated with it.

We can try to make the best of this and measure an amount of energy W by completely dissipating it and then determining how much entropy $S_g = W/T$ is generated at a given temperature T. The devices used for this are called "calorimeters," and we have already seen examples of them (Sect. 3.7). However, at this point we must be careful to neither lose any of the generated entropy nor to allow any addition from other sources. This is often the one viable method for measuring energy in chemical changes because it is nearly the only way of overcoming the ever-present inhibitions. We will come back to this later on. For now, we will begin with a mechanical example.

Let us suppose that we want to determine how much energy W is necessary to raise an object a distance h from the floor (Fig. 1.5). Instead of measuring W while the object is being lifted, we can find W while the object is being lowered, which we might do by drawing the rope over a braked hoisting drum connected to a calorimeter. An ice calorimeter could be used for this where the entropy generated (S_g) and the energy released ($W = T \cdot S_g$) in the brake shoe can be determined from the amount of melted ice. Theoretically, the energy released by expanding a spring, the impact of a thrown stone, the outflow of a compressed gas, or burning of a candle can be measured in this way.

Unfortunately, there is a hitch: latent heat, or rather, *latent entropy*. When an object is affected by compressing, stretching, electrifying, magnetizing, or by chemical alteration, it can become warm or cold even when no entropy is generated. Because of temperature differences, entropy begins to flow out into the environment or into the object from its environment making the amount of entropy in the object change. This process continues until temperatures are equalized again between the object and its environment. Such isothermal changes of entropy are called "latent entropies" ΔS_ℓ. The term "latent" was coined for caloric effects of this type in the eighteenth century. We will take a closer look at this concept in Sect. 8.7.

Every additional entropic effect interferes with measuring S_g. In mechanics, we have learned to overlook these effects because they appear to be totally meaningless. An example will make us realize that this impression is wrong. If we stretch a steel wire, it becomes colder, and when it is released, it warms up again. The change of temperature ΔT is small, only -0.5 K, even when the wire is extended to its limits. The wire needs to absorb entropy from its environment in order to retain its temperature. When the wire is allowed to shorten, the entropy flows back into the environment. In this case, the latent entropy is negative, $\Delta S_\ell < 0$. Energy W must be expended in order to expand the wire. We might determine W by allowing the expanded wire to snap back to its relaxed state in a calorimeter and measuring the generated entropy $S_g = W/T$. However, latent entropy greatly interferes with this because ΔS_ℓ is of approximately the same magnitude as S_g. If the stretching is small, it is even the dominant effect. In the calorimeter we measure the effects combined: $S_g - \Delta S_\ell = S_g + |\Delta S_\ell|$. Therefore, the procedure is only useful if it is possible to determine the latent entropy alongside the sum of the terms. In this example, it is easy to do because we can expand and relax the wire without noticeably generating entropy, so that $S_g \approx 0$ and therefore ΔS_ℓ can be determined using the same calorimeter.

In mechanics, energies are hardly ever measured directly, and certainly never calorimetrically. They are almost always calculated indirectly from measured or imagined forces and displacements. This is the preferred method because it is simpler to use and gives more exact results. In chemistry, though, things are different because to a large extent, one generally depends upon calorimetry. This gives the reverse impression that caloric effects are characteristics of transformations of substances and these types of processes cannot be properly described or understood without them. Fortunately, this impression is also false. We will return to caloric effects in Chap. 8.

3.13   Heat Pumps and Heat Engines

A *heat pump* like the one represented for example by the refrigerator described in Sect. 3.6 is a device that conveys entropy from a body of lower temperature T_1 to a body with a higher temperature T_2. The energy needed to transfer an amount of entropy $S_{t(ransfer)}$ can be easily found. It equals the energy $W_2 = T_2 \cdot S_t$ that is needed to press the entropy into the warmer body, minus the energy $W_1 = T_1 \cdot S_t$ that is gained when the entropy is removed from the colder body (Fig. 3.31):

$$W_t = (T_2 - T_1) \cdot S_t. \tag{3.27}$$

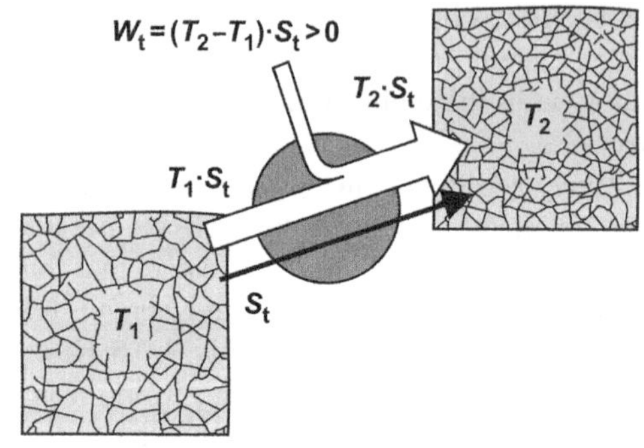

Fig. 3.31 Flow diagram of energy and entropy in an ideal heat pump (*gray circle*).

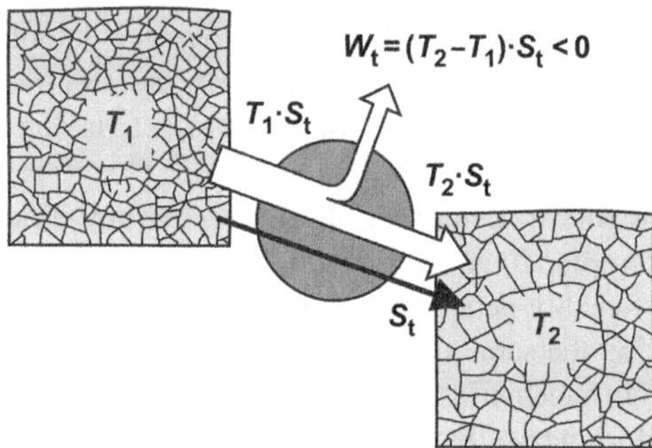

Fig. 3.32 Flow diagram of energy and entropy in an "ideal" thermal motor (heat engine) (*gray circle*).

Friction and other processes always generate some extra entropy in either smaller or larger amounts and this takes extra energy. The total amount of energy W_{total} becomes greater. The *efficiency* η of the device is expressed as follows:

$$\eta = \frac{W_t}{W_{total}}. \tag{3.28}$$

A *heat engine* or "thermal motor" (as an engine of this kind could be called following the language use in electricity) is the reverse of a heat pump. Energy is gained during the transfer of entropy out of a warmer body at temperature T_1 into a colder one with the temperature T_2 (Fig. 3.32). This energy can be calculated with the same equation that is used for finding the amount of energy needed for a heat pump. The only difference is that W_t is now negative because of $T_2 < T_1$. This means that W_t does not represent expended energy but energy gained, so-called *useful energy*.

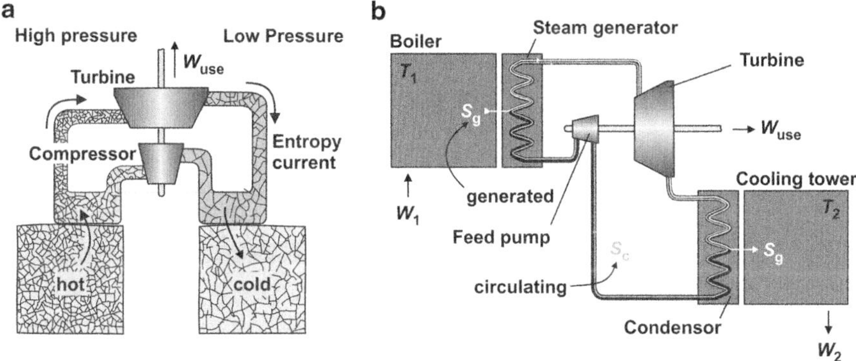

Fig. 3.33 (**a**) Possible inner setup of an "ideal" heat engine, (**b**) Simplified schematic diagram of a thermal power plant.

Figure 3.33 presents the possible inner setup of an "ideal" heat engine in more detail (Fig. 3.33a) as well as the strongly simplified schematic diagram of a thermal power plant (Fig. 3.33b). In the case of such a plant, the energy W_t ($= W_{use}$) is used which is gained during the transfer of entropy from the steam boiler to the cooling tower. The entropy itself is generated in the boiler by consumption of energy W_1.

When we use up energy W_1 to generate entropy S_g, we know that this is a one-way street with no return. Even so, we do not need to consider W_1 as completely lost. As we have seen in our example, if S_g is generated at the higher temperature T_1, it is actually possible to regain at least a part of W_1. A heat engine might transfer the entropy from a temperature of T_1 down to T_2, and ideally return energy $W_t = S_g \cdot (T_2 - T_1) < 0$ to us. In this case the quantity is counted as negative since energy is *released*. Entropy S_g cannot be destroyed so it must end up in some repository. If T_2 is the temperature of such a repository, then $W_2 = S_g \cdot T_2$ describes the amount of energy needed for this transfer, quasi the "fee" for use of the repository. Only W_2 can be considered to be lost but not W_1.

If it were possible to find a repository with a temperature of $T_2 \approx 0$, then $W_2 \approx 0$ and we would be able to recover W_1 almost completely. This energy would then be available for any type of use. It remains "undamaged" and retains its value while it is distributed over many atoms. It is neither really *lost* nor is it really *devaluated*. For this reason we will avoid using these expressions for their undesirable associations. The term "*burnt energy*" is actually much more exact because it simultaneously refers to two important aspects: the waste of useful energy and the heating associated with it.

A water mill works in exactly the same way as a thermal motor when water flows from a higher level to a lower one (Fig. 3.34). In this case, the entropy corresponds to the mass m of the water and the temperature corresponds to the term $g \cdot h$.

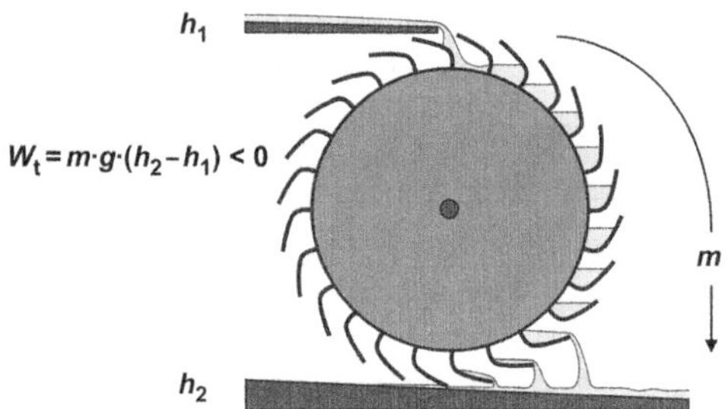

Fig. 3.34 Energy production by a water mill.

Experiment 3.11 *Magnetic heat engine*: When it heats up in the flame, the wheel flange (made of a CuNi alloy) loses in this part its ferromagnetism due to the low Curie temperature. (The Curie temperature is the temperature at which a ferromagnetic material loses its magnetism and becomes paramagnetic.) A force results which keeps the wheel in motion after a push-start. Because the left hot part of the heated wheel flange is less "magnetic" than the right cold part, the wheel flange is pulled from right to left in the area of the pole shoes.

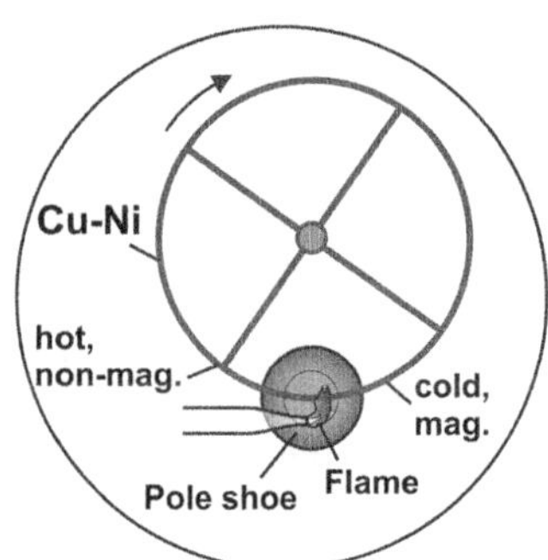

Another example would be a turbine operating between two water containers with different hydrostatic pressures. In electrodynamics the electric motor fulfills this role.

We try to direct the processes in nature in such a way that energy is left over which is freely available. By directing for example a rivulet over a mill wheel we cannot only grind corn but also pump water or drive a generator. *Free* means that the use is not predetermined. We will only say that energy is *set free* or *released* if we have the freedom to use it, even if it is just the freedom to "burn" it.

Some entropy is also generated in a thermal motor, a water mill etc., as a result of friction and other processes. This costs some of the energy W_t, so that the actual usable energy is smaller.

Finally, we will look at two examples of heat engines. Let us begin with the *magnetic heat engine* (Experiment 3.11).

The *rubber band heat engine* (Experiment 3.12) represents an alternative.

Experiment 3.12 *Rubber band heat engine*: While the wheel is centrally borne, the rubber bands pull at an eccentrically positioned wheel boss. Because the spokes are tauter after they are heated, the wheel begins to rotate and that from the right to the left in the *lower region*.

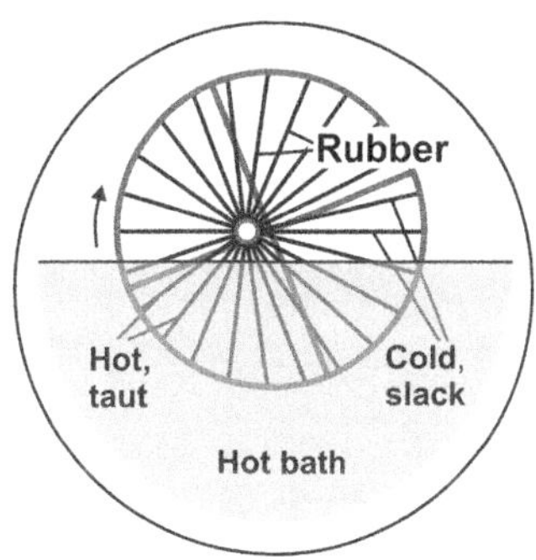

3.14   Entropy Generation in Entropy Conduction

We consider the flow of entropy through a conducting connection that we will call a "conducting segment," from a body with the higher temperature T_1 to another with the lower temperature T_2 (Fig. 3.35). We might imagine a rod made up of a material that conducts entropy well. It is insulated along its long side. One end is heated by a Bunsen burner, while the other is cooled by water. For the transfer of the amount S of entropy the energy $W = (T_2 - T_1) \cdot S$ is necessary. Its value is negative because of $T_2 < T_1$, meaning that the energy is released and has not to be expended. But where is the released energy? It cannot be stored anywhere, so it must have been used to generate entropy, it is "burnt," $W_b = -W$. The entropy S_g generated in the conducting connection must also flow—permanently increasing—down the temperature gradient to arrive finally at the cooler body with the temperature T_2. The amount S_g can be calculated from the released energy W_b:

$$S_g = \frac{W_b}{T_2} \tag{3.29}$$

with

$$W_b = -W = -(T_2 - T_1) \cdot S = (T_1 - T_2) \cdot S. \tag{3.30}$$

In the process of being conducted through a temperature gradient, entropy will increase according to set laws. This is a surprising but inevitable result of our considerations. The energy flowing to the cooler body is calculated according to

$$T_2(S + S_g) = T_2 \cdot S + T_2 \cdot \left[\frac{(T_1 - T_2)S}{T_2} \right] = S \cdot T_1. \tag{3.31}$$

Thus, it is exactly as much as the value $S \cdot T_1$ that is released by the hotter body. While the amount of entropy increases during conduction, the energy current remains constant. W_b represents the energy used up ("burnt") along the conducting

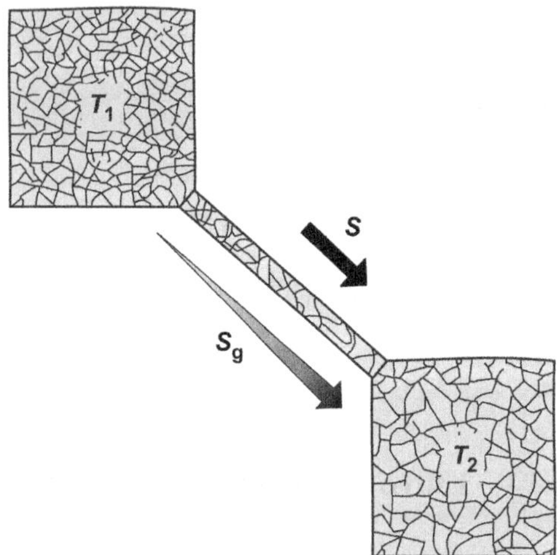

Fig. 3.35 Entropy generation related to the flow of entropy through a temperature gradient.

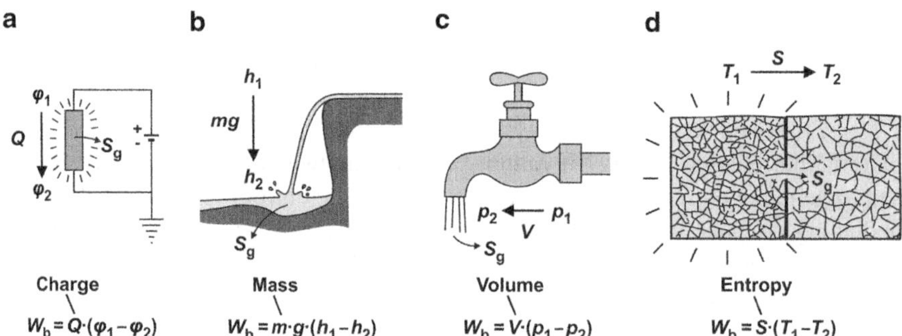

$$W_b = Q \cdot (\varphi_1 - \varphi_2) \qquad W_b = m \cdot g \cdot (h_1 - h_2) \qquad W_b = V \cdot (p_1 - p_2) \qquad W_b = S \cdot (T_1 - T_2)$$

Fig. 3.36 Energy release and entropy generation for (**a**) a potential drop of charge, (**b**) a mass falling from a height, (**c**) a pressure drop of a volume, (**d**) a temperature drop of entropy.

segment. If an ideal heat engine were interposed here instead of the conducting segment, this energy would be useful energy. Here, this energy is not used and it becomes devalued while entropy increases.

Entropy conduction (Fig. 3.36d) can be compared to electric conduction (Fig. 3.36a). If an electric charge Q is forced through an electric resistor—from a higher to a lower potential φ—the resistor will become warm. This is a simple way of generating entropy which we encountered with the immersion heater in Experiment 3.6. The energy W_b released and completely "burnt" in this case results from a substance-like quantity that is pushed through the "conducting segment"—here the electric charge—and the drop of a potential—here the electric potential φ:

$$W_b = -(\varphi_2 - \varphi_1) \cdot Q = (\varphi_1 - \varphi_2) \cdot Q. \tag{3.32}$$

The entropy generated is calculated as W_b/T_2, where T_2 is the temperature of the segment. Using analogical reasoning, we can interpret the generation of entropy in

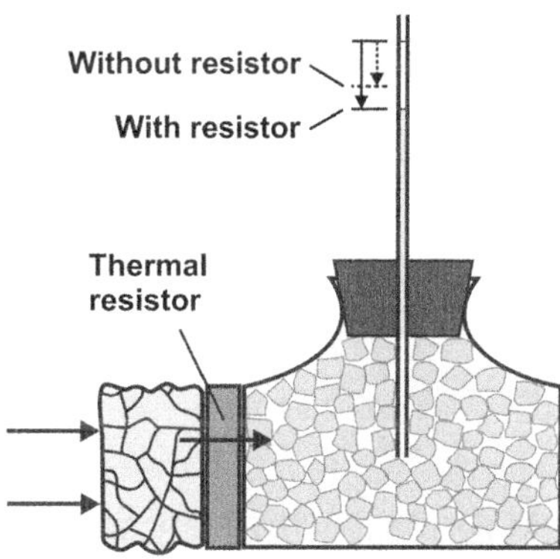

Fig. 3.37 Entropy generation by entropy exchange through a resistor.

entropy conduction as the result of the forcing of entropy through a "thermal" resistor. The temperature plays the role of a "thermal potential" and the entropy that of a "thermal charge" [Eq. (3.30)]:

$$W_b = (T_1 - T_2) \cdot S.$$

A vivid comparison is also that of a waterfall (Fig. 3.36b) where the released and "burnt" energy is found from the water mass m involved and the height of the drop, or more exactly, the drop of the "gravitational potential" $\psi = \psi_0 + g \cdot h$ where h represents the height above sea level:

$$W_b = (\psi_1 - \psi_2) \cdot m = m \cdot g \cdot (h_1 - h_2). \tag{3.33}$$

The amount of entropy generated can be calculated from the quotient W_b/T_2 where T_2 is the temperature of the effluent water. At last, let us mention an example from hydraulics, an opened water tap (Fig. 3.36c). Here, the pressure p acts as potential:

$$W_b = (p_1 - p_2) \cdot V. \tag{3.34}$$

There are two distinguishable steps that these processes all have in common:

1. Release of energy by a drop of a flowing "something" (characterized by a substance-like quantity) from a higher to a lower potential.
2. "Burning" of energy thereby generating entropy.

When entropy is conducted (Fig. 3.36d), this relation becomes a bit blurred because flowing and generated quantities have the same nature.

This type of entropy generation by forcing entropy through a resistor can be demonstrated experimentally (here as a thought experiment) (Fig. 3.37).

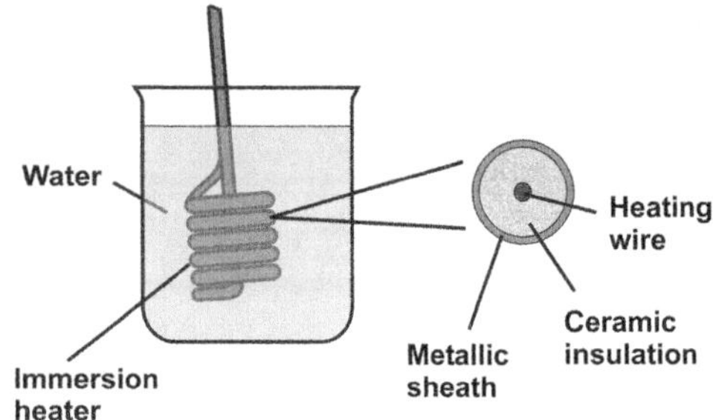

Fig. 3.38 Immersion heater in water. Magnified cross section (simplified) on the *right*.

- *Entropy flow without resistance*: If the auxiliary body is compressed, it remains cold because the entropy escapes into the bottle. The ice melts there, and the level in the capillary falls.
- *Entropy flow through a resistor*: If the same auxiliary body is compressed exactly as before, it will become warm because the entropy can only escape slowly through the resistor. It gradually seeps into the bottle and the capillary level falls even lower than before! Although the auxiliary body releases the same amount of entropy in both cases, the bottle shows more this time.

In closing, let us have a look at a concrete example, a 700 W immersion heater in water (Fig. 3.38). The heating wire should have a temperature T_1 of 1,000 K. Hence, in 1 s an amount of entropy S_g' of

$$S_g' = \frac{W}{T} = \frac{P \cdot \Delta t}{T} = \frac{700\,\mathrm{J\,s^{-1}} \times 1\,\mathrm{s}}{1,000\,\mathrm{K}} = 0.7\,\mathrm{J\,K^{-1}} = 0.7\,\mathrm{Ct}$$

is generated by the wire [see Eq. (3.8)]. At the surface, however, the immersion heater has the same temperature as the surrounding water. We suppose that the water temperature is $T_2 = 350$ K. Along the short path taken by the entropy S $\left(= S_g'\right)$ from the heating wire to the surface of the heater, an amount of entropy S_g of

$$S_g = \frac{W_b}{T_2} = \frac{(T_1 - T_2) \cdot S}{T_2} = \frac{(1,000\,\mathrm{K} - 350\,\mathrm{K}) \times 0.7\,\mathrm{Ct}}{350\,\mathrm{K}} = 1.3\,\mathrm{Ct}$$

is generated [see Eqs. (3.29) and (3.30)]. Therefore, an amount of entropy equal to $S_{\mathrm{total}} = 2.0$ Ct flows into the water per second.

Chapter 4
Chemical Potential

The chemical potential μ is used as a measure of the general tendency of a substance to transform. Only a few properties are necessary for a complete phenomenological description of this new quantity. They are easy to grasp and can be illustrated by everyday examples. It is possible to derive quantitative scales of μ values (initially at standard conditions) by using these properties, and after choosing a convenient reference level. A first application in chemistry is predicting whether or not reactions are possible by comparing the sum of potentials of the initial and the final states. This is illustrated by numerous experimental examples. The quantitative description can be simplified by defining a "chemical drive" $\mathcal{A}$ as the difference of these sums. In this context, a positive value of $\mathcal{A}$ means that the reaction proceeds spontaneously in the forward direction. As a last point, a direct and an indirect measuring procedure for the chemical potential are proposed.

4.1 Introduction

The Greek philosopher Heraclitus concluded from observations of his environment already in the fifth century before Christ that "Everything flows—Nothing stands still ($\pi\acute{\alpha}\nu\tau\alpha\ \rho\varepsilon\ddot{\iota}$)." Creation and decay are well known in the living world but also in inanimate nature the things around us change more or less rapidly. A lot of such processes are familiar to us from everyday life (Experiment 4.1):

- Objects made of iron rust when they come into contact with air and water.
- Bread dries out if one leaves it at air for a longer time.
- Rubber bands or hoses become brittle.
- Paper turns yellow.
- Butter or fat becomes rancid.
- Copper roofs turn green (so-called patina).
- Even the seemingly stable rocks ("solid as a rock") weather.
- Conversely, mud or also wood petrifies.

© Springer International Publishing Switzerland 2016

G. Job, R. Rüffler, *Physical Chemistry from a Different Angle*,

DOI 10.1007/978-3-319-15666-8_4

Experiment 4.1 *Changes in the world of substances*: (**a**) Rusted tin can, (**b**) Dried-out bread, (**c**) Embrittled rubber hose, (**d**) Yellowed and brittle pages of a book, (**e**) Quartz sand from eroded granite, (**f**) Petrified mud.

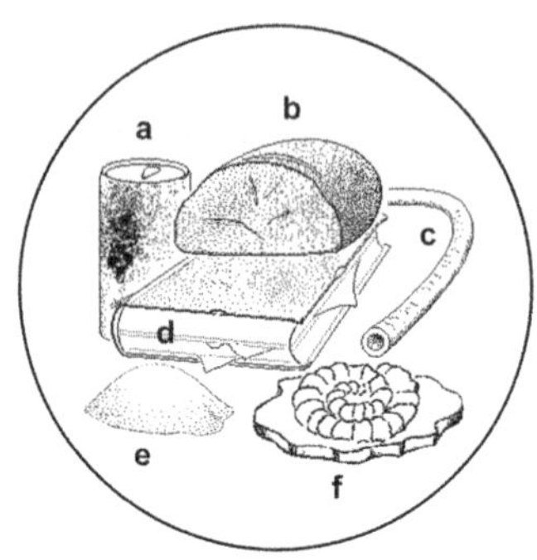

Experiment 4.2 *Aging of acrylic acid*: Acrylic acid as pure substance is a water-clear liquid strongly smelling of vinegar. If left to stand alone in a completely sealed container, it will change by itself after some time into a colorless and odorless rigid glass.

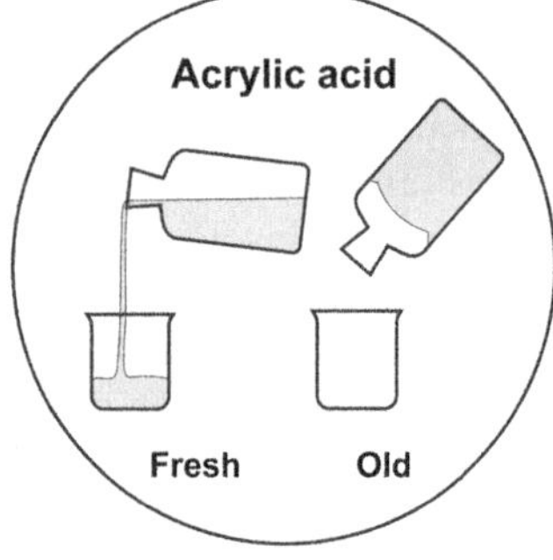

This list could be continued as long as one likes.

It would be possible to consider extraneous effects as the cause. For example, iron would not rust if oxygen were kept from it. However, this is not the point, because substances that are separated from the environment also change. For example, these objects "age" by themselves:

- Bread in a plastic bag,
- Tinned food in an unopened can,
- Chemicals in a sealed bottle such as acrylic acid (propenoic acid) (Experiment 4.2).

The hardening is caused by polymerization, meaning the small acrylic acid molecules combine to form long chains:

$$...+ CH_2{=}CH \atop } + CH_2{=}CH +... \rightarrow ...{-}CH_2{-}CH{-}CH_2{-}CH{-}... .$$

$$\underset{COOH}{|} \quad \underset{COOH}{|} \qquad \underset{COOH}{|} \quad \underset{COOH}{|}$$

The transformation of pure substances such as the weathering of natron ($Na_2CO_3 \cdot 10\ H_2O$) and Glauber's salt ($Na_2SO_4 \cdot 10\ H_2O$) in ambient air where the large colorless crystals become covered with a white powdery crust as they lose water,

$$Na_2CO_3 \cdot 10\,H_2O \rightarrow Na_2CO_3 \cdot 7\,H_2O + 3\,H_2O,$$

$$Na_2SO_4 \cdot 10\,H_2O \rightarrow Na_2SO_4 + 10\,H_2O,$$

the slow transition of the almost colorless monoclinic β-sulfur into the yellow rhombic α-sulfur, or that of the low-molecular white phosphorus into the high-molecular red phosphorus,

$$S|\beta \rightarrow S|\alpha,$$

$$P|white \rightarrow P|red,$$

all show that it is not an interaction between reaction partners that is the motor for the change of substances, but that the substances tend to transform by themselves. This means that each and every individual substance has a *"tendency to transform."* This inherent tendency to transform is certainly not the same for all substances, and it has no particular "goal." One might say that all substances are "driven to transform" to one extent or another. They use every opportunity that comes up to follow this "drive," or tendency. A somewhat casual but catchy way to express this would be that they somehow want to "sneak off." Most substances known to us only survive over a longer period of time because many of the transformation processes are inhibited, and not because the drive for them does not exist.

From the transition of the white into the red phosphorus mentioned above, it can be concluded that the white type has the stronger tendency to transform and forces the formation of the red type against its own tendency to transform. Similarly, we can imagine that iron sulfide is formed because the starting substances iron and sulfur together have a stronger tendency to transform than the product FeS. When various metal powders (such as magnesium, zinc, iron, copper, and gold) react with sulfur, the differences are very pronounced. For example, magnesium, when mixed with sulfur and ignited, explodes violently. In contrast, the last metal powder, gold, does not react with sulfur at all:

$$Mg \text{————} Zn \text{————} Fe \text{————} Cu \text{————} Au\,!$$
$$\text{explosive} \quad \text{glaring} \quad \text{glowing} \quad \text{glimmering} \quad \text{nothing.}$$

Here, it becomes immediately obvious that the suggested tendency to transform of the various metal sulfides (compared to the elements from which they are formed) is totally differently strong. On the basis of the violence of reaction, we arrive at the following sequence:

$$MgS < ZnS < FeS < CuS < \text{"AuS."}$$

Obviously, magnesium sulfide is the easiest to produce, meaning it has the weakest tendency to transform. Gold sulfide, on the other hand, seems to have a relatively strong tendency to transform. It is possible, however, to obtain various compounds of gold and sulfur by indirect means, but they all tend to decompose into the elements involved. We can, therefore, confidently assume that AuS is not produced because its tendency to transform exceeds that of Au + S combined.

We will now go more deeply into the meaning of tendency to transform and its quantitative description with the help of the so-called chemical potential.

4.2 Basic Characteristics of the Chemical Potential

Before we attempt to quantify this new concept we will create an overview of what it means, what it is good for, and how it can be dealt with. In order to do this, we compile the most important characteristics of the chemical potential into a short outline, a kind of "wanted poster," which we will subsequently go into more deeply.

- The tendency of a substance

    - To decompose or to *react* with other substances,
    - To undergo a *transition* from one type of state to another,
    - To *redistribute* in space

    can be expressed by one and the same quantity—its chemical potential μ.
- The strength of this tendency, meaning the numerical value of μ, is not unchangeable but

    - Is determined by the *nature* of the substance,
    - By its *milieu*,

    but *neither* by the nature of reaction partners *nor* the resulting products.
- A *reaction, transition, redistribution*, etc., can only proceed spontaneously if the tendency for the process is more pronounced in the initial state than in the final state.

We can assume that any substance, let us call it B, has a more or less pronounced tendency to *transform*. This means a tendency to *decompose* into its elementary (or other) components, to *rearrange* itself into some isomer, $B \rightarrow B^*$, or to *react* with other substances B', B'', $\ldots$,

$$B + B' + \ldots \rightarrow \ldots .$$

Even less drastic *transformations* of substance B, such as changing the state of aggregation, the crystalline structure, the degree of association, etc., which can be symbolized for example as follows:

$$B|\alpha \rightarrow B|\beta$$

are driven by the same tendency to transform. This also holds for the tendency of a substance to redistribute in space. This means its tendency to migrate to another location or to move into a neighboring region:

$$B|\text{location } 1 \rightarrow B|\text{location } 2.$$

The *chemical potential* μ is a measure of the strength of this tendency. We write μ_B or $\mu(B)$ to signify the potential of substance B. The greater the μ, the more active or "bustling" the substance. The smaller the μ, the more passive or "phlegmatic" it is.

As was mentioned earlier, the strength of the inherent tendency to transform, and with it the numerical value of μ_B, fundamentally depends upon the nature of the substance. In this context, we see the nature of a substance being determined by its chemical composition, characterized by its content formula, but also by its state of aggregation, its crystalline structure, etc. Hence, liquid water and water vapor as well as diamond and graphite will exhibit different chemical potentials under otherwise identical conditions and therefore need to be treated as different substances. In addition, the strength of the tendency to transform also depends upon the *milieu* in which the substance is located. By milieu we mean the totality of parameters such as temperature T, pressure p, concentration c, the type of solvent S, type and proportions of constituents of a mixture, etc., which are necessary to clearly characterize the environment of B. In order to express these relations, we may write

$$\mu_B(T, p, c, \ldots, S, \ldots) \qquad \text{or} \qquad \mu(B, T, p, c, \ldots, S, \ldots).$$

Experiment 4.3 illustrates how a substance reacts to a changed milieu. In this case, it is the change of solvent S. Obviously, iodine prefers ether as milieu compared to water. The tendency to transform and thereby the chemical potential of iodine is higher in the water than in the ether—under otherwise identical conditions. We will discuss the influence of the milieu in more detail in the following chapters.

An important characteristic of a substance's tendency to transform is that it is *not* dependent upon the partner it reacts with or what products result. μ is a characteristic of a single substance and not of a combination of substances. This reduces

Experiment 4.3 *Iodine in different milieu*: Iodine dissolved in water (*left side*) separates out when it is shaken with ether (*right side*). The ether floats on top of the specifically heavier, now colorless layer of water. The brown color of the dissolved iodine allows us to easily see where it is.

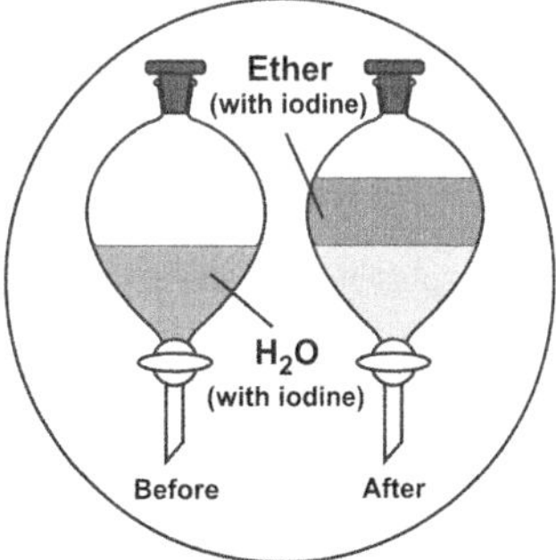

dramatically the amount of data necessary because the number of possible combinations is much, much larger than the number of individual substances itself.

4.3 Competition Between Substances

When a substance disappears, one or even several substances are produced from it, or the substance reappears in another location. The produced substances, however, also show a tendency to transform just like the reactants, so the direction in which a certain process will run depends upon which side has the stronger tendency. Therefore, chemical processes resemble a competition between the substances on either side of the conversion formula.

An image commonly used for this competition is the relationship between things on the right and left pans of an equal-arm balance scale (or seesaw) (Fig. 4.1). The direction in which the scale tips depends solely upon the sum of the weights G on each side of it. Even negative weights are allowed if the objects floating upward (maybe balloons) are attached to the scale.

This behavior can also be expressed mathematically: The left side wins, i.e., the objects B, B$'$, ... on the left side of the balance scale or seasaw are successful against the objects D, D$'$, ... on the right side in their attempt to sink downward if

$$G(\text{B}) + G(\text{B}') + \ldots > G(\text{D}) + G(\text{D}') + \ldots .$$

Equilibrium is established when the sums of the weights on the left and right side of the scale are just equal,

$$G(\text{B}) + G(\text{B}') + \ldots = G(\text{D}) + G(\text{D}') + \ldots .$$

The statements made here for weights correspond completely to the role of chemical potentials in transformations of substances. It makes no difference whether it is a reaction between several substances or a transition of a substance from one state to

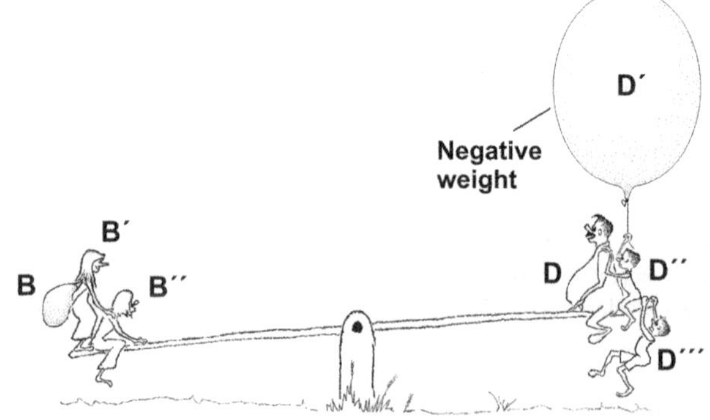

Fig. 4.1 Seasaw as model for the interaction between starting substances on the one side and final products on the other side during a transformation, whereby the weights of the objects correspond to the chemical potentials of the substances.

another, or just a change of location. The direction in which such a process progresses, for example, the reaction

$$B + B' + \ldots \rightarrow D + D' + \ldots ,$$

depends solely upon the sums of the chemical potentials μ of all the substances on either side. The substances on the left side prevail in their attempt to react if

$$\mu(B) + \mu(B') + \ldots > \mu(D) + \mu(D') + \ldots ,$$

(see e.g. Fig. 4.2). Equilibrium is established when the sum of the chemical potentials on both sides is the same and therefore no particular direction preferred:

$$\mu(B) + \mu(B') + \ldots = \mu(D) + \mu(D') + \ldots .$$

For example, a candle burns because the starting substances combined [in this case, atmospheric oxygen and paraffin wax, formula $\approx(CH_2)$] have a higher chemical potential than the final products (in this case, carbon dioxide and water vapor):

$$3\,\mu(O_2) + 2\,\mu((CH_2)) > 2\,\mu(CO_2) + 2\,\mu(H_2O).$$

Therefore, every feasible reaction may be viewed as representing a kind of balance scale that enables us to compare potential values or their sums, respectively. However, the measurement often fails because of inhibitions in the reactions; in other words, the scale is "stuck." In the case of a drop in the chemical potential from the left to the right side, this means that in principle the process *can* proceed in this direction; however, it does not mean that the process will actually run. Therefore, this drop is a necessary but not sufficient condition for the reaction considered. This is not really surprising. An apple tends to fall downward, but it will not fall as long as it hangs from its stem. The coffee in a cup does not flow out over the table although the tendency to do so is there. The porcelain walls of the cup inhibit it from

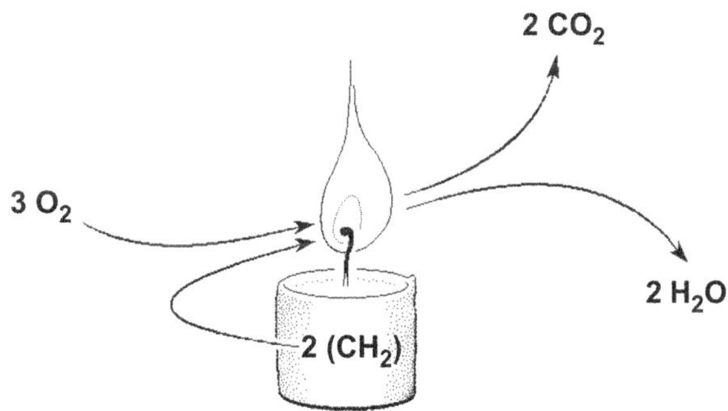

Fig. 4.2 Burning candle as example of a reaction that proceeds spontaneously.

doing so. We do not even have to bore a hole in the cup for the barrier to be overcome. A bent straw that acts as siphon is already enough. When candle wax and air are put together, no fire occurs. The candle wick and flame work as a kind of "siphon" which helps to overcome the inhibitions. Inhibitions are an important part of our world. Without them, we would end up as carbon dioxide, water, nitrogen, and ashes in the sea of oxygen in which we live.

If a transformation tends to run in one direction, this does not mean that the opposite direction is impossible, it just does not happen *spontaneously*. By itself, sand always trickles downward. A mole can shovel it upward, though, just as a harsh desert wind can pile it up into high dunes, but these processes do not occur spontaneously. Hydrogen and oxygen exhibit a strong tendency to react to form water. The reverse process never runs by itself at room conditions, but can be forced to do so in an electrolytic cell. Predicting substance transformations based upon chemical potentials always presupposes that there are no inhibitions to the process and that no "outside forces" are in play. We will gradually go into what this exactly means and what we need to look out for.

The adjoining cartoon concludes this section. Despite its anthropomorphic viewpoint, it is useful as an image of the general behavior of substances:

More active, more "bustling" substances are transformed into more passive, more "phlegmatic" substances. They migrate from "busier" places (with a lot of "activity") to "quieter" places (with weak "activity"). In short, matter aspires to a state of maximum "laziness."

4.4 Reference State and Values of Chemical Potentials

Reference Level Up to now, what we have been missing in order to make concrete predictions are the μ values of the substances we have been dealing with. The chemical potential can be assigned an absolute zero value, just as temperature can. In principle, the absolute values could be used, but they are enormous. It would mean that in order to work with the tiny differences in potentials common in chemical and biological reactions, at least 11 digits would be necessary (the ratio between the potential differences and the absolute values is around one to one billion!). This alone would lead to numbers that are much too unwieldy, not to mention that the absolute values are not known accurately enough for this to be feasible.

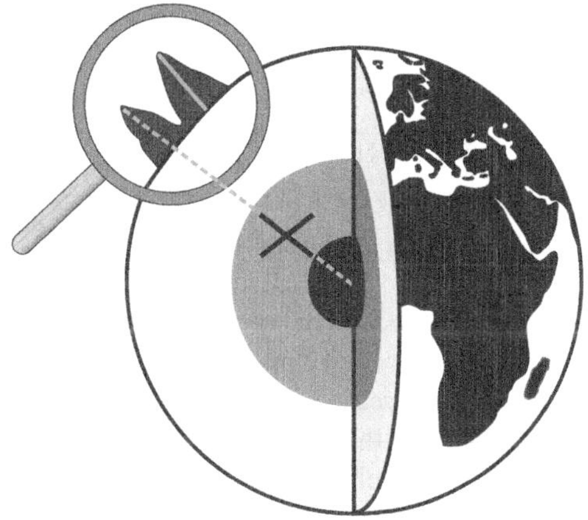

Fig. 4.3 Determining geographical elevations as an example for the selection of an appropriate reference level.

But there is a simple way out of this dilemma. The heights of mountains are not referred to the geocenter but to the sea level (Fig. 4.3). Everyday temperatures are not referred to absolute zero, but are given as Celsius temperatures based upon the freezing point of water. (The zero point of Daniel Gabriel Fahrenheit's original scale was determined by placing the thermometer in brine, here a mixture of ice, water, and ammonium chloride.)

It is similarly practical to choose a convenient level of reference for the values of the chemical potential because differences of μ can be determined much more precisely than absolute values. Moreover, because we only need to compare potential values or their sums, it does not matter what the unit is at first. The μ values could be expressed in various scales similarly to how temperature can be expressed (Celsius, Fahrenheit, Kelvin, Reaumur, etc.). We will use the SI coherent unit "*Gibbs*," abbreviated to G. This name has been proposed by the German chemist Egon Wiberg (Wiberg E (1972) Die chemische Affinität. De Gruyter, Berlin, p 164) to honor Josiah Willard Gibbs (1839–1903) who first introduced the concept of chemical potential. For use in chemistry, the unit kilo-Gibbs (kG), which corresponds to 1,000 Gibbs, is even handier.

Elements Used for "Zero Levels" Next, we will turn to the question of what reference states are suitable for measuring potential differences. It is useful to refer to the conventional basic substances in chemistry, the elements, as long as we limit the transformations of substances to chemical reactions in the broadest sense and exclude nuclear reactions. The values of the chemical potentials of substances are related to the chemical potentials of the elements they are composed of and can be determined experimentally by means of chemical reactions. Because it is not possible to transform one element into another by chemical means, the values of the various elements themselves are not related to each other. This means that in principle one could arbitrarily determine the reference level for each basic substance, i.e., for every element. Because in the case of chemical reactions the elements are preserved, i.e., an equal number of chemical symbols appears on both sides of a conversion formula, this has no effect upon the potential differences that are being observed and measured. Let us take a closer look at the synthesis of ammonia from nitrogen and hydrogen as an example:

$$\underline{N_2 + 3\,H_2 \quad \rightarrow\ 2\,NH_3}$$

$\mu(kG)$:   0 3×0 $2 \times (-16)$ $\Rightarrow \mu(N_2) + 3\,\mu(H_2) - 2\,\mu(NH_3) = +32\,kG$

　　　　　 0 $3 \times 2{,}000$ $2 \times 2{,}984$ $\Rightarrow \mu(N_2) + 3\,\mu(H_2) - 2\,\mu(NH_3) = +32\,kG$

N appears two times on the left as well as on the right side of the conversion formula; H, however, appears six times. Therefore, if the chemical potential of a substance is increased by a fixed, although arbitrary summand (say 1,000 kG, as shown above in the third line) for every H appearing in its content formula, this added value cancels when we compute the potential difference and we end up with the same result as in the second line above. The same holds for nitrogen. This means that the reference level for any element could in principle be chosen arbitrarily as mentioned earlier. For the sake of simplicity, the value 0 is assigned to the chemical potential of all elements.

Additionally, one has to consider the following for the specification of a reference state: The state of an element depends upon its temperature and pressure. It also depends upon whether, for instance, hydrogen appears in atomic or molecular form, carbon in the form of graphite or diamond, or oxygen as O, O_2, or O_3, etc. As an easily reproducible reference state, we will choose the state of the most stable modification of a particular element in its "pure form" and in its natural isotope composition at *standard conditions* (meaning 298 K and 100 kPa, as discussed in Chap. 3). For example, in the case of carbon graphite is used as reference state. An exception to this is phosphorus where the more accessible white (in some tables it is also the red) modification is preferred to the more stable, but very difficult to produce, black modification. In general, we will use the symbol $\mu^{\ominus}$ to designate μ values at standard conditions. Thus, it follows that (if E represents any arbitrary element in its most stable modification):

$$\mu^{\ominus}(E) = 0. \tag{4.1}$$

For elements E such as H, N, O, Cl, etc., which, at standard conditions, usually appear as diatomic gases, 1 mol E simply means $\tfrac{1}{2}$ mol E_2 and $\mu(E)$ correspondingly $\tfrac{1}{2} \cdot \mu^{\ominus}(E_2)$.

Just as the average sea level serves as the zero level for geographical altitude readings, the state of matter where the substances are decomposed into their elements at standard conditions represents the "zero level" of the potential scale. Analogously, Celsius temperature readings usually used in everyday life can replace those of absolute temperature if melting ice is chosen as reference state.

Substances of All Kinds The chemical potential μ of an arbitrary pure substance itself depends upon temperature and pressure (and possibly other parameters), $\mu(T, p, \ldots)$. Therefore, it is usual in chemistry to tabulate the potentials of substances (referred to the elements that form them) in the form of *standard values* $\mu^{\ominus}$, i.e., for 298 K and 100 kPa. In Table 4.1 we find such standard values for some common substances.

Note that the potential value 0 for iron does not mean that iron has no "tendency to transform," but only that we have used its potential as the zero level to base the values of the potential of other iron-containing substances upon.

The selection of substances in the table shows that not only well-defined chemicals are referred to when speaking about chemical potential, but everyday substances as well. In the case of marble, certain impurities are responsible for its colors, but these substances have almost no effect upon the chemical potential of its main component, $CaCO_3$. However, in order to specify the potential μ of a substance, an appropriate content formula has to be assigned to it which shows how it is composed of the elements, and which would then be binding for all calculations. This is why this formula must be present in such a table. But the chemical potential of a pure substance also depends on its state of aggregation, its crystal structure, etc. For example, liquid water and water vapor exhibit different chemical potentials at the same temperature and pressure; the same is valid for example for diamond and graphite. In order to arrive at unambiguous specifications, we once again call attention to the relevant additions |s (for solid), |l (for liquid), and |g (for gaseous) to the symbol of a substance (cp. Sect. 1.6); modifications can be characterized by Greek letters |α, |β, ... or the full name such as |graphite, |diamond, ... etc.

Table 4.1 Chemical potentials of several selected substances at standard conditions (298 K, 100 kPa, dissolved substances at 1 kmol m^{-3}).

Substance	Formula	$\mu^{\ominus}$(kG)
Pure substances		
Iron	Fe|s	0
Graphite	C|graphite	0
Diamond	C|diamond	+3
Water	H_2O|l	−237
Water vapor	H_2O|g	−229
Table salt	NaCl|s	−384
Quartz	SiO_2|s	−856
Marble	$CaCO_3$|s	−1,129
Cane sugar	$C_{12}H_{22}O_{11}$|s	−1,558
Paraffin wax	≈(CH_2)|s	+4
Benzene	C_6H_6|l	+125
Acetylene (Ethyne)	C_2H_2|g	+210
In water		
Cane sugar	$C_{12}H_{22}O_{11}$|w	−1,565
Ammonia	NH_3|w	−27
Hydrogen(I)	H^+|w	0
Calcium(II)	Ca^{2+}|w	−554

Because our immediate goal here is a first basic knowledge of the chemical potential, we will for the time being consider the μ values of substances as given, just as we would consult a table when we are interested in mass density or electric conductivity of a substance. Some measuring methods will be discussed in the concluding Sects. 4.7 and 4.8.

Dissolved Substances The potential of a substance B changes if it is brought into another milieu for example by dissolving it. Depending on the type of solvent S in question we obtain different values for the chemical potential. This type of state can be characterized in general by the addition |d (for dissolved) or more precisely by |S—if one would like to specify also the type of solvent. In the by far most common case, substances dissolved in water, we use the symbol |w. But, what matters in this context is not only the type of the solvent but also the concentration of B. Therefore, the concentration c of a dissolved substance, for which the tabulated value will be valid, must be specified in addition to p and T. The usual reference value is 1 kmol m^{-3} $(= 1 \text{ mol L}^{-1})$. There exist some peculiarities concerning the determination of these standard values (such as in the case of gases), but we will discuss them in Sect. 6.2.

We can summarize:

$$\mu^{\ominus} = \mu(T^{\ominus}, p^{\ominus}) \qquad \text{for pure substances} \qquad T^{\ominus} = 298 \text{ K}$$
$$\mu^{\ominus} \approx \mu(T^{\ominus}, p^{\ominus}, c^{\ominus}) \quad \text{for dissolved substances} \quad p^{\ominus} = 100 \text{ kPa}$$
$$c^{\ominus} = 1 \text{ kmol m}^{-3}$$

$T^{\ominus}, p^{\ominus}, c^{\ominus}$ indicate *standard temperature, standard pressure*, and *standard concentration*.

Zero-Order Approximation As long as the temperature does not vary by more than ± 10 K, and pressure and concentration do not vary more than a power of ten, the changes of potential of substances of low-molecular mass remain about ± 6 kG in general. Therefore, we can consider the μ values to be constant, at least very roughly. This precision is often sufficient for us so that we can use the $\mu^{\ominus}$ values found in tables. It is unnecessary to worry about temperature, pressure, and concentration dependencies of the potentials at the moment. We will only start dealing in more detail with these influences in the following chapters. The approximation used here is a kind of zero-order approximation.

Charged Substances Just like a substance, an assembly of ions can be assigned a chemical potential. When ions of a certain type are decomposed into their elements, there is a positive or negative amount n_e of electrons left over along with the neutral elements, for example,

$$CO_3^{2-} \rightarrow C + \tfrac{3}{2} O_2 + 2 e^-.$$

The electrons appear here as a kind of additional element (cf. Sect. 1.6) that, like all elements, can be assigned the value $\mu^{\ominus} = 0$ in a certain reference state. However, electrons in a free state play no role in chemistry. Therefore, a value for $\mu^{\ominus}(e^{-})$ has been arbitrarily chosen so that the most commonly appearing type of ion H^{+} (in an aqueous solution and at standard conditions) receives the $\mu^{\ominus}$ value of zero:

$$\mu^{\ominus}(H^{+}|w) \ = \ 0. \tag{4.2}$$

At first, this seems surprising because we know that the chemical potential of an element at standard conditions is zero, i.e., $\mu^{\ominus} = 0$. This is of course also valid for hydrogen, $\mu^{\ominus}(H_{2}|g) = 0$. That is why we expect that other states of hydrogen would show divergent $\mu^{\ominus}$ values. But let us have a look at the system hydrogen gas/hydrogen ion, which is capable of providing electrons without major inhibitions under suitable conditions (the symbol := should be read as "equal by definition"):

$$H_{2}|g \rightleftarrows 2\,H^{+}|w + 2\,e^{-}$$

with

$$\underbrace{\mu^{\ominus}(H_{2}|g)}_{:= 0} = 2\,\underbrace{\mu^{\ominus}(H^{+}|w)}_{0} + 2\,\underbrace{\mu^{\ominus}(e^{-})}_{:= 0}.$$

When H_{2} and H^{+} are present at standard conditions and *equilibrium* has been established, the chemical potential of the electrons, $\mu^{\ominus}(e^{-})$, is supposed to be zero. (The electron potential $\mu(e^{-})$, abbreviated μ_{e}, will be discussed in more detail in Chap. 22.) Because $\mu^{\ominus}(H_{2}|g)$ disappears by definition, it follows necessarily that in a state of equilibrium, $\mu^{\ominus}(H^{+}|w)$ has to be zero as well.

4.5   Sign of the Chemical Potential

If we use values of chemical potentials in the following, they are valid for room conditions and for dissolved substances of concentrations of 1 kmol m^{-3} ($= 1$ mol L^{-1}) where water is the usual solvent. Elements in their usual, stable state receive, as agreed, the value $\mu^{\ominus} = 0$ (see also Table 4.3 at the end of this chapter or Sect. A.2.1 in the Appendix). This is for example valid for molecular hydrogen $\mu^{\ominus}(H_{2}|g) = 0$, while atomic hydrogen has a rather high positive potential $\mu^{\ominus}(H|g) = +203\,kG$. This means that its tendency to transform into H_{2} is very strong.

   A look at Table 4.3 and Sect. A.2.1 in the Appendix shows something remarkable. Most of the potential values are *negative*. A substance with negative chemical potential can be produced spontaneously from the elements because it has a weaker tendency to transform than the elements it is produced from. However, this also means that most substances do not tend to decompose into their elements but, in

Experiment 4.4 *Decomposition of S_4N_4*: A small amount of tetrasulfur tetranitride explodes (like a cap for use in toy guns) when hit lightly with a hammer.

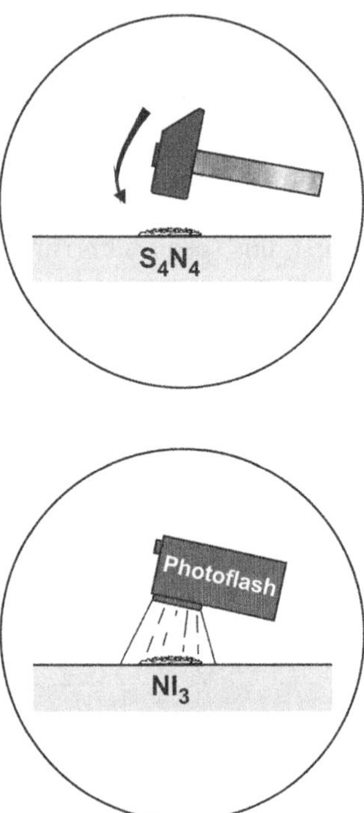

Experiment 4.5 *Decomposition of NI_3*: Nitrogen triiodide decomposes in a dry state if touched by a feather or irradiated by a flash of light thereby producing a sharp explosive sound. The produced iodine can be easily identified by the cloud of violet smoke.

contrast, tend to be produced from them. Therefore, most of the substances we deal with are *stable*; they do not decompose.

If, on the other hand, the potential is positive, the substance will tend to decompose into its elements. Such a substance is *unstable*, thus eluding preparation or is *metastable* at best, i.e., in principle a spontaneous decomposition is possible, but there exists an inhibition. If the inhibition can be overcome, e.g., by supplying energy or by making use of a catalyst, it is very common for the substance to react violently, especially when the value of μ is very large.

This behavior can be demonstrated quite impressively by means of the attractive orange crystals of tetrasulfur tetranitride S_4N_4 ($\mu^{\ominus} \approx +500$ kG) (Experiment 4.4).

Further examples would be heavy metal azides such as lead azide $Pb(N_3)_2$ (used in detonators to initiate secondary explosives) or silver azide AgN_3. Furthermore, the easily produced black nitrogen triiodide NI_3 ($\mu^{\ominus} \approx +300$ kG) tends to decompose into its elements as well (Experiment 4.5).

However, a positive μ does not always mean that the substance must be explosive. For example, benzene remains rather stable in spite of its $\mu^{\ominus}$ value of $+125$ kG. As discussed in Sect. 4.3, a positive μ value is a necessary but not sufficient condition for a spontaneous decomposition of a substance into its elements to take place. Therefore, we cannot simply assume that just because there is a possibility of transformation, it will take place within a certain span of time, be it years, millennia, or millions of years.

Comparing analogous substances shows best how the level of the chemical potential affects their properties. Here are three examples:

	CO_2|g	NO_2|g	ClO_2|g
$\mu^{\ominus}$ (kG)	-394	$+52$	$+123$

The gas CO_2 with its strongly negative $\mu^{\ominus}$ value is stable and spontaneously produced from carbon and oxygen, i.e., carbon is "combustible." NO_2 with positive $\mu^{\ominus}$ is not formed spontaneously from N_2 and O_2, but is so stable that it is not dangerous to handle. Finally, ClO_2 has an even higher chemical potential and is extremely explosive.

A similar consideration can be used for solid oxides:

	Al_2O_3|s	Fe_2O_3|s	Au_2O_3|s
$\mu^{\ominus}$ (kG)	$-1{,}582$	-741	$+78$

Aluminum and iron combine with oxygen to form their stable oxides, while solid Au_2O_3 must be handled carefully so that no oxygen separates from it.

The category of metal sulfides also contains similarly composed substances that are appropriate for comparison:

	MgS|s	ZnS|s	FeS|s	CuS|s	"AuS"|s
$\mu^{\ominus}$ (kG)	-344	-199	-102	-53	>0

The sequence deduced in Sect. 4.1 from the violence of the reactions of formation actually runs parallel with the values of the chemical potentials. However, be careful: A vague characteristic such as the violence of reaction that is dependent upon different factors can only be considered evidence under comparable conditions.

4.6   Applications in Chemistry and Concept of Chemical Drive

Concept of Chemical Drive   The most important application for the chemical potential μ is that it enables us to predict whether a transformation of substances can happen spontaneously or not. As we have seen, a chemical reaction

$$B + B' + \ldots \rightarrow D + D' + \ldots$$

is possible when the following is valid:

$$\mu(B) + \mu(B') + \ldots > \mu(D) + \mu(D') + \ldots.$$

If we wish to find out if a previously unknown process can run spontaneously, it is enough to find the corresponding μ values in appropriate tables and then to compare the potentials on the right and left side of the conversion formula. Spontaneously, processes only run "downhill," meaning from left to right, when the sum of the μ values on the left is greater than on the right.

The condition for a spontaneous process results in

$$\mu(B) + \mu(B') + \ldots - \mu(D) - \mu(D') - \ldots > 0$$

after rearrangement of the formula above. The summation of the variables can be presented in a shorter form by using the sigma sign, $\sum$. We summarize:

$$\text{reactants} \rightarrow \text{products is spontaneously possible if } \sum_{\text{initial}} \mu_i - \sum_{\text{final}} \mu_j > 0$$

That means that how a reaction runs has less to do with the levels of the potentials themselves than with the potential difference between the substances in their initial and final state. Therefore, it is convenient to introduce this difference as an independent quantity. We will call the quantity

$$\mathcal{A} = \sum_{\text{initial}} \mu_i - \sum_{\text{final}} \mu_j \tag{4.3}$$

the *chemical drive* $\mathcal{A}$ of the process (reaction, phase transition, redistribution, etc.), in short, the *drive*, when it is clear that no nonchemical influences are participating. The unit for drive is "Gibbs," as can be easily seen in the above definition.

Internationally, the quantity $\mathcal{A}$ is usually called *affinity*. The origin of this name reaches back into antiquity. However, it is, unfortunately, a bad indicator of the characteristic it describes (see below). The symbol recommended by IUPAC (International Union of Pure and Applied Chemistry) is A. So as to avoid confusion with other quantities labeled by the letter A, such as area, we shall use another font (like $\mathcal{A}$).

The name *chemical tension* for $\mathcal{A}$ would be appropriate as well when taking into consideration that the quantities electric potential φ and electric tension U (voltage),

$$U = \varphi_{\text{initial}} - \varphi_{\text{final}},$$

are similarly related both conceptually and formally as chemical potential and drive. U describes the (electric) drive for a charge transfer between two points. The simplest case of this would be between the input and output of a two-terminal electronic component (lightbulb, resistor, diode, etc.).

The quantity $\mathcal{A}$ has a centuries-old history under the name *affinity*. The first table with values of this quantity was compiled by Louis-Bernard Guyton de Morveau in 1786. This was hundred years before the concept of chemical potential was created.

At that time, people had very different ideas about the causes of substance transformations. The closer the "relationship (affinity)" of two substances, the stronger the driving force for them to bond. This was the main reason for using this name. Substance A might displace substance B from a compound BD, if it had a closer relationship or affinity to D than B. It might also occur that if A was already loosely bound to a partner C, it would then be free for a new partnership: $AC + BD \rightarrow AD + BC$. The German writer and polymath Johann Wolfgang von Goethe was inspired by this idea in his novel "The Elective Affinities" of 1809 in which he transferred this concept to human relationships.

A positive drive, $\mathcal{A} > 0$, "drives" a transformation as long as there are reactants available. A negative, $\mathcal{A} < 0$, leads to a reaction in the opposite direction of the reaction arrow. $\mathcal{A} = 0$ means no drive, therefore, a standstill where equilibrium is established. Here are some examples:

Decomposition of a Substance into Its Elements We have already encountered one type of reaction, namely the decomposition of a compound $A_\alpha B_\beta C_\gamma\ldots$ into the elements that make it up, A, B, C, $\ldots$:

$$A_\alpha B_\beta C_\gamma \ldots \rightarrow \nu_A A + \nu_B B + \nu_C C + \ldots ,$$

in which case the conversion number ν_A is numerically equal to α, ν_B to β, etc. For the strength of the tendency to decompose—the "drive to decompose"—we then obtain:

$$\mathcal{A} = \mu_{A_\alpha B_\beta C_\gamma\ldots} - \left[\nu_A \mu_A + \nu_B \mu_B + \nu_C \mu_C + \ldots\right].$$

Having arbitrarily set the potentials of the elements (in their most stable modification) under standard conditions equal to zero the expression in brackets disappears and the drive to decompose corresponds to the chemical potential of the substance:

$$\mathcal{A} = \mu_{A_\alpha B_\beta C_\gamma\ldots} - \underbrace{\left[\nu_A \cdot \mu_A^\ominus + \nu_B \cdot \mu_B^\ominus + \nu_C \cdot \mu_C^\ominus + \ldots\right]}_{0} = \mu_{A_\alpha B_\beta C_\gamma\ldots}.$$

These circumstances were already anticipated and taken into consideration by our discussion in Sect. 4.5. As a concrete example, we will consider the decomposition of ozone O_3. This tends to transform into oxygen gas O_2, which we can see easily by comparing the potentials:

$$O_3|g \rightarrow \tfrac{3}{2}\,O_2|g$$
$$\mu^\ominus(kG):\quad 163 \quad > \quad \tfrac{3}{2}\times 0 \quad \Rightarrow \mathcal{A}^\ominus = +163 \text{ kG}.$$

In this case $\mathcal{A}^\ominus$ refers to the drive to decompose under standard conditions. The decomposition process is so slow, however, that ozone can be technically used despite its limited stability. We just have to produce it fast enough to compensate for its decomposition.

Here is an anomaly that one can easily stumble over: We obtain different values for the drive to decompose of ozone depending upon which formula is being used to describe the process:

$$\mathcal{A}^{\ominus}(2\,O_3 \rightarrow 3\,O_2) = +326\;\text{kG},$$

$$\mathcal{A}^{\ominus}(O_3 \rightarrow \tfrac{3}{2}\,O_2) = +163\;\text{kG}.$$

Basically, only the sign of $\mathcal{A}$ matters, and it is the same in both cases. Still it seems strange that there appear to be different values of the drive for the same process. The first process, however, differs from the second one in the same way that a harnessed team of two horses differs from just one harnessed animal. We expect that the team will be twice as strong as the single one. This is also true for reactions. Just as with the ξ values (Sect. 1.7), it is always important to indicate the conversion formulas that one is referring to.

Transitions Another simple case is the transition of one substance into another one:

B $\rightarrow$ D is spontaneously possible if $\mu_B > \mu_D$, i.e., $\mathcal{A} > 0$.

A suitable substance for an example is mercury iodide HgI_2, which appears in beautiful red and yellow modifications with somewhat different chemical potentials:

$$\underline{HgI_2|\text{yellow} \rightarrow HgI_2|\text{red}}$$
$$\mu^{\ominus}(\text{kG}): \;\; -101.1 \quad\quad > \;\; -101.7 \quad \Rightarrow \mathcal{A}^{\ominus} = +0.6\;\text{kG}.$$

Because of the yellow modification's higher (not as strongly negative) tendency to transform, it must change into the red form. That this is actually the case is shown by Experiment 4.6.

Phase transitions such as melting and vaporization of substances can be treated in the same way. Such processes can also be formulated like reactions. An example of this is melting of ice:

$$\underline{H_2O|s \;\;\; \rightarrow \;\; H_2O|l}$$
$$\mu^{\ominus}(\text{kG}): \;\; -236.6 \;\; > \;\; -237.1 \quad \Rightarrow \mathcal{A}^{\ominus} = +0.5\;\text{kG}.$$

We have used the tabulated values valid for a temperature of 298 K or 25 °C. Therefore, a positive drive can be expected that allows ice to melt under these conditions. For given conditions, the phase with the lowest chemical potential is stable.

Therefore, diamond should undergo a transition into graphite because it has a higher chemical potential:

$$\underline{C|\text{diamond} \rightarrow C|\text{graphite}}$$
$$\mu^{\ominus}(\text{kG}): \;\; +2.9 \quad\quad > \;\; 0 \quad\quad\quad \Rightarrow \mathcal{A}^{\ominus} = +2.9\;\text{kG}.$$

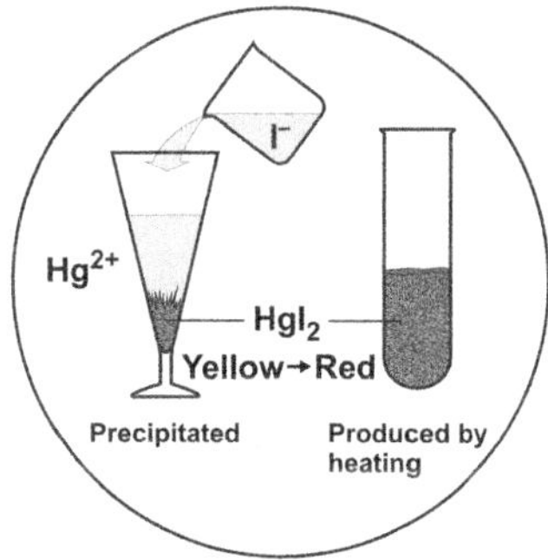

Experiment 4.6 *Change of modification of HgI₂*: Within an hour, a spoonful of yellow HgI_2 powder (produced by heating the red form in an oil bath or drying oven to over 125 °C) becomes spotted with red. These spots get larger and grow together to become uniformly red (*right*, in the figure). The process takes place within seconds when the poorly soluble HgI_2, precipitated out of a Hg^{2+} solution by addition of I^-, is used. At first, the precipitate is sallow yellow, which immediately turns to orange and finally to deep red (*left*, in the figure).

However, this does not happen at room temperature because the process is much too inhibited. The reason is that for the carbon atoms to form a graphite lattice, the very strong bonds of the carbon atoms in the diamond must be broken and this is just about impossible at room temperature. In this context let us once again recall that a potential drop from the left to the right side and therewith a positive value of the chemical drive $\mathcal{A}$ merely tells us that the process *can* proceed spontaneously in this direction in principle, but it does not signify that the process will actually run. Changes in the states of aggregation, gas $\rightarrow$ liquid $\rightarrow$ solid, take place largely without inhibition and therefore almost immediately due to a high mobility of the individual particles in participating gases or liquids, as soon as the potential gradient has the necessary sign for a spontaneous process. On the other hand, an unstable state in a solid body can be "frozen" and stay like that for thousands or even millions of years.

Reactions of Substances in General When several substances participate in a reaction, the decision about whether or not a process can take place is not more difficult to make.

We can take the reaction of marble with hydrochloric acid, an aqueous solution of hydrogen chloride, HC1, as first example (Experiment 4.7).

Therefore, we conclude that the reaction drive has to be positive. Indeed, we arrive at this result if we calculate the value of the drive by using the tabulated potential values (assuming an acid concentration of 1 kmol m^{-3}). In doing so we have to consider that HC1 is a strong acid and is entirely dissociated into hydrogen and chloride ions, H^+ and Cl^-. The H^+ ions are responsible for the reaction, while the Cl^- ions remain more or less inactive.

$$CaCO_3|s + 2\,H^+|w \;\rightarrow\; Ca^{2+}|w \;+\; CO_2|g + H_2O|l$$

$\mu^{\ominus}$(kG):	$-1{,}129$	2×0		-554	-394	-237

$$\underbrace{-1{,}129}\qquad\qquad > \qquad\qquad \underbrace{-1{,}185}\qquad\qquad \Rightarrow \mathcal{A}^{\ominus} = +56\ kG.$$

Another example is the development of hydrogen chloride gas when concentrated sulfuric acid reacts with table salt:

Experiment 4.7
Dissolution of marble in hydrochloric acid: If a few pieces of marble are put in hydrochloric acid, a strong effervescence of carbon dioxide can be observed.

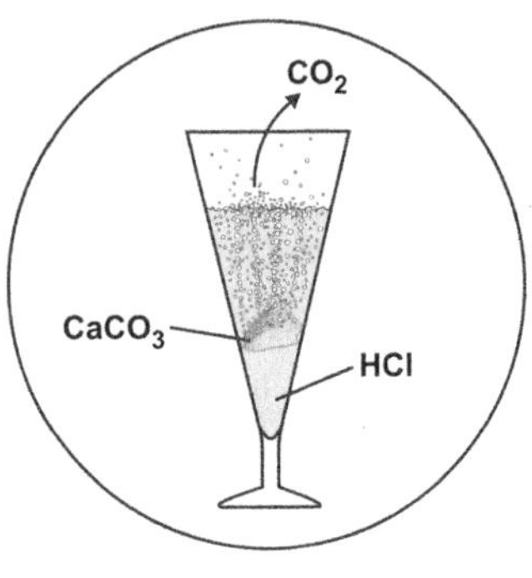

Experiment 4.8
Blackening of CuSO$_4$ by H$_2$S: If gaseous hydrogen sulfide is made to flow over anhydrous, white copper sulfate, black copper sulfide is produced. This let us observe the reaction easily.

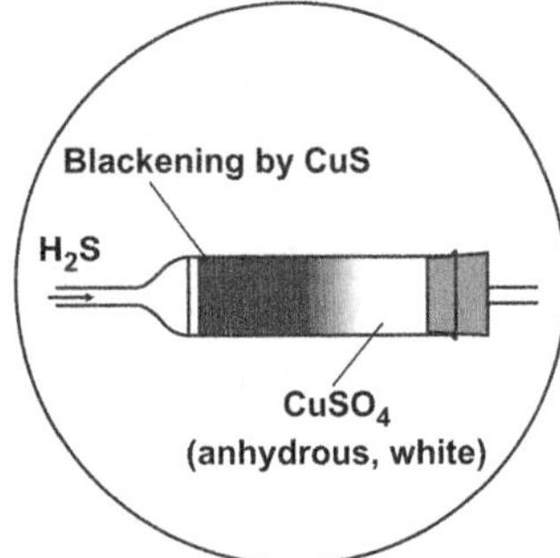

$$NaCl|s \ + \ H_2SO_4|l \ \rightarrow \ HCl|g \ + \ NaHSO_4|s$$

$\mu^{\ominus}(kG)$: $\underbrace{-384 \qquad -690}_{-1{,}074} \qquad > \qquad \underbrace{-95 \qquad -993}_{-1{,}088} \qquad \Rightarrow \mathcal{A}^{\ominus} = +14 \text{ kG}.$

For lack of better criteria, it is common to explain the fact that hydrogen chloride can be obtained from table salt and concentrated sulfuric acid by use of a rule that states that a less volatile acid displaces a higher volatile acid from its salts. In the case of dissolving marble in hydrochloric acid, also a stronger acid displaces a weaker one. These rules are often satisfied, but they are less than reliable. Experiment 4.8 shows an example that contradicts both rules:

$$CuSO_4|s \ + \ H_2S|g \ \rightarrow \ CuS|s \ + \ H_2SO_4|l$$

$\mu^{\ominus}(kG)$: $\underbrace{-661 \qquad -33}_{-694} \qquad > \qquad \underbrace{-53 \qquad -690}_{-743} \qquad \Rightarrow \mathcal{A}^{\ominus} = +49 \text{ kG}.$

In this case, the weak, volatile acid hydrogen sulfide displaces the strong, low volatile sulfuric acid from its salt.

It is also easy to predict the production of a low soluble precipitate from its ionic components when two solutions are combined:

$$Pb^{2+}|w + 2\,I^-|w \rightarrow PbI_2|s$$

$\mu^{\ominus}(kG):$ $\underbrace{-24 \qquad 2\times(-52)}_{-128}$ -174

$-128 \qquad > -174 \qquad \Rightarrow \mathcal{A}^{\ominus} = +46\ kG.$

Lead iodide must precipitate out of an aqueous solution containing Pb^{2+} and I^- ions. Many other precipitation reactions can be predicted according to the same pattern. When solutions containing Pb^{2+}, Zn^{2+}, or Ba^{2+} are mixed with those that contain CO_3^{2-}, S^{2-}, or I^- ions, precipitation can be expected (based on calculations similar to that for lead iodide) only in those instances that are marked with a plus sign in Table 4.2.

To save some calculation, the chemical potential of the possible precipitates and the combined potential of the ions forming them are included in Table 4.3 at the end of the chapter. The predicted result can be easily proven in a demonstration experiment, for example, by using S^{2-} (Experiment 4.9). Reactions with CO_3^{2-} or I^- can be carried out correspondingly. Because ionic reactions in particular are hardly inhibited in solutions and therefore usually proceed quickly if the potential gradient has the correct sign, they are especially well suited for comparing predictions with experimental results.

As discussed, a reaction always runs in the direction of a potential drop. This might give the impression that substances with a positive potential cannot ever be produced by normal reactions of stable substances (with negative μ). The production of ethyne (acetylene) with a high positive chemical potential from calcium

Table 4.2 Prediction of precipitation reactions.

	CO_3^{2-}	S^{2-}	$2I^-$
Pb^{2+}	+	+	+
Zn^{2+}	+	+	−
Ba^{2+}	+	−	−

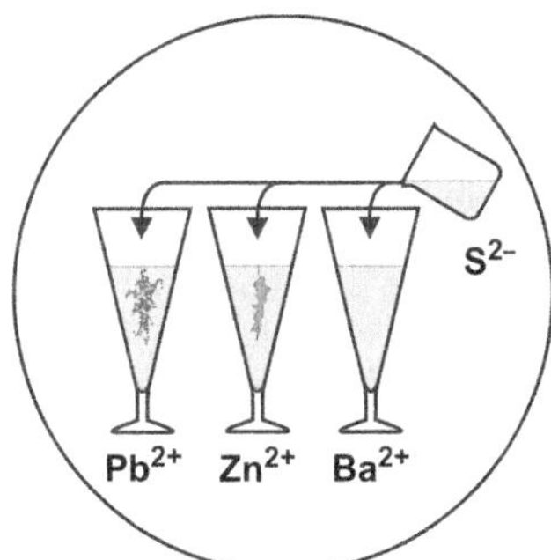

Experiment 4.9 *Precipitation of sulfides*: When a solution containing S^{2-} ions is mixed with solutions that contain Pb^{2+}, Zn^{2+}, or Ba^{2+} ions, precipitation only takes place in the first two cases.

carbide and water, both substances with a negative potential, shows that this is not the case (Experiment 4.10).

$$CaC_2|s + 2\,H_2O|l \;\rightarrow\; Ca(OH)_2|s + C_2H_2|g$$

$$\mu^{\ominus}(kG): \quad \underbrace{-65 \qquad\quad 2\times(-237)}_{-539} \quad > \quad \underbrace{-898 \qquad\quad +210}_{-688} \qquad \Rightarrow \mathcal{A}^{\ominus} = +149\ kG.$$

The very low chemical potential of calcium hydroxide on the product side makes sure that the chemical drive $\mathcal{A}$ is generally positive, even though μ(ethyne) is > 0. In earlier times, the gas extracted from this reaction was used to power miners' lamps and bicycle lights because of its bright flame. It is still used today for welding because of its high combustion temperature.

Dissolution Processes Dissolving substances in a solvent can also be described with the help of the concept of potentials. Whether a substance dissolves easily or not in water, alcohol, benzene, etc., is a result of the difference of its chemical potential in the pure and dissolved state. A first impression of the dissolution behavior of substances is all that will be given in this section. Chapter 6 will discuss how solubility can actually be calculated or estimated.

For dissolving for example cane sugar in water (more exactly: in a solution which already contains 1 kmol m^{-3} of sugar, which is about 340 g per liter!) we obtain:

$$C_{12}H_{22}O_{11}|s \;\rightarrow\; C_{12}H_{22}O_{11}|w$$

$$\mu^{\ominus}(kG): \quad -1{,}558 \qquad > \quad -1{,}565 \qquad\qquad \Rightarrow \mathcal{A}^{\ominus} = +7\ kG.$$

$\mathcal{A}^{\ominus} > 0$ means that the sugar dissolves by itself even in such a concentrated solution. Sugar dissolves easily, as we know from using it every day (Experiment 4.11).

Table salt also dissolves easily in water, as we know. The reason for this is that in an aqueous medium (even at a concentration of 1 kmol m^{-3}), the chemical potential

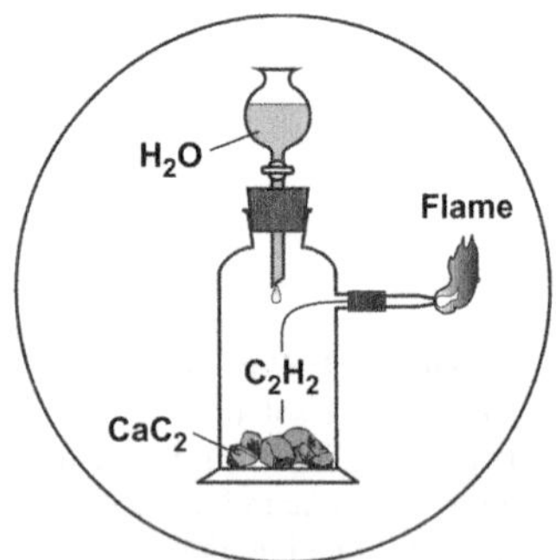

Experiment 4.10 *Carbide lamp*: When water is dripped onto some lumps of calcium carbide a vigorous generation of gas can be observed. The produced gaseous ethyne burns with a bright and sooty flame.

of the Na^+ and Cl^- ions together is noticeably lower than when it is a salt in solid form.

$$NaCl|s \rightarrow Na^+|w + Cl^-|w$$
$$\mu^{\ominus}(kG): \quad -384 \qquad -262 \quad -131$$
$$-384 \quad > \quad -393 \qquad \Rightarrow \mathcal{A}^{\ominus} = +9\ kG.$$

For contrast, let us consider the dissolution behavior of iodine.

$$I_2|s \rightarrow I_2|w$$
$$\mu^{\ominus}(kG): \ 0 \quad < \quad +16 \quad \Rightarrow \mathcal{A}^{\ominus} = -16\ kG.$$

The chemical drive is strongly negative so the process can only run backward spontaneously. Solid iodine would precipitate from a solution with a concentration of $1\ kmol\ m^{-3}$. However, this does not mean that iodine is not at all soluble in water. Increasing the dilution decreases the potential of iodine in water so that the drive can become positive when the dilution is high enough. More about this in Chap. 6.

Even the dissolution behavior of gases can be easily described in this way. For our first example, we choose ammonia as the gas and water as the solvent:

$$NH_3|g \rightarrow NH_3|w$$
$$\mu^{\ominus}(kG): \ -17 \quad > \quad -27 \qquad \Rightarrow \mathcal{A}^{\ominus} = +10\ kG.$$

Consequently, ammonia is very easily dissolved in water. An impressive way of showing this excellent solubility is with a so-called fountain experiment (Experiment 4.12).

The situation is different with carbon dioxide, which is much less soluble in water.

Experiment 4.11 *Dissolution of a sugar cube*: The process becomes noticeable by the shrinking of the sugar cube in a glass of tea even when it is not touched. An even more impressive version of this process is stacking sugar cubes into a tower on a shallow plate and then pouring some water colored with a few drops of food dye onto the plate. The water immediately begins to move up the tower and make it collapse after a short while.

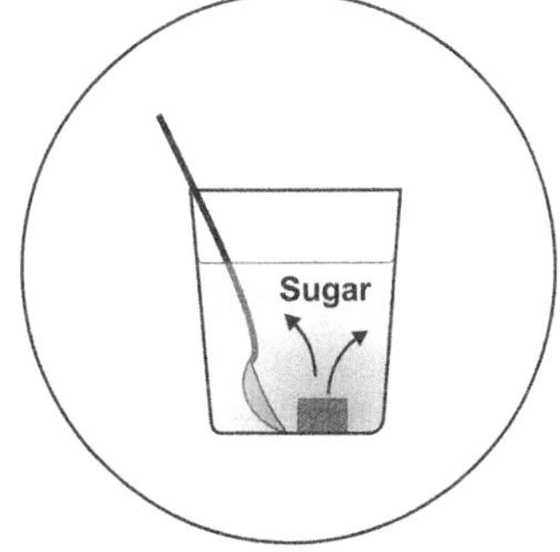

Experiment 4.12 *Ammonia fountain*: NH_3 gas dissolves so readily in water that just a few drops are enough to decrease the pressure in the flask filled with gas so drastically that more water is drawn upward into it in a strong jet. If a few drops of the acid–base indicator phenolphthalein are added to the water, the solution turns pink just as soon as it enters the flask (more in Chap. 7).

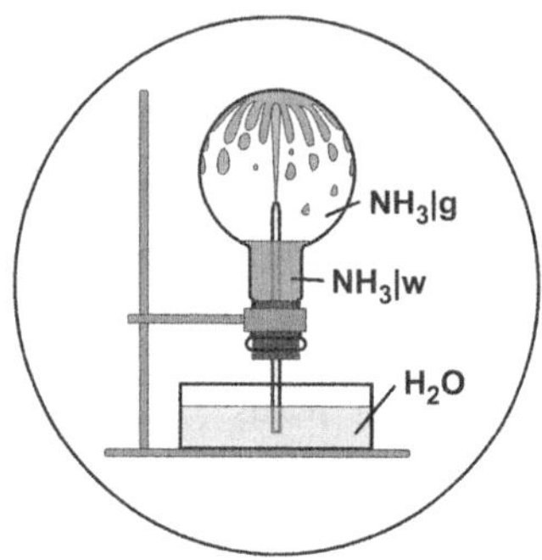

Experiment 4.13 *Effervescence of carbon dioxide*: Carbonated liquids such as champagne or mineral water are filled into bottles under excess pressure. When the pressure is reduced such a liquid releases carbon dioxide bubbles.

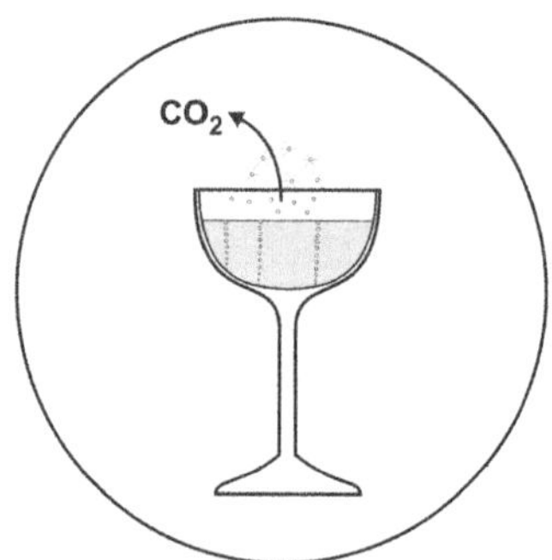

$$\begin{array}{ccc} CO_2|g & \rightarrow & CO_2|w \\ \end{array}$$
$$\mu^{\ominus}(kG): \quad -394 \quad > \quad -386 \qquad \Rightarrow \mathcal{A}^{\ominus} = -8 \text{ kG}.$$

Therefore, it tends to bubble out of an aqueous solution (Experiment 4.13).

Ammonia and carbon dioxide are both very voluminous in their gaseous states, so their appearance or disappearance in dissolving or escaping is very noticeable.

Potential Diagrams Rather than merely comparing numerical values we gain an even clearer picture of the process of substance transformations if we enter the $\mu^{\ominus}$ values into a diagram that charts the potentials, a so-called *potential diagram*. Such a diagram lets us see the drop in the potential that "drives" the process particularly well if in each case we add up the values of the chemical potentials $\mu^{\ominus}$ of the reactants and products. Let us take a closer look at the reaction of copper sulfate with hydrogen sulfide as an example (Fig. 4.4).

Up until now, we have considered the chemical potential in the roughest (zero-order) approximation as a constant. In doing so, we have neglected the dependencies of temperature, pressure, concentration, etc. We will deal with these influences in the next chapters and discuss the consequences for the behavior of substances. But for now we will discuss how the tendency of substances to transform can be quantified.

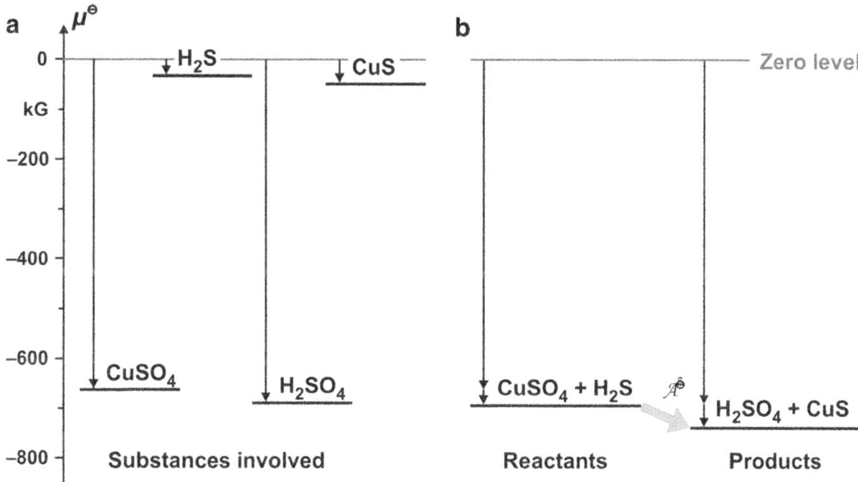

Fig. 4.4 Potential diagram for the reaction $CuSO_4 + H_2S \rightarrow H_2SO_4 + CuS$ under standard conditions: (**a**) Levels of the potential for the substances involved, (**b**) Adding of the values of the potential in the initial and the final state of reaction.

4.7   Direct Measurement of Chemical Drive

The usual methods do not measure the chemical potentials of substances themselves, but only the difference between the sums of the potentials of the initial substances and the final products, i.e., the drives $\mathcal{A} = \sum \mu_{\text{initial}} - \sum \mu_{\text{final}}$ of chemical transformations. In other words, $\mathcal{A}$ is the basic quantity from which we derive the chemical potential μ. This is also true for electric circuits where only the difference of electric potentials φ between two points, i.e., the voltage $U = \varphi_{\text{initial}} - \varphi_{\text{final}}$, can be measured, and not the potentials themselves. If an arbitrary zero point is selected, the potentials can be derived.

The chemical drive $\mathcal{A}$ can be measured both directly and indirectly just as other quantities can. However, the direct method has the advantage of not being dependent upon other physical quantities. This means that the meaning of the quantity $\mathcal{A}$ can be comprehended directly. A disadvantage is that some reference standard, a well-reproducible process that represents the unit $\mathcal{A}_1$ of the drive, must be chosen. Reference standards (etalons) for the units of length and mass are, for example, the original meter and original kilogram made of platinum or a platinum alloy which are kept in Paris. Values of chemical drives initially measured as multiples of $\mathcal{A}_1$ must then be converted to standard units.

Data in a SI coherent unit are desirable. G (Gibbs) is an example which has already been presented here. There is a trick that can be used so that preliminary values do not have to be remembered. We do not assign a value of 1 to the drive $\mathcal{A}_1$ of the process which has been chosen as the unit of the chemical drive. Rather, we take a value which comes as close as possible to the value in Gibbs. For instance,

Fig. 4.5 Cell representing a fixed value of chemical drive $\mathcal{A}$.

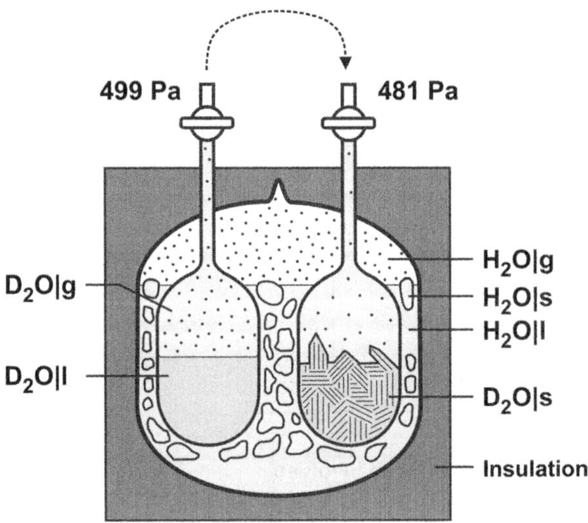

the temperature unit K (Kelvin) has been defined in this way and the temperature span of 1 K as closely approximated to the older unit 1 °C as possible. In this case, the temperature of a "triple point cell" (a cell where liquid water, water vapor, and solid ice are coexisting) is given the exact value $T = 273.16$ K.

The cell in Fig. 4.5 represents a fixed value of chemical drive just as the original meter and the original kilogram in Paris represent fixed length and mass values. This example shows the solidification of supercooled heavy water (freezing point 276.97 K),

$$D_2O|l \rightarrow D_2O|s,$$

which is embedded in airless light water and whose temperature is brought to 273.16 K. The transformation happens spontaneously if the D_2O vapor is allowed to move from the container on the left to the one on the right. Expressed in SI coherent units, the drive is

$$\mathcal{A}_l = 84 \text{ G}.$$

As we have already seen in the example of weight (Sect. 1.3), there are basically three agreements necessary for metricization. These are agreements about

(a)   Sign,
(b)   Sum,
(c)   Unit

of the quantity $\mathcal{A}$ which serves as the measure of the drive of a chemical transformation. We have just discussed how to introduce a unit (point c) in detail. Quite a bit was also said about the sign (point a) in Sect. 4.5: A process that runs forward spontaneously receives a positive value of drive $\mathcal{A} > 0$, one that tends to run backward against the direction of the reaction arrow, receives a negative value $\mathcal{A} < 0$, and a process that does neither is in equilibrium and has the value $\mathcal{A} = 0$.

Now we need only think about creating sums (point b). Two or more transformations with the drives $\mathcal{A}, \mathcal{A}', \mathcal{A}'' \ldots$ are coupled to each other—it does not matter how—so that they have to take place together. We make the agreement that the drive $\mathcal{A}_{\text{total}}$ of the entire process, i.e., of the sequence of the coupled processes, is the sum of the drives of these processes:

$$\mathcal{A}_{\text{total}} = \mathcal{A} + \mathcal{A}' + \mathcal{A}'' + \ldots .$$

There exist a number of methods for achieving a coupling of two or more chemical processes. Here are some of them:

(a) *Chemically* through shared intermediate substances,
     special case: *enzymatically* through enzyme substrate complexes,
(b) *Electrically* through electrons as the intermediate substance,
(c) *Mechanically* through cylinders, pistons, gears, etc.

Chemical coupling is the most common kind of coupling. Almost all reactions are made up of such coupled sub-steps. A strict synchronization and a close coupling are forced when, under the chosen experimental conditions, the *intermediate substance* Z no longer freely appears in noticeable quantities, i.e., just as Z is formed it is consumed by the next reaction:

$$A + B + \ldots \rightarrow C + D + \ldots + \boxed{Z}$$
$$\boxed{Z} + \ldots + F + G \rightarrow H + I + \ldots .$$

Both processes can only take place simultaneously or they have to rest simultaneously, i.e., the substance Z couples them rigidly like wheels in a clock or a gear. The short-lived intermediate substances are usually not noticed, so we can only guess what they might be. They can be quite exotic and we should not necessarily give them the status of proper substances. A simple example of a sequence of chemically coupled reactions for which the intermediate substances are well known is the precipitation of limestone from lime water that occurs when we blow exhaled air in it that contains carbon dioxide. In the process, the first two reactions are coupled by dissolved CO_2, the next by HCO_3^-, and the last by CO_3^{2-}.

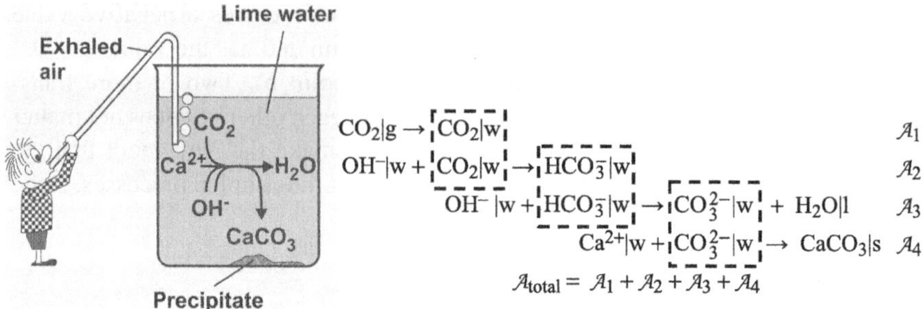

$$CO_2|g \rightarrow CO_2|w \qquad\qquad \mathcal{A}_1$$
$$OH^-|w + CO_2|w \rightarrow HCO_3^-|w \qquad\qquad \mathcal{A}_2$$
$$OH^-|w + HCO_3^-|w \rightarrow CO_3^{2-}|w + H_2O|l \qquad \mathcal{A}_3$$
$$Ca^{2+}|w + CO_3^{2-}|w \rightarrow CaCO_3|s \quad \mathcal{A}_4$$
$$\mathcal{A}_{total} = \mathcal{A}_1 + \mathcal{A}_2 + \mathcal{A}_3 + \mathcal{A}_4$$

Enzymatic coupling is an important special case of chemical coupling. This process has been developed to a high degree of perfection in biochemical reactions. For example, the innumerable reactions taking place in living cells are connected in this way which leads to the metabolism driving all other processes. Thereby reactions are interlocked like the wheels of a clock so that one transformation can drive many others.

Unfortunately, it is difficult to imitate the procedure with chemical equipment, and laboratory chemistry does not offer much room for systematic interlocking of various reactions. The coupling of a reaction with the chosen unit reaction required for measuring a drive is fundamentally possible but very difficult to realize with chemical methods.

Electrical coupling which makes use of reversible galvanic cells presents a much more flexible example. Theoretically, any chemical transformation can be used to transport electric charge from one terminal to another in a galvanic cell. After all, practically all substances contain electrons and therefore allow for dividing each transformation into a partial process which supplies electrons and another which consumes electrons. This can be accomplished in many different ways.

Let us select a general reaction

$$B + C \rightarrow D + E.$$

Theoretically, we can divide the reaction into two spatially separated partial processes where a sufficiently mobile ion B^+ is to act as the shared reaction partner. So as to keep the electrons from migrating along with the B^+ ions, we place a wall between them that is permeable only for ions. Gauze electrodes on both sides of the wall which will not hinder the migration of B^+ are used for the conduction of the electrons. In the simplest case, the wall is solid and the dissolved substances are located in a suitable trough (Fig. 4.6).

In order for substance B to go from left to right, it must be stripped of its surplus electrons,

$$B \rightarrow B^+ + e^-,$$

which accumulate on the electrode on the left, while they are in short supply on the one on the right because they are being consumed there,

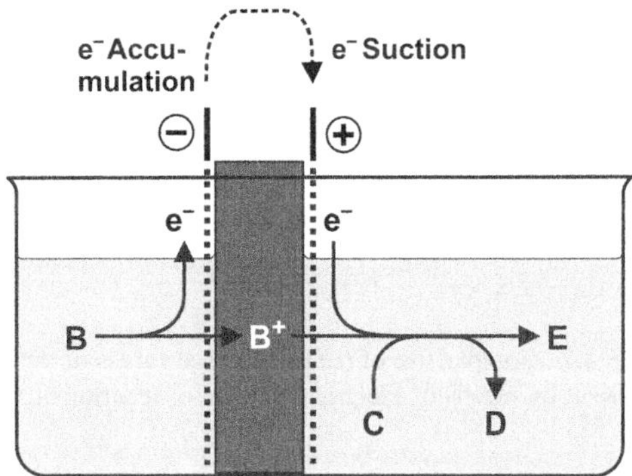

Fig. 4.6 Coupling of two reactions that are spatially separated by electrons that serve as shared reaction partner.

$$e^- + B^+ + C \rightarrow D + E.$$

As a result, an electric "tension," a voltage, is developed between the electrodes. The experimental arrangement thus represents nothing but a galvanic cell in which the entire reaction can only proceed when the electrons are allowed to flow from the cell's terminal on the left to the one on the right. Chapter 23 will go more deeply into how such cells are constructed.

Ideally, transport of charge and chemical transformation are closely coupled. By connecting two or more such cells in series, the reactions in the cells are coupled so that they only run forward or backward as a unit. Their drives add up. For simplicity's sake, it is assumed that the reactions are formulated so that the conversion number of electrons is $v_e = 1$. When the terminals of a cell in such a series connection are switched, the drive of this cell is given a negative sign—like a weight on the opposite side of a scale.

It is also possible to couple reactions *mechanically*. However, this method succeeds only when conducted as a thought experiment and therefore we will not discuss it any further here.

Drive $\mathcal{A}$ of a transformation can be measured by the same procedure that we explained when discussing weights. All we need to do is couple m specimens of the reaction to be measured inversely to as many specimens n of the unit reaction (or a reaction with a known drive) required to achieve equilibrium. In other words, the drive of the entire process is made to disappear:

$$\mathcal{A}_{\text{total}} = m \cdot \mathcal{A} + n \cdot \mathcal{A}_l = 0 \ \text{ or } \ \mathcal{A} = -(n/m) \cdot \mathcal{A}_l. \tag{4.4}$$

By applying this method it is in principle possible to measure the quantity $\mathcal{A}$ with as much precision as we desire. We can illustrate the procedure with the example of oppositely coupled vehicles (Fig. 4.7a). Like those vehicles, it is possible to inversely couple m galvanic cells which represent a particular reaction with

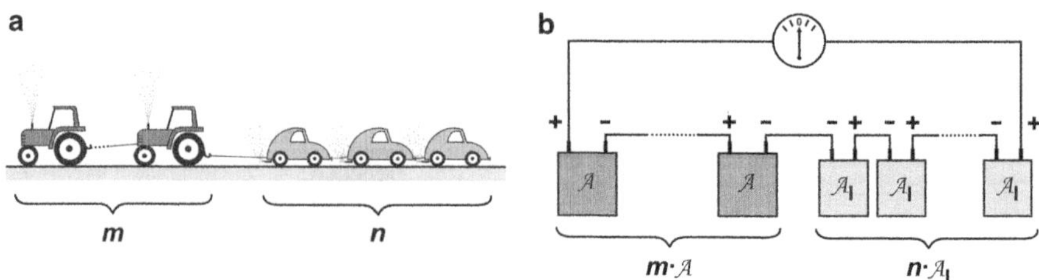

Fig. 4.7 Comparison of (**a**) the tractive forces of vehicles with (**b**) the measurement of chemical drive $\mathcal{A}$ by means of electric coupling of reactions.

unknown $\mathcal{A}$ to n cells based on a reaction with known drive such as the unit reaction ($\mathcal{A}_{\mathrm{I}}$) so that equilibrium is established and the electric current in the circuit is zero (Fig. 4.7b). As mentioned earlier, we achieve the inverse coupling through reverse poling, i.e., by interchanging the positive and the negative terminals.

The procedure can be simplified considerably. For example, it is possible to calibrate a sufficiently sensitive highly resistive voltmeter directly in the unit $\mathcal{A}_{\mathrm{I}}$. For this purpose one merely needs to connect the instrument to the two open terminals of various cell chains which consist of an increasing number of "unit cells." The pointer deflections are marked and in this way we construct a scale suitable for the measurement of unknown $\mathcal{A}$ values. The procedure is similar to the calibration of a spring balance by utilizing a number of different weights or even to the calibration of the ice calorimeter directly in entropy units (Sect. 3.7).

On the chosen scale, the chemical potential μ is nothing else than the drive of a substance to decompose into its elements. Therefore, μ can be measured using analogous methods if the reaction is chosen suitably.

In addition to the *direct* methods for determining chemical drives and potentials, respectively, there are numerous *indirect* methods that are more sophisticated and therefore more difficult to grasp, yet more universally applicable. These include chemical (using the mass action law) (Sect. 6.4), calorimetric (Sect. 8.8), electro-chemical (Sect. 23.2), spectroscopic, quantum statistical, and other methods to which we owe almost all of the values that are available to us today. Just as every relatively easily measured property of a physical entity that depends upon temperature (such as its length, volume, electrical resistance, etc.) can be used to measure T, every property (every physical quantity) which depends upon μ can be used to deduce μ values.

4.8 Indirect Metricization of Chemical Potential

In order to increase our understanding, we will consider a method which allows—in principle—the μ values of substances to be determined rather directly and in a way that approaches the way most commonly used. Figure 4.8 shows a theoretically

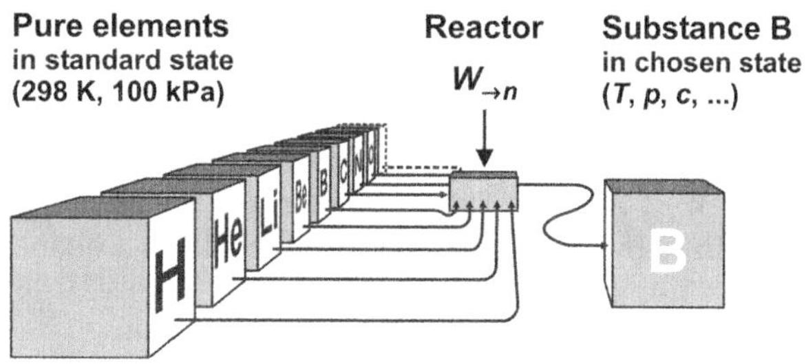

Fig. 4.8 Hypothetical arrangement for measuring chemical potentials μ.

possible setup for measuring μ which gives the values directly in the scale used by us. This method is indirect because the energy $W_{\to n}$, which is used for forming a small amount n of the substance B, is measured. Almost everything we are doing is associated with some kind of energy "turnover," so it is not easy to separate the energy contribution $W_{\to n}$, which serves exactly this purpose, from the other energy contributions which only accompany the process.

The containers on the left in the figure contain the elements in their normal stable states at 298 K and 100 kPa. In order to produce substance B, the correct proportions of these elements are supplied to a continuously working reactor. They are transformed there (the details of this process are not necessary for a first understanding) and then sent to the container on the right in the desired form of B (solid or liquid, warm or cold, pure or dissolved, etc.). One might say that the reactor transports substance B from a state on the left, where it is broken down into its elementary components with the potential 0, to a state on the right with a potential of μ_B. Whereas the matter on the left exists in a state which is identical for all substances to be formed, the matter on the right appears in a *specific form* and in a *specific milieu*. The form is determined by the selection of B with its fixed constituents and their arrangement; the milieu on the other hand is defined by temperature, pressure, concentration, type of solvent, etc. Energy is required to transform matter; as a rule the more complex and complicated the rearrangement the more energy is needed. We might say that matter will "resist" such change. This results in a more or less strong tendency to return to the old or even another state while releasing the consumed energy.

Let us recapitulate: The stronger the "drive" of a substance B to transform, here in particular the drive of the substance to decompose into the elements (in their standard state),

- The more difficult it is for the substance to be formed against its "drive,"
- The greater the amount of $W_{\to n}$ necessary to achieve this.

$W_{\to n}$ grows in proportion to the amount of substance n formed (as long as n remains small), so $W_{\to n}$ itself is not to be used as a measure of the tendency to transform and therefore of the chemical potential μ, but rather $W_{\to n}$ divided by n:

$$\mu = W_{\to n}/n. \tag{4.5}$$

The accumulation of substance B in a container gradually changes the milieu of this substance and therefore its potential. For this reason, it is required that amount n and energy $W_{\to n}$ are kept small in order to keep the disturbance small. This can be symbolized by dn and $dW_{\to n}$. μ itself results as a quotient of both quantities:

$$\mu = dW_{\to n}/dn. \tag{4.6}$$

It is of course necessary to avoid or subtract all energy contributions due to side effects (e.g., as a result of friction, lifting, entropy transfer, acceleration, production of other substances, solvents, or mixing partners, etc.). If, on the other hand, the process (the transport of substance B) runs spontaneously from left to right, it releases energy. In this case, $W_{\to n}$ and therefore μ are negative but apart from that our considerations remain essentially the same.

The unit for chemical potential which results from the equation $\mu = dW_{\to n}/dn$ is $J\,mol^{-1}$. Since we constantly deal with values of the chemical potential, we are justified in giving this unit its own name, "Gibbs," abbreviated G, in a manner analogous to "Volt" for the electric potential difference as we have done in Sect. 4.5:

$$1\ Gibbs(G) = 1\ J\,mol^{-1}.$$

Naturally, the energy of the portions of the elements taken from the left is to be found in the substance B formed from them. We do not have to worry about these contributions since they drop out when calculating the drive of a transformation of substances. This is so because, as is always the case in chemistry, the elements are conserved (see Sect. 4.4). Only the additional quantity $dW_{\to n}$ which we can identify with μdn, matters. Together with substance B, it is added to the system on the right and so increases its energy. We could use the increase dW of W to infer the value of $dW_{\to n} = \mu dn$, even if we did not know anything of the existence of the reactor or even if it did not exist at all.

Like the volume of water in a bathtub (Sect. 1.6), the energy W may change as a result of different processes, such as transfer or generation of entropy (see Sect. 3.11), increase or decrease of volume (Sect. 2.5), or by taking in other substances $B', B'', \ldots$ etc.:

$$dW = \underbrace{- pdV}_{dW_{\to V}} + \underbrace{TdS}_{dW_{\to S}} + \underbrace{\mu dn}_{dW_{\to n}} + \underbrace{\mu' dn'}_{dW_{\to n'}} + \underbrace{\mu'' dn''}_{dW_{\to n''}} + \ldots .$$

To avoid interferences caused by the different paths for energy transfer, we require $S, V, n', n'', \ldots$ to be held constant, i.e., $dS, dV, dn', dn'', \ldots = 0$, so that the related energy contributions vanish:

$$dW = \underbrace{-pdV}_{0} + \underbrace{TdS}_{0} + \underbrace{\mu dn}_{dW_{\to n}} + \underbrace{\mu'dn'}_{0} + \underbrace{\mu''dn''}_{0} + \ldots = (dW)_{S,V,n',n''\ldots}.$$

If we introduce $\mu = dW_{\to n}/dn$ into this equation, this yields

$$\mu = \frac{(dW)_{S,V,n',n''}}{dn} = \left(\frac{dW}{dn}\right)_{S,V,n',n''\ldots} = \left(\frac{\partial W}{\partial n}\right)_{S,V,n',n''\ldots}. \tag{4.7}$$

This equation already shows some similarity to Gibbs's approach. When Josiah Willard Gibbs introduced the quantity μ in 1876 which we now call chemical potential, he addressed experts in his field. Students who are not used to work with such expressions may feel repelled by equations of this type. The expression in the middle means that we should consider W a function of n, $W = f(n)$. We can consider this function to be given by a computational formula with W as the dependent variable and n the independent variable. V, S, n' n'' ... are constant parameters.

Here is an example from school mathematics. In the equation of a parabola $y = ax^2 + bx + c$, W corresponds to y, n to x. V, S, n', n'', ... correspond to the parameters a, b, c. To find μ, we must take the derivative of $W = f(n)$ with respect to n, just as we take the derivative of $y = ax^2 + bx + c$ with respect to x in order to find the slope of the function, i.e., $y' = 2ax + b$.

On the other hand, the expression on the right using the symbol ∂ assumes that W is to be considered a function of *all* the variables in the denominator and the index, $W = g(V, S, n, n', n'', \ldots)$. Since all these quantities, except for the one in the denominator, are kept constant when we take the derivative, there is no difference to the result.

Remember the formula which we used in Sect. 1.3 to indirectly determine the weight G of a body via the energy:

$$G = \left(\frac{dW}{dh}\right)_{v} = \frac{(dW)_{v}}{dh}.$$

Here, dW is the energy used to lift the object a small distance dh. The index v means that the velocity is to be kept constant. When we considered this example, we neglected the fact that the energy W of the body could also vary with its entropy S (such as through friction) or with the amount n of one of the substances it is made out of. To exclude these possibilities, we write

$$G = \left(\frac{dW}{dh}\right)_{p,S,n},$$

where, for consistency's sake, we have changed the variable v to the momentum $p = mv$. Written in this form, we must gain the impression that the weight G is a quantity that cannot be grasped and dealt with without the benefit of advanced mathematics and thermodynamics. The same happens when we try to understand the chemical potential as the partial derivative of special energy forms. That is why we prefer to introduce the chemical potential by characterizing it phenomenologically and by direct metricization. Once we have understood what the quantity μ means and which properties it has, it should be possible to follow the definition given by Gibbs. In concluding, we should once more bring to mind that μ like G is not an energy but rather corresponds to a "force," more precisely a "force-like" quantity in the sense of Helmholtz (see Sect. 2.7).

Table 4.3 Chemical potential μ (and its temperature and pressure coefficients α and β which we will get to know in the next chapter) under standard conditions (298 K, 100 kPa, dissolved substances at 1 kmol m^{-3}).

Substance	$\mu^{\ominus}$ (kG)	α (kG K^{-1})	β (μG Pa^{-1})	Substance	$\mu^{\ominus}$ (kG)	α (kG K^{-1})	β (μG Pa^{-1})
Fe|s	0	-0.027	7.1	HCl|g	-95		
NaCl|s	-384	-0.072		H$_2$SO$_4$|l	-690		
NaHSO$_4$|s	-993	-0.113		Na$_2$SO$_4$|s	$-1,270$		
SiO$_2$|s	-856	-0.041		CuSO$_4$|s	-661		
CaCO$_3$|s	$-1,129$	-0.093		CuS|s	-53		
C$_{12}$H$_{22}$O$_{11}$|s	$-1,558$	-0.392		H$_2$S|g	-33		
C$_2$H$_2$|g	210	-0.201	24.8×10^3	CaC$_2$|s	-65		
Ca^{2+}|w	-554	0.053	-17.8	OH$^-$|w	-157		
Element	0	Most stable form		Pb^{2+}|w	-24		
H$_2$|g	0			Zn^{2+}|w	-147		
H|g	203			Ba^{2+}|w	-561		
O$_2$|g	0			CO$_3{}^{2-}$|w	-528		
O$_3$|g	163			S^{2-}|w	86		
C|graphite	0			I$^-$|w	-52		
C|diamond	2.9					Substances	$\Sigma\mu$
NI$_3$|s	300			PbCO$_3$|s	-625	Pb^{2+} + CO$_3{}^{2-}$	-552
S$_4$N$_4$|s	500			ZnCO$_3$|s	-731	Zn^{2+} + CO$_3{}^{2-}$	-675
C$_6$H$_6$|l	125			BaCO$_3$|w	$-1,135$	Ba^{2+} + CO$_3{}^{2-}$	$-1,089$

(continued)

Table 4.3 (continued)

Substance	$\mu^{\ominus}$ (kG)	α (kG K^{-1})	β (µG Pa^{-1})	Substance	$\mu^{\ominus}$ (kG)	α (kG K^{-1})	β (µG Pa^{-1})
$CO_2|g$	-394	-0.214	Combustible	$PbS|s$	-98	$Pb^{2+}+S^{2-}$	62
$NO_2|g$	52	-0.240	Incombust.	$ZnS|s$	-199	$Zn^{2+}+S^{2-}$	-61
$ClO_2|g$	123	-0.257	Incombust.	$BaS|s$	-456	$Ba^{2+}+S^{2-}$	-475
		parent element		$PbI_2|s$	-174	$Pb^{2+}+2I^-$	-128
$Al_2O_3|s$	$-1{,}582$	-0.051	Oxidizes	$ZnI_2|s$	-209	$Zn^{2+}+2I^-$	-251
$Fe_2O_3|s$	-741	-0.087	Rusts	$BaI_2|s$	-602	$Ba^{2+}+2I^-$	-665
$Au_2O_3|s$	78	-0.130	Stable	$C_{12}H_{22}O_{11}|w$	$-1{,}565$		
$MgS|s$	-344	-0.050		$Na^+|w$	-262		
$ZnS|s$	-199	-0.059		$Cl^-|w$	-131		
$FeS|s$	-102	-0.060		$I_2|s$	0		
$CuS|s$	-53	-0.066		$I_2|w$	16		
$AuS|s$	>0			$NH_3|g$	-16		
$HgI_2|red$	-101.7	-0.180		$NH_3|w$	-27		
$HgI_2|yellow$	-101.1	-0.186		$CO_2|w$	-386		
$H^+|w$	0			$H_2O|s$	-236.6	-0.045	19.7
				$H_2O|l$	-237.1	-0.070	18.1
				$H_2O|g$	-228.6	-0.189	24.8×10^3

Chapter 5
Influence of Temperature and Pressure on Transformations

The chemical potential can be regarded as constant only in a first approximation. Frequently, temperature and pressure have a decisive influence on the chemical potential and therefore on the course of chemical processes. Water freezes in the cold and evaporates in the heat. Ice melts under the blades of ice skates and butane gas (the fuel of a cigarette lighter) becomes liquid when compressed. Therefore, a more detailed approach has to consider the temperature and pressure dependence of μ. Often linear approaches to these dependencies suffice. If the corresponding temperature and pressure coefficients are given, it is easily possible to predict the behavior of the substances when they are heated, compressed, etc. The melting, sublimation points, etc., can be calculated, but also the minimum temperature needed for a particular reaction. Only the pressure coefficient of gases shows a strong pressure dependence itself; therefore, the linear approach is only valid in a small pressure range. For wider application, a logarithmic approach has to be used.

5.1 Introduction

Until now, the tabular values we have used were the so-called standard values based upon room temperature and standard pressure (298 K and 100 kPa). For dissolved substances, the standard concentration is 1 kmol m^{-3}. Up to this point, our statements about the possibility of a transformation have been valid for these conditions only.

However, temperature and pressure often have a decisive influence on the chemical potential and therefore on the course of chemical processes. Water freezes in the cold and evaporates in the heat. Cooking fat melts in a frying pan and pudding gels while cooling, ice melts under the blades of ice skates, and butane gas becomes liquid when compressed. The chemical potential μ is not a material constant, but depends upon temperature, pressure, and a number of other parameters.

© Springer International Publishing Switzerland 2016
G. Job, R. Rüffler, *Physical Chemistry from a Different Angle*,
DOI 10.1007/978-3-319-15666-8_5

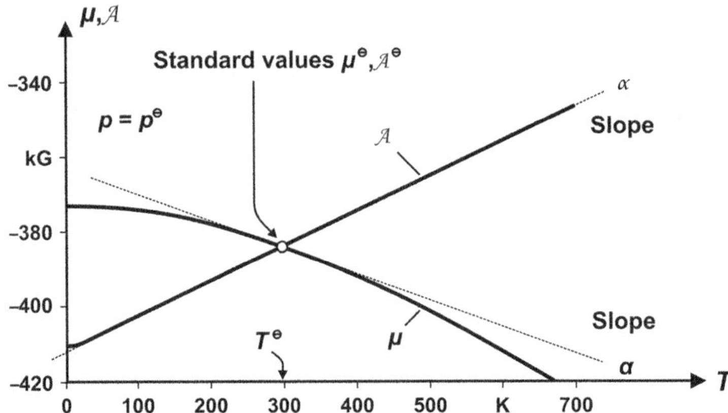

Fig. 5.1 Chemical potential of table salt and chemical drive to decompose according to $NaCl \rightarrow Na + \frac{1}{2} Cl_2$ depending upon temperature (at constant pressure $p^{\ominus}$).

5.2 Temperature Dependence of Chemical Potential and Drive

Introduction To begin, let us consider as a typical example the change with temperature in the chemical potential of table salt $\mu(NaCl)$ (Fig. 5.1). For comparison, the graphic also shows the temperature dependence of the chemical drive of table salt to decompose into the elements $\mathcal{A}(NaCl \rightarrow Na + \frac{1}{2}Cl_2)$.

It is striking that the chemical potential *falls* more and more steeply with *increasing temperature*. Except for a very few exceptions of dissolved substances (e.g., $Ca^{2+}|w$), all substances exhibit this behavior. The tendency of a substance to transform generally decreases when it is put into a warmer environment.

The chemical drive $\mathcal{A}(T)$, which is calculated from the temperature-dependent potentials, exhibits a noticeably more linear gradient than the $\mu(T)$ curves. Both curves intersect at the standard temperature $T^{\ominus}$ because the chemical potential of a substance at standard conditions corresponds to the drive to decompose into the elements (here sodium and chlorine).

The drop of potential appears, at first glance, to contradict the observation that reactions progress more readily and more quickly at higher temperatures than at lower ones. But it should be noted that a higher rate of reaction does not necessarily mean a stronger chemical drive. This can also be caused by a smaller or even vanishing inhibition as is actually often the case in chemical reactions. The strong decrease of inhibition resulting from an increase of warming masks the mostly weak change to the drive $\mathcal{A}$. Moreover, it should be remembered that $\mathcal{A}$ is determined by the difference of the chemical potentials of the starting substances and the final products, and *not* by the absolute levels of potentials. Since the potentials of the starting substances as well as of the final products decrease as a result of an increase in temperature, the potential difference which is alone responsible for the reaction drive does not necessarily decrease. It can remain constant or even increase, as in our example.

Temperature Coefficient In order to describe the drop of potential with increasing temperature, we will be content with a simple approach at first. For example, if one wishes to show how the length l of a rod changes with temperature, this can be done with the help of a *temperature coefficient* which tells us by how much the length increases when its temperature is changed by 1 K. The increase in length for a temperature increase from an initial value of T_0 to a value of T can be described by a *linear* equation as long as $\Delta T = T - T_0$ is not too large:

$$l = l_0 + \varepsilon \cdot (T - T_0). \tag{5.1}$$

The initial value of the length is represented by l_0 and ε represents the temperature coefficient.

To indicate the change of chemical potential as a result of warming, we proceed exactly in the same manner:

$$\mu = \mu_0 + \alpha \cdot (T - T_0). \tag{5.2}$$

Here, μ_0 characterizes the initial value of the chemical potential. This represents a value at arbitrarily chosen values of temperature T_0, pressure p_0, and concentration c_0 (in contrast to the standard value $\mu^\ominus$). However, standard values often serve as the initial values of a calculation, so that in special cases, $\mu_0 = \mu^\ominus$, but this is not necessarily the case. The *temperature coefficient* α represents the slope of the function $\mu(T)$ at the point $(T_0; \mu_0)$ (it is therefore strictly valid only for the reference temperature T_0), and is therefore almost always *negative*, as we have seen.

For the temperature dependence of the chemical drive $\mathcal{A}$ of a transformation

$$B + B' + \ldots \rightarrow D + D' + \ldots$$

we obtain analogously:

$$\mathcal{A} = \mathcal{A}_0 + \alpha \cdot (T - T_0). \tag{5.3}$$

The temperature coefficient α of the drive can be calculated by the same easy to remember procedure as the drive itself:

$$\alpha = \alpha(B) + \alpha(B') + \ldots - \alpha(D) - \alpha(D') - \ldots \tag{5.4}$$

$$(\text{Remember} : \mathcal{A} = \mu(B) + \mu(B') + \ldots - \mu(D) - \mu(D') - \ldots).$$

If we take room conditions as the starting point, the error is about 1 kG for low-molecular substances for ΔT values of about ± 100 K. This approximation remains useful for rough estimates up to $\Delta T \approx 1{,}000$ K and above, although $\mu(T)$ falls sharply with rising temperature. This remarkable and (for applications) important circumstance is based upon the fact that it is not the potentials that are decisive in chemical processes, but the drives. When taking the difference $\mathcal{A} = \sum \mu_{\text{reactants}} - \sum \mu_{\text{products}}$, the progressive contributions of the functions $\mu(T)$ largely cancel.

If higher precision is desired, the approach can be easily improved by adding more terms to the equation:

$$\mu = \mu_0 + \alpha \cdot \Delta T + \alpha' \cdot (\Delta T)^2 + \alpha'' \cdot (\Delta T)^3 + \dots . \tag{5.5}$$

Of course, there are other possible approaches; reciprocals for instance, or logarithmic terms. However, we do not wish to go into mathematical refinements of this type here because it is astounding how far one can actually go with the linear approximation. It is our goal here to show this.

Table 5.1 shows the chemical potential $\mu^{\ominus}$ as well as its temperature coefficient α for some substances. Along with the already mentioned basic rule which states that the temperature coefficient α is (almost) always negative, another rule (which almost all substances follow) becomes apparent when the α values are compared for changes of state of aggregation. The temperature coefficient α of the chemical potential of a substance B becomes increasingly negative when the phase changes from the solid to the liquid and finally to the gaseous state. The jump corresponding to the second transition (represented by the sign $\ll$) is considerably greater than the one corresponding to the first one. For a substance in an aqueous solution, α is mostly similar to that of the liquid state. The values scatter more strongly, though, so that we cannot easily fit $\alpha(B|w)$ into the other α values:

Table 5.1 Chemical potential μ and its temperature coefficient α for some selected substances at standard conditions (298 K, 100 kPa, dissolved substances at 1 kmol m^{-3}).

Substance	Formula	$\mu^{\ominus}$ (kG)	α (G K^{-1})
Iron	Fe|s	0	−27.3
	Fe|l	5.3	−35.6
	Fe|g	368.3	−180.5
Graphite	C|graphite	0	−5.7
Diamond	C|diamond	2.9	−2.4
Iodine	I$_2$|s	0	−116.1
	I$_2$|l	3.3	−150.4
	I$_2$|g	19.3	−260.7
	I$_2$|w	16.4	−137.2
Water	H$_2$O|s	−236.6	−44.8
	H$_2$O|l	−237.1	−70.0
	H$_2$O|g	−228.6	−188.8
Ammonia	NH$_3$|l	−10.2	−103.9
	NH$_3$|g	−16.5	−192.5
	NH$_3$|w	−26.6	−111.3
Calcium(II)	Ca^{2+}|w	−553.6	+53.1

$$\alpha(\mathrm{B|g}) \ll \alpha(\mathrm{B|l}) < \alpha(\mathrm{B|s}) < 0.$$

$$\longleftarrow \alpha(\mathrm{B|w}) \longrightarrow$$

For clarification, we will single out the values for iodine at standard conditions given in G K^{-1} from Table 5.1:

$$-260.7 \ll\; < -150.4 < -116.1 < 0.$$

$$-137.2$$

(As we will see in Sect. 9.3, the temperature coefficient α corresponds to the negative molar entropy S_m, i.e., $\alpha = -S_\mathrm{m}$. Anticipating this can help us to remember the two rules above more easily: First, in Chap. 3, we demonstrated that the molar entropy is always positive; the negative sign of the temperature coefficient easily results from this (the rare exceptions mentioned above will be discussed in detail in Sect. 8.4). Second, the fact that the molar entropy of a liquid is greater than that of a solid, and the molar entropy of a gas is much greater than that of a liquid (see Sect. 3.9), leads to the sequence above.)

Phase Transition   The chemical potential of gases therefore decreases especially fast with increase in temperature. Their tendency to transform decreases most strongly so that, by comparison to other states, the gaseous state becomes more and more stable. This simply means that, as a result of temperature increase, all other states must eventually transform into the gaseous state. At high temperatures, gases possess the weakest tendency to transform and therefore represent the most stable form of matter.

We will use water to take a closer look at this behavior. Under standard conditions, the chemical potential of ice, water, and water vapor has the following values:

	$H_2O|s$	$H_2O|l$	$H_2O|g$
$\mu^{\ominus}(kG)$	-236.6	-237.1	-228.6

One sees here that under these conditions, ice melts, and water vapor condenses because water in its liquid state has the lowest chemical potential and therefore the weakest tendency to transform. However, this changes if the temperature is raised or lowered sufficiently. For easy calculation, we will consider a temperature change of ± 100 K. The following results are obtained using the linear approach:

	$H_2O|s$	$H_2O|l$	$H_2O|g$
$\alpha(\mathrm{G\,K^{-1}})$	-45	-70	-189
$\mu(398\,\mathrm{K})(kG)$	-241	-244	-248
$\mu(198\,\mathrm{K})(kG)$	-232	-230	-210

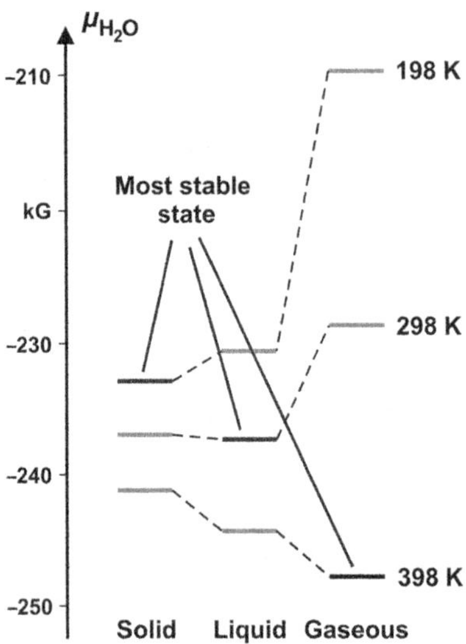

Fig. 5.2 Chemical potential of water in various states at 198, 298, and 398 K.

We see that at 398 K (125 °C), the chemical potential of water vapor has the smallest value and that water vapor must result from the other forms, while at 198 K (−75 °C), ice must develop. This result is represented graphically in Fig. 5.2.

Phase Transition Temperatures How to calculate the *phase transition temperatures* now appears obvious: If a substance like lead is solid at room temperature, this is because its chemical potential has its lowest value in the solid state. The potential of liquid lead has to exceed that of solid lead; otherwise, at room temperature, it would be liquid like mercury. We will now visualize this in a diagram (Fig. 5.3).

$\mu(Pb|s)$ as potential of an element at room temperature (and standard pressure) is equal to zero since this value has been arbitrarily chosen as the zero point of the μ scale. Under these conditions, $\mu(Pb|l)$ must lie above this. The chemical potentials decrease with warming. This happens more quickly in the liquid state than in the solid (according to the sequence presented above: $\alpha(B|l) < \alpha(B|s) < 0$). For this reason, the curves must intersect at some point, say at the temperature T_{sl}. This T_{sl} is the *melting temperature* (melting point) of lead because below T_{sl}, the most stable state of lead is the solid state; above T_{sl}, however, the most stable state is the liquid state. In order to indicate the phase transition in question, the symbols for the corresponding states of aggregation are inserted as indices (see also the comment in Sect. 3.9).

We can calculate the temperature T_{sl}. In order to do this we have to consider the melting process

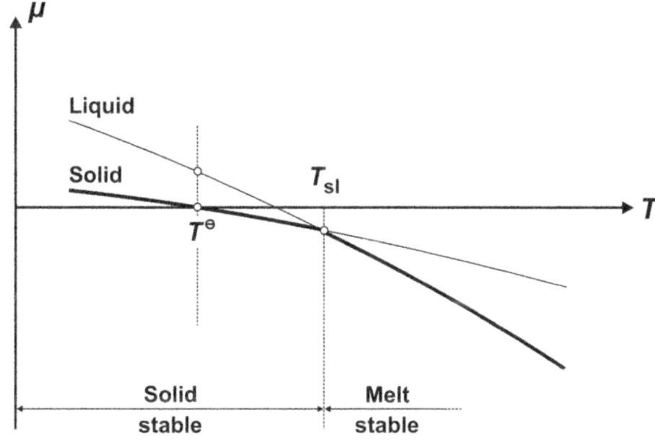

Fig. 5.3 Temperature dependence of the chemical potential of the solid and liquid phase of a substance (The lowest chemical potential for each is highlighted.).

$$\text{Pb}\,|\,\text{s} \longrightarrow \text{Pb}\,|\,\text{l}.$$

T_{sl} is the temperature at which the chemical potentials of solid and liquid phase are just equal,

$$\mu_s = \mu_l. \tag{5.6}$$

At this temperature, the two phases are in equilibrium. The temperature dependence of μ is expressed by the linear approximation:

$$\mu_{s,0} + \alpha_s \cdot (T_{sl} - T_0) = \mu_{l,0} + \alpha_l \cdot (T_{sl} - T_0).$$

By transforming this, we obtain

$$\mu_{s,0} - \mu_{l,0} = -(\alpha_s - \alpha_l) \cdot (T_{sl} - T_0)$$

and finally

$$T_{sl} = T_0 - \frac{\mu_{s,0} - \mu_{l,0}}{\alpha_s - \alpha_l} = T_0 - \frac{\mathcal{A}_0}{\alpha}. \tag{5.7}$$

The derivation is somewhat shortened when the equivalent of Eq. (5.6), $\mathcal{A} = \mu_s - \mu_l = 0$, is used as a starting point for the existence of a state of equilibrium. If the temperature dependence of the chemical drive is taken into account [Eq. (5.3)], we have

$$\mathcal{A}_0 + \alpha \cdot (T_{sl} - T_0) = 0$$

and therefore in the end as above

$$T_{sl} = T_0 - \frac{\mathcal{A}_0}{\alpha}.$$

Of course, strictly speaking, this mathematical relationship is not completely accurate because our formula for temperature dependence is only an approximation. The smaller ΔT ($:= T_{sl} - T_0$) is, the more exact the calculated value will be. The melting point of lead is actually 601 K. Based on the tabulated standard values (Sect. A.2.1 in the Appendix), our calculation yields

$$T_{sl} = 298\,\mathrm{K} - \frac{0 - 2220}{(-64.8) - (-71.7)} \frac{\mathrm{G}}{\mathrm{G\,K^{-1}}} = 620\,\mathrm{K}.$$

The result is surprisingly good for the rather rough approximation.

We will now complete the $\mu(T)$ diagram above by adding the chemical potential of lead vapor (Fig. 5.4). At room temperature, the chemical potential of vapor lies much higher than that of the liquid phase. However, with rising temperature, $\mu(\mathrm{Pb}|\mathrm{g})$ falls rather steeply, as is usual in all gases. At some temperature T_{lg} the potential of lead vapor intersects with that of liquid lead. When this temperature is exceeded, the melted lead undergoes a transition to vapor because now vapor is the most stable state. T_{lg} is nothing other than the *boiling temperature* (boiling point) of lead melt. The boiling temperature can be calculated in the same manner as the melting temperature, only now the potentials and their temperature coefficients for liquid and gaseous states will be used.

There are substances for which the chemical potential of the vapor is relatively low compared to that of the melt. The potential of the vapor can then intersect that of the solid below the melting point. This means that there is no temperature (for a given pressure) at which the liquid phase exhibits the lowest chemical potential and

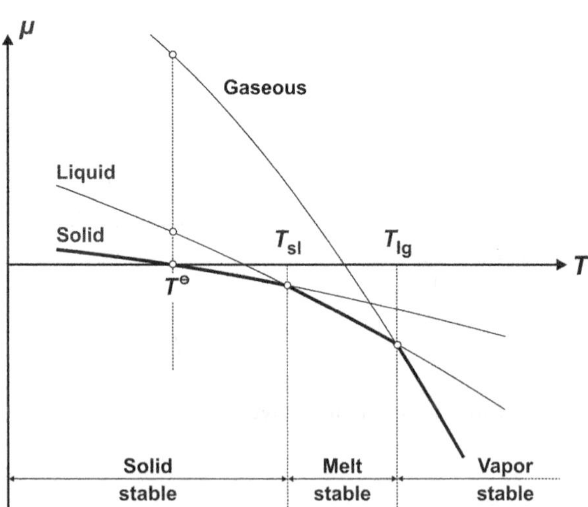

Fig. 5.4 Temperature dependence of the chemical potentials of a substance as solid, melt, or vapor.

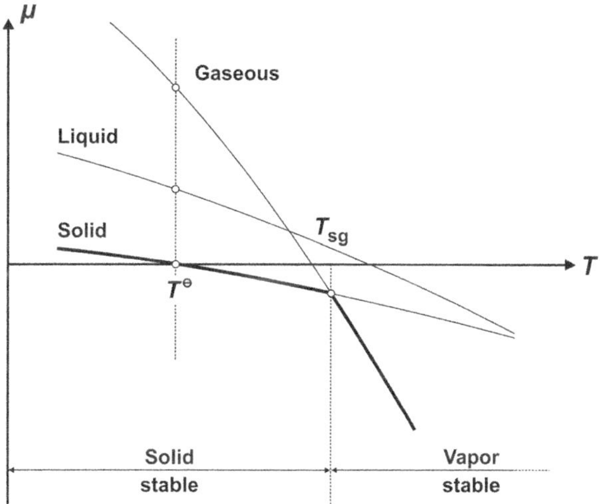

Fig. 5.5 Chemical potentials of all phases as a function of temperature in the case of sublimation.

is therefore stable. Such substances do not melt when warmed but transform immediately into the vapor state (Fig. 5.5). This phenomenon is called *sublimation*.

An excellent example of such a substance is frozen carbon dioxide which has the characteristic of vaporizing without melting. Because of this it is also called "dry ice." *Sublimation temperatures* (sublimation points) T_{sg} can be calculated based on the same procedure as used for melting and boiling temperatures, respectively.

Other transitions can be dealt with in the same way. A good object for demonstration is the already mentioned mercury iodide (cf. Sect. 4.6):

	HgI_2|yellow	HgI_2|red
$\mu^{\ominus}$(kG)	-101.1	-101.7
α(G K^{-1})	-186	-180

When heated, the temperature coefficient of the yellow form decreases more quickly than that of the red one because $\alpha(HgI_2|\text{yellow}) < \alpha(HgI_2|\text{red}) < 0$, so that above a certain temperature, $\mu(HgI_2|\text{yellow})$ falls below $\mu(HgI_2|\text{red})$, making the yellow form the more stable modification. The transition temperature (about 398 K or 125 °C) can be calculated just like the melting temperature of lead and can be easily verified by experiment (Experiment 5.1). The property of some substances to change color due to a change in temperature is called *thermochromism*.

Reaction Temperatures   Chemists are mostly interested in "real" chemical reactions. Because the temperature changes in gases have the strongest effect on their potentials, they are what shapes the behavior of reactions. Processes which produce more gas than is used up (so-called *gas forming* reactions) benefit from the strongly negative temperature coefficients α of gases when the temperature rises. In contrast, the chemical drive of a *gas binding* reaction is weakened by a rise in temperature. Consider the example of thermal decomposition of silver oxide:

Experiment 5.1 *Thermo-chromism of HgI$_2$*: A test tube containing red-orange mercury(II) iodide is slowly heated in a glycerine bath. At 398 K the iodide undergoes phase transition from the red-orange to the pale yellow modification.

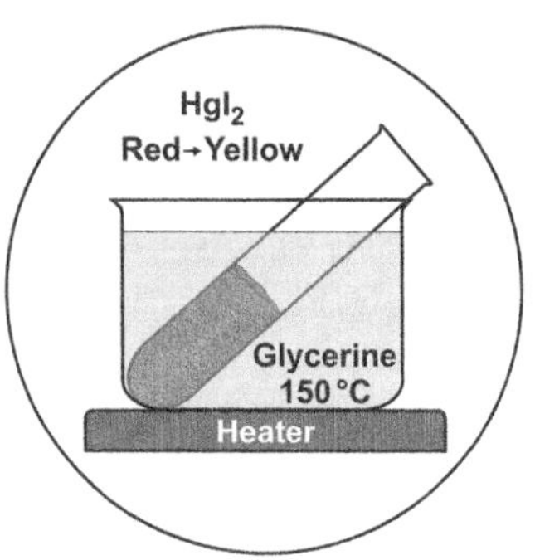

Experiment 5.2 *Heating of silver oxide*: When the blackish brown silver oxide is heated by a burner, the generation of a gas is detectable by the slow blowing up of the balloon. Subsequently, the gas can be identified as oxygen with a glowing splint. White shiny silver metal remains in the test tube.

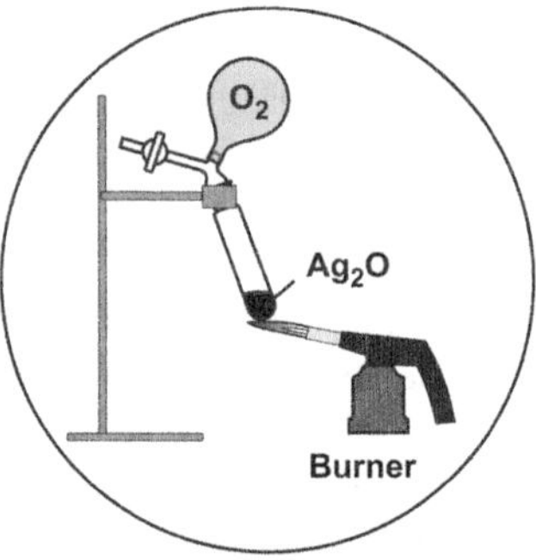

$$2\,Ag_2O|s \;\rightarrow\; 4\,Ag|s \;+\; O_2|g$$

$\mu^{\ominus}$(kG): $2\times(-11.3)$ 4×0 0 $\Rightarrow \mathcal{A}^{\ominus} = -22.6\;kG$

α (G K^{-1}): $2\times(-121)$ $4\times(-43)$ -205 $\Rightarrow \alpha = +135\;G\,K^{-1}$

The decomposition does not take place at room temperature due to the negative drive. However, since a gas should be formed, we expect that this process begins at a high enough temperature (Experiment 5.2). The minimum temperature T_D for the decomposition of Ag$_2$O is obtained from the condition that the combined chemical potentials of the initial and final substances must be equal or alternatively the chemical drive $\mathcal{A} = 0$:

$$\mathcal{A} = \mathcal{A}_0 + \alpha \cdot (T_D - T_0) = 0.$$

In analogy to Eq. (5.7), we obtain for the decomposition temperature

$$T_D = T_0 - \frac{\mathcal{A}_0}{\alpha}.$$

Based on the initial values $T_0 = T^{\ominus}(= 298\,K)$ and $\mathcal{A}_0 = \mathcal{A}^{\ominus}$, we obtain by inserting the $\mathcal{A}^{\ominus}$ and α values calculated above a decomposition temperature $T_D \approx 465\,K$ (i.e., 192 °C).

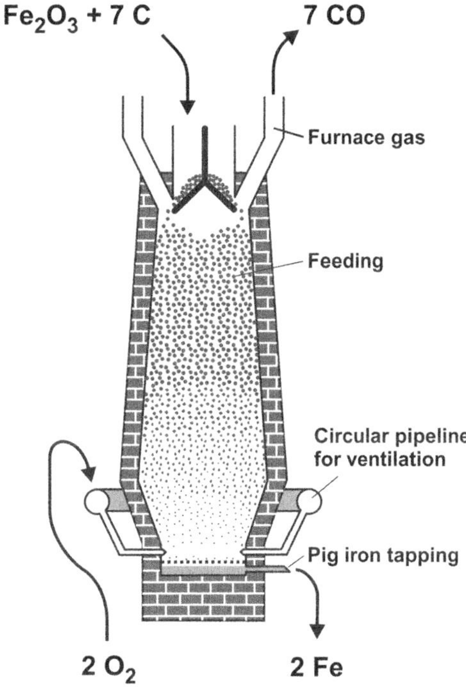

Fig. 5.6 Schematic of a blast furnace.

The same procedure can be used, for example, to calculate how strongly a compound containing crystal water must be heated in a drying oven in order to dehydrate it. Industrially important processes such as smelting of iron ore in a blast furnace (Fig. 5.6) can also be captured descriptively.

If the technical details are left out, a blast furnace can be considered a chemical reactor where iron ore, coal, and oxygen are introduced and furnace gas and pig iron exit. If this process uses the minimum amount of coal (in the conversion formula simplistically represented by carbon $C|s$ ($\approx$ graphite) it cannot take place at room temperature due to its negative chemical drive.

$$Fe_2O_3|s + 3\,C|s \;\rightarrow\; 2\,Fe|s \;+\; 3\,CO|g$$

	Fe₂O₃|s	3 C|s	2 Fe|s	3 CO|g	
$\mu^{\ominus}$(kG):	−741.0	3×0	2×0	3×(−137.2)	$\Rightarrow \mathcal{A}^{\ominus} = -329.4$ kG
α (G K^{-1}):	−87	3×(−6)	2×(−27)	3×(−198)	$\Rightarrow \alpha\; = +543$ G K^{-1}

However, a gas is formed, so we expect that the reaction should be possible at higher temperatures. If one wishes to find out if the 700 K in the upper part of the shaft of the furnace is hot enough, the drive must be approximated for this temperature according to $\mathcal{A} = \mathcal{A}_0 + \alpha \cdot (T - T_0)$ [Eq. (5.3)]. With a value of -111 kG, the drive is noticeably less negative, i.e., the potential difference between the reactants and products has become smaller, but the reaction still cannot take place. Again, the minimum temperature T_R needed for the reaction can be approximated by an equation which appears very familiar to us [entirely equivalent to Eq. (5.7)]:

$$T_R = T_0 - \frac{A_0}{\alpha}.$$

We therefore obtain a value for T_R of ≈ 900 K. Extra coal is needed for the furnace to reach this temperature. The chemical drive of the whole blast furnace process, beginning and ending with all substances at room temperature, is strongly positive because of the additional consumption of carbon:

$$Fe_2O_3|s + 7\,C|s + 2\,O_2|g \;\rightarrow\; 2\,Fe|s + 7\,CO|g$$

$\mu^\ominus$(kG): -741.0 7×0 2×0 2×0 $7\times(-137.2)$ $\Rightarrow A^\ominus = +219.4$ kG

Of course, all of these calculations depend upon access to the necessary data.

5.3   Pressure Dependence of Chemical Potential and Drive

Pressure Coefficient   As previously stated, the value of the chemical potential of a substance depends not only upon temperature but upon pressure as well. Moreover, the potential generally *increases* when the *pressure increases* (Fig. 5.7).

In a small range of pressures, all the curves can be approximated as linear, comparable to the way in which we described the influence of temperature:

$$\mu = \mu_0 + \beta \cdot (p - p_0). \tag{5.8}$$

μ_0 is the starting value of the chemical potential for the initial pressure p_0. The pressure coefficient β is almost always *positive*.

Analogously, the pressure dependence of the chemical drive A of a transformation

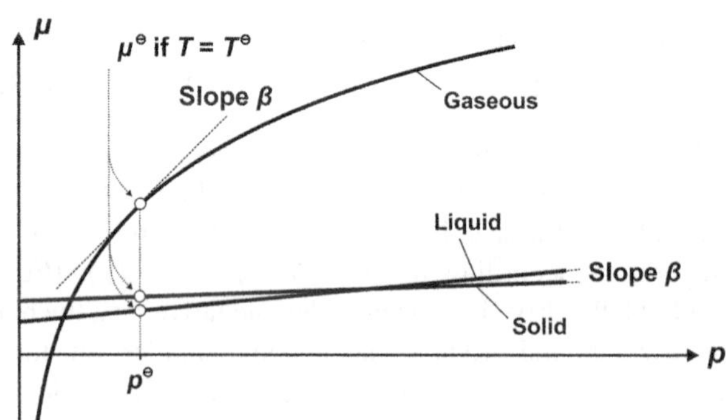

Fig. 5.7 Pressure dependence of the chemical potentials of a substance in solid, liquid, or gaseous state.

$$B + B' + \ldots \rightarrow D + D' + \ldots$$

results in

$$\mathcal{A} = \mathcal{A}_0 + \beta \cdot (p - p_0), \tag{5.9}$$

where the pressure coefficient β is:

$$\beta = \beta(B) + \beta(B') + \ldots - \beta(D) - \beta(D') - \ldots . \tag{5.10}$$

The linear approximation is useful for solid, liquid, as well as dissolved substances and for the drives of the corresponding transformations up to $\Delta p \approx 10^5$ kPa ($= 1{,}000$ bar). For obtaining general approximations, it is useful even up to 10^6 kPa ($= 10{,}000$ bar). In the case of gases and the drives of transformations in which gases participate, $\Delta p / p < 10\%$ is considered acceptable because the slope β of the corresponding curve changes relatively strongly with pressure. For greater ranges of pressure Δp, another approach must be applied to which we will be introduced later on (Sects. 5.5 and 6.5).

Table 5.2 shows the β values for the substances of Table 5.1. A rule similar to the one for temperature coefficients α is valid for pressure coefficients β. It is very useful for qualitative considerations:

$$0 < \beta(B|s) < \beta(B|l) <<<< \beta(B|g) .$$

$$\longleftarrow \quad \beta(B|w) \longrightarrow$$

Table 5.2 Chemical potential μ and its pressure coefficient β for some selected substances at standard conditions (298 K, 100 kPa, dissolved substances at 1 kmol m^{-3}).

Substance	Formula	$\mu^{\ominus}$ (kG)	β (μG Pa^{-1})
Iron	Fe|s	0	7.1
	Fe|g	368.3	24.8×10^3
Graphite	C|graphite	0	5.5
Diamond	C|diamond	2.9	3.4
Nitrogen	N_2|g	0	24.8×10^3
Iodine	I_2|s	0	51.5
	I_2|l	3.3	60.3
	I_2|g	19.3	24.8×10^3
	I_2|w	16.4	≈ 50
Water	H_2O|s	-236.6	19.8
	H_2O|l	-237.1	18.1
	H_2O|g	-228.6	24.8×10^3
Ammonia	NH_3|l	-10.2	28.3
	NH_3|g	-16.5	24.8×10^3
	NH_3|w	-26.6	24.1
Calcium(II)	Ca^{2+}|w	-553.6	-17.7

Experiment 5.3 *Boiling of lukewarm water at low pressure*: A suction flask is filled to one-third with lukewarm water, closed, and subsequently evacuated with a water aspirator. The water begins to boil.

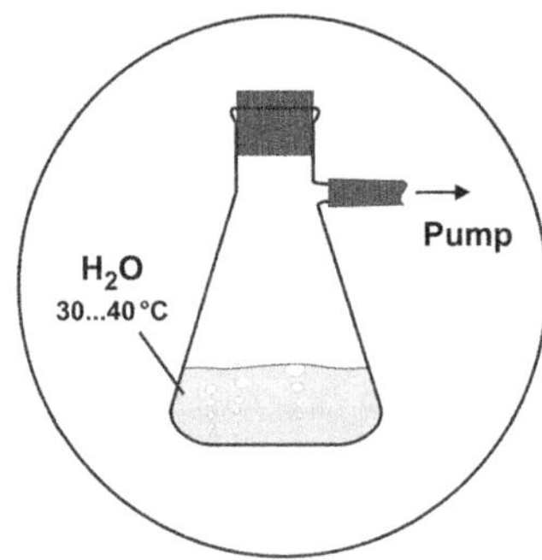

To make this clear, we will again single out the values for iodine at standard conditions, this time given in μG Pa^{-1}:

$$0 < 51.5 < 60.3 <\!\!<\!\!<\!\!< 24.8{\times}10^3$$

$$\approx 50$$

Like any rule, this one has exceptions. For instance, β for some ions in an aqueous solution is negative and sometimes—as in the case of water—β in the solid state is greater than in the liquid state. This is exactly the opposite from what the rule would lead us to expect.

(In this case, as well, there is a relation to a molar quantity, namely the molar volume V_m: We have $\beta = V_m$ (compare Sect. 9.3). Because all molar volumes are basically positive, the pressure coefficient always has a positive sign. (The very few exceptions and their cause will be discussed in detail in Sect. 8.2.) The molar volume of a gas is far greater (by a factor of 1,000) than that of the condensed phases (liquid and solid). On the other hand, the molar volume of a liquid phase is usually greater than that of the solid phase so that the sequence above results.)

Phase Transition Raising the pressure generally causes the chemical potential to increase although, as already stated, the increase varies for the different states of aggregation. In the solid state, it is smallest and in the gaseous state, greatest. As a rule, the higher the pressure is, the more stable the solid state is compared to the others and the greater the tendency of the substance to undergo a transition to the crystalline state. Conversely, a pressure reduction results in a preference for the gaseous state.

Let us once more consider the behavior of water from this new viewpoint. The following table summarizes the necessary chemical potentials and pressure coefficients:

	H$_2$O|s	H$_2$O|l	H$_2$O|g
$\mu^{\ominus}$(kG)	−236.6	−237.1	−228.6
$\beta(10^{-6}\,\mathrm{G\,Pa^{-1}})$	19.8	18.1	24.8×10^3

One sees that lukewarm water can boil at low pressure (Experiment 5.3), although, at room conditions, $\mu(H_2O|l) < \mu(H_2O|g)$, meaning liquid water is the

Experiment 5.4 *Causing warm water to boil by cooling*: Ice water is poured over a round-bottomed flask only filled with warm water and water vapor. The water begins to boil also in this case.

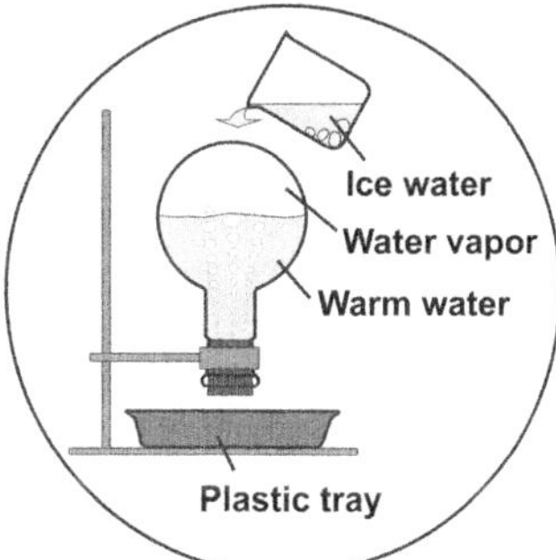

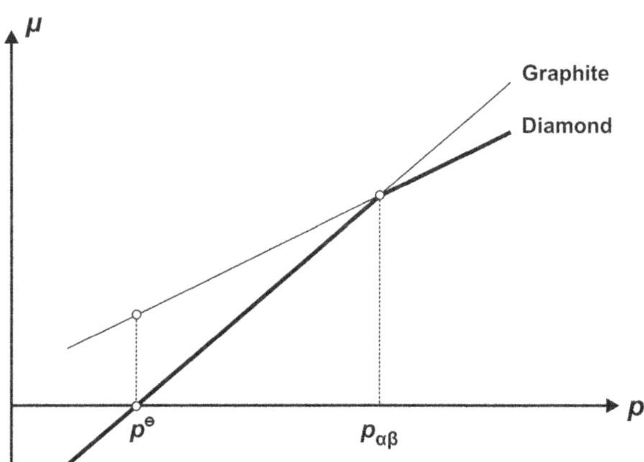

Fig. 5.8 Pressure dependence of the chemical potentials of graphite and diamond (the lowest chemical potential for each is again highlighted).

stable phase. But if the pressure is lowered enough by pumping the air above the water out of a closed container, $\mu(H_2O|g)$ will at some point fall below $\mu(H_2O|l)$, because β is especially great for the gaseous state. The reduction of pressure becomes noticeable by a strong decrease of chemical potential and the water begins to transform into water vapor by boiling.

But low pressure can also be created by cooling down water vapor which is in equilibrium with liquid water in a closed flask (Experiment 5.4). In the process, a part of the vapor condenses, leading to a decrease in pressure.

Phase Transition Pressure We shall take a closer look at a further example of the transition of a substance under pressure. Diamond is a high-pressure modification of carbon which should never appear at normal pressure. The most stable modification of carbon, the one with the lowest chemical potential, is graphite which we know from pencils. A characteristic of graphite is that its chemical potential increases more strongly with pressure than the potential of diamond so that, at one point, $\mu(C|graphite)$ should exceed $\mu(C|diamond)$ making it possible for diamond to form (Fig. 5.8).

At normal pressure and room temperature, $\mu(C|graphite)$ equals zero because this value has been arbitrarily set as the zero point of the μ scale. The $\mu(p)$ curve is steeper for graphite than for diamond. Therefore, the two curves must intersect at a pressure $p_{\alpha\beta}$, which we will call the transition pressure. The index $\alpha\beta$ indicates that the transition of one modification α (here graphite) into another modification β (here diamond) is considered. Below $p_{\alpha\beta}$, graphite is more stable; above it, diamond is more stable.

The pressure $p_{\alpha\beta}$ can be calculated because $p_{\alpha\beta}$ is the pressure for which

$$\mu_\alpha = \mu_\beta. \tag{5.11}$$

The pressure dependence of μ is expressed by a linear relation,

$$\mu_{\alpha,0} + \beta_\alpha \cdot (p_{\alpha\beta} - p_0) = \mu_{\beta,0} + \beta_\beta \cdot (p_{\alpha\beta} - p_0),$$

resulting in

$$\mu_{\alpha,0} - \mu_{\beta,0} = -(\beta_\alpha - \beta_\beta) \cdot (p_{\alpha\beta} - p_0)$$

and finally

$$p_{\alpha\beta} = p_0 - \frac{\mu_{\alpha,0} - \mu_{\beta,0}}{\beta_\alpha - \beta_\beta} = p_0 - \frac{\mathcal{A}_0}{\beta}. \tag{5.12}$$

The expression shows a great formal similarity to the one for determining a transformation temperature whether it applies to a phase transition, a decomposition, or something else.

Inserting the tabulated values results in $p_{\alpha\beta} \approx 14 \times 10^5$ kPa $(= 14{,}000$ bar$)$. Strictly speaking, this result cannot be accurate because the linear relations only represent approximations. However, as a general tool for orientation, it is quite useful.

5.4 Simultaneous Temperature and Pressure Dependence

There is nothing stopping us from expanding our ideas to transformations in which temperature *and* pressure change simultaneously. In this case the chemical potential can be expressed as follows:

$$\mu = \mu_0 + \alpha \cdot (T - T_0) + \beta \cdot (p - p_0). \tag{5.13}$$

Correspondingly, the chemical drive takes the form

$$\mathcal{A} = \mathcal{A}_0 + \alpha \cdot (T - T_0) + \beta \cdot (p - p_0). \tag{5.14}$$

The dependence of transition temperatures upon pressure can be described by these equations as well. Here is a familiar example representative of many others. Ice

Experiment 5.5 *Ice melting under pressure*: A wire loop with a heavy weight hanging from it slowly "melts" its way through a block of ice. The process is supported by the high entropy conductivity of the steel wire (see Sect. 20.4). The water formed below the wire under high-pressure flows around the wire and freezes again above it because of the lower pressure there. The ice block will remain intact even after the wire passes completely through.

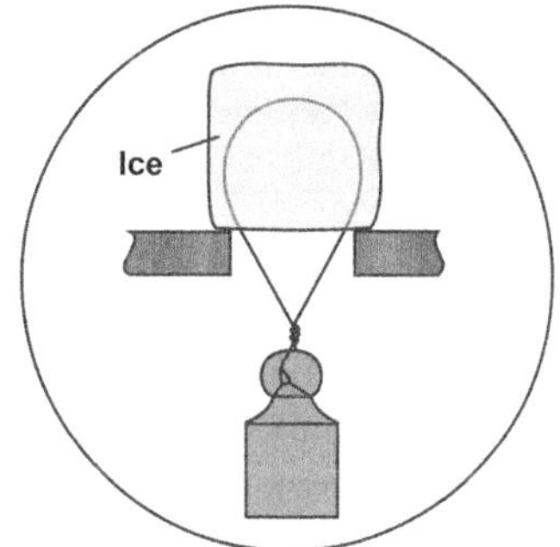

melts under high pressure [if it is not too cold]. The chemical potential of ice is the same as that of ice water ($\mu(\mathrm{H_2O|s}) = \mu(\mathrm{H_2O|l})$) at 273 K (0 °C) and standard pressure. However, because of $\beta(\mathrm{H_2O|s}) > \beta(\mathrm{H_2O|l})$, the value of $\mu(\mathrm{H_2O|s})$ increases above that of $\mu(\mathrm{H_2O|l})$ as the pressure increases, and the ice begins to melt (Experiment 5.5).

As mentioned, water is among the few exceptions where β in the solid state is greater than in the liquid state. This special characteristic of ice is responsible for the ability of a glacier to flow downward a few meters per day in a mountain valley like slow moving dough. Where the ice is under especially high pressure, it melts and becomes pliable so that it gradually moves around obstacles.

But a block of ice does not totally melt when compressed because it cools down during melting. The reason for the drop in temperature is that the entropy required for the phase transition solid $\rightarrow$ liquid is not supplied from outside (cf. Sect. 3.5). It has to be provided by the system itself, leading to a lowering of temperature. The chemical potentials increase because of the negative temperature coefficients α. Because of $\alpha(\mathrm{H_2O|l}) < \alpha(\mathrm{H_2O|s}) < 0$, the effect is stronger in water than in ice. The potential difference due to excess pressure is compensated and the process of melting stops. Again, there is equilibrium between the solid and the liquid phase, but this time at a lower freezing point. Only when the pressure is further increased does the ice continue to melt until additional cooling balances the potentials again.

To illustrate this, let us take a look at Fig. 5.9. If the pressure is increased, the chemical potentials of the solid and the liquid phase increase, but this increase is much more pronounced for the solid than for the liquid phase [because of $0 < \beta(\mathrm{B|l}) < \beta(\mathrm{B|s})$]. Thus the intersection point of the curves (T'_{sl}) shifts to the left, i.e., the freezing point is lowered by ΔT_{sl}.

It is easy to calculate the lowering of temperature in compressed ice, i.e., the freezing-point depression of water under pressure. The condition for equilibrium $\mu_{\mathrm{s}} = \mu_{\mathrm{l}}$ takes the following form:

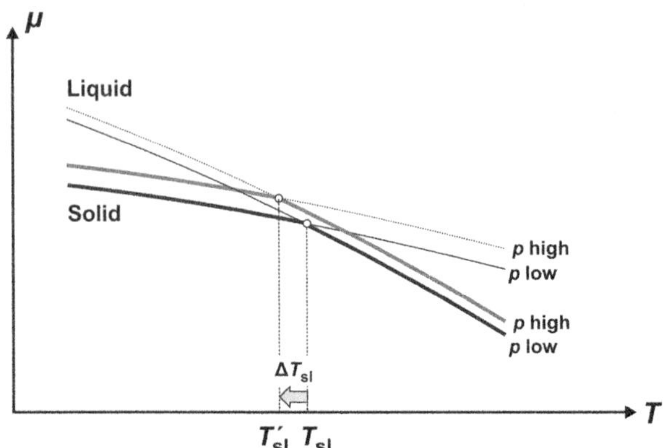

Fig. 5.9 Temperature dependence of the chemical potentials of a solid and a liquid phase at different pressures (in case of $0 < \beta_l < \beta_s$). The intersection point of the $\mu(T)$ curves at the pressure considered and hence the freezing-point shifts with increasing pressure to lower temperatures (lowering of freezing point).

$$\mu_{s,0} + \alpha_s \cdot (T - T_0) + \beta_s \cdot (p - p_0) = \mu_{l,0} + \alpha_l \cdot (T - T_0) + \beta_l \cdot (p - p_0)$$

or shortened:

$$\mu_{s,0} + \alpha_s \cdot \Delta T + \beta_s \cdot \Delta p = \mu_{l,0} + \alpha_l \cdot \Delta T + \beta_l \cdot \Delta p.$$

If the freezing point of water at standard pressure ($T_0 = 273$ K) is chosen as the initial value, then $\mu_{s,0}$ and $\mu_{l,0}$ are equal and drop out of the expression. The following relation remains with the change in temperature ΔT as the only unknown:

$$\Delta T = -\frac{\beta_s - \beta_l}{\alpha_s - \alpha_l} \Delta p = -\frac{\beta}{\alpha} \Delta p. \tag{5.15}$$

For $\Delta p = 10^4$ kPa (100 bar), the lowering of the freezing point due to pressure results in $\Delta T = -0.67$ K (calculated with the numerical values for ice and liquid water from Tables 5.1 and 5.2).

However, for most substances, the melting temperature increases with increased pressure [because of $0 < \beta(B|s) < \beta(B|l)$] (see Fig. 5.10). Correspondingly, the shifts in potentials cause higher pressure to raise the boiling point and lower pressure to lower the boiling point [because of $0 < \beta(B|l) <<<< \beta(B|g)$]. This is also valid for water as we have seen in Experiments 5.3 and 5.4. Again, the change ΔT can be approximated with the formula derived above. The value of β for boiling is roughly 10^4 times greater than for melting, whereas the α values do not vary so drastically. Therefore, even small changes of pressure are enough to noticeably shift the boiling point. To achieve a comparable change of the freezing point, much higher pressures are necessary. A pressure increase of about 10 kPa (0.1 bar) already results in a shift of the boiling point of water of about +2.0 K, while for a comparable change of the freezing point ($\Delta T = -2.0$ K), a pressure increase of more than 3×10^4 kPa (300 kbar) is necessary.

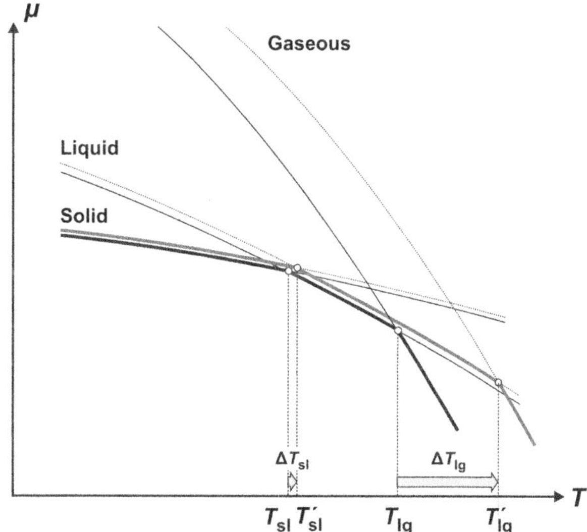

Fig. 5.10 Temperature dependence of the chemical potentials of a substance in solid, liquid, and gaseous states at low pressure (*lower curves*) and at high pressure (*upper curves*) (in case of $0 < \beta_s < \beta_l <<<< \beta_g$). The intersection points of the $\mu(T)$ curves and hence the freezing- and boiling-point shift with increasing pressure to higher temperatures (raising of freezing and boiling point).

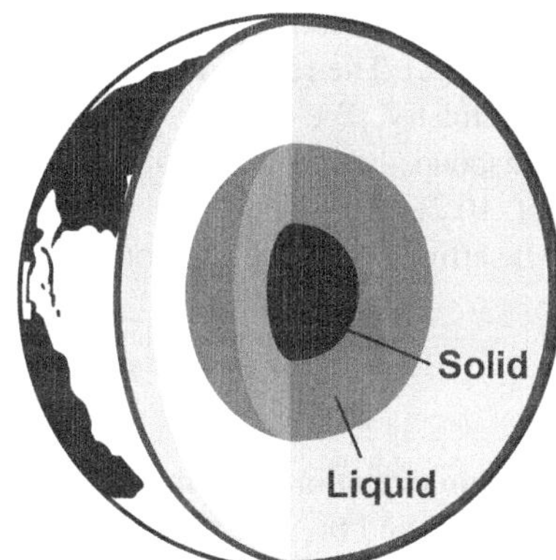

Fig. 5.11 Opposing effects of temperature and pressure in the Earth's interior: The temperature increase toward the middle of the Earth ($> 5{,}000$ K) causes the iron core to melt. The pressure, which grows to 3.6×10^8 kPa, turns it into a solid again at the very center (standard melting and boiling points of iron are about 1,809 and 3,340 K, respectively).

We will close this section with a look at our "home planet." It represents a very good example for the effect of increasing pressure and temperature upon the chemical potential and therefore the melting and freezing of substances (Fig. 5.11).

5.5   Behavior of Gases Under Pressure

As already stated, the chemical potential of gases is especially sensitive to changes of pressure. For this reason, the pressure coefficient β is greater by several powers of ten than those of solid or liquid substances. At the same time, β itself is strongly dependent upon pressure. For these reasons, the linear approximation is only applicable to a very narrow range of pressures $(\Delta p/p < 10\,\%)$. This is far too limiting for most applications so a formula must be sought that spans a much wider range of pressures. A look at the tabulated values shows that β has not only a large value but the same value for all gases at standard conditions. Apparently, the pressure coefficient β of gases is a *universal* quantity. For given T and p, it is the same for all gases in all milieus. Moreover, it is directly proportional to the absolute temperature T and indirectly proportional to the pressure p of the gas in question. This remarkable fact can be expressed as follows:

$$\beta = \frac{RT}{p} \quad \text{where } R = 8.314\ \mathrm{G\,K^{-1}}. \tag{5.16}$$

R is a fundamental constant and is the same for all substances. It is called the "*(general) gas constant*" because it was first discovered in a law valid for gases (Sect. 10.2). The relation above is based upon the phenomenon called *mass action* in chemistry. We will go more deeply into this in the next chapter. (Note: β corresponds here to the molar volume of a so-called ideal gas, as we will see in Sect. 10.2).

Inserting β into the relation (5.8) yields the following equation:

$$\mu = \mu_0 + \frac{RT}{p} \cdot (p - p_0). \tag{5.17}$$

Those proficient in mathematics immediately see that there is a logarithmic relation between μ and p:

$$\mu = \mu_0 + RT\ln\frac{p}{p_0}. \tag{5.18}$$

The pressure coefficient β of gases is nothing other than the derivative of the function $\mu(p)$ with respect to p. If we take the derivative with respect to p of the function above, we retrieve indeed Eq. (5.16).

For those interested in mathematics: Equation (5.17) can be transformed to result in

$$\mu - \mu_0 = \frac{RT}{p} \cdot (p - p_0) \quad \text{or} \quad \Delta\mu = \frac{RT}{p} \cdot \Delta p.$$

For very small (infinitesimal) changes, the relation is

$$\mathrm{d}\mu = \frac{RT}{p}\,\mathrm{d}p.$$

If we wish to calculate the change of the chemical potential from the initial value μ_0 to the final value μ for a change of pressure from p_0 to p, we must integrate both sides. (The concept of integration will be described in more detail in Sect. A.1.3 in the Appendix.) The following elementary indefinite integral will serve well for this:

$$\int \frac{1}{x}\,\mathrm{d}x = \ln x + \text{constant}.$$

Inserting the limits results in:

$$\int_{\mu_0}^{\mu} \mathrm{d}\mu = RT \int_{p_0}^{p} \frac{1}{p}\,\mathrm{d}p$$

and finally,

$$\mu - \mu_0 = RT\ln\frac{p}{p_0}.$$

We tend to consider a logarithm as unfamiliar and therefore complicated. This is unjustified. Basically, the relation is as simple as a linear one (cf. Sect. A.1.1 in the Appendix). In contrast to the linear approximation, this still relatively simple logarithmic formula spans the much wider range of pressures from between 0 and 10^4 kPa (100 bar). The range of validity will be discussed in more detail in Sect. 6.5.

Let us take a closer look at these relations using the example of butane (the fuel in a gas lighter) (Fig. 5.12). The $\mu(p)$ curve of gaseous butane shows the expected logarithmic relationship [cf. Eq. (5.18)]. Furthermore, we can see from the figure that, when compressed at room temperature, butane turns into liquid relatively easily. The so-called boiling pressure p_{lg}, i.e., the intersecting point of the potentials for the liquid and the gaseous phase, lies only a little above 200 kPa. This intersecting point characterizes the state of butane in a lighter at room temperature. However, further important information can be derived from the figure: The $\mu(p)$ curve for a liquid is an almost horizontal line. (Its slope is very small.) For this reason, the chemical potential of condensed phases (liquids and solids) can be considered nearly independent of pressure in most cases when they are present together with a gas. Furthermore, the chemical potential of a gas continues to decrease with falling pressure. The μ value approaches negative infinity if the pressure approaches zero.

This leads to the following remarkable conclusions. We can infer, for example, that calcium carbonate $CaCO_3$ cannot be stable if the CO_2 pressure in the surroundings falls to zero. In this case, the chemical potential of CO_2 would have the value $-\infty$. The reaction

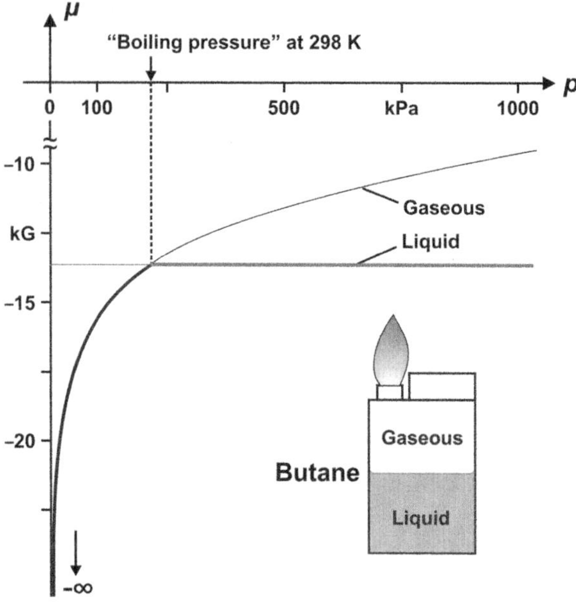

Fig. 5.12 Pressure dependence of the chemical potential of butane in liquid and gaseous states at room temperature (298 K).

$$\begin{array}{lcccl}
 & CaCO_3|s & \rightarrow & CaO|s & + \; CO_2|g \\
\end{array}$$

	$CaCO_3|s$	$\rightarrow$ $CaO|s$	$+ \; CO_2|g$	
$\mu^{\ominus}(kG)$:	-1128.8	-603.3	-394.4	$\Rightarrow \mathcal{A}^{\ominus} = -131.1 \; kG$
$\alpha \, (G\,K^{-1})$:	-93	-38	-214	$\Rightarrow \alpha \; = +159 \; G\,K^{-1}$,

which cannot take place at standard conditions, would have a positive drive. The sum of potentials on the left would be higher than on the right. However, decomposition produces CO_2, so that the CO_2 pressure must rise in a *closed* system. The process continues until the CO_2 pressure has reached a value for which the chemical potentials on the left and right sides balance. This CO_2 pressure is called the *decomposition pressure* of calcium carbonate.

The decomposition pressure can be easily calculated. If the chemical potentials satisfy

$$\mu_{CaCO_3} = \mu_{CaO} + \mu_{CO_2}$$

we have equilibrium. We ignore the pressure dependence of solid substances because, in comparison to gases, it is smaller by three orders of magnitude. We only take the dependence for CO_2 into account, and that at first for $T = T_0$:

$$\mu_{CaCO_3,0} = \mu_{CaO,0} + \mu_{CO_2,0} + RT\ln\frac{p}{p_0}.$$

Fig. 5.13 Dependence of CO_2 pressure upon temperature during decomposition of calcium carbonate (comparison of the calculated curve with measured values).

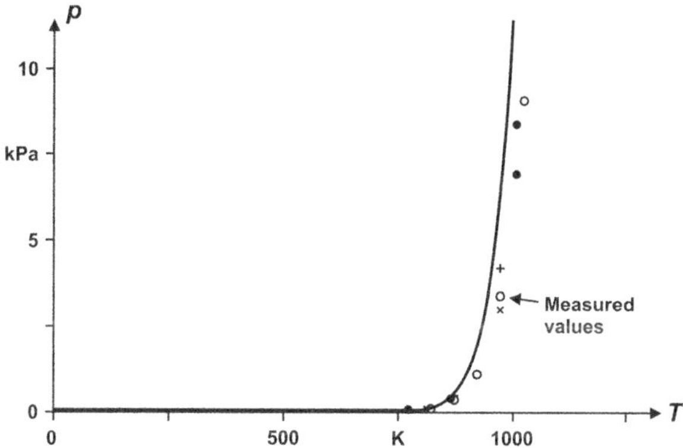

This results in

$$\underbrace{\mu_{CaCO_3,0} - \mu_{CaO,0} - \mu_{CO_2,0}}_{\mathcal{A}_0} = RT\ln\frac{p}{p_0}$$

as well as

$$\exp\frac{\mathcal{A}_0}{RT} = \exp\left(\ln\frac{p}{p_0}\right)$$

and finally in the following exponential relation:

$$p = p_0\exp\frac{\mathcal{A}_0}{RT}. \tag{5.19}$$

In order to calculate the decomposition pressure for a temperature different from the initial temperature T_0, the μ values in the exponents only need to be converted to the new temperature. The linear formula for temperature dependence used so far is generally good enough for this:

$$p = p_0\exp\frac{\mathcal{A}_0 + \alpha(T - T_0)}{RT}.$$

With the help of corresponding data, i.e., the standard values and the corresponding temperature coefficients from Sect. A.2.1 in the Appendix, the $p(T)$ curve can be determined (Fig. 5.13). This curve gives the decomposition pressure of calcium carbonate as a function of temperature:

$$p = 100\,\text{kPa} \cdot \exp\frac{-1.311 \times 10^5 + 159 \cdot (T/\text{K} - 298)}{8.314 \cdot T/\text{K}}.$$

Chapter 6
Mass Action and Concentration Dependence of Chemical Potential

The concept of mass action, its relation to the concentration dependence of the chemical potential (mass action equations), and subsequently, its relation to the chemical drive are discussed. An important application in the case of chemical equilibria is the derivation of the "mass action law." But we also examine some further consequences such as the solubility of ionic solids and gases in liquids, preferably in water. The former leads us to the concept of solubility product, the latter to Henry's law. With the help of Henry's law, we can, for example, estimate the oxygen content in bodies of water, a parameter of prime importance for biological processes. Another relevant application results in Nernst's distribution law which governs the distribution of a solute between two immiscible liquids. Distribution equilibria play a significant role in separating the substances in a mixture by the process of extraction or by partition chromatography. The last section of this chapter illustrates how the concept of mass action can be visualized with the help of potential diagrams.

6.1 The Concept of Mass Action

It has been known for a long time that the amounts of reacting substances can play an important role in the drive of chemical reactions. In 1799, the French chemist Claude-Louis Berthollet was the first to point out this influence and discuss it using many examples. Contrary to the prevailing concept of that time, he stated that it is not necessarily the case that a reaction must completely take place whenever a substance B displaces another substance C in a compound,

$$B + CD \rightarrow C + BD,$$

even if there is a great excess of B, but that an *equilibrium* is established which is dependent upon the amounts of the substances involved. The stronger B is bonded

© Springer International Publishing Switzerland 2016
G. Job, R. Rüffler, *Physical Chemistry from a Different Angle*,
DOI 10.1007/978-3-319-15666-8_6

to D, and the greater the amounts of unbound B in the reaction chamber compared to substance C, the more BD should form at the cost of CD and vice versa.

Based upon this finding, we can conclude that the tendency μ of substances to transform is not only dependent upon the types of those substances but also upon their amounts n: The greater the amount of a substance (or the mass proportional to it) in the reaction chamber, the higher its expected potential μ should be. Closer scrutiny of this effect which is known as *mass action* shows that, in this case, the quantity n itself is unimportant. It is n in relation to the volume V in which a substance is distributed, meaning its concentration $c = n/V$, that is important. If B or C or both participate as pure substances in a reaction, meaning at fixed concentrations, their amounts n_B and n_C have no influence upon the state of equilibrium and, therefore, upon the amounts of BD and CD formed. How much or how little of a substance is present, in this case, is apparently not decisive but rather how densely or loosely it is distributed in the space. This means that the more cumulative and concentrated the application, the more intense the effect (illustrated by the cartoon). In other words, the *mass* of a substance is not decisive for mass action, but its "massing," its "*density*" in space: not the *amount*, but the *concentration*. Cato Maximilian Guldberg and Peter Waage of Norway called attention to this in the year 1864.

Thus, the tendency to transform and therefore the chemical potential of substances increase according to how strongly concentrated they are. Conversely, the chemical potential goes down when the concentration of a substance decreases. We will use an example from everyday life to illustrate this. According to the values of the chemical potentials, pure water vapor must condense at room conditions:

$$H_2O|g \;\; \rightarrow \;\; H_2O|l$$
$$\mu^{\ominus}(kG): \; -228.6 \;\; > \;\; -237.1 \;\;\;\; \Rightarrow \mathcal{A}^{\ominus} = +8.5 \text{ kG}.$$

However, if the vapor is diluted by air, the value of its potential decreases below that of liquid water. It can then undergo a phase transition to the gaseous state. It *evaporates*. $\mu(H_2O|g) < \mu(H_2O|l)$ is required for wet laundry (Fig. 6.1), wet dishes, and wet streets to dry (if no other causes such as direct sunlight play a role).

6.2   Concentration Dependence of Chemical Potential

Concentration Coefficient The influence of concentration c upon the tendency μ of a substance to transform can basically be described by a linear relation like it was

Fig. 6.1 Influence of concentration upon the chemical potential, shown by the example of drying wet laundry from everyday life.

done in the last chapter to describe the influence of temperature T and pressure p. As long as $\Delta c = c - c_0$ is small enough we have:

$$\mu = \mu_0 + \gamma \cdot (c - c_0) \quad \text{for } \Delta c \ll c. \tag{6.1}$$

Mass action is an effect which superposes with other less important influences which we will address later and all of which contribute to the *concentration coefficient* γ: $\gamma = \overset{\times}{\gamma} + \gamma' + \gamma'' + \cdots$. The symbol $\times$ above a term will be used here and in the following to denote the quantities dependent upon mass action when it is desirable to distinguish them from similar quantities of different origins. Mass action appears most noticeably at small concentrations where the other influences recede more and more until they can be totally neglected, $\overset{\times}{\gamma} \gg \gamma', \gamma'', \ldots$. If one wishes to investigate this effect as directly as possible, experiments should be carried out with strongly diluted solutions $c \ll c^{\ominus} (= 1 \text{ kmol m}^{-3})$.

The temperature coefficient α and the pressure coefficient β (except for the latter in the case of gases) are not only different from substance to substance but also vary according to type of solvent, temperature, pressure, concentration, etc. In short, they depend upon the overall consistency of the *milieu* the substance is in. In contrast, the concentration coefficient $\overset{\times}{\gamma}$ related to mass action is a *universal* quantity. At the same T and c, it is the same for all substances in every *milieu*. It is directly proportional to the absolute temperature T and inversely proportional to concentration c of the substance in question and has the same basic structure as the pressure coefficient β of gases:

$$\gamma = \overset{\times}{\gamma} \equiv \frac{RT}{c} \quad \text{for } c \ll c^{\ominus} \text{ where } R = 8.314 \text{ GK}^{-1}. \tag{6.2}$$

Because the quantity T is in the numerator, we can conclude that the mass action gradually loses importance with a decrease of temperature and eventually disappears at 0 K.

Mass Action Equations If we insert Eq. (6.2) into Eq. (6.1), we obtain the following relation:

$$\mu = \mu_0 + \frac{RT}{c} \cdot (c - c_0) \quad \text{for } \Delta c \ll c \ll c^{\ominus}. \tag{6.3}$$

Analogous to the considerations about the pressure coefficient β of gases (Sect. 5.5), a logarithmic relation between μ and c is the result:

$$\mu = \mu_0 + RT\ln\frac{c}{c_0} \quad \text{for } c, c_0 \ll c^{\ominus} \text{ (mass action equation 1).} \tag{6.4}$$

We will return to the term "mass action equation" below.

As already mentioned, precise measurements show that the relation (6.4) is not strictly adhered to. At higher concentrations, values depart quite noticeably from this relation. If we gradually move to lower concentrations, the differences become smaller. The equation here expresses a so-called *limiting law* which strictly applies only when $c \to 0$. For this reason, we have added the requirement for a small concentration $(c, c_0 \ll c^{\ominus})$ to the equation.

In practice, the relation (6.4) serves as a useful approximation up to rather high concentrations. In the case of electrically *neutral* substances, deviations are only noticeable above 100 mol m^{-3}. For *ions*, deviations become observable above 1 mol m^{-3}, but they are so small that they are easily neglected if accuracy is not of prime concern. For practical applications, let us remember that:

$$\mu \approx \mu_0 + RT\ln\frac{c}{c_0} \quad \text{for } c < \begin{cases} 100 \text{ mol m}^{-3} & \text{for neutral substances,} \\ 1 \text{ mol m}^{-3} & \text{for ions.} \end{cases}$$

However, it is precisely for *standard concentration* $c^{\ominus} = 1,000$ mol m^{-3} $\left(= 1 \text{ mol L}^{-1}\right)$ (the usual reference value) that the logarithmic relation is not satisfied for any substance. Still, this concentration is used as the usual starting value for calculating potentials, and we write:

$$\mu = \overset{\circ}{\mu} + RT\ln\frac{c}{c^{\ominus}} = \overset{\circ}{\mu} + RT\ln c_{\mathrm{r}} \quad \text{for } c \ll c^{\ominus} \text{ (mass action equation 1').} \tag{6.5}$$

Basic Value Here, $c_{\mathrm{r}}(= c/c^{\ominus})$ is the *relative concentration*. $\overset{\circ}{\mu}$, intended as the *basic value* (at fixed concentration $c^{\ominus}$), has been chosen so that the equation gives the right results at low values of concentration. In contrast to the mass action equation 1, the initial value of μ is no longer real, but fictitious. Note that the basic value μ of a dissolved substance B depends upon pressure p and temperature T, $\overset{\circ}{\mu}_B(p,T)$. This distinguishes it from the *standard value* $\mu_B^\ominus \equiv \overset{\circ}{\mu}_B(p^\ominus, T^\ominus)$ already known to us. $\mu_B^\ominus$ represents the value usually tabulated. Even this standard value as particular basic value is, therefore, *not* a real value measured at standard concentration but a fictitious one. This fictitious value which mostly deviates only slightly from the real value is easier to handle for calculations. The residual term $RT\ln c_r = \overset{\times}{\mu}$ is also called the mass action term.

In its first or second version, this law describes mass action formally as a characteristic of the chemical potential. We assign to these equations a name in order to refer to them more easily. In fact, we will assign the same name to all relations of this type (there are several more of them), "*mass action equations.*" As expected, the tendency of a substance to transform increases with its concentration. This does not happen linearly, though, but logarithmically. We obtain the following graph for the dependence of the chemical potential of a dissolved substance upon its concentration (Fig. 6.2).

For small concentrations, the measured curve approximates the dotted logarithmic function very well, while, at higher concentrations, the two diverge noticeably. Depending upon the type of substance, the actual curves can run above or below the logarithmic ones. Keep in mind that the basic value $\overset{\circ}{\mu}$ of the chemical potential of the dissolved substance coincides with the logarithmic approximation and not with the measured function!

Decapotential The logarithmic initial part of the $\mu(c)$ curve, which theoretically extends to $-\infty$, is the same for all substances in every type of milieu. If the concentration increases one *decade* (a factor of ten), the chemical potential always increases by the same amount μ_d, the so-called *decapotential* (which is still dependent upon temperature T):

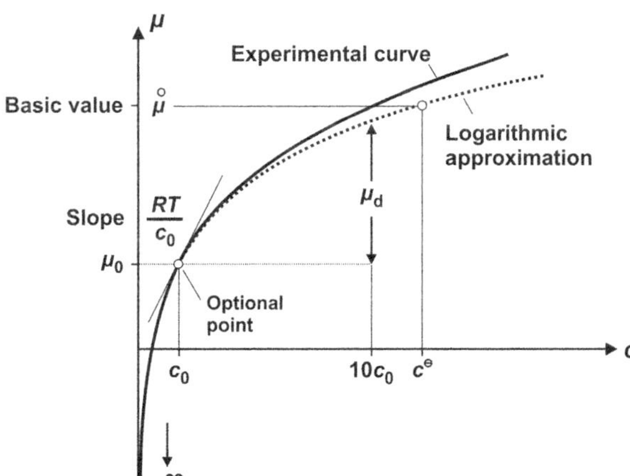

Fig. 6.2 Dependence of the chemical potential upon concentration.

$$\mu \rightarrow \mu + \mu_d \quad \text{for } c \rightarrow 10c, \text{ as long as } c \ll c^\ominus.$$

In order to calculate the μ_d value at room temperature, we only need to go back to the first of our mass action equations [Eq. (6.4)] and to insert $c = 10c_0$:

$$\mu = \mu_0 + RT \ln \underbrace{\frac{10c_0}{c_0}}_{\mu_d = RT \ln 10 = 8.314 \text{ G K-1} \times 298 \text{ K} \times \ln 10 = 5.705 \text{ kG}}.$$

It is helpful to remember the value $\mu_d \approx 5.7$ kG (approximately ≈ 6 kG) in order to quickly estimate the influence of a change of concentration of a substance upon the level of its potential or vice versa.

In summary, when the concentration c of a substance increases to 10 times its initial value at room temperature, its chemical potential μ increases by 6 kG. It does not matter,

- What substance it is,
- What it is dissolved in,
- How often this step is repeated (as long as the concentration remains small enough).

Practical Examples To obtain a better impression of the order of magnitude of the potentials, we will look at a concrete example. We have chosen $\mu(c)$ for ethanol in water (Fig. 6.3). The basic value of the potential of dissolved ethanol has been added to the graph. This value lies only about 0.1 kG above the actual μ value at standard concentration. The basic value of pure ethanol has also been included. (The peculiarities in the case of basic values of (nearly) pure substances will be discussed later.)

Using the newly derived relations, we will again take a closer look at evaporation. When the vapor is diluted by air, say, by a factor of 100 (by two orders of magnitude), its potential goes down by around 2×5.7 kG $= 11.4$ kG to about -240.0 kG. At that point, $\mu(H_2O|g)$ actually lies below the value for liquid water

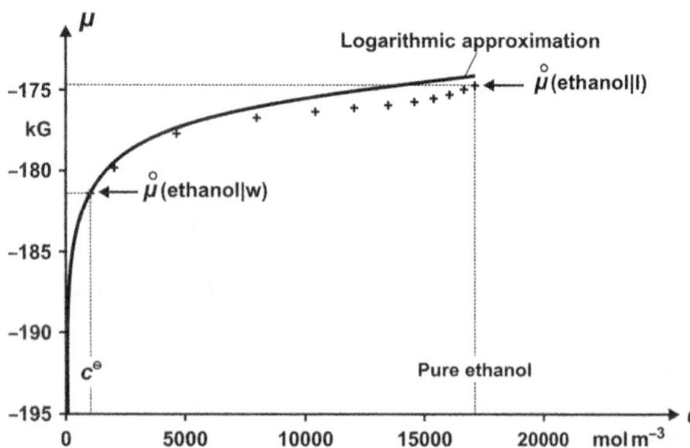

Fig. 6.3 Dependence upon concentration of the chemical potential of ethanol in water at 298 K.

(−237.1 kG) and evaporation takes place. At a concentration of $\frac{1}{30}$, the air is already so moist that it cannot absorb any more water. It is said to be saturated. A concentration of $\frac{1}{30}$ means about 1.5 orders of magnitude below the concentration of pure vapor. Therefore, the water vapor potential lies about $1.5 \times 5.7\,\text{kG} = 8.6\,\text{kG}$ below the value for pure vapor. At $-237.2\,\text{kG}$, it is at about the same level as that of liquid water and the drive to evaporate disappears. Even a little higher concentration leads to condensation, and the excess water precipitates as dew.

6.3   Concentration Dependence of Chemical Drive

We can now use what we have learned to easily show how shifts in concentration affect the chemical drive to react. Observe the following reaction

$$B + B' + \ldots \rightarrow D + D' + \ldots$$

between dissolved substances, meaning in a homogeneous solution. The drive results in

$$\mathcal{A} = [\mu_B + \mu_{B'} + \ldots] - [\mu_D + \mu_{D'} + \ldots].$$

If all the substances are present in small concentrations, we can apply the mass action equation for all of them:

$$\mathcal{A} = \left[\overset{\circ}{\mu}_B + RT\ln c_r(B) + \overset{\circ}{\mu}_{B'} + RT\ln c_r(B') + \ldots\right]$$
$$- \left[\overset{\circ}{\mu}_D + RT\ln c_r(D) + \overset{\circ}{\mu}_{D'} + RT\ln c_r(D') + \ldots\right].$$

The terms of the equation can be sorted a bit

$$\mathcal{A} = \underbrace{\left[\overset{\circ}{\mu}_B + \overset{\circ}{\mu}_{B'} + \ldots - \overset{\circ}{\mu}_D - \overset{\circ}{\mu}_{D'} - \ldots\right]}_{\overset{\circ}{\mathcal{A}}} + RT\left[\ln c_r(B) + \ln c_r(B') + \ldots\right.$$

$$\left. -\ln c_r(D) - \ln c_r(D') - \ldots\right].$$

The logarithmic terms can be summarized by using the logarithm rules $\ln x + \ln y = \ln(x \cdot y)$ and $\ln x - \ln y = \ln(x/y)$ [Eqs. (A1.1) and (A1.2)]:

$$\mathcal{A} = \overset{\circ}{\mathcal{A}} + RT\ln \frac{c_r(B) \cdot c_r(B') \ldots}{c_r(D) \cdot c_r(D') \ldots}. \tag{6.6}$$

The result is $\overset{\circ}{\mathcal{A}} = \overset{\circ}{\mu}_B + \overset{\circ}{\mu}_{B'} + \ldots - \overset{\circ}{\mu}_D - \overset{\circ}{\mu}_{D'} - \ldots$ for the basic value $\overset{\circ}{\mathcal{A}}$ of the chemical drive. It expresses the (hypothetical) drive when all reaction partners have the standard concentration of 1 kmol m^{-3}, and all other influences (except of mass action) are negligible.

$$RT\ln\frac{c_r(B)\cdot c_r(B')\cdot\ldots}{c_r(D)\cdot c_r(D')\cdot\ldots}=\overset{\times}{\mathcal{A}},\text{ in turn, represents the mass action term.}$$

The mass action term $\overset{\times}{\mathcal{A}}$ summarizes the deviations from the basic value of the individual substances which are caused by mass action.

We will explain the influence of concentration shifts upon the chemical drive using the example of cleavage of cane sugar,

$$\mathrm{Suc}\big|\mathrm{w}+\mathrm{H_2O}\big|\mathrm{l}\rightarrow\mathrm{Glc}\big|\mathrm{w}+\mathrm{Fru}\big|\mathrm{w}.$$

Suc is the abbreviation for cane sugar (sucrose, $C_{12}H_{22}O_{11}$), and Glc and Fru represent the isomeric monosaccharides grape sugar (glucose, $C_6H_{12}O_6$) and fruit sugar (fructose, $C_6H_{12}O_6$). From the chemical potentials, we obtain the following for the drive $\mathcal{A}$:

$$\mathcal{A}=\mu(\mathrm{Suc})+\mu(\mathrm{H_2O})-\mu(\mathrm{Glc})-\mu(\mathrm{Fru}),$$

$$=\overset{\circ}{\mu}(\mathrm{Suc})+RT\ln\frac{c(\mathrm{Suc})}{c^{\ominus}}+\overset{\circ}{\mu}(\mathrm{H_2O})-\overset{\circ}{\mu}(\mathrm{Glc})-RT\ln\frac{c(\mathrm{Glc})}{c^{\ominus}}-\overset{\circ}{\mu}(\mathrm{Fru})-RT\ln\frac{c(\mathrm{Fru})}{c^{\ominus}},$$

$$=\underbrace{\overset{\circ}{\mu}(\mathrm{Suc})+\overset{\circ}{\mu}(\mathrm{H_2O})-\overset{\circ}{\mu}(\mathrm{Glc})-\overset{\circ}{\mu}(\mathrm{Fru})}_{\overset{\circ}{\mathcal{A}}}+RT\ln\frac{c(\mathrm{Suc})\cdot c^{\ominus}}{c(\mathrm{Glc})\cdot c(\mathrm{Fru})}.$$

We cannot apply the mass action equation to water because its concentration lies far outside the equation's range of validity, $c(\mathrm{H_2O})\approx 50$ kmol m^{-3}. The potential curves for high c values are very flat, and the $c(\mathrm{H_2O})$ value in dilute solutions does not differ significantly from the concentration for pure water, so it is not only possible to replace the actual $c(\mathrm{H_2O})$ value with that of pure water but that is what we must do. We will indicate the potential for the pure solvent (in this case water) similarly to the basic potentials of dissolved substances with the index $^{\circ}$ placed above the symbol: $\overset{\circ}{\mu}(\mathrm{H_2O})$. In general, solvents in dilute solutions can be approximated well by pure substances. Therefore, we obtain for the drive of the cleavage of cane sugar at standard conditions:

$$\mathrm{Suc}|\mathrm{w}\ +\ \mathrm{H_2O}|\mathrm{l}\ \rightarrow\ \mathrm{Glc}|\mathrm{w}+\mathrm{Fru}|\mathrm{w}$$

$$\mu^{\ominus}(\mathrm{kG}):\ -1565\quad-237\qquad-917\qquad-916\quad\Rightarrow\mathcal{A}^{\ominus}=+31\ \mathrm{kG}.$$

A brief comment about how to write arguments and indexes: $\mu(\mathrm{H_2O})$, $c(\mathrm{H_2O})\ldots$ and $\mu_{\mathrm{H_2O}}$, $c_{\mathrm{H_2O}}\ldots$ are treated as equivalent forms. In the case of long names of substances or substance formulas with indexes (such as H_2O) or an accumulation of indexes, the preferred way of writing is the first one, otherwise, for the sake of brevity, the second.

The generalization on reactions where not all conversion numbers are just +1 (for products) or -1 (for reactants) is not difficult. Because we would like to retain the usual representation of the conversion formula with the starting substances on the left side of the reaction arrow, we choose the following notation:

$$|v_B|B + |v_{B'}|B' + \ldots \rightarrow v_D D + v_{D'} D' + \ldots .$$

Because the conversion numbers of reactants are negative, their absolute values characterized by two vertical lines have to be applied. The drive of the process can be calculated by using the same procedure as above:

$$\mathcal{A} = [|v_B|\mu_B + |v_{B'}|\mu_{B'} + \ldots] - [v_D\mu_D + v_{D'}\mu_{D'} + \ldots].$$

If the concentration dependence of the chemical potential for dissolved substances is applicable to all of them, the calculation results in

$$\mathcal{A} = \left[|v_B|\overset{\circ}{\mu}_B + |v_B|RT\ln c_r(B) + |v_{B'}|\overset{\circ}{\mu}_{B'} + |v_{B'}|RT\ln c_r(B') + \ldots\right]$$
$$- \left[v_D\overset{\circ}{\mu}_D + v_D RT\ln c_r(D) + v_{D'}\overset{\circ}{\mu}_{D'} + v_{D'}RT\ln c_r(D') + \ldots\right].$$

We can rearrange

$$\mathcal{A} = \left[|v_B|\overset{\circ}{\mu}_B + |v_{B'}|\overset{\circ}{\mu}_{B'} + \ldots - v_D\overset{\circ}{\mu}_D - v_{D'}\overset{\circ}{\mu}_{D'} - \ldots\right]$$
$$+ RT\left[|v_B|\ln c_r(B) + |v_{B'}|\ln c_r(B') + \ldots - v_D\ln c_r(D) - v_{D'}\ln c_r(D') - \ldots\right]$$

and obtain finally

$$\mathcal{A} = \overset{\circ}{\mathcal{A}} + RT\ln\frac{c_r(B)^{|v_B|} \cdot c_r(B')^{|v_{B'}|} \cdot \ldots}{c_r(D)^{v_D} \cdot c_r(D')^{v_{D'}} \cdot \ldots}. \tag{6.7}$$

Here, additionally the logarithm rule (A1.3) has been applied: $\log(x^a) = a \cdot \log x$.

Let us now take another look at a concrete example. For this, we choose the reaction of Fe^{3+} ions with I^- ions:

$$2\,Fe^{3+}\big|w + 2\,I^-\big|w \rightarrow 2\,Fe^{2+}\big|w + I_2\big|w.$$

Therefore, the conversion numbers are $v(Fe^{3+}) = -2$, $v(I^-) = -2$, $v(Fe^{2+}) = +2$, and $v(I_2) = +1$, the absolute values of the reactants in either case being 2. Insertion into Eq. (6.7) results in

$$\mathcal{A} = \overset{\circ}{\mathcal{A}} + RT\ln\frac{c_r(\mathrm{Fe}^{3+})^2 \cdot c_r(\mathrm{I}^-)^2}{c_r(\mathrm{Fe}^{2+})^2 \cdot c_r(\mathrm{I}_2)}. \tag{6.8}$$

However, the concentrations do not remain constant during a reaction. They change in the course of the reaction. If there is only one substance at the beginning, its concentration decreases continuously to the benefit of the product. Using the simplest reaction possible, we will discuss the transformation of a substance B into a substance D:

$$B \rightarrow D.$$

An example would be the transition of α-D-glucose into the isomeric β-D-glucose in aqueous solution. These are two stereoisomers of glucose $C_6H_{12}O_6$, i.e., the isomers do not differ in the structure (unlike the structural isomers described in Sect. 1.2) but only in the spatial placement of substituents with the same bond structure. In our case, the position of the OH group at the first C atom (characterized by *) is different. This OH group was formed by the ring closure thereby making the first C atom chiral as well, since its four bonds lead to four different groups. α-D-glucose and β-D-glucose only differ at this newly formed stereocenter.

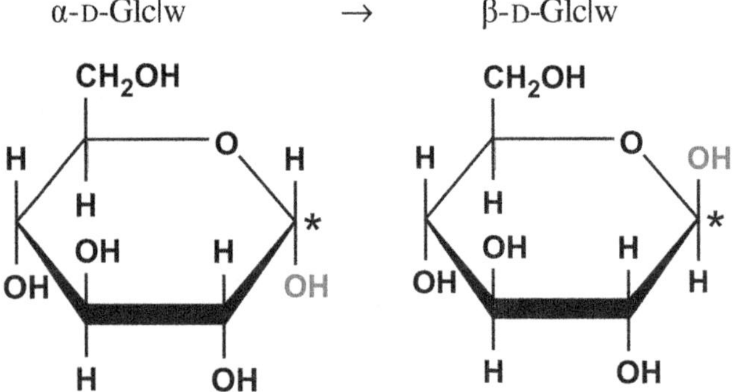

(The transition takes place via the open-chain aldehyde form, but its concentration is so small that it can be ignored.)

We obtain for the drive of the transition at standard conditions:

α-D-Glc|w → β-D-Glc|w	
$\mu^{\ominus}$(kG): −914.54 −915.79	$\Rightarrow \mathcal{A}^{\ominus} = +1.25$ kG.

The two substances are optically active, i.e., the plane of linearly polarized light is rotated when the light passes through their solutions. Pure α-D-glucose shows an angle of rotation of +112°, pure β-D-glucose, however, one of +18.7°. Therefore, a polarimeter (a scientific instrument used to measure the angle of rotation) can be used to observe the change in the solution's angle of rotation. When crystals of pure α-D-glucose are dissolved in water, the specific rotation of the solution decreases gradually from an initial value of $+112°$ to a value of $+52.7°$.

In Sect. 1.7, the extent of conversion (extent of reaction) ξ was introduced as a measure of the progress of a reaction. The amounts n_i and therefore the concentrations $c_i = n_i/V$ of the substances involved change with increasing ξ. Starting with the initial values $n_{i,0}$ and $c_{i,0}$, respectively, we obtain:

$$n_i = n_{i,0} + v_i\xi \quad \text{and} \quad c_i = c_{i,0} + v_i \cdot \xi/V. \tag{6.9}$$

The drive $\mathcal{A}$ of a reaction changes obligatorily along with the concentrations c_i. If one assumes a concentration of c_0 of the starting substance B as well as an absence of the final product D at the beginning of the reaction, the concentration of D after a certain period of time will be ξ/V, that of B $c_0 - \xi/V$. Therefore, the following relation for the dependence of the drive upon the extent of conversion is obtained:

$$\mathcal{A} = \overset{\circ}{\mathcal{A}} + RT\ln\frac{(c_0 - \xi/V)/c^{\ominus}}{(\xi/V)/c^{\ominus}} = \overset{\circ}{\mathcal{A}} + RT\ln\frac{c_0 - \xi/V}{\xi/V}. \tag{6.10}$$

For the sake of simplicity, we will now relate the extent of conversion ξ to the maximum possible value ξ_{max}. This value is reached when one of the starting substances (here only one) is entirely consumed, i.e., its concentration disappears. In this case, the numerator $c_0 - \xi/V$ equals 0, a fact that results in $c_0 = \xi_{max}/V$. Dividing numerator and denominator by c_0 yields:

$$\mathcal{A} = \overset{\circ}{\mathcal{A}} + RT\ln\frac{1 - \xi/\xi_{max}}{\xi/\xi_{max}}. \tag{6.11}$$

Because we will use the quotients ξ/V and ξ/ξ_{max} frequently, we assign to them their own symbol and name:

$$c_\xi := \frac{\xi}{V} \text{ "density of conversion,"} \quad \alpha_\xi := \frac{\xi}{\xi_{max}} \text{ "degree of conversion." } \tag{6.12}$$

The index ξ is added for clarification to avoid confusion with the molar concentration c_i and the temperature coefficient α_i of the chemical potential. For the drive, we obtain correspondingly:

$$\mathcal{A} = \overset{\circ}{\mathcal{A}} + RT\ln\frac{c_0 - c_\xi}{c_\xi} \quad \text{or} \quad \mathcal{A} = \overset{\circ}{\mathcal{A}} + RT\ln\frac{1 - \alpha_\xi}{\alpha_\xi}. \tag{6.13}$$

With a standard value of $\mathcal{A}^{\ominus} = 1.25$ kG for the transition of α-D- into β-D-glucose at room temperature, a characteristic S-shaped curve is obtained for the dependence of chemical drive upon degree of conversion α_ξ (Fig. 6.4).

Fig. 6.4 Dependence of chemical drive $\mathcal{A}$ upon the degree of conversion α_ξ for the transition of α-D-glucose into β-D-glucose in an aqueous solution at room temperature.

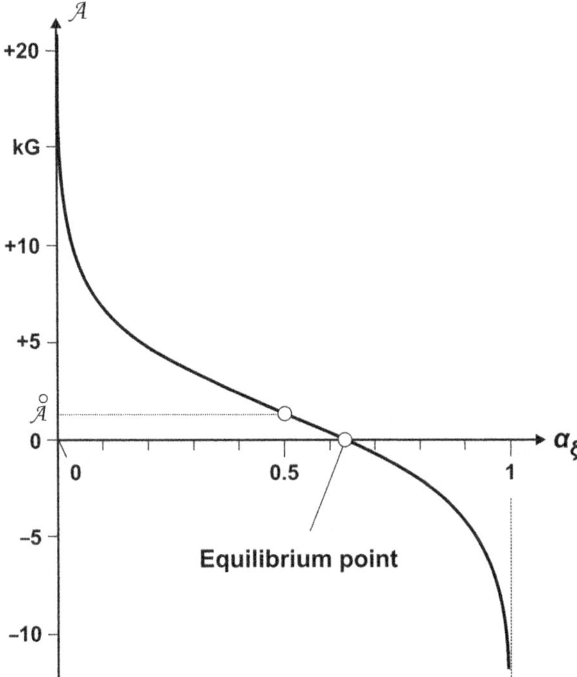

At the beginning of the reaction, meaning for $\alpha_\xi = 0$, $\mathcal{A}$ equals $+\infty$. $\mathcal{A}$ decreases during the reaction and, at an $\overset{\circ}{\mathcal{A}}$ dependent α_ξ value, finally reaches a value of zero (equilibrium point). When all the starting substance has been used up, meaning at $\alpha_\xi = 1$, $\mathcal{A}$ takes a final value of $-\infty$.

The mathematical relation becomes rather complicated for reactions with a more complex stoichiometry such as the reaction of Fe^{3+} ions with I^- ions mentioned above. For better overview, it would be advisable in this case to create a kind of table (Table 6.1). For each substance involved, the table lists in the first row the standard value of its chemical potential for calculating the drive of the reaction under standard conditions. In the following, we assume that at the beginning of the reaction, Fe^{3+} and I^- both have a concentration of c_0 and that Fe^{2+} and I_2 are absent. The concrete values of the initial concentrations of the substances follow in the next row. Finally, the formulas for the concentrations at an arbitrary time t are listed which can be calculated by using the stoichiometry of the reaction. c_ξ is the density of conversion mentioned above.

If we insert the expressions for the concentrations c_i into Eq. (6.8), we obtain for the drive of the reaction:

Table 6.1 Collection of data relevant for a concrete reaction

$$2\,\mathrm{Fe^{3+}|w}\ +\ 2\,\mathrm{I^-|w}\qquad \rightarrow 2\,\mathrm{Fe^{2+}|w}\ +\ \mathrm{I_2|w}$$

$\mu^{\ominus}$(kG):	$2\times(-4.7)$	$2\times(-51.6)$	$2\times(-78.9)$	16.4	$\Rightarrow \mathcal{A}^{\ominus} = +28.8\,\mathrm{kG}$
$c_{i,0}(\mathrm{kmol\,m^{-3}})$	0.001	0.001	0	0	
c_i	$c_0 - 2c_\xi$	$c_0 - 2c_\xi$	$2c_\xi$	c_ξ	

$$\mathcal{A} = \overset{\circ}{\mathcal{A}} + RT\ln \frac{[(c_0 - 2c_\xi/c^{\ominus})]^2 \cdot [(c_0 - 2c_\xi)/c^{\ominus}]^2}{[(2c_\xi)/c^{\ominus}]^2 \cdot [c_\xi/c^{\ominus}]}$$

$$= \overset{\circ}{\mathcal{A}} + RT\ln \frac{(c_0 - 2c_\xi)^4}{4c_\xi^3 \cdot c^{\ominus}}. \tag{6.14}$$

The density of conversion c_ξ reaches its maximum value if $c_0 - 2c_\xi$ in the numerator vanishes, meaning $c_0 = 2c_{\xi,\mathrm{max}}$. If we also consider the relationship $\alpha_\xi = c_\xi/c_{\xi,\mathrm{max}}$ ($= \xi/\xi_{\mathrm{max}}$), the equation above can be written as follows:

$$\mathcal{A} = \overset{\circ}{\mathcal{A}} + RT\ln \frac{2(1 - \alpha_\xi)^4 c_0}{\alpha_\xi^3 \cdot c^{\ominus}}. \tag{6.15}$$

Despite the more complex stoichiometry of the reaction and the correspondingly more complicated calculations, we obtain again the typical S-shaped curve (Fig. 6.5).

Let us remind ourselves of the criteria for a transformation which we were introduced to in Chap. 4: A reaction takes place spontaneously as long as drive $\mathcal{A}$ is positive. At $\mathcal{A} = 0$, there is equilibrium. A negative drive forces a reaction backward against the direction the reaction arrow points in.

Here are some important consequences for the reaction process:

- Every homogeneous reaction begins spontaneously (if the concentrations of the reaction products are equal to zero at the beginning, we start with $\mathcal{A} = +\infty$).
- At a certain extent of conversion, the reaction ceases to change the ratio of reactants to products. We can also say that equilibrium is established.
- Equilibrium can be reached from both sides, meaning from the side of the starting substances as well as from the side of the reaction products.

In equilibrium, neither the forward reaction nor the backward reaction takes place spontaneously. Macroscopically speaking, there is no more transformation and the composition of the reaction mixture remains constant. However, forward and backward reactions do continue to occur at the microscopic level between the particles. These happen at identical rates though, so that the transformations in the two directions compensate for each other. In this case, one speaks traditionally of a *dynamic equilibrium*, an equality of forward and backward "forces" although one means a *kinetic equilibrium*, an equality of forward and backward reaction rates. We will go into this in more detail later on in Sect. 17.2.

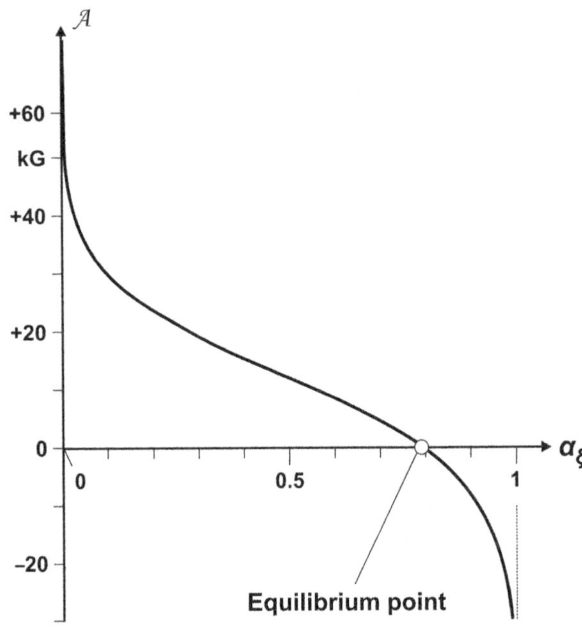

Fig. 6.5 Dependence of drive $\mathcal{A}$ upon the degree of conversion α_ξ for the reaction $2Fe^{3+} + 2I^- \rightarrow 2Fe^{2+} + I_2$ in an aqueous solution at room temperature (at initial concentrations of Fe^{3+} and I^- of 0.001 kmol m^{-3}).

6.4   The Mass Action Law

What is commonly known as the *mass action law* as defined by Guldberg and Waage is a consequence of a combination of the mass actions of individual substances participating in a reaction. Let us, once again, consider a reaction in a homogeneous solution

$$B + B' + \ldots \; \rightleftarrows \; D + D' + \ldots.$$

Equilibrium is established when there is no longer any potential gradient and the drive $\mathcal{A}$ disappears. Therefore:

$$\mathcal{A} = \overset{\circ}{\mathcal{A}} + RT\ln\frac{c_r(B) \cdot c_r(B') \cdot \ldots}{c_r(D) \cdot c_r(D') \cdot \ldots} = 0. \tag{6.16}$$

By solving the equation for $\overset{\circ}{\mathcal{A}}$ and applying $-\ln(x/y) = \ln(y/x)$, we obtain

$$\overset{\circ}{\mathcal{A}} = -RT\ln\frac{c_r(B) \cdot c_r(B') \cdot \ldots}{c_r(D) \cdot c_r(D') \cdot \ldots} = RT\ln\frac{c_r(D) \cdot c_r(D') \cdot \ldots}{c_r(B) \cdot c_r(B') \cdot \ldots}.$$

When we divide by RT and take the antilogarithm, we finally obtain

$$\overset{\circ}{\mathcal{K}} = \left(\frac{c_{\mathrm{r}}(\mathrm{D}) \cdot c_{\mathrm{r}}(\mathrm{D}') \cdot \ldots}{c_{\mathrm{r}}(\mathrm{B}) \cdot c_{\mathrm{r}}(\mathrm{B}') \cdot \ldots} \right)_{\mathrm{eq.}} \tag{6.17}$$

with

$$\overset{\circ}{\mathcal{K}} := \exp \frac{\overset{\circ}{\mathcal{A}}}{RT}. \tag{6.18}$$

Equation (6.17) characterizes the relationship between concentrations in *equilibrium* which has been referred to by the index eq. and shows a possible form of the mass action law for the reaction. The quantity $\overset{\circ}{\mathcal{K}}$ which is called the *equilibrium constant* of the reaction is so named because it does not depend upon the concentration of the substances. A more precise name, however, would be *equilibrium number* because $\overset{\circ}{\mathcal{K}}$ is a number and it is not constant but dependent upon temperature, pressure, solvent, etc. The index $^{\circ}$, which is actually superfluous, has been inserted in order to emphasize that $\overset{\circ}{\mathcal{K}}$ is to be formed from $\overset{\circ}{\mathcal{A}}$ (and not from $\mathcal{A}$!).

For the more general reaction between dissolved substances

$$|\nu_{\mathrm{B}}|\mathrm{B} + |\nu_{\mathrm{B}'}|\mathrm{B}' + \ldots \rightleftarrows \nu_{\mathrm{D}}\mathrm{D} + \nu_{\mathrm{D}'}\mathrm{D}' + \ldots$$

we obtain entirely appropriately

$$\overset{\circ}{\mathcal{K}} = \left(\frac{c_{\mathrm{r}}(\mathrm{D})^{\nu_{\mathrm{D}}} \cdot c_{\mathrm{r}}(\mathrm{D}')^{\nu_{\mathrm{D}'}} \cdot \ldots}{c_{\mathrm{r}}(\mathrm{B})^{|\nu_{\mathrm{B}}|} \cdot c_{\mathrm{r}}(\mathrm{B}')^{|\nu_{\mathrm{B}'}|} \cdot \ldots} \right)_{\mathrm{eq.}} . \tag{6.19}$$

Conventionally, the relative concentrations c_{r} are replaced by $c/c^{\ominus}$ and the fixed standard concentration $c^{\ominus}$ is combined with the equilibrium number $\overset{\circ}{\mathcal{K}}$ to form the new equilibrium constant $\overset{\circ}{K}$ (κ "*dimension factor*"):

$$\overset{\circ}{K} = \kappa \overset{\circ}{\mathcal{K}}, \quad \text{where } \kappa = (c^{\ominus})^{\nu_c} \text{ with}$$

$$\nu_c = \nu_{\mathrm{B}} + \nu_{\mathrm{B}'} + \cdots + \nu_{\mathrm{D}} + \nu_{\mathrm{D}'} + \cdots . \tag{6.20}$$

ν_c is the sum of the conversion numbers of dissolved substances, more precisely, those substances which show a dependence of the chemical potential upon concentration c; this is indicated by the index c. If the chemical potentials of all substances involved are concentration dependent, this fact can be emphasized by writing $\overset{\circ}{\mathcal{K}}_c$ and $\overset{\circ}{K}_c$. As stated above, $\overset{\circ}{\mathcal{K}}$ is always a number while the conventional constant $\overset{\circ}{K}$ has the unit $(\mathrm{mol}\ \mathrm{m}^{-3})^{\nu_c}$. Only when ν_c happens to be 0, are $\overset{\circ}{\mathcal{K}}$ and $\overset{\circ}{K}$ identical. $\overset{\circ}{K}$ is the more convenient quantity for formulating the mass action law,

$$\overset{\circ}{K} = \overset{\circ}{K}_c = \left(\frac{c(\mathrm{D})^{\nu_\mathrm{D}} \cdot c(\mathrm{D}')^{\nu_{\mathrm{D}'}} \cdot \ldots}{c(\mathrm{B})^{|\nu_\mathrm{B}|} \cdot c(\mathrm{B}')^{|\nu_{\mathrm{B}'}|} \cdot \ldots} \right)_{\mathrm{eq.}}, \tag{6.21}$$

while, for general considerations, $\overset{\circ}{K}$ is preferred since its dimension is the same for all reactions.

Here is an example of what has been said. With the help of the Table in Sect. A2.1 of the Appendix we will determine the *acidity constant* of acetic acid (CH_3COOH), abbreviated to HAc, in an aqueous solution, i.e., the equilibrium constant for the *dissociation* of acetic acid in water

$$\mathrm{HAc|w} \ \rightleftarrows \ \mathrm{H^+|w} + \mathrm{Ac^-|w}$$

$\mu^{\ominus}(\mathrm{kG})$: $-396.46 \qquad 0 \qquad -369.31 \quad \Rightarrow \ \mathcal{A}_1^{\ominus} = -27.15\,\mathrm{kG}$

or the equilibrium constant for the *proton transfer* from acetic acid to water in Brønsted's sense (this will be discussed in more detail in the next chapter)

$$\mathrm{HAc|w} \ + \ \mathrm{H_2O|l} \qquad \rightleftarrows \ \mathrm{H_3O^+|w} + \mathrm{Ac^-|w}$$

$\mu^{\ominus}(\mathrm{kG})$: $-396.46 \quad -237.14 \qquad -237.14 \quad -369.31 \quad \Rightarrow \ \mathcal{A}_2^{\ominus} = -27.15\,\mathrm{kG}.$

In both cases, the chemical drive has the same value meaning that the equilibrium numbers calculated by using this value are identical: $\mathcal{K}^{\ominus} = \exp(\mathcal{A}^{\ominus}/RT) = 1.74 \times 10^{-5}$. Because the dimension factors $\kappa = (c^{\ominus})^{\nu_c}$ are equal ($\nu_{c,1} = -1 + 1 + 1 = +1 = \nu_{c,2}$, since the solvent water is treated as pure substance and therefore does not appear in the sum $\nu_{c,2}$ of conversion numbers), this is also valid for the equilibrium constants $K^{\ominus} = 1.74 \times 10^{-5}\ \mathrm{kmol\ m^{-3}}$. In the first case, the mass action law is written numerically as

$$\overset{\circ}{\mathcal{K}}_1 = \frac{c_\mathrm{r}(\mathrm{H^+}) \cdot c_\mathrm{r}(\mathrm{Ac^-})}{c_\mathrm{r}(\mathrm{HAc})} \quad \text{and} \quad \overset{\circ}{K}_1 = \frac{c(\mathrm{H^+}) \cdot c(\mathrm{Ac^-})}{c(\mathrm{HAc})} = \kappa \, \overset{\circ}{\mathcal{K}}_1 .$$

In the second case, when we take into account that the solvent water has to be treated as a pure substance, we obtain for the mass action law

$$\overset{\circ}{\mathcal{K}}_2 = \frac{c_\mathrm{r}(\mathrm{H_3O^+}) \cdot c_\mathrm{r}(\mathrm{Ac^-})}{c_\mathrm{r}(\mathrm{HAc})} \quad \text{and} \quad \overset{\circ}{K}_2 = \frac{c(\mathrm{H_3O^+}) \cdot c(\mathrm{Ac^-})}{c(\mathrm{HAc})} = \kappa \, \overset{\circ}{\mathcal{K}}_2 .$$

We have to keep in mind that $\mathrm{H^+|w}$ and $\mathrm{H_3O^+|w}$ are merely two different notations for the same kind of particle. From now on, we will avoid the rather cumbersome index eq. as long as it is clear from the relation, as in this example, that we are dealing with the equilibrium composition.

The mass action law's range of validity is the same as that of the mass action equations (from which it is derived). The smaller the concentrations, the more strictly the law applies. At higher concentrations, deviations occur as the result of molecular or ionic interactions.

The magnitude of the equilibrium number determined unambiguously by $\overset{\circ}{\mathcal{A}}$ according to Eq. (6.18) or the equivalent relation

$$\overset{\circ}{\mathcal{A}} = RT\ln \overset{\circ}{\mathcal{K}} \tag{6.22}$$

is a good qualitative indication for how a reaction proceeds. The more strongly positive $\overset{\circ}{\mathcal{A}}$ is, the greater $\overset{\circ}{\mathcal{K}}$ ($\overset{\circ}{\mathcal{K}} \gg 1$) is. In this case, the final products dominate in the equilibrium composition. Because of the logarithmic relation, even small changes to $\overset{\circ}{\mathcal{A}}$ lead to noticeable shifts in the position of equilibrium. On the other hand, if $\overset{\circ}{\mathcal{A}}$ is strongly negative, $\overset{\circ}{\mathcal{K}}$ approaches zero ($\overset{\circ}{\mathcal{K}} \ll 1$) and the starting substances dominate in the equilibrium composition. At the same time, this also means that, even for negative $\overset{\circ}{\mathcal{A}}$, very small amounts of the starting substances are still converted into the final products because $\overset{\circ}{\mathcal{K}}$ has a small yet finite value. When $\overset{\circ}{\mathcal{A}} \approx 0$ and therefore $\overset{\circ}{\mathcal{K}} \approx 1$, the starting substances and final products are present in comparable amounts in equilibrium. (Keep in mind, however, that in all three cases discussed, $\mathcal{A} = 0$! since we have equilibrium in any case.)

With the help of the equilibrium number or the conventional equilibrium constant, the equilibrium composition of a mixture which has formed by spontaneous transformation of given amounts of starting substances can be quantitatively determined. If, for example, pure α-D-glucose at the concentration $c_0 = 0.1$ kmol m^{-3} is dissolved in water, one can use a polarimeter to observe the continuous change to the angle of rotation until a constant value is finally achieved. This can be ascribed to the partial transition of α-D-glucose into β-D-glucose (remember the discussion further above). If we indicate the concentration of β-D-glucose in equilibrium by the density of conversion c_ξ (which corresponds in this case to the concentration of β-D-glucose) we obtain in equilibrium

$$\overset{\circ}{K} = \kappa \cdot \exp\frac{\overset{\circ}{\mathcal{A}}}{RT} = \frac{c_\xi}{c_0 - c_\xi}.$$

The dimension factor κ equals 1 because of $v_c = 0$. The equilibrium constant at room temperature can be calculated by use of the standard value $\mathcal{A}^{\ominus} = 1.25$ kG (see Sect. 6.3):

$$K^{\ominus} = \exp\frac{1.25 \times 10^3 \text{ G}}{8.314 \text{ GK}^{-1} \times 298 \text{ K}} = 1.66.$$

Solving for c_ξ results in:

$$c_\xi = \frac{K^\ominus \cdot c_0}{K^\ominus + 1} = \frac{1.66 \times 0.1 \ \text{kmol} \ \text{m}^{-3}}{1.66 + 1} = 0.0623 \ \text{kmol} \ \text{m}^{-3}.$$

According to this, the state of equilibrium shows that 37.7 % of all dissolved molecules are α-D-glucose molecules and 62.3 % are β-D-glucose molecules.

The mathematical relations become rather complicated for reactions with a more complex stoichiometry. If we like to determine for example the equilibrium composition in the case of the reaction

$$2 \ Fe^{3+}\big|w + 2 \ I^-\big|w \rightarrow 2 \ Fe^{2+}\big|w + I_2\big|w$$

characterized by Table 6.1, we obtain

$$\overset{\circ}{K} = \kappa \cdot \exp\frac{\overset{\circ}{\mathcal{A}}}{RT} = \frac{4c_\xi^{\,3}}{(c_0 - 2c_\xi)^4}$$

with the dimension factor $\kappa = \left(c^\ominus\right)^{-1} = 1 \ \text{kmol}^{-1} \ \text{m}^3$ (because of $v_c = -2 - 2 + 2 + 1 = -1$). Because the value of $\mathcal{A}^\ominus$ is positive and relatively high ($\mathcal{A}^\ominus = +29 \ \text{kG}$), we have $K^\ominus \gg 1$, i.e., we can expect that the final products dominate in the equilibrium composition. For more detailed data we have to solve the equation above for c_ξ. Because a higher degree polynomial is involved, a numerical technique using appropriate mathematical software or a graphical approach is advisable. The equilibrium point can be determined from the figure in Sect. 6.3: $\alpha_\xi = c_\xi/c_{\xi,\text{max}} \approx 0.79$. The concentrations of the substances in the equilibrium mixture are therefore $c(Fe^{3+}) \approx 0.21 \ \text{mol} \ \text{m}^{-3}$ and $c(I^-) \approx 0.21 \ \text{mol} \ \text{m}^{-3}$ for the starting substances and $c(Fe^{2+}) \approx 0.79 \ \text{mol} \ \text{m}^{-3}$ and $c(I_2) \approx 0.39 \ \text{mol} \ \text{m}^{-3}$ for the final products, respectively.

We can also use Eq. (6.22) the other way around to experimentally determine the basic drive $\overset{\circ}{\mathcal{A}}$ of a reaction. In order to do this, it suffices to first calculate the equilibrium number $\overset{\circ}{\mathcal{K}}$ and then to derive $\overset{\circ}{\mathcal{A}}$ from it. At first glance, this looks amazingly easy, but the reaction can be so strongly inhibited that the concentrations being determined do not correspond to equilibrium values. This obstacle can be overcome, though, by adding a catalyst (compare to Sect. 19.2). As long as the added amount remains small, the position of equilibrium does not change and we can directly use the equilibrium values obtained in this way in Eq. (6.19). Once the basic drive is known, the drive for any other concentration can be calculated provided that its values lie within the range of validity of the mass action equations.

6.5 Special Versions of the Mass Action Equation

Until now, we have described mass action by using functions in which the concentrations c or, more exactly, the ratios c/c_0 or $c/c^{\ominus}$ appear as arguments. Instead of c, it would be possible to introduce any other measure of composition as long as it is proportional to concentration. This is almost always the case at small c values. We will highlight two of these measures here because they are of greater importance.

When the pressure of a gas is increased, the concentration of the gas particles also increases because they are compressed into a smaller volume. If the temperature remains unchanged, the concentration grows proportionally to the pressure: $c \sim p$, or

$$\frac{c}{c_0} = \frac{p}{p_0}. \tag{6.23}$$

As a result, the concentration ratio in the mass action equation for gases can be replaced by the pressure ratio:

$$\mu = \mu_0 + RT\ln\frac{p}{p_0} \quad \text{for } p, p_0 \ll 10p^{\ominus} \text{ (mass action equation 2)}. \tag{6.24}$$

This equation is precise enough to be applied to pressures up to about 10^2 kPa (1 bar). It also lends itself to estimates up to 10^3 or even 10^4 kPa. In anticipation of this, we have applied the equation above to treating the pressure dependence of the chemical potential of gases [cf. Eq. (5.18)].

The mass action equation 2 can be generalized somewhat. In the case of gaseous mixtures, one imagines that each component A, B, C, ... produces a *partial pressure* which is independent of its partners in the mixture. This corresponds to the pressure that the gaseous components would have if they alone were to fill up the available volume. The total pressure p of the gaseous mixture is simply equal to the sum of the partial pressures of all the components present (Dalton's law):

$$p_{\text{total}} = p_A + p_B + p_C + \dots \quad (\text{as well as } c_{\text{total}} = c_A + c_B + c_C + \dots). \tag{6.25}$$

If a gas is compressed, the concentrations of all the components and the partial pressures increase. This is exactly as if the gases were separate from each other. The formula $c \sim p$ is valid even when p represents only a partial pressure of a gas and not the total pressure. Hence, the equation $c/c_0 = p/p_0$, as well as the mass action equation, remains correct if we take c to be the partial concentration and p the partial pressure of a gas in a mixture.

Normally, the *standard pressure* $p^{\ominus} = 100$ kPa is chosen as the initial value for pressure although at this pressure, the chemical potential μ already deviates somewhat from the value the mass action equation yields. In order to have the results remain correct at low pressures, the true μ value at standard pressure cannot be inserted. Instead, a fictitious value which varies from it somewhat must be used.

(This is analogous to the procedure used for concentrations.) This fictitious value that is valid for standard pressure can be found in tables and then used to calculate the potentials at any other, not too high pressure. This special value is also called the *basic value* $\overset{\circ}{\mu}$ which should be indicated again by the index $^{\circ}$ placed above the symbol:

$$\mu = \overset{\circ}{\mu} + RT\ln \frac{p}{p^{\ominus}} = \overset{\circ}{\mu} + RT\ln p_r \quad \text{for } p \ll p^{\ominus} \text{ (mass action equation 2')}, \quad (6.26)$$

where p_r is the relative pressure. In contrast to this, all the μ values in the mass action equation 2 are real.

Another much used measure of composition is *mole fraction* x. As long as the content of a substance in a solution is small, concentrations and mole fractions are proportional to each other: $c \sim x$ for small c. In turn, this means

$$\frac{c}{c_0} = \frac{x}{x_0}. \quad (6.27)$$

Hence, x/x_0 can replace the concentration ratio c/c_0 in the mass action equation:

$$\mu = \mu_0 + RT\ln \frac{x}{x_0} \quad \text{for } x, x_0 \ll 1 \text{ (mass action equation 3).} \quad (6.28)$$

6.6 Applications of the Mass Action Law

Disturbance of Equilibrium One way to disturb a preexisting equilibrium would be to add a certain amount of one of the starting substances to the reaction mixture. Gradually, a new equilibrium would be established where the new equilibrium concentrations differ from the original ones. However, in all, the relations (6.19) and (6.21), respectively, remain fulfilled.

As an example, we consider the equilibrium in aqueous solution between iron hexaquo complex cations and thiocyanate anions on the one hand and the blood red iron thiocyanate complex on the other which can be described in the following simplified manner:

$$[Fe(H_2O)_6]^{3+}|w + 3\,SCN^-|w \rightleftarrows [Fe(H_2O)_3(SCN)_3]|w + 3\,H_2O|l.$$

If the blood red solution is diluted with water, the concentration and therefore also the chemical potential of the dissolved substances decrease for all substances by the same amount. This is indicated by an arrow placed above the formulas of the substances:

Experiment 6.1 *Iron(III) thiocyanate equilibrium*: If Fe^{3+} or SCN^- solutions are added to the pale orange dilute iron thiocyanate solution, it will turn red again in both cases.

$$\overset{\downarrow}{\left[Fe(H_2O)_6\right]^{3+}}\Big|w \;+3\;\overset{\downarrow\downarrow\downarrow}{SCN^-}\Big|w \rightleftarrows \overset{\downarrow}{\left[Fe(H_2O)_3(SCN)_3\right]}\Big|w +3\;H_2O\Big|l.$$

The potential of water remains nearly unchanged because of its high excess. At the beginning, the sum of potentials on both sides is equal. Afterward, the equilibrium is disturbed because the decrease on the left side is four times stronger than that on the right side. The complex decomposes and its potential decreases, whereas the potential of the substances on the left side increases. The reaction takes place (easily observable by the fading of the red color) until the equilibrium is again established. The pale orange color of the resulting solution is caused by the iron hexaquo complex. But also this new equilibrium can be displaced again (Experiment 6.1).

By adding excess iron (III) ions (1st case), their concentration in the solution and therefore also their potential increase; the reaction runs backwards and correspondingly the color deepens again to blood red. Adding excess thiocyanate ions shows the same result: the pale orange solution turns red again (2nd case). Both cases can be illustrated again by arrows above the chemical formulas.

$$\overset{\uparrow(1st\;case)}{\left[Fe(H_2O)_6\right]^{3+}}\Big|w \;+3\;\overset{\uparrow\uparrow\uparrow(2nd\;case)}{SCN^-}\Big|w \rightleftarrows \left[Fe(H_2O)_3(SCN)_3\right]\Big|w + 3\;H_2O\Big|l.$$

But one should keep in mind that together with the dissolved substances, water is also added which results in dilution. For achieving the desired effect, the added solutions of Fe^{3+} and SCN^- should therefore not be too thin for achieving the desired effect.

We can get the same results in a somewhat more cumbersome manner if we write down the mass action law and take into consideration that numerator and denominator must increase or decrease by the same factor in order to preserve equilibrium (water as solvent is treated as pure substance; therefore, it does not appear in the formula):

$$\overset{\circ}{\mathcal{K}} = \frac{c_r\left(\left[\text{Fe(H}_2\text{O)}_3\text{(SCN)}_3\right]\right)}{c_r\left(\left[\text{Fe(H}_2\text{O)}_6\right]^{3+}\right) \cdot c_r(\text{SCN}^-)^3}.$$

Diluting with water for example lowers the concentration of the complex but also the concentrations of the free ions. Therefore, the denominator would decrease much faster than the numerator. Because the quotient has to remain constant and equal to $\overset{\circ}{\mathcal{K}}$, the numerator must decrease as well: the equilibrium is displaced toward the reactant side.

Homogeneous Gas Equilibria To obtain homogeneous gas equilibria

$$|v_\text{B}|\text{B} + |v_{\text{B}'}|\text{B}' + \ldots \rightleftarrows v_\text{D}\,\text{D} + v_{\text{D}'}\text{D}' + \ldots$$

we can derive the equilibrium number analogously to the homogeneous solution equilibria, but instead of the mass action equation $1'$, we refer to the mass action equation $2'$. To indicate that partial pressures p replace concentrations c usually the symbol $\overset{\circ}{\mathcal{K}}_p$ is used instead of simply $\overset{\circ}{\mathcal{K}}$:

$$\overset{\circ}{\mathcal{K}}_p = \frac{p_r(\text{D})^{v_\text{D}} \cdot p_r(\text{D}')^{v_{\text{D}'}} \cdot \ldots}{p_r(\text{B})^{|v_\text{B}|} \cdot p_r(\text{B}')^{|v_{\text{B}'}|} \cdot \ldots}. \tag{6.29}$$

To convert into the conventional equilibrium constant $\overset{\circ}{K}_p$ a dimension factor must again be taken into account:

$$\overset{\circ}{K}_p = \kappa\,\overset{\circ}{\mathcal{K}}_p, \quad \text{where } \kappa = (p^{\ominus})^{v_p} \text{ with}$$

$$v_p = v_\text{B} + v_{\text{B}'} + \ldots + v_\text{D} + v_\text{D}' + \ldots. \tag{6.30}$$

As an example, let us consider the synthesis of ammonia:

$$\text{N}_2|\text{g} + 3\,\text{H}_2|\text{g} \rightleftarrows 2\,\text{NH}_3|\text{g}$$

$\mu^{\ominus}(\text{kG})$: 0   3×0   2×(−16.5) $\Rightarrow \mathcal{A}^{\ominus} = +33.0$ kG.

With the standard value of the chemical drive of +33 kG, the corresponding equilibrium number at room temperature is

$$\mathcal{K}_p^{\ominus} = \frac{p_r(\text{NH}_3)^2}{p_r(\text{N}_2) \cdot p_r(\text{H}_2)^3} = \exp\frac{\mathcal{A}^{\ominus}}{RT} = \exp\frac{33 \times 10^3 \ \text{G}}{8.314\text{G} \ \text{K}^{-1} \times 298 \ \text{K}} = 6.1 \times 10^5.$$

Because of $v_p = -1 - 3 + 2 = -2$ the conventional equilibrium constant is

$$K_p^{\ominus} = \frac{p(\mathrm{NH_3})^2}{p(\mathrm{N_2}) \cdot p(\mathrm{H_2})^3} = \underbrace{(p^{\ominus})^{v_p}}_{\kappa} \cdot \overset{\circ}{\mathcal{K}}_p = 6.1 \times 10^5 \times (100 \ \mathrm{kPa})^{-2} = 61 \ \mathrm{kPa}^{-2}.$$

In general, we notice that, depending upon the quantity used to describe the composition and its standard value ($c^{\ominus} = 1$ kmol m^{-3}, $p^{\ominus} = 100$ kPa ...), the same substance yields different basic values $\overset{\circ}{\mu}_c$, $\overset{\circ}{\mu}_p$, ... and therefore different equilibrium numbers $\overset{\circ}{\mathcal{K}}_c$, $\overset{\circ}{\mathcal{K}}_p$, ... for the same reaction. This is indicated by the varying indexes. Normally, we use $\overset{\circ}{\mu}_c$ for dissolved substances and $\overset{\circ}{\mu}_p$ for gases. If there is no index, we simply imagine what it is supposed to be.

Decomposition Equilibria As yet we have only considered homogeneous equilibria, i.e., chemical reactions in which all participating substances are in the same phase. Next we will discuss *heterogeneous equilibria* in which the substances are in different phases. Heterogeneous gas reactions in which solid phases are involved will be our first topic.

   We start with a process which is also industrially important, the calcination of limestone. In the case of the decomposition reaction of calcium carbonate in a *closed* container described by

$$\underline{\mathrm{CaCO_3|s} \ \rightleftarrows \ \mathrm{CaO|s} \ + \ \mathrm{CO_2|g}}$$

$$\mu^{\ominus}(\mathrm{kG}){:} \ -1128.8 \qquad -603.3 \quad -394.4 \qquad \Rightarrow \mathcal{A}^{\ominus} = -131.1 \ \mathrm{kG},$$

two pure solid phases ($\mathrm{CaCO_3}$ and CaO) and a gas phase are in equilibrium. The mass action equation 2′ is applied for the gas carbon dioxide. But how can we take pure solid substances (or pure liquids) B into account? In the case of these substances, the mass action term $RT\ln c_{\mathrm{r}}(\mathrm{B})$ is omitted, i.e., $\mu(\mathrm{B}) = \overset{\circ}{\mu}(\mathrm{B})$; the pure solid substance does not appear in the mass action law. In a dilute solution, this is also valid for the solvent which can be treated as a pure substance (see Sect. 6.3). The equilibrium number at standard temperature, $\overset{\circ}{\mathcal{K}}_p(T^{\ominus}) = \mathcal{K}_p^{\ominus}$, for the decomposition of carbonate is therefore equal to

$$\overset{\circ}{\mathcal{K}}_p = p_{\mathrm{r}}(\mathrm{CO_2}) \ \text{with} \ \mathcal{K}_p^{\ominus} = \exp(\mathcal{A}^{\ominus}/RT^{\ominus}) = 1.1 \times 10^{-23} \ \text{at 298 K.} \qquad (6.31)$$

The conventional equilibrium constant $\overset{\circ}{K}_p = \kappa \overset{\circ}{\mathcal{K}}_p$ with $\kappa = (p^{\ominus})^{v_p}$ results in

$$\overset{\circ}{K}_p = p(\mathrm{CO_2}) \ \text{with} \ K_p^{\ominus} = p^{\ominus} \cdot \mathcal{K}_p^{\ominus} = 100 \ \mathrm{kPa} \times 1.1 \times 10^{-23}$$

$$= 1.1 \times 10^{-21} \ \mathrm{kPa} \qquad (6.32)$$

because of $v_p = 1$. The equilibrium constant is identical to the decomposition pressure, i.e., the pressure of carbon dioxide at equilibrium, and hence not dependent on the amounts of the solid substances. Even though the pure solid substances do not appear in the mass action law, they have to be present for establishing equilibrium. The decomposition pressure is very low at room temperature, but it still depends (like the equilibrium constant) on temperature (see also Sect. 5.5 and Fig. 5.13). Only at temperatures considerably above 800 K, a noticeable pressure of carbon dioxide exists.

When the calcium carbonate is, however, heated in an *open* furnace like a lime kiln, the gas escapes into the surroundings, equilibrium is not established, and all of the carbonate decomposes.

The decomposition of crystalline hydrates, etc., can be described in the same way.

Phase Transitions The approach can also be applied to transitions where the state of aggregation changes. The evaporation of water represents an example for such a phase transition where a gas participates. The equilibrium number $\overset{\circ}{\mathcal{K}}_p$ for the equilibrium between liquid water and water vapor in a closed system,

$$\frac{\mathrm{H_2O|l} \quad \rightleftarrows \mathrm{H_2O|g}}{\mu^{\ominus}(\mathrm{kG}):\ -237.1 \qquad -228.6 \quad \Rightarrow \mathcal{A}^{\ominus} = -8.5\ \mathrm{kG},}$$

results in

$$\overset{\circ}{\mathcal{K}}_p = p_{\mathrm{r}}(\mathrm{H_2O|g}) \ \text{ with } \ \mathcal{K}_p^{\ominus} = 3.24 \times 10^{-2} \text{ at } 298\ \mathrm{K}. \tag{6.33}$$

Liquid water as pure liquid does not appear in Eq. (6.33). The corresponding conventional equilibrium constant is

$$\overset{\circ}{K}_p = p(\mathrm{H_2O|g}) = p_{\mathrm{lg}}(\mathrm{H_2O}). \tag{6.34}$$

Hence, the equilibrium constant represents the (saturation) vapor pressure of water, i.e., the pressure of water vapor in equilibrium with liquid water at the temperature considered. We will discuss phase transitions in more detail in Chap. 11.

Solubility of Ionic Solids Our next topic is heterogeneous *solution equilibria*. A substance submerged in a liquid will generally begin to dissolve. The extremely low chemical potential μ of this substance in the pure solvent rises rapidly—remember that $\mu \to -\infty$ for $c \to 0$—with increasing dissolution and therefore concentration. The process stops when the chemical potential of the substance in the solution is equal to that of the solid, i.e., equilibrium is established. We then refer to the solution as *saturated*, i.e., the solution contains as much dissolved material as possible under given conditions (temperature, pressure, such as standard conditions).

If the substance dissociates when dissolved, such as a salt in water

$$AB|s \rightleftharpoons A^+|w + B^-|w,$$

the products of the dissociation compensate together for the "tendency to transform" $\mu(AB)$ of the salt AB:

$$\mu(AB) = \mu(A^+) + \mu(B^-).$$

The mass action law for the dissolution process, here the transition from solid state to dissolved, s→d, is

$$\overset{\circ}{\mathcal{K}}_{sd} = c_r(A^+) \cdot c_r(B^-), \tag{6.35}$$

with

$$\overset{\circ}{\mathcal{K}}_{sd} = \exp\frac{\overset{\circ}{\mathcal{A}}_{sd}}{RT} = \exp\frac{\overset{\circ}{\mu}(AB) - \overset{\circ}{\mu}(A^+) - \overset{\circ}{\mu}(B^-)}{RT},$$

as long as some undissolved AB is present. Here the mass action term for a pure solid and therefore also the denominator in Eq. (6.35) have been omitted. Thus, the product of the relative concentrations of the ions in a saturated solution is constant under given conditions. In chemistry, the value $\overset{\circ}{\mathcal{K}}_{sd}$ usually receives its own name, *solubility product*. (This is commonly indicated by the index sp ($\overset{\circ}{\mathcal{K}}_{sp}$), but for the sake of consistency, we will use sd.) If a substance dissociates into several ions, then $\overset{\circ}{\mathcal{K}}_{sd}$ consists of the corresponding number of factors.

If the concentration of one of the ions, e.g., $c(A^+)$, increases, the concentration of the second, $c(B^-)$, must decrease in order to maintain equilibrium. This means that the substance AB precipitates from the solution. As an example, we consider a saturated table salt solution in which solid NaCl is in equilibrium with its ions in the solution (Experiment 6.2):

$$NaCl|s \rightleftharpoons Na^+|w + Cl^-|w$$

$$\mu^{\ominus}(kG):\ -384.1 \qquad -261.9 \qquad -131.2 \quad \Rightarrow \mathcal{A}^{\ominus} = +9.0\ kG.$$

Consequently, the equilibrium number $\mathcal{K}_{sd}^{\ominus}$ at room temperature is 37.8. The heterogeneous equilibrium can be described by the solubility product:

$$\overset{\circ}{\mathcal{K}}_{sd} = c_r(Na^+) \cdot c_r(Cl^-).$$

For the concentration $c_{sd} = c(Na^+) = c(Cl^-) = c^{\ominus}\sqrt{\mathcal{K}_{sd}^{\ominus}}$ of the saturated solution (*saturation concentration*) at 298 K, we obtain the value 6.1 kmol m^{-3} that corresponds—better than expected—to the measured value of 5.5 kmol m^{-3}.

Experiment 6.2 *Solubility product of table salt*: When Na^+ ions (in the form of concentrated sodium hydroxide solution) or Cl^- ions (in the form of concentrated hydrochloric acid) are added to the saturated table salt solution, NaCl precipitates.

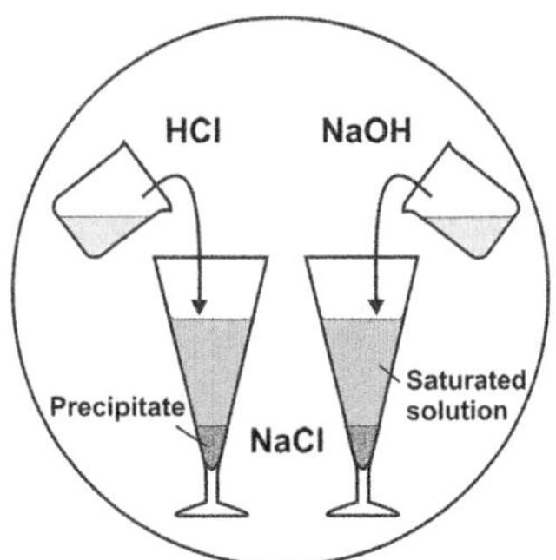

Consequently, adding compounds which have an ionic species in common with the salt under consideration, affects its solubility.

In the case of readily soluble salts we can only indicate general trends due to the strong ion–ion interaction in concentrated electrolyte solutions especially if polyvalent ions are involved. However, quantitative statements can be made in the case of slightly soluble compounds. As an example, let us consider lead (II) iodide:

$$PbI_2|s \quad \rightleftarrows \quad Pb^{2+}|w + 2\,I^-|w$$

$$\mu^{\ominus}(kG): -173.6 \qquad -24.4 \qquad 2\times(-51.6) \quad \Rightarrow \mathcal{A}^{\ominus} = -46.0\ kG,$$

with $\overset{\ominus}{\mathcal{K}}_{sd} = 8.6 \times 10^{-9}$ at room temperature. The mass action law for the solution equilibrium is

$$\overset{\circ}{\mathcal{K}}_{sd} = c_r\left(Pb^{2+}\right) \cdot c_r\left(I^-\right)^2.$$

We can calculate the saturation concentration c_{sd} of this salt from the numerical value for the solubility product at 298 K. In our example, the stoichiometry of equilibrium results in two I^- ions being produced for one Pb^{2+} ion. In contrast to the case of table salt, we now have

$$c\left(Pb^{2+}\right) = c_{sd} \quad \text{and} \quad c\left(I^-\right) = 2c_{sd}.$$

Insertion results in

$$\overset{\circ}{\mathcal{K}}_{sd} = \left(c_{sd}/c^{\ominus}\right) \cdot \left(2c_{sd}/c^{\ominus}\right)^2 = 4c_{sd}^{\,3}/\left(c^{\ominus}\right)^3$$

and the saturation concentration at 298 K is

$$c_{sd} = \sqrt[3]{\mathcal{K}^{\ominus}_{sd}/4}\, c^{\ominus} = \sqrt[3]{8.6 \times 10^{-9}/4}\ \text{kmol m}^{-3} = 1.3 \times 10^{-3}\ \text{kmol m}^{-3}.$$

It is now possible to estimate the effect of adding one type of the ions. Let us add enough of a concentrated NaI solution to the saturated lead iodide solution so that the I^- concentration will be equal to 0.1 kmol m^{-3}. Now, with the extra iodide ions present, the concentration of lead ions results in:

$$c\left(Pb^{2+}\right) = \frac{\mathcal{K}^{\ominus}_{sd}}{(c(I^-)/c^{\ominus})^2}\, c^{\ominus} = \frac{8.6 \times 10^{-9}}{0.01}\ \text{kmol m}^{-3} = 8.6 \times 10^{-7}\ \text{kmol m}^{-3}.$$

As expected, adding I^- drastically decreases the lead concentration. But "too much of a good thing" can be bad because a too big excess of I^- results in the formation of soluble complexes, e.g., $PbI_2 + I^- \rightarrow PbI_3^-$. This effect too could be calculated with the help of our approach.

The solubility of certain slightly soluble compounds can be controlled by different means as well, for example, by the pH value. We shall conclude this section by discussing the following interplay between the precipitation and the dissolution processes: When a potassium dichromate solution is added to a barium chloride solution (Experiment 6.3), a yellow precipitate of slightly soluble barium chromate is formed according to

$$2\, Ba^{2+}\big|w + Cr_2O_7^{2-}\big|w + H_2O\big|l \rightleftarrows 2\, BaCrO_4\big|s + 2\, H^+\big|w.$$

To be more precise, the concentrations of the starting substances decrease, i.e., barium chromate precipitates until equilibrium is established.

The precipitate dissolves after addition of hydrochloric acid. The concentration of H^+ ions and therefore their chemical potential increase. For reestablishing the equilibrium, $BaCrO_4$ has to dissolve:

Experiment 6.3 *Precipitation of barium chromate*: When a $K_2Cr_2O_7$ solution is added to a $BaCl_2$ solution, a yellow precipitate of $BaCrO_4$ is formed which gradually settles down at the bottom of the goblet. Addition of hydrochloric acid dissolves the precipitate again (*on the left*). If the solution is filtrated instead and solid sodium acetate is added to the clear filtrate, more $BaCrO_4$ precipitates (*on the right*).

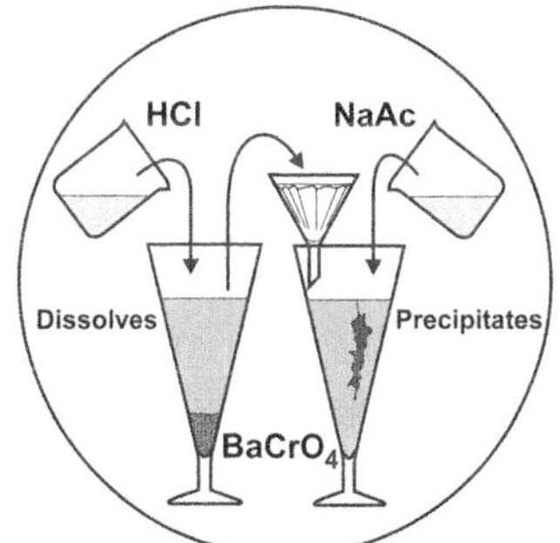

$$2\,\mathrm{Ba}^{2+}\big|\mathrm{w} + \mathrm{Cr_2O_7^{2-}}\big|\mathrm{w} + \mathrm{H_2O}\big|\mathrm{l} \rightleftarrows 2\,\mathrm{BaCrO_4}\big|\mathrm{s} + 2\,\overset{\uparrow\uparrow}{\mathrm{H}^+}\big|\mathrm{w}\,.$$

If, on the other hand, the H^+ ions which are released during precipitation are removed in the filtrate by adding sodium acetate ($\mathrm{Ac}^- + \mathrm{H}^+ \rightarrow \mathrm{HAc}$), their chemical potential decreases. Consequently, more $\mathrm{BaCrO_4}$ has to precipitate.

The same result can be obtained by considering the mass action law:

$$\overset{\circ}{\mathcal{K}} = \frac{c_\mathrm{r}(\mathrm{H}^+)^2}{c_\mathrm{r}(\mathrm{Ba}^{2+})^2 \cdot c_\mathrm{r}(\mathrm{Cr_2O_7^{2-}})}\,.$$

By adding H^+ the numerator increases so that the denominator must also increase in order for the quotient to remain constant and equal to $\overset{\circ}{\mathcal{K}}$. However, the concentrations $c_\mathrm{r}(\mathrm{Ba}^{2+})$ and $c_\mathrm{r}(\mathrm{Cr_2O_7^{2-}})$ can only increase when the precipitated $\mathrm{BaCrO_4}$ dissolves.

Solubility of Gases Next we will discuss the dissolution behavior of gases. If a gas B is brought into contact with a liquid (or solid), it diffuses within it, $\mathrm{B}\big|\mathrm{g} \rightleftarrows \mathrm{B}\big|\mathrm{d}$, until the chemical potential of the gas inside is as high as outside, $\mu(\mathrm{B}\big|\mathrm{g}) = \mu(\mathrm{B}\big|\mathrm{d})$. Consequently, the chemical drive $\mathcal{A}_{\mathrm{g}\rightarrow\mathrm{d}} = \mu(\mathrm{B}\big|\mathrm{g}) - \mu(\mathrm{B}\big|\mathrm{d})$ of the diffusion process has disappeared. $\mu(\mathrm{B}\big|\mathrm{d})$ is described by the mass action equation $1'$, $\mu(\mathrm{B}\big|\mathrm{g})$, however, by the mass action equation $2'$. The equilibrium number $\overset{\circ}{\mathcal{K}}_{\mathrm{g}\rightarrow\mathrm{d}}$ or, more briefly, $\overset{\circ}{\mathcal{K}}_{\mathrm{gd}}$ turns out to be

$$\overset{\circ}{\mathcal{K}}_{\mathrm{gd}} = \frac{c_\mathrm{r}(\mathrm{B}\big|\mathrm{d})}{p_\mathrm{r}(\mathrm{B}\big|\mathrm{g})} \quad \text{with} \quad \overset{\circ}{\mathcal{K}}_{\mathrm{gd}} = \exp\frac{\overset{\circ}{\mathcal{A}}_{\mathrm{gd}}}{RT} = \exp\frac{\overset{\circ}{\mu}(\mathrm{B}\big|\mathrm{g}) - \overset{\circ}{\mu}(\mathrm{B}\big|\mathrm{d})}{RT}\,. \qquad (6.36)$$

Taking into account that in this case, $\overset{\circ}{\mathcal{K}}$ corresponds to neither $\overset{\circ}{\mathcal{K}}_c$ nor $\overset{\circ}{\mathcal{K}}_p$, but represents a so-called mixed constant $\overset{\circ}{\mathcal{K}}_{pc}$. Like the mass action equations themselves, this relation is only valid as long as the concentration c in the solution and the pressure p outside it are small. Written in the conventional way, the equilibrium constant is

$$\overset{\circ}{K}_{\mathrm{gd}} = \frac{c(\mathrm{B}\big|\mathrm{d})}{p(\mathrm{B}\big|\mathrm{g})} \quad \text{with} \quad \overset{\circ}{K}_{\mathrm{gd}} = \overset{\circ}{\mathcal{K}}_{\mathrm{gd}} \cdot \frac{c^\ominus}{p^\ominus}\,. \qquad (6.37)$$

The solubility of a gas at constant temperature is therefore proportional to its partial pressure above the solution, $c(\mathrm{B}) \sim p(\mathrm{B})$. This relation was already discovered empirically in 1803 by the English chemist William Henry (*Henry's law*). Therefore, $\overset{\circ}{K}_{\mathrm{gd}}$ is also known as the Henry constant $\overset{\circ}{K}_\mathrm{H}$.

Let us take a look at an example, the solubility of oxygen in water, a quantity of prime importance for biological processes in bodies of water:

$$O_2|g \rightleftharpoons O_2|w$$

$$\mu^{\ominus}(kG): \quad 0 \qquad\qquad +16.4 \quad \Rightarrow \mathcal{A}^{\ominus} = -16.4 \text{ kG.}$$

In this case, the standard value of the chemical drive is -16.4 kG. The equilibrium number at room temperature is equal to

$$\overset{\circ}{\mathcal{K}}{}^{\ominus}_{\mathrm{gd}} = \frac{c_{\mathrm{r}}\left(O_2\big|w\right)}{p_{\mathrm{r}}\left(O_2\big|g\right)} \quad \text{with} \quad \overset{\circ}{\mathcal{K}}{}^{\ominus}_{\mathrm{gd}} = \exp\frac{-16,400 \text{ G}}{8.314 \text{ G K}^{-1} \times 298 \text{ K}} = 1.3 \times 10^{-3},$$

while the conventional equilibrium constant results in

$$\overset{\circ}{K}{}^{\ominus}_{\mathrm{gd}} = \frac{c\left(O_2\big|w\right)}{p\left(O_2\big|g\right)} \quad \text{with} \quad \overset{\circ}{K}{}^{\ominus}_{\mathrm{gd}} = 1.3 \times 10^{-3} \times \frac{1 \text{ kmol m}^{-3}}{100 \text{ kPa}}$$

$$= 1.3 \times 10^{-5} \text{ mol m}^{-3} \text{ Pa}^{-1}.$$

The partial pressure of O_2 in air, for example, is about 20 kPa; the concentration of O_2 in air-saturated water at 298 K therefore results in

$$c\left(O_2\big|w\right) = \overset{\circ}{K}{}^{\ominus}_{\mathrm{gd}} \cdot p\left(O_2\big|g\right) = 1.3 \times 10^{-5} \text{ mol m}^{-3} \text{ Pa}^{-1} \times 20 \times 10^{3} \text{ Pa}$$

$$= 0.26 \text{ mol m}^{-3}.$$

Distribution Equilibria Relations that can be dealt with in a theoretically similar way would be, for example, systems where a third substance B (possibly iodine) is added to two practically immiscible liquids such as water/ether. Substance B should be soluble in both liquid phases $(')$ and $('')$. It disperses then between these phases until its chemical potential is equal in both. The equilibrium number $\overset{\circ}{\mathcal{K}}_{\mathrm{d}\to\mathrm{d}}$ or, more briefly, $\overset{\circ}{\mathcal{K}}_{\mathrm{dd}}$ is then:

$$\overset{\circ}{\mathcal{K}}_{\mathrm{dd}} = \frac{c_{\mathrm{r}}\left(B\big|d''\right)}{c_{\mathrm{r}}\left(B\big|d'\right)} \quad \text{with} \quad \overset{\circ}{\mathcal{K}}_{\mathrm{dd}} = \exp\frac{\overset{\circ}{\mathcal{A}}_{\mathrm{dd}}}{RT} = \exp\frac{\overset{\circ}{\mu}\left(B\big|d'\right) - \overset{\circ}{\mu}\left(B\big|d''\right)}{RT}. \tag{6.38}$$

Conventionally, we obtain

$$\overset{\circ}{K}_{\mathrm{dd}} = \frac{c\left(B\big|d''\right)}{c\left(B\big|d'\right)} \quad \text{with} \quad \overset{\circ}{K}_{\mathrm{dd}} = \overset{\circ}{\mathcal{K}}_{\mathrm{dd}}. \tag{6.39}$$

In the case of small concentrations, the ratio of equilibrium concentrations (or mole fractions, etc.) of the dissolved substance in two liquid phases is a (temperature dependent) "constant" (*Nernst's distribution law*). The constant $\overset{\circ}{K}_{dd}$ is also called Nernst's distribution coefficient K_N.

Distribution equilibria play a significant role in separating the substances in a mixture by the process of *extraction*. The laboratory procedure called "extraction by shaking" (extracting a substance from its solution by using another solvent in which the substance dissolves much better) is based on such equilibria. This method can be used to completely remove iodine from water by repeatedly extracting it with ether. *Partition chromatography* is based upon the same principle. A solvent acts as the stationary phase in the pores of a solid carrier material (paper for example) and a second solvent (with the substance mixture to be separated) flows past it in the form of a mobile phase. This is known as a mobile solvent. The more soluble a substance is in the stationary phase, the longer it will remain there and the more strongly its movement along this phase will slow down. Eventually, a separation occurs in the mixture originally applied at a point.

Influence of Temperature   The equilibrium numbers (and constants) we have considered so far are valid only under certain conditions (mostly standard conditions at 298 K and 100 kPa). If the value of $\overset{\circ}{\mathcal{K}}$ at an arbitrary temperature is of interest, then the RT term as well as the temperature dependence of chemical drive $\mathcal{A}$ need to be taken into account. We refer here to the linear approximation introduced in Sect. 5.2:

$$\mathcal{A} = \mathcal{A}_0 + \alpha(T - T_0).$$

Insertion into Eq. (6.18) yields the following result for the equilibrium number at a temperature T:

$$\overset{\circ}{\mathcal{K}}(T) = \exp \frac{\overset{\circ}{\mathcal{A}}(T_0) + \alpha(T - T_0)}{RT}. \tag{6.40}$$

When the temperature is increased ($\Delta T > 0$), $\overset{\circ}{\mathcal{K}}(T)$ can increase or decrease relative to the initial value $\overset{\circ}{\mathcal{K}}(T_0)$ depending upon the values of $\overset{\circ}{\mathcal{A}}(T_0)$ and α which are typical for a particular reaction. In the first case, the equilibrium composition shifts to benefit the products, and in the second case, it shifts to benefit the starting substances. The equilibrium constant can be influenced by the choice of temperature. This is of potentially great importance for large-scale technical reactions as well as for environmentally relevant ones.

6.7   Potential Diagrams of Dissolved Substances

Energy must be used in order to transfer matter from a state of low μ value to a state of high μ value. Therefore, the potential μ can be regarded as a kind of energy level the matter is on. This is why matter with a high chemical potential is often called *energy rich* and matter with a low potential, *energy poor*. These terms are not to be considered absolute in themselves but only in relation to other substances with which the substance in question can reasonably be compared.

When the amount n of a dissolved substance in a given volume is continuously increased, the potential μ of the substance also increases. While at first, small changes Δn in the amount of substance are enough to cause a certain rise in potential $\Delta \mu$, later on increasingly large amounts are necessary for this. As long as the concentration is not too high, the mass action equation remains valid. This means that the concentrations (or when the volume remains constant, the amount) must always increase by the same factor β if μ is to increase by the same amount. n therefore increases exponentially along with the chemical potential μ.

Relative to the same increase of potential $\Delta \mu$, the solution takes up the more of the substance to be dissolved the more it already contains. Hence, the "capacity" for this substance increases with the amount already present—perhaps somewhat different from what one might expect. The capacity B of a substance, the so-called *matter capacity*, is defined by the following equation:

$$B = \frac{\mathrm{d}n}{\mathrm{d}\mu}. \tag{6.41}$$

A well-known example of the quantity B is the so-called *buffering capacity*, meaning the capacity B_p of a given amount of a solution for hydrogen ions. We will go into greater detail in Sect. 7.6. If the region being dealt with is homogeneous, B can logically be related to the volume:

$$b = \frac{B}{V}. \tag{6.42}$$

In contrast to B, we will call b the *matter capacity density*.

If the matter capacity B of a finite volume V of solution is plotted against the chemical potential μ (Fig. 6.6a), the area under the $B(\mu)$ curve from $-\infty$ to the actual potential μ represents the amount n of the substance in this volume. The relation becomes clearer if the axes are exchanged and the curve then appears as a two-dimensional outline of a "container" filled to the level μ with the amount of substance n (Fig. 6.6b). Finally, if $\sqrt{B/\pi}$ is plotted instead of B, the resulting curve can be thought of as the outline of a rotationally symmetrical goblet (Fig. 6.6c). In this case as well, n is the volume of the container filled up to the level μ. In the following we will make use of this image because it is not only vivid, but also has the advantage of reducing the width of otherwise very wide curves, making them easier to draw.

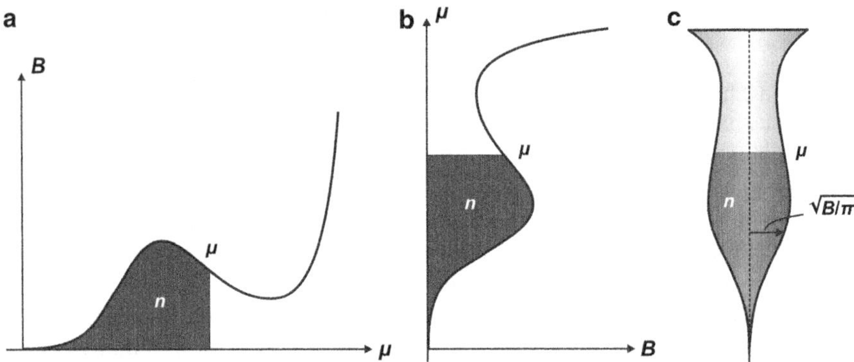

Fig. 6.6 (**a**) Plotting the matter capacity B as a function of chemical potential μ, (**b**) Exchanging axes, (**c**) Solid of revolution having the same content.

The area under the curve of a continuous function $f(x)$ in the interval [a;b] is calculated by

$$F = \int_a^b f(x)\,dx,$$

(cf. Sect. A1.2). The volume of a body created by rotating the area in the diagram about the x axis can be calculated analogously:

$$V = \pi \int_a^b f(x)^2\,dx.$$

Such bodies are also called *solids of revolution*.

As long as the mass action equations are valid, it is easy to express n as a function of μ. To do so, we solve the equation $\mu = \overset{\circ}{\mu} + RT\ln(c/c^{\ominus})$ with $c = n/V$ for n and obtain:

$$\frac{\mu - \overset{\circ}{\mu}}{RT} = \ln\frac{n}{Vc^{\ominus}} \quad \text{or} \quad n = Vc^{\ominus} \cdot \exp\frac{\mu - \overset{\circ}{\mu}}{RT}. \tag{6.43}$$

The matter capacity is the derivative with respect to μ, for constant $Vc^{\ominus}$ and $\overset{\circ}{\mu}$:

$$B = \frac{dn}{d\mu} = \frac{Vc^{\ominus}}{RT} \cdot \exp\frac{\mu - \overset{\circ}{\mu}}{RT} \quad \text{or} \quad \text{short} \quad B = \frac{n}{RT}. \tag{6.44}$$

Here a few hints concerning the calculation: If one writes—such as in school mathematics—y instead of n and x instead of μ and abbreviates the constant terms with a, b, c, the calculation is nothing else than taking the derivative of the function $y = a \cdot \exp(bx + c)$ with respect to x with $a = Vc^{\ominus}$, $b = 1/RT$, and $c = -\overset{\circ}{\mu}/RT$. Obviously, the function $z = g(x) = bx + c$ is nested in the function $y = f(z) = a \cdot \exp z$: $y = f(g(x))$. One takes the derivative of such functions by using the *chain rule* [rule (A1.13) in Sect. A1.2)]:

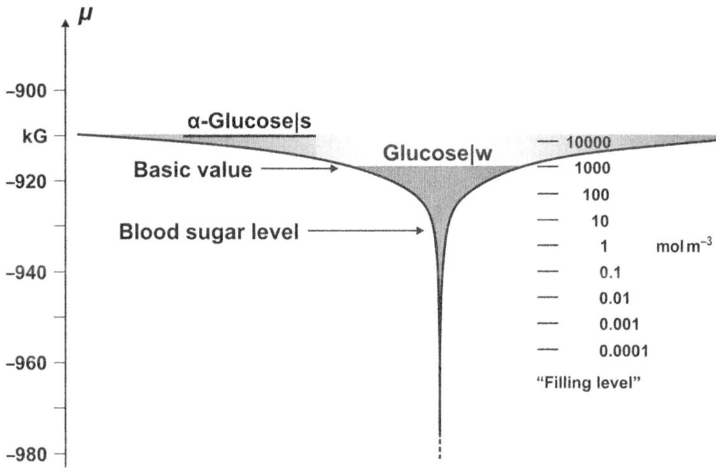

Fig. 6.7   Potential diagram of glucose.

$$y' = f'(z) \cdot g'(x) = a \cdot \exp z \cdot b = a \cdot b \cdot \exp(bx + c).$$

Additionally, we have used the following calculation rules: the exponential function is equal to its own derivative [rule (A1.7)], a constant factor is preserved when taking the derivative [rule (A1.9)], a constant summand, however, disappears. If we now replace the variables and constants by the corresponding quantities and expressions, we finally obtain:

$$B = \frac{Vc^{\ominus}}{RT} \cdot \exp\left(\frac{1}{RT}\mu - \frac{\overset{\circ}{\mu}}{RT}\right).$$

As a consequence, B, like n, depends exponentially upon μ. In this case, the container whose curve we are interested in has the form of an "exponential horn" which is open at the top.

The matter capacity density b can be easily calculated from B:

$$b = \frac{B}{V} = \frac{c^{\ominus}}{RT} \cdot \exp\frac{\mu - \overset{\circ}{\mu}}{RT} \quad \text{or short} \quad b = \frac{c}{RT}. \tag{6.45}$$

Like B and n, b depends exponentially on μ.

We will take a closer look at this approach using the example of glucose (Fig. 6.7). In its solid and pure state, glucose has a chemical potential which is not subject to mass action. For this reason, it is represented as a horizontal line in the potential diagram. Because, as previously stated, glucose occurs in two forms, α and β, two potential levels lying close together should actually be drawn in. However, for the sake of simplicity, only one is represented here.

In the dissolved state and depending upon concentration or amount, we have an entire band of potential values. Instead of the band, we will use the $B(\mu)$ curve as it is described for the general case in Fig. 6.6 to express this dependence. Alternatively, we can use the $b(\mu)$ curve, which looks identical to it. The radius of the rotationally symmetrical goblet equals $\sqrt{b/\pi}$. Therefore, the content up to a chosen level corresponds to the quantity of glucose present there relative to the volume of solution. This means it is equal to the total concentration of glucose. At small concentrations, the radius increases exponentially with rising μ; for high concentrations, this is only approximate. We do not need to distinguish between the α and the β forms because an equilibrium rather quickly develops between the two isomers. The basic value of the potential (at a concentration of 1,000 mol m^{-3}) applies to this equilibrium mixture. The content of the goblet has been drawn to this arbitrarily chosen potential level. We will also generally choose this fill level for other substances. The value in a living cell would be considerably lower, though.

If the amount of dissolved glucose were to be continuously increased, and the fill level of the goblet raised to the level of the solid glucose, the glucose would begin to crystallize. The glucose would begin to run over the rim of the goblet, so to speak. If, on the other hand, glucose were present as excess solid, it would need to dissolve for as long as it would take for all the crystals to disappear or until the potential in the solution increased to the level of the goblet rim in the drawing. One might say that in this state, the glucose solution is *saturated* relative to the solid.

Chapter 7
Consequences of Mass Action: Acid–Base Reactions

The concept of mass action can be applied to any transformation of substances. In the case of matter dynamics, it does not matter how we imagine the process in question working at the molecular level: Whether it is by formation or cleavage of chemical bonds, rearranging crystal lattices, migration of particles, transfer of electrons or whole groups of atoms from one type of particle onto the other, etc. In this chapter we will concentrate upon one important example for chemical transformations, namely acid–base reactions, in order to demonstrate that the chemical potential is well suited to describing very specialized and differentiated fields. Acid–base reactions are central to chemistry and its applications; for their quantitative description we introduce the "proton potential" μ_p as a measure of the strength of an acid–base pair. The level equation and the protonation equation are used to describe the behavior of weak acid–base pairs. Subsequently, one important application for acid–base equilibria, the analytic method called acid–base titration, is presented. Finally, the mode of reaction of buffers and indicators is discussed. Buffers play also a significant role in living organisms because even small shifts in the proton potential can there result in disease and death.

7.1 Introduction

The approach used up to now can be applied in the same way for any transformation of substances. In the case of matter dynamics, it does not matter how we imagine the process in question working at the molecular level: Whether it is by formation or cleavage of chemical bonds, rearranging crystal lattices, migration of particles, transfer of electrons or whole groups of atoms from one type of particle onto the other, etc. We will concentrate upon one important example here, namely acid–base reactions, in order to demonstrate that the chemical potential is well suited to describing very specialized and differentiated fields.

© Springer International Publishing Switzerland 2016
G. Job, R. Rüffler, *Physical Chemistry from a Different Angle*,
DOI 10.1007/978-3-319-15666-8_7

Before introducing the topic in more detail we have to discuss shortly the problems concerning the denomination of hydrogen ions. Hydrogen has three naturally occurring isotopes, called *protium* ^1H, *deuterium* ^2H, and *tritium* ^3H. The corresponding cations are called *proton* (denoted $^1H^+$ or shortly p), *deuteron* ($^2H^+$ or d), and *triton* ($^3H^+$ or t). *Hydron* is the name (recommended by IUPAC in 1988) for positive hydrogen ions H^+ without regard to nuclear mass, especially for the mixture of isotopes formed by natural hydrogen. Traditionally in acid–base chemistry, the term "proton" is used instead of "hydron." But the differences are minute and may be neglected, because more than 99.98 % of the naturally occurring hydrons H^+ are protons p anyway. In this chapter we follow the traditional practice.

7.2　The Acid–Base Concept According to Brønsted and Lowry

According to Johannes Nicolaus Brønsted and Thomas Lowry, an *acid* is a substance or more generally a type of particle (be it neutral or ionic) that tends to release protons p (H^+ ions). It represents a *proton donor* for which we will use the abbreviations HA or BH^+,

$$HA \rightarrow A^- + H^+, \quad \text{and} \quad BH^+ \rightarrow BH + H^+, \quad \text{respectively.}$$

The residual left over from separation is called the *corresponding* or *conjugated base*. The base abbreviated with A^- or BH acts as *proton acceptor*. For general purposes, it is more convenient to use symbols without charge numbers. We will use the abbreviation Ad for the acid (from the Latin acidum) and Bs for the base [(from the Greek βάσις (basis)]:

$$Ad \rightarrow Bs + \nu_p\, p.$$

Ad and Bs form an *acid–base pair* abbreviated by Ad/Bs. ν_p describes the *proticity*. If $\nu_p = 1$, one speaks of a monoprotic acid–base pair. However, if $\nu_p > 1$, it is called polyprotic. Simple examples for monoprotic pairs are HCl/Cl^- and H_3O^+/H_2O:

$$HCl \rightarrow Cl^- + H^+, \quad H_3O^+ \rightarrow H_2O + H^+.$$

In the first case the neutral acid hydrogen chloride HCl tends to release a proton by forming the anionic base Cl^-; in the second the cationic acid H_3O^+ does the same under formation of the neutral base H_2O.

More than one substance can appear in place of the simple one Ad and/or Bs. If we allow the abbreviations Ad and Bs to signify a combination of substances, then the generalized process reads as follows:

$$\overbrace{Ad' + Ad'' + Ad''' + \ldots}^{Ad} \rightarrow \overbrace{Bs' + Bs'' + Bs''' + \ldots}^{Bs} + \nu_p\, p.$$

In the following example, a combined acid appears:

$$CO_2 + H_2O \rightarrow HCO_3^- + H^+.$$

An acid-base pair Ad/Bs can be considered as a kind of chemical reservoir for protons that is

- Totally full (completely protonated) in the Ad state,
- Totally empty (completely deprotonated) in the Bs state.

Because even the smallest release of protons allows a base to be formed, we can assume from the start that both base and acid are always present to a greater or lesser degree. The separated protons usually do not appear as such. In a following reaction, they are bonded immediately to other particles that function as bases (such as H_2O). In an aqueous solution, the protons appear in the form of oxonium ions H_3O^+. In the case of the dissociation of an acid HA (such as HCl) in water according to $HA \rightarrow A^- + H^+$ protons are directly transferred from the HA molecules to the H_2O molecules (example of a so-called *acid–base reaction*). The following equilibrium is established immediately (within 10^{-8} s):

$$HA|w + H_2O|l \rightleftarrows A^-|w + H_3O^+|w.$$

Correspondingly, in a reaction of a base B (such as NH_3) with water, protons are transferred from the H_2O molecules to the B molecules:

$$B|w + H_2O|l \rightleftarrows BH^+|w + OH^-|w.$$

In this case, the water functions as the acid.

We have already seen an impressive example for such an acid–base reaction, namely the ammonia fountain (Experiment 4.12). The formation of an alkaline (basic) milieu in the flask according to

$$NH_3|w + H_2O|l \rightleftarrows NH_4^+|w + OH^-|w$$

is demonstrated by the change of color of the added indicator (phenolphthalein) from colorless to red violet. We will take a closer look at indicators and their mode of functioning in Sect. 7.7.

7.3 Proton Potential

Basic Idea Obviously, the protons in acid–base reactions are simply transferred from one base B to another B*:

$$BH^+ + B^* \rightleftharpoons B + B^*H^+.$$

This is also referred to as a *proton transfer reaction*. The older but ambiguous name *protolytic reaction* is not recommended. According to the general conversion formula, the reaction between gaseous hydrochloric acid and ammonia vapor produces a "sal-ammoniac fog," a mist of finely distributed ammonium chloride crystals, $NH_4Cl|s$ (Experiment 7.1):

$$HCl|g + NH_3|g \rightarrow \left[NH_4^+\right]\left[Cl^-\right]|s.$$

This means that the definitions given here also hold when no solvent is present.

Most acid–base reactions, however, take place in aqueous solutions. In this case, the second acid–base pair is provided by the solvent water itself (H_3O^+/H_2O or H_2O/OH^-). Despite this, it is possible to assign a chemical potential to the bonded yet exchangeable protons. This is the value established in the equilibrium of the reaction $Ad \rightleftharpoons Bs + \nu_p\,p$, where p denotes protons from any external source. In this particular case we have

$$\mu_{Ad} = \mu_{Bs} + \nu_p\mu_p$$

or

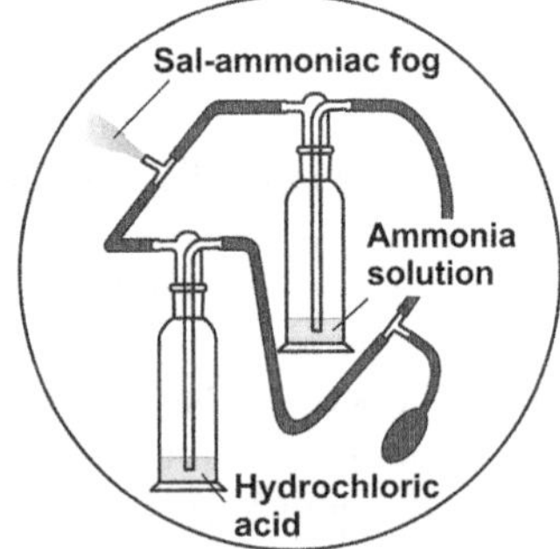

Experiment 7.1 *Formation of "sal-ammoniac fog" from vapors of hydrochloric acid and ammonia*: Concentrated hydrochloric acid is poured in one of the gas-washing bottles; concentrated ammonia solution is poured in the other. By pressing the rubber ball both gases (HCl and NH_3) are unified and the resulting "sal-ammoniac" fog emerges from the small tube.

$$\mu_p = \underbrace{\frac{1}{\nu_p}[\mu_{Ad} - \mu_{Bs}]}_{\mu_p(\text{inside})}, \tag{7.1}$$

respectively. Protons move from outside into the "reservoir" Ad/Bs if $\mu_p > \mu_p(\text{inside})$ and in the opposite direction if $\mu_p < \mu_p(\text{inside})$. The value of $\mu_p(\text{inside})$ is uniquely defined by the chemical potentials μ_{Ad} and μ_{Bs}, meaning it is a characteristic quantity of the acid–base pair under consideration. We express this by adding the name of the pair as argument or subscript:

$$\mu_p(\text{inside}) := \mu_p(\text{Ad/Bs}) := \mu_{p,\,Ad/Bs}.$$

The *proton potential* $\mu_p(\text{Ad/Bs})$ indicates the strength with which the acid–base pair Ad/Bs tends to transfer protons to any other reservoir. Hence, it is a measure for what is conventionally called the "strength of acid Ad," or more accurately the *acidic strength* of the pair Ad/Bs. (The tendency to release protons is determined equally by the chemical potential of the acid and that of the corresponding base.) For comparison of acidic strengths of various Ad/Bs pairs, the conditions must be specified. Table 7.1 shows some numerical values $\mu_p^{\ominus}(\text{Ad/Bs})\left[= \nu_p^{-1}\left(\mu_{Ad}^{\ominus} - \mu_{Bs}^{\ominus}\right)\right]$ for *standard conditions* that means 298 K, 100 kPa, and a concentration of

Table 7.1 Standard values of proton potentials of some acid–base pairs (298 K, 100 kPa, 1 kmol m^{-3} in an aqueous solution).

Ad $\rightleftarrows$ Bs $+ \nu_p\, p$	$\mu_p^{\ominus}$ (kG)
$HClO_4|w \rightleftarrows ClO_4^-|w + H^+$	$+57$
$HCl|w \rightleftarrows Cl^-|w + H^+$	$+34$
$H_2SO_4|w \rightleftarrows HSO_4^-|w + H^+$	$+17$
$HNO_3|w \rightleftarrows NO_3^-|w + H^+$	$+8$
$H_3O^+|w \rightleftarrows H_2O|l + H^+$	0
$HSO_4^-|w \rightleftarrows SO_4^{2-}|w + H^+$	-11
$H_2CO_3|w \rightleftarrows HCO_3^-|w + H^+$	-21
$C_2H_4OHCOOH|w \rightleftarrows C_2H_4OHCOO^-|w + H^+$	-22^a
$CH_3COOH|w \rightleftarrows CH_3COO^-|w + H^+$	-27^b
$CO_2|w + H_2O|l \rightleftarrows HCO_3^-|w + H^+$	-36
$NH_4^+|w \rightleftarrows NH_3|w + H^+$	-53
$HCO_3^-|w \rightleftarrows CO_3^{2-}|w + H^+$	-59
$\left[Ca(H_2O)_6\right]^{2+}|w \rightleftarrows Ca(OH)_2|s + 4H_2O|l + 2H^+$	-65
$H_2O|l \rightleftarrows OH^-|w + H^+$	-80
$NH_3|w \rightleftarrows NH_2^-|w + H^+$	-130
$OH^-|w \rightleftarrows O^{2-}|w + H^+$	-165
$H|g \rightleftarrows e^-|g + H^+$	-231
$HCl|g \rightleftarrows Cl^-|g + H^+$	-289
$HF|g \rightleftarrows F^-|g + H^+$	-441

[a]Lactic acid/lactate
[b]Acetic acid/acetate

1 kmol m^{-3} in aqueous solution for both acid and base. For determining such values, for example, the value $\mu_p^{\ominus}(\text{HAc}/\text{Ac}^-)$ of the pair acetic acid/acetate (the abbreviation Ac is used for the acetate group CH$_3$COO), one first selects the chemical potentials of the acid and the corresponding base at standard conditions from the Table in Sect. A.2.1 in the Appendix or a comparable one. Subsequently, the difference is calculated ($v_p = 1$):

$$\mu_p^{\ominus}(\text{HAc}/\text{Ac}^-) = \frac{1}{1}[\mu^{\ominus}(\text{HAc}) - \mu^{\ominus}(\text{Ac}^-)] = (-396.5\ \text{kG}) - (-369.3\ \text{kG})$$

$$= -27.2\ \text{kG}.$$

Even if the (fictitious) values of acid and base in aqueous solution at a concentration of 1 mol m^{-3} (cp. Sect. 6.2) are used, the proton exchange with the solvent water should be completely inhibited. That means that the acetic acid HAc in the solution in question should only exist as molecule (and not in the ionized form); the acetate ion Ac$^-$, however, should only exist as ion.

An exception concerning the standard conditions are the pairs H$_3$O$^+$/H$_2$O and H$_2$O/OH$^-$ where the water that functions as the acid or base (as the case may be) also represents the solvent and is therefore treated as a pure liquid (cf. Sect. 6.3).

We expect that any substance that has protons "impressed" upon it can again release them. Therefore, it follows that in its protonated form, it should be called an acid, and in its original form, its corresponding base. In this regard, protonated water H$_3$O$^+$ with its base H$_2$O is shown in the table as well as the acid H$_2$O with its base OH$^-$. This also holds for the hydrogensulfate anion HSO$_4^-$ (protonated H$_2$SO$_4$, deprotonated SO$_4^{2-}$) and for ammonia molecules NH$_3$ (protonated NH$_4^+$, deprotonated NH$_2^-$). Substances that function not only as acids but also as bases are called *amphoteric*.

Acid–base pairs with positive standard value $\mu_p^{\ominus}$ (or more generally positive basic value $\overset{\circ}{\mu}_p$) are considered *strongly acidic* and ones with $\mu_p^{\ominus} < 0\,\text{kG}$, *weakly acidic*. This rough classification may be refined as shown in Table 7.2. Talking about strong or weak acids or bases instead of acid–base pairs is common practice,

Table 7.2 Classification of acid–base pairs in aqueous solution according to their acidic strength.

Ad/Bs may be called	For $\overset{\circ}{\mu}_p$(kG) in the range
Very strong acidic	>+20
Strong acidic	0 … +20
Moderate acidic	−20 … 0
Weak acidic	−40 … −20
Weak alkaline	−60 … −40
Moderate alkaline	−80 … −60
Strong alkaline	−100 … −80
Very strong alkaline	<−100

but not very opportune. The acidic or alkaline (basic) strength is a property of the pair as a whole and not only of a part of it. (To avoid any possibility of confusion with the term "basic value" ("basic" in the sense of "fundamental") we will use only "alkaline" in the following.)

When two acid–base pairs are present in a solution, the "stronger acidic" pair forces protons upon the "weaker acidic" one. Because acid–base equilibrium occurs almost instantly, a uniform proton potential is established through the exchange of protons even when different acids and bases are present:

$$\mu_p = \mu_p(\text{Ad/Bs}) = \mu_p(\text{Ad*/Bs*})$$

$$= \ldots \qquad (\text{"Proton potential equalization"}). \qquad (7.2)$$

Equation (7.2) expresses an important fact: The proton potential μ_p represents a property of a chemical system which has a universal meaning comparable to that of pressure p and temperature T.

Strong Acid–Base Pairs In aqueous solutions, strong acidic pairs such as perchloric acid or hydrochloric acid almost completely lose their protons to the water (H_3O^+/H_2O pair). One might say they totally "dissociate," for example

$$HClO_4|w + H_2O|l \rightarrow ClO_4^-|w + H_3O^+|w$$

or

$$HCl|w + H_2O|l \rightarrow Cl^-|w + H_3O^+|w.$$

In both cases, the acid of the weaker acidic pair (H_3O^+) replaces that of the stronger acidic one ($HClO_4$ or HCl, respectively). Since the deprotonation is almost complete, the same acid H_3O^+ is present in both cases and therefore the pair $H_3O^+/$ H_2O with the highest possible acidic strength in not too concentrated aqueous solutions. In the process, the proton potential decreases from rather high and different positive values (+57 and +35 kG, respectively) to the same value of (just about) zero. Hence, strong acidic pairs cannot develop their full power in an aqueous solution. In this case, one speaks of the *leveling effect* of the solvent water. For this reason, the chemical potentials $\mu(\text{Ad})$ of undissociated strong acids (which are necessary for determining individual proton potentials) are measured in nonaqueous solutions and the results transferred as approximations to the solvent water.

If one wishes to find the proton potential of a strong acidic pair at arbitrary dilution, it is enough to just consider the acid–base pair H_3O^+/H_2O:

$$\mu_p = \mu_p(H_3O^+/H_2O) = \frac{1}{1}[\mu(H_3O^+) - \mu(H_2O)].$$

When the concentrations are small, the mass action equation can be applied.

However, there is such an excess of water (the solvent) present that its concentration shows almost no change during the reaction. As stated, it can be treated as a pure liquid (compare Sect. 6.3):

$$\mu_p = \overset{\circ}{\mu}(H_3O^+) + RT\ln c_r(H_3O^+) - \overset{\circ}{\mu}(H_2O)$$
$$= \left[\overset{\circ}{\mu}(H_3O^+) - \overset{\circ}{\mu}(H_2O)\right] + RT\ln c_r(H_3O^+).$$

The difference in parentheses corresponds to the basic value of the proton potential of the pair H_3O^+/H_2O, meaning

$$\mu_p = \overset{\circ}{\mu}_p(H_3O^+/H_2O) + RT\ln c_r(H_3O^+). \tag{7.3}$$

If, for example, we are interested in the proton potential of hydrochloric acid with a concentration of 0.01 kmol m^{-3}, the concentration of H_3O^+ corresponds to the stated HCl concentration due to the total dissociation. If we continue to take into account that $\overset{\circ}{\mu}_p(H_3O^+/H_2O) = \mu_p^{\ominus}(H_3O^+/H_2O) = 0$ (compare Table 7.1), we obtain for 298 K (and 100 kPa):

$$\mu_p = 0\ \text{G} + 8.314\ \text{G K}^{-1} \times 298\,\text{K} \times \ln 0.01 = -11\,\text{kG}.$$

The proton potential is noticeably smaller than the basic value which is valid for a H_3O^+ concentration of 1 kmol m^{-3} in an aqueous solution.

The bases of strong alkaline pairs $\left(\overset{\circ}{\mu}_p < -80\,\text{kG}\right)$, such as amide ions NH_2^-, suffer a similar fate where, due to the higher $\overset{\circ}{\mu}_p$ value, the water (H_2O/OH^- pair) forces the protons upon them:

$$NH_2^-|w + H_2O|l \rightarrow NH_3|w + OH^-|w.$$

The amide ion is a strong proton acceptor that is almost totally protonated when present in an excess of water. The concentration of the hereby produced base OH^- determines the proton potential μ_p, whereas the concentration and thus the chemical potential of its corresponding acid H_2O remains almost unaltered. In aqueous solutions of alkaline acid–base pairs, therefore, μ_p cannot fall far below -80 kG. Traditionally, this statement is expressed saying OH^- is the strongest base to be found in water.

Determining the proton potential of a dilute strong alkaline pair follows the same pattern as for a dilute strong acidic pair, although in this case, the pair H_2O/OH^- (instead of H_3O^+/H_2O) must be considered:

$$\mu_{\mathrm{p}} = \mu_{\mathrm{p}}(\mathrm{H_2O/OH^-}) = \tfrac{1}{1}[\mu(\mathrm{H_2O}) - \mu(\mathrm{OH^-})].$$

By taking into account the special situation of water as solvent as well as the mass action equation for $\mathrm{OH^-}$, one obtains

$$\mu_{\mathrm{p}} = \overset{\circ}{\mu}(\mathrm{H_2O}) - \left[\overset{\circ}{\mu}(\mathrm{OH^-}) + RT\ln c_{\mathrm{r}}(\mathrm{OH^-})\right]$$

and finally

$$\mu_{\mathrm{p}} = \overset{\circ}{\mu}_{\mathrm{p}}(\mathrm{H_2O/OH^-}) - RT\ln c_{\mathrm{r}}(\mathrm{OH^-}). \tag{7.4}$$

Thus, at 298 K (and 100 kPa), a strong alkaline pair with a concentration of $0.1\ \mathrm{kmol\ m^{-3}}$ in water has a proton potential of

$$\mu_{\mathrm{p}} = -80 \times 10^3\,\mathrm{G} - 8.314\,\mathrm{G\,K^{-1}} \times 298\,\mathrm{K} \times \ln 0.1 = -74\,\mathrm{kG},$$

which is noticeably higher than the standard value of -80 kG.

A proton potential considerably lower than -80 kG cannot be maintained in normal aqueous solutions because the water would continuously decompose due to loss of protons. The same is true for a potential considerably above 0 kG because here the $\mathrm{H_2O}$ molecules are destroyed by the production of $\mathrm{H_3O^+}$. In both cases, the water would largely disappear so that the term "aqueous solution" would no longer apply. Conditions of this type dominate in concentrated solutions of mineral acids or alkali hydroxides.

Weak Acid–Base Pairs Acids of a weak acidic pair such as acetic acid HAc, on the other hand, can be deprotonated to very different degrees in an aqueous solution. (As mentioned, Ac is used as abbreviation for the acetate group $\mathrm{CH_3COO}$.) If the acid–base pair is largely deprotonated, its proton reservoir is just about empty. However, if it is hardly deprotonated, meaning almost fully protonated, the proton reservoir is almost full. If one wishes to calculate the proton potential of a weak acidic pair such as $\mathrm{HAc/Ac^-}$ in a diluted aqueous solution, due to the incomplete proton transfer to the pair $\mathrm{H_3O^+/H_2O}$, both pairs must be taken into account according to

$$\mathrm{HAc|w + H_2O|l \rightleftarrows Ac^-|w + H_3O^+|w}.$$

We start with the following equation for "proton potential equalization,"

$$\mu_{\mathrm{p}} = \mu_{\mathrm{p}}(\mathrm{Ad/Bs}) = \mu_{\mathrm{p}}(\mathrm{H_3O^+/H_2O}),$$

a special case of Eq. (7.2). When the mass action equation is applied and the special situation of water as solvent is taken into account, the result is

$$\overset{\circ}{\mu}_{p}(Ad/Bs) + RT\ln c_{r}(Ad) - RT\ln c_{r}(Bs)$$

$$= \overset{\circ}{\mu}_{p}(H_{3}O^{+}/H_{2}O) + RT\ln c_{r}(H_{3}O^{+}). \tag{7.5}$$

It immediately follows from the conversion formula that the concentration of the base formed by the proton transfer from the acid to the water is equal to that of the resulting oxonium ions, $c_{r}(Bs) = c_{r}(H_{3}O^{+})$. If we furthermore assume that the acid of the weak acidic pair is only slightly dissociated, the undissociated portion $c_{r}(Ad)$ can, in first approximation, be equated to the initial concentration c_{0}, $c_{r}(Ad) \approx c_{0,r}$. When both terms are inserted into Eq. (7.5) and this equation is solved for $RT \ln c_{r}(H_{3}O^{+})$, we obtain

$$\overset{\circ}{\mu}_{p}(Ad/Bs) + RT\ln c_{0,r} - RT\ln c_{r}(H_{3}O^{+}) = \overset{\circ}{\mu}_{p}(H_{3}O^{+}/H_{2}O) + RT\ln c_{r}(H_{3}O^{+})$$

and finally

$$RT\ln c_{r}(H_{3}O^{+}) = \tfrac{1}{2}\cdot\left[\overset{\circ}{\mu}_{p}(Ad/Bs) - \overset{\circ}{\mu}_{p}(H_{3}O^{+}/H_{2}O) + RT\ln c_{0,r}\right].$$

Insertion of this term into Eq. (7.3) results in a relation for the proton potential:

$$\mu_{p} = \tfrac{1}{2}\cdot\left[\overset{\circ}{\mu}_{p}(Ad/Bs) + \overset{\circ}{\mu}_{p}(H_{3}O^{+}/H_{2}O) + RT\ln c_{0,r}\right]. \tag{7.6}$$

According to this, the proton potential at 298 K (and 100 kPa) of an acetic acid solution with a concentration of 1 kmol m^{-3} is:

$$\mu_{p} = \tfrac{1}{2} \times \left[(-27 \times 10^{3}\,G) + 0 + 8.314\,GK^{-1} \times 298\,K \times \ln 1\right] = -13.5\,kG.$$

This is noticeably higher than the standard value of -27 kG (see Table 7.1) and can be explained by the fact that the acetate concentration is negligibly small compared to the concentration of undissociated acetic acid. (On the other hand, the standard value is valid for a concentration ratio of $c(HAc):c(Ac^{-})$ of 1:1; compare Sect. 7.4.)

   The proton potential of a weak alkaline pair in aqueous solution can be derived analogously. The result is comparable to Eq. (7.6):

$$\mu_{p} = \tfrac{1}{2}\cdot\left[\overset{\circ}{\mu}_{p}(Ad/Bs) + \overset{\circ}{\mu}_{p}(H_{2}O/OH^{-}) - RT\ln c_{0,r}\right]. \tag{7.7}$$

Accordingly, a solution of ammonia in water with the corresponding conversion formula

$$NH_3|w + H_2O|l \rightleftharpoons NH_4^+|w + OH^-|w$$

has a proton potential of

$$\mu_p = \frac{1}{2} \times \left[\left(-53 \times 10^3\,G\right) + \left(-80 \times 10^3\,G\right) - 8.314\,G\,K^{-1} \times 298\,K \times \ln 0.1\right]$$
$$= -64\,kG,$$

when a concentration of 0.1 kmol m^{-3} at 298 K (and 100 kPa) is considered.

Acid–Base Disproportionation of Water Finally, we will take a look at the *acid–base disproportionation* of water. It has already been demonstrated that amphoteric water can function as an acid as well as a base. For this reason, proton transfer between the water molecules can take place where oxonium and hydroxide ions are formed even when there are no other acids or bases present:

$$\underbrace{H_2O|l}_{Bs} + \underbrace{H_2O|l}_{Ad^*} \rightleftharpoons \underbrace{H_3O^+|w}_{Ad} + \underbrace{OH^-|w}_{Bs^*}.$$

The proton potential of the acid–base pair H$_3$O$^+$/H$_2$O results from Eq. (7.3) and that of the pair H$_2$O/OH$^-$ from Eq.(7.4). Both potentials have to be equal in the same solution:

$$\mu_p = \overset{\circ}{\mu}_p(H_3O^+/H_2O) + RT\ln c_r(H_3O^+) = \overset{\circ}{\mu}_p(H_2O/OH^-) - RT\ln c_r(OH^-).$$

Pure water is electrically neutral, so the H$_3$O$^+$ concentration must be equal to the OH$^-$ concentration. Replacing $c_r(OH^-)$ by $c_r(H_3O^+)$ and solving the last equation for $c(H_3O^+) = c^\ominus \cdot c_r(H_3O^+)$ yields

$$c(H_3O^+) = c^\ominus \cdot \exp\frac{\overset{\circ}{\mu}_p(H_2O/OH^-) - \overset{\circ}{\mu}_p(H_3O^+/H_2O)}{2RT}.$$

Taking the standard values in Table 7.1 at 298 K (and 100 kPa) results in

$$c(H_3O^+) = 1\,kmol\,m^{-3} \times \exp\frac{-80 \times 10^3\,G - 0\,G}{2 \times 8.314\,G\,K^{-1} \times 298\,K} = 10^{-7}\,kmol\,m^{-3}.$$

If this concentration is inserted into Eq. (7.3), one obtains the corresponding proton potential. This is −40 kG and is called the *neutral value*. An aqueous solution with a proton potential above −40 kG is called acidic; one with a proton potential under −40 kG is called alkaline (basic).

Relationship to Other Measures of Acidity In closing, we will take a look at how the proton potential relates to other common measures used for acidity. There is a close relationship between the standard value of the proton potential $\mu_\mathrm{p}^{\ominus}(\mathrm{Ad/Bs})$ of an acid–base pair [more generally stated, its basic value $\overset{\circ}{\mu}_\mathrm{p}(\mathrm{Ad/Bs})$], and the acidity exponent or pK value. For this purpose, we will look again at proton transfer equilibrium,

$$\mathrm{HA}|\mathrm{w} + \mathrm{H_2O}|\mathrm{l} \rightleftarrows \mathrm{A}^-|\mathrm{w} + \mathrm{H_3O^+}|\mathrm{w},$$

where HA corresponds to an acid Ad and A^- to a base Bs. Equation (7.5) can be converted to

$$\overset{\circ}{\mu}_\mathrm{p}(\mathrm{Ad/Bs}) - \overset{\circ}{\mu}_\mathrm{p}(\mathrm{H_3O^+/H_2O}) = RT\ln\frac{c_\mathrm{r}(\mathrm{Bs}) \cdot c_\mathrm{r}(\mathrm{H_3O^+})}{c_\mathrm{r}(\mathrm{Ad})}. \qquad (7.8)$$

As already stated, we have $\overset{\circ}{\mu}_\mathrm{p}(\mathrm{H_3O^+/H_2O}) = 0$ (at standard pressure) for the proton potential. It represents the "reference level," so to speak.

The equilibrium "constants" of the reaction (see Sect. 6.4),

$$\overset{\circ}{\mathcal{K}}_\mathrm{a}(\mathrm{Ad/Bs}) = \frac{c_\mathrm{r}(\mathrm{Bs}) \cdot c_\mathrm{r}(\mathrm{H_3O^+})}{c_\mathrm{r}(\mathrm{Ad})} \quad \text{and} \quad \overset{\circ}{K}_\mathrm{a}(\mathrm{Ad/Bs}) = \frac{c(\mathrm{Bs}) \cdot c(\mathrm{H_3O^+})}{c(\mathrm{Ad})},$$

are both called *acidity constants*. Adding Ad/Bs as argument behind $\overset{\circ}{\mathcal{K}}_\mathrm{a}$ and $\overset{\circ}{K}_\mathrm{a}$ would be more elucidative, but is quite uncommon. If it is clear from the context which pair is meant or if the kind of pair does not matter we will in the future omit the argument Ad/Bs. These constants can vary by many orders of magnitude depending upon the chemical structure of the acids or bases. Therefore it is useful to introduce a logarithmic scale. Generally, one uses the negative decadic logarithm (logarithm to base 10) of the acidity constant, the acidity exponent or pK_a *value* (or simply the pK *value*):

$$pK_\mathrm{a} = -\lg\overset{\circ}{\mathcal{K}}_\mathrm{a}.$$

In this case, the numerical acidity constant $\overset{\circ}{\mathcal{K}}_\mathrm{a}$ needs to be used instead of the conventional $\overset{\circ}{K}_\mathrm{a}$ because the argument of the logarithm must be a number.

If we convert the natural logarithm to decadic logarithm in Eq. (7.8), by using $\ln x = \ln 10 \cdot \lg x$ [Eq. (A.1.5) in the Appendix], we obtain

$$\overset{\circ}{\mu}_{\mathrm{p}} = -RT\ln 10 \cdot \mathrm{p}K_{\mathrm{a}}.$$

The term $RT\ln 10$ corresponds to the decapotential μ_{d} (compare to Sect. 6.2), meaning that

$$\overset{\circ}{\mu}_{\mathrm{p}} = -\mu_{\mathrm{d}} \cdot \mathrm{p}K_{\mathrm{a}}. \tag{7.9}$$

In general, the basic value $\overset{\circ}{\mu}_{\mathrm{p}}(\mathrm{Ad/Bs})$ of the proton potential and the $\mathrm{p}K_{\mathrm{a}}$ value are proportional. If we place the acids in a sequence according to their strength, we get the same result independent of which of the two measures is being used.

A very similar looking equation relates the proton potential μ_{p} in the solution after proton potential equalization to the so-called hydrogen ion exponent or *pH value* that represents a special measure of acidity expressed by the H_3O^+/H_2O pair. We should call to mind that μ_{p} in a homogeneous solution has the same value for all acid–base pairs present [Eq. (7.2)]. Which pair one uses for measuring μ_{p} is not essential. The term pH value (derived from the Latin "pondus hydrogenii," meaning as much as "weight of hydrogen") was introduced by the Danish scientist Søren Peder Lauritz Sørensen in 1909. Originally, it was the "exponent of the power to the base 10" of the "concentration of hydrogen ions in water, measured in mol/L," $\{c(H^+|w)\}_{\mathrm{mol/L}}$. The negative sign in the exponent was dropped. The curly brackets signify that it is a pure numerical value. This means if we omit the supplement "|w" for the sake of simplicity:

$$\{c(H^+)\}_{\mathrm{mol/L}} = 10^{-\mathrm{pH}} \quad \text{or} \quad c(H^+)/c^{\ominus} = 10^{-\mathrm{pH}}.$$

The pH value as negative decadic logarithm of the relative concentration of (hydrated) hydrogen ions or, more exactly, (hydrated) oxonium ions—H^+, $H^+|w$, H_3O^+, $H_3O^+|w$ are here only different notations for the same kind of particles!—

$$\mathrm{pH} := -\lg\frac{c(H^+)}{c^{\ominus}} = -\lg c_{\mathrm{r}}(H^+) \quad \text{or} \quad \mathrm{pH} := -\lg\frac{c(H_3O^+)}{c^{\ominus}} = -\lg c_{\mathrm{r}}(H_3O^+)$$

represents (as does the $\mathrm{p}K$ value) a handier numerical value because the concentrations can vary by many orders of magnitude. For reasons of convenience, the specification $\{\ \}_{\mathrm{mol/L}}$ or $c^{\ominus}$ is generally dropped. However, this can easily lead to errors when other units of concentration are also being used and should, for this reason, be avoided.

Sørensen's equation can be contrasted with the proton potential μ_{p} of Eq. (7.3),

$$\mu_{\mathrm{p}} = \overset{\circ}{\mu}_{\mathrm{p}}(H_3O^+/H_2O) + RT\ln c_{\mathrm{r}}(H_3O^+).$$

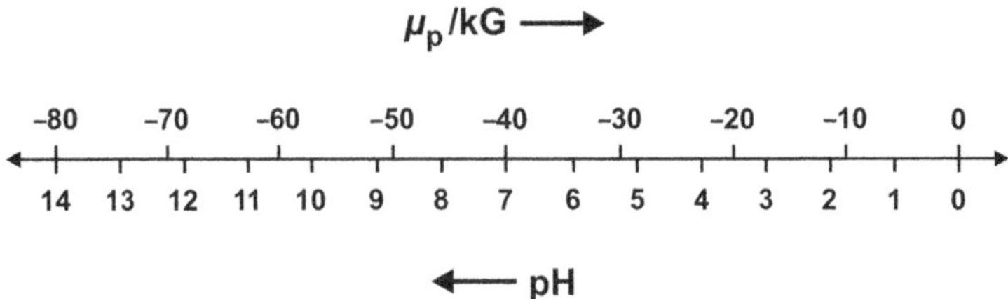

Fig. 7.1 Relation between proton potential μ_p and pH value in an aqueous solution at 298 K and 100 kPa.

$c_\mathrm{r}(\mathrm{H_3O^+})$ is the (relative) concentration of oxonium ions that is reached after proton potential equalization. We again observe that $\overset{\circ}{\mu}_\mathrm{p}(\mathrm{H_3O^+/H_2O}) = 0$ (at standard pressure) and convert the natural into the decadic logarithm. We find

$$\mu_\mathrm{p} = -RT\ln 10 \cdot \mathrm{pH}$$

or finally because of $RT\ln 10 = \mu_\mathrm{d}$

$$\mu_\mathrm{p} = -\mu_\mathrm{d} \cdot \mathrm{pH}. \tag{7.10}$$

This is a simple but important result.

According to Sørensen's definition, this relation is only valid for small concentrations c. These days, however, the pH value is defined so that this equation is *always* valid. Figure 7.1 clarifies the relation between proton potential μ_p and pH value in an aqueous solution at 298 K.

The pH value is just a variation of proton potential—in other words μ_p in a specialized scale. Differences of pH values ultimately mean differences in chemical potential and therefore, differences in the drive of the chemical reactions in which the protons take part.

In matter dynamics, the description of the proton potential μ_p as a measure of the strength of an acid–base pair that we have just been introduced to offers several advantages not present in conventional measures of acidity (pK and pH values):

- At the conceptual level, acid–base reactions and other transfer reactions, especially redox reactions, can be treated uniformly (cf. Sect. 22.4).
- Strengths of acids and numerical values of the measure of acidity run parallel (and do not, as with pK values, run opposite to each other).
- The proton potential also indicates the strength of an acid–base pair with respect to the concentration dependence (the pK value as a logarithm of an equilibrium constant cannot do this, so that in this case, another measure, the pH value is used instead).

7.4   Level Equation and Protonation Equation

Monoprotic Acid–Base Pairs The extent of protonation of an acid–base pair, briefly its "fill level," can be described by the *degree of protonation* Θ. In the case of a monoprotic pair, this is understood to be the portion of base molecules that are protonated in relation to the total concentration of acid and base molecules:

$$\Theta = \frac{c_{\mathrm{Ad}}}{c_{\mathrm{Bs}} + c_{\mathrm{Ad}}}. \tag{7.11}$$

The degree of protonation depends upon the proton potential μ_{p} in the solution. Based upon the condition of equilibrium, $\mu_{\mathrm{p}} = \frac{1}{1}[\mu_{\mathrm{Ad}} - \mu_{\mathrm{Bs}}]$, and the mass action equation,

$$\mu_{\mathrm{p}} = \left(\overset{\circ}{\mu}_{\mathrm{Ad}} + RT\ln(c_{\mathrm{Ad}}/c^{\ominus})\right) - \left(\overset{\circ}{\mu}_{\mathrm{Bs}} + RT\ln(c_{\mathrm{Bs}}/c^{\ominus})\right)$$

one obtains a relation that corresponds to the Henderson–Hasselbalch equation ($\mathrm{pH} = \mathrm{p}K_{\mathrm{a}} + \log(c_{\mathrm{Bs}}/c_{\mathrm{Ad}})$),

$$\mu_{\mathrm{p}} = \overset{\circ}{\mu}_{\mathrm{p}} + RT\ln\frac{c_{\mathrm{Ad}}}{c_{\mathrm{Bs}}} \quad \text{("Level equation"),} \tag{7.12}$$

where μ_{p} represents the hydrogen ion exponent or pH value, and $\overset{\circ}{\mu}_{\mathrm{p}}$ stands for the acidity exponent or $\mathrm{p}K_{\mathrm{a}}$ value. This is the so-called level equation. The name comes from the fact that μ_{p} describes how high the "acidity level" is in the Ad/Bs reservoir, or in other words, how strongly acidic the Ad/Bs pair reacts.

The quotient $c_{\mathrm{Ad}}/c_{\mathrm{Bs}}$ can be expressed using the degree of protonation Θ so that we have

$$\mu_{\mathrm{p}} = \overset{\circ}{\mu}_{\mathrm{p}} + RT\ln\frac{\Theta}{1 - \Theta}. \tag{7.13}$$

According to Eq. (7.11) we have

$$\Theta(c_{\mathrm{Bs}} + c_{\mathrm{Ad}}) = c_{\mathrm{Ad}} \quad \text{or} \quad \Theta \cdot c_{\mathrm{Bs}} = (1 - \Theta) \cdot c_{\mathrm{Ad}}$$

and finally

$$c_{\mathrm{Ad}}/c_{\mathrm{Bs}} = \Theta/(1 - \Theta).$$

$\overset{\circ}{\mu}_p$ vividly represents the "half-potential," meaning the proton potential present at a degree of protonation equal to $\Theta = \frac{1}{2}$. This is when the concentrations of acid and base are equal. In the case of $\mu_p > \overset{\circ}{\mu}_p$, the acid–base pair is present mostly in the protonated form, and for $\mu_p < \overset{\circ}{\mu}_p$ it is primarily present in the deprotonated form. When the relation (7.13) is solved for Θ, the following equation results:

$$\Theta = \frac{1}{1 + \exp\dfrac{\overset{\circ}{\mu}_p - \mu_p}{RT}} \quad (\text{"Protonation equation"}), \qquad (7.14)$$

which we will encounter repeatedly in a similar form.

We transform Eq. (7.13) into

$$\frac{\Theta}{1 - \Theta} = \exp\frac{\mu_p - \overset{\circ}{\mu}_p}{RT} = a.$$

With the abbreviation a and two intermediate steps,

$$a^{-1} = \frac{1 - \Theta}{\Theta} = \Theta^{-1} - 1 \ \text{ and } \ \Theta = \frac{1}{1 + a^{-1}},$$

we obtain Eq. (7.14).

Figure 7.2 graphically presents the relation $\Theta = f(\mu_p)$ described by the protonation equation (7.14). It shows that the curves of different monoprotic acid–base pairs have the same form. They are simply shifted along the μ_p axis.

Taking this aspect into account, let us reconsider our examples from Sect. 7.2. We had determined a proton potential of -13.5 kG for an aqueous solution of acetic acid with a concentration of 1 kmol m^{-3}. Inserting this into the protonation

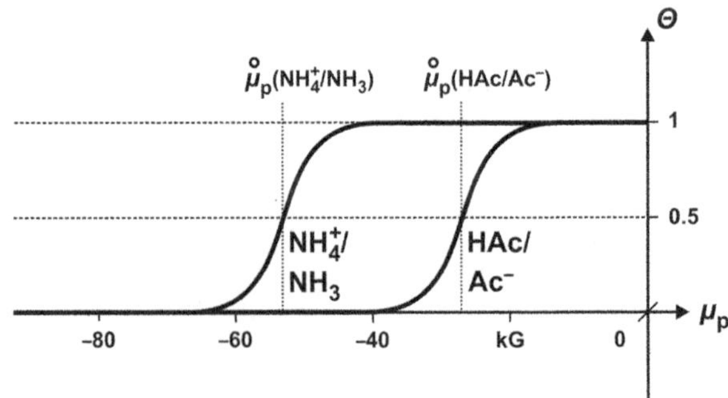

Fig. 7.2 Degree of protonation Θ of the acid–base pairs HAc/Ac$^-$ and NH$_4^+$/NH$_3$ at 298 K (and 100 kPa), as a function of the proton potential μ_p.

equation, and using the standard value of $\mu_p^{\ominus}(\mathrm{HAc}/\mathrm{Ac}^-) = -27\,\mathrm{kG}$ at 298 K (and 100 kPa), yields a degree of protonation of

$$\Theta = \frac{1}{1 + \exp\dfrac{-27 \times 10^3\,\mathrm{G} - \left(-13.5 \times 10^3\,\mathrm{G}\right)}{8.314\,\mathrm{G\,K^{-1}} \times 298\,\mathrm{K}}} = 0.996.$$

This means that 99.6 % of the amount of acid used exists in protonated form (in the form of acetic acid molecules). Only 0.4 % exists in deprotonated form as acetate ions. Our original assumption that acids of weak acidic acid–base pairs are present in dissociated form only to a very small proportion is therefore justified. (This is the case as long as the dissociated fraction remains smaller than about 5 %; since the degree of protonation is also dependent upon the initial concentration of c_0, via μ_p, the validity of the simplification must be ascertained from case to case.)

We can proceed analogously for an ammonia solution with a concentration of $0.1\,\mathrm{kmol\,m^{-3}}$ and obtain a degree of protonation of 0.012. This means only 1.2 % of the ammonia molecules have been protonated.

Let us return once again to our image of acid–base pairs as reservoirs for protons. We can consider the degree of protonation to be the relation of the amount of protons n_p in the reservoir to the maximum amount of protons $n_{p,\max}$ that can be stored:

$$\Theta = \frac{n_p}{n_{p,\max}}. \tag{7.15}$$

When this expression is inserted into Eq. (7.14) and solved for n_p, we obtain a variation of the protonation equation:

$$n_p = \frac{n_{p,\max}}{1 + \exp\dfrac{\overset{\circ}{\mu_p} - \mu_p}{RT}}. \tag{7.16}$$

The protonation equation quasi shows the "fill level" in the proton reservoir. The graphic representation corresponds to Fig. 7.2, just that the curve approaches $n_{p,\max}$ instead of a value of 1.

Water as a Special Case   Let us now consider a special case of the acid–base pairs where the solvent water is the reaction partner and where the pairs limit the range of potentials of the weakly and moderately acidic and alkaline pairs from that of the strong ones. For the acid–base pair H_3O^+/H_2O with $\overset{\circ}{\mu}_p(H_3O^+/H_2O) = 0$ (at standard pressure), we obtain a level equation corresponding to Eq. (7.3):

$$\mu_p = \overset{\circ}{\mu}_p(H_3O^+/H_2O) + RT\ln c_r(H_3O^+).$$

If we replace the relative concentration in this equation by the following expression:

$$c_r(H_3O^+) = \frac{c(H_3O^+)}{c^\ominus} = \frac{n(H_3O^+)}{V \cdot c^\ominus} = \frac{n_p}{V \cdot c^\ominus},$$

and solve for n_p, the corresponding protonation equation results:

$$n_p = V \cdot c^\ominus \cdot \exp\left(-\frac{\overset{\circ}{\mu}_p(H_3O^+/H_2O) - \mu_p}{RT}\right). \tag{7.17}$$

According to Eq. (7.4), the level equation for the pair H_2O/OH^-, with $\overset{\circ}{\mu}_p(H_2O/OH^-) = -80\,kG$, is:

$$\mu_p = \overset{\circ}{\mu}_p(H_2O/OH^-) - RT\ln c_r(OH^-).$$

It is now possible to describe the deprotonation of water as a deficit of protons, meaning it can be described by negative n_p values:

$$c_r(OH^-) = \frac{c(OH^-)}{c^\ominus} = \frac{n(OH^-)}{V \cdot c^\ominus} = \frac{-n_p}{V \cdot c^\ominus}.$$

By inserting the result in the equation above and solving for n_p, we obtain for the protonation equation

$$n_p = -V \cdot c^\ominus \cdot \exp\left(+\frac{\overset{\circ}{\mu}_p(H_2O/OH^-) - \mu_p}{RT}\right). \tag{7.18}$$

If there are several acid–base pairs Ad/Bs, Ad*/Bs*, … in a solution, the amounts of protons in the individual reservoirs add up to a total "fill amount" $n_{p,total}$:

$$n_{p,total} = n_p(Ad/Bs) + n_p(Ad*/Bs*) + \dots. \tag{7.19}$$

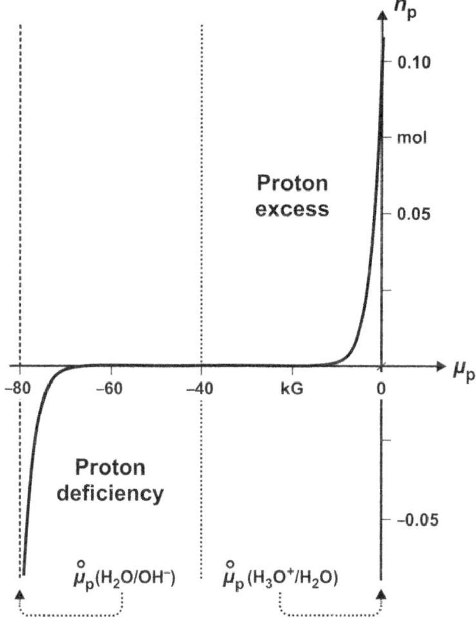

Fig. 7.3 Total amount $n_{p,\text{total}}$ in the storage medium water as a function of the proton potential μ_p in a volume of 100 mL at 298 K.

In the case of water, the pairs H_3O^+/H_2O and H_2O/OH^- must be taken into account due to the amphoteric character of the water. The resulting total amount is

$$n_{p,\text{total}} = n_p(H_3O^+/H_2O) + n_p(H_2O/OH^-).$$

If pure water is considered, the slight amount of protons that appear due to disproportionation is compensated for by the lack of protons so that the total "fill amount" equals zero. Figure 7.3 shows the cumulative curve that combines the two branches of the curve described by Eqs. (7.17) and (7.18).

How can the "fill level" in the proton reservoir—as a function of the proton potential—be determined in the case of a weak acid–base pair dissolved in water? For this, the contribution of the pair in question and that of the water add up to a total curve (Fig. 7.4).

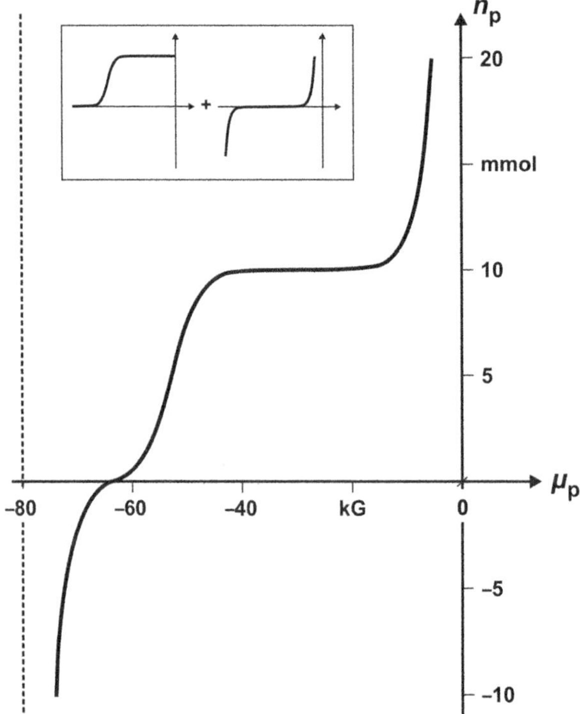

Fig. 7.4 "Fill level" of the protons as a function of proton potential for an aqueous solution of a monoprotic acid–base pair using the example NH_4^+ / NH_3 (10 mmol in 100 mL solution) at 298 K.

7.5   Acid–Base Titrations

An important application for acid–base equilibria is the analytic method called *titration*. With the help of titration, it is possible to investigate the composition of an initial solution using the *equivalence point*. Moreover, it can also be used to determine the basic value $\overset{\circ}{\mu}_p$(Ad/Bs) for various acid–base pairs. In order to do this, a solution of precisely known concentration, the so-called *titrator* (titrant), is added successively from a burette to the *analyte* (titrand), the solution with unknown concentration being titrated, while the pH value is continuously measured. The proton potential μ_p is then determined from this data (Experiment 7.2). Data can be collected on a computer and directly processed. In a further development of titration, the addition of the titration solution can be automatically controlled.

If the proton potential is plotted as a function of the added volume of the titrator (or any other quantity dependent upon the added amount, such as the amount of protons), one obtains a so-called *titration curve*.

Experiment 7.2 *Acid–base titration*: One possibility would be for example the titration of a sodium hydroxide solution with hydrochloric acid. A suitable sensor for these and other aqueous solutions would be a glass electrode that will be discussed in more detail in Sect. 22.7.

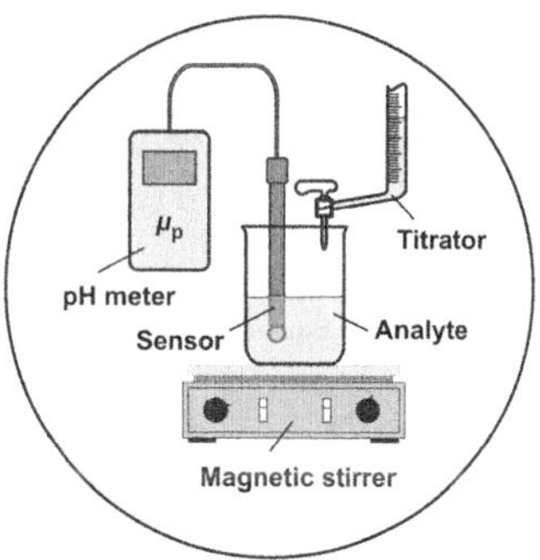

To begin with, let us consider the titration of the base of a strong alkaline pair with the acid of a strong acidic pair, for example, the titration of 100 mL of sodium hydroxide solution ($c(\text{NaOH}) = 0.1$ kmol m^{-3}) with hydrochloric acid ($c(\text{HCl}) = 1$ kmol m^{-3}). Sodium hydroxide in an aqueous solution is almost totally dissociated, so the behavior of the solution being titrated is determined by the pair H_2O/OH^-. At 298 K (and 100 kPa), the proton potential of the initial solution is -74 kG. This was already calculated in Sect. 7.3 using Eq. (7.4) which generally holds for strong bases of this concentration. In the hydrochloric acid solution, however, the proton potential is determined by the pair H_3O^+/H_2O. The original proton deficit of $n_p = -0.1$ kmol m$^{-3} \times 0.1 \times 10^{-3}$ m$^3 = -0.01$ mol $= -10$ mmol is depleted by 1 mmol per mL of added hydrochloric acid. If the titrator is present in a much higher concentration than the substance in the solution to be titrated, the increase of water due to the inflowing titration solution can be ignored. When titrating bases of strong alkaline pairs with acids of strong acidic pairs, generally only the pairs H_2O/OH^- and H_3O^+/H_2O have to be considered independent of the kind of strong acid–base pairs used (because of the leveling effect of the solvent water discussed in Sect. 7.3). The progress of titration results from the curve showing the total amount of protons in the storage medium water for a given volume of 100 mL (Fig. 7.3); or rather, it results from a section of this representation specified by the test conditions (Fig. 7.5a). At the beginning of titration, we have the pure sodium hydroxide solution (black point). As more hydrochloric acid is added, one moves in the direction of the arrow along the curve. If the proton potential is plotted as a function of the amount of protons added, the result is the corresponding titration curve (Fig. 7.5b).

At first, the proton potential changes only slightly as titrator is continuously added. However, as the point is approached where a stoichiometrically equivalent amount of hydrochloric acid (in this case 10 mmol) has been added to the sodium hydroxide solution, a drastic increase of proton potential occurs. At the *equivalence point*, there is no proton deficiency anymore, and the proton reservoir is completely filled. There is only an aqueous solution of Na$^+$ and Cl$^-$ ions that has almost no influence upon the proton potential which is then equal to the neutral value of -40 kG of pure water. If we continue to add hydrochloric acid to the neutralized

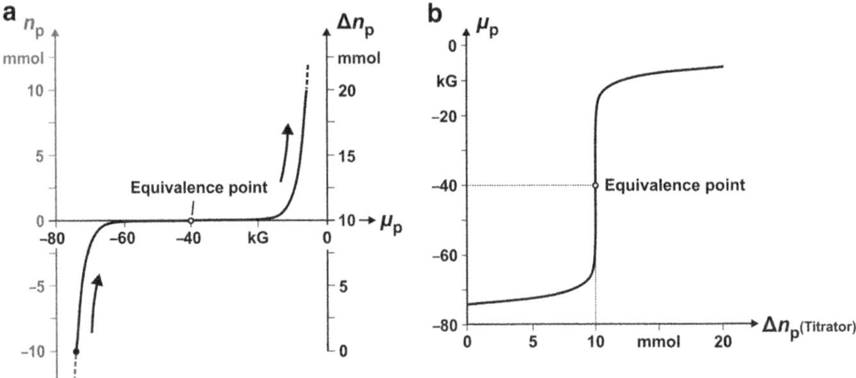

Fig. 7.5 (**a**) Total amount of protons $n_{p,\text{total}}$ in 100 mL water as a function of the proton potential μ_p, at 298 K, assuming a proton deficit of -10 mmol (black point), (**b**) Corresponding titration curve.

solution, the protons cannot be stored in the proton reservoir H_2O/OH^- already completely filled and the acid–base pair H_3O^+/H_2O of the titrator system determines the proton potential. Beyond the equivalence point, the increase of the proton potential is initially steep but starts leveling off very quickly.

Let us now turn to the titration of the base of a weak alkaline pair with the acid of a strong acidic pair using the example of titration of 100 mL of a 0.1 M ammonia solution with the standard solution of hydrochloric acid already used above. At first, the proton potential is -64 kG, which we have already calculated in Sect. 7.3 with the help of Eq. (7.7). The very low proton "fill level" of 1.2 % in the reservoir NH_4^+/NH_3 is just compensated for by the proton deficiency (which is caused by the OH^- ions produced by the proton transfer according to $NH_3 + H_2O \rightleftarrows NH_4^+ + OH^-$) so that the total "fill level" in the aqueous solution equals zero. The relation in Fig. 7.4, or more exactly, a section of it (Fig. 7.6a), is now what determines the form of the titration curve.

As hydrochloric acid continues to be added, we again move in the direction of the arrows along the curve. In the beginning, the form of the curve is determined by the acid–base pair NH_4^+/NH_3 and the corresponding protonation equation, meaning it is essentially this proton reservoir which is filled up. Halfway to the equivalence point (when half of the maximum amount of protons the pair can store have been added) the proton potential reaches the basic value of $\overset{\circ}{\mu}_p\left(NH_4^+/NH_3\right) = -53\,\text{kG}$. The fractions of base (NH_3) and corresponding acid $\left(NH_4^+\right)$ are now equal and the reservoir is just half full. Before this point, the pair appears primarily in deprotonated form and, afterward, in the protonated form. At the equivalence point, exactly as many protons are added with the standard solution so that the entire base is protonated. This means that it has completely transformed into its corresponding acid (except from some traces). At this point, we have an aqueous solution of an acid that has the same concentration as the original base. For this

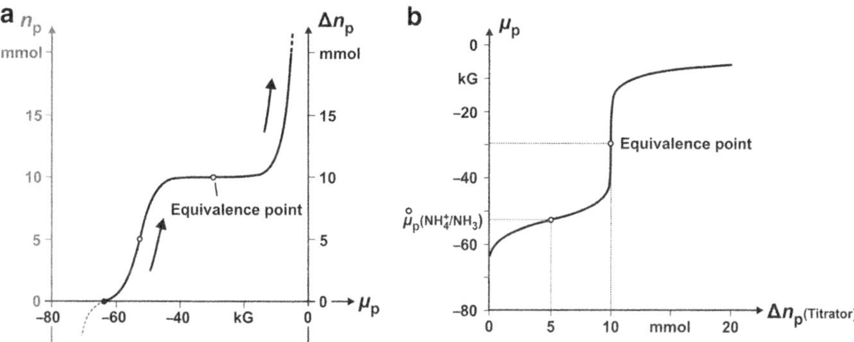

Fig. 7.6 (**a**) "Fill level" of the proton reservoir as a function of the proton potential for an aqueous solution of the acid–base pair NH_4^+/NH_3 (10 mmol in 100 mL solution) at 298 K, (**b**) Corresponding titration curve of an ammonia solution of equivalent concentration with the acid of a strong acidic pair.

reason we expect a proton potential that lies noticeably above the neutral point. It can be calculated according to Eq. (7.6) (the Cl^- ions that are also present have almost no influence upon the proton potential):

$$\mu_p = \tfrac{1}{2} \times \left(-53 \times 10^3\,G\right) + \tfrac{1}{2} \times 0\,G + \tfrac{1}{2} \times 8.314\,G\,K^{-1} \times 298\,K \times \ln 0.1$$
$$= -29\,kG.$$

With continued addition of acid, the form of the curve is now determined by the acid–base pair H_3O^+/H_2O of the titrator system.

Figure 7.6b shows the corresponding titration curve. It is striking that after a small initial rise, the proton potential changes only very slowly until shortly before the equivalence point. We will go into the great importance of this fact in the next section. It also becomes clear that the basic value $\overset{\circ}{\mu}_p$ of a weak acid–base pair can be determined from the measured data by just reading the potential value halfway to the equivalence point. Again, the equivalence point makes itself felt by a sudden change of proton potential, just like it does in titration of exclusively strong acid–base pairs (but now it is less pronounced).

The changes of potential value in the titration of the acid of a weak acidic pair (e.g., acetic acid) with the base of a strong alkaline pair (such as sodium hydroxide solution) proceed similarly in principle and can be derived analogously. Again, the starting point for consideration is the "fill level" for an aqueous solution of an acid–base pair. The example used here is the pair HAc/Ac^- (10 mmol in 100 mL solution). At first, the acetic acid is almost fully protonated. The slight amount that is deprotonated is just compensated for by the H^+ ions that are produced so that the total "fill level" equals 10 mmol (Fig. 7.7a, black point). Adding sodium hydroxide solution slowly empties the proton reservoir and one moves in the direction of the arrow along the curve. At the equivalence point, a solution of the base of the weak alkaline pair is present in the deprotonated form of Ac^-. The corresponding proton potential shows a value noticeably below the neutral value.

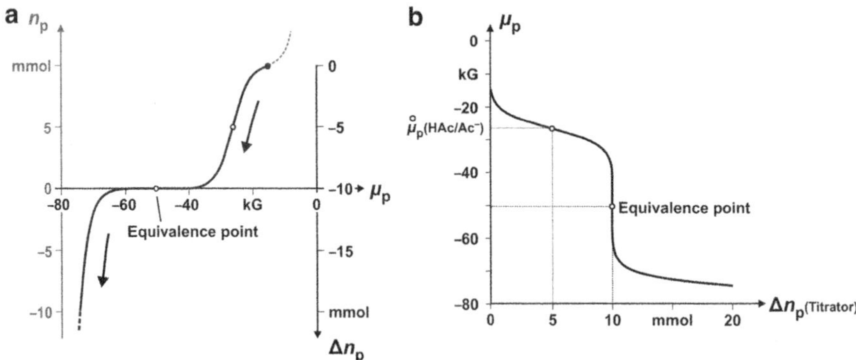

Fig. 7.7 (**a**) "Fill level" of a proton reservoir as a function of the proton potential for an aqueous solution of the acid–base pair HAc/Ac⁻ (10 mmol in 100 mL solution) at 298 K, (**b**) Corresponding titration curve of an acetic acid solution of equivalent concentration with the base of a strong alkaline pair.

Subsequently, the pair H_2O/OH^- determines the shape of the function. Figure 7.7b shows the corresponding titration curve.

If, however, the titrated system and the titrator are present in comparable concentrations, the initial approximation is no longer valid and the increase of the amount of water in the process of titration can no longer be ignored. The titration curve must then be calculated point for point by inserting the changing concentrations, which result from each addition of titrator, into the above relations. This is a lot of work. Since the basic form of the curves does not change much, however, we will leave it at a general understanding accepting the limitations mentioned.

7.6 Buffers

When a weak pair composed of approximately equal but large amounts of acid and its corresponding base is present in a solution, this pair determines the proton potential. Acids of weaker acidic pairs which are added cannot have any effect upon μ_p anyway (if we disregard effects of dilution, etc.), whereas acids of stronger acidic pairs—as long as they are at a shortfall—are robbed of their protons and so become ineffective. The same holds for adding a base. Our acid–base pair can absorb or *buffer* small outside disturbances without considerably changing μ_p of the solution. Hence, a solution in which μ_p reacts insensitively to the addition of slight amounts of acids or bases is called a *buffer*. Only when the acid of a stronger acidic pair is added to excess, for example, can the now almost totally protonated base no longer hinder the buildup of a higher proton potential. As a result, μ_p climbs to the value corresponding to the stronger acid–base pair.

The best way to show how a buffer works is in a potential diagram. Similar to matter capacity (see Sect. 6.7), we can introduce the *buffer capacity* B_p,

$$B_p = \frac{\mathrm{d}n_p}{\mathrm{d}\mu_p},$$

by taking the derivative of the protonation equation (7.16) with respect to the proton potential μ_p. The corresponding result

$$B_p = \frac{n_{p,\max}}{2RT \cdot \left(1 + \cosh\dfrac{\overset{\circ}{\mu}_p - \mu_p}{RT}\right)} \tag{7.20}$$

is not complicated but unfamiliar. The function cosh x is the so-called *hyperbolic cosine* whose values are simply the mean values of e^x and e^{-x}: $\cosh x = (e^x + e^{-x})/2$. Of course, we can directly insert the expression $(e^x + e^{-x})/2$ into Eq. (7.20), but the relation then appears less clearly.

We will make do with a qualitative discussion of the form of the function (Fig. 7.8a). At the point of inflection on the curve described by the protonation equation (7.16) (the point where the proton potential corresponds to the basic value $\overset{\circ}{\mu}_p$ of the acid–base pair), the function $B_p(n_p)$, being the derivative of the protonation equation, has, as expected, an extremum or, more precisely, a maximum value. As described in Sect. 6.7, the function of a buffer becomes even clearer when the axes are exchanged, $\sqrt{B/\pi}$ is calculated, and the curve is interpreted as the outline of a rotationally symmetric container (Fig. 7.8b). In contrast to the "exponential horn" that we have already seen, this container is rounded in form with the widest part in the area of the proton potential $\mu_p = \overset{\circ}{\mu}_p(\mathrm{Ad/Bs})$. If protons are added to the "reservoir" formed by the acid–base system, the "level" along with the proton potential changes very little in the range of the "wide-bellied container."

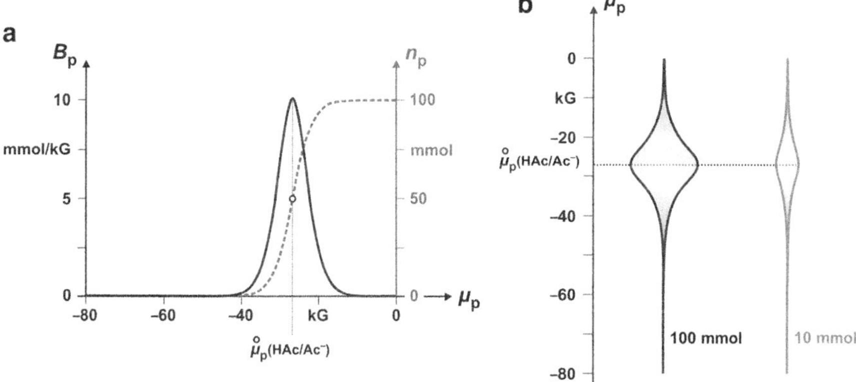

Fig. 7.8 (a) Plot of the stored amount of protons n_p (*gray dotted line*) and the corresponding buffer capacity B_p (*black continuous line*) as a function of proton potential μ_p at 298 K, using the example of the acid–base pair HAc/Ac⁻ (100 mmol), (b) Potential diagram of the buffer HAc/Ac⁻ for various total amounts of substance, $n = n(\mathrm{HAc}) + n(\mathrm{Ac}^-)$.

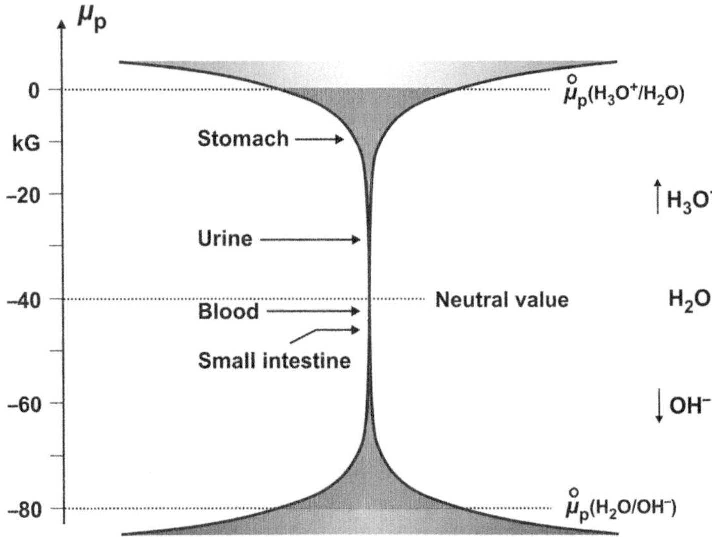

Fig. 7.9 Potential diagram for protons in water (by way of illustration, the proton potentials are shown in some bodily fluids).

The change of level is smaller the closer one gets to the most rounded out area of the container. Correspondingly, large amounts of protons can be added when the reservoir is only half full without the proton potential changing noticeably. The "proton reservoir" has the greatest capacity at this point. The size of the "proton reservoir" is determined by the total amount of substance $\mu_{\mathrm{p,max}}$; the greater the total amount of substance, the greater the buffer capacity. The size and shape of the "container" is the same for all acid–base pairs when the amount of substance remains the same. Only the position of the "container" shifts relative to the μ_{p} axis.

In the case of water, things look very different (Fig. 7.9). Within the range of $\mu_{\mathrm{p}} = -20\ldots-60$ kG, water has only a very small buffer capacity B_{p} or buffer capacity density $\beta_{\mathrm{p}} = B_{\mathrm{p}}/V$, so that even tiny amounts of protons going into or out of a water sample can strongly change its proton potential. Starting from the neutral point, the amount of protons bound to the H_3O^+ as well as the protons "lost" according to $H_2O \rightarrow OH^- + H^+$ is extremely small at first but grows exponentially. The upper "fill level" indicates the basic value of the proton potential in an acid–base system H_3O^+/H_2O, and the lower one shows the corresponding basic value in the H_2O/OH^- system.

In biological systems, the proton potential μ_{p} is often adjusted to certain values: in human blood it is fairly close to (-42.2 ± 0.3) kG, in stomach acid at around -10 kG, in urine at about -30 kG, and in the small intestine at about -50 kG. There must be buffer systems present that compensate for the water's lack of buffer capacity. Graphically speaking, the two "containers" in Figs. 7.8b and 7.9 are

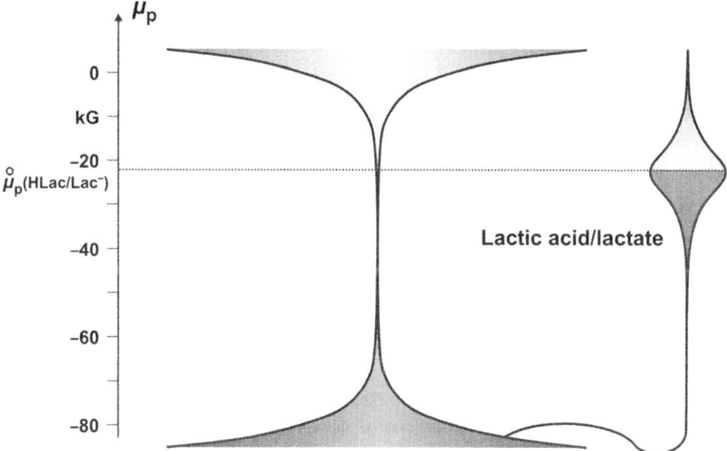

Fig. 7.10 Potential diagram for protons in the buffer system lactic acid/lactate in an aqueous solution (100 mL, 1 kmol m^{-3}, 298 K).

connected (Fig. 7.10). Expressed mathematically, this means that the buffer capacities of several acid–base pairs in a solution are added up:

$$B_{p,\text{total}} = B_p(\text{Ad}/\text{Bs}) + B_p(\text{Ad*}/\text{Bs*}) + \ldots$$

The lack of buffer capacity in water can be compensated for by adding a suitable buffer system of an approximately equimolar mixture of the acid and its corresponding base because B_p as well as b_p will then be greatest. Figure 7.10 illustrates the relation using the example of the pair lactic acid (CH_3–$CHOH$–$COOH$, abbreviated HLac)/lactate (Lac). As already discussed, the "proton reservoir" produced by such pairs has the greatest capacity when it is half full. This occurs in this case in the range of proton potential where the buffer capacity of the water is extremely small.

When there is enough excess quantity of lactate Lac$^-$ present together with its protonated form HLac (lactic acid) compared to the other dissolved pairs Ad/Bs, Ad*/Bs*, …, this *buffer system* determines the value of the proton potential. The bases are protonated or the acids deprotonated until the same proton potential is present everywhere. This is essentially determined by the level equation (7.12) of the buffer system lactic acid/lactate. When the lactate is half protonated, i.e., if $c(\text{Lac}^-) = c(\text{HLac})$, then we have:

$$\mu_p = \overset{\circ}{\mu}_p(\text{HLAc}/\text{Lac}^-) = -22 \text{ kG}.$$

This is why it is possible to buffer a solution at a given proton potential by choosing an acid–base pair whose basic value lies somewhere around the desired proton potential. In biology, the most important buffer is the carbon dioxide/hydrogen carbonate system with a basic value of -36.4 kG: $CO_2|g + H_2O|l \rightleftarrows HCO_3^-|w +H^+|w$. It is by far the most important part of the buffer system in the blood which maintains the proton potential of blood rather precisely between -42.5 and -41.9 kG (at 37 °C) and balances deviations caused by metabolism. A constant proton potential is indispensable to life because values greater than -38 kG or less than -46 kG often result in death.

As a result of the assumed excess and the consequent large buffer capacity, proton gain or loss leads to only slight μ_p shifts in the buffer system. For clarity, we will look more closely at the example of the lactic acid/lactate system. A buffer solution that contains lactate as well as lactic acid at a concentration of 0.1 kmol m^{-3} shows a proton potential μ_p of -22 kG. If 1 cm^3 of hydrochloric acid at a concentration of 1 kmol m^{-3} is added to one liter of this solution, it adjusts to a new proton potential μ_p'. Adding 0.001 mol HCl to the original 0.1 mol lactate in 1 L reduces the lactate by about 0.001 mol according to $Lac^- + H^+ \rightarrow HLac$, while the lactic acid increases by that much. Applying the level equation results in a new proton potential μ_p':

$$\mu_p' = \overset{\circ}{\mu}_p(HLac/Lac^-) + RT\ln\frac{c(HLac)'}{c(Lac^-)'}, \text{meaning}$$

$$\mu_p' = -22 \times 10^3\,G + 8.314\,GK^{-1} \times 298\,K \times \ln\frac{0.1 - 0.001}{0.1 + 0.001} = -21.95\,kG.$$

Adding the acid has only changed the proton potential by 0.05 kG.

If, however, the same amount of 1 cm^3 of hydrochloric acid is added to pure water, the proton potential μ_p'' will be

$$\mu_p'' = \overset{\circ}{\mu}_p(H_3O^+/H_2O) + RT\ln c_r(H_3O^+)$$

[according to Eq. (7.3)] and therefore

$$\mu_p'' = 0\,G + 8.314\,GK^{-1} \times 298\,K \times \ln 0.001 = -17\,kG.$$

By adding the acid, the proton potential has shifted by 23 kG in relation to that of pure water with -40 kG (compared to a change of only 0.05 kG ! in the case of the buffer solution).

Having introduced the concept of buffers, and having made it graphically clear in potential diagrams, we can now understand why the proton potential changes only very slowly until shortly before the equivalence point when the base of a weak alkaline pair (such as ammonia solution) is titrated with the acid of a strong acidic pair (Fig. 7.11). If the "communicating containers" are slowly filled with protons during titration, the level and the proton potential along with it will, at first, be largely determined by the "wide-bellied" proton reservoir of the buffer system

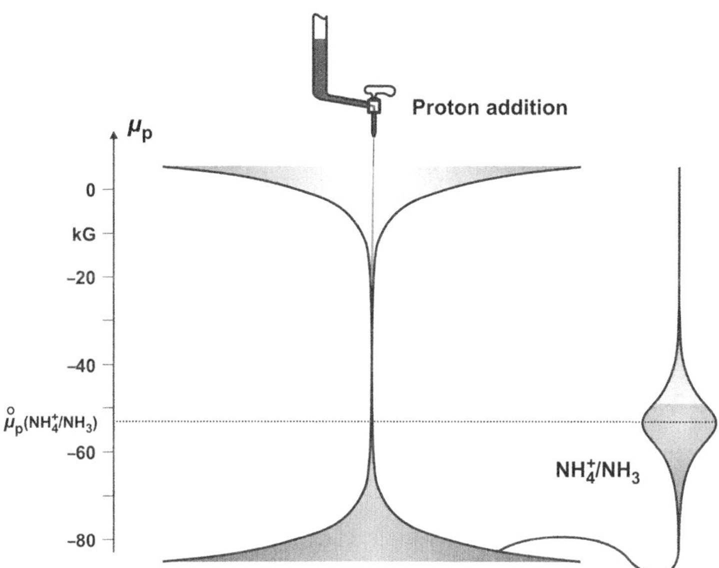

Fig. 7.11 Illustration of titration, using a potential diagram, of the base of a weak alkaline pair with the acid of a strong acidic pair.

$\mathrm{NH_4^+/NH_3}$. Greater amounts of protons can be added especially in the range that is most "rounded out" without the level changing noticeably. However, if the proton reservoir is completely full (equivalence point), a drastic change of proton potential occurs when more protons are added. However, this change slows down in the "funnel area" of the "exponential horn."

7.7   Acid–Base Indicators

Acid–base pairs with strongly contrasting colors are also interesting. Normally, these are large, water-soluble organic molecules. In tiny amounts, they are used as *indicators*. By themselves and in equal amounts in a solution, the members of these pairs produce a certain color mixture and a certain proton potential $\overset{\circ}{\mu}_{\mathrm{p}}(\mathrm{HInd/Ind^-})$, which is characteristic of the indicator system because indicator acids (HInd) and bases (Ind$^-$) obey the level equation (7.12) as well:

$$\mu_{\mathrm{p}} = \overset{\circ}{\mu}_{\mathrm{p}}(\mathrm{HInd/Ind^-}) + RT\ln\frac{c(\mathrm{HInd})}{c(\mathrm{Ind^-})}. \tag{7.21}$$

When the proton potential in a solution is raised by adding an excess quantity of the acid of a stronger acidic pair, for example, the indicator base disappears due to

Table 7.3 Standard values of some acid–base indicators and the corresponding changes of color (acid color—base color).

Indicator	$\mu_p^{\ominus}$ (kG)	Color change
Thymol blue	−10	Red—yellow
Bromophenol blue	−22	Yellow—blue
Bromocresol green	−28	Yellow—blue
Methyl red	−29	Red—yellow
Bromothymol blue	−41	Yellow—blue
Phenol red	−45	Yellow—red
Thymol blue	−51	Yellow—blue
Phenolphthalein	−54	Colorless—pink
Alizarin yellow	−64	Yellow—red

Experiment 7.3 *Acidity effect of mineral water*: If the indicator bromocresol green is put into a bottle of very cold mineral water, the solution turns yellow. When the bottle is opened at room temperature or the content heated, a large portion of the carbon dioxide escapes and the indicator color changes to green and finally to an intense blue.

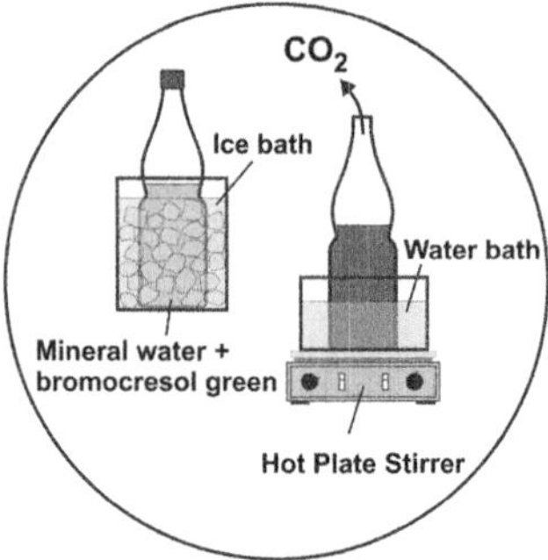

protonation, leaving only the color of the indicator acid visible. Just the opposite occurs when the proton potential is lowered. In this case, the acid is eliminated and the pure color of the base appears. Therefore, the color indicates whether μ_p is greater, smaller, or equal to $\overset{\circ}{\mu}_p$ (HInd/Ind$^-$). Table 7.3 shows the standard values of some acid–base indicators.

An interesting example for the change of an indicator in color due to a change of proton potential is the experiment about the acidity effect of sparkling water (Experiment 7.3). The reason for this acidity effect is the hydrogen carbonate that is produced when carbon dioxide dissolves in water. When it is cold and under pressure, it provides enough protons p which form oxonium ions with water:

$$CO_2|g + H_2O|l \rightleftarrows CO_2|w,$$

$$CO_2|w + 2\ H_2O|l \rightleftarrows H_3O^+|w + HCO_3^-|w.$$

The indicator's change of color can also be used to indicate the end point of acid–base titrations. This is possible because the proton potential climbs strongly while

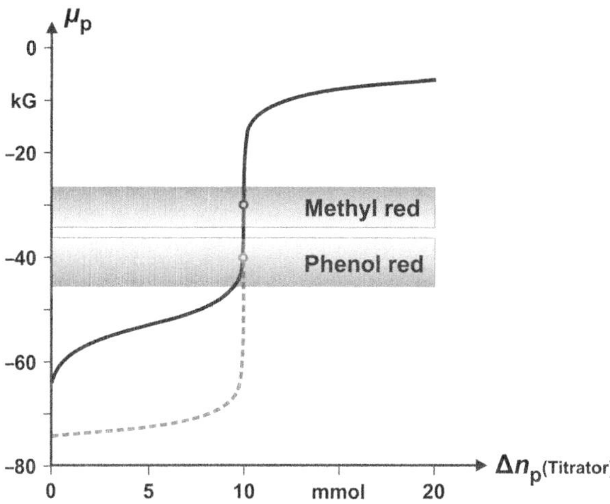

Fig. 7.12 Titration curves of weak alkaline pair–strong acidic pair (*black solid line*) and strong alkaline pair–strong acidic pair (*gray dotted line*) as well as the transition intervals of two indicators.

an acid of a strong acidic pair is being added, as we have seen in a previous section. This happens exactly when the base, whose concentration is not known, has been used up. Therefore, the indicator must be chosen so that its characteristic proton potential $\overset{\circ}{\mu}_p(\text{Hind}/\text{Ind}^-)$ lies between that of the acid–base pairs in the unknown solution and in the standard solution. This should correspond as well as possible to the proton potential at the equivalence point. For this reason, the indicator methyl red is suitable for the titration of the base of a weak alkaline pair with the acid of a strong acidic pair, but the indicator phenol red is not (Fig. 7.12). The latter can, however, be used in the titration of the base of a strong alkaline pair with the acid of a strong acidic pair. In this case, the changes of potential are so drastic that even with indicators of more strongly deviating $\overset{\circ}{\mu}_p$ values such as phenolphthalein, precise results can be obtained.

Indicators are, themselves, acids (HInd) or bases (Ind$^-$) so they also use up the standard solution when turning their color. They are employed in very small concentrations in order to keep possible errors to a minimum.

Chapter 8
Side Effects of Transformations of Substances

Transformations of substances like chemical reactions, phase transitions, distribution in space, etc., are often accompanied by very striking side effects, such as glowing and flashing, fizzling and cracking, bubbling and rising of smoke. These side effects which make chemistry so fascinating are primarily based upon changes of volume that can cause violent explosions and implosions, exchange and generation of entropy, which is responsible for effects like glowing and heating up, and energy exchange that we use in muscles, motors, and batteries. The goal of this chapter is to understand and quantitatively describe these phenomena, and to sensibly make use of them. For this purpose, the so-called partial molar properties such as the (partial) molar volume or the (partial) molar entropy of a dissolved substance are introduced. For describing the changes of volume and entropy associated with transformations, we will use the quantities molar reaction volume and molar reaction entropy. The special role of entropy makes a further differentiation into latent, generated, and exchanged reaction entropy necessary. We will also learn how the chemical drive of a reaction, the corresponding exchange of energy, and eventually the generated entropy are interrelated. In closing, this relationship will be used for determining the chemical drive with the help of a calorimeter.

8.1 Introduction

In the following, we will be concerned with transformations of substances of the most varied kind. Among these will be

- Absorption or release of substances,
- Spreading or aggregation,
- Mixing and dissolving processes,
- Phase transitions and chemical reactions.

© Springer International Publishing Switzerland 2016
G. Job, R. Rüffler, *Physical Chemistry from a Different Angle*,
DOI 10.1007/978-3-319-15666-8_8

All of these processes are accompanied by numerous side effects. Sometimes these are almost imperceptible, but more often they are very noticeable: something is glowing or flashing, fizzling or cracking, it is bubbling or smoke rises. These sorts of accompanying phenomena, which make chemistry so interesting, are based upon

- Changes of volume that can cause violent explosions and implosions,
- Exchange and generation of entropy, which is responsible for effects like glowing and heating up,
- Energy exchange that we use in muscles, motors, and batteries.

The goal of this chapter is to understand and quantitatively describe these phenomena, and to sensibly make use of them.

8.2   Volume Demand

Pure Substances We will begin with the simplest case, namely *change of volume* in transformations of substances. Every substance needs a certain amount of space. How much this is depends upon how much space is needed by its atoms and the gaps in between them. The volume taken up is greater, the more of the substance there is. In order to compare the volume needed by different substances (Experiment 8.1), one relates the volume to amount of substance. This so-called *molar volume* V_m then serves as the measure of the space needed by a pure substance:

$$V_m = \frac{V}{n} \quad molar \ volume \ of \ a \ pure \ substance.$$

The names or formulas of a substance can take the place of the index $_m$, for example, $V_{H_2O} = 18.07 \ \text{cm}^3 \ \text{mol}^{-1}$ or $V(H_2O) = 18.07 \ \text{cm}^3 \ \text{mol}^{-1}$ for the molar volume of (liquid) water.

The volume demand of a substance is by no means constant, but also depends upon its milieu. Substances are compressible to some extent and can expand when

Experiment 8.1 *Volume demand of various pure substances*: How different the volume demand, i.e., the space taken up by various pure substances can be, is easy to show. Cylindrical blocks all representing an amount of substance of 1 mol are placed side by side.

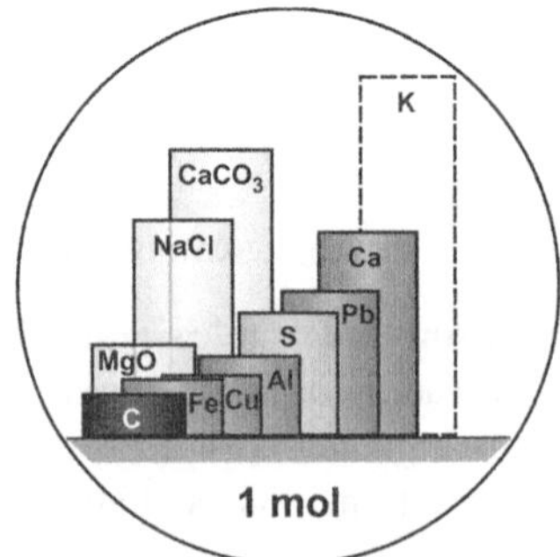

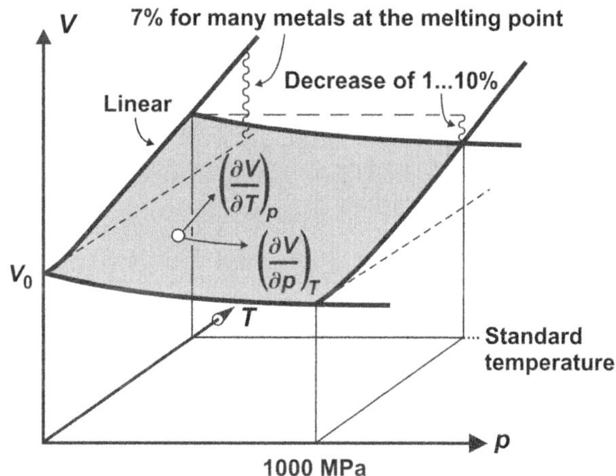

Fig. 8.1 Dependence of the volume of a solid substance upon pressure and temperature.

heated. Volume and molar volume depend upon pressure p as well as temperature T as demonstrated for a solid in Fig. 8.1.

When pressure increases, V generally decreases steeply at first and then more gradually. However, for solid substances, hundreds of MPa are necessary for attaining a noticeable change of volume. Gases, on the other hand, need much less pressure (some tens of kPa) for this. The $V(p, T)$ surface often rises in nearly a straight line in the T direction. For many metals, the increase of volume from 0 K up to the melting point is about 7 % (Grüneisen's rule). Toward lower temperatures, the tangent to the surface becomes horizontal.

Also in the case of molar volume the standard value is the value at "room conditions," meaning 298 K and 100 kPa. As before, we add the symbol $^\ominus$ as upper index to the symbol of the quantity, for example,

$$V_m^{\ominus}(\mathrm{H_2O}) = 18.07\,\mathrm{cm^3\,mol^{-1}} \quad \text{at}\quad 298\,\mathrm{K}\quad \text{and}\quad 100\,\mathrm{kPa}.$$

The standard values for some pure substances are summarized in Table 8.1. Molar volume also depends upon the state of aggregation, as the example of water shows.

The lowest molar volume at standard conditions is found for diamond with $3.4\ \mathrm{cm^3\ mol^{-1}}$. The values for solids and liquids are usually of the order of $10\ \mathrm{cm^3\ mol^{-1}}$ (where 1 mol refers to 6×10^{23} atoms of any type). In contrast, gases display a considerably greater molar volume of just a little less than $25\ \mathrm{L\ mol^{-1}}$. Why this is will be discussed in Sect. 10.2.

If the volume is known at a point p, T (e.g., for standard conditions), it is possible to approximately calculate the values in the vicinity of this point. For this we need the gradients of the surface in the direction of the p and T axes, $(\partial V/\partial p)_T$ and $(\partial V/\partial p)_p$ (see Fig. 8.1). The first coefficient measures the compressibility of the substance; the second measures its thermal expansion. The molar volume V_m for other p and T values can be calculated analogously to the method used for the

Table 8.1 Molar volumes of some pure substances under standard conditions (298 K, 100 kPa) as well as temperature and pressure coefficients (for corresponding reference values).

Substance	Formula	$V_{\mathrm{m}}^{\ominus}$ $(\mathrm{cm}^3\,\mathrm{mol}^{-1})$	$(\partial V_{\mathrm{m}}/\partial T)_p^{\ominus}$ $(\mathrm{cm}^3\,\mathrm{mol}^{-1}\,\mathrm{K}^{-1})$	$(\partial V_{\mathrm{m}}/\partial p)_T^{\ominus}$ $(\mathrm{cm}^3\,\mathrm{mol}^{-1}\,\mathrm{kbar}^{-1})$
Graphite	C|graphite	5.5	0.00004	−0.017
Diamond	C|diamond	3.4	0.00001	−0.001
Iron	Fe|s	7.1	0.00025	−0.004
Lead	Pb|s	18.3	0.00161	−0.045
Water ice	H_2O|s	19.7	[0.0010]	[−0.6]
Water	H_2O|l	18.1	0.0046	−0.836
Water vapor	H_2O|g	24.8×10^3	83.1	-25×10^7

The value for water ice was extrapolated linearly from 273 to 298 K. The values in brackets are valid for 273 K

Experiment 8.2 *Reduction of volume by mixing water and ethanol*: A test tube is half-filled with water (colored with an appropriate dye). Subsequently, it is filled to the top with ethanol and closed with a rubber stopper. After inverting the tube repeatedly, the formation of a gas bubble meaning a decrease in volume of the mixture can be observed.

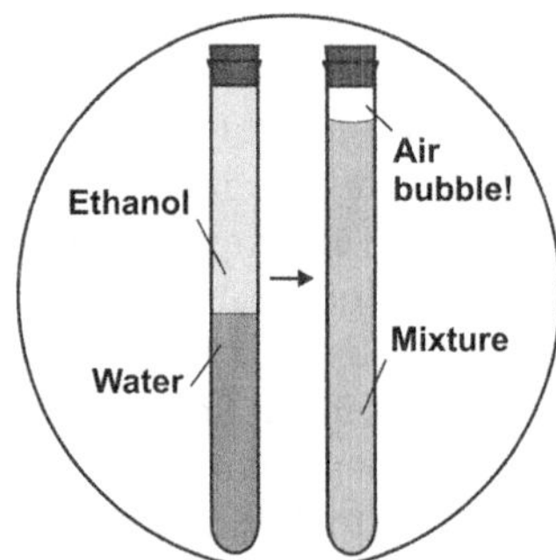

chemical potential by applying the appropriate pressure coefficient $(\partial V_{\mathrm{m}}/\partial p)_T$ or temperature coefficient $(\partial V_{\mathrm{m}}/\partial p)_p$, respectively.

Dissolved Substances It is noteworthy that the volume demand for a substance also depends upon what kind of a chemical environment it is in. Consider this example. 1 mol of pure water with a volume of about 18 cm^3 is stirred into 1 m^3 of concentrated sulfuric acid, and then the warmed mixture is cooled back down to the initial temperature. One finds that the entire volume has increased by only 8.5 cm^3 and not by 18 cm^3 as might have been expected. Obviously, water dissolved in sulfuric acid requires less space and the molar volume is smaller in this milieu:

$$V_{\mathrm{m}}^{\ominus}(\mathrm{H_2O\ in\ conc.\ H_2SO_4}) = 8.5\ \mathrm{cm}^3\,\mathrm{mol}^{-1}.$$

If sulfuric acid of half this concentration is used, one finds 17.5 cm^3 mol^{-1}. A similar but much smaller reduction of volume of about 4 % can be observed by mixing equal parts of water and ethanol (Experiment 8.2).

The volume demand for some substances in certain solvents can even be *negative*. The volume shrinks when such a substance is dissolved. An example of this is the solution of sodium hydroxide in water:

$$V_m^{\ominus}(\text{NaOH in } H_2O) = -6.8 \text{ cm}^3 \text{ mol}^{-1}.$$

When 1 mol of sodium hydroxide in form of pellets is dissolved in 1 m^3 of water, the volume of the solution shrinks by 6.8 cm^3, as long as the temperature and pressure are kept constant (Experiment 8.3). This contraction is caused by the H_2O molecules (which are rather loosely packed when in pure water) being concentrated more densely in the hydration shells of the Na^+ and OH^- ions.

As we have seen, the molar volume for a pure substance can be easily defined and calculated. How should we proceed, though, when we want to find the volume demand of a substance distributed inside another material environment?

Consider a body that absorbs a small amount of a substance. It will generally expand somewhat (Fig. 8.2). The volume grows because the substance now inside the body needs space, and by taking this space, the particles loosen the body's atomic structure. An example would be the volume demand of water penetrating a more or less moist block of wood causing the wood to expand more. The measure of

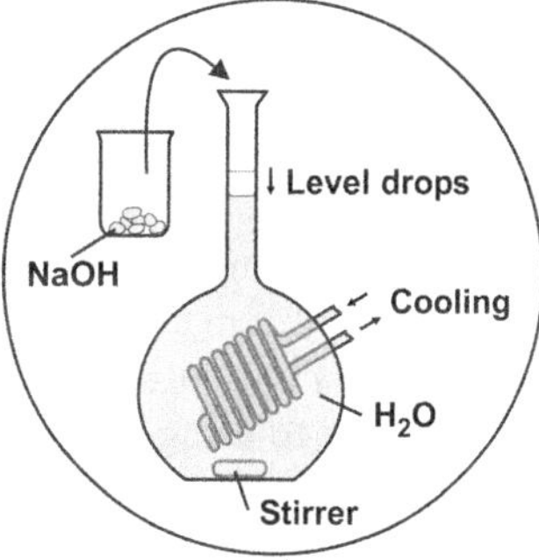

Experiment 8.3 *Negative volume demand of NaOH in water*: The flat-bottomed flask is filled with colored water up to the mark (with the cooling water running). Subsequently, as many pellets of sodium hydroxide as possible are put into the flask. After dissolution of the sodium hydroxide and cooling down of the resulting solution, the water level is considerably lower than before.

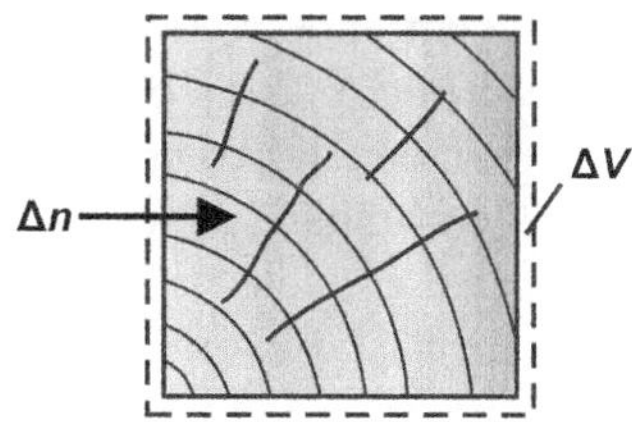

Fig. 8.2 Increase of volume ΔV of a block of wood when a small amount Δn of a substance (e.g., water) is added (strongly simplified description; in reality, the dimensional change with changing moisture content is anisotropic because of the inhomogeneity of wood).

volume demand of the substance added is the small observed change of volume ΔV, relative to the small amount of substance Δn added:

$$V_\mathrm{m} \approx \frac{\Delta V}{\Delta n} \quad \text{for a small amount } \Delta n. \tag{8.1}$$

In our example, this is of the order of 15 $\mathrm{cm}^3\,\mathrm{mol}^{-1}$.

To be precise, the added amount Δn (added to an amount n_0 possibly already present) should be kept as small as possible to keep the body from changing too much. The volume demand by water in a dry block of wood is different from that in wood which is already moist. We can express the changeover to infinitesimally small amounts of substance by replacing the difference quotient by the differential quotient. Naturally, the pressure p, temperature T, and the amounts n', n'', ... of all other substances inside the body must be kept constant during the entire process. This is necessary so that no changes occur during volume measurements due to mechanical compression, thermal expansion, and/or compositional changes while the substance is being added. In short, the milieu has to remain fixed. This can be expressed by adding the symbols of these quantities as indices to the differential quotient (see Sect. A.1.2 in the Appendix):

$$V_\mathrm{m} \equiv \left(\frac{\partial V}{\partial n}\right)_{p,T,n',n'',\ldots} \qquad \textit{(partial) molar volume of a dissolved substance.}$$

$$\tag{8.2}$$

More graphically: The molar volume of a substance corresponds to the change of volume which occurs when 1 mol of a substance is put into a very large sample of a given composition. The great excess ensures that the composition of the sample, as requested, will remain practically unchanged when the substance is added. This method is not only useful for mixtures, but for single substances as well. We can therefore forego using the epithet "partial."

We use the value of V_m at infinite dilution ($c \to 0$) as the *basic value* $\overset{\circ}{V}_\mathrm{m}$ of the molar volume of a dissolved substance, meaning the volume demand of the substance in the practically pure solvent. The basic value at standard temperature and pressure is, again, called *standard value* $V_\mathrm{m}^{\ominus} = \overset{\circ}{V}_\mathrm{m}(T^{\ominus}, p^{\ominus})$. For example, the molar volume of water strongly diluted in ethanol is not 18.1 $\mathrm{cm}^3\,\mathrm{mol}^{-1}$ but only about 14 $\mathrm{cm}^3\,\mathrm{mol}^{-1}$. As a result of the different packing densities of the molecules, the basic value of a dissolved substance also depends upon what kind of solvent it is in. Thus, the molar volume of water in concentrated sulfuric acid decreases to only about 9 $\mathrm{cm}^3\,\mathrm{mol}^{-1}$, as we have seen.

Depending upon the overall composition of a mixture, the molar volume of a substance can take very different values. The values can vary between the extremes of the pure state and the state of infinite dilution. Figure 8.3 shows how the molar volume of water depends upon the mole fraction of ethanol in an ethanol–water mixture at 298 K. The molar volume of ethanol is also dependent upon the

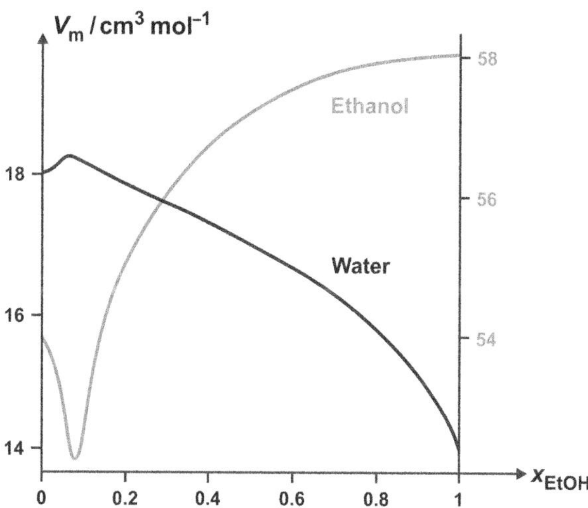

Fig. 8.3 Molar volume of water and ethanol in water–ethanol mixtures as a function of the ethanol content at 298 K. [Note the differing scales for water (*left*) and ethanol (*right*).]

composition. The minimum of the ethanol curve and the maximum of the water curve are found at the same mole fraction.

If the molar volumes V_A and V_B in a mixture of A and B are known for a certain composition, the volume of a portion of this mixture results from the amounts and volume demands of the components:

$$V = n_A \cdot V_A + n_B \cdot V_B. \tag{8.3}$$

(Here we have abbreviated $V_{m,A}$ or $V_{m,B}$ to V_A and V_B.) In order to derive this relation, we consider the increase of volume dV when we add small amounts dn_A and dn_B to a mixture of substances A and B while keeping pressure p and temperature T constant:

$$dV = \underbrace{\left(\frac{\partial V}{\partial n_A}\right)_{p,T,n_B} dn_A}_{dV_{\to n_A}} + \underbrace{\left(\frac{\partial V}{\partial n_B}\right)_{p,T,n_A} dn_B}_{dV_{\to n_B}} = V_A dn_A + V_B dn_B.$$

We have chosen to use the arrow in the index to make clear that we have increases in the directions of different variables. However, as mentioned in Sect. 3.11, this way of writing is optional. The equation itself looks more complicated than it actually is. The quantities p and T do not change here, so they can be left out, making the expressions seem somewhat simpler. We can imagine the graph of the function $V(n_A, n_B)$ as a mountainside with different slopes in the n_A und n_B directions (see Sect. A.1.2 in the Appendix). The first differential quotient shows the gradient $m_{\to n_A}$ in the n_A direction (imagine it as going east) and the second the gradient $m_{\to n_B}$ in the n_B direction (going northward). The product $m_{\to n_A} dn_A$ is then simply the increase $dV_{\to n_A}$ of "altitude" V, when one proceeds a small distance dn_A

in the n_A direction. The same holds true for movement in the n_B direction. The total increase dV for a change in both directions is simply the sum of both of these. Hopefully, the above equation is now clear.

Let us now imagine adding substances A and B at constant proportions, and right from the beginning when n_A and n_B are still both zero. The composition of the growing regions will then remain unchanged over the entire process, as will the volume demands V_A and V_B of both substances. The contribution of each individual substance to the total volume V is then simply the product of $V_A \cdot n_A$ or $V_B \cdot n_B$ and V itself is the sum of these.

In closing, a reminder that molar volume is not a measure of the volume which the molecules themselves fill, but only a measure of the space they lay claim to. This space can be much greater. For instance, in its gaseous state at room conditions, a substance requires about a thousand times the volume that it would in its condensed state. However, it can be much smaller (even negative), for instance when it causes the molecules of the substance it is mixed with to move more closely together, as we have seen happen with NaOH in a dilute aqueous solution. Naturally, the salt in the solution must have a positive volume, but if the volume contraction due to hydration is greater than the proper volume of the added ions, the total volume decreases.

8.3   Changes of Volume Associated with Transformations

The changes of volume observed during a chemical reaction are essentially the result of the volume demands of reactants and products. We will consider the reaction of pure or dissolved substances in order to calculate the effect at an arbitrary extent ξ of the reaction,

$$B + B' + \ldots \rightarrow D + D' + \ldots .$$

The starting as well as the final substances may be present concurrently in large or small amounts and in pure or dissolved states. We assume the pressure and temperature to remain constant during the whole process in order to avoid unwanted effects caused by compressibility and thermal expansion. We also require that no other reaction runs in parallel, i.e., ξ', ξ'', ... are constant. A *small extra* conversion $\Delta\xi$ then results in the following change of volume (instead of $V_{m,B}$ for a substance B, we use, as mentioned, the abbreviated V_B, etc.):

$$\Delta V = V_D \cdot \Delta\xi + V_{D'} \cdot \Delta\xi + \ldots - V_B \cdot \Delta\xi - V_{B'} \cdot \Delta\xi - \ldots . \qquad (8.4)$$

This follows from $\Delta n_B = \Delta n_{B'} = \ldots = -\Delta\xi$ and $\Delta n_D = \Delta n_{D'} = \ldots = +\Delta\xi$, respectively.

Every product requires an additional volume $V_m \cdot \Delta\xi$, while every reactant releases a volume $V_m \cdot \Delta\xi$. V_m denotes the required space of a given substance at extent ξ of the reaction. In order for the volume demands to have definite values, concentrations may not change noticeably. This can be achieved by allowing only small conversions. However, when all participating substances are in pure states, this limitation is unnecessary.

Because the change of volume ΔV is proportional to the conversion $\Delta\xi$ (at least as long as $\Delta\xi$ remains small), it is more useful to relate information of this kind to the conversion. Instead of ΔV, we use the quantity

$$\Delta_R V \equiv \frac{\Delta V}{\Delta\xi} = V_D + V_{D'} + \ldots - V_B - V_{B'} - \ldots \quad \text{for small } \Delta\xi;\ p,T,\xi',\xi'',\ldots\text{const.}$$

$$(8.5)$$

$\Delta_R V(\xi)$ is the *molar reaction volume* which is the measure of how strongly the transformation of the substances taking place changes the volume at a particular extent ξ of reaction. The index $_R$ refers to "reaction" and serves to differentiate the molar reaction volume (unit $\text{m}^3\ \text{mol}^{-1}$) from a change of volume ΔV (unit m^3).

The expression on the right can be made somewhat easier to understand with the help of the conversion numbers v_i. So far, for the sake of simplicity, we have chosen $v_B = v_{B'} = \ldots = -1$ and $v_D = v_{D'} = \ldots = +1$ in the conversion formula. Thus, only plus and minus signs have appeared in Eq. (8.5). In the more general case

$$|v_B|B + |v_{B'}|B' + \ldots \rightarrow v_D D + v_{D'} D' + \ldots$$

we obtain

$$\Delta_R V = \frac{\Delta V}{\Delta\xi} = v_B V_B + v_{B'} V_{B'} + \ldots + v_D V_D + v_{D'} V_{D'} + \ldots = \sum_i v_i V_i. \quad (8.6)$$

The conversion numbers for the reactants are negative and they are positive for the products. Therefore, the expression in the middle can be read as a difference, which explains the Δ in the symbol used for the quantity $\Delta_R V$.

In the limit, we require the $\Delta\xi$ to be infinitesimally small in Eq. (8.5). This is again expressed formally by using the symbol ∂ instead of the difference Δ. If we now introduce all the quantities that are to be kept constant as indices of the differential quotient, the equation takes the following form:

$$\Delta_R V \equiv \left(\frac{\partial V}{\partial\xi}\right)_{p,T,\xi',\xi'',\ldots} = \sum_i v_i V_i. \quad (8.7)$$

It should not be difficult to transfer this to other types of transformations such as phase transitions, dissolving processes, etc., that can be considered special cases of reactions. Depending upon the process (change of modification, melting, sublimation, dissolving, $\ldots$), one may write $\Delta_{\alpha\beta} V, \Delta_{sl} V, \Delta_{sg} V, \Delta_{sd} V, \ldots$ or in more detail

$\Delta_{\alpha\to\beta}V, \Delta_{s\to l}V, \Delta_{s\to g}V, \Delta_{s\to d}V, \dots$. Conversely, $\Delta_{\to}V$ can be used when the type of transformation is unimportant.

A few numerical values are useful for orientation. Volume increases generally around 3 % during melting. Water ice, whose volume actually decreases during melting is a well-known but rare exception. Evaporation volume is determined almost solely by the volume demand of the vapor with 25 L mol^{-1} at room conditions. (Compared to that, the volume required by the same substance in its condensed state is so small that it can be ignored.)

8.4 Entropy Demand

Pure Substances A substance that contains no entropy is absolutely cold. In order to bring it up to room temperature at standard pressure, a certain amount of entropy is necessary. This can be generated internally or added from outside. The amount of entropy necessary varies from substance to substance. It is proportional to the amount of substance, so we relate the entropy required by a substance to the amount needed for 1 mol of substance. This quantity, which we were introduced to in Sect. 3.9, is called *molar entropy*:

$$S_\mathrm{m} = \frac{S}{n} \quad molar\ entropy\ for\ pure\ substances.$$

Both entropy and molar entropy depend upon pressure and temperature. Therefore, if the temperature is kept constant, a solid body exposed to a pressure of 1,000 MPa loses about 1 % … 10 % of its entropy. In an ideal cooling process to 0 K, S decreases to $S_0 = 0$ Ct. Figure 8.4 illustrates the dependence of the entropy of a solid substance upon p and T. In the case of an ideal solid substance, the S surface originates at the p axis with a horizontal tangent and transforms into a rather

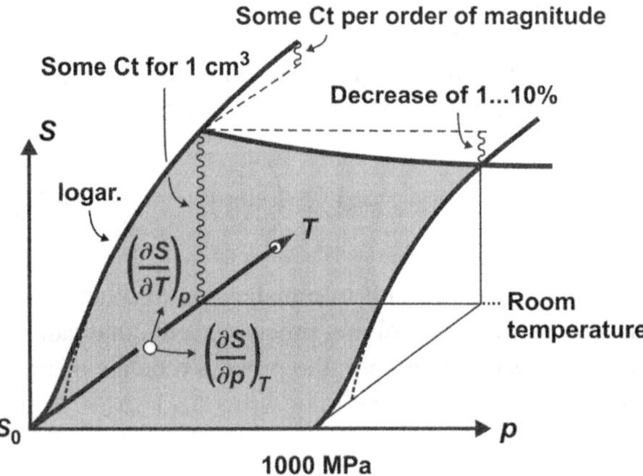

Fig. 8.4 Dependence of the entropy of a solid substance upon pressure and temperature.

logarithmically increasing slope. In this range, S increases a few Ct for 1 cm^3 of matter, if the temperature increases by one order of magnitude. The decrease of entropy in the direction parallel to the p axis is similar to the decrease of volume with rising pressure. Gases do not behave much differently within their range of existence. However, the entropy density is about a thousand times smaller at room conditions than it is in liquid or solid substances. So as the pressure rises, the decrease of entropy is very slight. It is only about 1 Ct for an increase of pressure of one order of magnitude, if one assumes 1 dm^3 of gas at standard conditions. The rise in the T direction is logarithmic such as that for solids, but steeper, some Ct per order of magnitude.

The standard value is indicated, as usual,

$$S_{\mathrm{m}}^{\ominus}(\mathrm{H_2O}) = 69.9 \ \mathrm{Ct \ mol^{-1}} \quad \text{at} \quad 298 \ \mathrm{K} \quad \text{and} \quad 100 \ \mathrm{kPa}.$$

The values of the entropy at a point p, T (such as at standard conditions) can be converted to other p and T values if the gradients of the surfaces in the p and T directions are known at the former point, i.e., if $(\partial S / \partial p)_T$ and $(\partial S / \partial p)_p$ are known. The first coefficient describes the substance's loss of entropy by increase of pressure. The second one corresponds to its entropy capacity $\mathcal{C}$, which we were introduced to in Sect. 3.9.

Dissolved Substances The entropy required by a substance distributed within a body differs from what it requires in its pure state. It is mostly considerably greater because the atomic disorder increases when atoms or molecules are scattered over a larger volume. $\mathrm{NaNO_3}$ in a 1 kmol m^{-3} aqueous solution at room conditions requires just about twice as much entropy as it does in its solid state. Therefore, when $\mathrm{NaNO_3}$ is dissolved in water, the solution cools down so strongly due to the salt extracting entropy from the water so that the glass it is in fogs up (Experiment 8.4). In order to keep the temperature constant, entropy must be absorbed from the surroundings. As is the case in almost every process, entropy is generated here too, but it is not enough to cover the high additional entropy demand of the salt.

Experiment 8.4 *Cooling during dissolving of NaNO$_3$ in water*: Solid sodium nitrate is poured all at once into the water and subsequently, one stirs vigorously with a glass rod. A strong decrease in temperature can be observed.

Distributing a tiny amount Δn of a substance inside a body results in a small change of entropy ΔS. The entropy change relative to the small amount of substance is used for defining a measure of the entropy demand of the substance:

$$S_\mathrm{m} \approx \frac{\Delta S}{\Delta n} \quad \text{for small } \Delta n. \tag{8.8}$$

To be exact, one must again deal with the limit of infinitesimally small additional amounts Δn of the substance, keeping pressure, temperature, and the amounts of all other substances constant in the process:

$$S_\mathrm{m} \equiv \left(\frac{\partial S}{\partial n}\right)_{p,T,n',n'',\ldots} \quad \textit{(partial) molar entropy of a dissolved substance.} \tag{8.9}$$

The molar entropy corresponds to the change of entropy required by the addition of a small amount of the component in order to keep the temperature constant at a given pressure, *extrapolated linearly* to 1 mol. Statements of this type can make things difficult at first, but actually come up a lot in everyday life. An example: when a car traveling at 50 km/h crosses some zebra stripes, it might take 1 s to do so. If we linearly extrapolate the width of the stripes by a factor of 3,600, we obtain a distance of 50 km. That's the meaning of "traveling at 50 km/h."

Like molar volume, molar entropy of dissolved substances can be negative. This occurs mostly with polyvalent ions in aqueous solution, for example, with Ca^{2+}. When such ions are put into water, they *organize* water molecules, previously distributed in an *unorganized* way in the liquid, into their hydration shells. The effect can be so great that a state emerges that is more strongly organized overall, although the arbitrary distribution of the ions in the solvent actually increases disorder. The entropy can then decrease. The decrease of entropy just described does not contradict the second law! The process does not *destroy* entropy, the entropy is pressed out of the water going into the hydration shells into the surrounding liquid, warming it. If the temperature is to be kept constant, this surplus entropy must be removed. The entropy of the solution decreases only as a result of this removal. The entropy demand is now smaller.

The molar entropy grows along with increasing dilution. It always grows by the same magnitude ΔS_m, if the concentration is multiplied by a fixed factor (<1) *no matter what the substance is and what milieu it is in.* Conversely, the molar entropy decreases when the concentration increases. If, for example, the concentration increases by a factor of 10, the molar entropy (at 298 K) decreases by

$$S_\mathrm{d} \approx -19\,\mathrm{Ct\,mol}^{-1} \quad \text{for } c \to 10c, \text{ as long as } c \ll c^{\ominus}.$$

We have seen a similar kind of behavior with the chemical potentials (remember the decapotential $\mu_\mathrm{d} = 5.7$ kG in Sect. 6.3). Indeed, the two patterns of behavior are closely related to each other. We will show this later on in Sect. 9.3 (S–n coupling). We conclude that S_m must depend logarithmically upon c:

$$S_\mathrm{m} = S_{\mathrm{m},0} - R\ln \frac{c}{c_0} \quad \text{for } c, c_0 \ll c^{\ominus}. \tag{8.10}$$

This resembles the mass action equation. If we use $c = 10c_0$, we recover the result for S_d mentioned above. We can use the value of 19 Ct mol^{-1} for rough estimates. This equation, which we accept for now on empirical grounds [and which we are going to derive later (Sect. 13.4)], holds strictly for low concentrations. Deviations occur at higher concentrations.

Unlike with molar volume, we cannot use the values for infinite dilution as the *basic values* $\overset{\circ}{S}_\mathrm{m}$ for molar entropy S_m because they would be infinite. A value at finite concentration, the standard concentration $c^{\ominus} = 1\,\mathrm{kmol\,m^{-3}}$, is used instead. The true values $S_\mathrm{m}(c^{\ominus})$ are not used but one extrapolates from the measured or calculated values for small concentrations to $c^{\ominus}$, according to the relation given above. Just like the basic values of the chemical potentials, those of molar entropies are fictive values. Written in terms of the basic value, the equation above is

$$S_\mathrm{m} = \overset{\circ}{S}_\mathrm{m} - R\ln \frac{c}{c^{\ominus}} \quad \text{for } c \ll c^{\ominus}. \tag{8.11}$$

In general, the *standard values* (the basic values at standard temperature and standard pressure) are tabulated, $S_\mathrm{m}^{\ominus} = \overset{\circ}{S}_\mathrm{m}(T^{\ominus}, p^{\ominus})$.

The amount of entropy contained in a portion of a mixture is obtained analogously to its volume from the amounts and entropy demands of its components, for example, components A and B:

$$S = n_\mathrm{A} \cdot S_\mathrm{A} + n_\mathrm{B} \cdot S_\mathrm{B}. \tag{8.12}$$

The derivation which we forgo here can be accomplished according to the same pattern used for volume [Eq. (8.3)].

8.5   Changes of Entropy Associated with Transformations

In a chemical reaction, the substances involved produce new ones with changed entropy demands. Here, we are interested in the amount of entropy ΔS which is added or removed for compensation when a reaction takes place at constant pressure and constant temperature. Let us consider the reaction of 0.1 mol of iron and 0.1 mol of sulfur forming 0.1 mol of iron sulfide, at room conditions:

$$\text{Fe} \quad + \text{S} \quad \rightarrow \text{FeS}$$

	Fe	S	FeS	
$S_{\mathrm{m}}^{\ominus}$ (Ct mol^{-1})	27	32	60	
S (Ct)	2.7	3.2	6.0	contained in each 0.1 mol.

$$2.7 + 3.2 = 5.9$$

We see that exactly $\Delta S = 0.1$ Ct is lacking. This is what is needed to cover the entropy demands of the FeS produced by a conversion of $\Delta\xi = 0.1$ mol. This amount of entropy ΔS must be introduced from outside if the iron sulfide is to be as warm at the end of the reaction as the iron and sulfur were before the process began. Without this added entropy, it would be colder. If the conversion is multiplied, the required entropy multiplies correspondingly.

Our example results in the following for an arbitrary conversion $\Delta\xi$:

$$\Delta S = S_{\text{FeS}} \cdot \Delta\xi - S_{\text{Fe}} \cdot \Delta\xi - S_{\text{S}} \cdot \Delta\xi,$$

where S_{FeS}, S_{Fe}, and S_{S} each represent the molar entropies of the corresponding substances. The additional demand ΔS is proportional to the conversion if temperature and pressure are kept constant during the reaction and no side reactions take place. Because of this proportionality, it makes sense to relate the additional requirement to the conversion:

$$\Delta_{\text{R}}S = \frac{\Delta S}{\Delta\xi} = S_{\text{FeS}} - S_{\text{Fe}} - S_{\text{S}} \quad \text{for small } \Delta\xi;\ p, T, \xi', \xi'', \ldots \text{const.}$$

This conversion-related quantity is called the *molar reaction entropy* $\Delta_{\text{R}}S$. In our example, the caveat "for small $\Delta\xi$" is unnecessary because only pure substances are participating in the reaction. However, if dissolved substances appear in the conversion formula, we can then only allow small additional conversions $\Delta\xi$ for any arbitrary extent ξ of the reaction. This is to ensure that the composition of the solution and, therefore, the entropy demands of the substances in it do not change noticeably.

For an arbitrary reaction between pure and dissolved substances,

$$|\nu_{\text{B}}|\text{B} + |\nu_{\text{B}'}|\text{B}' + \ldots \rightarrow \nu_{\text{D}}\text{D} + \nu_{\text{D}'}\text{D}' + \ldots,$$

we can calculate the molar reaction entropy according to the pattern used for many other conversion-based "extensive" (often substance-like) quantities, which we were introduced to through the example of molar reaction volume:

$$\Delta_{\text{R}}S = \nu_{\text{B}}S_{\text{B}} + \nu_{\text{B}'}S_{\text{B}'} + \ldots + \nu_{\text{D}}S_{\text{D}} + \nu_{\text{D}'}S_{\text{D}'} + \ldots = \sum \nu_i S_i. \tag{8.13}$$

The molar reaction entropy is the change of entropy—based upon the conversion— at constant p and T. It equals the sum of the molar entropies of the reaction partners weighted by the conversion numbers. Using our iron-and-sulfur example, we have $\Delta_R S = S_{FeS} - S_{Fe} - S_S = 1\,Ct\,mol^{-1}$ because of $v_{Fe} = v_S = -1$ and $v_{FeS} = +1$.

The conditions added to the equation above can be expressed as for the case of molar reaction volume by replacing the Δ by ∂ in the difference quotient and adding the quantities which will be kept constant to the index:

$$\Delta_R S = \left(\frac{\partial S}{\partial \xi}\right)_{p,T,\xi',\xi'',\ldots} = \sum v_i S_i. \tag{8.14}$$

If the standard values for the molar entropies of all the reaction participants are applied, one obtains $\Delta_R S^{\ominus}$ as the standard value.

The following considerations can be helpful for estimating reaction entropy: In the case of liquids and especially in the case of gases, the values of molar entropies are generally far above those of solid substances. The algebraic sign and the absolute value of $\Delta_R S$ are therefore primarily determined by how much the number of liquid or gaseous molecules changes during a reaction. The more the number of liquid or gas molecules increases in a reaction, the more positive the molar reaction entropy will be. In a net consumption of gas or liquid molecules, the reaction entropy decreases so that in the reaction

$$2\,H_2|g + O_2|g \rightarrow 2\,H_2O|l$$

we have a strong decrease of entropy of $\Delta_R S^{\ominus} = -327\,Ct\,mol^{-1}$ which results from the formation of a liquid from two gases.

As in the case of volume (see end of Sect. 8.3), it should be easy to relate this to other kinds of transformations of substances such as phase transitions, dissolving processes, etc. We can write $\Delta_{\rightarrow} S$ instead of $\Delta_R S$ if we wish to emphasize that we mean any type of transformation and not only reactions. Egon Wiberg (Wiberg E (1972) Chemical Affinity, 2nd edn. de Gruyter, Berlin, New York, p. 103) suggested calling processes with positive transformation entropy, $\Delta_{\rightarrow} S > 0$, *endotropic*, and those with negative $\Delta_{\rightarrow} S < 0$, *exotropic*.

However, in contrast to the case of volume, the effects caused by the differing entropy demands of substances can be masked by others because energy is released in many processes which then generates entropy. In order to better understand the consequences of this circumstance, we will now deal with energy released or used by these processes.

Again, it would be good to note some reference values. For monoatomic substances, the increase of entropy in a melting process is about $10\,Ct\,mol^{-1}$ (Richards's rule), while for all substances boiling at normal pressure, it is around $100\,Ct\,mol^{-1}$ (Pictet–Trouton's rule).

8.6 Energy Conversion in Transformations of Substances

As we have seen in the introduction to the chemical potential (Sect. 4.8), in order to increase the amount n of a substance B, against its own "tendency to transform" μ, we need an amount of energy equal to

$$W_{\to n} = \mu \cdot \Delta n \quad \text{for small} \quad \Delta n \quad \text{or} \quad dW_{\to n} = \mu dn \qquad (8.15)$$

(Fig. 8.5). Whether the growth is caused from production inside or by addition from outside makes no difference. If the substance is produced inside it only means that at the same time a certain amount of one or more substances participating in the creation of B either disappear or are produced. This can be included separately by the corresponding expressions $dW_{\to n'} = \mu' dn'$, $dW_{\to n''} = \mu'' dn''$, ... and can therefore be ignored here.

If we want to obtain the contribution $W_{\to n}$ from the change ΔW of the energy content W of the system, we have to make sure that no other substances, entropy, or similar quantities are added as well. Furthermore, the volume must be kept constant. This can be achieved by fixing the amounts of all other substances, the entropy, the volume, etc., of the body

$$W_{\to n} = (\Delta W)_{S,V,n',n'',\ldots} \quad \text{or} \quad dW_{\to n} = (dW)_{S,V,n',n'',\ldots}.$$

This is not different from what we learned in the example of the bathtub (compare Sect. 1.6). The way water flows in and out of the bathtub over various paths is also valid here for energy. The energy content W of the area is a state variable, but the amounts of energy $W_{\to V}$, $W_{\to S}$, $W_{\to n}$, $W_{\to n'}$, ... that are exchanged with the surroundings over different paths are not. They are so-called process quantities whose combined effect can best be imagined by considering a process taking place in time t. $P_V(t) = dW_{\to V}/dt$, $P_S(t) = dW_{\to S}/dt$, etc., represent the energy currents flowing over various paths, while $\dot{W} = dW/dt$ tells us how quickly the amount of energy increases as a result:

$$\dot{W} = P_V + P_S + P_n + P_{n'} + \ldots \quad \text{"Continuity equation."}$$

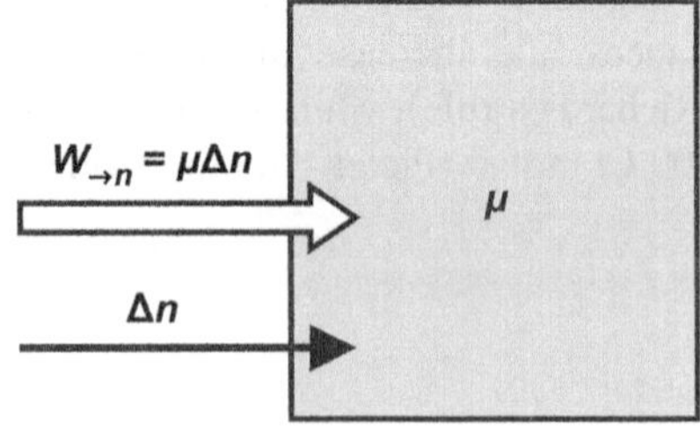

Fig. 8.5 The energy needed to increase the amount n of a substance inside a material system (where the potential equals μ) by adding or producing a small amount of Δn.

The energy current P_n flowing over path n can be formally described by $\dot{W}$ if we imagine all other pathways to be blocked, $P_n = \left(\dot{W}\right)_{S,V,n',n'',\ldots}$. When we use this method to calculate the energy $\mathrm{d}W_{\to n}$ flowing in during a short time span $\mathrm{d}t$, we obtain the equation we started with:

$$\mathrm{d}W_{\to n} = P_n \cdot \mathrm{d}t = \left(\dot{W} \cdot \mathrm{d}t\right)_{S,V,n',n'',\ldots} = (\mathrm{d}W)_{S,V,n',n'',\ldots}.$$

The following fact is remarkable and important. If entropy S_g is generated (possibly by friction) in an area at temperature T, this costs an additional amount of energy $W_b = T \cdot S_g$. The condition that S must be kept constant just means that S_g and therefore W_b cannot remain in the area but must be removed. Whether or not energy is "burnt" and entropy is generated in the procedure does not affect the result and can, therefore, be ignored.

We arrive at the chemical potential μ if we divide $\mathrm{d}W_{\to n}$ in the equation above by $\mathrm{d}n$, which leads us back to the following equation we know from Sect. 4.8:

$$\mu = \left(\frac{\partial W}{\partial n}\right)_{S,V,n',n'',\ldots}. \tag{8.16}$$

The energy W_t necessary for *transferring* a small amount n_t of a substance from a body 1 with the chemical potential μ_1 to another body 2 with the potential μ_2 (Fig. 8.6) results from the effort $\mu_2 \cdot n_t$ for supplying the amount n_t to body 2, less the profit $\mu_1 \cdot n_t$ resulting from the removal from body 1:

$$W_t = (\mu_2 - \mu_1) \cdot n_t = \Delta\mu \cdot n_t \quad \text{for small } n_t. \tag{8.17}$$

If $\mu_2 > \mu_1$, and therefore the substance is pumped "up the potential hill," W_t is positive, and increasingly so, the greater the potential lift is.

Conversely, energy can be gained when a substance goes from higher μ to lower μ, and W_t becomes negative. Similar to thermal engines where the temperature fall

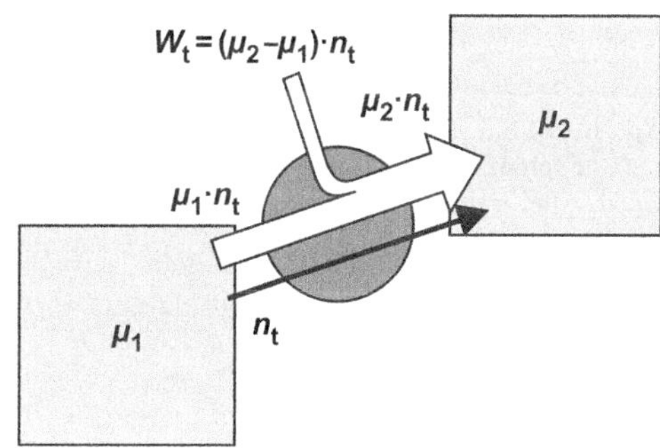

Fig. 8.6 Flow diagram for energy and amount of substance of an "ideal substance pump" (*gray circle*). In reality, entropy is constantly generated—possibly by friction—which costs extra energy.

of entropy is used, corresponding *"chemical engines"* can be constructed. Such chemical engines are to be found all over in nature in the form of muscles and the flagella of single-celled organisms. A simple apparatus of this type that makes use of the potential difference of liquid water in a glass and water vapor in the air of the room is the toy known as the "drinking duck" (Experiment 8.5).

Chemical reactions can be described using the same paradigm. It makes no difference whether or not the substances all appear in the same homogeneous area, or if they are distributed over various areas. For a reaction

$$B + B' + \ldots \rightarrow D + D' + \ldots$$

the total effort $W_{\rightarrow\xi}$ for an increase of the conversion by a small $\Delta\xi$ is simply the sum of the positive or negative contributions due to changes of amount $\Delta n_B = -\Delta\xi$, $\Delta n_{B'} = -\Delta\xi$, ... of the reactants and $\Delta n_D = \Delta\xi$, $\Delta n_{D'} = \Delta\xi$, ... of the products:

$$W_{\rightarrow\xi} = \mu_B \Delta n_B + \mu_{B'} \Delta n_{B'} + \ldots + \mu_D \Delta n_D + \mu_{D'} \Delta n_{D'} + \ldots .$$

The sums of the chemical potentials can be combined to give the chemical drive of the reaction:

$$W_{\rightarrow\xi} = -(\mu_B + \mu_{B'} + \ldots - \mu_D - \mu_{D'} - \ldots) \cdot \Delta\xi = -\mathcal{A} \cdot \Delta\xi.$$

Generalizing this for the case where the conversion numbers are not only $+1$ or -1 is just a formality. The energy $W_{\rightarrow\xi}$ needed for the general reaction

$$0 \rightarrow v_A A + v_B B + v_C C + \ldots \qquad \text{or} \qquad 0 \rightarrow \sum_i v_i \mathcal{B}_i$$

(compare Sect. 1.7) with $\Delta n_A = v_A \Delta\xi$, $\Delta n_B = v_B \Delta\xi$, ... or, alternatively, $\Delta n_i = v_i \Delta\xi$, equals

Experiment 8.5 *Drinking duck*: First, the felt of the duck's head is wetted. After a short while, the duck begins to "drink," i.e., it slowly swings back and forth, finally dips its beak into the water, and comes back up nodding. After a number of oscillations, the process starts anew.

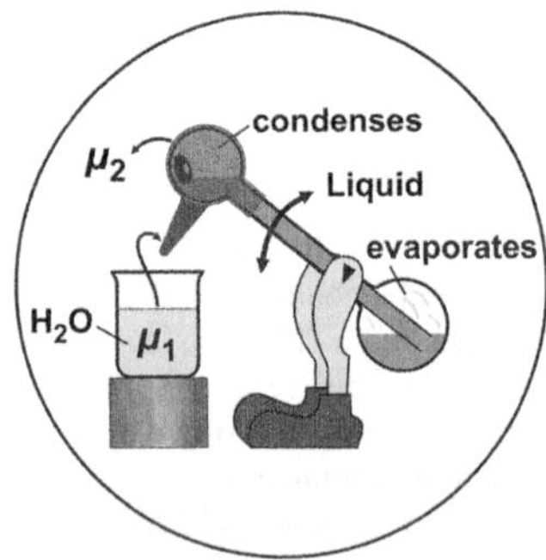

$$W_{\to\xi} = \mu_A \Delta n_A + \mu_B \Delta n_B + \mu_C \Delta n_C + \dots$$

and so

$$W_{\to\xi} = (\nu_A \mu_A + \nu_B \mu_B + \nu_C \mu_C + \dots) \cdot \Delta\xi = \left(\sum_i \nu_i \mu_i\right) \cdot \Delta\xi = -\mathcal{A} \cdot \Delta\xi.$$

The final formula is very simple and remains the same for both cases. The more "negative" drive $\mathcal{A}$ (the greater $-\mathcal{A}$) and the greater the conversion $\Delta\xi$ are, the more the energy needed increases:

$$W_{\to\xi} = -\mathcal{A} \cdot \Delta\xi \quad \text{for small } \Delta\xi. \tag{8.18}$$

One might call $-\mathcal{A}$ a measure of "reverse force," meaning how strongly a reaction resists when it is forced to proceed against its own drive. This is not different from what we know from mechanics when a spring is stretched or a weight is lifted.

As a reaction progresses, the composition of the mixture changes and along with it, the chemical drive. In order for Eq. (8.18) to remain generally valid, the conversion $\Delta\xi$ (and consequently, the energy $W_{\to\xi}$) must stay small enough. This is usually expressed by using differentials:

$$\mathrm{d}W_{\to\xi} = -\mathcal{A} \cdot \mathrm{d}\xi. \tag{8.19}$$

In the case of spontaneous reactions, where the drive $\mathcal{A}$ is positive, we have $W_{\to\xi} < 0$, and energy is therefore released. This released energy is usually "burnt," thereby generating entropy. This generally leads to a heating up of the reaction mixture (Sect. 8.7). The energy may, however, be available for lots of other uses. For instance, muscles use the energy released by glucose oxidation. This does not happen perfectly, but with a higher efficiency as if the glucose were burnt in a thermal power plant:

$$\underline{C_6H_{12}O_6|w + 6\,O_2|g \to 6\,CO_2|w + 6\,H_2O|l}$$

$\mu^{\ominus}\,(kG)$:   –915 6×0 $6\times(-386)$ $6\times(-237)$ $\Rightarrow$ $\mathcal{A}^{\ominus} = +2821$ kG.

For a conversion of $\Delta\xi = 1$ mol, the result is an energy requirement of $W_{\to\xi} = -\mathcal{A} \cdot \Delta\xi$, meaning a gain of $-W_{\to\xi} = \mathcal{A} \cdot \Delta\xi = 2,821\,\mathrm{kG}$ in the ideal case. Flashlight batteries, for example, use the energy released by zinc oxidation through MnO_2.

8.7   Heat Effects

Preliminary Remarks In the introduction to entropy in Chap. 3, we mentioned that this quantity describes exactly what we consider "amount of heat" in everyday life. Phenomena are called *caloric* when the *amount* of heat is the primary characteristic and called *thermal* when temperature is most significant. It is easiest to understand the caloric effects that accompany all types of transformations of substances through entropy and not through energy. Entropy is the characteristic substance-like quantity for effects of this type. Energy also plays its role, not only here but in (almost) all other effects as well. It is important but nonspecific. If one is paying too much attention (or is only paying attention) to energy, the wrong parameter is being emphasized.

"Great things cast long shadows" is a saying we all know. One can find out a lot from shadows alone, such as the form, motion, and behavior of whatever is producing the shadow. However, some things get lost such as colors or the actual shape and form of the body in question. Everything that happens in nature leaves behind traces at the level of energy. In many cases, these shadow images suffice and are even very good for describing an important aspect: conservation of energy—but this is not the only important aspect. It is like the floor plan of a house that does not give a sufficiently clear picture of its habitability.

An energy "shadow" belongs to every change of entropy dS in a system no matter if it occurs as a result of exchange with the surroundings (dS_e) or through generation inside (dS_g), or both:

$$\begin{aligned} dS &= dS_e + dS_g & \text{level of entropy,} \\ TdS &= TdS_e + TdS_g & \text{level of energy.} \end{aligned} \qquad (8.20)$$

In order to understand the basics, it suffices if, for now, we limit ourselves to systems that do not exchange substances with their surroundings. As long as changes are small or the temperature T remains constant, the shadow is a good image of what happens at the entropic level: $T\Delta S = T\Delta S_e + T\Delta S_g$. If this condition is breached in any way, a distorted image emerges. It may be distorted only slightly, but just as easily it can be deformed to the point of being unrecognizable. Only TdS_e (or the sum of such contributions, $\int_{\text{initial}}^{\text{final}} T\,dS_e$), is usually called "heat" (compare Sect. 3.11). We will return to this at the end of this section.

Many of the characteristic quantities that describe caloric effects hold for isothermal conditions. In this case, T represents a fixed scale factor between the original entropic image and the secondary image at the energy level. Luckily, we are only dealing with isothermal effects in this section. Therefore, it hardly makes any difference whether we discuss things at the entropic level or rewrite them in terms of energy quantities. Although it is unusual, we will choose the first approach because it better clarifies the essentials.

Interaction of Two Effects Let us consider a simple system with uniform pressure p and temperature T which conform to the values p^* and T^* in the surroundings. Moreover, only one reaction may occur inside this system. Now there are three pathways over which the energy content W of the system can be changed:

$$\mathrm{d}W = -p\mathrm{d}V + T\mathrm{d}S - \mathcal{A}\mathrm{d}\xi. \tag{8.21}$$

In order to direct the energy to take a particular path, the system as well as the surroundings must be adjusted to this. We might need heat conducting walls and an entropy reservoir outside to absorb or deliver energy thermally. Cylinders, pistons, and transmission gears are used in order to make the energy mechanically useful. Finally, electrodes, diaphragms, ion conductors, and the like are employed to use the energy made available chemically. It is unimportant here how these devices are constructed.

Let us now imagine a system and its surroundings equipped for energy to be exchanged via all three paths. When the reaction progresses by a small $\mathrm{d}\xi$, the amounts of substances change and their volume and entropy demands change along with them:

$$\mathrm{d}V = \Delta_\mathrm{R}V \cdot \mathrm{d}\xi \quad \text{and} \quad \mathrm{d}S = \Delta_\mathrm{R}S \cdot \mathrm{d}\xi. \tag{8.22}$$

The increases $\mathrm{d}V$ and $\mathrm{d}S$ occur at the cost of the surroundings, i.e., $\mathrm{d}V = -\mathrm{d}V^*$ and $\mathrm{d}S = -\mathrm{d}S^*$. This is the case as long as no entropy is generated, which we will assume for the present. There is an exchange of energy connected with this. Because we have assumed pressure and temperature to be the same inside and outside, $p = p^*$ and $T = T^*$, no energy is released. The word *released* means that it is then available for other purposes. In this case, what is released on one side is used on the other side so that nothing is left over: $p\mathrm{d}V + p^*\mathrm{d}V^* = 0$ and $T\mathrm{d}S + T^*\mathrm{d}S^* = 0$; therefore, nothing is released. This type of energy exchange in which energy is passed on for an *intended purpose* cannot be tapped into and is therefore uninteresting.

The situation is different for the third path. The energy $-\mathrm{d}W_{\to\xi} = \mathcal{A}\mathrm{d}\xi$ released in a spontaneous reaction (with $\mathcal{A} > 0$) cannot, as a rule, be used completely. To express this, we introduce an efficiency $\eta < 1$. The *useful* energy which the surroundings (Index $*$) receives, is then $\mathrm{d}W_\mathrm{use}^* = \eta\mathcal{A}\mathrm{d}\xi$. The rest $\mathrm{d}W_\mathrm{b}^* = (1 - \eta)\mathcal{A}\mathrm{d}\xi$ is "burnt," thereby generating entropy. With $\mathrm{d}S_\mathrm{g} = \mathrm{d}W_\mathrm{b}^*/T$, we have:

$$\mathrm{d}S_\mathrm{g} = \frac{\mathrm{d}W_\mathrm{b}^*}{T} = \frac{(1 - \eta)\mathcal{A}\mathrm{d}\xi}{T}. \tag{8.23}$$

The "burnt" energy is transferred thermally together with the contribution $-T\mathrm{d}S$ (e.g., through rigid but thermally conducting walls) into the surroundings (Fig. 8.7):

Fig. 8.7 Energy flow diagram for a material system when pressure and temperature are the same inside and outside. The energy released inside during a small conversion $\mathrm{d}\xi$ is used with an efficiency of η.

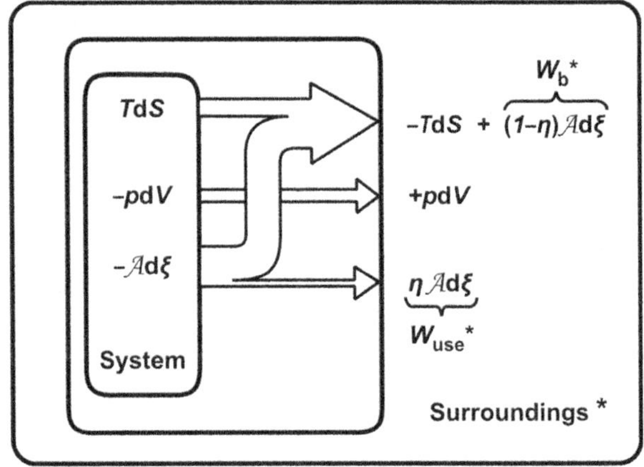

$$\mathrm{d}W = -p\mathrm{d}V + T\mathrm{d}S - \mathcal{A}\,\mathrm{d}\xi \qquad\qquad \text{system,}$$

$$\mathrm{d}W^* = \underbrace{+p\mathrm{d}V}_{-p\mathrm{d}V^*} \underbrace{-T\mathrm{d}S +(1-\eta)\mathcal{A}\,\mathrm{d}\xi}_{T\mathrm{d}S^*} \underbrace{+\eta\,\mathcal{A}\,\mathrm{d}\xi}_{\mathrm{d}W_{\mathrm{use}}^*} \qquad \text{surroundings *.}$$

One could say that the third path is "leaky," meaning that a part of the useful energy transported along this path may get lost and finally dissipates in the surroundings. The other paths are not impervious to loss, either. It is just the lack of pressure and temperature gradients that hinder them being tapped.

When dealing with only one system and its surroundings, there are stipulations for the signs: Inflows to the system receive positive signs and outflows receive negative ones. Whatever happens in the surroundings is described accordingly. The entropy content S of a system can, as we know, change through exchange or generation (compare Sect. 3.6):

$$\mathrm{d}S = \mathrm{d}S_\mathrm{e} + \mathrm{d}S_\mathrm{g}$$

or rearranged:

$$\mathrm{d}S_\mathrm{e} = \mathrm{d}S - \mathrm{d}S_\mathrm{g}. \tag{8.24}$$

While $\mathrm{d}S$ describes the entropic effect inside, $\mathrm{d}S_\mathrm{e} = -\mathrm{d}S^*$ indicates the effect that can be noticed outside. If $\mathrm{d}S_\mathrm{e} < 0$, the process is called *exothermic* and if $\mathrm{d}S_\mathrm{e} > 0$, it is called *endothermic*. In contrast, the sign of $\mathrm{d}S$ is the deciding factor for the terms *exotropic* and *endotropic* introduced in Sect. 8.5:

$$dS < 0 \quad \text{exotropic,} \qquad dS > 0 \quad \text{endotropic,}$$
$$dS_e < 0 \quad \text{exothermic,} \qquad dS_e > 0 \quad \text{endothermic.}$$

dS_e is *not* a simple quantity. According to Eq. (8.24), it is made up of two dissimilar contributions. Along with the process itself, we will call these contributions exothermic or endothermic depending upon their signs. Both are proportional to the conversion $d\xi$ but have different origins and depend upon very different parameters:

$$dS = \Delta_R S \cdot d\xi \quad [\text{according to Eq.(8.22)}]$$

- Can be either positive or negative, providing consequently an endothermic or exothermic contribution.
- Has nothing directly to do with the chemical drive $\mathcal{A}$ of the process.
- Is independent of whether or not and how the released energy is used or wasted.

$$dS_g = (1 - \eta)\mathcal{A}d\xi/T \quad [\text{according to Eq.(8.23)}]$$

- Is always positive and therefore always provides an exothermic contribution.
- Is directly proportional to the chemical drive $\mathcal{A}$ of the process.
- Varies between 0 and 100 % depending upon the type of energy usage.

Because of these differences it is a good idea to discuss the two contributions always separately and not to try to combine them as is usual in chemistry. dS and dS_g are proportional to the conversion $d\xi$ so it makes sense relating them to the conversion, especially when giving concrete values:

$$\left(\frac{dS}{d\xi}\right)_{p,T} = \Delta_R S \text{ and } \frac{dS_g}{d\xi} = \frac{(1 - \eta)\mathcal{A}}{T}. \tag{8.25}$$

We have inserted into the first expression the condition that p and T are to be kept constant, which is unnecessary in the second expression because it is independent of this condition. Another difference emerges that makes it advisable to keep the two effects separate. When the entropy S is given as a function of p, T, and ξ, $S = f(p, T, \xi)$, then $\Delta_R S$ can be calculated as the derivative of f with respect to ξ at constant p and T (according to the usual mathematical rules). Formally, this is expressed by replacing the straight d by the rounded ∂: $\Delta_R S = (\partial S/\partial \xi)_{p,T}$.

Latent Entropy The consideration above is correspondingly valid for other types of transformations of substances. In phase transitions such as melting, boiling, etc., the drive disappears at the corresponding transition points, $\mathcal{A} = 0$. This is analogously valid for dissolving processes at saturation and generally for transformations at equilibrium. The second expression becomes zero in each case and the entropic effects (and the caloric ones along with them) in the system and its surroundings become equal, $dS = dS_e$ (or $TdS = TdS_e$). In this special case, it has long been customary to use the term "latent heat," namely for the infinitesimal quantity $T\mathrm{d}S$ and, respectively, its integral $T\Delta S$.

The term "latent heat" originated in the eighteenth century when people had only vague conceptions about the nature of heat and its characteristics. The most common belief was that heat was an entity contained in objects to greater or lesser degree that was movable and could be exchanged between them, comparable to a substance. When something felt warm or hot, it was because there was more of this entity contained within it, and when something felt cold, it meant there was less of it. This describes almost exactly the characteristics we have attributed to entropy, with the difference that it was assumed back then that the entity could be neither generated nor destroyed.

Already in the eighteenth century it was clear that heat had to be added to vaporize water, and heat had to be withdrawn if steam was to condense. Even though the produced steam takes much more heat than is needed to heat the cold water to the boiling point, the temperature of the steam is not higher than the boiling temperature. It is as if the heat is "hiding"; it has become *latent* according to the term used back then. When the steam condenses to its liquid, the latent (hidden) heat is released and becomes *sensible*.

We have taken over this name for lack of a better one in order to distinguish between the effects caused by differences of entropy demands and those caused by entropy generation. In Eq. (8.25), the expression on the left represents the *latent molar reaction entropy* and the one on the right, the *generated molar reaction entropy*.

Balance of Entropy Only the difference of these two effects is noticeable outside due to $\mathrm{d}S_e = \mathrm{d}S - \mathrm{d}S_g$:

$$\left(\frac{\mathrm{d}S_e}{\mathrm{d}\xi}\right)_{p,T} = \left(\frac{\mathrm{d}S - \mathrm{d}S_g}{\mathrm{d}\xi}\right)_{p,T} = \left(\frac{\mathrm{d}S}{\mathrm{d}\xi}\right)_{p,T} - \frac{\mathrm{d}S_g}{\mathrm{d}\xi}$$

or

$$\underbrace{\left(\frac{\mathrm{d}S_e}{\mathrm{d}\xi}\right)_{p,T}}_{} = \underbrace{\Delta_R S}_{} - \underbrace{\frac{(1-\eta)\mathcal{A}}{T}}_{} \qquad (8.26)$$

$$\text{exchanged} = \text{latent} - \text{generated}\quad\text{molar reaction entropy.}$$

Naturally, formal symbols corresponding to the various reaction entropies can be introduced, which are more or less self-explanatory. This allows us to abbreviate the three molar entropies introduced above, and their integral counterparts, to:

$$\Delta_R S_e = \Delta_R S_\ell - \Delta_R S_g \quad\text{and}\quad \Delta S_e = \Delta S_\ell - \Delta S_g \quad \text{(balance of entropy)} \quad (8.27)$$

The index ℓ for "latent" is only inserted for the sake of clarity. We assume pressure and temperature to be constant, as stated at the beginning. Remember that merely $\Delta_R S_\ell (\equiv \Delta_R S)$ is determined only by the state parameters p, T, ξ used here. $\Delta_R S_g$ and

$\Delta_R S_e$ along with it depend upon the efficiency η and therefore upon other parameters such as "resistances" of various types inside or upon the devices that make exchange of energy between system and surroundings possible.

$\Delta_R S_e$ is mostly negative because the absolute value of latent entropy in chemical reactions is in general much smaller than the generated entropy, $|\Delta_R S| \ll \Delta_R S_g$. This results in most chemical reactions being exothermic, so that

- The entropy must flow off for the temperature and pressure to be retained or
- The temperature will rise if the outflow of entropy is hindered.

Let us return to the reaction of iron with sulfur with a conversion of $\Delta\xi = 0.1$ mol (compare Sect. 8.5). In order for the resulting iron sulfide not to emerge sub-cooled from the reaction, we noticed that 0.1 Ct of entropy was lacking, which needed to be absorbed from the surroundings. However, we know that iron sulfide actually forms as a brightly glowing product which must emit quite a lot of entropy in order to attain the original temperature of the starting substances (see Experiment 3.5). Where does all this excess entropy come from?

The energy $\mathcal{A} \cdot \Delta\xi$ is released during the reaction. It is easy to calculate $\mathcal{A}$ with the help of the chemical potentials. This energy can be used for anything we want, for example, if we manage to couple the reaction process in a galvanic cell with an electric current, which itself drives a motor, light bulb, electrolysis cell, etc. If, however, a mixture of iron and sulfur powder is ignited in the open, the released energy is used to generate entropy. This entropy is deposited in the end at temperature T which lies around 300 K. Therefore, according to $\mathcal{A} \cdot \Delta\xi = T \cdot \Delta S_g$, the amount of entropy *finally* generated, i.e., S_g, equals

$$\Delta S_g = \frac{\mathcal{A} \cdot \Delta\xi}{T} = \frac{(\mu(\text{Fe}) + \mu(\text{S}) - \mu(\text{FeS}))\Delta\xi}{T}$$
$$= \frac{\left(0 + 0 - \left(-102 \times 10^3\,\text{G}\right) \times 0.1\,\text{mol}\right)}{300\,\text{K}} = 34\,\text{Ct}.$$

The large value of the excess entropy mentioned above comes from the *generated entropy* S_g. What the temperatures in the interim are, or how the entropy actually comes to be generated, makes no difference to the final result.

The general rule is that all released energy remains unused, meaning it is used up to generate entropy. However, if the energy involved is not simply "burnt," but is used with an efficiency of η, say of 70 % (possibly by a galvanic cell driving a motor), then only $(1 - \eta)\mathcal{A}\Delta\xi$ is available for generating entropy:

$$S_g = \frac{(1 - \eta)\mathcal{A}\Delta\xi}{T} = \frac{(1 - 0.7) \times 102 \times 10^3\,\text{G} \times 0.1\,\text{mol}}{300\,\text{K}} = 10\,\text{Ct}.$$

In the ideal case of a complete use of energy ($\eta = 1$), the term S_g would disappear and the previous exothermic reaction would become endothermic. Galvanic cells, which allow such usage, were not actually developed for the reaction between iron and sulfur, but for the reaction of sulfur with sodium. We will discuss the design of such cells in electrochemistry, later on.

Accompanying Exchange of Energy We mentioned above that until the middle of the nineteenth century, certain characteristics were attributed to heat that correspond well to the properties attributed to entropy discussed in Chap. 3, with the exception that heat was considered a quantity that cannot be generated. Discoveries during the nineteenth century increasingly contradicted this assumption and finally led to a restructuring of the whole intellectual edifice. Since then, heat has been considered a special form of energy transfer. Formally, this rededication leads to the differential $T\mathrm{d}S_e$ or, more precisely, the integral $\int T\mathrm{d}S_e$ being called heat Q (rather than S):

$$\mathrm{d}Q = T\mathrm{d}S_e \ \text{ or } \ Q = \int_{\text{initial}}^{\text{final}} T\mathrm{d}S_e.$$

While the step from the quantity S to Q is comparatively easy, reversing this is very difficult. One reason is that we cannot see anymore on which path the energy arrived once it is in the system, similar to how we cannot tell the way a person arrived at work, whether on foot, by bicycle, in a car, etc., once he or she is in the office. Formally, this means that Q (differently from S) is not a state variable; it is not determined by the state of the system.

Another reason is that the (*thermodynamic*) *temperature* T can be defined easily based on entropy, but only in a temporary form and awkwardly without this concept. Without entropy, this definition is commonly achieved via the thermal expansion of gases. However, one usually forgoes demonstrating that the temperature θ defined this way actually corresponds to the thermodynamic T. Let us overlook this difficulty! When the equation $\mathrm{d}Q = T\mathrm{d}S_e$ above is solved for $\mathrm{d}S_e$ and then integrated, the result is:

$$\mathrm{d}S_e = \frac{\mathrm{d}Q}{T} \ \text{ and } \ \Delta S_e = \int_{\text{initial}}^{\text{final}} \frac{\mathrm{d}Q}{T}.$$

We have not reached our goal yet. What we are looking for is $\Delta S = \Delta S_e + \Delta S_g$ and not ΔS_e. Therefore, we must also make sure that $\Delta S_g = 0$. In order to assure this, the whole process must be *reversible* and this is indicated by the index $_{\text{rev}}$ for Q:

$$\mathrm{d}S = \frac{\mathrm{d}Q_{\text{rev}}}{T} \ \text{ and } \ \Delta S = \int_{\text{initial}}^{\text{final}} \frac{\mathrm{d}Q_{\text{rev}}}{T}.$$

The effects we are dealing with at the moment occur at constant temperature so that the relation between exchanged entropy ΔS_e and the accompanying transfer of energy or heat $Q = T\Delta S_e$ is very simple. If one wishes to discuss the accompanying energy exchange, it is enough in this case to multiply the corresponding entropies by T. Taking it a step further, and multiplying the balance of entropy (8.27) by T, we obtain the following expression:

$$\underbrace{T\Delta S_e}_{Q_e} = \underbrace{T\Delta S_\ell}_{Q_\ell} - \underbrace{T\Delta S_g}_{Q_g}.$$

$$Q_e = Q_\ell - Q_g \qquad \text{"exchanged"} = \text{"latent"} - \text{"generated heat."}$$

If we were to abide by the usual rules of terminology, we would only call the first element heat and use the symbol Q for it (without the index $_e$), even though the other contributions cause the same effects. We will overlook this limitation, though, because it means nothing to us and only creates an unnecessary obstacle.

Gibbs–Helmholtz Equations One of these equations (of which there are different versions) will serve here as an example for applying the balance of entropy discussed above. The most common version of this equation describes the relation between drive $\mathcal{A}$ and the heating effect that is observed when a reaction runs *freely* at constant p and T, meaning that the released energy $\mathcal{A} \cdot \Delta\xi$ is "burnt" without any use ($\eta = 0$), so that $\mathcal{A} \cdot \Delta\xi = T \cdot \Delta S_g = Q_g$. Again, we assume $\Delta\xi$ to be small. The fame of these equations can be traced to an error, the idea common at that time, that the emitted heat $-Q_e$ or the "heat tone" as it was called was a measure of the drive $\mathcal{A}$ of a reaction (Berthelot's principle, 1869). $Q_g = -Q_e + Q_\ell$ would have been correct, instead of $-Q_e$ alone. The error was not immediately seen because the latent heat Q_ℓ is generally considerably smaller than the observable heat effects. The attractiveness of this approach was that there existed simple calorimetric methods to measure Q_e. This gave people at that time a much simpler procedure for determining or at least estimating the drive $\mathcal{A}$ than had been possible until then. Before that, one had to be content with simply ranking the drives for certain reactions.

Therefore, there was good reason to measure these heats and to collect the data. The error was only gradually dealt with by the work of Josiah Willard Gibbs, Hermann von Helmholtz, and Jacobus Henricus van't Hoff who all showed that the "heat tone" $-Q_e$ itself did not represent the correct measure for the drive, but that a positive or negative contribution corresponding to the latent heats Q_ℓ (such as was known from phase transitions) had to be added:

$$\mathcal{A} \cdot \Delta\xi = \underbrace{-Q_e + Q_\ell}_{Q_g} \qquad \text{(a version of the Gibbs–Helmholtz equation)}.$$

8.8 Calorimetric Measurement of Chemical Drives

The concept is simple: In a reaction, the released energy $\mathcal{A} \cdot \Delta\xi$ is "burnt" during a certain conversion $\Delta\xi$ at given values of p and T and the generated entropy ΔS_g is then determined calorimetrically. Because $\mathcal{A} \cdot \Delta\xi = T \cdot \Delta S_g$, $\mathcal{A}$ is then easy to calculate from the measured data.

Unfortunately, the latent entropy becomes a problem because depending upon the type of substances, the entropy demands S_m and therefore the entropy content ΔS of the sample change in the reactor. A positive ΔS becomes noticeable as a negative contribution $-\Delta S$ in the calorimeter (Index *) so that not ΔS_g but $\Delta S^* = \Delta S_\mathrm{g} - \Delta S = -\Delta S_\mathrm{e}$ is measured there.

This value is only useful if it is possible to also determine ΔS in some way. This step is easy to imagine. One measures the entropy content S_1 of the sample before the reaction and then the value S_2 afterwards. The value being sought is the difference $\Delta S = S_2 - S_1$. In Sect. 3.9, we indicated how such a measurement might look. The fact that the sample must be cooled down to a temperature of just about absolute zero causes such entropy measurements to be technically complicated, comprising the greatest obstacle to this method. In fact, when possible, such entropy values are measured separately for the individual substances and then tabulated as molar entropies. The missing value for ΔS is then calculated from these tabular values.

How, then, is the first part of the measurement carried out, meaning how is $\Delta S^* = -\Delta S_\mathrm{e}$ determined? We might make use of one of the "ice calorimeters" described in Sect. 3.7 and deduce the entropy from the amount of melted water. More often, though, a calorimeter is used that determines the exchanged entropy through the small temperature changes in a water bath (or in a metal block). It is essentially made up of a container in which the reaction takes place and the water bath (or metal block) mentioned above as well as a sensitive thermometer (Fig. 8.8). The entire device is completely thermally insulated from the surroundings. The calorimeter must be calibrated before (or perhaps after) the actual measurement. For this, the measuring assembly receives a well-determined amount of entropy and the changes of temperature associated with this are measured. The easiest way to put entropy at a certain place is to generate it in an electric heating coil directly

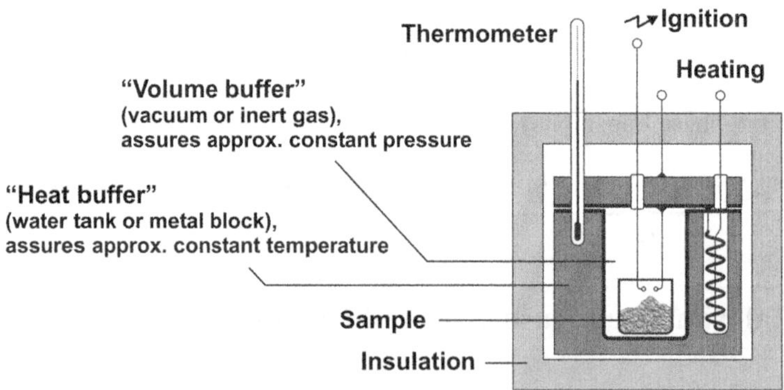

Fig. 8.8 Calorimeter.

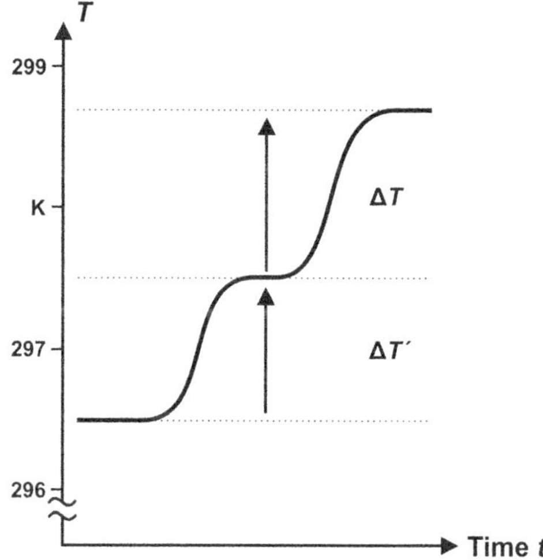

Fig. 8.9 Increases of temperature $\Delta T'$ and ΔT in the calorimeter as a result of the electrically generated entropy for calibrating $\Delta S'_g$ and the entropy $\Delta S_e = \Delta S_g - \Delta S$ emitted subsequently during the reaction.

where it is wanted. The expended electric energy W_b can be easily calculated from the current I, voltage U, and the power-on time Δt, according to

$$W_b = I \cdot U \cdot \Delta t. \tag{8.28}$$

The "burnt" energy, divided by the measured temperature T, results in the increase of entropy $\Delta S'_g$ that leads to the small temperature rise of $\Delta T'$ (Fig. 8.9). If we start a reaction in the container, it is possible to calculate the entropy $\Delta S^* = -\Delta S_e$ released by the sample, from the observed changes of temperature ΔT.

To make this procedure somewhat clearer, let us return one last time to our example of the reaction of iron and sulfur into iron sulfide at room conditions. Again, 0.1 mol each of iron and sulfur are used:

$$\text{Fe}|s + \text{S}|s \rightarrow \text{FeS}|s \quad \text{at } T = T^{\ominus} \text{ (influence of pressure is negligible)}.$$

- *Determining the latent entropy ΔS by measuring the entropy in the sample in its initial state (Fe + S) and final state (FeS):*
  Method: Cooling the sample down to approximately 0 K and measuring the entropy S needed for heating up to T, both before and after the reaction (or instead, calculating from the tabulated data obtained via the same procedure).
  Result: $\Delta S = 6.0$ Ct $- 2.7$ Ct $- 3.2$ Ct $= 0.1$ Ct.
- *Measuring the entropy $-\Delta S_e$ emitted by the sample during the reaction:*
  Calibration: Warming by $\Delta T' = 1.0$ K by electrically generated entropy $\Delta S'_g = 28.0$ Ct. A heater with a power of $P = 60$ W needs to be operated for a time span of $\Delta t = 139$ s for this: $\Delta S'_g = P \cdot \Delta t/T^{\ominus} = 60 \times 139/298$ Ct $= 28.0$ Ct.

<u>Measurement:</u> Warming by $\Delta T = 1.2$ K due to the entropy $-\Delta S_e$ emitted by the sample during the reaction.

<u>Analysis:</u> $\Delta T / \Delta T' = \Delta S^* / \Delta S^{*\prime} = -\Delta S_e / \Delta S'_g$ is valid for small changes; therefore,

$$\Delta S_e = -\frac{\Delta T}{\Delta T'} \Delta S'_g = -\frac{1.20\,\text{K}}{1.00\,\text{K}} 28.0\,\text{Ct} = -33.6\,\text{Ct}.$$

- *Summing up of the calorimetric partial results*:

  The entropy generated during the reaction, calculated from the balance of entropy $\Delta S = \Delta S_e + \Delta S_g$:

$$\Delta S_g = \Delta S - \Delta S_e = 0.1\,\text{Ct} - (-33.6\,\text{Ct}) = 33.7\,\text{Ct}.$$

Chemical drive, calculated from the relation $\mathcal{A} \cdot \Delta\xi = T \cdot \Delta S_g$:

$$\mathcal{A} = \frac{T \cdot \Delta S_g}{\Delta\xi} = \frac{298\,\text{K} \times 33.7\,\text{Ct}}{0.1\,\text{mol}} = 100\,\text{kG}.$$

Historically, this purely calorimetric method represents the first accessible way to determine chemical drives. At the same time, our example shows just how small the contribution of latent entropy ($\Delta S_\ell \equiv$) $\Delta S = 0.1$ Ct is, compared to the generated entropy $\Delta S_g = 33.7$ Ct, so it is no surprise that this small amount was overlooked at the beginning.

However, the situation changes as soon as gases are consumed or produced. We will look a little closer at an example, the reaction of oxyhydrogen (a mixture of hydrogen and oxygen gases):

$$
\begin{array}{llll}
& \text{H}_2|\text{g} \; + \; \tfrac{1}{2}\text{O}_2|\text{g} \;\; \to \;\; \text{H}_2\text{O}|\text{l} & \quad \text{at } T = T^\ominus,\; p = p^\ominus \\
\mu^\ominus(\text{kG}): & \quad 0 \qquad \tfrac{1}{2}\times 0 \qquad\; -237 & \quad \Rightarrow \mathcal{A}^\ominus \;\;\; = +237\,\text{kG}. \\
S^\ominus(\text{Ct mol}^{-1}): \; 131 & \quad \tfrac{1}{2}\times 205 \quad\; 70 & \quad \Rightarrow \Delta_R S^\ominus = -164\,\text{Ct mol}^{-1}.
\end{array}
$$

The following values for generated, latent, and exchanged entropies are the result of a conversion of $\Delta\xi = 0.1$ mol at standard conditions:

$$\Delta S_g = \frac{\mathcal{A} \cdot \Delta\xi}{T} = \frac{237 \times 10^3\,\text{G} \times 0.1\,\text{mol}}{298\,\text{K}} \qquad = +80\ \text{Ct},$$

$$\Delta S_\ell = \Delta_R S \cdot \Delta\xi = -164\,\text{Ct mol}^{-1} \times 0.1\,\text{mol} \quad = -16\ \text{Ct},$$

$$\Delta S_e = \Delta S_\ell - \Delta S_g \qquad\qquad\qquad\qquad\qquad = -96\ \text{Ct}.$$

In this case the latent entropy plays a significant role.

Chapter 9
Coupling

As we have seen in previous chapters, it is possible to act on a material system mechanically (by expansion and compression ...), thermally (by heating and cooling ...), and chemically (by addition or reaction of substances). All these actions are accompanied by changes of energy. These energy changes can be combined into a single equation, the so-called *main equation*.

The quantities appearing in this equation (such as p, V, S, T, $\mathcal{A}$, ξ), the *main quantities*, depend on each other in many ways. These dependencies can be described quantitatively by different coefficients such as $(\partial S/\partial T)_{p,\xi}$. Between the coefficients, there exist plenty of cross relations (couplings) which can be found in two ways: First, by using energy balances for appropriate cyclic processes and second as a direct result of a mathematical operation we call "flipping." Important couplings like the equivalence of the temperature coefficient of μ and the negative molar entropy are easily deduced by special "flip rules." We also discuss the relationship between coupling of main quantities and Le Chatelier–Braun's principle.

9.1 Main Equation

Introduction Homogeneous regions, in which pressure, temperature, and composition are the same throughout, form the basic building blocks of the systems that matter dynamics deals with. A region of this type is called a *phase* (Sect. 1.5). When two parts of such a phase are combined, the quantities such as volume V, entropy S, amount of substance n_i, etc., add up while pressure p, temperature T, the chemical potential μ_i, etc., remain the same. The first type of quantities are called extensive quantities and the second intensive quantities (Sect. 1.6). Not every quantity fits into one of these categories. For instance, V^2 and $\sqrt{S}$ are neither one nor the other and there are others like them.

© Springer International Publishing Switzerland 2016
G. Job, R. Rüffler, *Physical Chemistry from a Different Angle*,
DOI 10.1007/978-3-319-15666-8_9

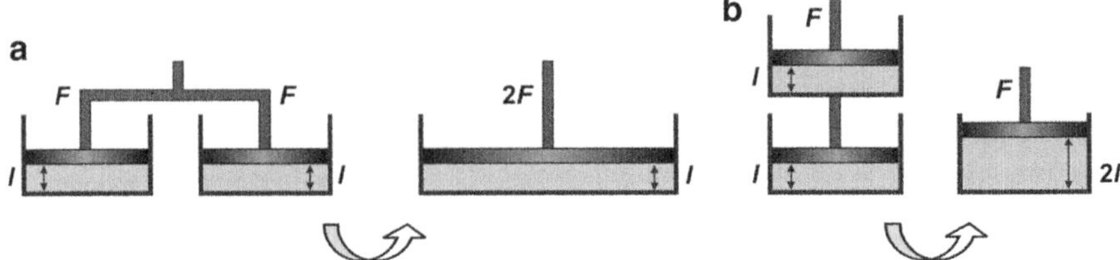

Fig. 9.1 Merging of two identical gas-filled regions (in *gray*), enclosed in cylinders with pistons: (**a**) Parallel connection: the force F is doubled, the length l remains constant, (**b**) Series connection: the force remains constant, the length is doubled.

Let us imagine a gas in a cylinder with a piston. The material system we observe is the gas and the concrete system we operate with is the gas-filled cylinder. Keeping the gas in the enclosure allows us to control the mechanical quantities p and V. We specify the force F upon the piston being pushed inside the cylinder, and a length l to quantify its position such as the distance between the piston and the bottom of the cylinder. We now imagine two identical gas-filled cylinders. How do the two quantities F and l behave when we merge the two systems into one? This is not initially clear and depends upon how this step is carried out (Fig. 9.1).

Similar problems appear in many other systems as well, such as in the galvanic cells which we will discuss in Chap. 23. In parallel and series connections, voltage U and transported charge Q behave like F and l in our example. The possibility of classifying quantities as either extensive or intensive is a distinctive feature of homogeneous systems and should not be thoughtlessly generalized.

Main Equation As we have seen in previous chapters, a material system can be affected mechanically (by expanding, compressing, ...), thermally (by heating, cooling, ...), chemically (by addition or reaction of substances...), etc. Each of these effects is related to changes of energy ΔW of the system involved. For example, energy is absorbed when a body is compressed and the more that it is absorbed, the greater the loss of volume $-\Delta V$ and the higher the pressure p (Sect. 2.5),

$$\Delta W = -p \cdot \Delta V.$$

Inflow of entropy also leads to an increase in the content of energy (Sect. 3.11) of

$$\Delta W = T \cdot \Delta S.$$

The same holds for the addition of a substance that results in absorption of energy according to

$$\Delta W = \mu \cdot \Delta n$$

(Sects. 4.8 and 8.6). As mentioned in Sect. 4.8, all of these changes of energy can be combined into a single equation, the so-called *main equation*. In a homogeneous region, where pressure p, temperature T, and the chemical potentials $\mu_1, \mu_2, \mu_3, \ldots$ are the same throughout, the following is valid for small changes of volume V, entropy S, and amounts of substance $n_1, n_2, n_3, \ldots$:

$$\Delta W = -p \cdot \Delta V + T \cdot \Delta S + \mu_1 \cdot \Delta n_1 + \mu_2 \cdot \Delta n_2 + \mu_3 \cdot \Delta n_3 + \ldots \quad \text{(Gibbs 1876)}. \tag{9.1}$$

The summation of the terms associated with the amounts of substance can be abbreviated with the help of the symbol for a sum, $\sum$:

$$\Delta W = -p \cdot \Delta V + T \cdot \Delta S + \sum_{i=1}^{n} \mu_i \cdot \Delta n_i. \tag{9.2}$$

This "main or fundamental equation" is the key to many important statements for systematically constructing the field of matter dynamics.

We are considering only loss-free systems that are able to fully release the energy put into them when the process is reversed. This means that no entropy may be generated inside the system, i.e., $S_g = 0$, so that the amount of entropy contained there can only vary by exchange from outside, $\Delta S = \Delta S_e$.

The quantities p, T, and n_i are considered easily measurable, so they are mostly chosen as independent variables. If a material system, possibly a body, is brought into contact with an electrical or magnetic field, or if it was charged or accelerated, etc., further terms would be added to the sum above; this we will ignore for the moment.

When no substances are exchanged with the surroundings but are only transformed inside the system, it is more advantageous to indicate its state by using the momentary extent ξ of a reaction in progress. The main equation then simplifies to

$$\Delta W = -p \cdot \Delta V + T \cdot \Delta S - \mathcal{A} \cdot \Delta \xi \quad \text{(De Donder 1920)}. \tag{9.3}$$

The condition of sufficiently small changes can be emphasized by using differentials. We would then obtain the following in the case of Eq. (9.3):

$$dW = -p \cdot dV + T \cdot dS - \mathcal{A} \cdot d\xi. \tag{9.4}$$

We already encountered another relation of this type, Eq. (8.21) in Sect. 8.7. There was a short discussion about concrete experimental conditions necessary for exchange of energy with the surroundings along all three paths. Unlike in Chap. 8, we will assume that it is possible for an exchange to occur without loss via the third path. This means that it always runs reversibly with no generation of entropy, i.e., the efficiency is $\eta = 1$. This also means that we can set the drive $\mathcal{A}$ and/or the extent ξ of the reaction in question by making use of appropriate aids.

This is similar to setting the pressure p through the force upon the piston or prescribing the volume V through its position in the cylinder. We consider $\mathcal{A}$ and ξ as adjustable parameters, like p and V or T and S, with whose help we can bring about certain states of a system and make changes over certain paths.

Main Quantities Certain mechanical $(-p, V)$, thermal (T, S), and chemical *main quantities* $(\mu, n$ or $-\mathcal{A}, \xi)$ appear on the right side of the main equations and are always pairs made up of an intensive quantity (without Δ- or d-symbol in front, and with proper signs $-p, T, \mu, -\mathcal{A}$) and a corresponding extensive, mostly substance-like quantity (with Δ- or d-symbols in front: V, S, n, ξ). One says that these two quantities are *conjugated* with respect to each other, or more precisely, they are energetically conjugated. The intensive quantity belonging to a substance-like quantity can also be considered the potential having an effect upon it. So understood, the chemical potential μ belongs to the amount of substance n, and the "thermal potential" T belongs to entropy S. We will encounter more such examples in the next chapters. We will see that calculations and descriptions are simple and clear when it is possible to formulate a problem using these main quantities. This is especially true in the case of heat effects. These are best calculated using entropies because it is the entropy accompanying an exchange of energy that actually qualifies it as "heat" (exchange of "disordered energy").

In order to simplify our terminology, we will allow volume V to be considered "substance-like." V represents an *improper* quantity of this kind with degenerate but simple characteristics: a conserved quantity with constant density equal to 1. Whatever volume a system loses, the surroundings gain. The negative pressure $-p$ would be the corresponding potential. The energy expended to increase a volume (volume "inflow") from the surroundings with a pressure of p_1 into the system with higher pressure p_2 is negative and amounts to $\Delta W = (-p_2 \cdot \Delta V) - (-p_1 \cdot \Delta V)$. The equation is completely analogous to those for transferring entropy or a substance:

$$\Delta W = [(-p_2) - (-p_1)] \cdot \Delta V, \quad \Delta W = [T_2 - T_1] \cdot \Delta S, \quad \Delta W = [\mu_2 - \mu_1] \cdot \Delta n.$$

We should not be too narrow in our choice of terminology if we want to discuss systems of more general type, i.e., those made up of more than just one phase, such as systems like the gas plus cylinder and piston, or a galvanic cell as a whole. If, for example, the term $-p \cdot dV$ is replaced by $-F \cdot dl$ or the term $-\mathcal{A} \cdot d\xi$ by $-U \cdot dQ$ (where U is the voltage and Q the charge), we obtain the following equation,

$$dW = -F \cdot dl + T \cdot dS - U \cdot dQ, \tag{9.5}$$

which is of the same nature as the one formulated before. It is the main equation for our new, expanded system. However, the terminology of extensive and intensive does not necessarily work with all the quantities in it now, at least not in the sense it is used today. We can refer to a term introduced by Hermann von Helmholtz for help (compare Sect. 2.7). He called the quantities on the right side of our main equation that represent differentials "(position) coordinates" and the factors in front "forces." Both were meant in a general sense. Referring back to this, we call the quantities l, S, Q in the main equation (9.5) "position-like," and F, T, and U, "force-like."

Main Effects When a main quantity is changed, this affects the conjugated quantities as well. The type of effect in which one observes the reciprocal dependency of two associated quantities is called *main effect*. The main effect of an increase

- Of volume V is a decrease of pressure, meaning an increase of $-p$ (note that it is not V and p that are conjugated, but V and $-p$ or $-V$ und p),
- Of entropy S is warming, meaning an increase of temperature T,
- Of an amount n of a (dissolved) substance is an increase of its chemical potential μ,
- Of the extent ξ of a reaction (in solutions) is a decrease of its drive, meaning there is an increase of $-\mathcal{A}$ (here, too, it is important to notice the sign).

Conversely, the effect of an increase of $-p$, T, μ, $-\mathcal{A}$ in the surroundings, is an increase of V, S, n, ξ in the system.

Coefficients quantitatively describe main effects. In the strict sense, such coefficients are differential quotients whose numerators and denominators contain conjugated main quantities. In order to refer to them more easily, we will call them and their offspring (which only differ by certain factors) *main coefficients*. If we allow only small changes, we can use difference quotients instead of differential quotients, which simplifies things. To give an example, the following coefficients, which we were introduced to in Sect. 3.9 as *entropy capacities*, can characterize the main effect of an increase of temperature on the De Donder system above:

$$\left(\frac{\partial S}{\partial T}\right)_{p,\xi}, \left(\frac{\partial S}{\partial T}\right)_{V,\xi}, \left(\frac{\partial S}{\partial T}\right)_{p,\mathcal{A}}, \ldots \quad \text{or} \quad \left(\frac{\Delta S}{\Delta T}\right)_{p,\xi}, \left(\frac{\Delta S}{\Delta T}\right)_{V,\xi}, \left(\frac{\Delta S}{\Delta T}\right)_{p,\mathcal{A}}, \ldots$$

The first coefficient describes the most common case, namely how much entropy ΔS flows in if the temperature outside and (also inside as a result of entropy flowing in) is raised by ΔT and the pressure p and extent ξ of the reaction are kept constant. In the case of the second coefficient, volume is maintained instead of pressure (this only works well if there is a gas in the system). In the case of $\mathcal{A} = 0$, the third coefficient characterizes the increase of entropy *during equilibrium*, for example when heating nitrogen dioxide (NO_2) (see also Experiment 9.3) or acetic acid vapor (CH_3COOH) (both are gases where a portion of the molecules are dimers). Multiplied by T, the coefficients represent *heat capacities* (the isobaric C_p at constant pressure, the isochoric C_V at constant volume, etc.). It is customary to relate the coefficients to the "size" of the system, possibly the mass or the amount of substance. The corresponding values are then presented in tables. In the case above, they would be tabulated as specific (mass related) or molar (related to amount of substance) heat capacities. The qualifier "isobaric" and the index $_p$ will only be used if necessary for clarity. As a rule, when these attributes are lacking, it is the isobaric coefficients that are being described:

$$C = T\left(\frac{\partial S}{\partial T}\right)_{p,\xi}, \qquad c = \frac{T}{m}\left(\frac{\partial S}{\partial T}\right)_{p,\xi}, \qquad C_{\mathrm{m}} = \frac{T}{n}\left(\frac{\partial S}{\partial T}\right)_{p,\xi}.$$

("global")   specific   molar   [isobaric] heat capacity.

Side Effects When a pair of conjugated main quantities is changed, there are always *side effects* to a greater or lesser degree affecting other main quantities. Almost all bodies tend to expand when entropy S is added to them. Volume V increases as long as the pressure is kept constant. If expansion is hindered, the pressure p inside will rise. In this case, S and V are *coupled in the same direction* which is expressed by $S \uparrow\uparrow V$. The relationship is reciprocal. When the volume is increased, the body also seeks to increase S by absorbing entropy from the surroundings. If this entropy absorption is obstructed, the temperature T will drop. Here, the material behaves similarly to a sponge in water as mentioned previously (Sect. 3.6). It "swells up" as it absorbs entropy and, when pressed, releases it again. There are very few exceptions to this. Ice water—meaning water between 0 and 4 °C—is the best example of these exceptions. The coupling here is *in the opposite direction, $S \uparrow\downarrow V$*. When ice water is pressed, it becomes even colder so that entropy begins to flow into it from the surroundings. When the pressure is released, the water becomes warmer and the entropy flows back out again.

This type of reciprocal relation between V and S is called a *mechanical–thermal* or in short a *V–S coupling*. All pairs of main quantities influence each other similarly. Change at one position almost always causes side effects somewhere else. A *coupling* of two processes in which one facilitates the other is said to go *in the same direction*. However, if one process impedes the other, we speak of a *coupling in the opposite direction*. It is always the behavior of the "position-like" partners in two main quantity pairs which is being compared and not the "force-like" ones.

Coefficients also can quantitatively describe side effects. However, in this case, the main quantities appearing in the numerator and denominator are not conjugated. Therefore, there is a large number of possible coefficients. For example, the first side effect mentioned above, the one having to do with S upon V, can be described by the following differential quotients in the upper line and the inverse one, V upon S, by the ones in the line below:

$$\left(\frac{\partial V}{\partial S}\right)_{p,n} \text{ or } \left(\frac{\partial p}{\partial S}\right)_{V,n}, \quad \text{but also} \quad \left(\frac{\partial V}{\partial T}\right)_{p,n} \text{ or } \left(\frac{\partial p}{\partial T}\right)_{V,n},$$

$$\left(\frac{\partial S}{\partial V}\right)_{T,n} \text{ or } \left(\frac{\partial T}{\partial V}\right)_{S,n}, \quad \text{but also} \quad \left(\frac{\partial S}{\partial p}\right)_{T,n} \text{ or } \left(\frac{\partial T}{\partial p}\right)_{S,n}.$$

There are numerous cross relations between these types of coefficients, which we will call *side coefficients*. We will deal with these relations in the following.

The law of conservation of energy requires the reciprocal influence to be symmetrical and certain coefficients describing these effects to be identical (*mechanical–thermal cross relation*). The same holds for all the side effects.

9.2   Mechanical–Thermal Coupling

Introductory Example   We will investigate the interactions between mechanical and thermal changes in a body. For the time being, we will ignore chemical changes (n or ξ, respectively, remain constant).

There are many cross relations. Here is one of the most important:

$$\left(\frac{\partial S}{\partial(-p)}\right)_{T,\xi} = \left(\frac{\partial V}{\partial T}\right)_{p,\xi} \quad \text{or} \quad \left(\frac{\partial S}{\partial p}\right)_{T,\xi} = -\left(\frac{\partial V}{\partial T}\right)_{p,\xi}. \tag{9.7}$$

This is one of the so-called *Maxwell relations*. The expression on the left lets us recognize its origin more easily. The one on the right shows the most common form. The coefficient on the far left of the upper line of equations describes entropy absorption during decompression (lowering of pressure); multiplied by T, this corresponds to the heat absorption. The one next to it on the right, divided by V, is the *cubic expansion coefficient γ*,

$$\gamma = \frac{1}{V}\left(\frac{\partial V}{\partial T}\right)_{p,\xi} \quad \text{``expansion coefficient,''} \tag{9.8}$$

which has been tabulated in many cases for pure substances. γ not only describes the relative volume increase of a body during heating, but also the amount of entropy emitted per unit of volume during compression as well.

Maxwell's Method   There are various methods for derivation of cross relations. We will look at the one that nineteenth-century Scottish mathematical physicist James Clerk Maxwell used. Although it is rather complicated, it best allows us to see the relation to the law of conservation of energy. To accomplish this, we will determine the balance of energy for the following cyclic process, by keeping all the changes small (the curves appearing as short straight lines) and inhibiting reaction (ξ constant):

1. Expansion of the body as a consequence of change of pressure by $\Delta(-p)$ at constant temperature, as required by the first coefficient, where the entropy $\Delta S = (\Delta S/\Delta(-p))_{T,\xi} \cdot \Delta(-p)$ is absorbed.
2. Heating of the body by ΔT at constant pressure $(p - \Delta p)$, as required by the second coefficient. The volume then increases by $\Delta V = (\Delta V/\Delta T)_{p,\xi} \cdot \Delta T$.
3. "Reversal" of step 1: Compression of the body by Δp, while the temperature is kept constant at $T + \Delta T$ and entropy ΔS is emitted.

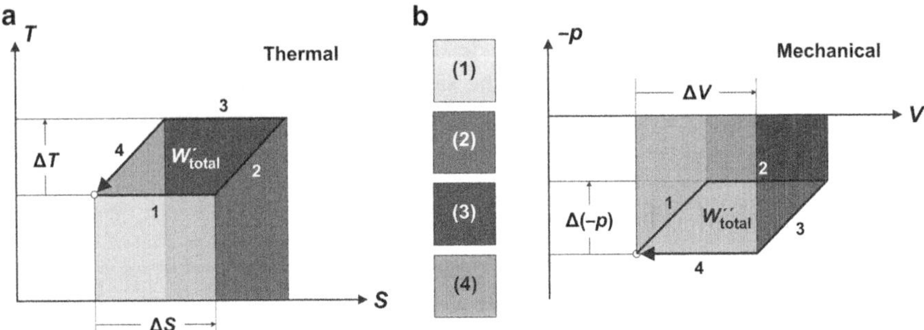

Fig. 9.2 (**a**) Cyclic process in a (T, S) or (**b**) in a $(-p, V)$ diagram. The individual steps are indicated by *various shades of gray*.

4. "Reversal" of step 2: Cooling of the body by ΔT at constant pressure, whereby V takes the original value again and the initial state is attained once more; the cycle is complete.

The process can now be schematically represented in a so-called (T, S) diagram where the temperature T and the entropy S are shown (Fig. 9.2a), as well as in a $(-p, V)$ diagram where negative pressure $(-p)$ and volume V appear (Fig. 9.2b).

The energy W' thermally added to the body at each step (left), as well as the energy W'' added mechanically (right), corresponds to the absolute value of the area below each piece of the curve through which the process runs. The amounts of energy are considered positive in motion in the direction of the x-axis, i.e., when S or V increases, and negative for movement in the opposite direction. The sign must be changed for areas below the x-axis. Let us use the diagram on the left to look more closely at this. The total amount of energy expended is made up of four contributions:

$$W'_{\text{total}} = W'_1 + W'_2 + W'_3 + W'_4.$$

W'_2 and W'_4 have the same absolute value, but different signs, so they cancel. Because of $W'_1 = T \cdot \Delta S$ and $W'_2 = -(T + \Delta T) \cdot \Delta S$, we have:

$$W'_{\text{total}} = T \cdot \Delta S - (T + \Delta T) \cdot \Delta S = -\Delta T \cdot \Delta S.$$

W'_{total} then corresponds to the area enclosed by the cycle in the (T, S) diagram, appearing as a parallelogram, which is to be considered negative here. Analogously, we can calculate the energy added mechanically, except that now W''_1 and W''_3 cancel each other out. We are left with the contribution in the second step that results in $(-p + \Delta(-p)) \cdot \Delta V$, as well as the one in the fourth step which equals $-(-p) \cdot \Delta V$:

$$W'_{\text{total}} = (-p + \Delta(-p)) \cdot \Delta V - (-p) \cdot \Delta V = \Delta(-p) \cdot \Delta V.$$

The thermally expended energy corresponds to $-\Delta T \cdot \Delta S < 0$, and the mechanically expended energy corresponds to $\Delta(-p) \cdot \Delta V > 0$. Because energy can neither be created nor destroyed, the sum of both contributions disappears, meaning that

$$-\Delta T \cdot \Delta S + \Delta(-p) \cdot \Delta V = 0 \quad \text{or written differently} \quad \Delta S/\Delta(-p) = \Delta V/\Delta T.$$

We can still add the indices—the pressure should be changed at constant temperature and the temperature should be changed at constant pressure, while ξ is kept constant—and replace the symbol Δ by either "d" or "∂," because we are only considering small changes. We finally obtain the sought-after Eq. (9.7):

$$\left(\frac{\Delta S}{\Delta(-p)}\right)_{T,\xi} = \left(\frac{\Delta V}{\Delta T}\right)_{p,\xi} \quad \text{and finally} \quad \left(\frac{\partial S}{\partial p}\right)_{T,\xi} = -\left(\frac{\partial V}{\partial T}\right)_{p,\xi}.$$

Flip Rule Cross relations can be found in two ways. First, by using energy balances for appropriate cyclic processes and second as a direct result of a mathematical operation we call "*flipping*." Naturally, the *flip rule* can be derived mathematically (see end of the Chapter), but we will give instructions here in the form of a kind of "recipe": Take the differential quotient in question (or difference quotient) you wish to transform and

1. Exchange numerator and denominator, simultaneously replacing the main quantities there with the corresponding partner in each case,
2. If the quantities in the numerator and denominator are of the same type, change the sign,
3. And insert as indices all the main quantities that appeared unpaired in the original expression (and the pairs of main quantities that were totally missing there).

The most important pairs of main quantities that change their partners in the first step are $-p \leftrightarrow V, T \leftrightarrow S, \mu_i \leftrightarrow n_i, -\mathcal{A} \leftrightarrow \xi, \ldots,$, where the algebraic sign should not be forgotten. It makes no difference here which partner it is assigned to. When we are talking about "main quantities of the same type" we mean that both are either "position-like" or that both are "force-like." "Unpaired" means that the corresponding partner is missing. The addition in parentheses in step 3 usually affects only rather exotic coefficients, so it can generally be ignored.

When applied to a side coefficient, the end result of this operation is a new differential quotient and therefore one of the cross relations of interest to us. Let us look at the approach using our concrete example. Again, we will start with the differential quotient $(\partial S / \partial (-p))_{T,\xi}$.

1. T is assigned to S and V is assigned to $(-p)$. Therefore, in the flipped differential quotient, T is in the denominator and V in the numerator.
2. The algebraic sign remains positive because T is a "force-like" quantity and V is a "position-like" quantity.
3. In the original expression, p and ξ were unpaired (because the corresponding partners V and $(-\mathcal{A})$ were missing) and therefore have to be inserted into the new index. Additional pairs should not be added because each pair of main quantities is represented by at least one quantity in the given coefficient.

$$\left(\frac{\partial S}{\partial(-p)}\right)_{T,\xi} \overset{1)}{\times} \left(\frac{\partial V}{\partial T}\right) \overset{2)}{\longrightarrow} \left(\frac{\partial V}{\partial T}\right) \overset{3)}{\longrightarrow} \left(\frac{\partial V}{\partial T}\right)_{p,\xi}.$$

The cross relation in question, which is one of Maxwell's relations, is here obtained directly through flipping. If a main coefficient is flipped, it is generally reproduced; there is no new information as we can easily see:

$$\left(\frac{\partial S}{\partial T}\right)_{p,\xi} \overset{1)}{\times} \left(\frac{\partial S}{\partial T}\right) \overset{2)}{\longrightarrow} \left(\frac{\partial S}{\partial T}\right) \overset{3)}{\longrightarrow} \left(\frac{\partial S}{\partial T}\right)_{p,\xi}.$$

9.3   Coupling of Chemical Quantities

The fact that pressure and temperature influence the chemical potential of substances and therefore the drive for transformations can be considered a result of coupling of mechanical, thermal, and chemical quantities. With rare exceptions, volume V or entropy S and the amounts of substance n_i are coupled in the same direction: $V \uparrow\uparrow n_i$ and $S \uparrow\uparrow n_i$. However, the couplings between V and ξ as well as between S and ξ exhibit no directional preference because starting and final substances can exchange their roles according to which conversion formula is chosen.

The most commonly used state variables in chemistry are T, p, n, or ξ. Therefore, the coefficients in which these quantities are used to characterize changes of state are the most important ones. We will discuss only these in the following. Cross relations will be written in two ways: In the first, they will be expressed with main quantities; in the second we will apply the previously used symbols.

***S–n* Coupling** Let us first turn to the temperature coefficient α of the chemical potential. We want to formulate it as a differential quotient: $(\partial\mu/\partial T)_{p,n}$. For practice, we will again show the process of "flipping" in detail:

1. n is assigned to μ and S is assigned to T. This means that in the flipped differential quotient, n is in the denominator and S is in the numerator.
2. The sign must change because both n and S are "position-like" quantities.
3. In the original expression, T and p are unpaired and therefore to be put into the new index:

$$\left(\frac{\partial\mu}{\partial T}\right)_{p,n} \quad\overset{1)}{\diagdown\!\!\!\diagup}\quad \left(\frac{\partial S}{\partial n}\right) \overset{2)}{\longrightarrow} -\left(\frac{\partial S}{\partial n}\right) \overset{3)}{\longrightarrow} -\left(\frac{\partial S}{\partial n}\right)_{T,p}.$$

This finally leads to:

$$\left(\frac{\partial\mu}{\partial T}\right)_{p,n} = -\left(\frac{\partial S}{\partial n}\right)_{T,p}. \tag{9.9}$$

The same procedure applied to a component (1) of a mixture of two substances (1 and 2, main equation $\mathrm{d}W = -p\mathrm{d}V + T\mathrm{d}S + \mu_1\mathrm{d}n_1 + \mu_2\mathrm{d}n_2$) results in:

$$\left(\frac{\partial\mu}{\partial T}\right)_{p,n_1,n_2} \quad\overset{1)}{\diagdown\!\!\!\diagup}\quad \left(\frac{\partial S}{\partial n_1}\right) \overset{2)}{\longrightarrow} -\left(\frac{\partial S}{\partial n_1}\right) \overset{3)}{\longrightarrow} -\left(\frac{\partial S}{\partial n_1}\right)_{T,p,n_2},$$

thus

$$\left(\frac{\partial\mu}{\partial T}\right)_{p,n_1,n_2} = -\left(\frac{\partial S}{\partial n_1}\right)_{T,p,n_2}. \tag{9.10}$$

These are further important cross relations analogous to those derived by James Clerk Maxwell. The expressions on the right are the molar entropies of the substance in a pure state S_m or in a mixture $S_{\mathrm{m},1}$. Positive S_m means that entropy tries to flow into a body along with any substance penetrating it, or that the temperature falls when the inflow of entropy is prevented. Since there are no walls that let a substance pass but not entropy, the effect cannot be observed directly. It becomes noticeable, though, when a substance is formed inside the body. This can only occur when another substance disappears, so there are always two or more effects which are superimposed additively or subtractively. Raising temperature not only increases S, but promotes absorption of substances, i.e., lowers the potential, as long as we do not allow matter to flow in. The conclusions of cross relations are simple, but possibly surprising. Along with the entropy demand of a substance, the molar entropy also describes the negative temperature coefficient α of its chemical potential:

$$\alpha = -S_\mathrm{m} \quad \text{or more generally} \quad \alpha_i = -S_{\mathrm{m},i}. \tag{9.11}$$

This relation easily results in the fact that α is always negative for pure substances and almost always negative for dissolved substances. This is because the molar entropy is just about always positive. Moreover, the molar entropy of a liquid is greater than for a solid and the molar entropy of a gas is much greater than that of a liquid, which leads to the ordering

$$\alpha(B|g) \ll \alpha(B|l) < \alpha(B|s) < 0$$

presented in Sect. 5.2. The form of the $\mu(T)$ curve in Fig. 5.1 can now be better justified by theoretical means. It begins with a horizontal tangent as long as the entropy content for $T = 0$ disappears like it should according to the third law of thermodynamics, and then falls more and more steeply. This drop is steeper for gases and substances in a dilute solution than for pure condensed substances (solids and liquids) because of the greater amount of entropy contained in them.

By means of the relation (9.10) we can also derive Eq. (8.10). The chemical potential of a dissolved substance is described by the mass action equation 1 [Eq. (6.4)] (or one of its variants):

$$\mu = \mu_0 + RT \ln \frac{c}{c_0}.$$

Its derivative with respect to T,

$$\left(\frac{d\mu}{dT}\right)_{p,n} = \left(\frac{d\mu_0}{dT}\right)_{p,n} + R \ln \frac{c}{c_0},$$

represents the negative molar entropy of the substance in question as long as the pressure and the amount of substance are held constant. Consequently, we obtain

$$-S_m = -S_{m,0} + R \ln \frac{c}{c_0} \quad \text{or} \quad S_m = S_{m,0} - R \ln \frac{c}{c_0}.$$

which corresponds to Eq. (8.10).

S–ξ Coupling Starting from the main equation (9.4), $dW = -pdV + TdS - \mathcal{A}d\xi$, we can use the flip rule to convert the temperature coefficient α of the drive of a reaction or any other type of transformation of substances:

$$\left(\frac{\partial(-\mathcal{A})}{\partial T}\right)_{p,\xi} = -\left(\frac{\partial S}{\partial \xi}\right)_{T,p} \quad \text{or} \quad \left(\frac{\partial \mathcal{A}}{\partial T}\right)_{p,\xi} = \left(\frac{\partial S}{\partial \xi}\right)_{T,p}. \tag{9.12}$$

The temperature coefficient α corresponds to the molar reaction entropy $\Delta_R S$ introduced in Sect. 8.5, or more generally, to the corresponding entropy $\Delta_{\rightarrow} S$ of a transformation:

Experiment 9.1 *Changing the position of equilibrium*:
If we mix equal weights of solid NaOH and water, the
resulting solution reaches about 100 °C. The process is
strongly exothermic. On the other hand, in a saturated
solution at 100 °C, the proportion of $n(\text{NaOH})$ to $n(\text{H}_2\text{O})$
is about three times that at 25 °C. The solubility of
NaOH strongly increases with rising temperature. While
the formation of a diluted or concentrated solution of
water and solid NaOH runs exothermically, dissolving
NaOH in an (almost) saturated sodium hydroxide
solution is an endothermic (endotropic) process. Only
this effect is responsible for changing the position of
equilibrium.

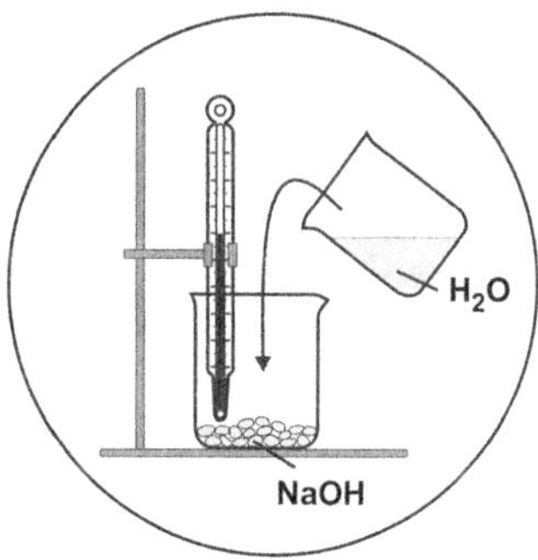

$$\alpha = \Delta_\text{R} S \ \text{ or } \ \alpha = \Delta_\rightarrow S. \tag{9.13}$$

This shows us that the drive of an endotropic reaction ($\Delta_\text{R} S > 0$) increases with rise
of temperature, $\alpha > 0$. If the reaction is in equilibrium, $\mathcal{A} = 0$, then a rise in
temperature leads to the drive $\mathcal{A}$ becoming positive and pushing the process further
forward in favor of the products. This is especially true for common phase transitions such as melting, boiling and sublimation, all of which are endotropic. The
opposite holds for exotropic reactions. If the initial drive is zero (equilibrium), it
will become negative as T increases and the equilibrium moves in the direction of
the starting substances.

The drive disappears when equilibrium is reached, so the terms endotropic and
endothermic as well as exotropic and exothermic all overlap (compare Sect. 8.7)
and become interchangeable. However, this only holds for the equilibrium state and
can lead to mistakes when applied elsewhere. Equilibrium constants for example
are calculated from data valid for (often idealized) states that are mostly nowhere
near equilibrium. It is rather amazing to find that the heat effects observed or
theoretically calculated under these circumstances actually result in useful, if
limited, conclusions about change of equilibrium conditions.

An explanatory example might be useful here (Experiment 9.1).

We could have used the concept of coupling without performing any calculations to correctly predict the *qualitative* behavior, i.e., to say whether heating causes
the drive to increase or decrease or the equilibrium to shift forward or backward.
Let us imagine an endotropic reaction. S and ξ would then be coupled in the same
direction ($S \uparrow\uparrow \xi$). If ξ increases, S does as well and when S increases, so does ξ. If
an equilibrium exists it will shift forward when heating takes place (increase of S).

The coupling is not rigid, though, and a change to the second quantity as a result
of change to the first can be overridden by other effects. A reaction can be blocked,
forcing ξ to remain constant. In this case, the addition of entropy (heating) will lead
to an increase of $\mathcal{A}$ so that, although ξ itself does not increase, the drive for this
change becomes stronger. We conclude that heating intensifies the drive of an
endotropic reaction.

Something similar holds for the reverse effect. Inflow of entropy can be blocked so that S cannot change. An increase of ξ would then become noticeable in a lowering of temperature because the missing entropy cannot be replaced.

V–n Coupling The pressure coefficient β of the chemical potential can be derived analogously to the temperature coefficient α. Using the main equation for a pure phase $dW = -pdV + TdS + \mu dn$, and by flipping, we obtain

$$\left(\frac{\partial \mu}{\partial(-p)}\right)_{T,n} = -\left(\frac{\partial V}{\partial n}\right)_{T,p} \tag{9.14}$$

or, based on the main equation for a mixture of two substances (1 and 2), $dW = -pdV + TdS + \mu_1 dn_1 + \mu_2 dn_2$, we have

$$\left(\frac{\partial \mu}{\partial(-p)}\right)_{T,n_1,n_2} = -\left(\frac{\partial V}{\partial n_1}\right)_{T,p,n_2}. \tag{9.15}$$

These are two forms of a further cross relation. If the volume of a body grows when a substance is added to it, the molar volume V_m $(= \partial V/\partial n)_{p,T}$ is positive. Raising the pressure will therefore impede the absorption of the substance, i.e., more energy will be needed for this. Correspondingly, the potential rises. V_m does not only indicate the volume demand of a substance but also the pressure coefficient of its potential:

$$\beta = V_m \quad \text{or more generally} \quad \beta_i = V_{m,i}. \tag{9.16}$$

The molar volumes of pure substances are fundamentally positive, and those of dissolved substances almost always are. Therefore, the pressure coefficient of the chemical potential is almost always positive. The molar volume of gases is, as we have discussed (Sect. 8.2), about a factor of 1,000 times greater than that of condensed phases (liquids and solids). The molar volume for the liquid phases of most substances is greater than that of the solid phase. This all results in the ordering discussed in Sect. 5.3:

$$0 < \beta(\text{B}|\text{s}) < \beta(\text{B}|\text{l}) <<<< \beta(\text{B}|\text{g}).$$

In regions of solid and liquid matter, where molar volume depends very little upon pressure ($V_m \approx$ const.), μ increases almost linearly with p. This is different from gases where the $\mu(p)$ curve runs almost logarithmically and much more steeply (Fig. 5.7).

V–ξ Coupling The pressure coefficient β of the drive $\mathcal{A}$ results from the same pattern as the temperature coefficient α. We start with the same main equation for both: $dW = -pdV + TdS - \mathcal{A}d\xi$. By flipping we obtain:

$$\left(\frac{\partial(-\mathcal{A})}{\partial(-p)}\right)_{T,\xi} = -\left(\frac{\partial V}{\partial \xi}\right)_{T,p} \quad \text{or} \quad \left(\frac{\partial \mathcal{A}}{\partial p}\right)_{T,\xi} = -\left(\frac{\partial V}{\partial \xi}\right)_{T,p}; \tag{9.17}$$

this means that the pressure coefficient β equals the negative molar reaction volume $\Delta_R V$,

$$\beta = -\Delta_R V. \tag{9.18}$$

Assume that the volume increases during a reaction ($\Delta_R V > 0$). Then its drive $\mathcal{A}$ weakens if we impede expansion by raising the pressure. If equilibrium has been established in a reaction of this type and therefore $\mathcal{A} = 0$, an increase of pressure causes $\mathcal{A}$ to become negative. The process then begins to run backward.

We can obtain the same result qualitatively and without calculations if we consider V and ξ to be coupled. If the coupling is in the same direction, i.e., $V\!\uparrow\!\uparrow\!\xi$, a rise in pressure will decrease the volume and ξ along with it, as long as the process is not blocked.

n–n Coupling   Basically, a coupling in the same or in opposite directions can exist between any two "position-like" quantities. Experiment 9.2 shows an effect based upon coupling of two amounts of substances n_1 and n_2, more precisely upon their coupling in the opposite direction ($n_1 \uparrow\downarrow n_2$).

An example of coupling in the same direction ($n_1 \uparrow\uparrow n_2$) would be a small amount of undissolved $PbCl_2$ in a beaker with water that dissolves when KNO_3 is added ("salting-in effect"). The first substance's rise in potential can be used to measure the strength of reciprocal action caused by the second substance. This is the so-called *displacement coefficient* $(\partial\mu_1/\partial n_2)_{T,p,n_1}$. The opposite effect, the displacement of the second substance by the first, which we describe by $(\partial\mu_1/\partial n_2)_{T,p,n_1}$ is just as great, as we can see by applying the flip rule:

Experiment 9.2
Precipitation of table salt resulting from the addition of acetone: When drops of acetone are added to an almost saturated aqueous solution of table salt, the salt begins to precipitate but as the acetone evaporates it begins dissolving again.

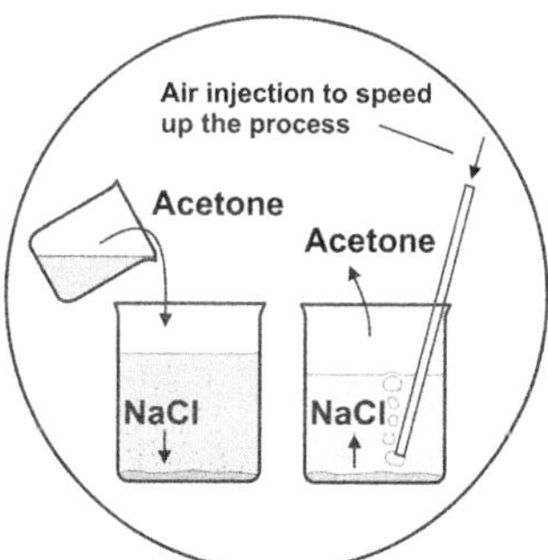

$$\left(\frac{\partial \mu_1}{\partial n_2}\right)_{T,\,p,n_1} = \left(\frac{\partial \mu_2}{\partial n_1}\right)_{T,\,p,n_2}. \tag{9.19}$$

The reverse effect can easily be shown using our first example: When table salt is added to a 1:1 mixture of acetone and water, the acetone separates into a second layer on top of the solution ("salting-out effect," compare Experiment 13.5).

We conclude that we can directly obtain all kinds of cross relations by flipping. The flip rule can be considered a memory aid for all such relations. Using it in this way is advantageous:

- Coefficients that are difficult to measure can be calculated from more easily accessible ones.
- Many coefficients do not need their own symbols; in the literature, we often find $\alpha, \beta, \alpha, \beta$ replaced by $-S_\mathrm{m}, V_\mathrm{m}, \Delta_\mathrm{R}S, -\Delta_\mathrm{R}V$ in the final results.

Le Chatelier–Braun's Principle or "Principle of Mobile Equilibrium" This "principle" was presented already at the end of the nineteenth century and is often mentioned in connection with our current discussion. Chemists use it to predict whether the extent ξ of a reaction in equilibrium will be shifted forward or backward when certain parameters are changed, especially pressure, temperature, amounts of reactants and products, etc. We have found our answers in what we have already discussed, so we can avoid using this "principle." For completeness sake, and because of the problems it causes even today, we will deal with it briefly. We will choose one of the various ways it can be formulated which is close to its original version, but we will avoid misunderstandings by using explanatory additions [in square parentheses]:

When a system at equilibrium is disturbed by changing one of the ["force-like"] equilibrium parameters [possibly by compressing, i.e., by an increase in pressure due to decreasing volume], the system responds with a change of state [$\Delta \xi$], in which the parameter changes in the opposite direction [this means that as long as the volume is kept constant during this step, the pressure goes down] and a new equilibrium is established.

This formulation of Le Chatelier–Braun's principle discusses how the pair of main quantities $(-\mathcal{A}, \xi)$ interacts with another such pair, here $(-p, V)$ in the special case of $\mathcal{A} = \text{const.}$ (or even $\mathcal{A} = 0$, which is unnecessarily special).

Let us consider the reaction between nitrogen gas and hydrogen gas to synthesize ammonia:

$$\mathrm{N_2|g} + 3\,\mathrm{H_2|g} \rightleftarrows 2\,\mathrm{NH_3|g}.$$

There are more gas molecules on the left-hand side of the conversion formula (4 in total) than on the right-hand side (2). Thus, compressing the system causes the equilibrium position to shift in the direction that leads to a reduction in the number of particles in the gas phase, as this tends to reduce the pressure and therefore to minimize the effect of compression. Consequently, a higher yield of ammonia is

Experiment 9.3 *Equilibrium between nitrogen dioxide and dinitrogen tetroxide*: A sealed tube containing a slightly brown mixture of NO_2 and N_2O_4 is submerged in a hot water bath. The color of the gas mixture becomes darker. However, if the tube is placed in an ice bath, the gas mixture becomes lighter in color.

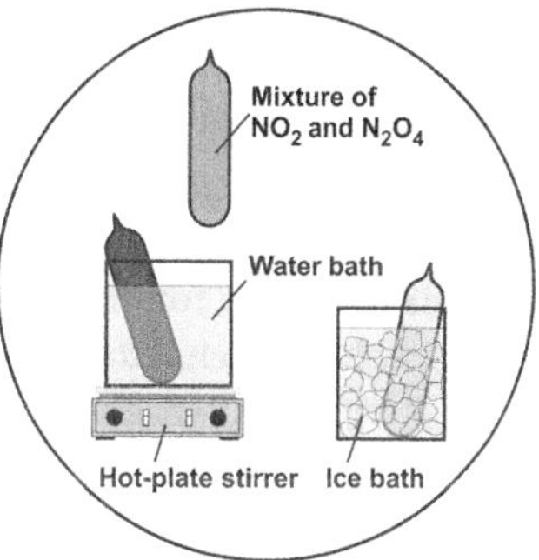

obtained. This effect was actually the solution to Fritz Haber's dilemma: Only when he carried out the synthesis at high pressure, he succeeded finally in 1909 in producing ammonia with sufficiently high yields.

If we imagine replacing the pair $(-p, V)$ by (T, S) or (μ, n), we could reformulate the corresponding parts of the text above:

When a system at equilibrium is disturbed by adding entropy to it so that the temperature rises, the system responds with a change of state $\Delta\xi$ where the temperature decreases (as long as no more entropy is allowed to be added).

When a system at equilibrium is disturbed by raising the chemical potential of one of the substances participating in the reaction by adding it from outside, the system responds with a change of state $\Delta\xi$ in which the potential decreases (as long as we impede that more substance comes in).

For the first case above, this means that raising the temperature shifts the equilibrium of an endotropic reaction in the direction of increased ξ. In the case of an exotropic reaction, this happens in the opposite direction. We have seen this before. But let us have again a look at an example, an equilibrium mixture between brown nitrogen dioxide and colorless dinitrogen tetroxide (Experiment 9.3) (see also Sect. 9.1):

$$2\,NO_2|g \rightleftarrows N_2O_4|g$$

The reaction is exotropic $\left(\Delta_R S^{\ominus} = -176\,Ct\,mol^{-1} < 0\right)$. If the temperature is raised, the equilibrium shifts in the direction of reduced ξ meaning in the direction of the reactant; a decrease in temperature, however, favors the formation of the final product.

The second case above [pair (μ, n)] does not show us anything we do not already know, either. As long as a reactant in a system is in a dissolved state, adding to it increases its chemical potential, driving the reaction forward. Adding more of a product to the system drives the reaction backward.

Recapulating, we can formulate in short: When a constraint is applied to a system in equilibrium, the equilibrium will shift so as to counteract the effect of the constraint. Constraint means here the change of a "force-like" (not a "position-like") parameter.

We are making only qualitative statements here, so it is naturally possible to replace one quantity with another one proportional to it or one that changes in the same direction. Pressure can be replaced by density, the chemical potential by concentration, the added entropy by the added heat or by a corresponding increase of internal energy or another appropriate state quantity. Trying to include all these possibilities makes formulating this principle difficult and the formulations become opaque. It has also led to phenomena being included that appear similar but do not belong here and to examples where the "principle" appears to fail.

Coupling of Several Pairs of Quantities Predicting behavior becomes more difficult when dealing with a coupling not just between two pairs of quantities, but maybe between three such pairs. When a dissolved substance is added, the solvent is also added at the same time so that the potentials of all other dissolved substances change. We saw such behavior in Sect. 6.6 with the formation of the red iron thiocyanate complex (Experiment 6.1):

$$\left[\mathrm{Fe(H_2O)_6}\right]^{3+}|\mathrm{w} \,+\, 3\,\mathrm{SCN^-}|\mathrm{w} \rightleftarrows \left[\mathrm{Fe(H_2O)_3(SCN)_3}\right]|\mathrm{w} \,+\, 3\,\mathrm{H_2O}|\mathrm{l}.$$

Adding the starting substances $\left[\mathrm{Fe(H_2O)_6}\right]^{3+}|\mathrm{w}$ and $\mathrm{SCN^-}|\mathrm{w}$ drives the reaction forward. Adding water drives it backward because the decrease of potential caused by dilution with water is four times as strong on the left than on the right. If a too strongly diluted solution of $\mathrm{Fe^{3+}}$ is added, the second effect can override the first so that the solution becomes paler red instead of deeper red, as would be expected.

The effect is similar in gas reactions when they are carried out at constant pressure. Addition of any kind of gas increases the volume, causing a diluting effect upon all the other gases. This diluting effect can override the increase of potential caused by addition of a reactant and the expected increase of the drive $\mathcal{A}$ along with it, driving the process backward and not forward. For example, in the case of ammonia synthesis,

$$\mathrm{N_2}|\mathrm{g} + 3\,\mathrm{H_2}|\mathrm{g} \rightleftarrows 2\,\mathrm{NH_3}|\mathrm{g},$$

addition of $\mathrm{N_2}$ leads to a higher yield of $\mathrm{NH_3}$, if the molar fraction of nitrogen in the reaction volume $x(\mathrm{N_2})$ is smaller than $1/2$, and leads to a decrease if $x(\mathrm{N_2})$ is greater than $1/2$.

9.4 Further Mechanical–Thermal Applications

In closing, we will apply what we have learned in this chapter to mechanical–thermal phenomena. We will consider only the simplest case, a body at rest at isotropic pressure p having a uniform temperature T, consisting of just one substance, such as a drop of water. The main equation for this is:

$$dW = -p \cdot dV + S \cdot dT. \tag{9.20}$$

Previously, we have discussed two coefficients from this field, two varieties of entropy capacity $\mathcal{C}$ (Sect. 3.9),

$$\mathcal{C} = \left(\frac{\partial S}{\partial T}\right)_p, \qquad \mathcal{C}_m = \frac{1}{n}\left(\frac{\partial S}{\partial T}\right)_p,$$

("global")   molar entropy capacity

and the "cubic expansion coefficient" γ (Sect. 9.2),

$$\gamma = \frac{1}{V}\left(\frac{\partial V}{\partial T}\right)_p. \tag{9.21}$$

We will not need the energetic version of entropy capacity, heat capacity $C = T \cdot \mathcal{C}$, which itself comes in different variants [Eq. (9.7)].

$\mathcal{C}$ is an extensive quantity that affects an entire body. $\mathcal{C}_m$ and γ are intensive quantities, properties of the substance making up the body. Another important property is *compressibility* χ,

$$\chi = -\frac{1}{V}\left(\frac{\partial V}{\partial p}\right)_T. \tag{9.22}$$

It describes how easy it is to compress a material. It is especially high for gases, which are easily compressed. Although it is, of course, possible to form other differential quotients, as well, the three coefficients $\mathcal{C}_m$, γ, and χ suffice for calculating all the first derivatives of the main quantities or coefficients made up of them.

Conversion of Differential Quotients   We will need some more computational rules for differential quotients. There are essentially four of which some will remind us of the usual rules of arithmetic with fractions and are therefore easy to remember. We will present them here briefly and without deriving them and show some examples of how they can be used.

(a)   *Inverting* a differential quotient:

$$\left(\frac{\partial p}{\partial q}\right)_{r\ldots} = 1 \Big/ \left(\frac{\partial q}{\partial p}\right)_{r\ldots}.$$

Numerator and denominator are exchanged, like with fractions, and the index remains unchanged.

(b) *Expanding* a differential quotient with a new quantity (here it is s):

$$\left(\frac{\partial p}{\partial q}\right)_{r...} = \left(\frac{\partial p}{\partial s}\right)_{r...} \cdot \left(\frac{\partial s}{\partial q}\right)_{r...}.$$

The differential quotient is expanded like a fraction with ∂s; the indices are the same in all expressions.

(c) *Insertion* of a quantity (here it is r) from the index of a differential quotient:

$$\left(\frac{\partial p}{\partial q}\right)_{r...} = -\left(\frac{\partial p}{\partial r}\right)_{q...} \cdot \left(\frac{\partial r}{\partial q}\right)_{p...}.$$

The differential quotient is expanded with ∂r, and the sign is changed; in the index the quantity missing from the complete set $p, q, r, \ldots$ is inserted.

(d) *Replacement* of a quantity in the index of a differential quotient with a new one:

$$\left(\frac{\partial p}{\partial q}\right)_{r...} = \left(\frac{\partial p}{\partial q}\right)_{s...} + \left(\frac{\partial p}{\partial s}\right)_{q...} \cdot \left(\frac{\partial s}{\partial q}\right)_{r...}.$$

In order to replace r with s in the index, the differential quotient is written with the changed index. The original expression that was expanded with ∂s is added as a "correction." The new quantities $s, q, \ldots$ appear as independent variables in the first term and the old $q, r, \ldots$ in the second term.

For these computational rules to be valid, all the differential quotients appearing on the left and right sides must make sense. This means that the quantities in the numerator must really be differentiable functions of the variables appearing in the denominator and index.

Of the four main quantities $-p, V, T, S$ mentioned above, $-p$ and T are most easily controlled ($-p$ or p because one often works in the laboratory with containers which are open to the atmosphere) so they are the preferred independent variables. These "preferred" quantities appear in the denominator or index of the differential quotients that, possibly multiplied by certain factors, can be found in Tables. The three coefficients $C_{\mathrm{m}}, \gamma, \chi$ mentioned above, are of this type. Consequently, we will attempt to convert a given coefficient so that only "preferred" quantities (in this case, p and T) appear in the denominator or index of the differential quotient, but never in the numerator. If we abbreviate all the other quantities with $a, a', \ldots$ and the "preferred" ones with $b, b', \ldots$ (here we have only two of these, but the method remains the same even when there are more), the given differential quotients are to be replaced by some of the type $(\partial a / \partial b)_{b',...}$.

First, attention should be paid to the quotient itself. This defines the first step:

$$\left(\frac{\partial a}{\partial b}\right)_{\dots}, \quad \left(\frac{\partial b}{\partial a}\right)_{\dots}, \quad \left(\frac{\partial a}{\partial a'}\right)_{\dots}, \quad \left(\frac{\partial b}{\partial b'}\right)_{\dots}.$$
$$\text{remains,} \quad \text{inverting,} \quad \text{expanding,} \quad \text{inserting.}$$

We "expand" with a quantity b which does not appear in the index, and "insert" a quantity a from the index. Two new differential quotients are created by this. The second of these is inverted, resulting in all the quotients now having the desired form $(\partial a / \partial b)\dots$. If an undesired quantity a is still in the index of one of the new expressions, it can be replaced by one of the "preferred" $b, b', \dots$. This process is then repeated until all the expressions have the form $(\partial a / \partial b)_{b',\dots}$.

Let us look at this approach using a concrete example, the relative pressure coefficient β_r, which is a measure of how steeply pressure rises when a body is heated at constant volume:

$$\beta_r = \frac{1}{p}\left(\frac{\partial p}{\partial T}\right)_V. \tag{9.23}$$

It can be expressed by the cubic expansion coefficient γ and the compressibility χ,

$$\beta_r = \frac{\gamma}{p\chi}, \tag{9.24}$$

where the following conversions led to this result:

$$\beta_r = \frac{1}{p}\left(\frac{\partial p}{\partial T}\right)_V = -\frac{1}{p}\left(\frac{\partial p}{\partial V}\right)_T\left(\frac{\partial V}{\partial T}\right)_p = -\frac{1}{p}\left(\frac{\partial V}{\partial T}\right)_p\left(\frac{\partial V}{\partial p}\right)_T^{-1} = -\frac{1}{p}\cdot\frac{V\gamma}{-V\chi}$$
$$= \frac{\gamma}{p\chi}.$$

A few words of explanation: The initial quotient has the form $(\partial b / \partial b')_a$. The undesired quantity a in the index is inserted into the quotient leading to a negative sign. The first of the two new quotients of the type $(\partial b / \partial a)\dots$ is inverted. All of the differential quotients then have the desired form. What now remains to be done is to replace the expressions with the usual coefficients [Eqs. (9.21) and (9.22)].

We will calculate the difference of the two entropy capacities $\mathcal{C}_p - \mathcal{C}_V$ as our second example. These are the usual isobaric $\mathcal{C} = \mathcal{C}_p$ (at constant pressure) and the more rarely used isochoric capacity $\mathcal{C}_V$ (at constant volume). The latter must be smaller than $\mathcal{C} = \mathcal{C}_p$ (as implied in Sect. 3.9), because the absorption of entropy is made more difficult if the change of volume (positive or negative) related to it is impeded. It makes no difference whether V and S are coupled in the same or in the opposite direction, i.e., $V \uparrow\uparrow S$ or $V \uparrow\downarrow S$, $\mathcal{C}_p \geq \mathcal{C}_V$ is always valid. To calculate the difference

$$C_p - C_V = \left(\frac{\partial S}{\partial T}\right)_p - \left(\frac{\partial S}{\partial T}\right)_V,$$

we have to replace the index $_V$ by $_p$ in the second differential quotient and to delete the terms that cancel:

$$C_p - C_V = \left(\frac{\partial S}{\partial T}\right)_p - \left[\left(\frac{\partial S}{\partial T}\right)_p + \left(\frac{\partial S}{\partial p}\right)_T \cdot \left(\frac{\partial p}{\partial T}\right)_V\right] = -\left(\frac{\partial S}{\partial p}\right)_T \cdot \left(\frac{\partial p}{\partial T}\right)_V.$$

We could stop the calculations at this point because both differential quotients are now known to us: According to Eqs. (9.7) and (9.8), $(\partial S/\partial p)_T$ corresponds to $-V \cdot \gamma$; on the other hand, according to Eqs. (9.23) and (9.24), $(\partial p/\partial T)_V$ corresponds to γ/χ. We obtain for the difference of the entropy capacities:

$$C_p - C_V = V\frac{\gamma^2}{\chi}. \tag{9.25}$$

We also could have just continued to obtain the following result in which all the differential quotients take the desired form $(\partial a/\partial b)_{b',\ldots}$:

$$C_p - C_V = -\left(\frac{\partial S}{\partial p}\right)_T \left(\frac{\partial V}{\partial p}\right)_T^{-1} \left(\frac{\partial V}{\partial T}\right)_p.$$

We know that some differential quotients can be converted into more easily measured forms by flipping them. We need only to check those that represent side coefficients quantifying some effects of coupling. Here, this would mean the first and the last quotients. Let us take a look at the results of this operation

$$\left(\frac{\partial S}{\partial p}\right)_T = \left(\frac{\partial(-V)}{\partial T}\right)_V \quad \text{and} \quad \left(\frac{\partial V}{\partial T}\right)_p = \left(\frac{\partial S}{\partial(-p)}\right)_T.$$

It is only worth flipping the differential quotient on the left but not the one on the right because this operation is just the inverse of the first one. The second-to-last step of this systematic operation for calculating a given differential quotient is to check whether or not the result can be improved by flipping. The very last step is then to replace the remaining differential quotients with the pertinent coefficients.

By the way, the flip rule itself can be derived by the four computational rules mentioned at the beginning. There are better ways of doing this, but it is not worth going into that here. Readers interested in the mathematics of this can take a look at Job G (1972) Neudarstellung der Wärmelehre. Akademische Verlagsgesellschaft, Frankfurt am Main, pp. 52–56, as well as Job G (1970) Zur Vereinfachung thermodynamischer Rechnungen. Das "Stürzen" einer partiellen Ableitung. Z. Naturforsch. 25a:1502–1508.

Chapter 10
Molecular-Kinetic View of Dilute Gases

In this chapter, we will deal with the special characteristics of dilute substances especially those of dilute gases. In this context, the term "ideal gas" will be introduced. Subsequently, the *general gas law*, one of the most cited equations in physical chemistry, is deduced from experimental observations made in the seventeenth and eighteenth century (Boyle–Mariotte's law, Charles's law, Avogadro's principle). Our understanding for these relationships will be deepened by an introduction to the kinetic theory of gases. We learn, for example, how this theory can be used to account for the pressure of a gas. In order to derive the distribution of particle velocities in a gas (Maxwell distribution), the concentration dependence (mass action equation) and additionally the energy dependence (excitation equation) of the chemical potential have to be considered. The last section of the chapter will show how we can glean the barometric formula and the Boltzmann distribution.

10.1 Introduction

In their diluted state, all dissolved and gaseous substances exhibit interesting commonalities of behavior. We will deal with some of these traits in this chapter. The behavior of gases is a prime example of the special characteristics of *dilute substances*. The more diluted gases are, the more clearly these properties emerge. In the limit of strong dilution, they take the character of strict laws. The most important of these is the so-called *general gas law*, or *gas law* for short, which concisely concentrates a whole series of important characteristics into one simple formula. For example, depending upon what kind it is, the volume of a certain amount of a solid substance changes more or less when heated or compressed. In contrast, all dilute gases behave the same way. At room conditions, the relative deviations of volume calculated with the help of the general gas law from the actual value are of the order of only 1 %. With decreasing pressure p, they tend toward

© Springer International Publishing Switzerland 2016
G. Job, R. Rüffler, *Physical Chemistry from a Different Angle*,
DOI 10.1007/978-3-319-15666-8_10

zero in proportion to p. The deviations for ambient air are of the order of tenths of a percent, making air a convenient model for dilute gases.

In the following, we will briefly summarize the uniformly valid experimental findings for all *dilute* gases and derive the general law of gases from them. Afterward, this will be interpreted from a molecular-kinetic viewpoint.

10.2   General Gas Law

In the seventeenth century, the experiments carried out by the Anglo-Irish scientist Robert Boyle were among the first to lead to fundamental results in physical chemistry. When rides in hot-air balloons became popular in the eighteenth century, these experiments became interesting again. The French scientists Jacques Charles and Joseph Louis Gay-Lussac studied the behavior of gases under various conditions in order to apply this knowledge to the new technology.

Boyle–Mariotte's Law   Robert Boyle investigated the pressure dependent changes in the volume of gases when temperature is kept constant. We want to model this investigation using a demonstration experiment (Experiment 10.1).

In 1664 and 1676, Robert Boyle and Edme Mariotte, independently found (and also we found in our experiment): The volume of a given amount of gas at constant temperature is inversely proportional to its pressure (Boyle–Mariotte's law):

$$V \sim \frac{1}{p} \quad \text{(at constant } T \text{ and } n\text{).} \tag{10.1}$$

For example, when the pressure is doubled, the volume will decrease by half. It is easy to compress gases because there is so much space between their particles.

If the volume is represented as a function of pressure (Fig. 10.1), one sees that the decrease of volume caused by an increase of pressure follows a hyperbolic curve. Such a curve is called an *isotherm* (from Greek ἴσος or *ísos* meaning "equal"

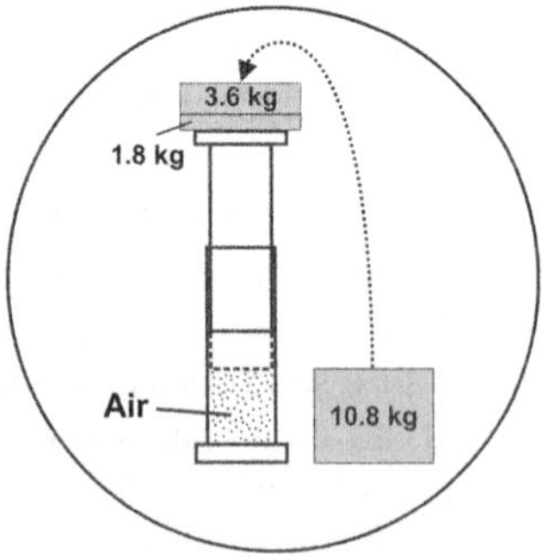

Experiment 10.1 *Boyle–Mariotte's law*: Starting with the lowest weight, different weights are placed successively upon a piston of a cylinder made of acrylic glass. If the cross section of the piston equals 1.8 cm^2, the pressure increases a factor of 2, 4, 10 due to the weights stacked upon each other, and the volume decreases to $\frac{1}{2}$, $\frac{1}{4}$, $\frac{1}{10}$ of the initial value.

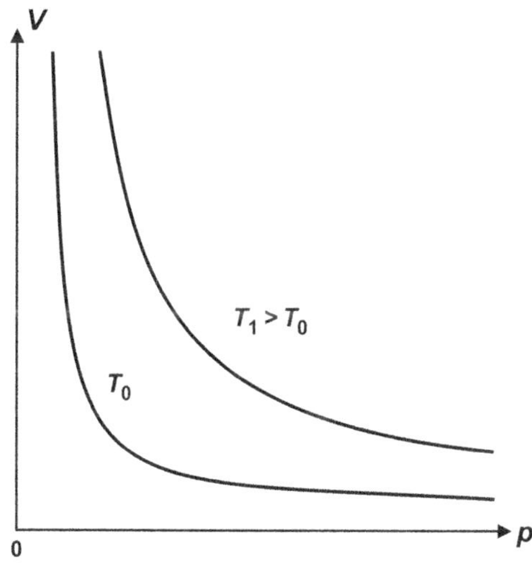

Fig. 10.1 Isotherms of a gas obeying Boyle–Mariotte's law, for two different temperatures $(T_1 > T_0)$.

and θέρμη or *thérmē* meaning "heat") because it describes the change of state (in this case, change of volume) at constant temperature. If this experiment is carried out at different temperatures, it yields differing isotherms but all follow Boyle–Mariotte's law and are therefore hyperbolas.

Law of Charles and Gay-Lussac If pressure as a parameter is kept constant, there is also a simple relation between the volume of a gas and its temperature. First investigations of this type were done by Jacques Alexandre César Charles (1787) and Joseph Louis Gay-Lussac (1802). They found that the volume of a certain amount of gas changes linearly with temperature. The mathematical expression for this linear relation is:

$$V = V_0 + \alpha_0 \vartheta \quad \text{(at constant } p \text{ and } n\text{)}. \tag{10.2}$$

V_0 is the initial volume of a certain amount of gas at, for example, a temperature of $0\,^\circ\text{C}$ (ice point), and ϑ is its temperature on the Celsius scale. A graphic or analytic extrapolation of the *isobars* $V(\vartheta)$ (Fig. 10.2) leads to an important conclusion: All the linear functions $V(\vartheta)$ belonging to different constant pressures, will intersect with the temperature axis at about $\vartheta = -267\,^\circ\text{C}$ (actually at $-273.15\,^\circ\text{C}$, as later measurements have shown), independent of the type of gas and amount of substance. The experiments carried out by Charles and Gay-Lussac were therefore a further indication of the existence of an *absolute zero point* of temperature. This had already been postulated in 1706 by Guillaume Amonton. It seemed, therefore, reasonable to introduce a new temperature scale and to measure temperature from this point because volume is never negative. This is how we arrive at the so-called *absolute temperature scale*. We were already introduced to this scale in Sect. 3.8 (but there we did not use the ice point but the more convenient triple point of water to fix the scale). Further, it also becomes clear that Eq. (10.2) is an example of a

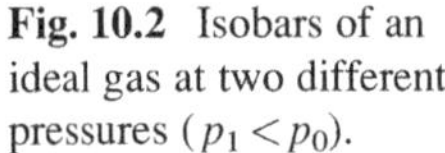

Fig. 10.2 Isobars of an ideal gas at two different pressures ($p_1 < p_0$).

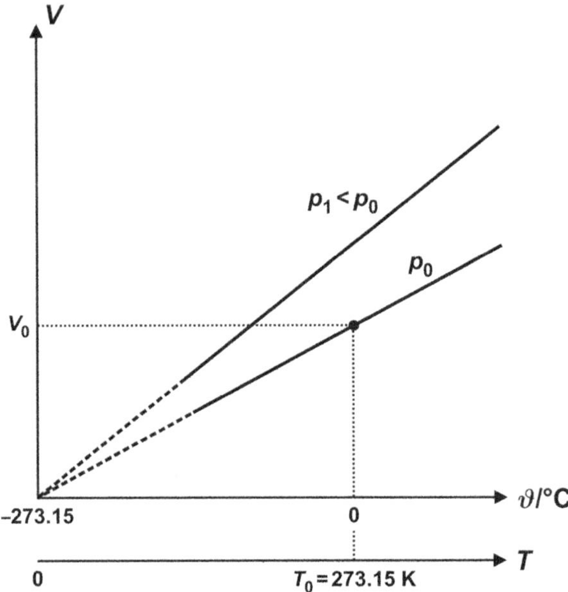

limiting law because it can only describe such gases across the entire span of temperature whose volumes actually do decrease to zero at $\vartheta = -273.15$ °C. This would only be possible if the gas particles themselves have no volume. Moreover, interactions between gas particles leading to the effects of condensation that produces a liquid and finally solidification at low temperatures will not be taken into account. A substance having such unreal characteristics is called an *ideal gas*.

When we replace the Celsius temperature ϑ by the absolute temperature ($\vartheta = T - T_0$), the gradient $\alpha_0 = V_0/T_0$ (which we can easily determine by using the boundary condition $V = 0$ for $T = 0$) helps us to obtain the following simple relation:

$$V = \frac{V_0}{T_0}T.$$

(10.3)

Since the ratio V_0T_0 is constant for a given amount of gas at unvarying pressure, we have

$$V \sim T \quad (\text{at constant } p \text{ and } n),$$

(10.4)

which means that the volume of a given amount of gas at constant pressure is proportional to the absolute temperature.

Doubling the temperature (given in Kelvin!) from, say, 298 K to 596 K (from 25 °C to 323 °C) leads to a doubling of the gas volume.

Avogadro's Principle   In closing, we will discuss *Avogadro's principle*. Lorenzo Romano Amedeo Carlo Avogadro's contribution to gas theory is the idea that the volume of a gas is a measure of the number of particles it is made up of, independent of the type of the particles. At given temperature and pressure, the volume of a gas is proportional to the amount of substance in question:

$$V \sim n \quad \text{(at constant } T \text{ und } p). \tag{10.5}$$

In mixtures of different gases B, C, D, ..., $n = n_B + n_C + n_D + \ldots$, i.e., n is equal to the sum of the amounts of substance of all the gases involved.

From Avogadro's principle it follows that the molar volume of a gas is independent of what type of gas it is and depends only upon temperature and pressure. At standard conditions ($T^\ominus = 298$ K, $p^\ominus = 100$ kPa), different gases yield experimentally determined V_m values that are almost identical at close to 25 L mol^{-1}.

General Gas Law   Combined, the three relations (10.1), (10.4) and (10.5) result in

$$V \sim n \cdot T / p, \tag{10.6}$$

a law that, after introducing a factor of proportionality R, can be written in the following form:

$$pV = nRT. \tag{10.7}$$

This is called the *general gas law* and is one of the most cited equations in physical chemistry. $R = 8.314$ G K^{-1} is the *general* (or *universal*) *gas constant* or *gas constant* in short, which we already know (Sect. 5.5). The general gas law describes the behavior of a (hypothetical) ideal gas. No existing gas is actually ideal, but the equation describes the behavior of most gases at pressures of 100 kPa and below rather well.

With the help of the general gas law, the molar volume of an ideal gas at any pressure and temperature values can be given. Transforming Eq. (10.7) yields:

$$V_m = \frac{V}{n} = \frac{RT}{p}. \tag{10.8}$$

The molar volume of an ideal gas at standard conditions ($T^\ominus = 298$ K, $p^\ominus = 100$ kPa) is therefore 24.79 L mol^{-1}, which can be shown easily by inserting the values.

Equation (10.8) has also another more far-reaching significance. In Sect. 9.3, we demonstrated that the molar volume of a substance corresponds to the pressure coefficient β of its chemical potential. For an ideal gas, we therefore have:

$$\beta = V_{\mathrm{m}} = \frac{RT}{p}. \tag{10.9}$$

This pressure coefficient for gases was introduced empirically in Sect. 5.5.

Based upon the general gas law, for example, according to

$$V(T, p) = \frac{nRT}{p} \tag{10.10}$$

some of the coefficients introduced in the last chapter can be calculated. The cubic expansion coefficient γ of an ideal gas turns out to be given by

$$\gamma = \frac{1}{V}\left(\frac{\partial V}{\partial T}\right)_p = \frac{p}{nRT} \cdot \frac{nR}{p} = \frac{1}{T}. \tag{10.11}$$

The compressibility χ, however, equals

$$\chi = -\frac{1}{V}\left(\frac{\partial V}{\partial p}\right)_T = -\frac{p}{nRT} \cdot -\frac{nRT}{p^2} = \frac{1}{p} \tag{10.12}$$

(cf. rule (A.1.6) for calculating derivatives in the Appendix). For an ideal gas, the difference between the entropy capacities $\mathcal{C}_p - \mathcal{C}_V$ ultimately results in

$$\mathcal{C}_p - \mathcal{C}_V = V\frac{\gamma^2}{\chi} = \frac{nRT}{p} \cdot \frac{p}{T^2} = \frac{nR}{T}. \tag{10.13}$$

10.3   Molecular-Kinetic Interpretation of the General Gas Law

Fundamentals Much can be understood about gases if they are assumed to be made up of a huge number of small particles in perpetual random motion at high velocities colliding elastically with each other. These molecules can be made up of just one atom, as they are in noble gases, or composed of several atoms, which is more often the case. When they collide with each other or with a wall, they bounce back like billiard balls. The particle density is so small that there is enough space for unhindered motion. Experiment 10.2 shows a good example of a "model gas."

These few assumptions are enough to make a series of useful conclusions. Energy is exchanged but not lost during collisions if there is no interference from outside so the molecular motion must go on indefinitely. The huge number of tiny collisions happening at every moment upon all bounding surfaces would appear to us as uniform and constant pressure. Some simple ideas can be used to make several statements about this pressure that can then be compared to the general gas law.

Experiment 10.2 *Model gas with steel beads*: A good example of a "model gas" would be a large number of small steel or glass beads being caused to move erratically by a rapidly oscillating piston, reflecting the behavior of gas particles.

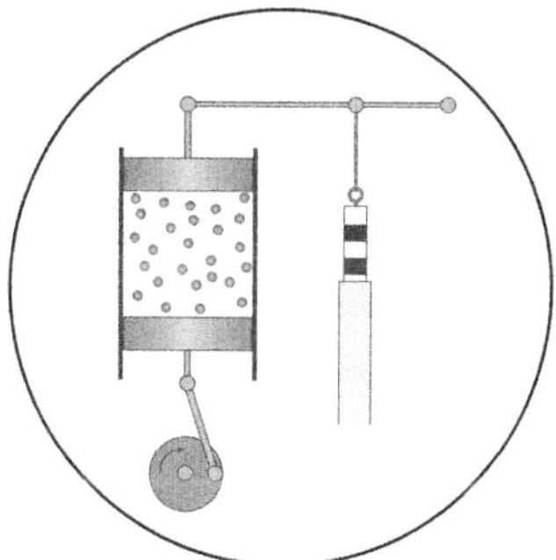

Assuming such a molecular structure of a gas, we immediately obtain the following without calculations:

$p \sim N$ Because double the amount of particles in the same container causes double the number of collisions and, therefore, twice the pressure, under otherwise identical conditions;

$p \sim 1/V$ Because halving the volume and keeping the number of particles constant is like a doubling of the number of particles in the original volume;

$p \sim v \cdot v$ Because at twice the velocity of all particles and at otherwise identical conditions

1. The number of collisions in the same time span is twice as high
2. And every collision is twice as strong (twice the momentum transferred);

$p \sim m$ Because at the same velocity, the impact of a particle with twice the mass equals the impacts of two particles of the original mass.

We combine the proportionalities into a single relation. An additional factor $\frac{1}{2}$ can be inserted but does not change the proportionality:

$$p \sim \frac{N \cdot \overline{\frac{1}{2}mv^2}}{V}. \tag{10.14}$$

The simultaneous averaging (indicated by the horizontal line above the formula) is required because neither the mass nor the velocity of the particles needs to be uniform. The expression

$$\overline{w}_{\text{kin}} := \overline{\tfrac{1}{2}mv^2} \tag{10.15}$$

obviously describes the average kinetic energy of a particle.

According to Sect. A.1.4 in the Appendix, the (arithmetic) mean of the values $m_i v_i^2$ of all N particles ($i = 1, 2, \ldots, N$) is

$$\overline{mv^2} = \frac{1}{N}\sum_{i=1}^{N} m_i v_i^2.$$

Before we compare the resulting relation

$$p \sim \frac{N \cdot \overline{w}_{\mathrm{kin}}}{V} \tag{10.16}$$

with the general gas law, we will give this equation a somewhat different form with the help of $n = N \cdot \tau$ [Eq. (1.2)] and $k_{\mathrm{B}} = R \cdot \tau$:

$$pV = N\tau \cdot \frac{k_{\mathrm{B}}}{\tau} \cdot T \quad \text{with} \quad \tau = 1.6606 \times 10^{-24}\,\mathrm{mol} \text{ and } k_{\mathrm{B}} = 1.3805 \times 10^{-23}\,\mathrm{J\,K^{-1}}.$$
$$\tag{10.17}$$

In this case, N is the particle number, τ is the elementary amount of substance that we already used in Sect. 1.4, and k_{B} is the Boltzmann constant (a natural constant just as R and τ). Accordingly, we obtain for pressure according to the (modified) gas law:

$$p = \frac{Nk_{\mathrm{B}}T}{V}. \tag{10.18}$$

Relation Between Average Kinetic Energy and Temperature If we compare the above with the gas law, the proportionality for p according to Eq. (10.16) yields

$$k_{\mathrm{B}}T \sim \overline{w}_{\mathrm{kin}}. \tag{10.19}$$

We see that the role played by temperature, which at first does not appear in our mechanical gas model, is taken over by a relatively simple mechanical quantity, the average kinetic energy of the gas particles. We can imagine in terms of this model that a gas appears hotter the faster its particles move. Doubling the velocity results in a quadrupling of the energy and a quadrupling of the temperature, along with it.

Up until now, our considerations have yielded the proportionality of temperature and kinetic energy but do not yield the proportionality factor itself. We will derive it in the following. For the sake of simplicity, we will imagine a dilute gas enclosed in a rectangular box. The N nearly point-like spherical particles elastically bounce off the smooth walls when they collide with them (Fig. 10.3). We also assume that when the particles graze the walls at an angle, they do not begin to spin. The gas is so diluted that collisions between its particles are very seldom. The particle motion can then be considered an overlapping of three *independent* motions in the x, y, and z directions.

When a particle collides with a side wall of the box (parallel to the x direction) the component of the velocity in the x direction, v_x, does not change (only v_y or v_z changes their sign) (Fig. 10.4). The particle continues its motion in the x direction as

Fig. 10.3 Box model for deriving gas pressure.

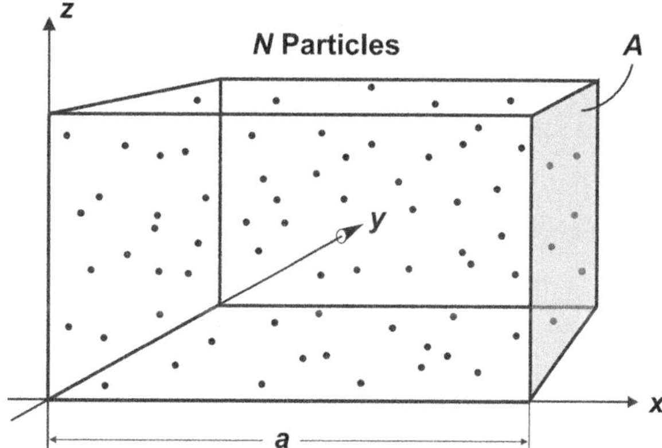

Fig. 10.4 Vectorial representation of the x and y components of particle velocity before and after colliding with a side wall.

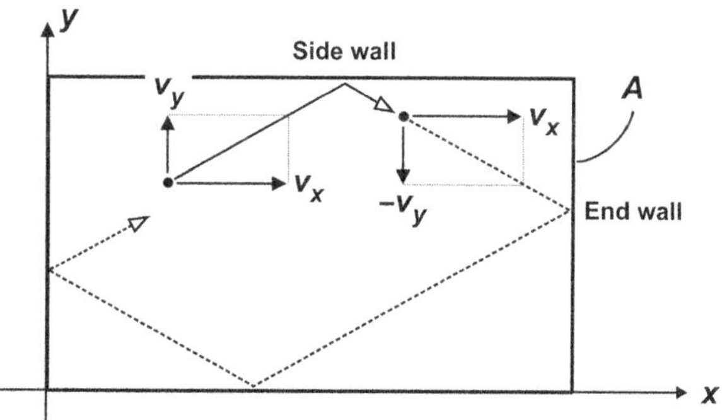

if nothing had happened at all. The transverse motion in the y and z directions has no effect upon the particle's movement in the x direction between the two end walls, which corresponds to a simple back and forth motion.

The *momentum p* of a particle is a measure for what is also called an "impact force" (compare Sect. 2.7). The heavier and faster the particle is, the greater its momentum. It has the character of a quantity that can be transported.

On the way toward the wall having a surface area A, the particle transports momentum mv_x, and after colliding with the wall, it transports momentum $-mv_x$ in the opposite direction (Fig. 10.5). The wall has absorbed the difference $2mv_x$ during the collision. Between two collisions with wall A, the particle travels the distance $2a$. In the time span Δt, it covers the distance $v_x \cdot \Delta t$. This means that in the time span Δt, the number of collisions is given by

$$\frac{v_x \cdot \Delta t}{2a}.$$

The product of momentum transferred per collision and the number of collisions, summed over the contributions from all the particles, yield the momentum p_{total} transferred to the wall in time Δt by all the particles:

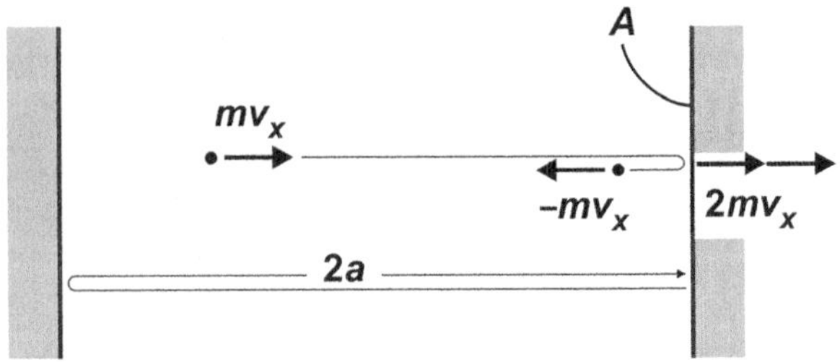

Fig. 10.5 Transfer of momentum during collision of a particle with the end wall A.

$$p_{\text{total}} = \sum_{i=1}^{N} 2m_i v_{x,i} \cdot \frac{v_{x,i} \cdot \Delta t}{2a} = \frac{2\Delta t}{a} \cdot \sum_{i=1}^{N} \tfrac{1}{2} m_i v_{x,i}^2. \tag{10.20}$$

The momentum current $J_p = p_{\text{total}}/\Delta t$ is the force $F = J_p$ acting upon the wall on the right as a result of the collisions (compare Sect. 2.7). The momentum current density j_p is simply the pressure p:

$$j_p = \frac{J_p}{A} = \frac{F}{A} = p.$$

With the help of Eq. (10.20), we obtain the following expression for the pressure:

$$p = \frac{p_{\text{total}}}{\Delta t \cdot A} = \frac{2 \cdot \sum_{i=1}^{N} \tfrac{1}{2} m_i v_{x,i}^2}{a \cdot A}. \tag{10.21}$$

The following is valid for the mean value of the kinetic energy in the x direction averaged over all the particles, the so-called ensemble average,

$$\overline{w}_{\text{kin},x} = \frac{\sum_{i=1}^{N} w_{\text{kin},x,i}}{N} = \frac{\sum_{i=1}^{N} \tfrac{1}{2} m_i v_{x,i}^2}{N}. \tag{10.22}$$

If we apply $V = a \cdot A$ and the gas law in the form of $p = N k_{\mathrm{B}} T / V$, we finally get:

$$\frac{2N \cdot \overline{w}_{\text{kin},x}}{V} = p = \frac{N k_{\mathrm{B}} T}{V}$$

or

$$\overline{w}_{\text{kin},x} = \tfrac{1}{2} k_{\mathrm{B}} T. \tag{10.23}$$

Degrees of Freedom and Law of Equipartition If we consider the kinetic energy of motion in *only* one direction, in this case the x direction, the missing proportionality factor in the relation (10.19) is $\tfrac{1}{2}$. We can find analogous equations for the

y and z directions using the same method. The average energy for every spatial direction in which the particles freely move is $\frac{1}{2}k_\mathrm{B}T$, so that

$$\overline{w}_{\mathrm{kin},x} = \overline{w}_{\mathrm{kin},y} = \overline{w}_{\mathrm{kin},z} = \tfrac{1}{2}k_\mathrm{B}T. \tag{10.24}$$

Because of the freedom of motion in all three spatial directions we also speak of (quadratic) *degrees of freedom*.

Correspondingly, due to

$$\overline{w}_{\mathrm{kin}} = \overline{w}_{\mathrm{kin},x} + \overline{w}_{\mathrm{kin},y} + \overline{w}_{\mathrm{kin},z} \tag{10.25}$$

the factor $\frac{3}{2}$ results for the free motion in all spatial directions:

$$\overline{w}_{\mathrm{kin}} = \tfrac{3}{2}k_\mathrm{B}T. \tag{10.26}$$

The average kinetic energy of a gas particle at room temperature ($T = 298$ K) is therefore

$$\overline{w}_{\mathrm{kin}} = \tfrac{3}{2} \times 1.3805 \times 10^{-23}\,\mathrm{J\,K^{-1}} \times 298\,\mathrm{K} = 6.17 \times 10^{-21}\,\mathrm{J}.$$

Relative to amount of substance (here $n = 1 \cdot \tau$) and according to

$$\overline{W}_{\mathrm{kin,m}} = \frac{\overline{w}_{\mathrm{kin}}}{\tau} = \tfrac{3}{2}\,\frac{k_\mathrm{B}}{\tau}T = \tfrac{3}{2}RT \tag{10.27}$$

an ideal gas at room temperature has an average molar kinetic energy of

$$\overline{W}_{\mathrm{kin,m}} = \tfrac{3}{2} \times 8.314\,\mathrm{G\,K^{-1}} \times 298\,\mathrm{K} = 3{,}716\,\mathrm{G} = 3.72\,\mathrm{kJ\,mol^{-1}}.$$

Equation (10.26) has a relatively far-reaching significance. It not only applies if we take the average for the entire ensemble of particles at a given moment, the so-called *ensemble average*, but also to each individual particle if we form the *time average*, which is the temporal average of the energy of motion of an individual gas particle. Even when the gas is so dense that the motion of its particles is strongly impeded or when it condenses into a liquid or crystal, this equation still applies to some extent. It reaches its limits when the quantum mechanical characteristics of molecules cannot be ignored any longer.

Equation (10.24) as well has a greater significance. Until now, we have only considered the translational motion of particles. However, if there are polyatomic molecules, rotations and oscillations appear as well, which add to the particles' energy. These forms of motion also have degrees of freedom. The previous result suggests that these degrees of freedom themselves have an average energy of $\frac{1}{2}k_\mathrm{B}T$, so we can say that generally:

The same average energy $\frac{1}{2}k_\mathrm{B}T$ is allotted to every degree of freedom (*law of equipartition of energy*).

Root Mean Square Speed of Gas Molecules Let us return once again to translational motion: The average speed of gas molecules is somewhat easier to understand than their average energies. (The term "speed" is used for the scalar magnitude of the vector quantity "velocity.") Since the average molar kinetic energy of a certain gas is given by

$$\overline{w}_{\text{kin,m}} = \frac{\overline{w}_{\text{kin}}}{\tau} = \frac{\overline{\tfrac{1}{2}mv^2}}{\tau} \tag{10.28}$$

and the relation between the molar mass M and the molecular mass m is

$$M = m/\tau, \tag{10.29}$$

we obtain with

$$\overline{w}_{\text{kin,m}} = \tfrac{1}{2}M\overline{v^2} \tag{10.30}$$

in combination with Eq. (10.27) for the mean square speed

$$\overline{v^2} = 3\frac{RT}{M} \tag{10.31}$$

and finally for the root mean square speed of gas molecules:

$$\sqrt{\overline{v^2}} = \sqrt{3\frac{RT}{M}}. \tag{10.32}$$

As expected, the speed of gas particles increases with a rise in temperature (proportionally to its square root). It also decreases correspondingly with increasing molar mass.

The root mean square speed of N_2 molecules ($M = 28.02$ g mol^{-1}) at 298 K can, for instance, be calculated as

$$\sqrt{\overline{v^2}} = \sqrt{3\frac{8.314\,\text{G}\,\text{K}^{-1} \times 298\,\text{K}}{0.02802\,\text{kg}\,\text{mol}^{-1}}} = 515\,\text{m}\,\text{s}^{-1}\left(= 1{,}854\,\text{km}\,\text{h}^{-1}\right).$$

This value is of the order of the speed of sound in air (346 m s^{-1} at 298 K). This makes sense because the sound waves, which are simultaneously density and pressure waves, propagate via molecular motion.

Entropy Capacities of Ideal Gases The entropy capacity $\mathcal{C}_V$ of an ideal gas at constant volume can be calculated using the law of equipartition. For the sake of simplicity, we will consider a monatomic gas because its particles cannot oscillate and have no rotational energy. (Because the mass lies upon the rotational axis, the moment of inertia as well as the rotational energy equals zero.) The energy of an ideal monatomic gas therefore corresponds to the average translational energy,

$$W = \tfrac{3}{2}nRT.$$

(10.33)

From the main equation

$$dW = -p \cdot dV + T \cdot dS$$

it follows for isochoric processes ($dV = 0$) that

$$dW = T \cdot dS$$

and therefore

$$\left(\frac{\partial W}{\partial T}\right)_V = T \cdot \left(\frac{\partial S}{\partial T}\right)_V.$$

(10.34)

If we take the derivative of the expression in Eq. (10.33) with respect to T, we obtain

$$\mathcal{C}_V = \left(\frac{\partial S}{\partial T}\right)_V = \frac{1}{T}\left(\frac{\partial W}{\partial T}\right)_V = \frac{3}{2}\frac{nR}{T}$$

(10.35)

for the entropy capacity at constant volume. Now, the entropy capacity at constant pressure can be found with the help of Eq. (10.13):

$$\mathcal{C}_p = \mathcal{C}_V + \frac{nR}{T} = \frac{3}{2}\frac{nR}{T} + \frac{nR}{T} = \frac{5}{2}\frac{nR}{T}.$$

(10.36)

10.4   Excitation Equation and Velocity Distribution

Introduction As mentioned earlier, the velocity of particles in a gas is not uniform. Speed (magnitude of the velocity vector) and direction change with each collision. Although basically every speed and every direction can be assumed to occur, they do not all occur with the same frequency but actually form a characteristic distribution. An impression of this distribution might be gained by imagining a small amount of gas, possibly about 1 μm^3, whose particles are allowed to escape into a vacuum by suddenly removing the walls enclosing it. The positions the particles have reached after about 30 μs are then marked producing an image as in Fig. 10.6a. The position reached by a particle simultaneously characterizes the direction and absolute value of the particle's velocity vector $\vec{v}$.

Fig. 10.6 (**a**) Density of points in three-dimensional velocity space as a spherical point cloud, (**b**) along any diameter of the cloud (values given for nitrogen at 298 K), (**c**) cubic grid having mesh length Δv (greatly magnified; center cell is shown where the molecular velocity is represented by *vectors*).

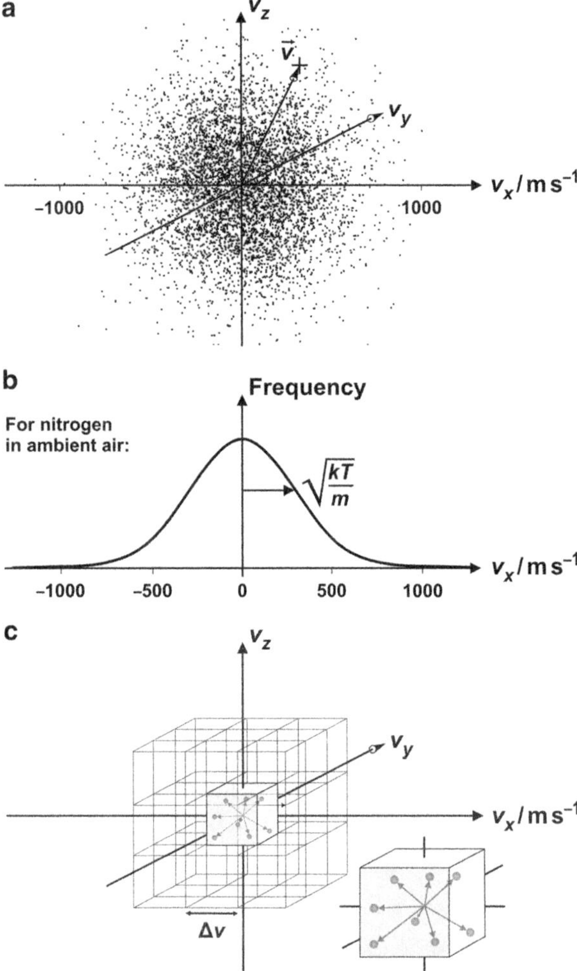

The density of points along any one of the diameters of the point cloud (e.g., along the v_x axis) is reflected in the *bell curve* in Fig. 10.6b.

In order to derive the particle velocity distribution of a gas we need a trick. We imagine all the particles with the same velocity $\vec{v}$ to be molecules of a substance $B(\vec{v})$ and the entire gas as a mixture of a large number of such substances. However, we run into a difficulty at this point. The number of particles having *exactly* the velocity $\vec{v}$ is, in the strict sense, zero. We therefore imagine the velocity space to be divided into a cubic grid with the mesh length Δv (Fig. 10.6c). Δv is assumed to be small compared to the width of the velocity distribution. For example, $\Delta v = 1\,\mathrm{m\,s}^{-1}$ would be a useful value for ambient air. All the particles whose velocity vectors end inside such a cube are then considered molecules of the same substance $B(\vec{v})$.

Assumptions To continue our derivation, we need the mass action equation valid for dilute gases

$$\mu = \mu_0 + RT \ln \frac{c}{c_0} \quad \text{(mass action equation 1)},$$

that we were introduced to in Sect. 6.2. The value μ_0 at the reference value of concentration, c_0, might be considered the basic value of the potential (in a generalized sense). In practice it is a good idea to limit the expression "basic value" (in a more limited sense) to the case most often encountered where the reference value c_0 of the concentration corresponds to the standard value, for example, $c^{\ominus} = 1\,\mathrm{kmol\,m^{-3}}$, and to use a special symbol, possibly the one introduced in Sect. 6.2, $\overset{\circ}{\mu}$, to characterize it.

We also need another important equation. When the molecules of a substance are excited and lifted to an energy state higher by w without otherwise changing them or their milieu (temperature, pressure, concentration, type of solvent, field strength, etc.), the potential of the substance will increase by the molar energy w/τ:

$$\mu(w) = \mu(0) + w/\tau \quad \text{(\textit{excitation equation}).} \tag{10.37}$$

$\mu(0)$ represents the chemical potential of the non-excited molecules, i.e., the chemical potential previously used. A simple example of a "purely energetic" excitation that leaves the molecules themselves unchanged might be a shift in an outer field such as a gravitational or electric field to a position with potential energy that is higher by w. By choosing an appropriate initial position, we may set $w = m \cdot g \cdot h$ in the case of the gravitational field. Because of the relation $M = m/\tau$ between the molar mass M and the molecular mass m, we obtain for the molar potential energy $M \cdot g \cdot h$ and correspondingly for the "gravitochemical" potential, $\mu(w) = \mu(0) + M \cdot g \cdot h$. The *electrochemical potential* can be defined analogously (Sect. 22.1).

Maxwell's Distribution of Speeds We will now return to our derivation of velocity distribution. The particles moving in different directions do not vary from each other chemically, so we can assign them the basic potential μ_0. The various energies at differing speeds $v = |\vec{v}|$ are taken into account by using the corresponding term $w/\tau = \tfrac{1}{2}\,mv^2/\tau = \tfrac{1}{2}Mv^2$, the *molar kinetic energy* of the substances:

$$\overset{\circ}{\mu}\left(\vec{v}\right) = \overset{\circ}{\mu}\left(\vec{0}\right) + \tfrac{1}{2}Mv^2. \tag{10.38}$$

Actually, this potential is not a purely chemical potential, but a *mechanochemical potential*.

The particles' changes of velocity caused by the multitude of collisions with each other or the walls appear as "transformations" of the following type:

$$B(\vec{v}) \rightarrow B(\vec{v}').$$

If we do not continuously disturb the gas by stirring or other means, all these processes will reach equilibrium after a short time. This means that the chemical potential $\mu(\vec{v})$ will be equal for all the substances $B(\vec{v})$, i.e., equal to $\mu(\vec{v}) = \mu$ for all $\vec{v}$. If the excitation equation as well as the mass action equation with the equilibrium values $c(\vec{v})$ of the concentrations are taken into account, we obtain

$$\mu = \overset{\circ}{\mu}(\vec{0}) + \tfrac{1}{2}Mv^2 + RT\ln\frac{c(\vec{v})}{c^{\ominus}} \quad \text{for all } \vec{v}. \tag{10.39}$$

Solving for $c(\vec{v})$ gives us the distribution we are looking for:

$$c(\vec{v}) = \underbrace{c^{\ominus}\cdot\exp\left(\frac{\mu - \overset{\circ}{\mu}(\vec{0})}{RT}\right)}_{c(\vec{0})} \cdot\exp\left(-\frac{\tfrac{1}{2}Mv^2}{RT}\right). \tag{10.40}$$

$c(\vec{0})$ therefore corresponds to the concentration of those particles that have the velocity $\vec{0}$ in equilibrium (stated more exactly, a velocity taken from the small cubic volume around zero). Inserting $\vec{v} = \vec{0}$ shows that we can really combine the terms in Eq. (10.40) to $c(\vec{0})$.

This result can be put into a more common form if we use the mass $m = M \cdot \tau$ of a molecule instead of the molar mass M and the Boltzmann constant $k_{\mathrm{B}} = R \cdot \tau$ instead of R. In addition, we use the relation $v^2 = v_x^2 + v_y^2 + v_z^2$ between the velocity $\vec{v}$ of a molecule in arbitrary direction and its components v_x, v_y, and v_z in the three spatial directions ("three-dimensional Pythagorean theorem"). We now arrive at

$$c(\vec{v}) = c(\vec{0}) \cdot \exp\left(-\frac{m \cdot \left(v_x^2 + v_y^2 + v_z^2\right)}{2k_{\mathrm{B}}T}\right). \tag{10.41}$$

If we represent $c(\vec{v})$ as density of points in the three-dimensional velocity space and apply $m = m(\mathrm{N_2})$ and $T = 298$ K, we obtain the spherical point cloud in Fig. 10.6.

We can make some statements about the pre-exponential factor $c(\vec{0})$ without needing calculations:

$c(\vec{0}) \sim 1/\sqrt{T}^3$ Because the root mean square speed [compare Eq. (10.32)] increases with rising temperature T. The point cloud therefore increases by $\sqrt{T}$ in all three spatial directions and the concentration of particles with velocity $\vec{0}$ decreases correspondingly.

$c(\vec{0}) \sim \sqrt{m}^3$ Because the root mean square speed decreases with an increase of mass m of the gas particles. The point cloud becomes compressed, resulting in an increase of particle concentration with velocity $\vec{0}$.

The density of points along the x-axis results from Eq. (10.41) for the special case of $v_y^2 = v_z^2 = 0$:

$$c\left(\vec{v}_x\right) = c\left(\vec{0}(x)\right) \cdot \exp\left(-\frac{mv_x^2}{2k_\mathrm{B}T}\right). \tag{10.42}$$

For the sake of simplicity, we will limit ourselves at first to such a one-dimensional distribution in velocity space. The volumes of all the cubic elements are the same so Eq. (10.42) can be also formulated using the number of particles:

$$N\left(\vec{v}_x\right) = N\left(\vec{0}\right) \cdot \exp\left(-\frac{mv_x^2}{2k_\mathrm{B}T}\right). \tag{10.43}$$

If the fact is taken into account that the number of all the particles distributed across the entire velocity space must equal N (*normalization*), it is then possible to calculate $N(\vec{0})$ yielding the following relation:

$$\frac{\mathrm{d}N\left(\vec{v}_x\right)}{N} = \sqrt{\frac{m}{2\pi k_\mathrm{B}T}} \exp\left(-\frac{mv_x^2}{2k_\mathrm{B}T}\right)\mathrm{d}v_x. \tag{10.44}$$

For readers interested in mathematics: The number $N(\vec{0})$ of the particles with velocity $\vec{0}$ can be expressed as a fixed fraction A of the total number N. However, in doing so, one must take into consideration that $N\left(\vec{v}_x\right)$ and $N(\vec{0})$ depend upon the width of the interval Δv_x. The greater Δv_x is, the more particles can be found in the chosen interval. Therefore:

$$N(\vec{0}) = A \cdot N \cdot \Delta v_x.$$

The mathematical expression for the normalization condition is then:

$$\sum_{v_x=-\infty}^{v_x=+\infty} A \cdot N \cdot \exp\left(-\frac{mv_x^2}{2k_BT}\right)\Delta v_x = N. \tag{10.45}$$

Principally, the velocity v_x can assume values between $-\infty$ and $+\infty$ where the minus sign indicates that the velocity vector is oriented in the negative x direction. If the width of the interval is made infinitesimally small, summation changes to integration (see Sect. A.1.3 in the Appendix):

$$\int_{v_x=-\infty}^{v_x=+\infty} A \cdot N \cdot \exp\left(-\frac{mv_x^2}{2k_BT}\right)dv_x = N. \tag{10.46}$$

Because A and N are independent of v_x the following is valid:

$$A \cdot N \cdot \int_{v_x=-\infty}^{v_x=+\infty} \exp\left(-\frac{mv_x^2}{2k_BT}\right)dv_x = N \tag{10.47}$$

or respectively

$$A = \frac{1}{\displaystyle\int_{v_x=-\infty}^{v_x=+\infty} \exp\left(-\frac{mv_x^2}{2k_BT}\right)dv_x}. \tag{10.48}$$

Using integral tables, we find for the integral

$$\int_{-\infty}^{+\infty} \exp\left(-ax^2\right)dx = 2\int_{0}^{\infty} \exp\left(-ax^2\right)dx$$

the value $\sqrt{\pi/a}$. Applied to Eq. (10.48) the result for A is:

$$A = \sqrt{\frac{m}{2\pi k_BT}}. \tag{10.49}$$

The familiar Eq. (10.44) can now be obtained for the one-dimensional velocity distribution of a gas made up of N particles.

The relationship to Fig. 10.6b can be directly seen in Eq. (10.44) because v_x appears only in the exponent of the exponential function and that quadratically. The distribution must therefore be symmetrical about the axis $(v_x = 0)$ and possess a maximum at this point. For large positive and negative v_x values, the function decreases exponentially with v_x^2.

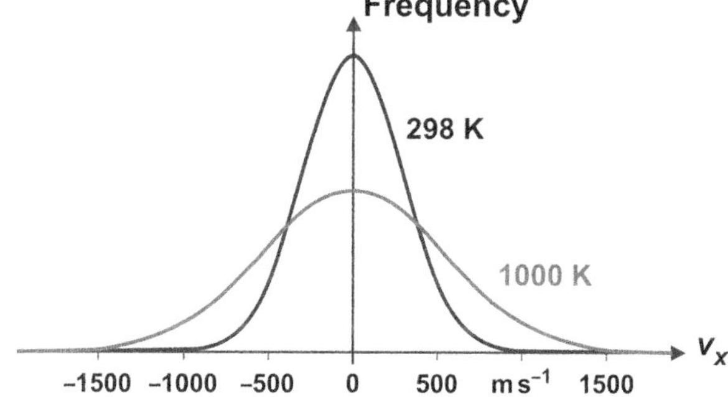

Fig. 10.7 One-dimensional velocity distribution of N_2 molecules at different temperatures.

Let us now look at the change of the distribution function of a certain gas as a function of temperature. We will use the example of nitrogen (Fig. 10.7).

We see that at higher temperatures, the distribution function becomes wider (the area under the curve remains the same since the total number N of particles does not change).

The concentration $c\left(\vec{v}_x\right)$ can be interpreted as a measure of the probability $p\left(\vec{v}_x\right)$ of finding a particle with velocity $\vec{v}_x$. It is therefore unsurprising that the density of points along an arbitrary diameter of the spherically symmetrical cloud (in this case, the x-axis) corresponds to the so-called *normal distribution* (*Gaussian distribution*) known from statistics (see Sect. A.1.4 in the Appendix). This becomes especially clear if Eq. (10.44) is somewhat reformulated:

$$\frac{dN\left(\vec{v}_x\right)}{N} = \frac{1}{\sqrt{2\pi} \cdot \sqrt{k_B T/m}} \exp\left(-\frac{v_x^2}{2\sqrt{k_B T/m}^2}\right) dv_x. \tag{10.50}$$

The "width" of the bell curve, in this case $\sqrt{k_B T/m}$ (Fig. 10.6b), meaning the distance of the inflection points from the center line (or standard deviation), is about $300\ \mathrm{m\ s^{-1}}$ for ambient air. This is just about the speed of sound.

We will now return to the three-dimensional overall distribution. The density of points inside a volume element is equivalent to the probability that the components of the velocity vector of the gas particles lie in the intervals of v_x to $v_x + dv_x$, v_y to $v_y + dv_y$, and v_z to $v_z + dv_z$. The probability of finding them simultaneously in the three intervals, i.e., in the volume element $dv_x \cdot dv_y \cdot dv_z$ of the velocity space, is given by the product of the individual probabilities:

$$\frac{\mathrm{d}N(\vec{v})}{N} = \left(\sqrt{\frac{m}{2\pi k_\mathrm{B} T}}\, \exp\left(-\frac{m v_x^2}{2 k_\mathrm{B} T} \right) \mathrm{d}v_x \right)$$
$$\cdot \left(\sqrt{\frac{m}{2\pi k_\mathrm{B} T}}\, \exp\left(-\frac{m v_y^2}{2 k_\mathrm{B} T} \right) \mathrm{d}v_y \right) \tag{10.51}$$
$$\cdot \left(\sqrt{\frac{m}{2\pi k_\mathrm{B} T}}\, \exp\left(-\frac{m v_z^2}{2 k_\mathrm{B} T} \right) \mathrm{d}v_z \right).$$

The resulting distribution function is (according to the product rule of exponents $a^n \cdot a^m = a^{n+m}$):

$$\frac{\mathrm{d}N(\vec{v})}{N} = \left(\sqrt{\frac{m}{2\pi k_\mathrm{B} T}} \right)^3 \exp\left(-\frac{m\left(v_x^2 + v_y^2 + v_z^2\right)}{2 k_\mathrm{B} T} \right) \mathrm{d}v_x \mathrm{d}v_y \mathrm{d}v_z. \tag{10.52}$$

As we argued before, there actually is a direct proportionality between the pre-exponential factor and the expression $\sqrt{m}^3$, as well as an inverse proportionality to $\sqrt{T}^3$.

It is customary to consider the distribution of the absolute values of the velocity, $v = |\vec{v}|$, rather than that of the velocity vectors $\vec{v}$:

$$v^2 = v_x^2 + v_y^2 + v_z^2 \quad \text{or} \quad v = \sqrt{v_x^2 + v_y^2 + v_z^2}. \tag{10.53}$$

Velocity vectors are organized according to their "length" v (speed), independent of their direction. Equation (10.53) is an analytic representation of the surface of a sphere with a radius of v (Fig. 10.8). All the vectors with values within the intervals v and $v + \mathrm{d}v$ end inside a thin spherical shell with a radius of v and a thickness of $\mathrm{d}v$

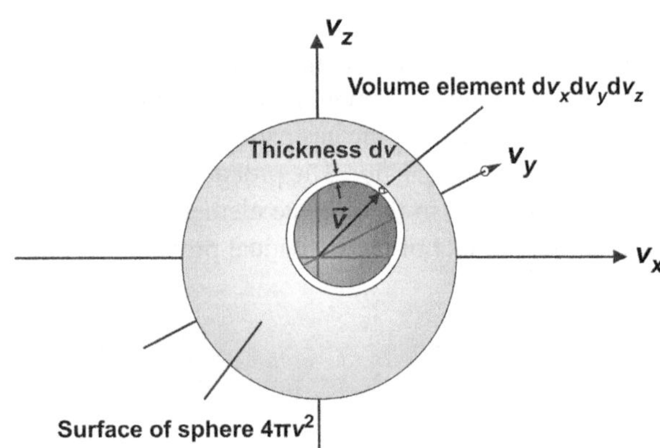

Fig. 10.8 Relation between the volume elements $\mathrm{d}v_x\mathrm{d}v_y\mathrm{d}v_z$ and $4\pi v^2 \mathrm{d}v$ (spherical shell).

and, therefore, a volume = surface $\times$ thickness = $4\pi v^2 \cdot dv$. The number $dN(v)$ of particles with those velocities results from the distribution (10.52) by summation over all the volume elements $dv_x \cdot dv_y \cdot dv_z$ inside the spherical shell.

The distribution within the shell is the same overall, so that in this step of the calculations, $4\pi v^2 dv$ takes the place of the volume element $dv_x \cdot dv_y \cdot dv_z$:

$$\frac{dN(v)}{N} = 4\pi \left(\sqrt{\frac{m}{2\pi k_\mathrm{B} T}} \right)^3 \exp\left(-\frac{mv^2}{k_\mathrm{B} T} \right) v^2 dv. \tag{10.54}$$

This expression is the so-called *Maxwell's speed distribution* of gas particles. A first impression can be gained in an experiment using glass beads as a "model gas" (Experiment 10.3).

Equation (10.54) may appear somewhat complicated at first sight, but important characteristics of the function are relatively easy to recognize:

- The curve increases parabolically close to the zero point (meaning for very small speeds) because the behavior is determined by the factor v^2 with which the exponential function is multiplied. For this reason, the fraction of gas particles with very small speeds is very small (Fig. 10.9).

- The exponential function of type e^{-ax^2} leads to a rather steep drop in the curve which, at high particle speeds, approaches the v axis asymptotically. Correspondingly, the proportion of gas particles with very high speeds must also be very small.

- For a given gas with constant molar mass, the expression $a = m/2k_\mathrm{B} T = M/2RT$ decreases with increasing temperature. This leads to a slower decrease of the exponential function with increasing v. At higher temperatures, there is a greater number of very fast gas particles. This fact has great importance for the kinetics of chemical reactions.

- Assuming constant temperature, the term a takes a large value for gases having high molar mass. As a result, the exponential function decreases more quickly (Fig. 10.10). Correspondingly, the probability of finding heavy particles with very high speed is very low.

Experiment 10.3 *Maxwell's speed distribution*: The flight paths of the beads that can only reach the outside through a chamber acting as a filter approximate those of horizontal projectile motion. The distance covered is a measure of the initial speed of a bead. The beads are then collected in the acrylic glass compartments of a registration chamber in order to determine how far they went. The "staircase curve" (histogram) resulting from the fill heights in the compartments gives us an idea of the speed distribution. The number of beads emerging from the filter is proportional to their speed so the distribution is distorted, but the qualitative impression is still correct.

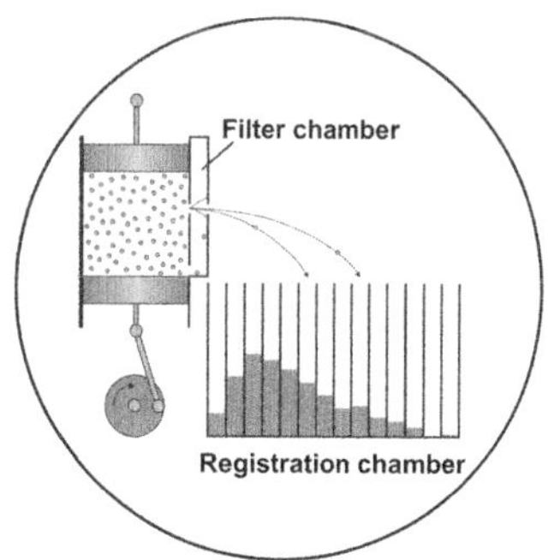

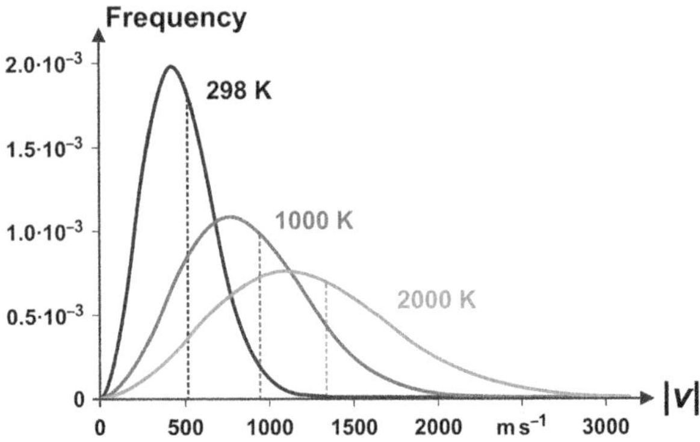

Fig. 10.9 Maxwell's speed distribution of N_2 molecules at different temperatures. [Included is the corresponding root mean square speed (*dashed line*).]

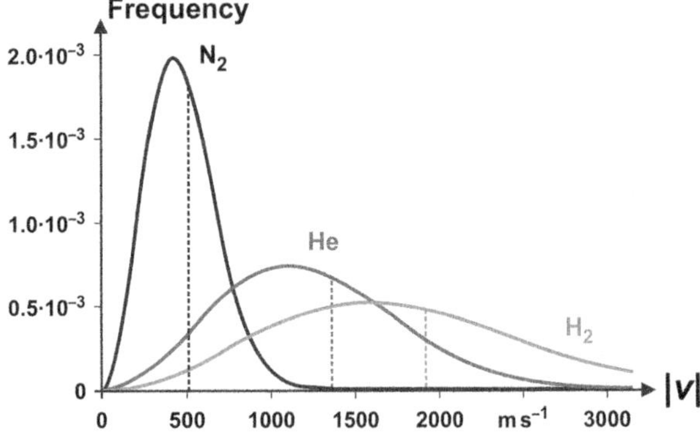

Fig. 10.10 Maxwell's speed distribution for gases with varying molar masses at 298 K.

The distribution is rather narrow for heavy molecules and most particles have a speed close to the mean speed. In contrast, when compared to heavier molecules under the same conditions, the light molecules (such as H_2) have both a noticeably higher mean speed as well as a much wider distribution.

10.5   Barometric Formula and Boltzmann Distribution

We can glean another well-known relation, the density distribution of a gas in a homogeneous gravitational field, by using the same procedure. The Earth's atmosphere whose density decreases just about exponentially with altitude is a good example of this. Similar to what we did in the last section, we will consider the particles at a given altitude h to be molecules of a substance B(h). The basic potentials $\overset{\circ}{\mu}(h)$ of the substances B(h) only differ from each other by the above-mentioned *molar potential energy* $w/\tau = M \cdot g \cdot h$ because they are chemically identical:

$$\overset{\circ}{\mu}\,(h) = \overset{\circ}{\mu}\,(0) + Mgh. \tag{10.55}$$

The exchange of particles between different altitudes can be described as a reaction of the following type:

$$B(h) \rightarrow B(h').$$

When temperatures are uniform, sooner or later equilibrium is established for all the substances so that the potential becomes the same for all altitudes, $\mu(h) = \mu$ for all h. Applying the mass action equation, we obtain

$$\mu = \overset{\circ}{\mu}\,(0) + Mgh + RT\ln\frac{c(h)}{c^{\ominus}} \qquad \text{for all } h \tag{10.56}$$

for the condition of equilibrium. Solving for $c(h)$ yields the equation

$$c(h) = \underbrace{c^{\ominus} \cdot \exp\left(\frac{\mu - \overset{\circ}{\mu}(0)}{RT}\right)}_{c(0)} \cdot \exp\left(-\frac{Mgh}{RT}\right) \tag{10.57}$$

or

$$c(h) = c(0) \cdot \exp\left(-\frac{Mgh}{RT}\right) \quad (\textit{barometric formula}). \tag{10.58}$$

RT/Mg represents the range of the exponential distribution. This is the altitude at which the gas concentration has fallen to $1/e = 36.8\,\%$ compared to the value $c(0)$ at sea level. For nitrogen as the main component of the atmosphere at 300 K, the range is

$$\frac{8.314\,\mathrm{kg\,m^2\,s^{-2}\,mol^{-1}\,K^{-1}} \times 300\,\mathrm{K}}{28 \times 10^{-3}\,\mathrm{kg\,mol^{-1}} \times 9.81\,\mathrm{m\,s^{-2}}} = 9,080\,\mathrm{m}.$$

The altitude at which the concentration has fallen to $\frac{1}{2}c(0)$, the so-called half-height h_{H}, is a factor of ln2 lower, $h_{\mathrm{H}} \approx 6,300$ m. This is a little more than the highest point of Mt. Kilimanjaro. The air around this peak has therefore about half the density of the air on the coast.

As the examples above have shown, the mass action and excitation equations $\mu = \overset{\circ}{\mu} + RT\ln(c/c^{\ominus})$ and $\overset{\circ}{\mu}\,(w) = \overset{\circ}{\mu}\,(0) + w/\tau$ together accomplish what *Boltzmann's distribution law* does. When put together, they seem to represent a special version of this law that is closer to chemistry. We obtain the usual form if we interpret the concentration $c(w)$ of the particle of type $B(w)$ as a measure of the probability $p(w)$ of encountering the B particles in a state with energy w, i.e., $p(w) \sim c(w)$. It is only necessary then to insert the second formula into the first one and to solve for $c = c(w)$, thereby obtaining:

$$c(w) = c^{\ominus} \cdot \exp\left(\frac{\mu - \overset{\circ}{\mu}(0)}{RT}\right) \cdot \exp\left(-\frac{w}{R_{\tau}T}\right), \tag{10.59}$$

or in other words:

$$p(w) \sim e^{-w/k_{\mathrm{B}}T} \quad \text{(Boltzmann's distribution law)}. \tag{10.60}$$

Chapter 11
Substances with Higher Density

If one changes from dilute (ideal) gases to real gases with higher density, the interaction between the particles and the phenomenon of condensation cannot be neglected any longer. The consideration of such effects results in the *van der Waals equation*, a modification of the general gas law. A closer look at the process of condensation leads us to the critical phenomena, meaning the unusual physical properties displayed by substances near their *critical points*. If we are interested to know how the phase transition liquid $\rightleftarrows$ gaseous can be influenced by factors such as temperature and pressure, we can use the T and p dependence of the chemical potential for calculating the *boiling pressure curve* (vapor pressure curve) of a given pure substance. This curve illustrates how the vapor pressure of the substance varies with temperature and is an example of a so-called *phase boundary*. The other phase transitions can also be represented in a $p(T)$ diagram in the form of phase boundaries, producing a complete *phase diagram*. Such a diagram is a kind of "map" which shows the conditions of temperature and pressure at which a certain phase is most stable and illustrates the ranges of existence of stable phases.

11.1 The van der Waals Equation

The *general gas law* is an approximation that becomes more exact the more diluted a gas is. If a gas becomes denser, deviations from the general gas law become increasingly noticeable. We will use an example to illustrate this: Oxygen can be obtained in steel cylinders at pressures up to 20 MPa (200 bar). Under these circumstances, the particles have almost no space to move in, their packing density is similar to a liquid. The characteristics of such a compressed gas are naturally different from those of a dilute gas.

The Dutchman Johannes Diderik van der Waals came up with an enlightening idea for understanding the behavior of gases at higher densities. He based it upon two very simple assumptions:

© Springer International Publishing Switzerland 2016
G. Job, R. Rüffler, *Physical Chemistry from a Different Angle*,
DOI 10.1007/978-3-319-15666-8_11

1. Every particle possesses a certain spatial extension and therefore occupies a
   certain volume. It excludes all other particles from this space.
2. The particles attract each other. The forces of attraction are weak but increase
   quickly as the particles move closer together.

These forces of attraction between uncharged particles with complete electron
shells in gases or liquids are called *van der Waals forces*. Among them are the
interactions between permanent electric dipole moments (also called dipole–dipole
forces or Keesom interaction, named after the Dutch physicist Willem Hendrik
Keesom), as well as the interactions between permanent moments and moments
induced by polarization (dipole-induced dipole forces) and the interactions between
induced moments only (instantaneous dipole-induced dipole forces or London
dispersion forces, named after the German-American physicist Fritz London). Interactions between electric dipoles are comparable to the more familiar ones between
magnetic dipoles. The dispersion forces mentioned last originate from the formation
of "temporary" dipole moments through fluctuations in the electron distribution
where for a short time, even nonpolar particles display an irregular distribution of
electron density that results in a positive or negative partial charge. This dipole,
which was created by "spontaneous polarization," can, itself, induce a "temporary"
dipole moment in a neighboring particle. The fact that even noble gases at low
enough temperatures will become liquid suggested the existence of such dispersion
forces. A characteristic of all these attractive interaction energies between uncharged
particles with permanent or temporary dipole moments is that they decrease with the
6th power of the distance between the molecules: $W_{\text{pot}} \sim -1/r^6$.

Among the van der Waals forces are also the strongly increasing repulsive forces
when the particles "touch" each other. The particles move so closely together that
their electron shells overlap and, as a result of Pauli's exclusion principle, they
begin to repel each other. For practical reasons, we often chose $W_{\text{pot}} \sim 1/r^{12}$ for the
repulsive interaction energy. Attracting and repulsing contributions to the interaction energy can be summed up by the Lennard-Jones (12-6) potential (Fig. 11.1):

$$W_{\text{pot}} = \frac{A}{r^{12}} - \frac{B}{r^6}.$$

The short-range repulsive interaction effect was taken into account by van der
Waals by the first of the assumptions above. This assumption implies that the gas
particles do not have the entire volume of a container available for motion. The
volume needed to be reduced by a contribution determined by the volume from
which the particles exclude each other. This volume unavailable for molecular
motion is called the *co-volume* (van der Waals volume) of a gas. The assumed
attraction in point 2 leads to the gas particles moving more closely together, just as
if there was additional pressure upon them. This "pressure" or "pull" (or "tensile
stress") caused by the forces of attraction is called the *internal pressure* or *cohesion
pressure* of a gas. Van der Waals assumed that the general gas law should continue
to be valid, except for the following two changes, a lessening of volume by the

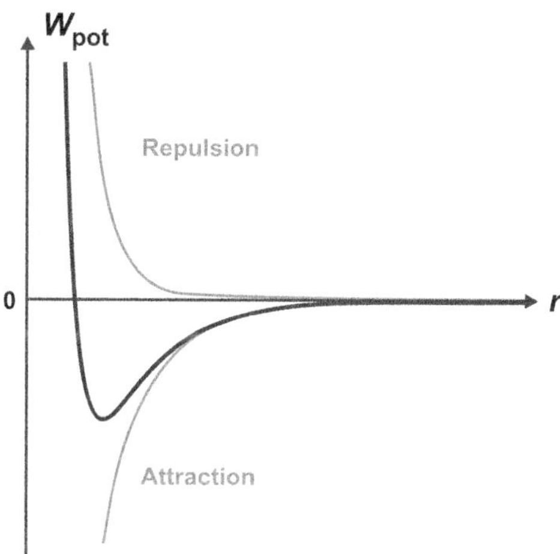

Fig. 11.1 Lennard-Jones potential (*black*) as the sum of the attracting and repelling parts (*gray*).

"unavailable" co-volume V_W and an increase of pressure by the internal pressure p_{int}:

$$(p + p_{int}) \cdot (V - V_W) = nRT. \tag{11.1}$$

We obtain the formula above by replacing, or rather "correcting," the pressure with $p + p_{int}$, and the volume with $V - V_W$ in the general gas law. The van der Waals volume V_W, from which the particles exclude each other, naturally grows along with the number of particles, meaning with the amount of gas n, so we can conclude that:

$$V_W \sim n. \tag{11.2}$$

However, internal pressure p_{int} increases with the square of the gas concentration $c = n/V$,

$$p_{int} \sim c^2 \sim \left(\frac{n}{V}\right)^2. \tag{11.3}$$

To derive this, let us visualize more exactly how "internal pressure" comes to be. We envision a plane section with the surface A at an arbitrary position in a gas (Fig. 11.2). We assume that each particle attracts every other one within the "range" l of the intermolecular forces. In particular, the particles on one side of the section attract the ones on the other side. This assumption has a simple consequence: The greater the number N_1 of particles on one side of the surface in the volume $V_1 = A \cdot l$, and the greater the particle number N_2 on the other side of it in the

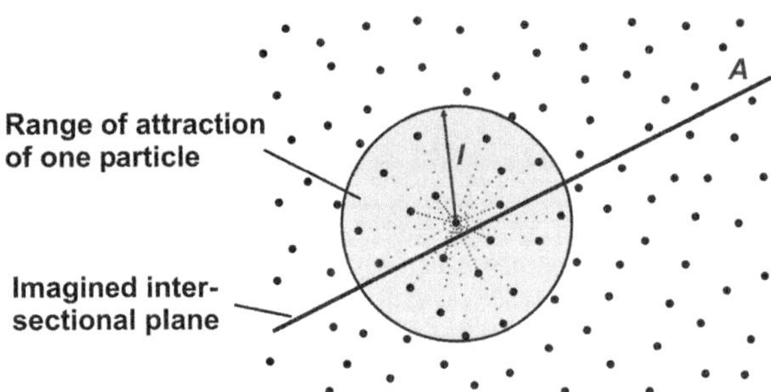

Fig. 11.2 Illustration of "internal pressure." A: imagined intersectional plane through the gas. The *gray circle* with the radius l, the "range" of the intermolecular forces, characterizes the "range of attraction" for an individual particle.

volume $V_2 \sim l^3$ within the range l, the stronger the total force of attraction, i.e., $F \sim N_1 \cdot N_2$. Because N_1 as well as N_2 are proportional to the particle density N/V and, therefore, to concentration $c = n/V$, the internal pressure (meaning the internally directed pull that is produced by the substance itself leading to a moving together of the particles) results in:

$$p_{\text{int}} = F/A \sim N_1 \cdot N_2/A \sim (V_1 \cdot c) \cdot (V_2 \cdot c)/A \sim A \cdot l \cdot l^3 \cdot c^2/A \sim n^2/V^2.$$

The dependencies for internal pressure and for co-volume will now be inserted into Eq. (11.1). In this form, it is known as the *van der Waals equation*:

$$\left(p + \frac{an^2}{V^2} \right) \cdot (V - bn) = nRT \qquad \text{van der Waals equation.} \tag{11.4}$$

It was derived by van der Waals in his dissertation of 1873.

The two substance-specific proportionality constants a and b are called *van der Waals constants*. In the case of water they are:

$$a(\text{H}_2\text{O}) = 0.55\,\text{Pa}\,\text{m}^6\,\text{mol}^{-2},$$

$$b(\text{H}_2\text{O}) = 2.7 \times 10^{-5}\,\text{m}^3\,\text{mol}^{-1} \approx\approx V_{\text{m}}(\text{H}_2\text{O}|\text{l}) (\approx\approx \text{ read: approximately equal}).$$

Further values of empirically determined constants for various gases can be found in Table 11.1.

The a constants for the various gases are quite different from each other due to strongly varying forces of interaction. In contrast, the b constants vary only slightly from each other. This means that the space required by the different particles is relatively similar.

If the van der Waals equation is transformed, the $p(V)$ isotherms can be calculated:

$$p = \frac{nRT}{V - bn} - \frac{an^2}{V^2}. \tag{11.5}$$

Table 11.1 Van der Waals constants for various gases (from: Lide D R (ed) (2008) CRC Handbook of Chemistry and Physics, 89th edn. CRC Press, Boca Raton).

Gas	a (Pa m^6 mol^{-2})	b (m^3 mol^{-1})
H_2	0.0245	2.7×10^{-5}
He	0.0035	2.4×10^{-5}
N_2	0.137	3.9×10^{-5}
O_2	0.138	3.2×10^{-5}
Cl_2	0.634	5.4×10^{-5}
Ar	0.136	3.2×10^{-5}
CO_2	0.366	4.3×10^{-5}
CH_4	0.230	4.3×10^{-5}
C_2H_2	0.452	5.2×10^{-5}
NH_3	0.423	3.7×10^{-5}
H_2O	0.554	3.1×10^{-5}

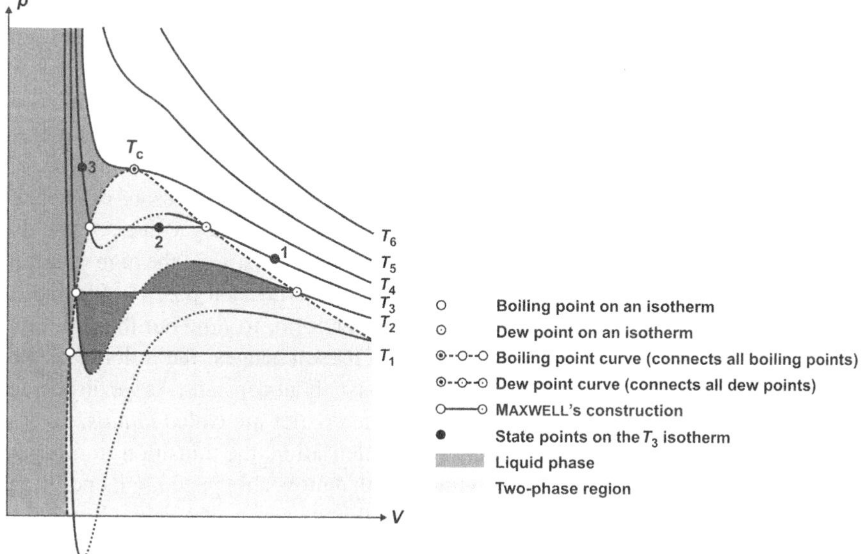

Fig. 11.3 $p(V)$ isotherms of a real gas according to van der Waals equation (with $T_1 < T_2 < \ldots < T_6$).

Figure 11.3 shows some of these isotherms for different temperatures $T_1 < T_2 < \ldots < T_6$. The graphic makes it clear that, at higher temperatures, the isotherms obtained by the van der Waals equation resemble the hyperbolas of the Boyle–Mariotte law. This is understandable because at high temperatures, the product nRT becomes so great that the second term in Eq. (11.5) can be ignored. In addition, at high volumes (and low pressures), $V \gg bn$, so that the van der Waals equation gives way to the general gas law, $p = nRT/V$.

Below a certain temperature (T_4), minima and maxima appear. These delineate a physically unrealistic part of the curve (which is therefore shown by a dashed line).

Volume would increase along with increase of pressure in this range, which contradicts experience. In order to understand the behavior of real gases at low temperatures, we must study *condensation*.

11.2 Condensation

When a gas is expanded, its internal pressure decreases quickly as a consequence of the quadratic concentration dependency. The internal pressure eventually becomes imperceptible and can be ignored when dealing with dilute gases. Conversely, internal pressure increases sharply when the density of a gas is increased. This happens so strongly that, at very high concentrations, a gas will become unstable and collapse. This is called *condensation*. Experiment 11.1 illustrates this phenomenon.

In the condensed state, particles adhere to each other. The short distances between them lead to their attraction being very strong and they hold together tightly. This is what causes the stability of a condensed state. The particles' speed is the same as it would be in a gas of the same temperature. At room temperature, this is a few hundred m s^{-1}, which is just about the speed of sound! This "frantic swarm of particles" can only hold together under the enormous internal pressure of several thousand bar. Although these particles adhere to each other, they will also quickly slide by each other, when the temperature is not too low. Because of the high speed at which they move, they change positions very fast. This makes it possible for such a dense swarm of particles to adapt to any container form, to *flow* out through tiny openings and to *flow* through pipes. During all these changes, the volume of the substance (of the ensemble of particles) remains just about constant. As we discussed in Sect. 1.6, substances with these kinds of characteristics are called *liquids*.

Let us take a closer look at the process of condensation, the transition from a gas to a liquid (Fig. 11.4). We imagine a cylinder with a moveable piston in it enclosing a certain amount of gas. The temperature should have a constant value during the entire process. It should therefore be possible for the excess entropy that would cause warming (or when missing, would cause supercooling) to escape through the

Experiment 11.1 *Using pressure to liquefy a gas*: If butane (the fuel in a gas lighter) is filled into a glass pressure cylinder with a piston and the piston is pressed down, it produces a visible amount of liquid.

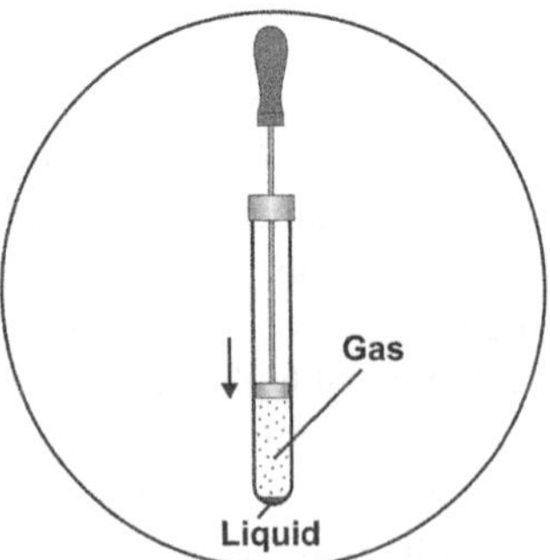

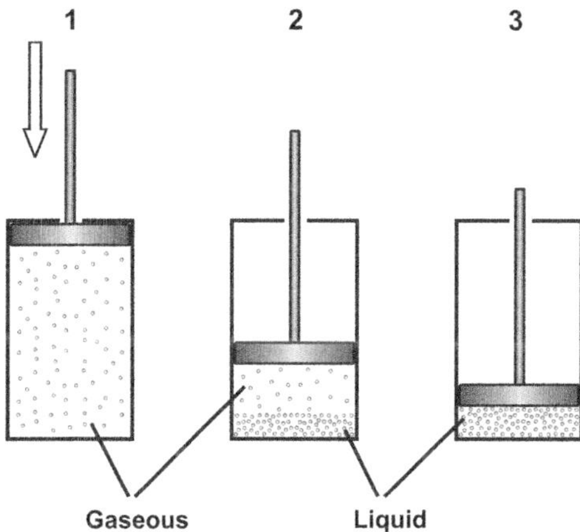

Fig. 11.4 Compressing a gas using a moveable piston (1), beginning condensation (2) and final compression of the liquid produced by condensation (3).

walls into the surroundings (or to be absorbed from there). Numbers 1, 2, and 3 refer to states indicated on the van der Waals isotherm at temperature T_3 (Fig. 11.3).

1. A gradual decrease of volume leads to a slow rise in "thermal" pressure, caused by molecular collisions, which drives the particles apart. Internal pressure simultaneously increases tending to concentrate the particles in a small space.
2. Because internal pressure increases more quickly than thermal pressure there exists a point at which the increase of thermal pressure cannot compensate for the increase of the internal pressure anymore. At this so-called *dew point*, the gas begins to "collapse," i.e., to condense.

   The gas does not condense all at once, but only a fraction of it does at first. In the process, a small number of particles disappear from the space occupied by the gas making the concentration and internal pressure decrease and causing the gas phase to restabilize.

   If the volume is further decreased, the processes repeat. More liquid forms while the rest of the gas stabilizes. If this is carried out further and further, the entire gas is eventually "squeezed" into a liquid state. Pressure will remain constant as long as temperature is unchanged during the condensation process. The gas above the liquid is usually called the liquid's *vapor*. The pressure in the gas compartment is called the *vapor pressure* of the liquid. Vapor pressure is independent of the amount of liquid it is associated with, whether it is one drop or a whole containerful.
3. Further compression leads to a strong increase of pressure because a liquid is much harder to compress than a gas.

We return once more to the van der Waals isotherms (Fig. 11.3): The van der Waals curves describe fairly exactly compression of gases up to the dew point. If the volume is decreased beyond this point, pressure does not increase, but condensation sets in. This occurs at constant pressure, meaning that the corresponding piece of the curve must be horizontal until the gas phase completely disappears at the *boiling point*. These lines have been constructed so that the areas enclosed by the van der Waals curves above and below the straight line are equal (Maxwell construction or equal area rule) (compare Fig. 11.3 where the two areas in the case of the T_2 isotherm are dark gray). Both gas and liquid exist simultaneously along these lines. The subsequent steep rise in pressure with further decrease of volume is characteristic of the low compressibility of liquids.

11.3 Critical Temperature

As temperature rises, it becomes harder to bring about a condensation process. The gas involved must be much more strongly compressed before it begins to condense and the two-phase region becomes more and more constricted. As we know, higher temperatures are related to faster particle motion. At higher velocities, higher internal pressure, i.e., higher compression is necessary to force the particles to collapse. If the temperature is continuously raised, making condensation more difficult, we eventually reach a critical value above which condensation is no longer possible. This limit is called the *critical temperature* T_c. Together with T_c, the corresponding values of critical pressure p_c and critical volume V_c make up the *critical point* of the substance in question. At this critical point, the corresponding van der Waals isotherm displays a saddle point, meaning a point of inflection with a horizontal tangent that simultaneously determines the maximum of the two-phase region (Fig. 11.3). Above this critical temperature, gases cannot be liquefied by compression. The critical temperatures for gases with weak attractive forces, such as helium, hydrogen, nitrogen, and oxygen, are far below room temperature so they remain gaseous even under high pressures. Gases like carbon dioxide, ammonia and water vapor, however, have higher critical temperatures and can be liquefied at room temperature if enough pressure is applied (Table 11.2).

If a gas is compressed at a temperature above the critical temperature, the result is a dense fluid whose characteristics are neither clearly liquid nor gaseous. Although this fluid's density is comparable to a liquid, making it possible to be used as a solvent, there is no surface visible between liquid and gas. Such a medium is called a *supercritical fluid*. It combines the positive characteristic of low viscosity of gases with the good solvent power of liquids, making it an attractive solvent for separating processes such as high-pressure extraction, polymer fractionation, and cleaning of monomers.

A further advantage is that it can be entirely removed from a product by expansion leaving no undesired, possibly poisonous, residue. Supercritical carbon

Table 11.2 Critical temperatures and pressures for some substances (from: Lide D R (ed) (2008) CRC Handbook of Chemistry and Physics, 89th edn. CRC Press, Boca Raton).

Substance	Critical temp. T_c (K)	Critical pressure p_c (MPa)
H_2	33.0	1.29
He	5.2	0.23
N_2	126.2	3.39
O_2	154.6	5.04
Cl_2	416.9	7.99
Ar	150.9	4.90
CO_2	304.1	7.38
CH_4	190.6	4.60
C_2H_2	308.3	6.14
NH_3	405.6	11.36
H_2O	647.1	22.06

dioxide is used to extract caffeine from tea or coffee. Supercritical water is used to dissolve quartz (and many other minerals) leading to hydrothermal solutions. The process of crystallizing on seed crystals produces high purity quartz monocrystals that can be cut into disks to be piezoelectric crystals in watches, for example. Hydrothermal solutions make also essential contributions to the formation of most vein deposits and ore stocks.

11.4 Boiling Pressure Curve (Vapor Pressure Curve)

We will consider the process of condensation from another point of view by using a $p(T)$ diagram (Fig. 11.5).

1. We begin with a state in which the gas is strongly expanded so its pressure is very low. We will keep the temperature at a constant value by conducting the entropy squeezed out by compression to the surroundings. A certain pressure will eventually be reached, the so-called dew point mentioned above, at which the gas begins to condense, or to "dew." Pressure remains constant during the condensation process. Only when the gas is completely condensed and the piston lies upon the surface of the liquid, does the pressure further increase. We conclude that the cylinder contains only gas up to the dew point and above the dew point, only liquid.
2. If this experiment is carried out again at a higher temperature, the process runs the same, but the dew point is shifted to a higher pressure (Fig. 11.6).
3. When all the dew points measured in this way are connected, a steep curve to the right is the result, the so-called *boiling pressure curve* or *vapor pressure curve* (Fig. 11.7). It gives the values of pressure and temperature at which gas and

Fig. 11.5 Representation of the dew point in a $p(T)$ diagram.

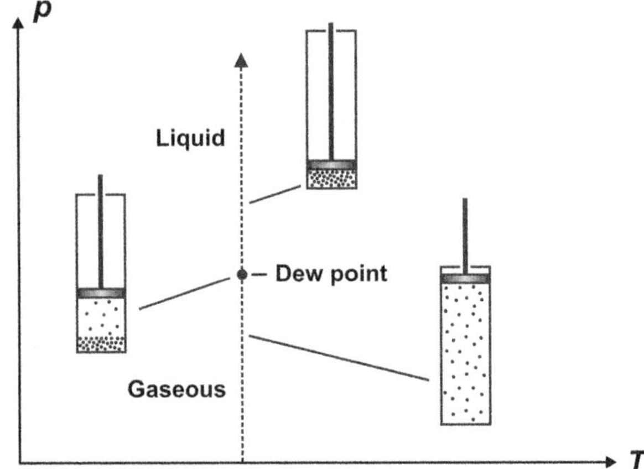

Fig. 11.6 Temperature dependency of dew points.

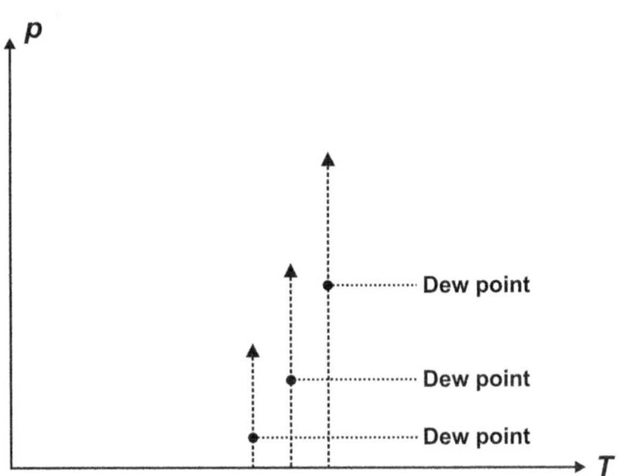

Fig. 11.7 Temperature dependency of a liquid's vapor pressure (boiling pressure curve or vapor pressure curve).

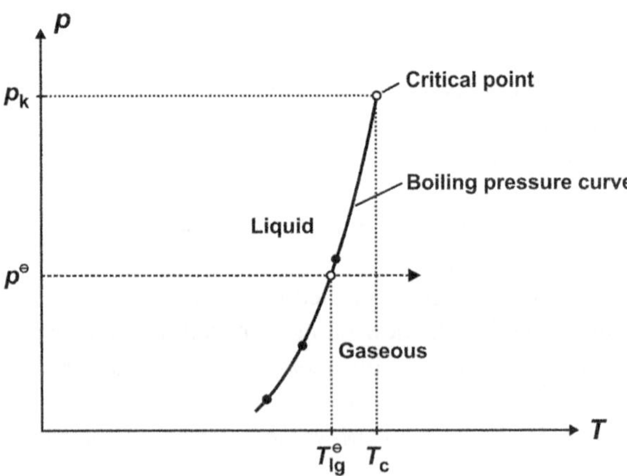

liquid are in equilibrium with each other. Only gas exists at the points below this curve and only liquid exists above it.

4. When a liquid condensate is heated at constant pressure (e.g., ambient air pressure $p^{\ominus}$ of 100 kPa), it will transform into a gaseous state and *boil* as it goes past the boiling pressure curve. The temperature at which this occurs is the *boiling temperature* (boiling point) associated with the pressure in question (e.g., $T_{lg}^{\ominus}$). The boiling point of a substance is not a constant but depends upon pressure. Tabulated values are based upon standard pressure of 100 kPa. In order to distinguish this special boiling point from other boiling points, we call it the *standard boiling point* and characterize it by $T_{lg}^{\ominus}$.

5. No dew point exists above the critical temperature T_c. This is the highest temperature at which a liquid can exist. For this reason, the boiling pressure curve ends at the critical temperature—in the critical point mentioned above.

How can we now quantitatively approximate the boiling pressure curve? To do this we will need to refer back to the chemical potential. As we have seen, every phase of a substance has its own chemical potential, which is dependent upon temperature and pressure. We can easily calculate these potentials for different temperatures and pressures. At an arbitrary condition, the most stable phase is the one with the lowest chemical potential. The stability range of the liquid phase is characterized by the chemical potential $\mu_l(p, T)$ being lowest there. Where the gaseous phase is stable, $\mu_g(p, T)$ is minimal.

In order to calculate the curve that separates the two stability ranges (the curve that reflects the equilibrium between two phases), we can refer back to the formula derived earlier for calculating the reaction pressure of gases (compare Sect. 5.5) because the boiling process can be considered a reaction:

$$B|l \rightarrow B|g.$$

If we proceed as we did in deriving Eq. (5.19), we obtain for the (saturation) vapor pressure p_{lg} in equilibrium in a closed, initially evacuated container:

$$p_{lg} = p_0 \exp \frac{\mathcal{A}_{lg,0}}{RT}. \tag{11.6}$$

The index $_{lg}$ (or in more detail, $_{l \rightarrow g}$) indicates the type of process. In this case, it is vaporization (the transition from liquid to gaseous state). p_0 is the arbitrarily chosen initial pressure and $\mathcal{A}_{lg,0} = \mu_{l,0} - \mu_{g,0}$ is the drive at the chosen initial state (T_0, p_0). The pressure dependency of the chemical potential of the liquid phase is ignored here. We are allowed to do this because it is generally smaller by several orders of magnitude than that of gases as long as the pressure p does not exceed standard pressure $p^{\ominus}$ by much. For this reason, our approach becomes invalid near the critical point where pressures are usually far above $p^{\ominus}$. In order to calculate the equilibrium pressures corresponding to various temperatures T, we need to use the chemical potentials and the corresponding drive at the chosen temperatures. We

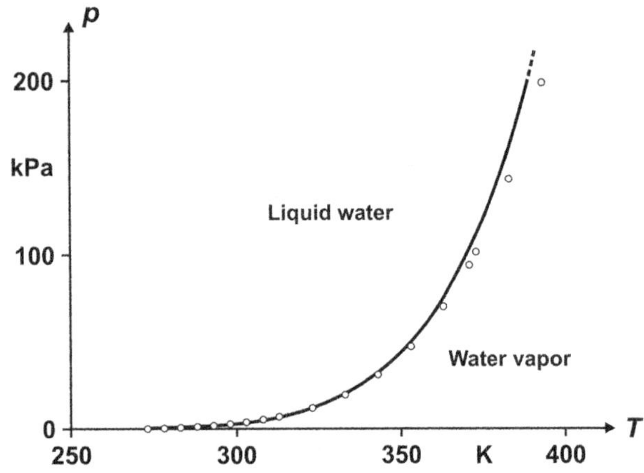

Fig. 11.8 Temperature dependency of saturation vapor pressure of water (comparison of the calculated curve with measured values).

will again make use of our linear approach by replacing $\mathcal{A}_{\mathrm{lg},0}$ by $\mathcal{A}_{\mathrm{lg},0} + \alpha_{\mathrm{lg}}(T - T_0)$ in Eq. (11.6). We obtain:

$$p_{\mathrm{lg}} = p_0 \exp\frac{\mathcal{A}_{\mathrm{lg},0} + \alpha_{\mathrm{lg}}(T - T_0)}{RT} \tag{11.7}$$

or, after transforming the equation and leaving out index $_{\mathrm{lg}}$ for simplicity:

$$\ln\frac{p}{p_0} = \frac{\mathcal{A}_0 + \alpha(T - T_0)}{RT} = \frac{(\mathcal{A}_0 - \alpha T_0)/R}{\vartheta + 273\,\mathrm{K}} + \frac{\alpha}{R}. \tag{11.8}$$

Equation (11.8) formally corresponds to the August vapor pressure formula. In 1862, Ernst Ferdinand August developed a formula based upon water vapor pressure, having the form $\lg\{p\} = -A/(\vartheta + C) + B$; the quantities A, B, C were parameters to be determined empirically, ϑ indicated the Celsius temperature, and $\{p\}$ the numerical value of the pressure.

In Fig. 11.8, the temperature dependency of vapor pressure is shown using water as the example, where the data for standard conditions (298 K, 100 kPa) were used, i.e.:

	$H_2O|l$	$\rightarrow$	$H_2O|g$	
$\mu^{\ominus}$ (kG):	-237.1		-228.6	$\Rightarrow \mathcal{A}^{\ominus} = -8.5$ kG
α (G K^{-1}):	-70		-189	$\Rightarrow \alpha = +119$ G K^{-1}

and therefore,

$$p_{\mathrm{lg}} = 100\,\mathrm{kPa} \times \exp\left(\frac{-8.5 \times 10^3 + 119 \times (T/\mathrm{K} - 298)}{8.314 \times T/\mathrm{K}}\right).$$

If the critical point is far enough away and if we limit ourselves to a relatively small temperature span, Eq. (11.7) will (despite the approximations used) yield a quite useful result. This can be seen in the comparison with experimental values.

As we have shown in Chap. 9, the temperature coefficient α of the drive corresponds to a molar reaction entropy. In this case, this is the change of entropy $\Delta_{lg}S = S_{m,g} - S_{m,l}$ due to the vaporization process at the initial point, meaning at temperature T_0 and pressure p_0. If we abbreviate to $\Delta_{lg}S_0$, we obtain:

$$p_{lg} = p_0 \exp\frac{\mathcal{A}_{lg,0} + \Delta_{lg}S_0 \cdot (T - T_0)}{RT}. \tag{11.9}$$

If instead of choosing an arbitrary initial state when calculating the boiling pressure curve (characterized by temperature T_0 and pressure p_0), we use the special case of an equilibrium state, meaning a known boiling point, e.g., the standard boiling point $T_{lg}^{\ominus}$, the relation (11.9) can be simplified: In equilibrium, the drive of the vaporization process equals zero, and we find

$$p_{lg} = p^{\ominus} \exp\frac{\Delta_{vap}S^{\ominus} \cdot \left(T - T_{lg}^{\ominus}\right)}{RT}. \tag{11.10}$$

If we set $y = \ln\left(p_{lg}/p^{\ominus}\right)$ and $x = T^{-1}$, the result is an equation of a straight line:

$$\underbrace{\ln\frac{p_{lg}}{p^{\ominus}}}_{y} = -\underbrace{\frac{\Delta_{vap}S^{\ominus} \cdot T_{lg}^{\ominus}}{R}}_{m} \cdot \underbrace{\frac{1}{T}}_{x} + \underbrace{\frac{\Delta_{vap}S^{\ominus}}{R}}_{b}. \tag{11.11}$$

Here, $\Delta_{vap}S^{\ominus}$ is the *molar entropy of vaporization* at the standard boiling point. Now when $\ln\left(p_{lg}/p^{\ominus}\right)$ is plotted as a function of $1/T$ for various substances, we notice that the intersection points b with the ordinate of the (extrapolated) straight lines lie rather close together for many nonpolar compounds. This means that the molar vaporization entropies are also similar, with an average of about 88 Ct K^{-1} mol^{-1}. This was already recognized by Frederick Thomas Trouton in 1884 through his work comparing measured values in Tables (*Pictet–Trouton's rule*). The reason for the approximate concordance of molar vaporization entropies is the comparably great increase of "disorder" for a transition from a relatively highly condensed phase to a gas with particles far away from each other. Great deviations from this rule can be explained by strong interactions between the molecules and a higher degree of order in the corresponding phase. For example, hydrogen bridge bonds form in liquid water (and other polar substances) resulting in higher entropy of vaporization $\left(\Delta_{vap}S^{\ominus}(H_2O) = 109.1\ \text{Ct K}^{-1}\ \text{mol}^{-1}\right)$.

11.5   Complete Phase Diagram

The other phase transitions of a pure substance can also be represented in a $p(T)$ diagram in the form of *phase boundaries*, producing a complete *phase diagram*. This diagram shows the conditions of temperature and pressure under which a certain phase is most stable and illustrates the ranges of existence of stable phases.

1. Below a certain temperature, a condensate forms as a solid and not a liquid. The direct transition from a gaseous to a solid state, or vice versa, is called *sublimation* and the vapor pressure curve of a solid substance, the *sublimation (pressure) curve* (Fig. 11.9). Gas and solid are in equilibrium with each other along this phase boundary. As long as pressure remains below this curve at constant temperature during the compression of the gas, only gas will be present in the cylinder. However, if the pressure lies above this, only a solid condensate will appear.
2. The almost vertical curve in the illustration (Fig. 11.10), called the *melting (pressure) curve*, separates the regions at which solid or liquid condensate is stable. The slope of this curve has been greatly exaggerated for the sake of clarity.
3. When a solid condensate is heated at constant pressure (e.g., at ambient air pressure of 100 kPa) starting at the absolute zero point, it will go into a liquid state as it exceeds the melting pressure curve: it *melts*. The temperature at which this occurs, where both solid and liquid phase are in equilibrium, is the *melting temperature* (melting point) corresponding to the pressure (e.g., $T_{sl}^{\ominus}$). When the substance is further heated, it finally reaches the boiling pressure curve and will begin to boil (compare Sect. 11.4).

   Like the boiling point of a substance, its melting point is not a constant but, to a much lesser extent, also depends upon pressure. The *standard melting point* $T_{sl}^{\ominus}$ is based upon a pressure of 100 kPa.

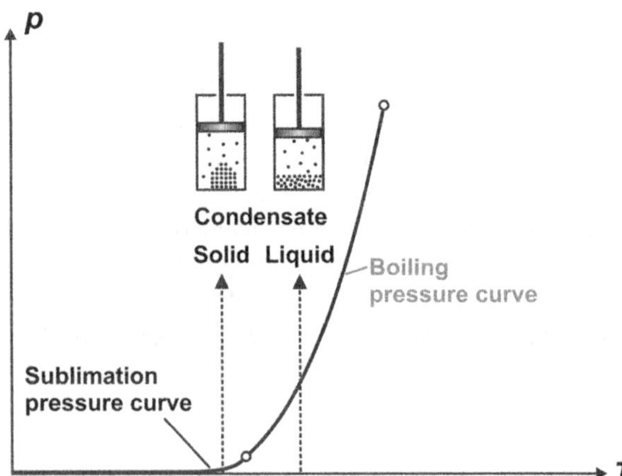

Fig. 11.9 Sublimation and boiling pressure curve of a substance.

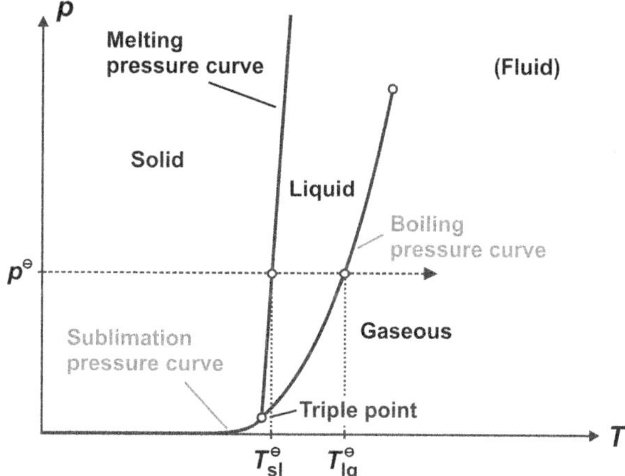

Fig. 11.10 Typical phase diagram of a pure substance.

4. The point at which the sublimation, boiling, and melting pressure curves all converge is called the *triple point* because at the conditions of temperature and pressure there, the substance is simultaneously in a solid, liquid, and gaseous state. At the triple point both pressure and temperature are characteristic properties of a pure substance. The triple point of water, for example, is at 273.16 K and 611 Pa. Only at this exact temperature and this exact pressure are ice, liquid water, and water vapor in equilibrium with each other. This triple point is used to define the unit called *Kelvin* (compare Sect. 3.8). If the pressure at the triple point lies noticeably above 100 kPa, the liquid state cannot exist no matter what the temperature, and only sublimation can be observed. An example of this is carbon dioxide (217 K, 518 kPa), which, when exposed to air, changes directly from solid to gas (so it is called "dry ice").

The other phase boundaries can also be calculated using the chemical potential. For example, the sublimation pressure curve can be described analogously to the boiling pressure curve. We only need the drive $\mathcal{A}_{sg,0} = \mu_{s,0} - \mu_{g,0}$ responsible for sublimation and the corresponding change of entropy, the molar sublimation entropy $\Delta_{sg}S_0$ (both at temperature T_0 and pressure p_0):

$$p_{sg} = p_0 \exp \frac{\mathcal{A}_{sg,0} + \Delta_{sg}S_0 \cdot (T - T_0)}{RT}. \tag{11.12}$$

When temperature increases, the sublimation pressure increases just as boiling pressure does, both reciprocally and exponentially $(\sim e^{-a/T})$.

A linear approach for both the temperature and pressure dependency of the chemical potential is sufficient for calculating the melting pressure curve (compare Sect. 5.4). For the process

$$B|s \rightarrow B|l$$

we obtain the following condition in the case of equilibrium ($\mu_s = \mu_l$):

$$\mu_{s,0} + \alpha_s \cdot \Delta T + \beta_s \cdot \Delta p = \mu_{l,0} + \alpha_l \cdot \Delta T + \beta_l \cdot \Delta p.$$

Because of $\mathcal{A}_{sl,0} = \mu_{s,0} - \mu_{l,0}$ as well as $\alpha_{sl} = \alpha_s - \alpha_l$ and $\beta_{sl} = \beta_s - \beta_l$, converting results in:

$$\mathcal{A}_{sl,0} + \alpha_{sl} \cdot \Delta T = -\beta_{sl} \cdot \Delta p$$

and therefore with $\Delta p = p_{sl} - p_0$ and $\Delta T = T - T_0$, if one solves for p_{sl}:

$$p_{sl} = p_0 - \frac{\mathcal{A}_{sl,0} + \alpha_{sl} \cdot (T - T_0)}{\beta_{sl}}. \tag{11.13}$$

$\mathcal{A}_{sl,0}$ represents the drive of the melting process at the chosen initial state (in this case, a temperature of T_0 and pressure of p_0). The temperature coefficient α_{sl} of the drive corresponds to the molar reaction entropy $\Delta_{sl}S_0$ of the melting process and the pressure coefficient β_{sl} corresponds to the negative molar reaction volume $-\Delta_{sl}V_0$ (compare Chap. 9), both at the initial state (T_0, p_0). This yields:

$$p_{sl} = p_0 + \frac{\mathcal{A}_{sl,0} + \Delta_{sl}S_0 \cdot (T - T_0)}{\Delta_{sl}V_0}. \tag{11.14}$$

The linear slope of the melting curve (gradient $\Delta_{sl}S_0/\Delta_{sl}V_0$) is positive for most substances (just as the reciprocally exponential gradient of the vapor pressure curves is) because $\Delta_{sl}S_0$ is always positive and $\Delta_{sl}V_0$ almost always is. There are very few substances—water is the best-known example (Fig. 11.11)—that contract during melting so that $\Delta_{sl}V_0$ becomes negative.

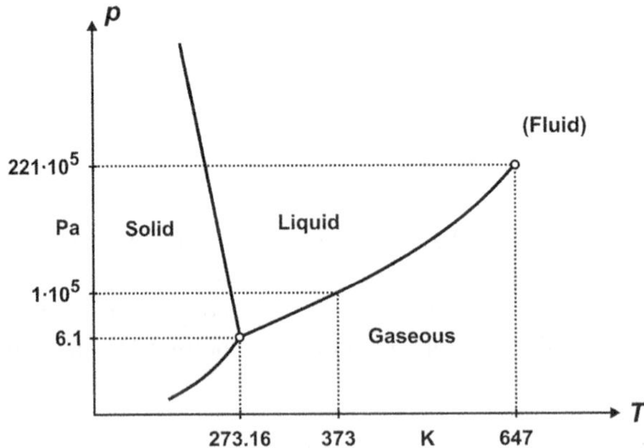

Fig. 11.11 Phase diagram of water (schematic).

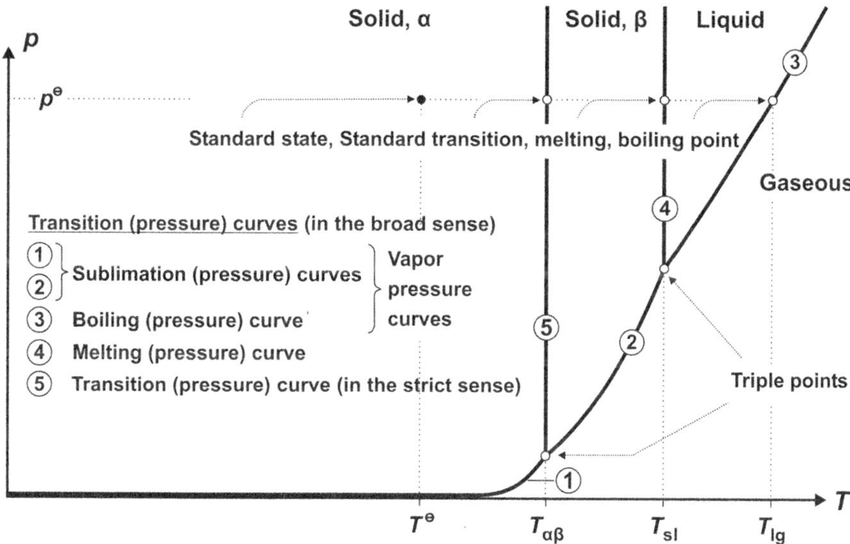

Fig. 11.12 Phase diagram of a substance with two modifications α and β at low pressures.

If we use the special case of an equilibrium point as the initial state, meaning a known melting point, perhaps the standard melting point $T_{sl}^{\ominus}$, then because the drive of the melting process equals zero, the relation (11.14) simplifies to

$$p_{sl} = p^{\circ} + \frac{\Delta_{fus}S^{\ominus}}{\Delta_{fus}V^{\ominus}} \cdot \left(T - T_{sl}^{\ominus}\right). \tag{11.15}$$

Here, $\Delta_{fus}S^{\ominus}$ is the molar entropy of fusion at the standard melting point and $\Delta_{fus}V^{\ominus}$ is the change in molar volume that occurs on melting.

While all substances in gaseous state and most substances in liquid state [except for certain compounds consisting of long-chain molecules which may flow like a liquid, but whose molecules may be oriented in a crystal-like way (so-called liquid crystals)] only form a single phase, almost all solid substances exist in various phases or *modifications* (compare Sect. 1.6). Because of these phenomena, which are known as *allotropy* for elements and as *polymorphism* for compounds, the phase diagram includes additional *transition pressure curves* (in the strict sense; Fig. 11.12). They separate the regions of existence of two different modifications (e.g.. α and β) and lead to new triple points. The actual single substance phase diagram for pressure values in the low-pressure range is an example of this.

At other pressures, especially in the high-pressure range, further solid phases γ, δ, ε … can appear (Fig. 11.13), whose regions of existence can be approximated using the same paradigm.

Fig. 11.13 Single substance phase diagram with further modifications at higher pressure.

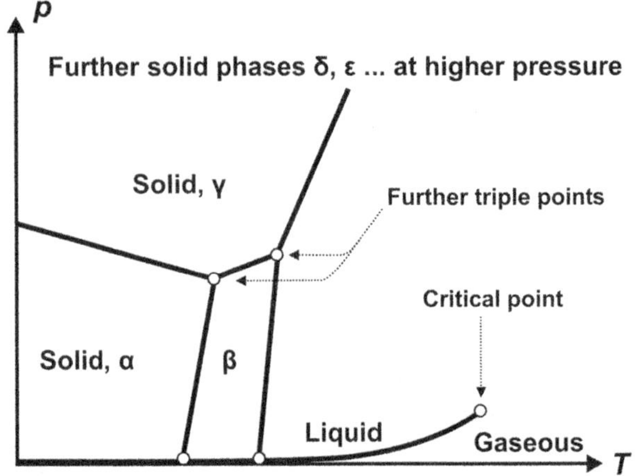

Fig. 11.14 Section of the phase diagram of water at higher pressure.

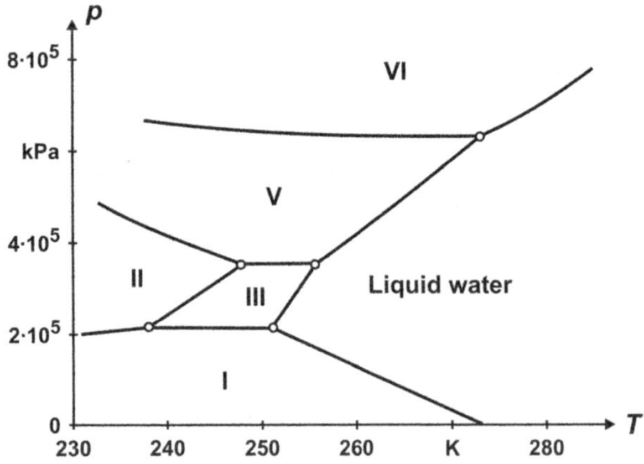

An example of this is water that, when under high enough pressure, forms not only the usual ice (I) but other solid phases as well (Fig. 11.14). These phases have different arrangements of H_2O molecules. In the figure, we see that phase IV is missing. This is simply because people originally believed that a new phase had been discovered which was later found to actually not exist at all. The numbering was kept, though for the sake of simplicity.

Chapter 12
Spreading of Substances

So far, the discussion of the chemical potential has concentrated primarily on chemical reactions and phase transitions. But another property of substances is also of great importance: their tendency to spread out or disperse in space. The phenomenon of diffusion will be explained in this context. The subject area of this chapter also includes the discussion of the effect of a small amount of solute on certain properties of a solution. The properties we have in mind are the lowering of vapor pressure of the solvent, the elevation of its boiling point, the lowering of its freezing point, and last but not least the origin of osmotic pressure. These phenomena are found everywhere, in households and in nature but also in engineering. In everyday life, a prime example for freezing-point depression is the melting effect of road salt. Or have you ever asked yourself why juice is drawn out of sugared strawberries but cherries swell up and burst after a long rain? Then have a look at Sect. 12.4 dealing with osmosis. For a quantitative discussion of all these phenomena, we first have to learn about indirect mass action and the corresponding colligative lowering of chemical potential.

12.1 Introduction

Until now, when we have considered the chemical potential we have primarily concentrated upon chemical reactions and phase transitions. However, there is another characteristic of substances that is almost as important. This is the tendency to spread out or disperse in space whether it is "empty" or filled with matter. This phenomenon can easily be illustrated by everyday processes. Mostly, substances migrate extremely slowly and in infinitesimally small amounts so that this migration remains imperceptible. However, there are many examples of spreading that

© Springer International Publishing Switzerland 2016
G. Job, R. Rüffler, *Physical Chemistry from a Different Angle*,
DOI 10.1007/978-3-319-15666-8_12

Experiment 12.1 *Redistribution of water between zwieback and bread*: A piece of zwieback stored for 2 or 3 days in a bread box or plastic bag together with fresh bread absorbs moisture and becomes soft and bendable, while a slice of bread in a bag full of zwieback becomes hard and brittle from losing moisture to the dry zwieback.

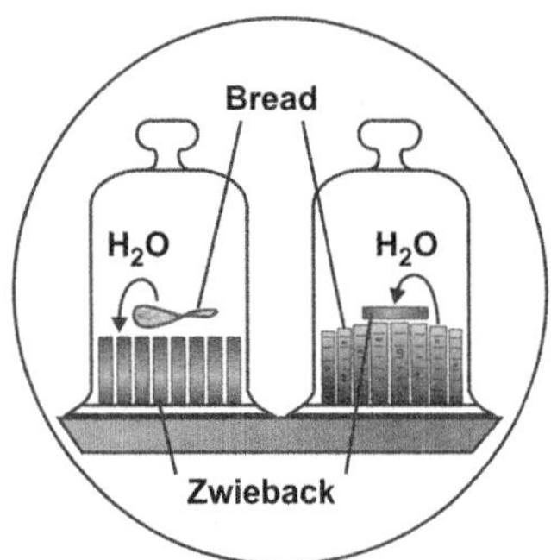

Experiment 12.2 *Spreading of KMnO₄ in agar gel* (view from above on a thin layer of gel in a Petri dish): A few small KMnO$_4$ crystals are cautiously distributed on the agar gel. Immediately, a kind of red violet "halo" is formed around each crystal. Because of its color, the spreading of the "halo" away from the source can be observed easily.

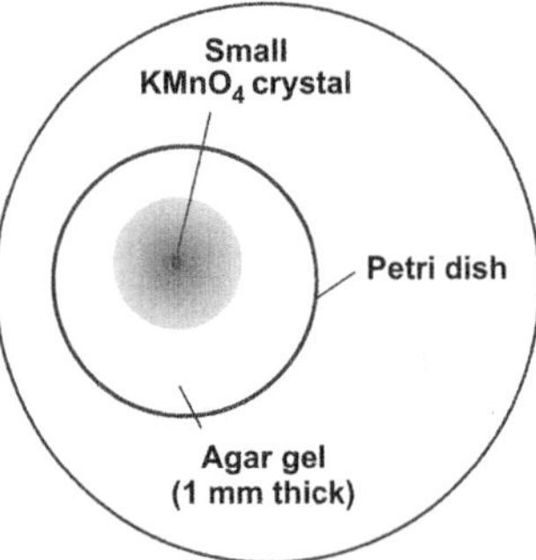

are quite noticeable. When the aroma of freshly ground, unpacked coffee escapes over the course of a few days, when the water in a rain puddle evaporates in a few hours, when glue from a tube congeals in one minute, or when the ink from a felt pen dries within seconds are all examples of how mobile and volatile some substances can be. Experiment 12.1 is an example that shows clearly how substances do not disappear but simply redistribute.

It is easy to follow the spreading of strong smelling or colored substances. The scent of a bouquet of lilacs, a peeled orange, or the pungent odor of potent cheese quickly fills a whole room. The spreading of colored low-molecular chemicals such as potassium permanganate in a liquid, or better, in a gel (to hinder convection) happens so fast that it can readily be observed (Experiment 12.2).

Gases spread out in the atmosphere even faster. This can be observed very easily in the case of reddish-brown bromine vapor (Experiment 12.3).

Even crystallized, compact bodies are not impenetrable. This can be illustrated by Experiment 12.4.

These examples may suffice to show that spreading of substances is a very general characteristic of matter.

Experiment 12.3 *Spreading of Br$_2$ in air*: The reddish-brown vapor from a drop of bromine in a gas jar (filled with air) spreads out quickly and fills the whole space inside it.

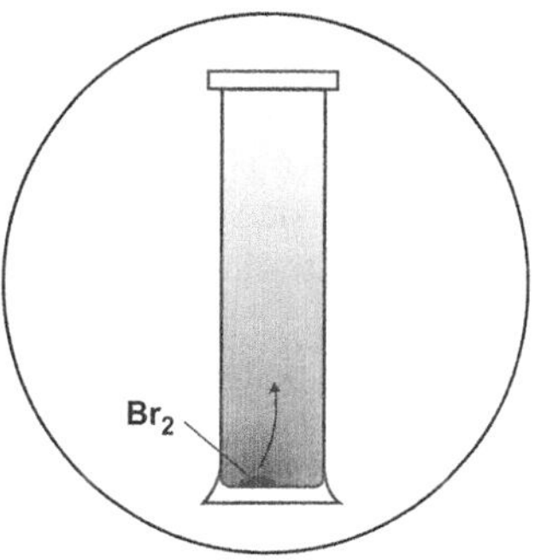

Experiment 12.4 *"Carbonizing" and "decarbonizing" of iron* (schematic representation): Iron can be "carbonized" by annealing in charcoal powder at approx. 1,000 °C. This means that the carbon atoms move into it. The changes of the grain structure at the border of the sample are visible to the naked eye as a dark area, but they can be examined in more detail with a microscope. The iron can also be "decarbonized" by heating in an oxidizing flame or in a furnace in air.

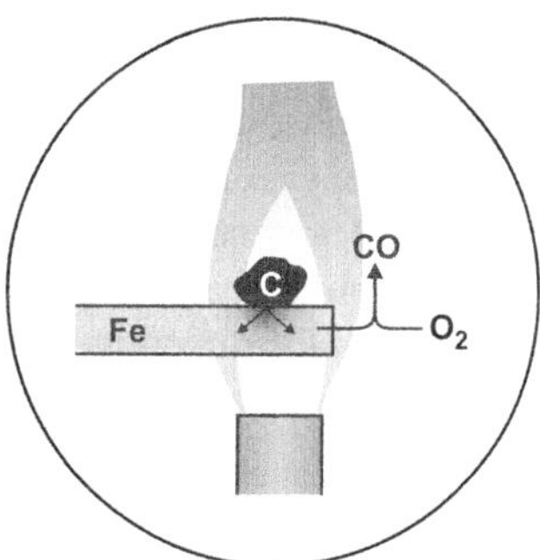

The migration of a substance from one place to another can be considered as a transformation

$$B|\text{Origin} \rightarrow B|\text{Destination},$$

so it is plausible that the chemical potential also controls this process. The transport of substances always follows the direction of the potential gradient. This means that the substance only moves spontaneously in one direction, the direction where the μ value at the starting point is greater than at the destination (if no other forces play a role like centrifugal forces in a centrifuge or electrical forces in an electrolytic cell). In this case, a characteristic of μ plays a decisive role that we have already been introduced to, namely its concentration dependency (compare Sect. 6.2): *The more diluted a substance is, the lower its chemical potential and the chemical potential can be lowered to any degree if the dilution is strong enough.*

In the following, we will apply this statement to the spreading of substances and other phenomena related to it. Chapter 13 will deal with the special features that result when the rule mentioned above no longer applies which is the case at higher concentrations.

12.2   Diffusion

If the substance is mobile enough, it must distribute in an otherwise homogeneous material or space uniformly over the entire area. This is because at locations of lower concentrations, its chemical potential is lower causing the substance to move from locations of higher concentration into areas of lower concentration (Fig. 12.1). When several substances migrate in an area at the same time, this holds for every one of them. The substances strive intrinsically for homogeneous distribution. This is called *diffusion*.

Although difference of concentration is by far the most important cause for diffusion, it is not the only cause. Other factors influencing the chemical potential can also play a role. It is entirely possible that at certain locations in inhomogeneous areas, a substance is enriched at the cost of neighboring areas. This characteristic is applied in microscopy in order to color areas that would tend to preferentially absorb certain dyes (staining in histology). A dye more or less evenly applied distributes unevenly even without outside influence.

We shall now take a closer look at transport of matter. As we have already seen, the determining quantity is the chemical potential. The substance moves spontaneously only in the direction in which the potential falls. If the transfer of a substance B from position x_1 to position x_2 (Fig. 12.2) is formulated as a reaction, then

$$B|x_1 \rightarrow B|x_2 \quad \text{occurs spontaneously if} \quad \mu_B(x_1) > \mu_B(x_2).$$

Fig. 12.1 Flow of substance caused by an inhomogeneous distribution of the substance in an otherwise homogeneous area.

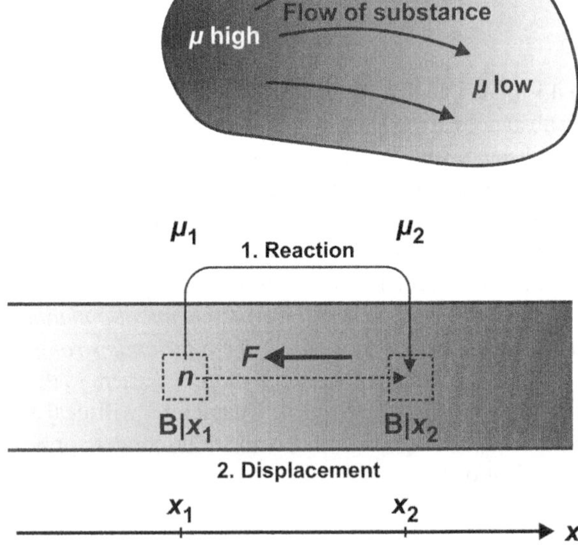

Fig. 12.2 Determination of force F upon a small sample of amount n of a substance B in a gradient of chemical potential.

The chemical drive $\mathcal{A}$ of this process is the corresponding potential difference,

$$\mathcal{A} = \mu_B(x_1) - \mu_B(x_2) = -\Delta\mu_B,$$

and the conversion $\Delta\xi$ is the amount of substance n_B transferred from position x_1 to position x_2. The potential gradient causes a force F that pushes the substance in the direction of the falling potential. F is easy to calculate if the energy W for transporting a small amount n_B from position x_1 to position x_2 (against the potential gradient) is expressed first as expenditure for the reaction and then as that required for the displacement against the force F. For this, let us consider that the potential μ_B increases in the x direction and that the substance is transported "uphill" opposite to the potential gradient. The energy W is then > 0, while the chemical drive $\mathcal{A}$ is negative. F also counts as negative because it is oriented opposite to the positive x-axis. In general, F depends upon the position x. If the segment $\Delta x = x_2 - x_1$ is made small enough, F can be considered constant. As a consequence, we obtain:

$$W = -\mathcal{A} \cdot \Delta\xi = \Delta\mu_B \cdot n_B \quad \text{and} \quad W = -F \cdot \Delta x \quad \text{for small} \quad \Delta x.$$

When both equations are combined, we find

$$F_B = -n_B \cdot \frac{\Delta\mu_B}{\Delta x} \quad \text{or more precisely} \quad F_B = -n_B \cdot \frac{d\mu_B}{dx}. \tag{12.1}$$

It is fitting to call F *diffusion force* because F is the driving force behind the kind of transport called diffusion. In Chap. 20, which deals with transport phenomena, we will discuss the velocity of substance transport and the laws pertaining to it (Sect. 20.2).

When we apply molecular kinetic considerations, diffusion tends to be conceptualized as a result of random motion of molecules and not as an effect of a directed force. When the boundaries of two areas of different concentrations of a substance B touch each other, a totally random movement of molecules occurs where, on average, more particles move from the concentrated area into the diluted area than vice versa. Random chance suffices to explain that B gradually moves from areas of higher concentration into ones of lower concentration until it is evenly distributed. No special driving force is necessary for this. Actually, there is no discrepancy between the two models. The greater number of molecular collisions occurring at higher concentrations of the diffusing substance B must lead to a directed force at the transition to the area of lower concentration. This is the diffusion force F calculated above.

Other phenomena already familiar to us also belong to the category of spreading. Water (Sect. 6.1) but also alcohol, ether, etc., evaporate into the air. They convert to a gaseous state although the chemical potential of liquid A is smaller below its boiling point than that of *pure* vapor. What makes this possible is the fact that the vapor is not pure but so diluted by air that $\mu(A|g) < \mu(A|l)$.

Any substance B dissolves in finite amounts in any other substance A (even if these amounts are immeasurably small) because when it is diluted enough, the chemical potential of the dissolved substance B falls below the fixed μ value of the non-dissolved solute of B, so that B begins to migrate away from it (Sect. 6.6).

Finally, the mass action law that is so important in chemistry can also be mentioned in this context as long as it deals with an exchange of substances taking place between spatially separate areas. Nernst's law of distribution, Henry's law of gas solubility, and even the vapor pressure formula of pure substances are all examples of this.

12.3 Indirect Mass Action

When a *small* amount n_B of a foreign substance is dissolved in a liquid A, the chemical potential μ of this liquid will decrease at constant p and T. In fact, it decreases in proportion to the mole fraction $x_B = n_B/(n_A + n_B)$ of the foreign substance, *independent of the kind of solute in question*:

$$\mu_A = \overset{\bullet}{\mu}_A - RT \cdot x_B \quad \text{for } x_B \ll 1 \quad \text{"colligative lowering of potential."} \quad (12.2)$$

(The term "colligative" is explained at the end of this section). $\overset{\bullet}{\mu}_A$ (pronounced "mu-A-pure") designates the potential of A in its pure state ($x_A = 1$). Until now, we have ignored the "moderating influence" that the addition of a small amount of foreign substance can have upon another substance's tendency to transform. An example would be the application of the mass action law when the solvent takes part in the reaction in question. The reason for this is that for dilute solutions, the contribution $-RTx_B$ is small compared to the concentration-dependent contributions $\overset{\times}{\mu} = RT\ln c_r$ of the dissolved substances. The latter tend to $-\infty$ for decreasing concentration. In the mass action law which applies in the limit of strong dilution, the contribution $-RTx_B$ disappears, but the mass action terms $\overset{\times}{\mu}$ do not; they actually increase without limit.

In order to describe the dependency of the chemical potential μ of any substance upon composition (concentration c, partial pressure p, mole fraction x, etc.), chemists generally separate the potential μ into two parts: a basic component $\overset{\circ}{\mu}$ independent of the composition and a residual that is dependent upon it (compare Sect. 6.2). In the sense explained here, $\overset{\bullet}{\mu}$ represents a particular basic value. Only when this needs to be emphasized will we use the notation $\overset{\bullet}{\mu}$; otherwise we will stay with $\overset{\circ}{\mu}$.

Equation (12.2) for the lowering of potential is valid as long as the foreign substance B or foreign substances F (there can be several different ones, B, C, D, ..., since their kind does not matter) dissolve molecularly but do not associate or dissociate, meaning they may not decompose into smaller components or form aggregates of several molecules. This remarkable relation, which is valid for all substances, is the indirect result of the mass action of the dissolved substances. Remember that mass action is independent of what kind of substances make up solvent A and solute B (see the mass action equations in Chap. 6).

Note that this new equation holds only in the *limit* of a small amount of a foreign substance being added to the solvent. Admittedly, this change of potential is small. However, because substance A is highly concentrated, it can have significant effects, which we will look into in the next sections.

For the mathematically interested: In order to derive Eq. (12.2), we will refer back to the cross relation discussed in Sect. 9.3 known as n–n coupling. When one substance tries to displace (or favor) another one, this happens reciprocally and with equal strength. The corresponding displacement coefficients are equal as can easily be shown by applying the flip rule (main equation $dW = -pdV + TdS + \mu_A dn_A + \mu_B dn_B$):

$$\left(\frac{\partial \mu_A}{\partial n_B}\right)_{p,T,n_A} = \left(\frac{\partial \mu_B}{\partial n_A}\right)_{p,T,n_B}.$$

Taking into account that μ_B is dependent upon c_B and c_B is dependent upon n_A, $\mu_B(c_B(n_A))$, which means that we have to apply the chain rule [see Eq. (A.1.13) in the Appendix)] in order to calculate the derivative on the right, the result is:

$$\left(\frac{\partial \mu_B}{\partial n_A}\right)_{p,T,n_B} = \left(\frac{\partial \mu_B}{\partial c_B}\right)_{p,T} \left(\frac{\partial c_B}{\partial n_A}\right)_{p,T,n_B}.$$

When we started examining the phenomenon called mass action, we chose the situation where the concentration coefficient $\overset{\times}{\gamma}$ of the chemical potential at low concentrations c is a universal quantity (Sect. 6.2). When applied to substance B, this means:

$$\left(\frac{\partial \mu_B}{\partial c_B}\right)_{p,T} = \overset{\times}{\gamma} = \frac{RT}{c_B}.$$

Now, $c_B = n_B/V$ is indirectly dependent upon n_A even at a constant n_B, because $V = n_A V_A + n_B V_B \approx n_A V_A$. Since $n_B \ll n_A$, the contribution $n_B V_B$ to the volume V can be ignored and V_A can be considered equal to the volume demand of the pure substance A and therefore independent of n_A. When we take the derivative of $c_B = n_B/(n_A V_A)$ with respect to n_A [by using rule (A.1.6) for calculating derivatives in the Appendix],

$$\left(\frac{\partial c_B}{\partial n_A}\right)_{p,T,n_B} = -\frac{n_B}{n_A{}^2 V_A} = -\frac{n_B}{n_A V} = -\frac{c_B}{n_A},$$

and insert the result further above, we obtain:

$$\left(\frac{\partial \mu_A}{\partial n_B}\right)_{p,T,n_A} = \frac{RT}{c_B} \cdot \frac{-c_B}{n_A} = -\frac{RT}{n_A}.$$

The derivative is not dependent upon n_B, meaning that, with increasing amounts n_B, μ_A decreases linearly with constant slope from the initial value $\overset{\bullet}{\mu}_A$:

$$\mu_A = \overset{\bullet}{\mu}_A - \frac{RT}{n_A} \cdot n_B \approx \overset{\bullet}{\mu}_A - RT \cdot x_B.$$

Figure 12.3 shows the chemical potential μ as a function of x within the entire range from $x=0$ to 1. The solid curve illustrates the logarithmic relation considered the ideal case. If the mole fraction is reduced by a power of ten each time, the chemical potential always decreases by the same value, the decapotential μ_d of 5.7 kG at room temperature. As already stated (and derived above), all $\mu(x)$ curves must show the same slope RT for $x \approx 1$ (that is $x \approx 0$ for the second component of the mixture). We will go more deeply into this subject in Sect. 13.2.

"Colligative lowering of potential" leads to several effects: development of osmotic pressure, lowering of vapor pressure of a solution (compared with the pure solvent), raising of its boiling point, and lowering of its freezing point. These effects are determined solely by the mole fraction of the foreign substances, i.e., the *number* of dissolved particles ("assemblies of atoms") and not their chemical nature, size, and form. For this reason, the term *colligative properties* is used (from the Latin colligare "to assemble"). An aqueous solution with a mole fraction of glucose of 0.001 corresponds quite well to a urea solution with the same mole fraction for all the properties just mentioned (osmotic pressure, vapor pressure, freezing point, and boiling point). However, each type of particle in the solution must be treated as an individual substance. For example, the cations and anions in an electrolyte solution need to be counted separately. For a solution of table salt with $x_{NaCl} = 0.001$, $x_F = x_{Na^+} + x_{Cl^-} = 2 \cdot x_{NaCl} = 0.002$; this is so because NaCl is fully dissociated into the Na^+ and Cl^- ions. In a calcium chloride solution, we even have $x_F = x_{Ca^{2+}} + x_{Cl^-} = 3 \cdot x_{CaCl_2} = 0.003$.

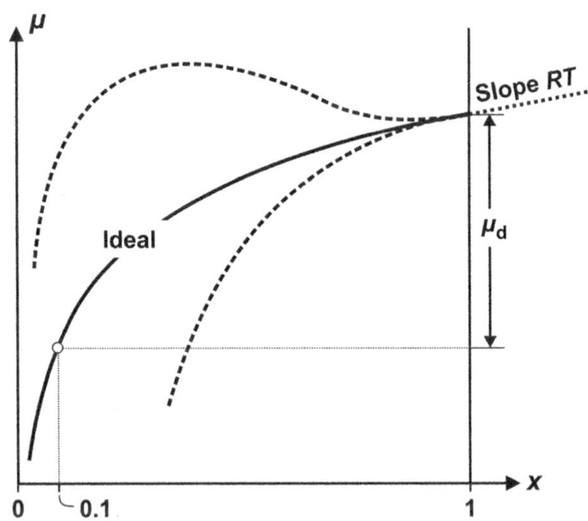

Fig. 12.3 Dependency of chemical potential upon mole fraction x.

12.4 Osmosis

When two solutions with different concentrations of a dissolved substance B are separated by a thin wall that only allows solvent A to pass through (a so-called *semipermeable* membrane, Fig. 12.4), solvent A will migrate through this membrane from the more diluted solution (with respect to B) to the one of higher concentration. In this case, one of the solutions can be composed of the pure solvent for which $c_B = 0$. In the solution having a higher concentration of B, substance A appears more strongly diluted due to its solution partner B. This means that the concentration of solvent is smaller, and because of this, its chemical potential μ_A is also smaller. Thus, the chemical drive for this process called *osmosis* is the difference of potential generated by different concentrations of foreign substances.

In the simple membrane model shown in Fig. 12.4, selective permeability results from the maximum pore size: Only the smaller solvent molecules manage to move through the membrane. But semipermeability can also occur through other mechanisms. The biological membranes surrounding living cells are also semipermeable. They allow water and molecules of comparable size through while holding back enzymes and proteins inside cells. In biology, osmotic exchange of water represents a ubiquitous phenomenon. It is responsible for the effect that juice is "drawn out" from strawberries that are sugared (Experiment 12.5) or that cherries swell up and burst after a long rain. In the first case, water migrates out through the peel into the concentrated, therefore water-poor, sugar solution. In the second case, water flows inward, because the water is more diluted there.

Another good example of the first effect would be salted slices of white radish (Experiment 12.6).

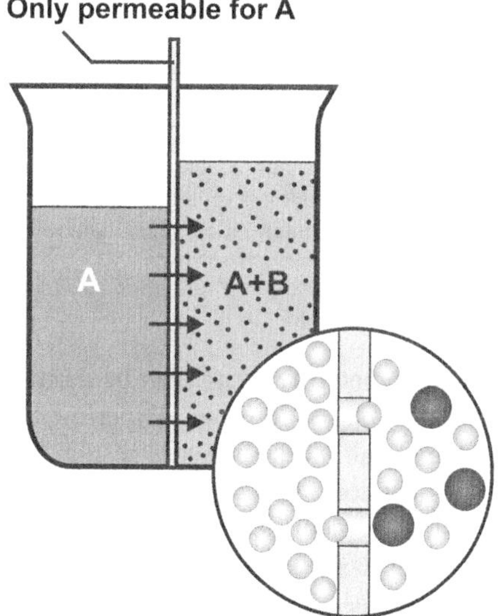

Fig. 12.4 Migration of solvent A in the direction opposite to a pressure gradient into a solution. The molecules of the dissolved substance B are shown as *dots*; for clarity, solvent A is shown as a continuum, where the slightly different *gray tones* indicate the differences of concentrations. The detail shows spheres symbolizing the molecules of the dissolved substance (*dark*) as well as the solvent (*light*).

Experiment 12.5 *Juice "extraction" from sugared fruit*: Two handfuls of strawberries (or slices of mandarins) are carefully dabbed dry with a paper towel. One half of the strawberries is filled in a goblet so that there is some space left at the bottom. A lot of dry sugar is sprinkled on the other part of the strawberries, and subsequently, they are cautiously filled in the same way in the second goblet. After some hours, a watery syrup begins to gather on the bottom of the glass beneath the sugared fruit; after 2 days, a volume of 30–40 mL can be obtained. However, there is no formation of a liquid in the case of the unsugared fruit.

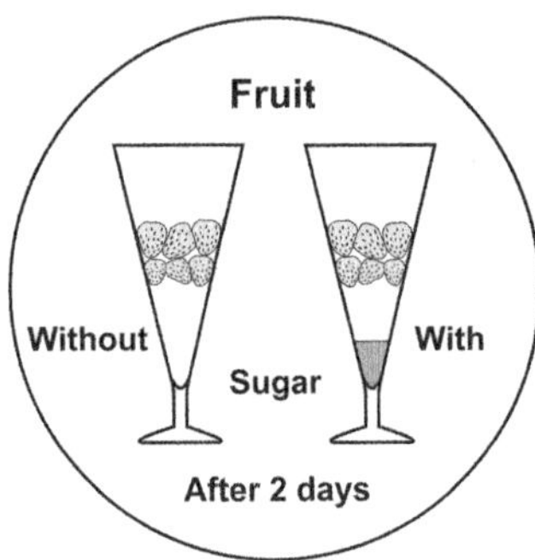

Experiment 12.6 *Juice "extraction" from slices of salted white radish*: The radish is cut into thin slices and these slices are piled in two stacks. The slices of one of the stacks are picked up in turn, salted very well, and piled up again. Subsequently, both stacks are speared on the wire. Immediately, juice begins to drip out of the stack with the salted slices. The measuring cylinder contains approx. 30 mL juice after 15 min.

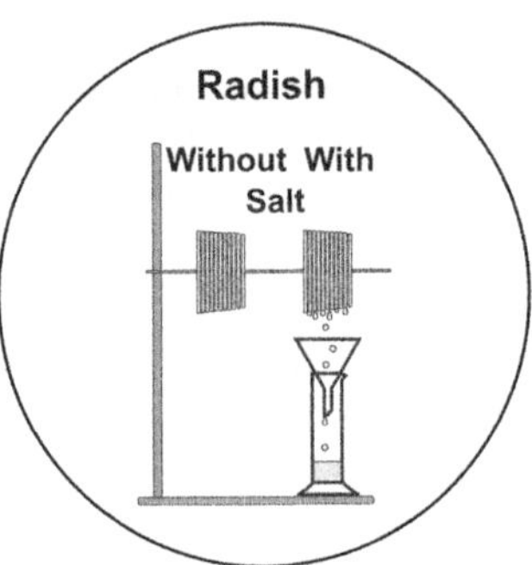

Experiment 12.7 *Swelling of a decalcified egg in water*: One of two raw eggs (as equal as possible in size) is placed in a beaker with hydrochloric acid (or vinegar) to dissolve the calcareous egg shell— without breaking the membrane surrounding the egg. Subsequently, each of the eggs is put in a separate beaker filled with water. After 2 days, the shell-less egg has grown visibly in size.

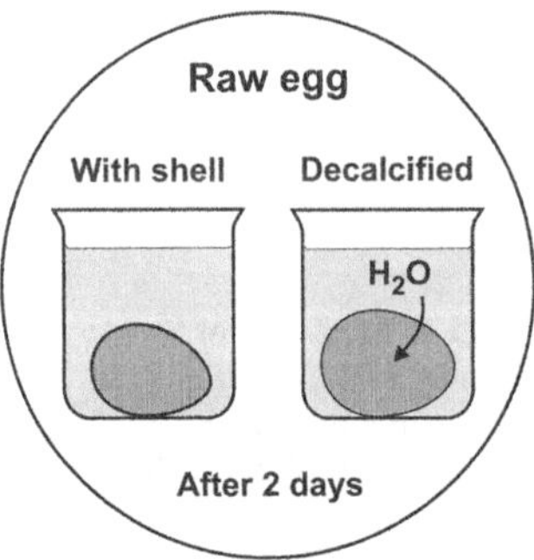

The second effect can also be demonstrated on a raw decalcified egg which is cautiously placed in water (Experiment 12.7).

A gradual excess pressure results from the flow of solvent A into the concentrated solution. The chemical potential μ_A also gradually increases so that the potential gradient decreases. The process stops when μ_A on the right and on the left of the wall becomes equal (or when the substance A completely disappears from one side). The resulting excess pressure is called *osmotic pressure*.

We will now take a closer look at osmosis. We consider a vessel containing an amount n_A of a liquid A. When a small amount n_B of a foreign substance is dissolved in it, the chemical potential μ_A of the solvent decreases ["colligative lowering of potential," Eq. (12.2)]:

$$\mu_A = \overset{\bullet}{\mu}_A - RT \cdot x_B \quad \text{for } x_B \ll 1. \tag{12.3}$$

Let us imagine the container connected to another one by a wall that is permeable only for the solvent (Fig. 12.5). The liquid in the second container is in its pure state. The potential gradient causes this liquid to flow through the wall into the solution. This flow can be suppressed by compensating for the loss of potential by raising the pressure on the solution. The chemical potential grows with increasing pressure (compare Sect. 5.3):

$$\mu_A = \overset{\bullet}{\mu}_A - RT \cdot x_B + \beta_A \cdot \Delta p, \tag{12.4}$$

where the pressure coefficient β_A corresponds to the molar volume V_A of the pure solvent (compare Sect. 9.3). The following is then valid for the osmotic equilibrium:

$$\overset{\bullet}{\mu}_A - RT \cdot x_B + V_A \cdot \Delta p = \overset{\bullet}{\mu}_A \text{ and respectively, } - RT \cdot x_B + V_A \cdot \Delta p = 0. \tag{12.5}$$

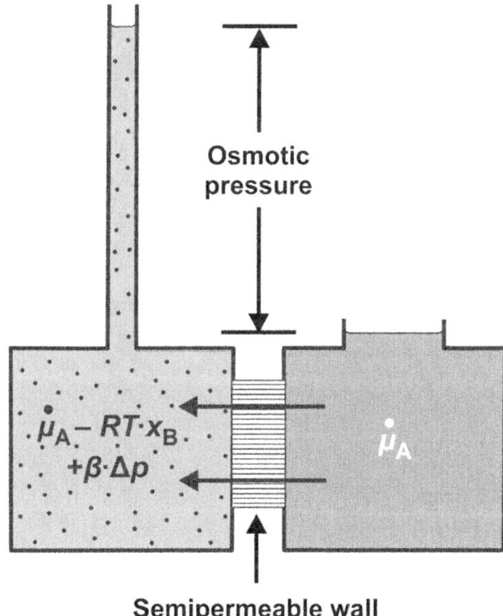

Fig. 12.5 Experiment illustrating osmotic pressure. The riser pipe on the *left* serves as manometer.

This means that the chemical potential of the solution again corresponds to that of the pure solvent. The excess pressure Δp necessary for establishing equilibrium serves as measure for the *osmotic pressure* p_{osm} in the solution.

The simple arrangement in Fig. 12.5 shows the pressure working against the flow of solvent into the solution as a result of the gravitational pressure of the column of solution in the riser pipe on the left. This pressure is produced by the osmosis itself as the pure solvent moves through the semipermeable wall into the solution. In the process, the difference of level between the two riser pipes gradually becomes greater until, eventually, the gravitational pressure compensates for the effect of osmotic pressure. This means that osmotic equilibrium has been established. The osmotic pressure $p_{osm} = \rho g h$ can be easily calculated from the resulting rise h of the column of liquid, the density ρ of the solution, and the gravitational acceleration g.

With the help of a carrot and a riser pipe with a funnel-shaped end, it is easy to construct an experiment to prove the existence of osmotic pressure (Experiment 12.8).

We have $x_B = n_B/(n_A + n_B) \approx n_B/n_A$ for a diluted solution because the amount n_B of solute is so small compared to the amount n_A of the solvent that it can be ignored. Multiplying Eq. (12.5) by n_A yields:

$$-RT \cdot n_B + n_A V_A \cdot p_{osm} = 0. \tag{12.6}$$

If $V \approx n_A \cdot V_A$ indicates the volume of liquid (ignoring the small amount $n_B \cdot V_B$ of foreign substance), then the osmotic pressure p_{osm} results in:

$$p_{osm} = n_B \frac{RT}{V} \quad \text{Van't Hoff's equation.} \tag{12.7}$$

At room temperature ($T = 298$ K), a solution of an arbitrary non-electrolyte with a concentration of 0.1 kmol m^{-3} results in an osmotic pressure of 250 kPa (2.5 bar). This would be enough to raise the column of liquid more than 25 m. Even at very small concentrations, osmotic pressure is of such a magnitude that it is easy to measure with sufficient precision.

Experiment 12.8 *Experimental demonstration of osmotic pressure*: The inside of the carrot is hollowed out in a cylindrical form and filled with a colored saturated calcium chloride solution. Then the riser pipe is attached. After a short time, one observes a continuous rise of the solution in the riser pipe. In this case, the cell membranes in the carrot act as the semipermeable wall.

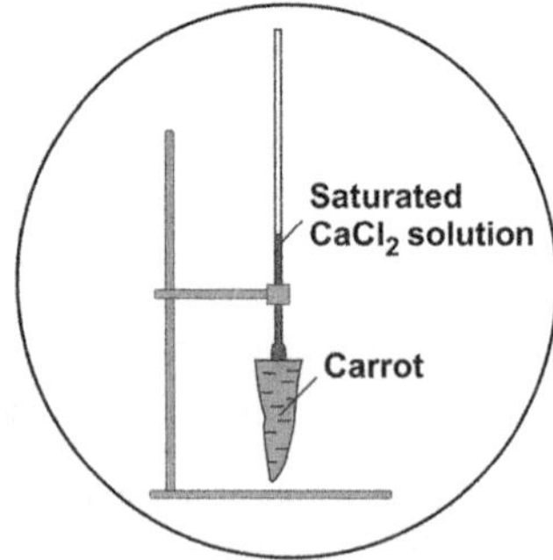

Van't Hoff's equation is very similar to the general gas law. In fact, both equations can be interpreted in the same way. Here we need to keep in mind that the forces of attraction between the A particles keep the liquid together (compare Sect. 11.1, keyword "cohesion pressure"). The contribution of the external pressure p is comparatively small. The B particles that drift far away from each other and scarcely influence each other cause a pressure like that of a dilute gas. However, in this case the pressure is not compensated by the container walls but by the cohesion of the A particles. When the osmotic pressure p_{osm} is higher than the external pressure—a condition that is often attained—the liquid A behaves as if it were under negative pressure. If we calculate the potential μ_A of the liquid for a pressure reduced by p_{osm} and keep in mind that for dilute solutions $V \approx n_A \cdot V_A$ and $n_B/n_A \approx x_B$, we again end up with Eq. (12.3). This demonstrates that both descriptions are equivalent:

$$\overset{\bullet}{\mu}_A - \beta_A \cdot p_{osm} = \overset{\bullet}{\mu}_A - V_A \frac{n_B RT}{V} = \overset{\bullet}{\mu}_A - x_B RT.$$

If the dilute solution contains several types of particles which cannot penetrate the membrane, we have

$$p_{osm} = n_F \frac{RT}{V} = c_F RT, \qquad (12.8)$$

where n_F as well as c_F are the sum of the amounts of substance and the sum of concentrations of all types of particles, respectively. c_F is called the *osmotic concentration* (formerly known as *osmolarity*) of the solution. Since the total number of dissolved particles must be taken into account when calculating the osmotic concentration, the number of ions that form an ionic substance must also be considered. For example, the osmotic concentration of an aqueous solution of the salt $CaCl_2$ with a concentration c is three times as great as this concentration. Correspondingly, the osmotic pressure of the solution of this salt is three times that of a solution of a non-electrolyte with concentration c.

As we have mentioned at the beginning, osmotic phenomena play an essential role in biological processes. They have great importance for the balance of water in living organisms and influence the shape of their cells. The osmotic concentration c_F of cell liquid in human red blood cells, for example, is approximately 300 mol m^{-3}. In this case, we can apply Van't Hoff's law only with reservations due to this relatively high concentration. However, for body temperature, an osmotic pressure of $p_{osm} = 300 \text{ mol m}^{-3} \times 8.3 \text{ J mol}^{-1} \text{ K}^{-1} \times 310 \text{ K} = 770 \text{ kPa}$ can be at least estimated. If red blood cells were to be suspended in pure water, they would have to withstand about 8 times normal atmospheric pressure. In fact, they swell up and burst long before this point (Fig. 12.6, left). On the other hand, if red blood cells are put into contact with an aqueous saline solution that has a much higher osmotic pressure than 770 kPa, the water in the cells will flow out and they will shrink (the membranes of red blood cells are almost impenetrable for Na^+, Fig. 12.6, right). Only if the osmotic pressure is the same inside the red blood cells

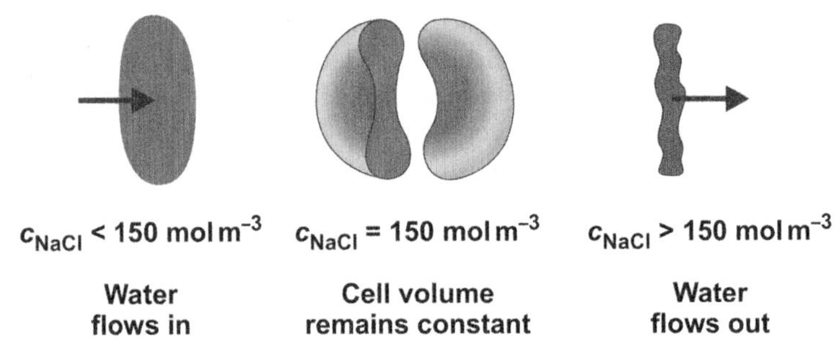

Fig. 12.6 Osmotic behavior of red blood cells in aqueous NaCl solutions at concentration c.

and the surrounding solution (which is the case for blood plasma) do the cells keep their normal shape (Fig. 12.6, middle).

Solutions where the water content of the cells remains constant [$\mu(H_2O)$ is the same inside and outside the cells] are described as *isotonic*. When giving intravenous infusions, it is important to be aware that in order not to damage the blood cells, only an isotonic solution matching the blood may be used. A physiological saline solution has a concentration of 150 mol m^{-3} and therefore an osmotic concentration of 300 mol m^{-3}. On the other hand, the cell damaging effects of concentrated saline solutions can be put to good use in order to preserve foods. One example is salting of meat (pickling) where the water is removed from possibly unhealthy microorganisms through osmosis. This hinders their cell functions and strongly reduces their reproduction.

In *reverse osmosis*, an external excess pressure is exerted upon the side of the concentrated solution that is higher than the osmotic pressure p_{osm} of this solution. This causes the solvent's molecules to be "forced" in the direction opposite to the osmotic effect. They are forced through the semipermeable membrane and into the more diluted solution where they are enriched. This procedure is used in desalinization of water as well as wine processing.

12.5 Lowering of Vapor Pressure

A pure liquid A is assumed to be in equilibrium with its vapor at a pressure of p_{lg} (initial situation: $\overset{\bullet}{\mu}_{A|l} = \overset{\bullet}{\mu}_{A|g}$). The *vapor pressure* of A is lowered if a low-volatile foreign substance B is dissolved in it (Fig. 12.7).

Qualitatively, this can be understood immediately. Adding B dilutes liquid A|l, thereby lowering its chemical potential $\mu_{A|l}$. The dissolved substance B should have low volatility so it contributes nothing to the vapor A|g and the potential $\mu_{A|g}$ remains unchanged. Because $\mu_{A|l}$ is now lower than $\mu_{A|g}$, the vapor has to condense on the surface of the solution, thereby causing the pressure to fall.

Fig. 12.7 Establishment of
equilibrium between a
solution of a foreign
substance B in a liquid A
and the pure vapor phase
of A.

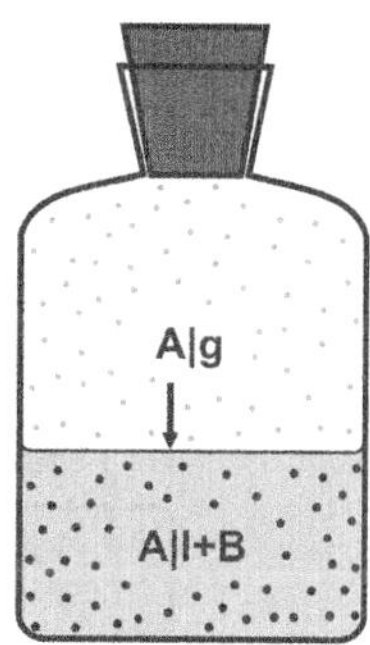

The quantitative discussion is also not difficult. The process continues until
equality of potentials is regained, $\mu_{A|g} = \mu_{A|l}$. For this purpose, reducing the
pressure by a small amount Δp is sufficient. The chemical potential $\mu_{A|g}$ of the
vapor falls steeply with decreasing pressure because of its high pressure coefficient,
$\beta_{A|g} >>> \beta_{A|l}$, whereas the change of the potential $\mu_{A|l}$ is so small that it can be
neglected. If we consider the "colligative potential lowering" $-RTx_B$ caused by B
we obtain for the equilibrium condition:

$$\mu_{A|g} = \overset{\bullet}{\mu}_{A|g} + \beta_{A|g} \cdot \Delta p = \overset{\bullet}{\mu}_{A|l} - RT \cdot x_B = \mu_{A|l}. \tag{12.9}$$

We act on the assumption that $\overset{\bullet}{\mu}_{A|l} = \overset{\bullet}{\mu}_{A|g}$; therefore these contributions cancel each
other out. Because we also have $\beta_{A|g} = V_{A|g} = RT/p$ (see Sect. 9.3), the equation
can be simplified according to:

$$RT\frac{\Delta p}{p} = -RT \cdot x_B. \tag{12.10}$$

In this context, p corresponds to the vapor pressure p_{lg} of the *pure* solvent. If it is
important to emphasize this fact, we add the symbol $\bullet$ and write $p_{lg}^{\bullet}$ instead of p_{lg}.
For the "lowering of vapor pressure" Δp_{lg}, we obtain a relation that was discovered
empirically in 1890 by the French Chemist François Marie Raoult:

$$\Delta p_{lg} = -x_B \cdot p_{lg}^{\bullet} \quad \text{Raoult's law.} \tag{12.11}$$

Let us have a look at Fig. 12.8 for illustration: At the intersection of the potentials
for pure solvent and pure vapor, there is equilibrium between the liquid and its
vapor phase at the vapor pressure p_{lg}. A dissolved substance of low volatility lowers
the chemical potential of the solvent by $-RTx_B$ ("colligative potential lowering"
corresponding to the distance between the almost horizontal straight lines), but

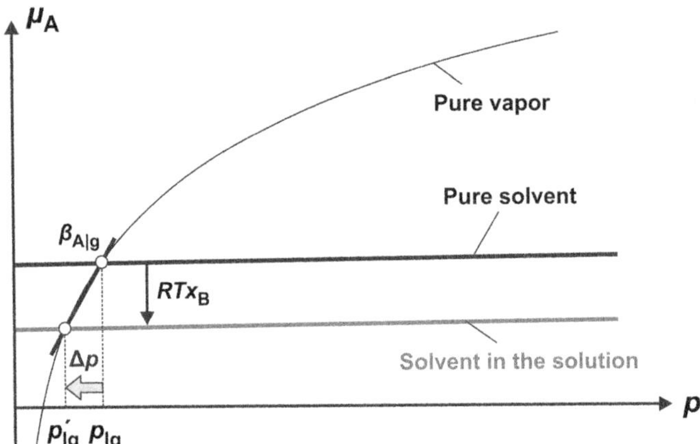

Fig. 12.8 Dependence of the chemical potential upon pressure, and lowering of vapor pressure.

Experiment 12.9 *Comparison of the vapor pressures of ether and an ether–oleic acid mixture*: Both gas-washing bottles are filled about one-fifth of their volumes with ether. After removal of most of the air with the help of a pump, the stop cock is closed. Subsequently, the oleic acid is added drop by drop. The vapor pressure of ether in the gas-washing bottle with added oleic acid decreases in comparison to that of pure ether. This is shown by the level of liquid in the manometer.

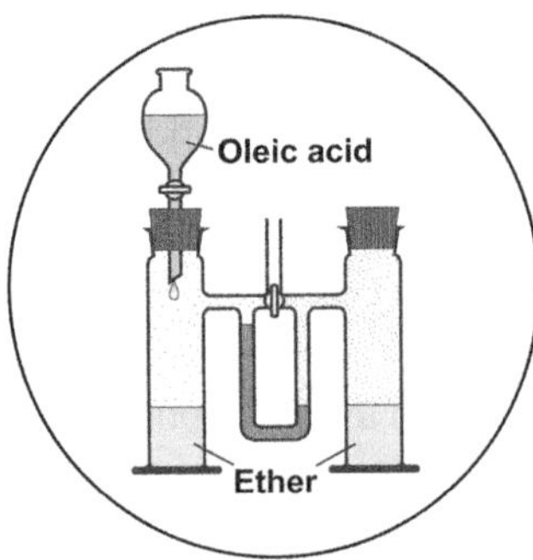

leaves that of the vapor unchanged. Thus, the intersection of the curves (axis of abscissas p'_{lg}) shifts to the left. This means that the vapor pressure is lowered by Δp ($\Delta p < 0$!).

With the aid of the "slope triangle" drawn in the figure, the slope $\beta_{A|g}$ of the potential curve for pure vapor results in

$$\beta_{A|g} = \frac{-RTx_B}{\Delta p}.$$

Considering that $\beta_{A|g} = V_{A|g} = RT/p$ and solving the equation for Δp results in Raoult's law. Whether one prefers the first pure mathematical or the second more geometrical derivation is a question of personal preference.

The simple setup shown in Experiment 12.9 illustrates the discussed effect.

12.6   Lowering of Freezing Point and Raising of Boiling Point

A frozen liquid A melts more easily when a substance B that is soluble in the liquid but not in the solid is added (Fig. 12.9). At the normal freezing point T_{sl} of the liquid A, the chemical potentials for the solid and the liquid state are equal ($\overset{\bullet}{\mu}_{A|s} = \overset{\bullet}{\mu}_{A|l}$). If a foreign substance is dissolved in the liquid phase, the chemical potential of this phase decreases so that it falls below that of the solid phase which then begins to melt. The entropy required for the phase transition solid $\rightarrow$ liquid is not added from outside but has to be brought up by the system itself. Therefore, the entire mixture cools down and the chemical potentials rise due to their negative temperature coefficients. However, because the temperature coefficient for a liquid is smaller than for a solid ($\alpha_{A|l} < \alpha_{A|s} < 0$), $\mu_{A|l}$ grows faster with decreasing temperature than does $\mu_{A|s}$. This causes the potential gradient to disappear again at a certain lower temperature and the melting process stops.

An illustration (Fig. 12.10) explains this phenomenon: When the chemical potentials for the pure solvent as well as for the pure solid are drawn as functions of temperature, the intersection of the two curves yields the freezing point T_{sl}. The dissolved substance lowers the chemical potential of the liquid but does not influence the solid phase. The intersection of the curves ($T_{sl'}$) is therefore shifted to the left. This means that the freezing point is lowered by ΔT_{sl} ($\Delta T_{sl} < 0$).

In the corresponding calculation, we again assume a reestablished equilibrium. This time it is between the liquid and solid phase, $\mu_{A|l} = \mu_{A|s}$. Because the changes in temperature T we are interested in are mostly small, we can assume a linear dependency of the chemical potential upon T. At constant pressure, we obtain the following result (keep in mind that in this case the freezing point T_{sl} is the reference point for the temperature coefficient):

$$\mu_{A|l} = \overset{\bullet}{\mu}_{A|l} - RT_{sl} \cdot x_B + \alpha_{A|l} \cdot \Delta T = \overset{\bullet}{\mu}_{A|s} + \alpha_{A|s} \cdot \Delta T = \mu_{A|s}. \qquad (12.12)$$

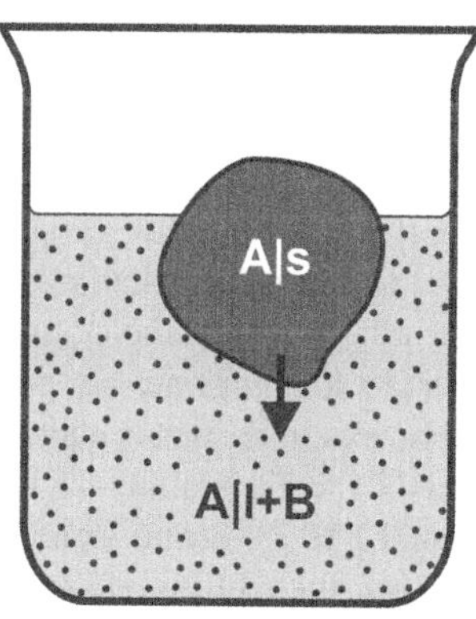

Fig. 12.9 Melting of a frozen liquid in a solution.

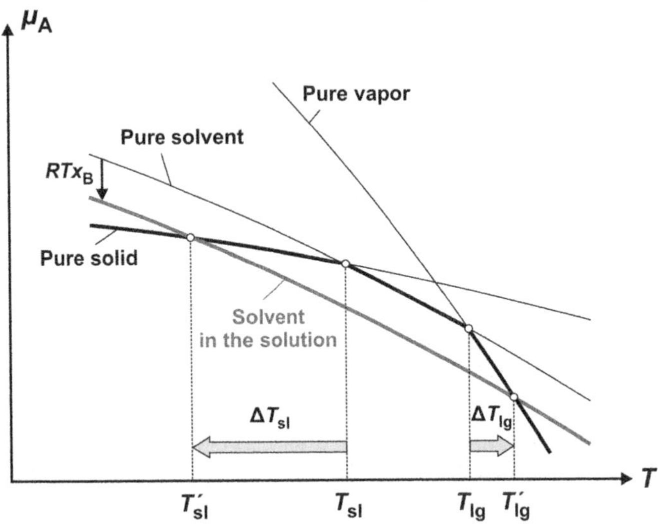

Fig. 12.10 Temperature dependence of chemical potentials, lowering of freezing point, as well as raising of boiling point. The lowering of potential of solvent A by $-RTx_B$ caused by the foreign substance B is compensated at the freezing point T_{sl} by a lowering and at the boiling point T_{lg} by a raising of temperature.

Again the basic values cancel each other out, $\overset{\bullet}{\mu}_{A|l} = \overset{\bullet}{\mu}_{A|s}$. When the equation is solved for ΔT (or rather ΔT_{sl}), the freezing-point depression results in:

$$\Delta T_{sl} = \frac{RT_{sl} \cdot x_B}{\alpha_{A|l} - \alpha_{A|s}}. \tag{12.13}$$

Here, the temperature coefficient corresponds to the negative molar entropy of the substance (compare Sect. 9.3). Subsequently, the difference $S_{A|l} - S_{A|s}$ can be summarized as the molar entropy of fusion $\Delta_{fus}S_A$ of the pure solvent (at the freezing point) (compare Sect. 11.5):

$$\Delta T_{sl} = -\frac{RT_{sl}^{\bullet} \cdot x_B}{\Delta_{fus}S_A}. \tag{12.14}$$

Like the vapor pressure lowering, the freezing-point depression is directly proportional to the mole fraction x_B of dissolved substance. The additional character $\bullet$, indicating that the labeled quantity applies to the pure substance, can be omitted if misunderstandings are not possible.

For example, in an aqueous non-electrolyte solution with a mole fraction of $x_B = 0.01$, ΔT_{sl} is about -1 K. An example from everyday life is shown in Experiment 12.10. A prime example of an application of freezing-point depression is, however, the melting effect of road salt.

Experiment 12.10 *Whisky "on the rocks"*: When rum, schnaps, whisky, or, alternatively, ethanol is poured over ice, it will become considerably colder than 0 °C.

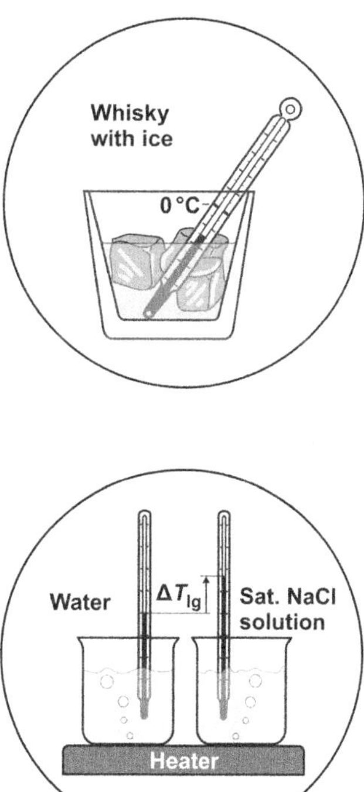

Experiment 12.11 *Raising of boiling point of a saturated solution of table salt*: A saturated solution of table salt begins to boil at a noticeable higher temperature than tap water.

In contrast to the freezing point, the boiling point of a solution is raised as demonstrated in Experiment 12.11. Correspondingly, the position of the equilibrium between liquid and vapor is shifted to a higher temperature in Fig. 12.10. However, this raising of boiling point is much smaller than the lowering of freezing point which can be attributed to the different slopes of the curves representing the potentials. These, in turn, are determined by the molar entropies, which naturally leads to the steepest drop for the gaseous state of the solvent.

A consideration similar to that for lowering the freezing point leads to the equation for raising the boiling point:

$$\Delta T_{\mathrm{lg}} = -\frac{RT_{\mathrm{lg}} \cdot x_{\mathrm{B}}}{\alpha_{\mathrm{A|g}} - \alpha_{\mathrm{A|l}}} \quad \text{and correspondingly,} \quad \Delta T_{\mathrm{lg}} = \frac{RT^{\bullet}_{\mathrm{lg}} \cdot x_{\mathrm{B}}}{\Delta_{\mathrm{vap}}S_{\mathrm{A}}}, \qquad (12.15)$$

where $\Delta_{\mathrm{vap}}S_{\mathrm{A}}$ now represents the molar entropy of vaporization $(S_{\mathrm{A|g}} - S_{\mathrm{A|l}})$ at the boiling point T_{lg}. In an aqueous solution of a non-electrolyte with a mole fraction of $x_{\mathrm{B}} = 0.01$, ΔT_{lg} is only about 0.3 K.

In summary, when a low-volatile foreign substance is dissolved in a liquid, the potential of the pure vapor, which is then higher than that of the solution, can be lowered in two ways. First, by lowering the pressure (Sect. 12.5), and second, by raising the temperature.

12.7   Colligative Properties and Determining Molar Mass

The four phenomena just described (osmosis, lowering of vapor pressure, lowering of freezing point, and raising of boiling point) have a common feature: they are all dependent upon the indirect mass action of dissolved substances, i.e., the lowering of the chemical potential by mixing in small amounts of foreign substances. These so-called colligative phenomena depend solely upon the mole fraction of these foreign substances and therefore the number of dissolved particles. However, what type of substances they are is unimportant.

Because of this peculiarity, the colligative properties can be used to determine the amount of substance n_B of a sample of an unknown substance B and, therefore, if the mass m_B of the sample is known, also the molar mass $M_B = m_B/n_B$. Let us take a quick look at this by considering the example of lowering of freezing point. $x_B \approx n_B/n_A$ is valid at high dilution, and because of $n_A = m_A/M_A$, we have $x_B \approx n_B \cdot M_A/m_A$. The quotient n_B/m_A corresponds to the molality b_B (compare Sect. 1.5). Inserting these expressions in Eq. (12.14) for the freezing-point depression results in

$$\Delta T_{sl} = -k_f \cdot \frac{n_B}{m_A} = -k_f \cdot b_B \quad \text{with} \quad k_f = -\frac{RT_{sl}M_A}{\Delta_{sl}S_A}, \tag{12.16}$$

a coefficient called "*cryoscopic constant*" which is only dependent upon pressure and type of solvent. k_f corresponds to the lowering of freezing point which is obtained from 1 mol of dissolved substance in 1 kg of solvent. At such high concentrations, the equation above can only be an approximation. For measuring the temperature changes at low concentrations with sufficient precision, it is advisable to use solvents with k_f values as high as possible. Table 12.1 shows the "cryoscopic constants" of some solvents.

There are analogous relations and applications for the raising of boiling point:

$$\Delta T_{lg} = +k_b \cdot \frac{n_B}{m_A} = +k_b \cdot b_B \quad \text{with} \quad k_b = \frac{RT_{lg}M_A}{\Delta_{lg}S_A}. \tag{12.17}$$

The "ebullioscopic constant" k_b that corresponds to k_f is positive, and because of its higher denominator ($\Delta_{vap}S_A > \Delta_{fus}S_A$), it is (in absolute terms) smaller than k_f

Table 12.1 Cryoscopic and ebullioscopic constants of some solvents (from: Lide D R (ed) (2008) CRC Handbook of Chemistry and Physics, 89th edn. CRC Press, Boca Raton).

Solvent	T_{sl} (K)	k_f (K kg mol^{-1})	T_{lg} (K)	k_b (K kg mol^{-1})
Water	273.2	1.86	373.2	0.51
Benzene	278.6	5.07	353.2	2.64
Cyclohexane	279.7	20.8	353.9	2.92
Cyclohexanol	299.1	42.2	434.0	3.5
Campfor	452.0	37.8	–	–

(compare Table 12.1). For this reason, the change of temperature here is smaller than for lowering of the freezing temperature and therefore more difficult to measure. Finally, the desired molar mass M_B results from the equations above by solving for n_B and calculating according to $M_B = m_B/n_B$.

Development of osmotic pressure can also be used in determining amounts of substances and thereby molar masses. The principle of this method, which is also known as *osmometry*, is to measure the osmotic pressure of a solution of known molality. The advantage of this method compared to the other methods using colligative properties is that it is much more sensitive. For instance, an aqueous solution of cane sugar with a concentration of 0.01 mol kg^{-1} exhibits a raising of boiling point of 0.005 K and a lowering of freezing point of 0.02 K. However, the osmotic pressure is 25 kPa (0.25 bar), which can be measured both easily and precisely. Because of its sensitivity, osmometry is useful particularly for the investigation of macromolecular substances such as synthetic polymers, proteins, or enzymes having molar mass between 10^4 and 10^6 g mol^{-1}. The lowering of the freezing point is commonly used in medicine for determining the total osmolality of aqueous solutions such as blood plasma or urine.

The gas law also belongs to the colligative properties, but often it is not mentioned in this context. It can be used for the same purposes. The vacuum appears here in the role of the solvent, whereas the gas pressure corresponds to the osmotic pressure.

Chapter 13
Homogeneous and Heterogeneous Mixtures

In chemistry but also in everyday life, we are very often confronted with mixtures, be they homogeneous or heterogeneous. Think for example of hard liquor, basically a homogeneous mixture of ethanol and water, but also of fog, a heterogeneous mixture of air and minuscule water droplets. First, we concentrate on mixtures made up of two liquid components. We discuss the behavior of the chemical potential of one component in such mixtures and the reason for spontaneous mixing or demixing. For an adequate quantitative description, the concept of chemical potential has to be extended on substances in real solutions by introducing an extra potential $\overset{+}{\mu}$.

For the characterization of mixing processes, it is useful to assign an (average) chemical potential to a mixture of two components A and B (with the mole fractions x_A and x_B), as is done for pure substances. Depending on whether the resulting mixture is homogeneous or heterogeneous the concentration dependence of this average potential is different. On this basis, concepts such as miscibility gap and lever rule are discussed.

13.1 Introduction

To begin with, let us take a look at mixtures made up of two liquid components. A homogeneous mixture of ethanol and water (like it is found in hard liquor) can be preserved over long periods of time; we always observe only one single *phase*. (We were introduced to the concept of phase for a homogeneous region of matter in Sect. 1.5.) However, if we let a hot mixture of phenol and water cool down, it will split up into two separate parts (Experiment 13.1) meaning that *demixing* (phase separation) occurs.

A similar situation occurs with ether and water (Experiment 13.2).

If only a small amount of ether—colored brown with iodine—is added to water, a homogeneous brown-colored solution results as we have seen. This is because the

© Springer International Publishing Switzerland 2016

G. Job, R. Rüffler, *Physical Chemistry from a Different Angle*,

DOI 10.1007/978-3-319-15666-8_13

Experiment 13.1 *Demixing of phenol/ water*: We let a hot mixture of phenol and water (at a ratio of 1:1) cool down in the air. After a while, demixing takes place thereby forming streaks. The phenol-rich phase settles to the bottom because of its higher density. The demixing becomes nicely visible when a tiny amount of methyl red is added to the original mixture. Because the dye is not soluble in water but is very soluble in phenol, it remains in the phase rich in phenol.

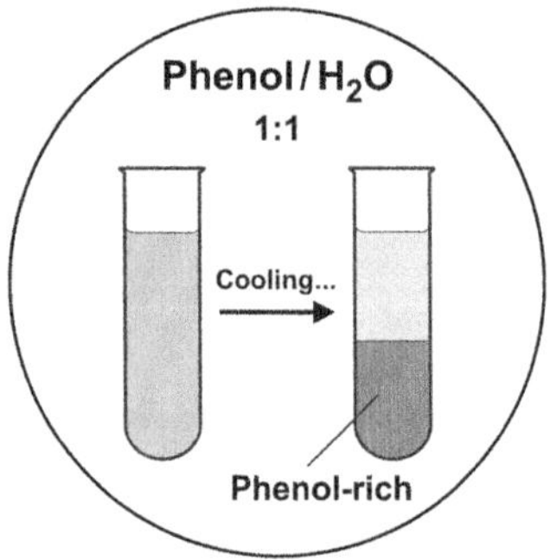

Experiment 13.2 *Mixing of ether with water*: We add a small amount of ether—colored brown with a bit of iodine—to water in a separating funnel (at a ratio of 10:1), and subsequently, we shake the funnel gently. A homogeneous brown-colored solution results. When the same amount of ether is added once more (ratio now 5:1) and we shake the funnel again, a large portion of the ether separates as a brown layer on top of the almost colorless water.

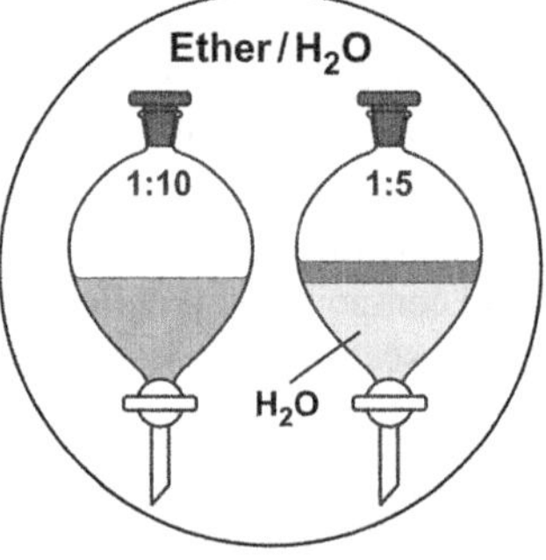

small amount of ether completely dissolves in the water and the iodine distributes through the water along with it. When the ratio of ether to water is 1:5, the ether separates as a brown layer on top of the water because water can only absorb about 10 % of its own volume in ether. Iodine can dissolve much better in ether than in water so it moves from the aqueous phase and collects in the layer of ether lying on top of it. This has been already discussed in Sects. 4.2 and 6.6.

Solid mixtures behave very similarly. For example, α-brass (an alloy of copper and up to 40 % zinc) can be preserved for just about any amount of time (Experiment 13.3).

On the other hand, carbon steel made up of iron with a maximum of 2 % carbon separates more or less quickly—when cooled from its melted state—into two homogeneous but intricately entangled areas (Experiment 13.4).

Processes of this type can be discussed in the same way as chemical reactions. The constituents of the mixture assume the role of elements as basic substances whose amounts are conserved during transformation (see Sect. 1.2). The homogeneous and heterogeneous mixtures themselves, however, correspond to chemical compounds. Therefore, the composition of these mixtures can be given by a content formula but with the peculiarity that the content numbers are not necessarily integer

Experiment 13.3 *Polished cross section of brass*: A polished cross section studied under a light microscope after etching shows well-distinguishable grain boundaries that separate the variously oriented but otherwise identical regions of matter.

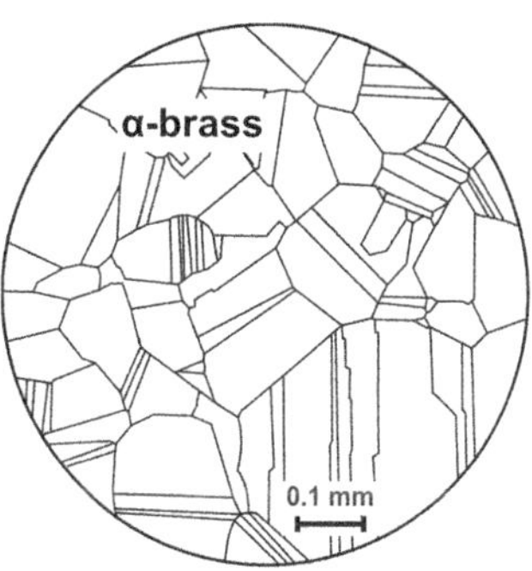

Experiment 13.4 *Polished cross section of carbon steel*: The *white areas* are ferrite (materials science term for α-Fe, i.e., (nearly) pure iron with a body-centered cubic crystal structure). The *dark gray areas*, however, are pearlite (a fine lamellar structure of ferrite and cementite, an iron–carbon compound with the formula Fe_3C).

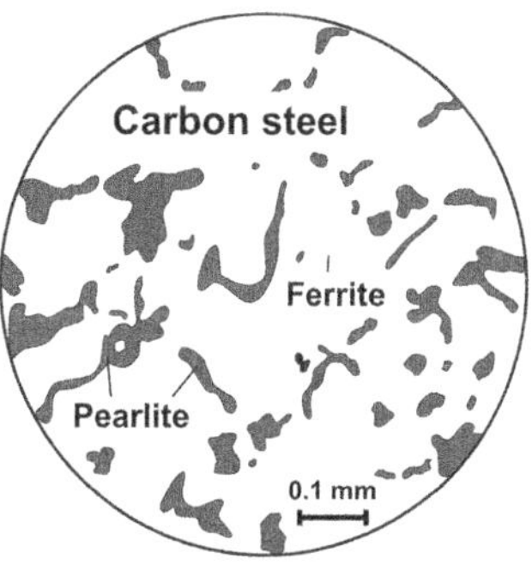

numbers but can also be real. Normally, one chooses the numbers in such a way that their sum equals 1. The simplest case would be the mixing of two substances A and B resulting in a mixture with defined composition, for example, with a mole fraction x_B of B and therefore $x_A = 1 - x_B$ of A:

$$v_A A + v_B B \rightarrow A_{x_A} B_{x_B}.$$

One glance is enough to recognize that the conversion numbers v_i on the left side have to be identical with the content numbers g_i in the formula on the right side and hence with the mole fractions x_i in the mixture: $v_A = g_A = x_A$ and $v_B = g_B = x_B$. The content formula on the right describes only the fractions of the components, but no information is provided if the resulting mixture is homogeneous (such as water and alcohol) or heterogeneous (like sugar and flour when preparing a cake dough). However, the meaning of the formula on the right is usually clear from the context. We will first discuss homogeneous mixtures; in this case, the content formula on the right is sufficient for characterization. The situation is more complex for heterogeneous mixtures because their constituents have not to be pure substances but can also be homogeneous or even heterogeneous mixtures. But also in this case we basically need no new means for capturing the essence.

13.2   Chemical Potential in Homogeneous Mixtures

Why do certain mixtures split up when others do not? How are phases formed? To answer these questions, we will again refer to the chemical potential μ. Until now we have considered the situation as follows: If a region is inhomogeneous, the substances move more or less quickly along the potential gradient until the chemical potential for every substance is equal everywhere in that region. Although a homogeneous region is what would be expected as the final result, this is obviously not always the case. For clarity we will consider the $\mu(x)$ curve.

To provide an example, we will now look at how the chemical potential μ of water depends upon its mole fraction x in various mixtures (Fig. 13.1).

On the right, all curves show the same slope RT in the vicinity of $x = 1$. On the left, they all approach negative infinity. The similar initial part of the curves on the right side is a consequence of indirect mass action which is the same for all substances (see Sect. 12.3). The similarities and differences for varying mixtures become even clearer when plotted logarithmically (Fig. 13.2).

In this representation, all curves end up on the left side in parallel straight lines with the slope RT. In Sect. 6.5 we set up the following equation to describe mass action [Eq. (6.28)]:

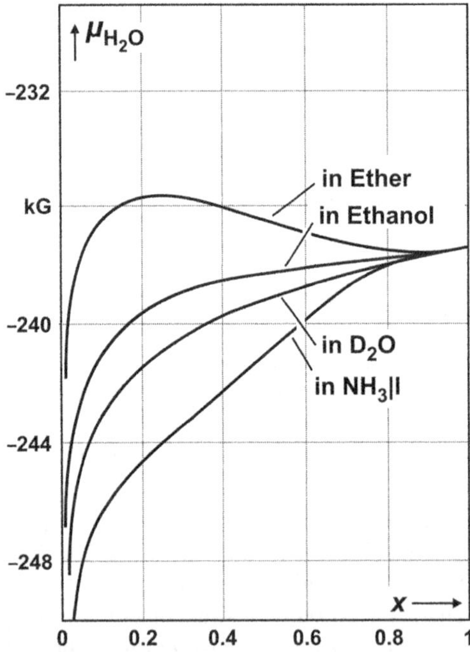

Fig. 13.1 Chemical potential of water in different mixtures as a function of its mole fraction (at a temperature of 298 K and a pressure of 100 kPa).

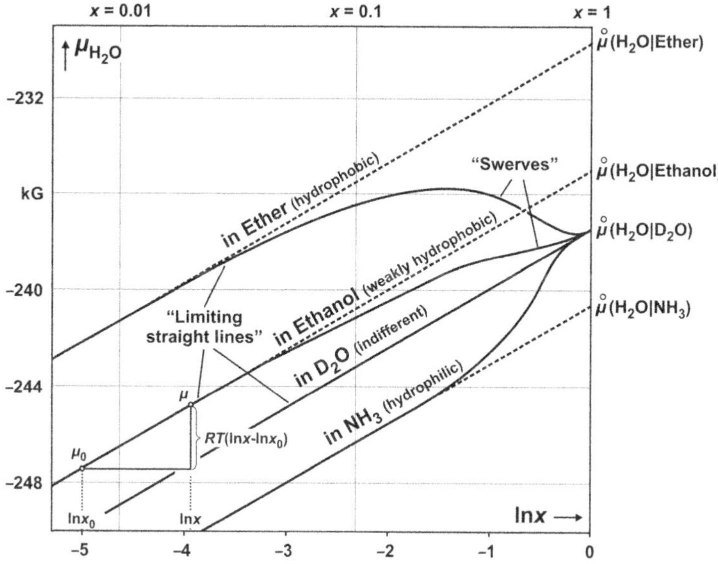

Fig. 13.2 Presentation of the chemical potential of water in different mixtures as a function of concentration on a logarithmic scale.

$$\mu = \mu_0 + RT\ln\frac{x}{x_0} \quad \text{for } x, x_0 \ll 1.$$

A logarithmic plot where μ is chosen to be the ordinate and $\ln x$ the abscissa, should result in a straight line with a slope of RT:

$$\mu = \mu_0 + RT(\ln x - \ln x_0).$$
$$y = y_0 + m \cdot (x - x_0) \qquad \text{linear equation}$$

(As a reminder, the well-known equation you learned at school for a straight line with a slope of m through the point $(x_0; y_0)$ is given.)

A mixture is called *ideal* if this relation is not only valid for low mole fractions but in the whole range $0 \leq x \leq 1$ and in particular for $x_0 = 1$. If we write again $\overset{\bullet}{\mu}$ for the chemical potential of the pure substance, the equation above simplifies to (cf. Sect. 12.3):

$$\mu = \overset{\bullet}{\mu} + RT\ln x \quad \text{for } 0 \leq x \leq 1 \text{ in the ideal case.} \tag{13.1}$$

When we take the derivative of this function with respect to x at constant T, we obtain the value of RT at $x = 1$ as we should expect because of the indirect mass action:

$$\left(\frac{\partial \mu}{\partial x}\right)_T = \frac{RT}{x} \quad \text{and therefore} \quad \left(\frac{\partial \mu}{\partial x}\right)_T = RT \quad \text{for } x = 1.$$

However, only very similar substances that are *indifferent* to each other such as a homogeneous mixture of light water (H_2O) and heavy water (D_2O) show an ideal behavior. (The molecular rearrangement $H_2O + D_2O \rightarrow 2HDO$ which occurs easily is considered to be suppressed.).

A similar approach can be used to describe the behavior at low mole fractions. The $\mu(x)$ curves appear in Fig. 13.2 as straight lines parallel to that for the ideal case; they only differ from each other in the intercepts on the y-axis (vertical line at the far right in the figure). $\overset{\circ}{\mu}_x$ represents a special basic value (such as already $\overset{\bullet}{\mu}$) meaning a "basic contribution" independent from the composition. The index $_x$ can be omitted if it is clear from the context that the usual basic value specified in the concentration scale (i.e., the value for 1 kmol m^{-3}) is not meant. To avoid confusion, we can write $\overset{\circ}{\mu}_c$ instead of simply $\overset{\circ}{\mu}$ as before.

$\overset{\circ}{\mu}_x(B|A)$ or abbreviated $\overset{\circ}{\mu}(B|A)$ represents the chemical potential for a hypothetical state of the "pure" substance B in question (here water) in which the interaction of the substance (B) and the solvent molecules (A) (here ether, ethanol, etc.) determine the outcome, and not the interactions of the B molecules with each other. Naturally, each solvent leads to a different value (Table 13.1). We will take a short look at this again in the next section.

The substances' individual characteristics recede near $x = 1$ as well as $x = 0$ and general laws that are broadly independent of substance-specific quantities become valid. However, the form of the functions varies noticeably between the limits mentioned. But characteristic for all curves is the "swerve" from the straight line on the left with slope RT to the parallel straight line on the right.

The potential difference between $\overset{\circ}{\mu}(B|A)$ and $\overset{\bullet}{\mu}(B)$ can serve as measure for the compatibility of B with A. The higher the value $\overset{\circ}{\mu}(B|A)$ lies above that of $\overset{\bullet}{\mu}(B)$ the stronger the tendency of B to separate from A, the worser the compatibility between the substances. As long as one of the substances is added in small or very small amounts, it will always be tolerated. The situation can become critical when both components are present in comparable amounts. We call the substances "lowly compatible" when they do not yet separate from each other and "incompatible"

Table 13.1 Basic values of chemical potential of substances in several mixtures (at 298 K and 100 kPa).	Substance/solvent		$\overset{\circ}{\mu}_x$ (kG)
	H_2O	Pure	−237.4
		In ether (hydrophobic)	−230
		In D_2O (indifferent)	−237.4
		In H_2SO_4 (hydrophilic)	−260
	Hg	Pure	0
		In H_2O (lowly compatible)	+40
		In benzene (lowly compatible)	+30
		In liquid Na (highly compatible)	−150
	Fe	Pure	0
		In Cu (lowly compatible)	+20

when they do. Also the opposite can occur meaning the value of $\overset{\circ}{\mu}$ (B|A) lies beneath that of $\overset{\bullet}{\mu}$(B). In this case, A and B are better compatible among each other than each of the substances alone. We call substances showing this behavior "highly compatible."

In the case of a mixture of two "highly compatible" substances such as H_2O and NH_3 (in this special case also called a *hydrophilic* or "water-loving" substance), a downward deviation from the continuous straight line for indifferent substances can be observed. "Lowly compatible" substances like H_2O and Ethanol, however, show an upward deviation. The curve for a mixture of "incompatible" substances (such as H_2O and *hydrophobic* ether) shows a "swerve" raised to a maximum.

The varying behaviors of these mixtures are due to the different interactions of the components A and B at their molecular levels. If the attraction between particles of different types A and B is about equal to the average attraction between particles of the same type (A and A or B and B), the substances will behave *indifferently*. This holds for mixtures of substances that are chemically closely related such as H_2O/D_2O, hexane/heptane, benzene/toluene, etc. Dilute gases also behave indifferently toward each other because the attraction is almost nonexistent due to the large distances between the particles. For this reason,

- Non or weakly polar liquids: hexane, ether, carbon tetrachloride or
- Substances that form hydrogen bonds: water, ammonia, methanol, glycerol

are among each other more or less compatible.

If the attraction between particles A and B is stronger than that between the different types of particles themselves, one speaks of *highly compatible* substances. If, however, the attraction is considerably weaker, the substances are called *incompatible*. Incompatible ones are

- Polar and nonpolar liquids such as water in combination with organic solvents like benzene, hexane, or carbon tetrachloride,
- Metallic and nonmetallic liquids such as Hg/H_2O, $Hg/benzene$.

We can qualitatively interpret the demixing (separation) of a mixture like water and ether having a ratio of 1:1 ($x_{H_2O} = 0.5$) as follows (Fig. 13.3): A tiny arbitrary accumulation of H_2O molecules at some location in the mixture lowers the

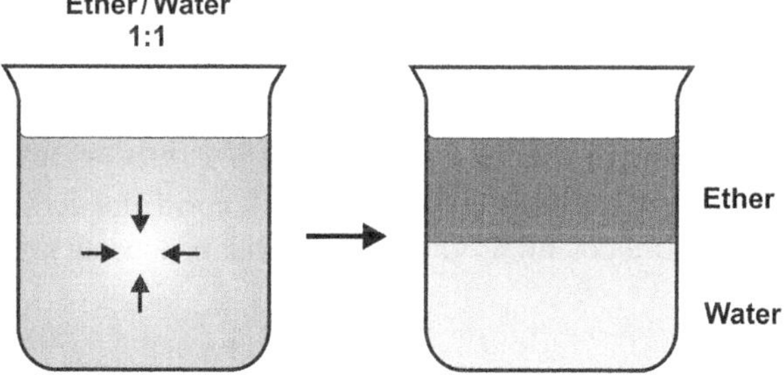

Fig. 13.3 Demixing of a mixture of ether and water into an upper ether-rich layer with 1 % water and a water-rich layer below with an ether fraction of 8 %.

Experiment 13.5 *Demixing of an acetone-salt water solution*: Acetone is colored with a little bit of methyl violet and the same amount of water is added. When table salt is added to the homogeneous mixture, a demixing takes place resulting in a deep violet acetone and a pale violet water layer.

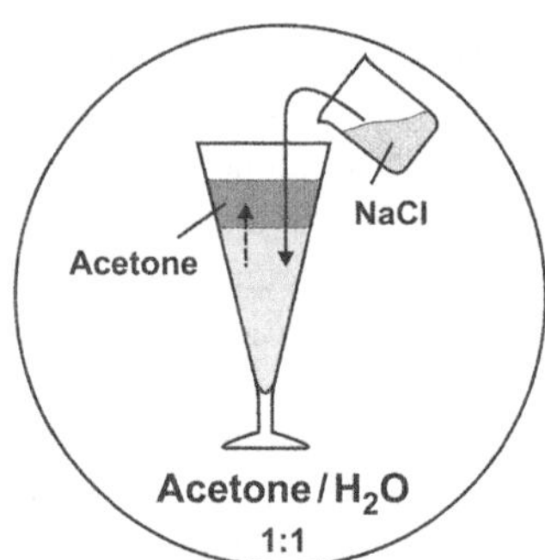

chemical potential μ of the water there because it decreases with an increase of the mole fraction (compare Fig. 13.1). As a result, additional H_2O particles migrate into this spot from the area around it and it becomes gradually larger and richer in H_2O. This process continues until the surroundings are so depleted of water that the chemical potential there drops. The final state is a water-poor, lighter layer on top and a water-rich and heavier layer below.

The compatibility of two components A and B can be influenced by the addition of a third substance and the corresponding change of the concentrations of the components. For example, when table salt is added to a homogeneous mixture of water and acetone, a demixing takes place (Experiment 13.5). The reason for this is the low compatibility of the components which increases with the salt content of the water.

13.3   Extra Potential

When we were introduced to the ideal case for the potential μ of a substance in a homogeneous mixture, we learned about the following behavior:

$$\mu(x) = \overset{\bullet}{\mu} + RT\ln x \quad \text{for } 0 \leq x \leq 1.$$

Deviations from this simple mass action equation can be explained by the fact that the chemical interactions between the particles upon each other are not taken into account. Corrections must therefore be made in order to describe the behavior correctly. This is most easily done by the addition of a correction term $\overset{+}{\mu}$, a so-called *extra potential*. This extra potential is not constant but is dependent upon the mole fraction x:

$$\mu(x) = \overset{\bullet}{\mu} + RT\ln x + \overset{+}{\mu}(x). \tag{13.2}$$

Substances such as light water and heavy water or dilute gases that are dissolved in each other follow the ideal curve, so the extra potential $\overset{+}{\mu}$ disappears. This means that $\overset{+}{\mu}(x) \equiv 0$. Otherwise, the extra potential would be either positive or negative.

For a strongly diluted substance, the change of $\overset{+}{\mu}$ with the mole fraction x is negligible compared to the strongly changing term $RT \cdot \ln x$, which in the limit tends toward $-\infty$ (compare Fig. 13.1). It is therefore possible to replace $\overset{+}{\mu}$ by the constant limiting value $\overset{+}{\mu}_0$ for "infinite" dilution meaning vanishingly low concentration (for better distinction we use a "slashed zero" as index):

$$\mu(x) = \underbrace{\overset{\bullet}{\mu} + \overset{+}{\mu}_0}_{\overset{\circ}{\mu}} + RT\ln x \quad \text{for small } x. \tag{13.3}$$

As we have seen in the previous section, when substances are indifferent to each other, both reference potentials are identical, $\overset{\circ}{\mu} = \overset{\bullet}{\mu}$. The extra potential vanishes in the entire range, $\overset{+}{\mu}(x) \equiv 0$, and so do all properties related to it.

In Chap. 9, which dealt with cross relations, we learned that the volume and entropy demands of a substance in a mixture can be derived from the pressure coefficient β and the temperature coefficient α of chemical potential, meaning the derivatives of μ with respect to p and T at constant composition:

$$\beta = \left(\frac{\partial \mu}{\partial p}\right)_{T,n} = V_{\mathrm{m}} \quad \text{and} \quad \alpha = \left(\frac{\partial \mu}{\partial T}\right)_{p,n} = -S_{\mathrm{m}}.$$

Here, V_{m} and S_{m} are the molar volume and molar entropy, respectively. If we use the equation for the chemical potential above, $\mu(x) = \overset{\bullet}{\mu} + RT\ln x + \overset{+}{\mu}(x)$, as our starting point and abbreviate the derivatives of the terms $\overset{\bullet}{\mu}$, and $\overset{+}{\mu}$ correspondingly, we obtain

$$V_{\mathrm{m}} = \overset{\bullet}{V}_{\mathrm{m}} + \overset{+}{V}_{\mathrm{m}} \tag{13.4}$$

and, respectively,

$$S_{\mathrm{m}} = \overset{\bullet}{S}_{\mathrm{m}} - R\ln x + \overset{+}{S}_{\mathrm{m}}. \tag{13.5}$$

In the case of volume, the term $RT \cdot \ln x$ drops out because it is not dependent upon pressure. We call $\overset{+}{V}_{\mathrm{m}}(x)$ "molar extra volume" and $\overset{+}{S}_{\mathrm{m}}(x)$ "molar extra entropy." The extra quantities disappear in homogeneous mixtures of indifferent substances. While the volume demand V_{m} is independent of composition in this special case, this is not true for the entropy demand S_{m} which increases continually with falling mole fraction x. It tends toward $+\infty$ for $x \rightarrow 0$, but "extremely slowly."

13.4   Chemical Potential of Homogeneous and Heterogeneous Mixtures

Potential of Homogeneous Mixtures   A process of mixing can be described as a "reaction" between two substances. When one-third ethanol and two-thirds water are mixed, the result is the mixed phase schnaps. This phase can be further put to use to produce a grog, for example (Experiment 13.6).

In order to describe these kinds of processes in the usual way, it is practical to assign an amount of substance and a chemical potential to a portion of a homogeneous mixture, too. The sum of the amounts n_A, n_B, n_C, ... of the pure substances A, B, C, ... that make up the homogeneous mixture M equals the amount of substance n_M of the mixture:

$$n_M = n_A + n_B + n_C + \dots . \tag{13.6}$$

The weighted average of the chemical potentials of the components—weighted with the mole fractions x_A, x_B, x_C ...—is the chemical potential of the mixture, μ_M:

$$\mu_M = x_A\mu_A + x_B\mu_B + x_C\mu_C + \dots . \tag{13.7}$$

μ_M is often called the *average* chemical potential (to keep in mind the averaging). It agrees nicely with the definition of chemical potential we learned earlier. We have seen that μ_i is the energy $dW_{\to n_i}$ (abbreviated dW_i) necessary for creating the substance i (no matter if the substance is pure or mixed with others; however, it is essential to avoid or subtract all energy contributions expended for any other purpose), per amount of substance n_i:

$$\mu_i = \frac{dW_{\to n_i}}{dn_i} = \frac{dW_i}{dn_i} .$$

The energy $dW_{\to n_M}$ (abbreviated dW_M) necessary for producing a homogeneous mixture M is simply the sum of the energies of formation of its components,

Experiment 13.6 *Mixing of a grog*: Two or three sugar cubes are put in a glass. Subsequently, we fill the glass halfway with boiling water, add the rum, and stir. A new mixed phase, the grog, has been produced from the two original mixed phases, the aqueous sugar solution and the rum.

$$dW_M = \mu_A dn_A + \mu_B dn_B + \mu_C dn_C + \ldots,$$
$$= \mu_A x_A dn_M + \mu_B x_B dn_M + \mu_C x_C dn_M + \ldots .$$

Dividing this expression by the amount of the mixture n_M results in a quantity that we appropriately call the chemical potential of M:

$$\mu_M = \frac{dW_M}{dn_M} = x_A \mu_A + x_B \mu_B + x_C \mu_C + \ldots .$$

How does the (average) chemical potential change with the composition of the mixed phase? Let us consider the simple case of a so-called binary mixture of two pure components A and B that are indifferent to each other. The chemical potential of component A in the mixed phase is

$$\mu_A = \overset{\bullet}{\mu}_A + RT \ln x_A .$$

The relationship for component B can be formulated correspondingly. The chemical potential for the mixed phase is then

$$\mu_M = x_A \mu_A + x_B \mu_B = x_A \overset{\bullet}{\mu}_A + x_B \overset{\bullet}{\mu}_B + RT(x_A \cdot \ln x_A + x_B \cdot \ln x_B) \tag{13.8}$$

$$= \underbrace{\left(\overset{\bullet}{\mu}_B - \overset{\bullet}{\mu}_A\right) \cdot x_B + \overset{\bullet}{\mu}_A}_{\text{straight line}} + \underbrace{RT((1 - x_B) \cdot \ln(1 - x_B) + x_B \cdot \ln x_B)}_{\text{“drooping belly”}} . \tag{13.9}$$

We replaced x_A by $1 - x_B$ in the second step. Figure 13.4 illustrates the curve of μ_M (shaped as a "drooping belly") as a function of x_B. At $x_B = 0$ as well as $x_B = 1$, the curve has vertical tangents, a fact that is not easy to recognize in the figure but that has important consequences which we will discuss later.

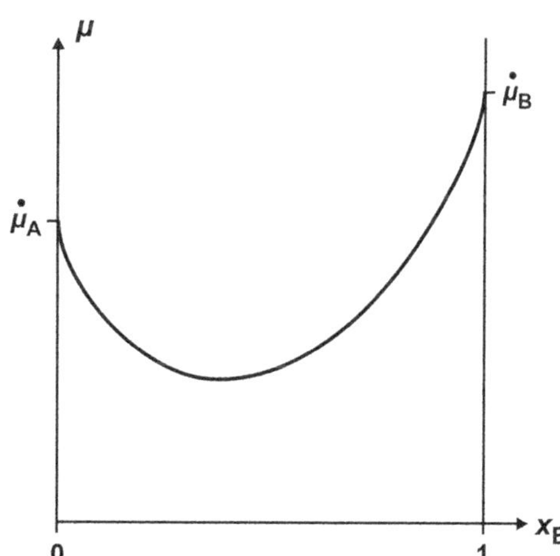

Fig. 13.4 The (average) chemical potential as a function of the composition of a homogeneous mixture of two indifferent substances A and B.

Potential of Heterogeneous Mixtures A portion of a heterogeneous mixture $\mathcal{M}$ of several immiscible components A, B, C, ... can also be assigned an amount of substance n and an (average) chemical potential μ according to the same pattern used for homogeneous mixtures:

$$n_{\mathcal{M}} = n_A + n_B + n_C + \ldots \tag{13.10}$$

and

$$\mu_{\mathcal{M}} = x_A\mu_A + x_B\mu_B + x_C\mu_C + \ldots . \tag{13.11}$$

There is, however, a fundamental difference: While in the case of homogeneous mixtures, the chemical potentials of the components are different in their mixed and unmixed state, μ_A, μ_B, μ_C ... always have the same values in heterogeneous mixtures whether A, B, C, ... are present in their mixed or unmixed states. In order to differentiate the chemical potential of a heterogeneous mixture from the chemical potential of a homogeneous mixture, we will label it with the index $\mathcal{M}$.

The following is then valid for a heterogeneous mixture of two pure components A and B:

$$\mu_{\mathcal{M}} = x_A\overset{\bullet}{\mu}_A + x_B\overset{\bullet}{\mu}_B. \tag{13.12}$$

Because $x_A + x_B = 1$, we obtain

$$\mu_{\mathcal{M}} = (1 - x_B)\overset{\bullet}{\mu}_A + x_B\overset{\bullet}{\mu}_B = \underbrace{\left(\overset{\bullet}{\mu}_B - \overset{\bullet}{\mu}_A\right) \cdot x_B + \overset{\bullet}{\mu}_A}_{\text{straight line}} . \tag{13.13}$$

If the potential $\mu_{\mathcal{M}}$ of a heterogeneous mixture is plotted as a function of the mole fraction x_B, a straight line with the slope $\left(\overset{\bullet}{\mu}_B - \overset{\bullet}{\mu}_A\right)$ and the y-intercept $\overset{\bullet}{\mu}_A$, which runs through points $(0; \overset{\bullet}{\mu}_A)$ and $(1; \overset{\bullet}{\mu}_B)$, appears in place of the "drooping" curve (Fig. 13.5).

"Lever Rule" An important fact to remember here is that the components themselves do not need to be pure substances but can be homogeneous mixtures of two (or more) components A and B. Let us assume that the entire system (homogeneous or heterogeneous mixture) characterized by $\blacktriangle$, with a mole fraction $x_B^{\blacktriangle}$ of B, is composed of two homogeneous mixtures M' and M'' of which one (x_B') is poorer in B and the other (x_B'') is richer. If the system as a whole is made up of an amount of substance $n^{\blacktriangle}$, the balance for component B according to the general formula $n_B = x_B \cdot n$, applied to each mixed phase, yields:

$$x_B'n' + x_B''n'' = x_B^{\blacktriangle} n^{\blacktriangle} .$$

Fig. 13.5 The (average) chemical potential as a function of the composition of a heterogeneous mixture (*solid line*) (For comparison, the dotted curve for a homogeneous mixture is also included in the graphic.).

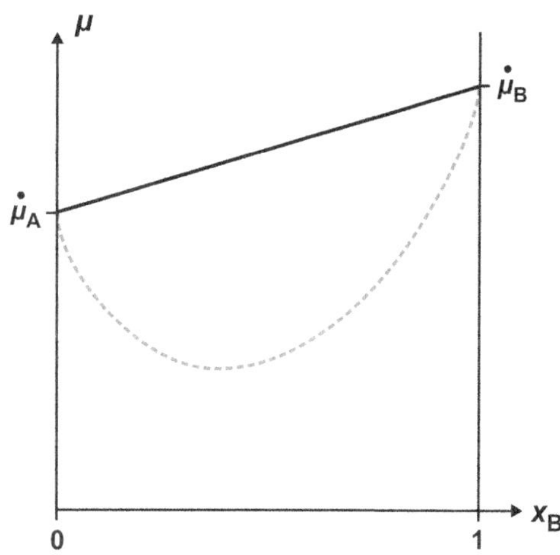

Because of $n' + n'' = n^{\blacktriangle}$, the following is valid:

$$x_B' n' + x_B'' n'' = x_B^{\blacktriangle} \cdot (n' + n'') \quad \text{and rearranged} \quad \left(x_B' - x_B^{\blacktriangle}\right) \cdot n' = \left(x_B^{\blacktriangle} - x_B''\right) \cdot n''.$$

Thus, the ratio of amounts of substance of the initial mixtures is

$$\frac{n'}{n''} = \frac{x_B'' - x_B^{\blacktriangle}}{x_B^{\blacktriangle} - x_B'}. \tag{13.14}$$

This is the so-called "*lever rule*" (of the amounts of different phases). The name is borrowed from mechanics. The form

$$n' \cdot \left(x_B^{\blacktriangle} - x_B'\right) = n'' \cdot \left(x_B'' - x_B^{\blacktriangle}\right)$$
$$\text{"load} \times \text{load arm} = \text{force} \times \text{force arm"}$$

brings a lever to mind that is supported at $x_B^{\blacktriangle}$ and upon whose ends the two phases hang like two "weights" n' and n''. In this case, too, we see that the shorter the lever arm is that is oriented toward its corresponding phase, the greater the "weight" needs to be. In this case, the weight is the amount of substance.

In closing, let us take a look at the graphic representation of the potential of a heterogeneous mixture $^{\blacktriangle}$ of two homogeneous mixtures $'$ and $''$ (Fig. 13.6). The total potential is again defined by linear variation of the starting values. In this case, it lies upon the line connecting the two points (x_B', μ') and (x_B'', μ'').

Volume Demand and Entropy Demand The temperature coefficient α_M and the pressure coefficient β_M of chemical potential μ_M of a homogeneous mixture M are obtained by taking the derivative μ_M with respect to T or p. Based on the approach for a mixture of substances A, B, C, ...

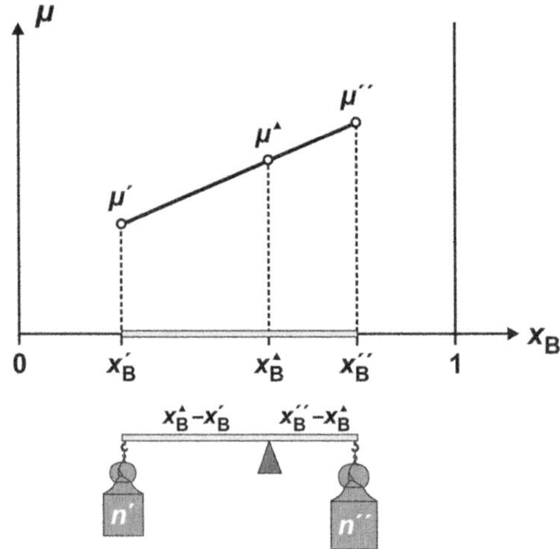

Fig. 13.6 Applying the "lever rule" to a heterogeneous mixture.

$$\mu_M = x_A \mu_A + x_B \mu_B + \dots \quad \text{with} \quad \mu_A = \overset{\bullet}{\mu}_A + RT\ln x_A + \overset{+}{\mu}_A, \dots,$$

this results in the desired equations for α_M and β_M. We forgo writing down these equations and select instead a more well-known version by replacing α with molar entropy S_m (but pay attention to the different algebraic sign), $\alpha = -S_m$, and β with molar volume V_m, $\beta = V_m$:

$$S_M = x_A S_A + x_B S_B + \dots \quad \text{with} \quad S_A = \overset{\bullet}{S}_A - R\ln x + \overset{+}{S}_A, \dots,$$

$$V_M = x_A V_A + x_B V_B + \dots \quad \text{with} \quad V_A = \overset{\bullet}{V}_A + \overset{+}{V}_A, \dots,$$

where S_A is the molar entropy of a substance A, S_B that of a substance B, ... The same applies to the molar volume.

The ellipses (three points) at the end of the last two lines signify the corresponding expressions for the substances B, C, ..., etc., which only differ by the index from the previous one.

13.5   Mixing Processes

Indifferent Substances The chemical potential of homogeneous mixtures can be applied so that reactions between mixed phases can be treated exactly like reactions between pure substances. As an example the chemical drive $\mathcal{A}_{mix}$ for the mixing process of two substances that are indifferent to each other should be determined. Because the conversion numbers ν_A and ν_B coincide with the mole fractions x_A and x_B, the conversion formula simplifies to

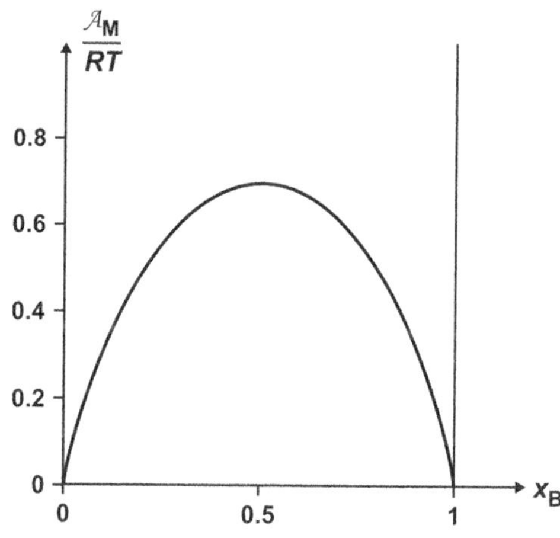

Fig. 13.7 Chemical drive of mixing $\mathcal{A}_{\mathrm{mix}}$ as a function of the composition of a homogeneous mixture of two indifferent substances.

$$x_A A + x_B B \rightarrow A_{x_A} B_{x_B}.$$

As usual, the chemical drive corresponds to the potential drop from reactants to products. When we calculate the potential μ_M of the homogeneous mixture $M = A_{x_A} B_{x_B}$ in the manner discussed in the last section [see Eq. (13.7)],

$$\mathcal{A}_{\mathrm{mix}} = x_A \overset{\bullet}{\mu}_A + x_B \overset{\bullet}{\mu}_B - \mu_M$$
$$= x_A \overset{\bullet}{\mu}_A + x_B \overset{\bullet}{\mu}_B - \left[x_A \overset{\bullet}{\mu}_A + x_B \overset{\bullet}{\mu}_B + RT(x_A \cdot \ln x_A + x_B \cdot \ln x_B) \right]$$

we obtain:

$$\mathcal{A}_{\mathrm{mix}} = -RT(x_A \cdot \ln x_A + x_B \cdot \ln x_B).$$

Figure 13.7 shows the "chemical drive of mixing" $\mathcal{A}_{\mathrm{mix}}$ as a function of the composition of the mixture. Notice that the drive for any arbitrary composition is invariably positive because the mole fractions x_A and x_B are always smaller than 1 so that the two logarithms are always negative ($\ln x < 0$ for $x < 1$). This means that two substances that are indifferent to each other mix spontaneously in any proportion.

Side Effects in the Ideal Case Changes of volume and entropy in mixing processes work just like those of chemical processes discussed in Chap. 8. Let us again consider a homogeneous mixture of two indifferent substances A and B. Because the extra quantities, in this case $\overset{+}{V}_m(x)$ and $\overset{+}{S}_m(x)$, disappear, the molar volume of mixing $\Delta_{\mathrm{mix}}V$ and the molar entropy of mixing $\Delta_{\mathrm{mix}}S$ turn out to be

$$\Delta_{\mathrm{mix}}V = V_M - x_A \overset{\bullet}{V}_A - x_B \overset{\bullet}{V}_B = \underbrace{(x_A V_A + x_B V_B)}_{\overset{\bullet}{V}_A \qquad \overset{\bullet}{V}_B} - x_A \overset{\bullet}{V}_A - x_B \overset{\bullet}{V}_B \qquad \text{meaning}$$

$$\Delta_{\mathrm{mix}}V = 0 \qquad\qquad\qquad (13.15)$$

and

$$\Delta_{\mathrm{mix}}S = S_M - x_A\,\overset{\bullet}{S}_A - x_B\,\overset{\bullet}{S}_B = \left(x_A S_A + x_B S_B\right) - x_A\,\underbrace{\overset{\bullet}{S}_A}_{\overset{\bullet}{S}_A - R\ln x_A} - x_B\,\underbrace{\overset{\bullet}{S}_B}_{\overset{\bullet}{S}_B - R\ln x_B} \qquad \text{meaning}$$

$$\Delta_{\mathrm{mix}}S = -R \cdot (x_A \cdot \ln x_A + x_B \cdot \ln x_B). \tag{13.16}$$

When substances that are indifferent to each other (such as dilute gases and light and heavy water) are mixed, the volume does not change. They neither try to absorb entropy from the surroundings nor emit it and their temperatures remain constant. This is just as if different parts of one and the same substance were mixed. For this reason, such mixtures are called "*ideal*." In fact, it is $\Delta_{\mathrm{mix}}V = 0$, but not $\Delta_{\mathrm{mix}}S$, because the term $-R(x_A \cdot \ln x_A + x_B \cdot \ln x_B)$, which is always positive because of $x < 1$, remains in the expression. The total required entropy has become greater so that the mixture would have to cool down if entropy cannot flow in from outside. However, this is not the case because energy is released as a result of the substances going through a potential difference during the mixing process. Exactly as much entropy is generated as is needed. The energy released by a small conversion $d\xi$ and subsequently dissipated results in $dW_b = \mathcal{A}_{\mathrm{mix}} \cdot d\xi$ and therefore the generated entropy is $dS_g = dW_b/T$ which is exactly equal to $\Delta_{\mathrm{mix}}S \cdot d\xi$:

$$dS_g = \frac{\mathcal{A}_{\mathrm{mix}}d\xi}{T} = \frac{-RT(x_A \cdot \ln x_A + x_B \cdot \ln x_B)}{T}d\xi = -R(x_A \cdot \ln x_A + x_B \cdot \ln x_B)d\xi.$$

Indifferent behavior occurs on the molecular level when the interactions between particles (such as in dilute gas mixtures) are nonexistent or when they are of the same size independently of the type of particles.

Real Mixtures Let us now turn to real mixtures where interactions cannot be ignored any longer and the extra potential $\overset{+}{\mu}$ must be taken into account. The (average) chemical potential of a homogeneous mixture of two components A and B will be equal to:

$$\mu_M = x_A\mu_A + x_B\mu_B = \underbrace{x_A\overset{\bullet}{\mu}_A + x_B\overset{\bullet}{\mu}_B}_{\overset{\circ}{\mu}_M} + \underbrace{RT(x_A \cdot \ln x_A + x_B \cdot \ln x_B)}_{\overset{\times}{\mu}_M} + \underbrace{x_A\overset{+}{\mu}_A + x_B\overset{+}{\mu}_B}_{\overset{+}{\mu}_M}.$$

When the chemical potential is plotted as a function of the composition of the mixture (characterized by the mole fraction x_B, Fig. 13.8), the three cases discussed above can again be distinguished: highly compatible, lowly compatible, and incompatible. For clarity, the graphic also includes the relation for indifferent behavior. The curve corresponding to highly compatible substances droops the most compared to the ideal case, while the curve corresponding to lowly compatible substances is more compressed. The curve for incompatible substances exhibits a noticeable "dent" upward.

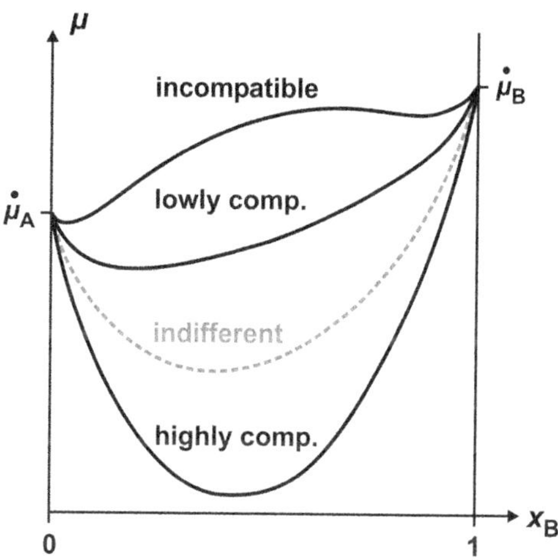

Fig. 13.8 (Average) chemical potentials for mixtures of two substances A and B of varying compatibility, as functions of concentration.

Just as in the case of a single substance, the chemical potential μ_M of a homogeneous mixture is divided into three contributions: "basic term" $\overset{\circ}{\mu}_M(x_B)$, "mass action term" $\overset{\times}{\mu}_M(x_B)$, and "extra term" $\overset{+}{\mu}_M(x_B)$. Figure 13.8 illustrates the effect of the three contributions. The first term results in a straight connection between the "pivot points" $(0, \dot\mu_A)$ and $(1, \dot\mu_B)$, and the second one is responsible for the formation of the "drooping belly" in between (dotted) which begins and ends with a vertical tangent. At last, the third term deforms the "belly," but the vertical tangents at the borders are always preserved no matter how strong this deformation in upward direction is. The $\overset{+}{\mu}_M(x_B)$ curve begins and ends in the "pivot points." The simplest approach would be a parabolic arc described by an equation such as

$$\overset{+}{\mu}_M(x_B) = a \cdot x_B(1 - x_B).$$

We have to choose an appropriate coefficient a which itself can be dependent on temperature. Negative a means high compatibility, positive a low compatibility, or even incompatibility when $a > 2\,RT$.

Demixing   In closing, we will deal in more detail with the behavior of incompatible substances. To do so, we will take a closer look at the process of demixing (the reverse process of mixing):

$$M^{\blacktriangle} \rightarrow \nu'M' + \nu''M''.$$

This means that the initial homogeneous mixture $M^{\blacktriangle}$ need not separate into the starting components A and B, but can also separate into two homogeneous mixtures M' and M'', of which one is richer in B than the initial mixture and the other one poorer. Because of $n'' = n^{\blacktriangle} - n'$, the balance for component B, for example, results in

$$x_B^{\blacktriangle} n^{\blacktriangle} = x_B' n' + x_B'' n'' = x_B' n' + x_B''(n^{\blacktriangle} - n') = x_B'' n^{\blacktriangle} - (x_B'' - x_B')n', \text{ meaning}$$

$$n' = \frac{x_B'' - x_B^{\blacktriangle}}{x_B'' - x_B'} \cdot n^{\blacktriangle} \quad \text{and finally} \quad v' = \frac{n'}{n^{\blacktriangle}} = \frac{x_B'' - x_B^{\blacktriangle}}{x_B'' - x_B'}.$$

v'' can be deduced accordingly or follows more simply from the condition $v' + v'' = 1$:

$$v'' = 1 - v' = \frac{x_B^{\blacktriangle} - x_B'}{x_B'' - x_B'}.$$

The demixing process occurs spontaneously when its chemical drive $\mathcal{A}$ is positive, i.e., $\mathcal{A} = \mu^{\blacktriangle} - v'\mu' - v''\mu'' > 0$ or, put another way, when

$$\mu^{\blacktriangle} > v'\mu' + v''\mu''.$$

Let us now consider the $\mu(x_B)$ curve for incompatible substances (Fig. 13.9). Mixture M$^{\blacktriangle}$ separates into two homogeneous mixtures M' and M'' if its chemical potential $\mu^{\blacktriangle}$ has a higher value than the chemical potential $\mu_{\mathcal{M}}$ of the heterogeneous mixture, which is made up of M' with the fraction v' and M'' with the fraction v''. The potential $\mu_{\mathcal{M}}$ lies on the gray straight line connecting the points (x_B', μ') and (x_B'', μ'') and is therefore noticeably lower than $\mu^{\blacktriangle}$.

The lever rule is again valid for the ratio of the two coexisting phases:

$$\frac{n'}{n''} = \frac{v'}{v''} = \frac{x_B'' - x_B^{\blacktriangle}}{x_B'' - x_B'} : \frac{x_B^{\blacktriangle} - x_B'}{x_B'' - x_B'} = \frac{x_B'' - x_B^{\blacktriangle}}{x_B^{\blacktriangle} - x_B'}.$$

Miscibility Gap Figure 13.9, however, does not yet represent the final situation. Further connecting lines that lie beneath the gray one—and therefore meaning lower potentials $\mu_{\mathcal{M}}$—are also conceivable. The lowest possible $\mu_{\mathcal{M}}$ value can be

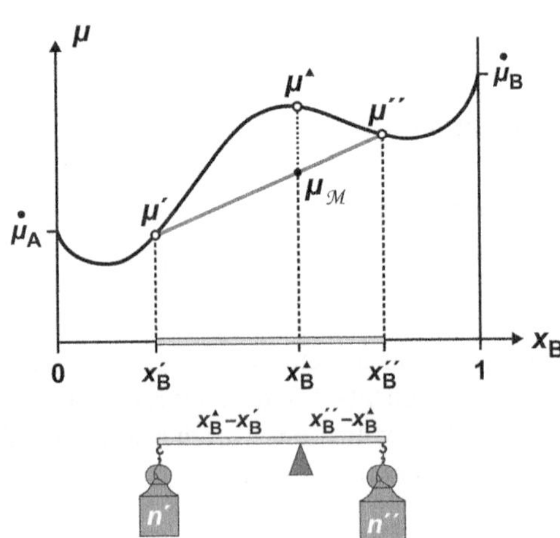

Fig. 13.9 Application of the "lever rule" to a mixture of two incompatible components A and B.

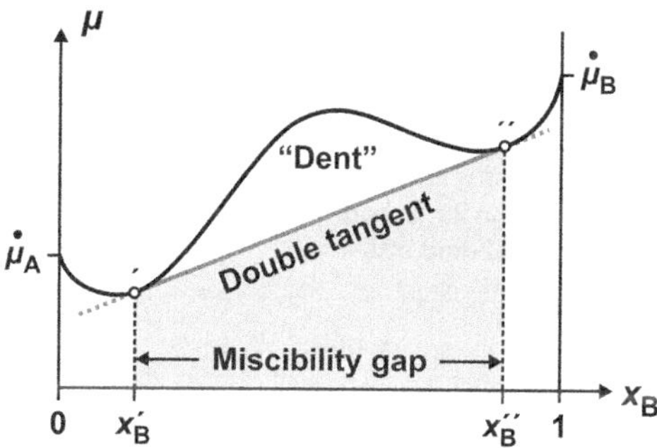

Fig. 13.10 Illustration of the double tangent rule as well as the appearance of a miscibility gap.

found by connecting the points of contact $'$ and $''$ of the common tangents on the "dented" curve, creating a double tangent (Fig. 13.10). These two points limit the so-called *miscibility gap*. For compositions $x_B^{\blacktriangle}$ which lie in the range of the gap, no homogeneous mixture is stable. Instead, it spontaneously separates more or less rapidly into a heterogeneous mixture $\mathcal{M}$ of the two homogeneous mixtures M$'$ and M$''$ with the compositions x_B' and x_B'' at the left and right border of the miscibility gap.

If the heterogeneous mixture is liquid, differences in density are responsible for a further separation which often takes place in the gravitational field. The mixture with higher density collects at the bottom, the one with lower density at the top. Such systems of substances which are still connected by shared interfaces and in which pressure and temperature are still uniform can be regarded as *heterogeneous mixtures in a wider sense* because they can be described in the same manner as the systems which are usually considered as heterogeneous mixtures. When a differentiation should be necessary we will call such unusual heterogeneous mixtures *degenerate*.

13.6 More Phase Reactions

In addition to mixing, there are a lot of other processes that can be considered "reactions" between phases and described using the (average) chemical potential. An example of this might be the solidification of magma into mica, feldspar, and quartz.

For every state of an A–B mixture (vapor, melt, and every form of crystal), there is a corresponding (average) chemical potential μ_M, just as there is a chemical potential μ_B for each state of an individual substance B. The most stable state of a phase is the one with the lowest chemical potential, be it a pure phase or a mixed phase. Let us consider as a simple example the behavior of two substances A and B, which are completely miscible in both their solid and liquid states (indifferent behavior, Fig. 13.11). In this case, a gap appears as well, meaning there is a two-phase area. However, this time there is a melt 1 with the composition x_B^l and a solid phase

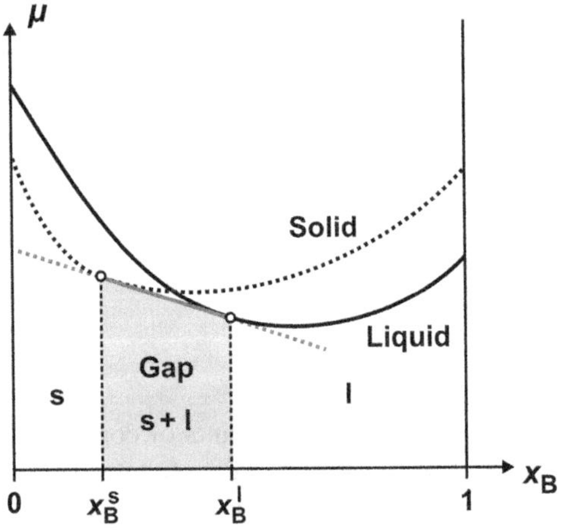

Fig. 13.11 The (average) chemical potential $\mu(x_B)$ as a function of concentration for mixtures of two components that show indifferent behavior in both their solid and liquid states.

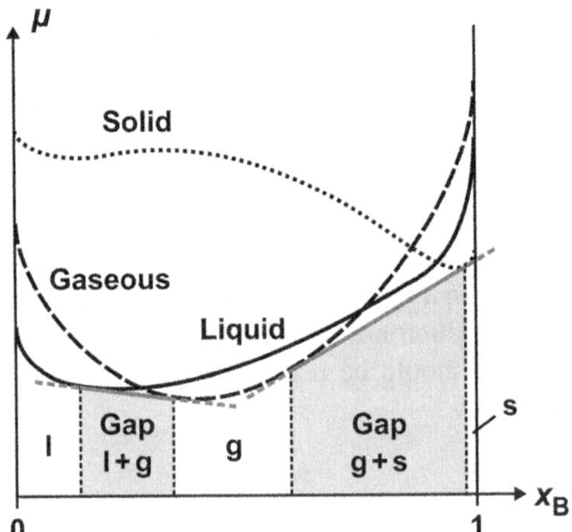

Fig. 13.12 The (average) chemical potential $\mu(x_B)$ as a function of concentration for mixtures of two components that show varying compatibility in their gaseous, liquid, and solid states.

s with the composition x_B^s coexisting with it. Between the points of contact, the double tangent lies everywhere below the $\mu_l(x_B)$ or $\mu_s(x_B)$ curves and therefore exhibits smaller values for the potential $\mu_{\mathcal{M}}$ of the heterogeneous mixture $\mathcal{M}$ which consists partly of mixed crystals M′ with the composition x_B^s and partly of a mixed melt M″ with the composition x_B^l. The separation of the phases is facilitated by the positive chemical drive.

Even more complex circumstances can be treated in this manner. Figure 13.12 shows $\mu(x_B)$ for two substances A and B, which

- As gases, are indifferent as usual,
- As liquids, are lowly compatible ("compressed drooping belly"),
- As solids, are incompatible ("dent upward").

In this case two gaps appear. Because each of the curves shows a vertical tangent on the left side at $x_B = 0$ and on the right side at $x_B = 1$, the gap can never exactly extend so far with the result that there is a narrow sometimes no longer recognizable region in which A and B are miscible.

Based on this, we can now construct phase diagrams just as we did in the case of single-component systems (Chap. 11). This will be discussed extensively in the next chapter.

Chapter 14
Binary Systems

The average chemical potential of a mixture depends not only upon the composition but also upon the temperature (and pressure). These dependencies and the fact that the phase with the lowest chemical potential at a given temperature (or pressure) will be stable can be used to construct the phase diagrams of different mixtures. First, we will discuss the temperature–composition diagrams of two liquid phases. With the help of these diagrams, we can judge under which conditions the two liquids are mutually miscible and under which they are not; the diagrams are therefore also called miscibility diagrams. Liquid–solid phase diagrams are used to identify the regions of temperature and composition at which solids and liquids exist in a two-component system. Such diagrams are of great commercial and industrial relevance; they play an important role in metallurgy but also in the manufacture of ceramics and semiconductors. In the last section, the phase diagrams of binary mixtures of two volatile components are discussed. This kind of diagram is important for understanding distillation, one of the most significant processes used in chemical laboratories and industry for separating liquid mixtures. It has been in use since ancient times to extract essential oils such as attar of roses. An important industrial application is distilling of petroleum in oil refineries that produce the heavy and light gasoline used to fuel engines.

14.1 Binary Phase Diagrams

In Chap. 11, we were introduced to the phase diagrams of pure substances. They can be used to find which phase is the most stable under given conditions (such as temperature or pressure). Analogous to these, phase diagrams for mixtures can be constructed. In the following, we will confine ourselves to two-component systems, meaning so-called *binary* mixtures of two components. In this case, the composition x of the mixture appears as third variable along with temperature T and pressure p. Hence, a complete description of the system is only possible using

© Springer International Publishing Switzerland 2016
G. Job, R. Rüffler, *Physical Chemistry from a Different Angle*,
DOI 10.1007/978-3-319-15666-8_14

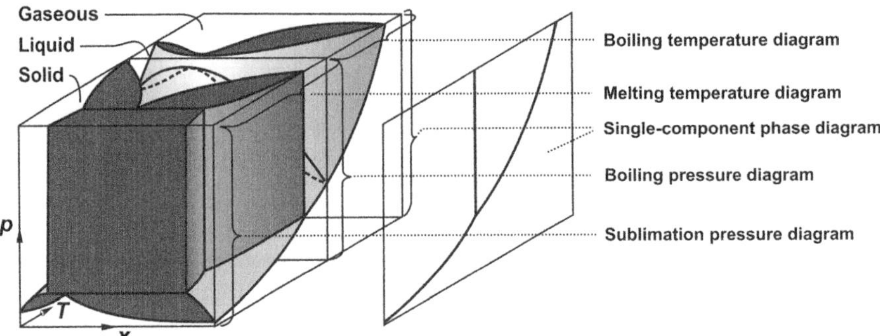

Fig. 14.1 Three-dimensional phase diagram.

three-dimensional phase diagrams. Figure 14.1 shows an example. The curved surfaces divide the diagram into single-phase and two-phase (dark colored) spatial regions. It is limited on the left ($x = 0$) and on the right ($x = 1$) by the already familiar plane phase diagrams of pure components (compare the diagram on the right with Fig. 11.10). The whole thing looks rather complicated at first, but don't worry, one variable is usually left out and either the temperature or pressure is kept constant, resulting in a phase diagram that is just two-dimensional. It gives us, for example, the most stable phase as a function of temperature and composition of the mixture. To this end, the temperature is plotted on one of the axes and the mole fraction of one of the components on the other. (Because it is a binary system, the mole fraction of the second component is then known as well.) Such a $T(x)$ diagram is equivalent to an isobaric cut through the three-dimensional diagram. Examples would be miscibility diagrams, and melting and boiling temperature diagrams which we will go into in more detail in the next sections. A $p(x)$ diagram is analogous to an isothermal cut. Examples of this would be boiling pressure and sublimation pressure diagrams.

14.2 Liquid–Liquid Phase Diagrams (Miscibility Diagrams)

Using what we learned from the section on mixing processes in the last chapter, we will now deal with mixtures of two liquid phases A and B. We saw that (at particular values of temperature and pressure) the substances can be indifferent, highly compatible, lowly compatible, or even incompatible with each other. However, this behavior can change with temperature (at constant pressure). For example, there are substances such as phenol and water or hexane and nitrobenzene that are quite compatible at high temperatures and incompatible at lower temperatures.

Fig. 14.2 Behavior of (average) chemical potential $\mu(x_B)$ in a mixture of two liquid components depending upon temperature (*top*) and the corresponding phase diagram at constant pressure with an upper critical solution point (*bottom*).

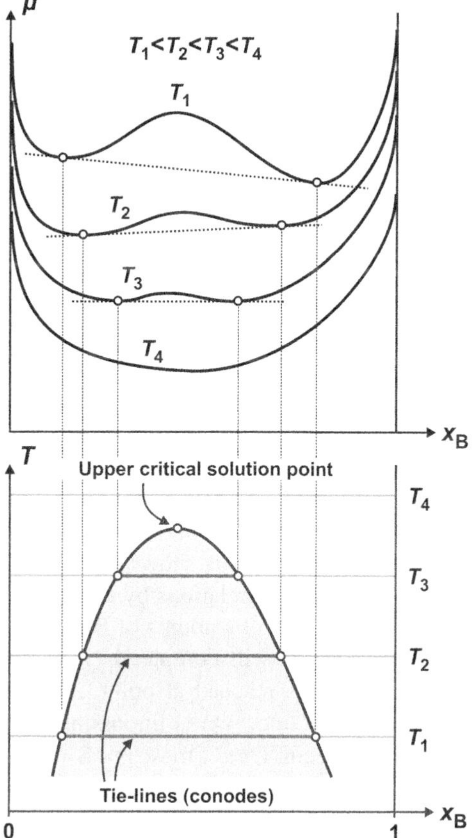

We can now determine $\mu(x)$ for every temperature (at a constant pressure, such as standard pressure of 100 kPa) as we learned to do in the last chapter (Fig. 14.2, above).

We see that the miscibility gap is greatest at the lowest temperature (T_1). Both the chemical potential and the contribution from the extra potential decrease as the temperature increases (T_2, T_3). The miscibility gap becomes increasingly smaller as the temperature goes up. The difference of the composition x_B in the two separate mixed phases becomes increasingly smaller as well. Finally, at the temperature T_4, there is only one phase left. It is possible to construct a $T(x)$ diagram (Fig. 14.2, below, a so-called *miscibility diagram*) from the $\mu(x)$ isotherms by plotting the double tangent's points of contact for every temperature and connecting them to form a curve. The horizontal line, which relates a pair of coexisting phases to each other, is called a *tie-line* or *conode* and their end points are called *nodes*. During heating, the tie-lines become continuously shorter until the end points finally coincide at the *upper critical solution point*. The corresponding temperature is the

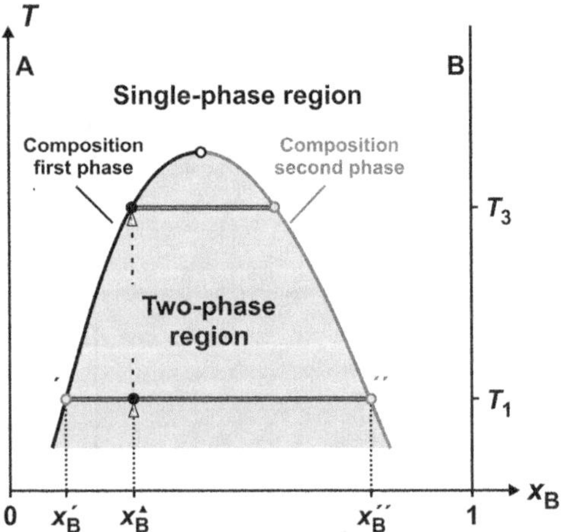

Fig. 14.3 Applying the "lever rule" using the example of a phase diagram with an upper critical solution point.

highest temperature at which a phase separation can take place. Above this *upper critical solution temperature,* both substances are completely miscible. The system phenol/water, for example, shows an upper critical solution temperature of 339 K.

Let us clarify these relations by using Fig. 14.3: For example, if we have a pure substance A, and small amounts of B are successively added to it at a temperature T_1, the two liquids will completely mix at first. This will be the case until the miscibility limit is reached at point $'$. However, a homogeneous mixture with a higher B content (e.g., $x_B^{\blacktriangle}$) is impossible to produce. Instead, two separate liquid mixed phases occur. One of these is a B-poor phase with the composition x_B' and the other a B-rich phase with the composition x_B''. The ratio of amounts of the two coexisting mixed phases results from the lever rule [analogous to the procedure in the case of $\mu(x)$ curves (Chap. 13)] from which the $T(x)$ diagram was constructed. If B continues to be added to the mixture, we still have two mixed phases $'$ and $''$. However, the amount of B-rich phase increases at the expense of the B-poor one because the corresponding lever arm shortens. When the phase boundary is crossed at point $''$, the two mixed phases finally merge into one.

On the other hand, if a sample with the mole fraction $x_B^{\blacktriangle}$ at temperature T_1 is continuously heated (vertical dotted line), the compositions of the liquid mixed phases that are in equilibrium with each other change. The B-poor phase becomes gradually richer in B (while the composition remains below $x_B^{\blacktriangle}$), whereas the B-rich phase loses some B. The ratio of the amounts of the two mixed phases changes according to the lever rule. The phase richer in B gradually disappears because as the temperature rises, the ratio of lever arms shifts in its favor and therefore the ratio of amounts in its disfavor. When the phase boundary line is finally crossed at temperature T_3, only one mixed phase with the composition $x_B^{\blacktriangle}$ is present.

Some systems exhibit a lower critical solution point (Fig. 14.4). At higher temperatures (and depending upon the composition), two phases can be present. At lower temperatures, the two substances are totally miscible. An example of this

Fig. 14.4 Phase diagram of a system with a lower critical solution point.

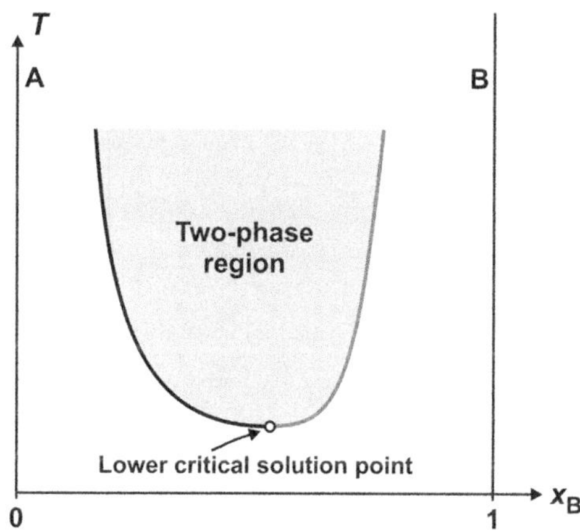

Experiment 14.1 *Demonstration of the presence of a miscibility gap with the help of the systems phenol/water and triethylamine/ water*: When heated, a heterogeneous mixture of phenol and water will become a homogeneous solution when the upper critical solution temperature (approx. 339 K) is exceeded. However, even at higher temperatures, a heterogeneous mixture of triethylamine and water remains separated, but when cooled with ice to below the lower critical solution temperature (approx. 292 K), it will become a homogeneous solution. The phenol–water mixture, however, continues to consist of two phases after cooling.

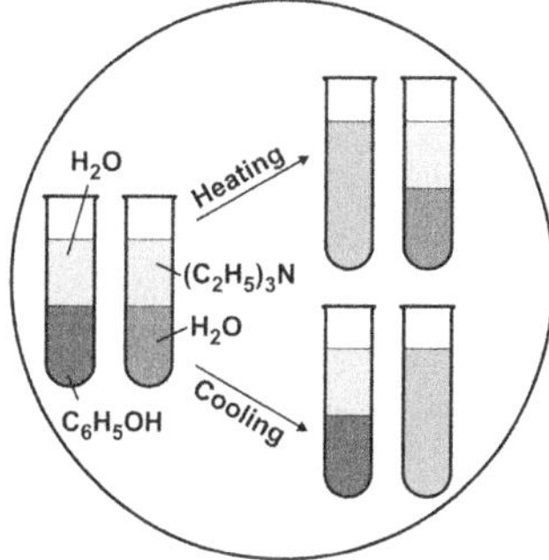

would be a system of triethylamine and water which has a lower critical solution temperature of 292 K.

Let us illustrate these relationships by a demonstration experiment using on the one hand a mixture of phenol (C_6H_5OH) (the phenol is colored with methyl red for visibility) and water and on the other hand a mixture of triethylamine ($(C_2H_5)_3N$) and water colored with orange G (Experiment 14.1).

Some systems have both an upper and a lower critical solution point (Fig. 14.5). These kinds of systems are mostly found at higher pressures. It is therefore plausible to assume that all systems having a lower critical solution point will also exhibit an upper critical solution point if the temperature and pressure are high

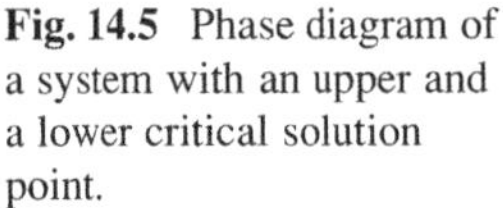

Fig. 14.5 Phase diagram of a system with an upper and a lower critical solution point.

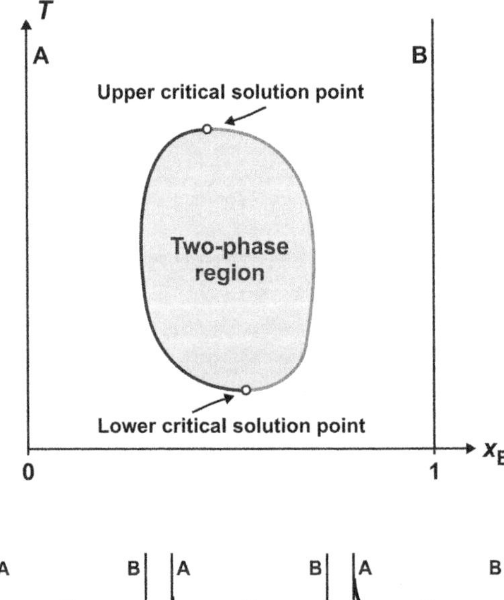

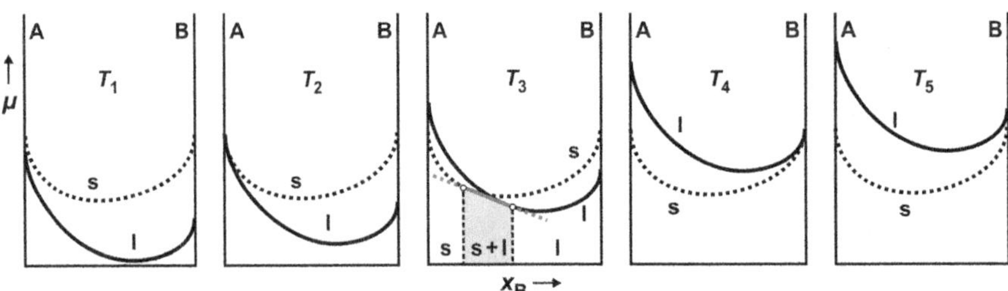

Fig. 14.6 Behavior of (average) chemical potential $\mu(x_B)$ for mixtures of two components that are indifferent both in their solid and liquid states for various temperatures ($T_1 > T_2 > T_3 > T_4 > T_5$).

enough. A well-known example is the system nicotine/water which exhibits a closed miscibility gap with the critical temperatures 334 and 483 K.

14.3   Solid–Liquid Phase Diagrams (Melting Point Diagrams)

Solid Solution Systems We will now take a look at phase diagrams containing both solid and liquid phases. They are also called *melting temperature diagrams* or *melting point diagrams* that, for example, play a big role in metallurgy. Especially simple melting point diagrams represent systems whose components A and B are infinitely soluble with each other both in the melted (1) as well as the solid states (s), meaning that they form *mixed crystals*. For constructing these diagrams, the $\mu(x)$ curves of the liquid and solid mixed phases must be considered as functions of temperature (Fig. 14.6).

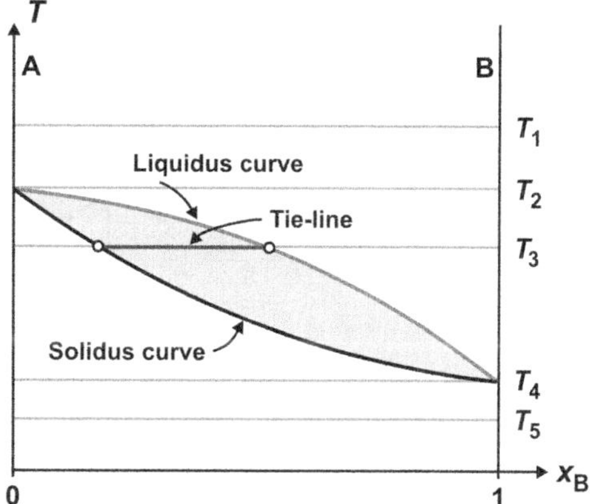

Fig. 14.7 Melting point diagram of a system of two substances showing indifferent behavior in both the liquid and solid states (constructed using Fig. 14.6).

At very high temperatures (T_1), μ_l has for any arbitrary composition a smaller value than μ_s, which means the melting process mixed phase|s $\rightarrow$ mixed phase|l always proceeds spontaneously. For this reason, the system is liquid independent of the composition. When the system is cooled, it will eventually reach the point (T_2) where the two $\mu(x)$ curves touch for the first time. This means that they exhibit the same value $(\mu_l = \mu_s)$. In our example, this is at $x_B = 0$ and therefore at the melting point of the pure substance A. At the melting temperature of A, a solid phase (in equilibrium with a liquid phase) is present only at a mole fraction of $x_B = 0$, while the phases with all other compositions are still completely liquid. When the temperature is further reduced (e.g., to T_3), separate regions of composition appear in which either the solid phase s or the melted phase l exhibits the smaller chemical potential. Between these regions, the smallest chemical potential is obtained through a heterogeneous mixture of melt and solid phases (l + s) as we have seen in the last section of the previous chapter. The compositions that are determined by the double tangent on the μ_l and μ_s curves define the solidus and liquidus points at a given temperature. Further cooling causes the tangent's points of contact to move. This means that the range of the two-phase region shifts. When the melting temperature of substance B (which melts at a lower temperature) is finally reached, the chemical potential of the solid phase is smaller than that of the liquid phase for all compositions except for $x_B = 1$. On the other hand, at $x_B = 1$ the chemical potentials are equal so that melt is still present. The solid state is present at any and all compositions below this temperature (e.g., at T_5).

It is possible to construct a phase diagram by applying these considerations consistently for as many temperatures as possible (Fig. 14.7). The result is a spindle-shaped two-phase region. The upper curve shows the composition of melt (*liquidus curve* or freezing curve, at which freezing begins as the mixture is cooled down) and the lower one shows that of the solid phase (*solidus curve* or melting curve, at which melting begins as the mixture is heated). Above the liquidus curve is only melt and beneath the solidus curve is only the solid phase. Between the two

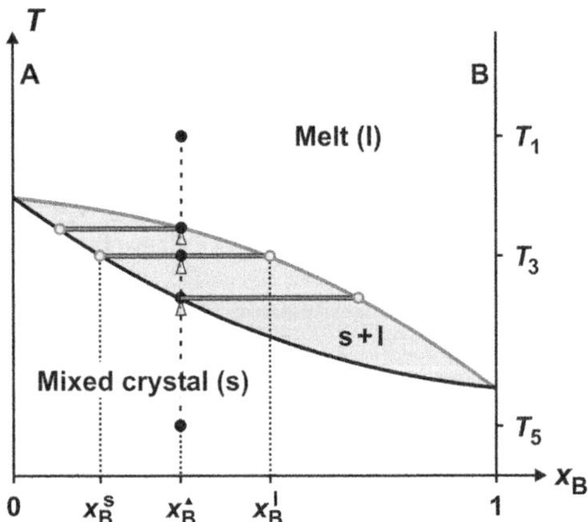

Fig. 14.8 Applying the "lever rule" in a melting point diagram for a system containing two substances that are indifferent in liquid and solid states.

curves a separation into melt and mixed crystals of the compositions given by the intersection points of the tie-line with the liquidus and the solidus curve takes place.

To illustrate this we will take a look at the solidification process of a homogeneous melt with the composition $x_B^▲$, starting at a temperature T_1 (Fig. 14.8, vertical dotted line). When the liquidus curve is reached, a rather B-poor mixed crystal begins to separate in miniscule amounts. Its composition results from the intersection point of the tie-line with the solidus curve. However, the melt exhibits a composition that is (almost) the same as the initial one. With a lowering of temperature, the melt becomes gradually richer in the component B (which melts at a lower temperature) because we are moving downward on the liquidus curve. If the melt remains in equilibrium with the mixed crystals, their B content increases again along with progressive crystallization. For example, a melt at temperature T_3 with the composition x_B^l can coexist with a mixed crystal having the composition x_B^s. The ratio of the amounts of the two phases can be found with the help of the now familiar lever rule. As mentioned before, the downward movement on the solidus curve means that the composition of the already crystallized fraction must change continuously during cooling (which is conceptualized as an equilibrium process). In reality, such changes of composition are not that simple since the diffusion of atoms in solids takes a very long time (see Sect. 20.2). If we cool the substance further very slowly, we finally reach the solidus curve. The intersection point of the tie-line with the liquidus curve shows the composition of the last tiny drop of melt. Below the solidus curve, the entire melt has solidified. All that is left is a mixed crystal having the same composition $x_B^▲$ as the original melt.

Melting point diagrams with complete miscibility only appear when the form and size of the particles allow them to be inserted into a common lattice. Examples of this are copper/nickel as well as the minerals fayalite (Fe_2SiO_4)/forsterite (Mg_2SiO_4).

More Complicated Melting Point Diagrams More complicated phase diagrams can be developed by considering the $\mu(x)$ curves of the participating components.

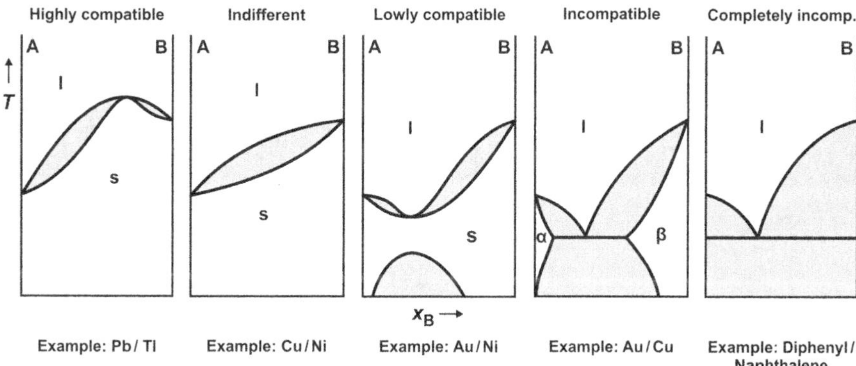

Fig. 14.9 Melting point diagrams for systems containing components having indifferent behavior in the liquid state and varying behaviors in the solid state.

While components A and B in their liquid states are mostly infinitely soluble with each other, they may be indifferent in their solid states, as just discussed, but also highly compatible, lowly compatible, or even incompatible or completely incompatible. We will take a short look at the corresponding melting point diagrams (Fig. 14.9).

α and β mean different mixed crystals. For example, in the case of Au/Cu, this would be a gold-rich and a copper-rich mixed crystal.

Although this may look somewhat complicated at first, a phase diagram can be compared to a map. It is difficult and time-consuming to create a map by topographical survey, but it is easy to use it when that has already been done and when some rules and conventions are known. It is just as difficult to calculate a phase diagram and very time-consuming to measure it. However, once it is known it can be used like a map assuming one observes also in this case some rules and conventions which we have already learned about to some extent. We will illustrate this using the melting point diagram for two components that are incompatible in their solid states (Fig. 14.10).

There are three two-phase regions $(\alpha + l, \beta + l, \alpha + \beta)$ along with the single-phase melt (l) and the mixed crystals (α, β).

Eutectic Mixture What happens when we cool a melt having a composition $x_B^{\blacktriangle}$? When the temperature is lowered and the boundary of the two-phase region solid/liquid is reached, a very B-rich mixed crystal β begins to separate from the melt. Further cooling causes more and more solid substance to crystallize whereby the ratio of melt to mixed crystal is determined by the lever rule. In the process, the melt is constantly depleted of B because almost pure B (mixed only with a bit of A) precipitates. When the system reaches the temperature T_e and thereby the horizontal line, the residual melt with the composition x_B^e solidifies. The $\mu(x)$ curves (Fig. 14.11) are applied again here for the purposes of illustration.

At temperature T_e, the tangent touches the $\mu(x_B)$ curves at three points (once the μ_l curve and twice the μ_s curve). This means that the melt is in equilibrium with the

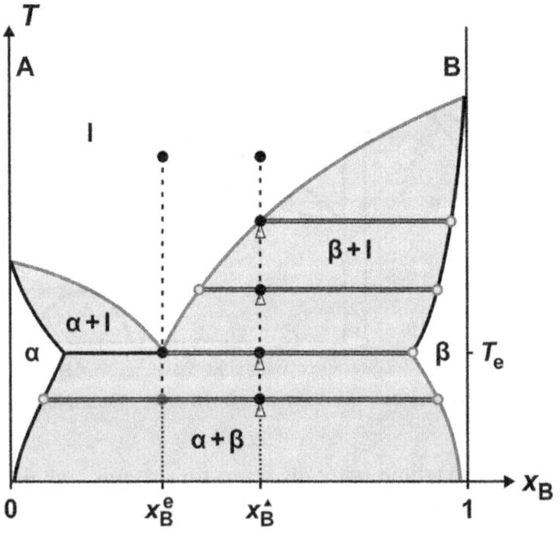

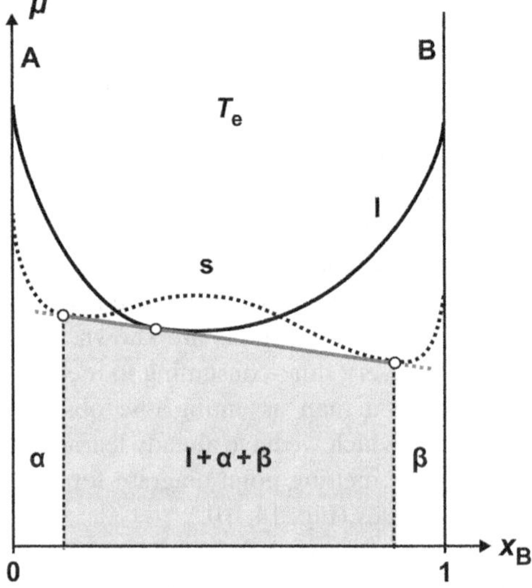

Fig. 14.10 Melting point diagram for a system of two components that show incompatible behavior in their solid states as well as application of the "lever rule".

Fig. 14.11 Behavior of (average) chemical potential $\mu(x_B)$ for a system of two substances that are indifferent in their liquid states and incompatible in their solid states, at temperature T_e.

mixed crystal α and the mixed crystal β. At this point, a tangent can be plotted for the last time on the μ_l curve because at lower temperatures, the chemical potential of the melt is always greater than that of the solid state so the curve increasingly retracts upwards. T_e is therefore the lowest temperature at which the melt can exist. This temperature is known as *eutectic* temperature; the term "eutectic" (abbreviated e) comes from the Greek eutektos, meaning "easily melted."

Let us now return to the phase diagram. Beneath the eutectic temperature T_e lies a two-phase system of a very A-rich mixed crystal α and a very B-rich mixed crystal β. Further cooling alters the composition of the mixed crystals. The α mixed crystal

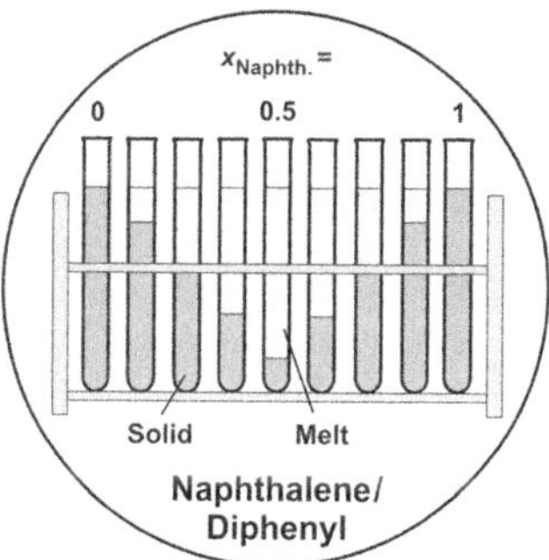

Experiment 14.2 *Melting point diagram of naphthalene and diphenyl*: The test tubes with the nine different mixtures are put in a boiling water bath in order to melt the mixtures. Subsequently, the mixtures are allowed to cool down in front of a black background. Starting with the test tubes on the ends the transparent melts begin to congeal and they become gradually white and more opaque. After a while, only the melts at the center are liquid. In the end, just about everything has solidified.

becomes continuously richer in A and the β mixed crystal becomes continuously richer in B. However, these changes in composition require an extremely long time.

Now what happens if a melt with a eutectic composition of x_B^e is cooled? A liquid with this composition will solidify as a whole at a single definite temperature (like a pure substance). This means that none of the components have separated out before. A heterogeneous mixture has been formed of simultaneously precipitated α and β mixed crystals that also exhibits a total composition of x_B^e (for simplicity, the lever rule is not drawn here). In contrast to all the others, the eutectic mixture does not need to be cooled slowly in order to have conditions for equilibrium and one obtains a uniform and very fine-grained structure (microcrystals).

We will look at an experiment showing what happens when we cool down different mixtures of liquid naphthalene ($C_{10}H_8$) and diphenyl (phenylbenzene, $C_{12}H_{10}$) (Experiment 14.2).

In their solid states, naphthalene and diphenyl are completely incompatible. This means that they crystallize when cooled into an (almost) pure state (diagram on the far right in Fig. 14.9). (The infinite gradient of the tangent showing the impossibility of the existence of a pure substance has been mentioned in the last chapter.) Finally, only the melts with nearly eutectic compositions, namely $x_{Naphthalene} = 0.45$, are liquid.

Thermal Analysis *Thermal analysis* has proven to be especially suited to investigating phase diagrams. Samples of varying compositions have been melted and then cooled down again for this. During cooling, a thermocouple measures the mixture's temperature as a function of time and a *cooling curve* is recorded (Experiment 14.3).

Supplementary micrographs and structural investigations are often used as well.

Experiment 14.3 *Cooling curve of a mixed melt (using the example of a lead–tin alloy):* The lead–tin alloy (40 wt% Sn; so-called soft solder) in the test tube is slowly melted over a Bunsen burner. Subsequently, the sample is allowed to cool down. Using a thermocouple, the temporal change of temperature is recorded by a plotter or a computer.

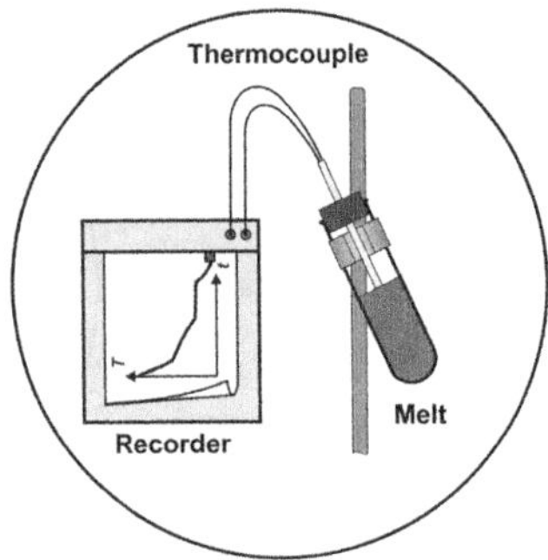

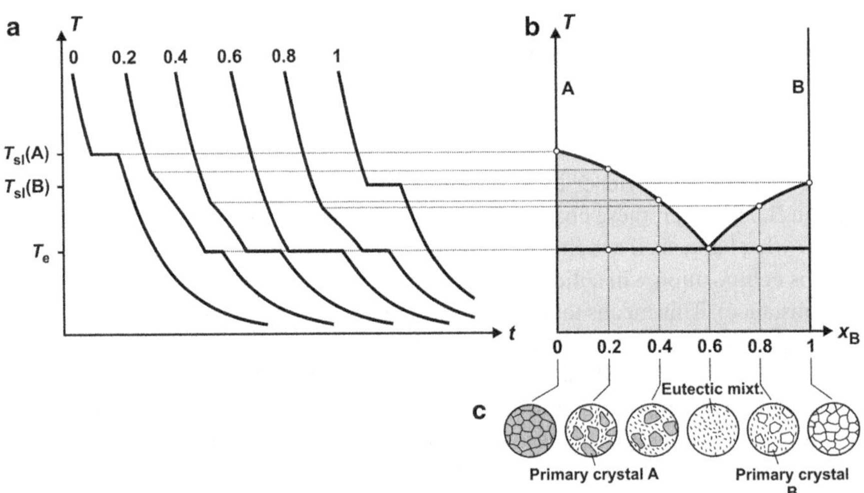

Fig. 14.12 (**a**) Cooling curves for a system of two substances A and B, which are completely incompatible in their solid states, (**b**) Melting point diagram constructed from the cooling curves, (**c**) Corresponding micrographs.

Figure 14.12 shows (idealized) cooling curves for various liquid mixtures of the two components A and B (which are completely incompatible in their solid states) as well as the schematic of the corresponding micrographs. These curves were used to construct the melting point diagram.

The pure substances (curves for $x_B = 0$ and $x_B = 1$) yield *arrest points* at the crystallization temperature T_{sl} because during the process of isothermal solidification, the cooling is "arrested" by the entropy emitted by crystallization. Only when the entire melt has solidified, does the temperature begin to drop again.

Let us now consider the process for a mixture with the mole fraction of $x_B = 0.2$. At first, this liquid system cools evenly until it reaches a temperature at which solid A begins to precipitate. During the continuous crystallization of A, meaning the phase transition liquid $\rightarrow$ solid, entropy is emitted steadily; therefore, the cooling of the system slows down considerably. Accordingly, an *inflection point* appears on the cooling curve where the phase field with solid A and the remaining melt is entered. When the system finally reaches the eutectic temperature T_e, A and B crystallize simultaneously. The rest of the sample solidifies without its composition changing. This is shown by an arrest point on the cooling curve comparable to what we have seen with pure substances. Initially, substance A crystallizes continuously during the cooling process and new substance grows on crystals already present, producing large primary A crystals. When the eutectic temperature is attained, the entire amount of B (as well as the remaining A) must freeze at once, thereby creating a great number of seed crystals. The micrograph of the solid mixture shows large A crystals imbedded in extremely fine-grained B and A crystals.

If the melt has eutectic composition from the start (in our example, $x_B = 0.6$), it will continuously cool until it reaches the eutectic freezing temperature. When the mixture's temperature falls below this temperature, a simultaneous precipitation of A and B takes place until the entire sample has solidified. Correspondingly, the temperature remains constant over a longer period of time compared to other mixtures. A micrograph shows a heterogeneous mixture of A and B microcrystals of approximately the same size.

If there are enough cooling curves of mixtures with different compositions available, it is possible to construct the corresponding phase diagram.

14.4 Liquid–Gaseous Phase Diagrams (Vapor Pressure and Boiling Temperature Diagrams)

Finally, we will deal with phase diagrams for mixtures of two volatile liquids where we will initially assume an indifferent behavior.

Vapor Pressure Diagrams At a certain temperature, and in equilibrium, a saturation vapor pressure of $p^{\bullet}_{\mathrm{lg,A}}(T)$ develops over an easily evaporating liquid A (compare Sect. 12.5). (The symbol $^{\bullet}$ indicates again that the quantity refers to a pure phase.) In order to avoid an unattractive piling up of indices, we will use $p^{\bullet}_A$ in the following. If an easily evaporating substance B is dissolved in A (Fig. 14.13), the chemical potential of A decreases as a result of dilution.

In Sect. 12.5, we were introduced to a similar situation that leads to lowering of vapor pressure over solutions. However, foreign substance B had low volatility so that the vapor phase was made up of only A (at least approximatively). In the case of two volatile components, reestablishment of equilibrium is caused by lowering of the <u>partial</u> pressure of A to p_A in the <u>mixed</u> vapor (above the liquid mixed phase):

Fig. 14.13 Forming mixed
vapor in equilibrium over a
mixture of two indifferent
liquid components A and B.

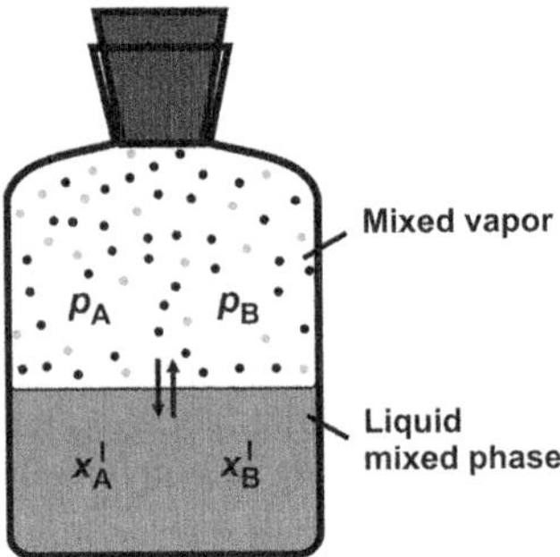

$$p_A = x_A^l \cdot p_A^\bullet. \tag{14.1}$$

The same is also valid for the second volatile component, the substance B:

$$p_B = x_B^l \cdot p_B^\bullet. \tag{14.2}$$

Here is a somewhat different formulation of Raoult's law: The partial pressure of
every component i in a mixed vapor equals the product of the vapor pressure of the
pure component and its mole fraction x_i^l in the liquid mixed phase. Ideal mixtures
are subject to Raoult's law independent of chemical composition. Conversely, this
law also represents a further experimental criterion for the indifferent behavior of
the two components relative to each other.

If we assume ideal behavior in the gas phase as well, then according to Dalton's
law, the total vapor pressure above the mixture is the sum of the partial pressures:

$$p = p_A + p_B. \tag{14.3}$$

The partial pressures

$$p_A = \left(1 - x_B^l\right) \cdot p_A^\bullet = p_A^\bullet - p_A^\bullet \cdot x_B^l \tag{14.4}$$

and

$$p_B = p_B^\bullet \cdot x_B^l \tag{14.5}$$

as well as the total pressure

$$p = p_A + p_B = p_A^\bullet - p_A^\bullet \cdot x_B^l + p_B^\bullet \cdot x_B^l = p_A^\bullet + \left(p_B^\bullet - p_A^\bullet\right) \cdot x_B^l \tag{14.6}$$

change linearly with the changing composition of the liquid mixture characterized
by the mole fraction x_B^l (Fig. 14.14).

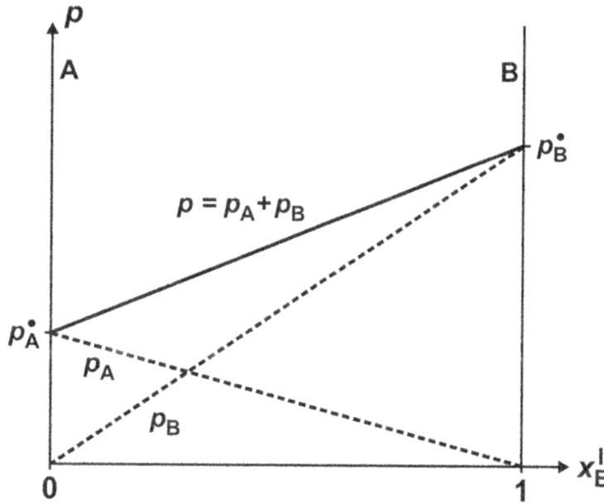

Fig. 14.14 Total pressure (*solid line*) and partial pressures (*dotted lines*) above a mixture of two indifferent components A and B at constant temperature.

The compositions of a liquid mixed phase and its corresponding mixed vapor need not be identical. In fact, we can expect the fraction of the more volatile component in the vapor to be higher. According to Dalton's law, the following is valid for the mole fraction x_B^g of component B in the vapor phase:

$$x_B^g = \frac{p_B}{p}. \tag{14.7}$$

Inserting Eq. (14.2) then yields

$$x_B^g = \frac{p_B^\bullet}{p} x_B^l. \tag{14.8}$$

If B in its pure state has a higher vapor pressure than A, then the following is valid: $p_B^\bullet/p > 1$ (compare Fig. 14.14) and therefore $x_B^g > x_B^l$. In fact, the vapor is enriched with the more volatile component B.

If the vapor pressure (as a function of the vapor composition) is inserted into a $p(x)$ diagram (*vapor pressure diagram* or boiling pressure diagram) along with the linear vapor pressure curve (function of the liquid's composition) (Fig. 14.15), the corresponding curve will always lie below the straight line. This is called the *dew point curve*, while the straight line is called the *boiling point curve*. Below the dew point curve, there is only the gas phase, and above the boiling point curve, there is only the liquid phase. Both curves delineate a two-phase region in which both vapor and liquid mixed phases are in equilibrium.

Tie-lines can again be used to determine the chemical compositions of coexisting phases. This can be done analogously to the approach used in $T(x)$ diagrams discussed earlier, but here the temperature is kept constant while the pressure is varied (Fig. 14.16). The lever rule can also be applied here to determine the ratio of amounts. For example, if we have a liquid mixture with the composition $x_B^\blacktriangle$ at pressure p_1, and slowly lower the pressure while keeping the temperature constant, it will begin to vaporize when it reaches the two-phase region at pressure

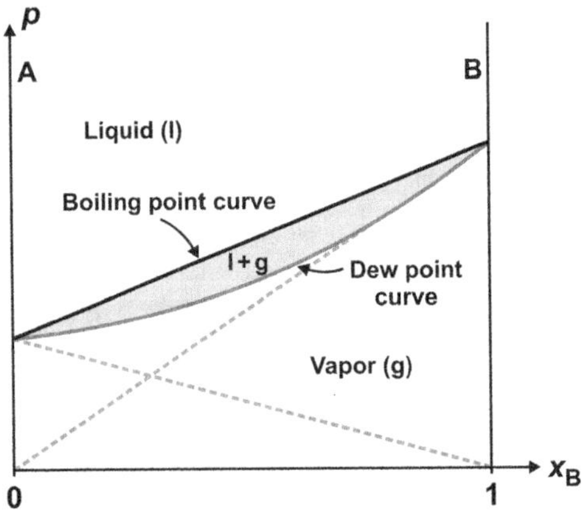

Fig. 14.15 Vapor pressure diagram of a system of two largely indifferent components A and B (at constant temperature).

Fig. 14.16 Applying the "lever rule" in a vapor pressure diagram.

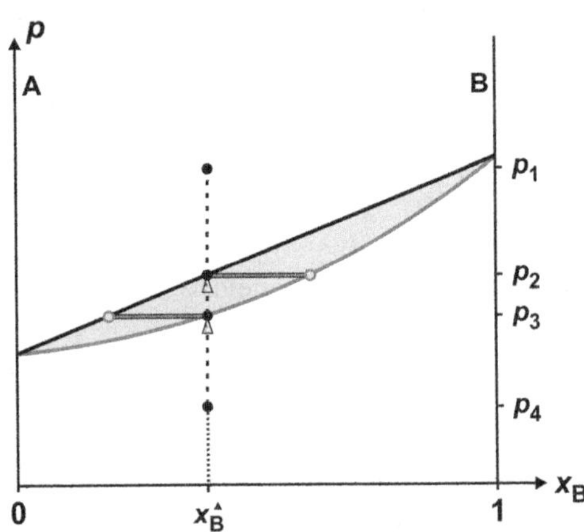

p_2. As we have already seen, the more volatile component is enriched in the vapor phase. As the pressure is lowered even more and the vaporizing continues, the liquid phase becomes increasingly depleted of this component. Finally, below pressure p_3, there is only vapor of the same composition as the initial liquid phase.

Boiling Temperature Diagrams Vaporization takes place much more often at constant pressure than at constant temperature. If we continue to assume the substances A and B to be indifferent to each other in the liquid as well as the gaseous phase, the corresponding $T(x)$ diagram (*boiling temperature diagram* or often also *boiling point diagram*) can be constructed analogously to the respective melting point diagram (Fig. 14.17). The *boiling point curve* in this case is nothing more than a plotting of the liquid mixture's boiling temperature (at constant pressure such as standard pressure) versus the mole fraction of one of the components. It delineates the homogeneous liquid phase's region of existence toward higher temperatures. The composition of the vapor phase, which is in equilibrium with the corresponding

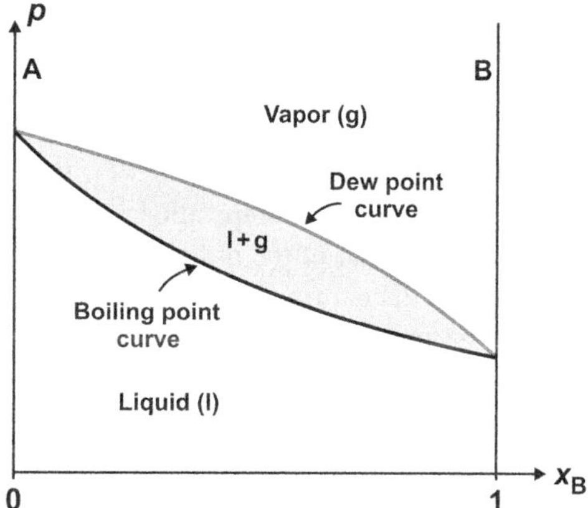

Fig. 14.17 Boiling point diagram of a system with indifferent components (at constant pressure).

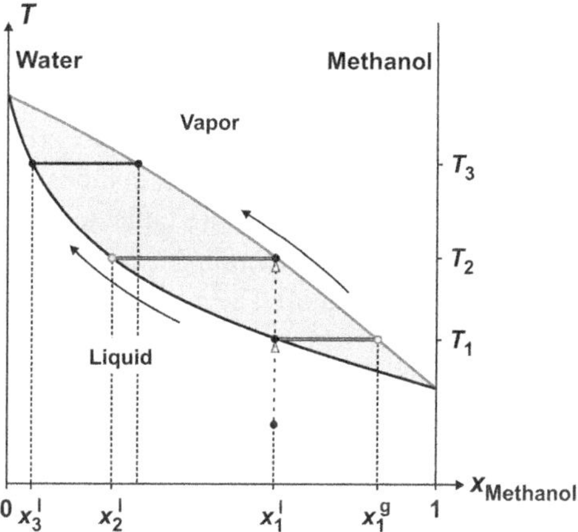

Fig. 14.18 Schematic boiling point diagram of the system water/methanol to show simple distillation.

liquid mixture at the respective boiling point, is indicated by the *dew point curve*. Above the dew point curve lies a homogeneous gas phase and between the two curves is the two-phase region. Usually, when there are two volatile liquids A and B, the one with the lower vapor pressure has the higher boiling temperature. Therefore, in the boiling point diagram the regions representing different states are exchanged compared to the ones in the vapor pressure diagram.

Distillation The differing compositions of the liquid mixed phase and its coexisting mixed vapor in the two-phase phase region can be utilized in separating the substances by means of *distillation*. First, though, we will take a look at the boiling process of a system of the (largely) indifferent components water and methanol (Fig. 14.18). When a mixture of composition x_1^l is heated at constant pressure until it boils, the vapor rising from it at boiling temperature T_1 will have the composition x_1^g. As a consequence, the methanol which has higher volatility (lower boiling temperature) becomes enriched in the vapor phase. Further heating causes a depletion of the methanol in the liquid which leads to a rise in boiling temperature: a slow raising of temperature changes it in the direction of T_2. At the same time, the methanol content in the vapor phase decreases and its composition shifts along the dew point curve in the same direction the liquid phase moves along the boiling point curve. Finally, at temperature T_2, the last drop of the (almost) completely vaporized liquid has the composition x_2^l, while the composition of the vapor phase is now the same as that of the initial mixture of the liquid.

Because of the enormous increase of volume during vaporization by a factor of about a thousand, this procedure is not practicable. The so-called *simple distillation* in which the liquid mixture is brought to a boil in a flask, the vapor condensed in a cooler, and the resulting condensate (distillate) collected in a receiver is the most common separation process in this context. Because methanol with its lower boiling temperature escapes preferably, the water with lower volatility is enriched in the flask. Therefore, the boiling temperature rises more and more—even over and above temperature T_2. When the distillation process is stopped near the boiling temperature T_3 of the component which has the higher boiling temperature, a mixture with composition x_3^l (water with little methanol) remains in the flask. The distillate in the receiver contains the more volatile component methanol with a reduced fraction of water in contrast to x_1^l (but higher compared to x_1^g).

The changes of the distillate in composition become more obvious when separate portions (fractions) of it are collected in interchangeable receivers (*fractional distillation*). The first fraction has, in fact, the composition x_1^g and is strongly enriched with methanol. The next fractions are increasingly poorer in methanol and therefore richer in water than the first one. The separation effect of fractional distillation can be improved by redistilling the individual fractions. In the process, the composition of the distillate moves along the dew point curve in the direction of pure methanol. After numerous repetitions, both components are just about pure. The disadvantage of this method is the low yield determined by the lever rule, which means that the distillation steps must be taken very often with continuously renewed initial mixture.

The tedious separate steps of vaporization and condensation are therefore combined in practice by the process of countercurrent distillation (rectification). In a *distillation column*, the ascending hotter vapor is flowing past the cooler *reflux*, i.e., a part of the condensate that is flowing back (Fig. 14.19). The close contact between the reverse flows favors a fast entropy and energy exchange. Temperature and composition are therefore close to the particular equilibrium (which depends upon the height in the column) but never reach it.

Let us take a closer look at this process using the water/methanol system as an example (Fig. 14.20). For describing the process, one imagines that the column is separated into (hypothetical) zones in which the equilibrium between liquid and vapor should have been established (*plate* or tray). When the initial liquid mixture with a mole fraction x_0 of methanol is heated in the flask, it begins to boil at temperature T_0. A part of the vapor condenses at the first plate, whereas the remaining vapor rises further to the next plate. At the same time, cooler condensate

Fig. 14.19 Sketch of a distillation column for use in laboratories.

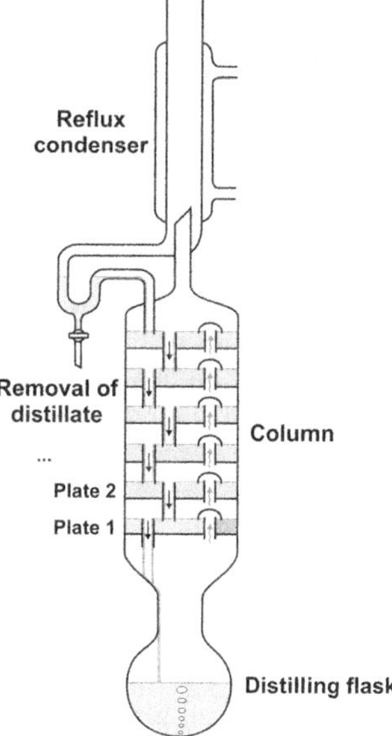

Fig. 14.20 Illustration of the process of rectification using the example of a mixture of water and methanol with an initial composition x_0.

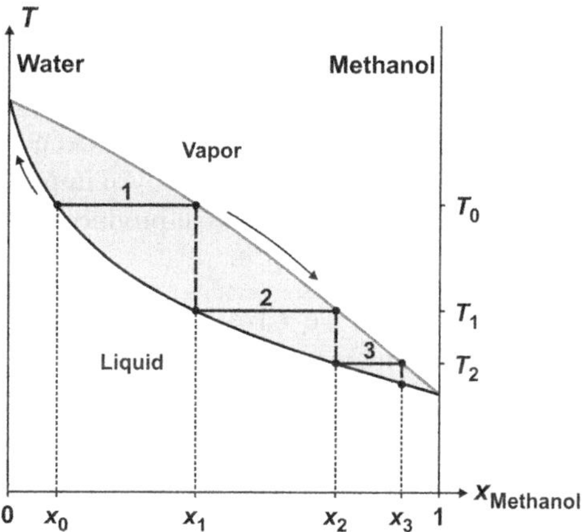

from the plate above flows into this plate and warmer excess condensate leaves it. A liquid with temperature T_1 and composition $x_1 > x_0$ accumulates on the plate. It fills the plate until it flows over. In dynamic equilibrium, inflow and outflow of each component have to compensate each other on each plate.

Each plate can be seen as an isolated distillation unit. The assumption that liquid and vapor are in equilibrium on each plate requires that the corresponding parts of the step curve appear in Fig. 14.20 as tie-lines and therefore horizontal lines. The vertical parts of the curve in between imply that the exchange of substances between the plates takes place at constant composition. In the process, the portion of the more volatile component in the liquid and correspondingly in the vapor, $x_i = x_i^{\mathrm{l}}$ and x_i^{g}, increases from one plate to the next higher one, $x_0 \rightarrow x_1 \rightarrow x_2 \rightarrow \ldots$. At the same time, the temperature in the column falls, $T_0 \rightarrow T_1 \rightarrow T_2 \rightarrow \ldots$, thereby approaching that of pure methanol. A "step" in the boiling point diagram (a combination of a horizontal and a vertical line) is called a *theoretical plate*. The number of theoretical plates (our example shows three) indicates the efficiency of the distillation column, even if it contains a packing material (such as small glass helices) instead of true plates.

When there are enough plates and the separation efficiency of the column is therefore sufficient, the component boiling at lower temperature will be just about pure in the distillate. Instead of separate plates, columns in laboratories are mostly packed with a material which has a large surface, such as rings, helices, or small spheres made of glass, for example. Successive vaporization and condensation steps take place on these surfaces at increasing heights in the column. The number of theoretical plates for separating a homogeneous mixture of a given pair of substances can be determined by inserting the possible "equilibrium steps" between the initial composition and the composition of the distillate in the boiling point diagram. However, in practice one has to assume a slightly higher number of plates.

Distillation is one of the most important processes used in chemical laboratories for separating liquid mixtures. It has been in use since ancient times to extract essential oils such as attar of roses. An important industrial application is distilling of petroleum in oil refineries that produce the heavy and light gasoline used to fuel engines.

Azeotropes What we have discussed so far in this section is only valid for mixtures of components that behave indifferently toward each other in both their liquid and vapor states. However, the liquid state often exhibits differing behavior. If the two components are highly compatible, the stronger interaction of the particles in the liquid mixture relative to the pure state hinders the transition to the vapor phase. The partial pressures of the components are smaller than in the case of indifferent behavior and the vapor pressure curves show a negative deviation from Raoult's law. Compared to the behavior of indifferent substances, the curves appear more or less distorted. As long as the disturbance is small, the behavior can be described in a similar manner as before.

A strong disturbance results in a *vapor pressure minimum* for the total pressure in the vapor phase and therefore in the vapor pressure curve (or boiling point curve) (Fig. 14.21a).

The dew point curve must again lie below the boiling point curve—exactly as it does with indifferent behavior. The two curves touch each other at the vapor pressure minimum. This means that they have a common tangent at this point,

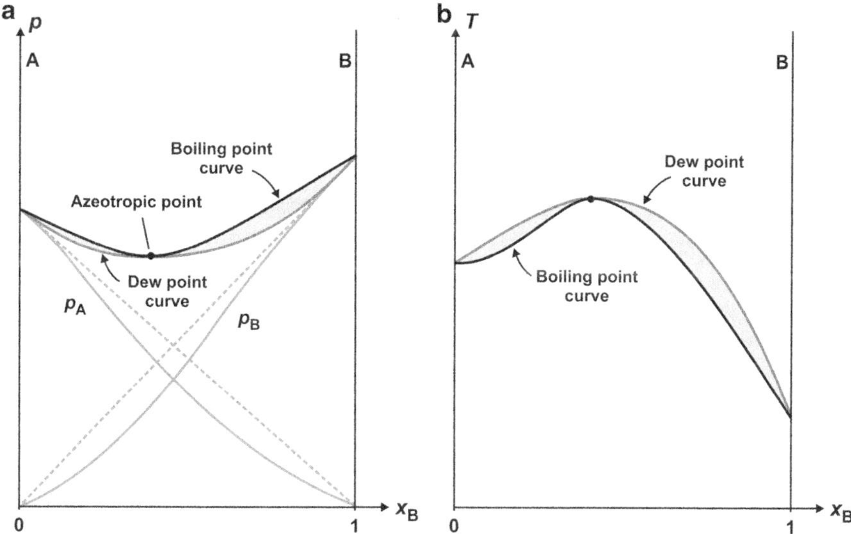

Fig. 14.21   (**a**) Vapor pressure diagram with azeotropic minimum of a binary system of two highly compatible substances, (**b**) Corresponding boiling point diagram with azeotropic maximum.

i.e., the liquid mixture is in equilibrium with a vapor of the same composition (!). This type of homogeneous mixture behaves just like a pure substance. This is called an *azeotropic mixture*, or an *azeotrope*. Its corresponding position in the vapor pressure diagram is called an *azeotropic point*. The word *azeotrope* is derived from the Greek words ζέειν (boil) and τρόπος (state) combined with the prefix α- (no) to give the overall meaning, "no change on boiling." In the boiling point diagram, not only the phase regions are inverted compared with the vapor pressure diagram, but the vapor pressure minimum also becomes a *boiling point maximum* (Fig. 14.21b). Therefore, these azeotropes are called maximum-boiling azeotropes or minimum-pressure azeotropes, sometimes also negative azeotropes. Systems whose components show highly compatible behavior in their liquid states are chloroform/acetone or hydrogen chloride/water (hydrochloric acid).

Lowly compatible behavior by the components in the liquid state with weaker particle interactions leads to a positive deviation from Raoult's law. Correspondingly, one can observe a *vapor pressure maximum* (Fig. 14.22a) as well as a *boiling point minimum* (Fig. 14.22b). Acetone/carbon disulfide and ethanol/water are examples of positive azeotropes (minimum-boiling or maximum-pressure azeotropes).

The appearance of azeotropic points has important consequences for the distillation of the mixtures concerned. First let us consider a system with a boiling point maximum (Fig. 14.23). A liquid mixture having the composition x_1^l boils at temperature T_1 and its corresponding vapor is enriched by the more volatile component B (x_1^g). If the vapor is removed continuously from equilibrium by simple distillation, meaning by condensation in a receiver, the composition of the

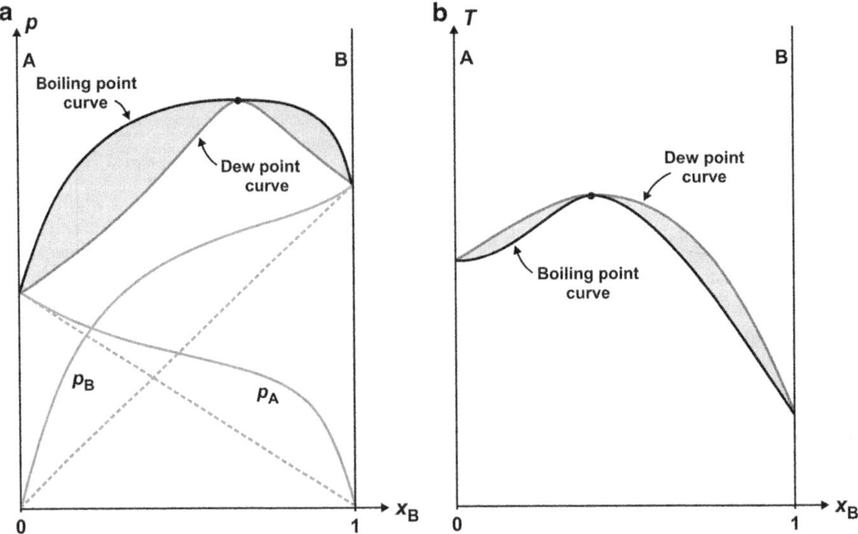

Fig. 14.22 (a) Vapor pressure diagram with azeotropic maximum of a binary system of two lowly compatible substances, (b) Corresponding boiling point diagram with azeotropic minimum.

Fig. 14.23 Simple distillation using the example of a boiling point diagram with an azeotropic maximum.

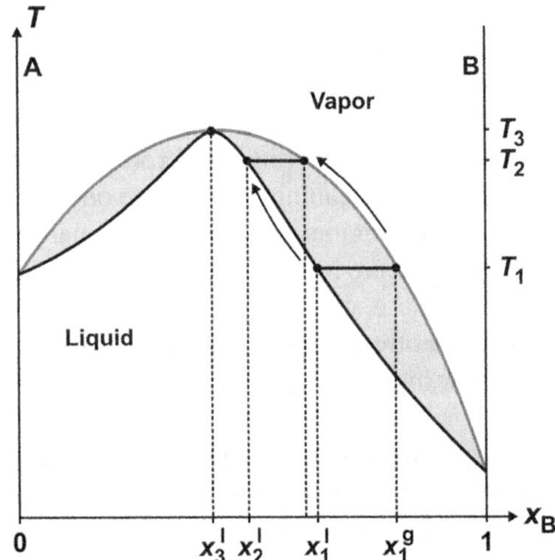

remaining liquid shifts along the boiling point curve to higher mole fractions of A (x_2^l). At the same time, the boiling temperature (T_2) rises and the difference in the composition of the liquid and vapor phases becomes noticeably smaller. By continuing this process of distillation, the residual liquid finally reaches the azeotropic composition x_3^l. The boiling liquid and the vapor (or condensate as the case may be)

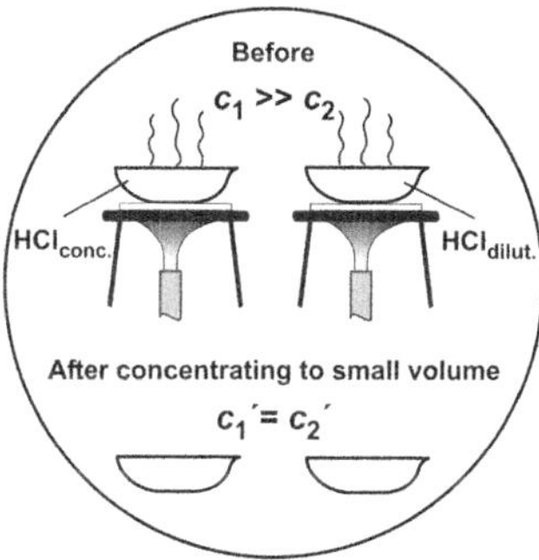

Experiment 14.4 *Azeotropic behavior of hydrochloric acid*: When diluted hydrochloric acid in a porcelain bowl is heated, it is primarily the water that evaporates until the residue reaches the azeotropic composition. Further separation is not possible any longer at this point, because only hydrochloric acid with 20 % HCl is distilled off. However, when concentrated hydrochloric acid is heated, it is mostly hydrogen chloride that evaporates until, again, the azeotropic point is reached. The residues in both cases exhibit identical concentrations of hydrochloric acid. This can be easily demonstrated by titration with sodium hydroxide solution.

Fig. 14.24 Fractional distillation using the example of a boiling point diagram with azeotropic minimum.

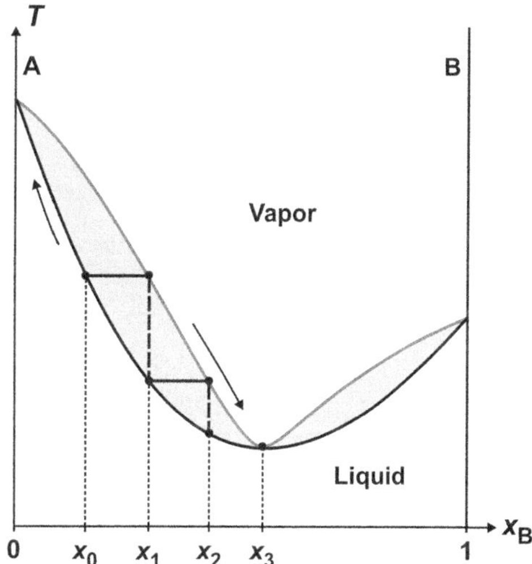

exhibit the same composition and further separation of the mixture is no longer possible.

An example of such a negative azeotrope is hydrochloric acid with a mass fraction $w_{HCl} = 20\ \%\,(= 200\ \mathrm{g\ kg^{-1}})$ that boils at 382 K (at standard pressure) without its composition changing (Experiment 14.4).

No matter what the initial composition of a mixture is, a complete separation by distillation is not possible but only one of the substances can be obtained in pure form together with the azeotropic mixture. This is valid as long as the distillation is done at normal pressure, as is commonly the case. Because the azeotropic point changes with pressure, it is possible to finally separate such mixtures as well but only with a considerable effort.

In closing we will discuss the azeotropic behavior of a system with a boiling point minimum (Fig. 14.24). Let us assume we have begun a fractional distillation of a homogeneous mixture with a composition of x_0 and follow the composition of the vapor in the column. The fraction of the component that boils at higher temperatures decreases along the dew point curve in the direction $x_1 \rightarrow x_2$, etc., until the azeotropic point is reached. This point may not be exceeded, meaning that at the top of the column there is always a condensate with the azeotropic composition x_3. A well-known and technically relevant example is the ethanol/water system that has an azeotrope with an alcohol content of $w = 96\,\%$ ($= 960$ g kg^{-1}) and a boiling temperature of 78 °C. The residue from this is almost pure water.

Chapter 15
Interfacial Phenomena

In this chapter, we will discuss how the chemical and physical properties of substances at interfaces differ from those in the bulk. For quantitative description, quantities like surface tension and surface energy have to be introduced. With the help of these quantities, phenomena known from everyday life like the lotus effect can be explained. However, perhaps you are more interested to learn how detergents clean? Then have a look at Sect. 16.3 which deals with the adsorption on liquid surfaces. The next section covers the adsorption on solid surfaces and the variation of the extent of coverage with pressure or concentration of the substance to be adsorbed. Langmuir's isotherm, the simplest description of such an adsorption process, is deduced by kinetic interpretation of the adsorption equilibrium. Alternatively, it can be derived by introducing the chemical potential of free and occupied sites and considering the equilibrium condition. In the last part of the chapter, some important applications such as surface measurement and adsorption chromatography are discussed.

15.1 Surface Tension, Surface Energy

The *interface* is defined as the layer separating two phases. The interface adjacent to a gas phase is also simply called a *surface*.

The particles at an interface between two phases, for example, those on the surface of a solid or liquid, are subject to different intermolecular forces than those inside a phase (Fig. 15.1). A particle inside a phase is attracted equally on all sides by identical neighboring particles. This means that intermolecular forces are in equilibrium and the net attractive pull on the particle in question equals zero. Particles at an interface such as between a solid and air or between a liquid and air are missing part of their neighbors. This leads to an imbalance of forces where (especially in the case of surfaces) a one-sided pull occurs toward the interior of the denser phase. Consequently, the neighboring particles in the interface will move

© Springer International Publishing Switzerland 2016
G. Job, R. Rüffler, *Physical Chemistry from a Different Angle*,
DOI 10.1007/978-3-319-15666-8_15

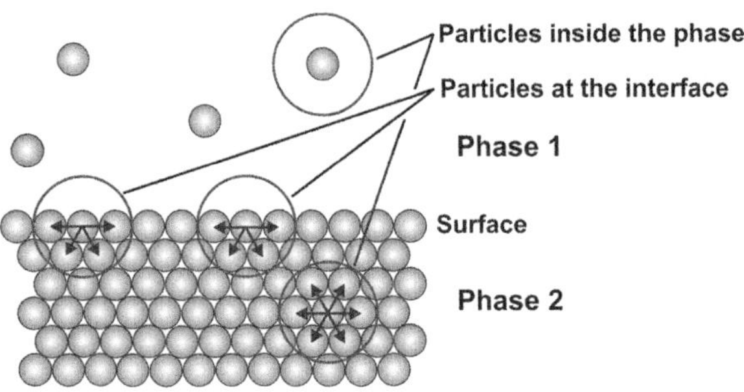

Fig. 15.1 Model of the surface of a condensed phase (solid or liquid).

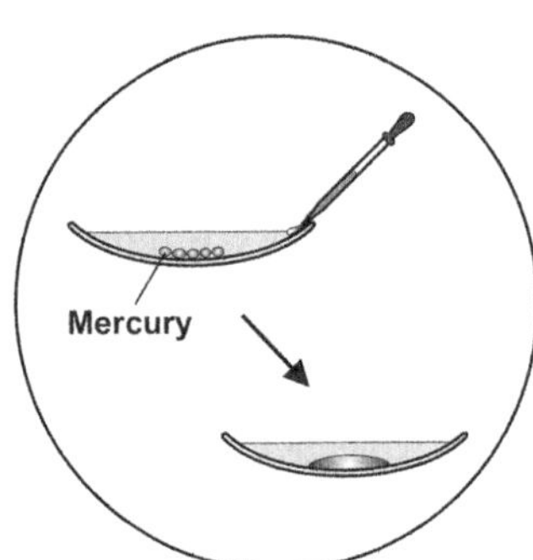

Experiment 15.1 *Merging of droplets of mercury*: Small droplets form when dripping mercury in a watch-glass filled with ethanol. They will gradually merge to form one large drop.

somewhat closer together and a tensile stress in the surface layer will appear (comparable to that in a stretched rubber membrane). This phenomenon is called *interfacial tension* or *surface tension* and is abbreviated by σ.

Surface tension results in a tendency for drops of liquid or gas bubbles to minimize their surface areas. If there are no other forces such as gravity at work, they will assume a spherical shape because a sphere has the smallest surface area for a given volume. Large drops grow at the cost of smaller ones because this also leads to a minimizing of the total surface area (Experiment 15.1).

The tensile forces F_σ that appear most noticeably at the boundary lines of the surface are proportional to the length l of such a contact line:

$$F_\sigma \sim l.$$

Therefore we define

$$\text{Surface tension} := \frac{\text{Tensile force}}{\text{Contact line}} \quad \text{or} \quad \sigma := \frac{F_\sigma}{l}. \tag{15.1}$$

The SI unit for surface tension σ is N m^{-1}.

To increase the surface area by ΔA, an amount of energy $W_{\to A}$ is needed because of the surface tension σ. Interpreted atomistically, molecules are transported against the tensile forces from inside the phase to its surface and this costs energy. The molecules at the surface of the phase therefore have an amount of energy that is higher by the *surface energy* $W_{\rightarrow A}$ than the energy of the molecules inside the phase. Surface tension can also be understood as surface energy density.

The concept of surface tension can be illustrated by a liquid film (such as a soap film) between a U-shaped wire frame and a slider (moveable piece of wire), comparable to a two-dimensional cylinder and piston (Fig. 15.2).

l is the total width of the surface on the front and the back of the liquid film. The slider with the weight hanging from it keeps the system in equilibrium. The weight F_G just compensates for the force F_σ that tries to contract the film.

In order to increase the liquid surface of the film by the small amount $dA = l \cdot ds$, the slider is shifted downward using only slightly more force than F_σ. The force needed is independent of the starting position of the slider and therefore of the size of the surface. This would be different when expanding a rubber membrane where the applied force increases with the elongation. The energy necessary results in

$$dW_{\rightarrow A} = F_\sigma ds = \sigma \cdot l \cdot ds \quad \text{or finally} \quad dW_{\rightarrow A} = \sigma dA. \tag{15.2}$$

After transforming the expression we arrive at what we wanted to show,

$$\sigma = \frac{dW_{\rightarrow A}}{dA}.$$

An interface can be understood to be a separate phase with an area A but not a volume ($V = 0$). We will not go into this right now. The main equation (Sect. 9.1) for such an "interface phase" where substances can be enriched from the neighboring phases, or can migrate into those phases, is:

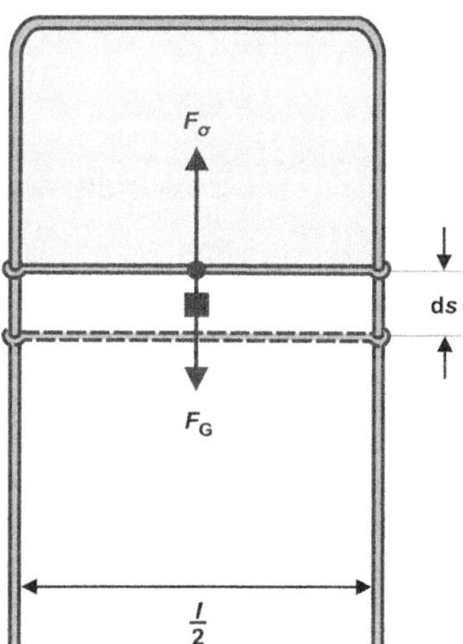

Fig. 15.2 Illustration of surface tension using a soap film in a wire frame with a slider.

Experiment 15.2 *Soap film*: When the slider is slowly pulled away from the end of the frame (see the *hand symbol*), the soap film expands. If we let go, the film contracts to its former size and the slider moves back to its original position (note the *arrow*).

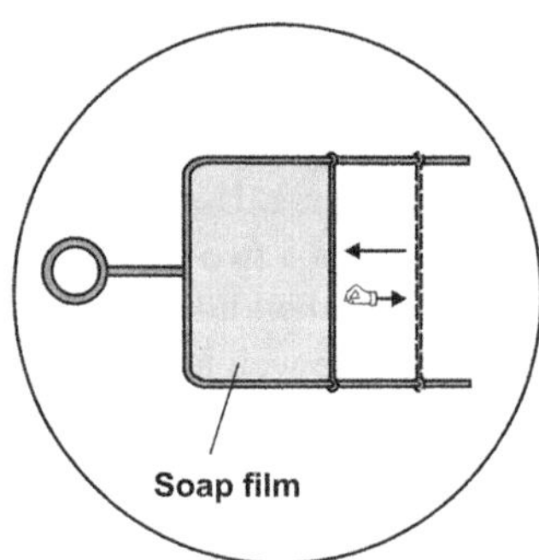

Table 15.1 Surface tensions of various liquids at 298 K (from: Lide D R (ed) (2008) CRC Handbook of Chemistry and Physics, 89th edn. CRC Press, Boca Raton).

Liquid	σ (mN m^{-1})
Diethyl ether	16.7
n-Hexane	17.9
Ethanol	22.0
Carbon tetrachloride	23.4
Acetic acid	27.1
Benzene	28.2
Water	72.0
Mercury	485.5

Table 15.2 Surface tension of water at various temperatures (from: Lide D R (ed) (2008) CRC Handbook of Chemistry and Physics, 89th edn. CRC Press, Boca Raton).

Temperature (K)	σ (mN m^{-1})
283	74.2
298	72.0
323	67.9
348	63.6
373	58.9

$$\mathrm{d}W = \sigma \mathrm{d}A + T\mathrm{d}S + \sum_i \mu_i \mathrm{d}n_i. \tag{15.3}$$

The sum $\sum \mu_i n_i$ can be left out if we are only dealing with a pure substance. The main equation then becomes especially simple.

The effects of surface tension can be observed concretely in an experiment (Experiment 15.2).

Surface tension is a substance-specific quantity. For many organic liquids, at around 298 K, it is between 15 and 30 mN m^{-1} (Table 15.1). The much higher value of $\sigma = 72$ mN m^{-1} for water is due to the high polarity of water molecules and the resulting relatively strong hydrogen bridge bonds between them. Surface tension of mercury is even six times higher than that of water. This is because of the metallic bonds between the atoms.

Surface tension decreases when temperature increases because the more intense motion of the molecules leads to a lessening of the intermolecular forces. It disappears when the critical point is reached. Table 15.2 shows the surface tension of water at various temperatures.

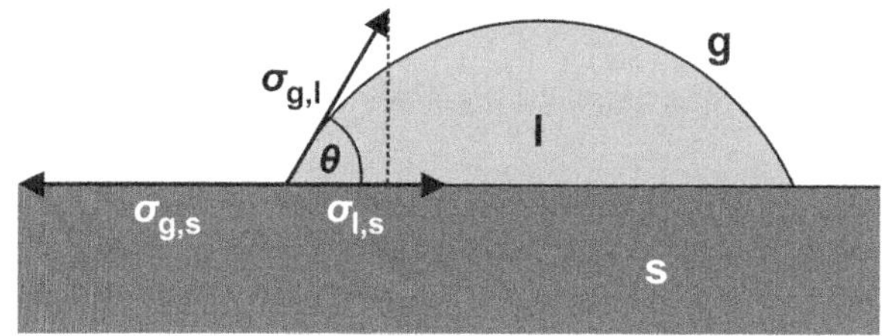

Fig. 15.3 Overlapping of different interface tensions and the corresponding contact angle (or wetting angle) θ of a liquid drop on a plane solid surface.

15.2   Surface Effects

Wetting   Wetting is defined as the complete covering of a solid surface by a liquid film caused by the attractive forces between different substances at a common interface.

If a drop of liquid is put upon a solid surface, three phases adjoin each other: gaseous (g), liquid (l), and solid (s) (Fig. 15.3).

Relating to a short section of length l of the three-phase contact line (which in this case is perpendicular to the plane of the drawing), we have to consider three interface tensions: $\sigma_{g,l}$, $\sigma_{g,s}$, and $\sigma_{l,s}$. As a result, three forces $\sigma_{g,l} \cdot l$, $\sigma_{g,s} \cdot l$, and $\sigma_{l,s} \cdot l$ act parallel to the arrows in Fig. 15.3. The contact line shifts right or left, and in doing so, the *contact angle* θ changes accordingly until equilibrium of forces is attained. A shift upward is impossible due to the rigid base, so we only need to consider the force components that are parallel to the solid surface:

$$\sigma_{g,s} \cdot l = \sigma_{l,s} \cdot l + \left(\sigma_{g,l} \cdot \cos\theta\right) \cdot l \quad \text{or} \quad \sigma_{g,s} = \sigma_{l,s} + \sigma_{g,l} \cdot \cos\theta. \tag{15.4}$$

If $\theta < 90°$, the liquid will spread across the solid surface; it *wets* the surface. Complete wetting exists when $\theta = 0°$ (or $\sigma_{g,s} > \sigma_{l,s} + \sigma_{g,l}$; in this case, an equilibrium of force is impossible). Water on greaseless glass shows a contact angle of $\approx 0°$.

If, however, $\theta > 90°$ (in the ideal case, 180°), no wetting will take place (Fig. 15.4) [Examples: mercury on glass, water on lotus leaves (lotus effect), water on polytetrafluoroethene fabric (Gore-Tex$^{®}$)].

Capillary Pressure   Capillary pressure corresponds to excess pressure p_σ in a gas bubble or a drop as a result of interface tension. Experiment 15.3 will serve to clarify this.

Apparently, capillary pressure decreases as the radius increases. How can this be explained? There is an excess pressure p_σ inside a bubble, which balances the interface tension. If the radius r grows by dr due to further inflating of the bubble, thereby increasing volume V by $dV = 4\pi r^2 dr$, the energy to be expended is

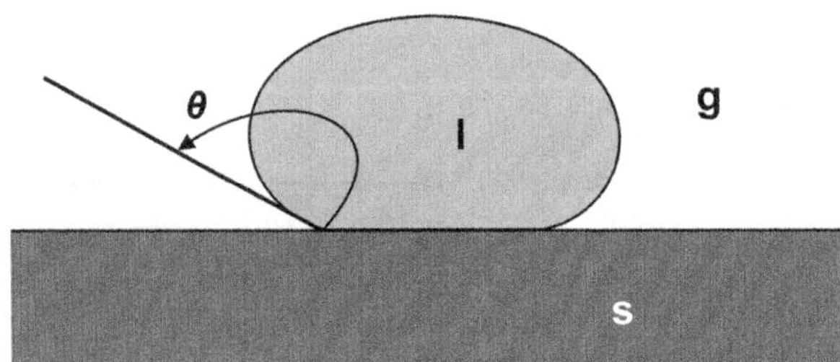

Fig. 15.4 Lack of wetting of the surface at a contact angle of $\theta > 90°$.

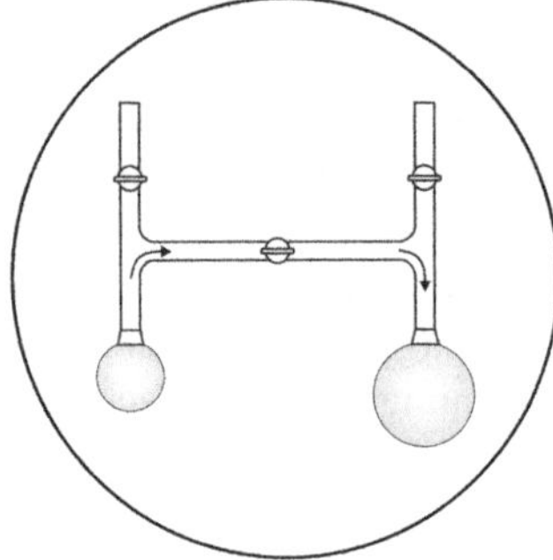

Experiment 15.3 *Connected soap bubbles*: Two soap bubbles of differing sizes are connected via a closed stopcock. When the stopcock is opened, the smaller bubble will "inflate" the larger one and disappear in the process.

$$dW = p_\sigma dV = p_\sigma \cdot 4\pi r^2 dr.$$

At the same time, the surface of the bubble grows by $dA = 8\pi r dr$ and the surface energy grows along with it. However, in the case of a soap bubble, there are both an inner and an outer surface to be dealt with. This means that, in total, the change of surface energy is:

$$dW_{\rightarrow A} = \sigma dA = \sigma \cdot 16\pi r dr.$$

The volume of a sphere is $\frac{4}{3}\pi r^3$ and its surface is $4\pi r^2$. By taking the derivative, we obtain

$$\frac{dV}{dr} = \frac{d}{dr}\left(\frac{4}{3}\pi r^3\right) = 4\pi r^2$$

and after rearranging

$$dV = 4\pi r^2 dr.$$

Analogously, ist results for the surface:

$$\frac{dA}{dr} = \frac{d}{dr}(4\pi r^2) = 8\pi r$$

so that

$$\mathrm{d}A = 8\pi r\mathrm{d}r.$$

In equilibrium, the following is valid:

$$p_\sigma \cdot 4\pi r^2 \mathrm{d}r = \sigma \cdot 16\pi r\mathrm{d}r.$$

The resulting capillary pressure p_σ in a soap bubble is then

$$p_\sigma = \frac{4\sigma}{r} \tag{15.5}$$

which is, as expected, inversely proportional to the radius of the bubble.

Instead of a soap bubble, let us now consider a gas "bubble" (or more precisely a gas-filled cavity) in a liquid such as the "bubbles" in champagne or, possibly, a drop of a liquid. Now there is only one interface to be taken into account. Correspondingly, the capillary pressure is

$$p_\sigma = \frac{2\sigma}{r}. \tag{15.6}$$

Capillary pressure disappears for plane surfaces ($r \to \infty$), but for very small drops, it is quite important. For example, a drop of water with a radius of 1 µm has the capillary pressure of 146 kPa.

Vapor Pressure of Small Drops A bulk liquid is subject to a saturation vapor pressure of $p_{\mathrm{lg},r=\infty}$. As a result of capillary pressure, the chemical potential of a drop of liquid is higher by

$$\Delta\mu_1 = p_\sigma \cdot \beta = \frac{2\sigma}{r} V_\mathrm{m}$$

than that of a bulk liquid. This means that a decrease in the size of the drop increases its tendency to evaporate. Equilibrium with the vapor occurs when its chemical potential has also grown by the same difference $\Delta\mu_\mathrm{g}$ due to a raise of pressure from $p_{\mathrm{lg},r=\infty}$ to $p_{\mathrm{lg},r}$:

$$\Delta\mu_\mathrm{g} = RT\ln\frac{p_{\mathrm{lg},r}}{p_{\mathrm{lg},r=\infty}} = \frac{2\sigma}{r} V_\mathrm{m} = \Delta\mu_1.$$

We obtain

$$\ln\frac{p_{\mathrm{lg},r}}{p_{\mathrm{lg},r=\infty}} = \frac{2\sigma V_{\mathrm{m}}}{rRT}$$

and, respectively,

$$p_{\mathrm{lg},r} = p_{\mathrm{lg},r=\infty}\exp(2\sigma V_{\mathrm{m}}/rRT) \quad \text{Kelvin equation.} \qquad (15.7)$$

Small drops have a higher vapor pressure $p_{\mathrm{lg},r}$ than that of a bulk liquid ($p_{\mathrm{lg},r=\infty}$). Table 15.3 illustrates the rise of vapor pressure as a function of the size of drops in the case of water drops.

Very small droplets are therefore quite unstable and the question arises of how condensation of water vapor in air can occur at all. There have to be "condensation nuclei," i.e., molecules, ions, dust particles, etc., with which even just a few water molecules can form stable aggregates that can continue to grow. If such nuclei or surfaces that water can precipitate onto are absent, supersaturated water vapor can exist for a very long time. Air for example can contain supersaturated water vapor even in a clear sky. This water vapor condenses on the particles left by the engine exhaust of an airplane and so-called condensation trails (contrails) appear behind the plane.

Capillary Action When a capillary tube is submerged in a wetting liquid, this liquid will rise in it up to a certain height. Experiment 15.4 is a good example for illustrating the dependency of capillary rise upon the diameter of the capillary tube.

Wetting of the inner wall of the capillary tube by the liquid film increases the liquid's surface. This is opposed by surface tension. A reduction of the surface can only occur if the liquid rises to a height h in the tube having a radius of r_{c} (Fig. 15.5). It forms a *meniscus* (the term for a curved surface in a narrow pipe). The meniscus of a completely wetting liquid, for example, water in a glass tube, assumes a hemispherical shape curved upward (concave surface). The water tries to cover as much of the glass surface as possible and the hemisphere is the smallest possible surface for the liquid. In the present case of a contact angle of $\theta \approx 0°$, the radius of curvature is equal to the capillary radius r_{c}. The liquid rises in the capillary until the weight $F_{\mathrm{G}} = mg = \rho V g$ of the liquid drawn up the tube just compensates for the force F_{σ} resulting from surface tension along the capillary circumference. We obtain

	Radius (nm)	Number of particles	$p_{\mathrm{lg},r}/p_{\mathrm{lg},r=\infty}$
Table 15.3 Increase of vapor pressure for varying sizes of water drops.	10^3	1.4×10^{11}	1.001
	10^2	1.4×10^8	1.011
	10	140,000	1.111
	1	140	2.88

Experiment 15.4 *Capillary tubes in action*: A communicating system of several capillary tubes with varying diameters is filled with colored water. The water rises in the tubes due to the capillary effect and reaches different levels. The narrower the tube, the higher the water rises.

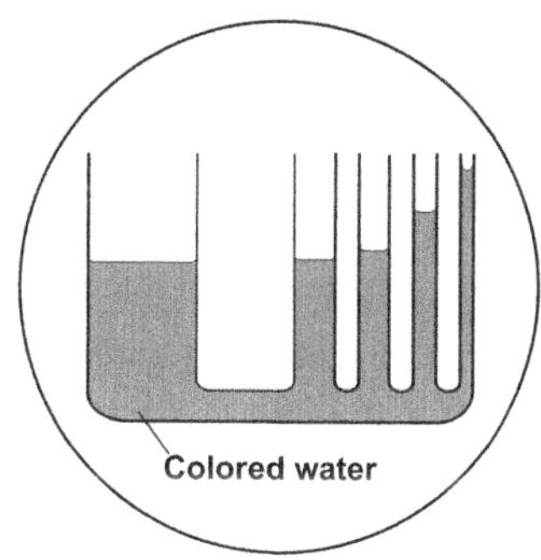

Fig. 15.5 A wetting liquid rising in a capillary tube (capillary rise).

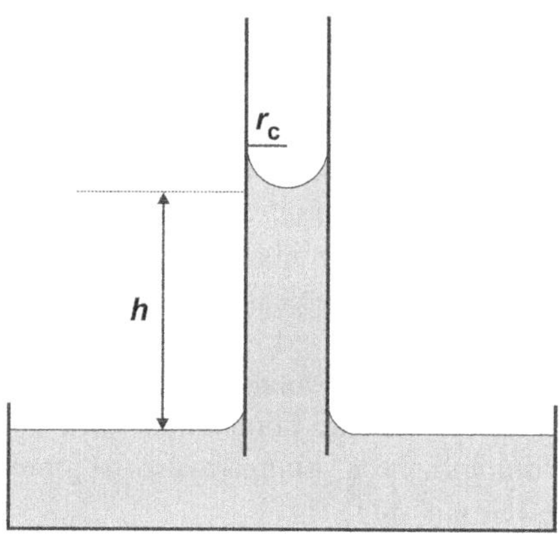

$$F_\sigma = 2\pi r_c \sigma = \rho \pi r_c^2 h g = F_G$$

or after rearranging the terms

$$h = \frac{2\sigma}{\rho r_c g}, \tag{15.8}$$

where ρ stands for the density of the liquid. The capillary rise of a liquid is inversely proportional to the capillary radius. For a water-filled glass tube in air at standard conditions ($\sigma = 0.072$ N m^{-1} at 298 K, $\rho = 1{,}000$ kg m^{-3}, and $g = 9.81$ m s^{-2}), the height of the water column is

$$h = \frac{1.47 \times 10^{-5}}{r_c}\,\text{m}.$$

Thus for a tube with a radius of 2 cm, the water would rise only 0.7 mm, but for one with a radius of 0.2 mm, it would already rise to a height of 70 mm.

Moreover, the capillary rise of a liquid is directly proportional to the surface tension. This relation can therefore be used to determine the surface tension of liquids.

The capillary force works in the opposite direction for non-wetting liquids: The level of the liquid is lowered and it displays a convex surface. In this case, one speaks of a "capillary depression" (Example: mercury in a glass tube as in thermometers and barometers).

15.3 Adsorption on Liquid Surfaces

Dissolved substances can influence interface tension by enriching themselves in the interface. This phenomenon is called *adsorption*. The forces of attraction A–B between the molecules of the dissolved substance B and those of the solvent A are smaller than A–A, so that the dissolved substance is forced out of the interior of the phase. The characteristics of the interface are modified due to enrichment of the dissolved substance. The surface energy and therefore the surface tension decrease so that the capillary action increases. These types of substances are called *capillary active, surface active,* or *interface active.* They are also called *surfactants.* In aqueous solutions, this characteristic is exhibited mainly by organic compounds with long hydrophobic hydrocarbon chains and hydrophilic head groups [hydroxyl group, carboxylate group (COO^-), or sulfonic acid group $\left(SO_3^-\right)$].

An experiment using a razor blade shows the influence upon surface tension by surfactants of the kind found in dish-washing liquid (Experiment 15.5).

The surfactant molecules in the solution of a dish-washing liquid or a laundry detergent slip between the water molecules and the hydrophobic residue of these

Experiment 15.5 *Floating razor blade*: When a razor blade is carefully laid upon a water surface, it will sink slightly (comparable to a weight on a tensed membrane), but continues to float. When a solution of dish-washing liquid is added, the razor blade sinks to the bottom of the container because it can no longer be supported by the surface tension.

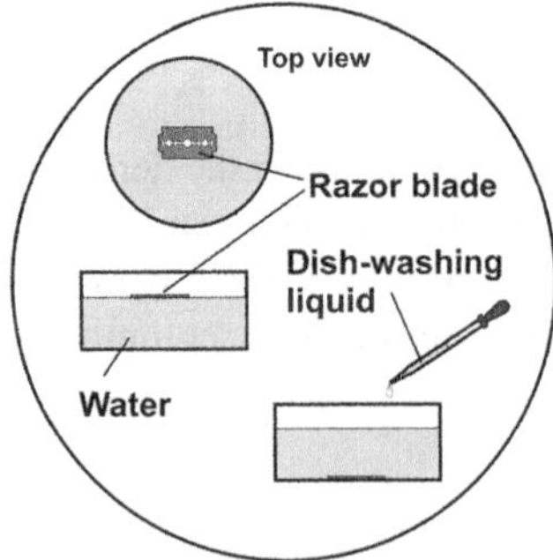

molecules then extend out of the water (Fig. 15.6). As a result, the attraction between the water molecules due to the strong hydrogen bridge bonds decreases along with surface tension.

If the concentration of the surfactant molecules continues to increase, the surface will finally become completely covered by a layer of molecules (monomolecular layer). If this concentration is exceeded, surfactant molecules will also be present in the interior of the liquid. However, they will be oriented so that the hydrophobic ends of the molecules are joined and shielded from the solution by the hydrophilic head groups. As a result, *micelles* are formed above a certain surfactant concentration, the so-called *critical micelle concentration* (CMC). Micelles are colloid-sized clusters.

The "*cleaning power*" of surfactants depends upon the hydrophobic hydrocarbon residues penetrating the dirt particles (drops of oil or grease, for example) and the textile fibers while the hydrophilic groups protrude into the water. The motion of the piece being washed separates the dirt particles from the fibers and solubilizes them. This means that their solubility in the solvent (in this case, water) is noticeably improved by adding a third chemical. Moreover, micelle formation is essential for the resorption of fat-soluble vitamins and the digestion of complicated lipids within the human body.

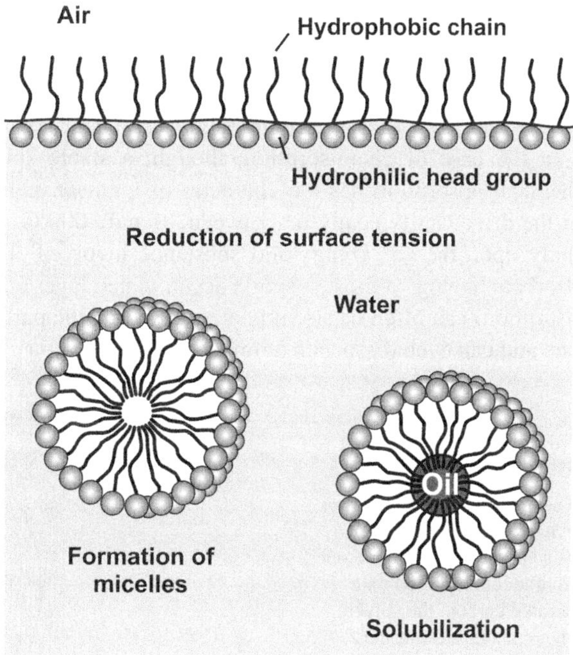

Fig. 15.6 Arrangement of surfactant molecules in a monolayer on the surface of water, formation of micelles, and emulsifying power of the surfactants.

15.4   Adsorption on Solid Surfaces

Physisorption and Chemisorption   The phenomena of adsorption on solid surfaces are much more varied than those on liquid surfaces. A good example for observing these phenomena is activated carbon, a highly porous carbon with a large specific surface (300 to 2,000 $m^2\ g^{-1}$ carbon) and possessing excellent adsorption capacity (Experiment 15.6).

The actual process of adsorption happens to lie between two extreme forms of adsorption: *physical adsorption* (*physisorption*) and *chemical adsorption* (*chemisorption*). These differ from each other primarily by the strength of the bonding of the *adsorptive* (free particles before adsorption, gas molecules for example) to the *adsorbent*, i.e., the molecules of the solid surface (for clarity, compare to Fig. 15.7). The particles that have accumulated on the solid surface are called *adsorbate*.

The term physisorption is used when the molecules of a gaseous or dissolved substance B that have accumulated on a solid adsorbent are loosely "physically" bound (Fig. 15.8, left) as might happen by Van Der Waals' forces:

$$\square + B_2 \rightarrow \boxed{B_2}.$$

The symbol $\square$ labels a site on the surface.

Physical adsorption with a chemical drive $\mathcal{A}$ of the order of 8 to 10 kG has the character of a *condensation*. The drive is determined almost exclusively by the type of substance being adsorbed. The adsorbed particles can attach into several layers, one on top of the other (multilayer adsorption), and essentially keep their structure. Noble gases, for example, would be physisorbed at low temperatures.

In the case of chemisorption, though, a stable "chemical" bond is formed. Chemical adsorption has the character of a *chemical reaction*, where the values for the drive can typically lie between 40 and 800 kG. The drive depends significantly upon the adsorbing solid substance involved. The molecular bond in the adsorbate having, at most, a single accumulated layer on the adsorbent (monolayer adsorption) can often be strongly altered so that the particles are in a very reactive state and can even dissociate (compare Fig. 15.8, right). The adsorptive bonding of

Experiment 15.6 *Adsorption on activated carbon*: When a dye solution is poured into one end of a column filled with activated carbon, the solvent will come out clear at the other end. This experiment can also be done with soft drinks containing food coloring, or even red wine.

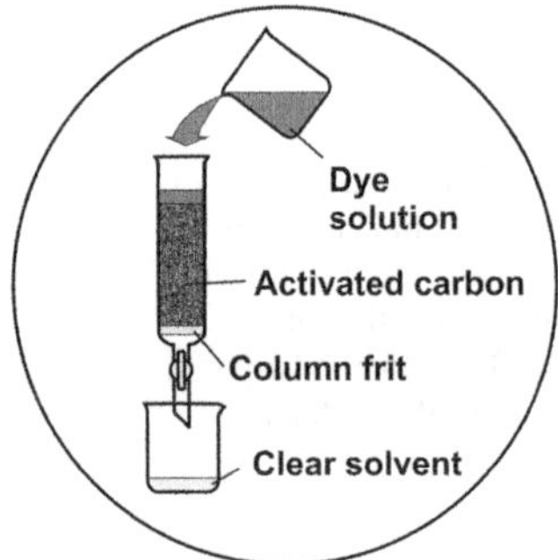

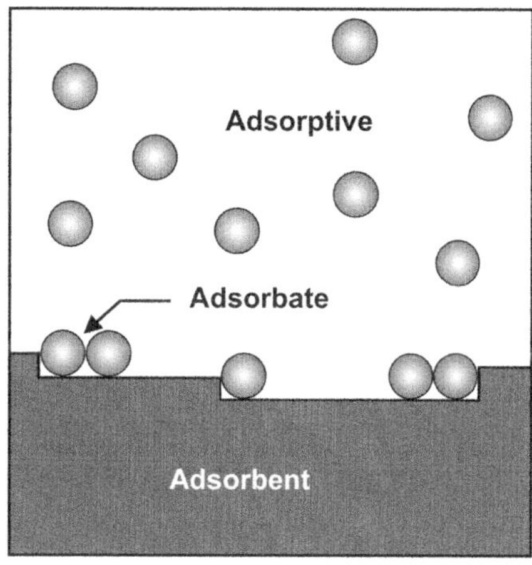

Fig. 15.7 Illustration of important terms describing adsorption.

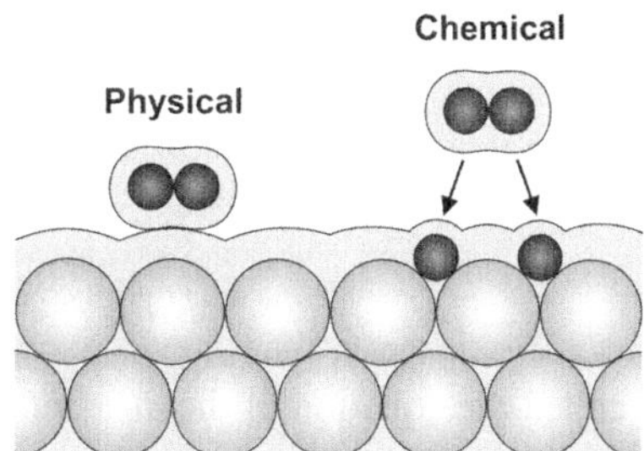

Fig. 15.8 Difference between physisorption and chemisorption.

hydrogen to surfaces of transition metals such as Pd or Fe—which is important for catalytic reactions—is a typical case of chemisorption (see also Sect. 19.4). In this case, hydrogen is not adsorbed in molecular form but in its atomic form:

$$2\,\square + B_2 \rightleftarrows 2\,\boxed{B}.$$

$\square$ indicates a site on the surface able for chemisorption.

Adsorption is accompanied by a "heat effect" (Experiment 15.7). As always in chemical reactions, there are actually two effects at work. Energy is released and dissipated whereby entropy is generated: $S_{\mathrm{g}} = \mathcal{A} \cdot \Delta\xi/T$. This exothermal contribution is complemented by the (usually) exothermal latent entropy $S_\ell = \Delta_\square S \cdot \Delta\xi$ (the symbol $\square$ replaces the index R which we have commonly used to indicate chemical reactions). The reason for the latter is the fact that adsorption on a solid surface limits the mobility of the particles, resulting in the release of entropy.

Adsorption Isotherm   We consider the adsorption of a substance B out of a gas or a solution on adsorption sites $\square$:

Experiment 15.7 *Rise of temperature in adsorption*: When acetone is poured over activated carbon, a noticeable rise in temperature occurs.

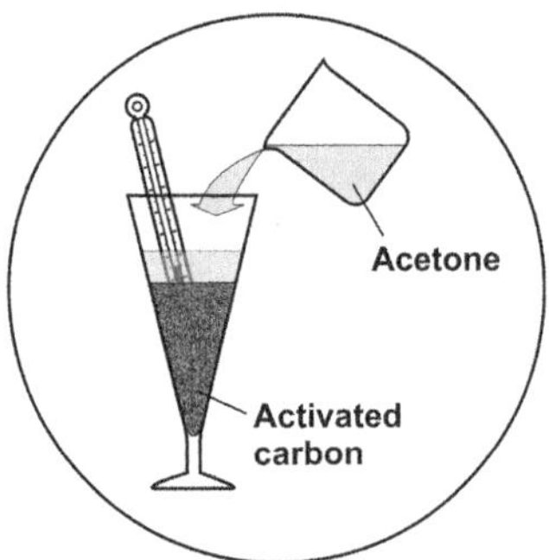

$$\square + B \rightleftarrows \boxed{B}.$$

If the temperature remains constant, an *adsorption equilibrium* will be established where the addition of particles B to and their elimination from the sites of adsorption both occur at the same rate. This is symbolized by a double arrow.

In the case of a gaseous adsorptive, the adsorbed amount n_B is dependent upon pressure p, and in the case of a dissolved adsorptive, it is dependent upon concentration c. The relation

$$n_B = f(p) \quad \text{or} \quad n_B = f(c) \qquad T = \text{const.}$$

is called the equation of the *adsorption isotherm*.

In the case of the formation of a *monomolecular adsorption layer*, the *fractional coverage Θ* is often used instead of the *adsorbed amount n_B* to symbolize the extent of adsorption. It indicates the portion of the surface that is coated:

$$\Theta = \frac{n_B}{n_{B,\text{mono}}} = \frac{m_B}{m_{B,\text{mono}}}. \tag{15.9}$$

The simplest theoretical description of an isotherm is the so-called *Langmuir isotherm* which is based upon a model of equivalent and independent adsorption sites upon a homogeneous surface whereby adsorption cannot proceed beyond monolayer coverage. Although we have mostly been dealing with static aspects of chemical dynamics so far, we will derive this adsorption isotherm first by using kinetic aspects (in anticipation of Chap. 16).

Adsorption equilibrium is established when the rate r_{ads} of adsorption equals the rate r_{des} of *desorption*, meaning the release of already adsorbed molecules.

According to the concepts of kinetics, the rate of adsorption is proportional to the product of the concentrations of the reaction partners. In this case, these are the adsorptive and the free sites on the surface. Partial pressure p or molar concentration c can be used as the measure of adsorptive concentration. The concentration of free sites, on the other hand, must be proportional to the fraction of surface that is not covered, $1 - \Theta$. In summary, we have

$$r_{\text{ads}} = k_{\text{ads}} \cdot p \cdot (1 - \Theta), \tag{15.10}$$

whereby the proportionality constant k_{ads} can also be called the rate coefficient of adsorption. (We will deal with rate coefficients in Chap. 16 and again in more detail in Chap. 18.)

The rate of desorption is proportional to the concentration of sites that are already occupied and, therefore, the fractional coverage Θ. We obtain

$$r_{\text{des}} = k_{\text{des}} \cdot \Theta \tag{15.11}$$

with k_{des} as the rate coefficient of desorption.

The following is valid for dynamical equilibrium:

$$k_{\text{ads}} \cdot p \cdot (1 - \Theta) = k_{\text{des}} \cdot \Theta.$$

Solving for Θ yields:

$$\Theta = \frac{k_{\text{ads}} \cdot p}{k_{\text{des}} + k_{\text{ads}} \cdot p}. \tag{15.12}$$

Using $\overset{\circ}{K} = k_{\text{ads}}/k_{\text{des}}$ we obtain Langmuir's adsorption isotherm:

$$\Theta = \frac{\overset{\circ}{K} \cdot p}{1 + \overset{\circ}{K} \cdot p} \quad \text{for} \ \ T = \text{const.} \tag{15.13}$$

$\overset{\circ}{K}$ can be interpreted as the equilibrium constant for the process of adsorption. Correspondingly, $\overset{\circ}{K}$ is dependent upon temperature (compare Chap. 6).

At low pressures, $\overset{\circ}{K} \cdot p \ll 1$ applies; the isotherm then rises proportionally to p. At high pressures $\left(\overset{\circ}{K} \cdot p \gg 1 \right)$, the fractional coverage asymptotically approaches the limiting value of 1 (Fig. 15.9).

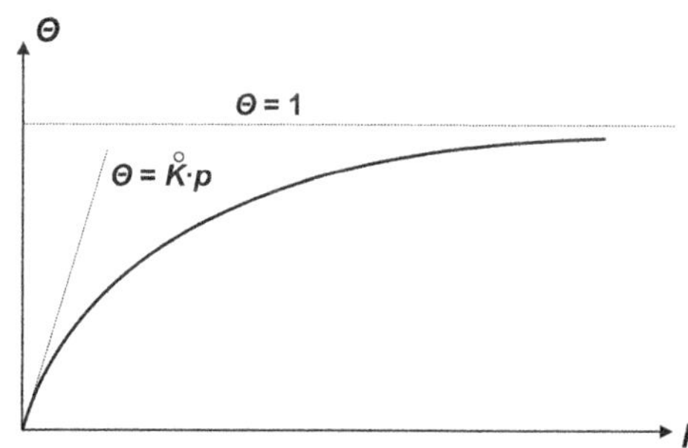

Fig. 15.9 Langmuir's adsorption isotherm.

We can also derive Langmuir's isotherm with the help of the chemical potential. In order to do this, we consider the following sequence of simple processes:

$Bs + H^+ \rightarrow BsH^+$,   Protonation of a base during a titration,

$E + B \rightarrow EB$,   Formation of an enzyme − substrate complex in a cell,

$\square + B \rightarrow \boxed{B}$,   Adsorption of a molecule at a surface site.

In Sect. 7.4, we dealt extensively with the first process and its description with the help of proton potential. A characteristic common to all three processes is that in each one, a certain type of particle occupies a certain type of site. There is obviously a smooth transition from the first type of process to the last one in the sequence because it is easy to insert additional intermediate links. For instance, the gap between the first *homogeneous* reaction and the third *heterogeneous* reaction can be bridged by the second which can be considered a bimolecular reaction between the dissolved substances E (enzyme) and B (substrate) as well as an adsorption of B at a site E. If we imagine the E-molecules to be combined into larger and larger two-dimensional complexes, we arrive in stages at a contiguous surface.

We therefore want to deal with the question of the chemical potential of *sites*. In order to do this, we will again consider adsorption of a substance B out of a gas or a solution on *independent* adsorption sites, but now from this new point of view:

$$\square + B \rightleftarrows \boxed{B}.$$

Adsorption equilibrium is determined by the free and occupied sites $\square$ and $\boxed{B}$, so it seems obvious to assign to them chemical potentials $\mu(\square)$ and $\mu\left(\boxed{B}\right)$. A comparison to the corresponding homogeneous reaction suggests an interesting idea:

$$A + B \rightleftarrows AB.$$

A particle A can be considered a *carrier* of a single adsorption site $\square$ for B. In order to keep the sites from influencing each other, the total concentration $c = c(A) + c(AB)$ of free and bonded A must remain low. This then allows us to apply the mass action equations for $\mu(A)$ and $\mu(AB)$ (compare Sect. 6.2). The condition for equilibrium $\mu(A) + \mu(B) = \mu(AB)$ therefore assumes the form

$$\overset{\circ}{\mu}(A) + RT\ln[c(A)/c^{\ominus}] + \mu(B) = \overset{\circ}{\mu}(AB) + RT\ln[c(AB)/c^{\ominus}]. \qquad (15.14)$$

In order to attain a description that is independent of whether a site sits upon separate particles or upon a continuous surface and is also independent of the parts of the carrier A that are inessential for adsorption, we slightly alter the condition for equilibrium. $c(A)/c \equiv \Theta(\square)$ is the *fraction of empty sites*, $c(AB)/c \equiv \Theta\left(\boxed{B}\right)$ is the *fraction of occupied sites*. We replace $c(A)$ and $c(AB)$ by $c \cdot \Theta(\square)$ and $c \cdot \Theta\left(\boxed{B}\right)$ and subtract $\overset{\circ}{\mu}(A) + RT\ln(c/c^{\ominus})$ from both sides:

$$\underbrace{\overset{\circ}{\mu}(\square) + RT\ln\Theta(\square) + \mu(B)}_{} = \underbrace{\overset{\circ}{\mu}\left(\boxed{B}\right) + RT\ln\Theta\left(\boxed{B}\right)}_{},$$

$$\mu(\square) + \mu(B) = \mu\left(\boxed{B}\right) \qquad \text{(condition for equilibrium).} \tag{15.15}$$

We consider $\overset{\circ}{\mu}\left(\boxed{B}\right) \equiv \overset{\circ}{\mu}(AB) - \overset{\circ}{\mu}(A)$ to be *the basic value of the chemical potential of occupied sites* $\boxed{B}$, i.e., the potential $\mu\left(\boxed{B}\right)$ at full occupancy, $\Theta\left(\boxed{B}\right) = 1$. The term $\overset{\circ}{\mu}(\square) \equiv 0$ is only inserted for the sake of uniformity. It functions as the *basic value of the chemical potential of empty sites* $\square$, meaning the potential $\mu(\square)$ for $\Theta(\square) = 1$.

As a result of the bond between A and B, both A and B are altered. In the case of larger molecules, the changes primarily affect the atoms near the place of bonding; atoms that are farther away are hardly touched. The definition above of the quantity $\overset{\circ}{\mu}\left(\boxed{B}\right)$ boils down to the fact that all changes to molecules A and B are formally assigned to the *adsorbed* particles B. The contribution of the unaltered parts of the carrier A is canceled, especially the contribution by all atoms of A that are outside the sphere of influence of the bonding site.

We can now use the mass action equations shown above for *independent* sites,

$$\mu(\square) = \overset{\circ}{\mu}(\square) + RT\ln\Theta(\square) \tag{15.16}$$

and

$$\mu\left(\boxed{B}\right) = \overset{\circ}{\mu}\left(\boxed{B}\right) + RT\ln\Theta\left(\boxed{B}\right), \tag{15.17}$$

to describe the adsorption of a substance B out of a dilute solution or a dilute gas on a solid surface having equivalent adsorption sites—independently of whether or not they are occupied. Taking the mass action equation for B, $\square$ and $\boxed{B}$ into account, as well as the equations $\Theta\left(\boxed{B}\right) = \Theta$ and $\Theta(\square) = 1 - \Theta$ with *fractional coverage* Θ, the condition for adsorption equilibrium is:

$$\overset{\circ}{\mu}(\square) + RT\ln(1 - \Theta) + \overset{\circ}{\mu}(B) + RT\ln(c/c^{\ominus}) = \overset{\circ}{\mu}\left(\boxed{B}\right) + RT\ln\Theta. \tag{15.18}$$

We subtract $\overset{\circ}{\mu}\left(\boxed{B}\right)$ from both sides, move the logarithmic terms to one side, divide by RT, take the exponential, and divide by $c^{\ominus}$. Because of $\overset{\circ}{\mu}(\square) = 0$, this leads to the relation

$$\underbrace{\frac{1}{c^{\ominus}} \cdot \exp\left(\frac{\overset{\circ}{\tilde{\mu}}(B) - \overset{\circ}{\tilde{\mu}}\left(\boxed{B}\right)}{RT}\right)}_{\overset{\circ}{K}} = \frac{\Theta}{(1 - \Theta) \cdot c}. \tag{15.19}$$

Here, $\overset{\circ}{K}$ is the equilibrium constant, for which (according to Sect. 6.4) the following is valid:

$$\overset{\circ}{K} = (c^{\ominus})^{\nu}\,\overset{\circ}{\mathcal{K}} = (c^{\ominus})^{-1}\exp\left(\frac{\overset{\circ}{\mu}(B) - \overset{\circ}{\mu}\left(\boxed{B}\right)}{RT}\right) \quad \text{with } \nu = -1.$$

Transformation of Eq. (15.19) finally results in the familiar equation for Langmuir's adsorption isotherm

$$\Theta = \frac{\overset{\circ}{K} \cdot c}{1 + \overset{\circ}{K} \cdot c}. \tag{15.20}$$

15.5   Applying Adsorption

Surface Measurement The specific surface of a porous solid material can be determined from the adsorbed amount of an adsorptive when its surface demand is known. The most reliable method for doing this is based upon the physisorption of gases (mostly nitrogen) near their boiling points. The so-called BET isotherm is used for the analysis. This isotherm developed in 1938 by Brunauer, Emmet, and Teller takes multilayer adsorption into account as well.

Separation of Substances Adsorption also plays a major role in separation of substances, especially in *adsorption chromatography*. The method is based upon the difference of adhesion probabilities of the substances being separated which, being in a mobile phase (liquid, gas), are passed along a stationary phase (solid material, for example, Al_2O_3, SiO_2). The greater the attraction of a substance for the stationary phase compared to the mobile phase, the slower this substance moves up. We differentiate between *gas–solid chromatography* (GSC) and *liquid–solid chromatography* (LSC). Depending upon the method in use, we also distinguish between column chromatography, paper chromatography, or thin-layer chromatography. A simple yet convincing example from everyday life is the chromatographic separation of felt pen ink (Experiment 15.8).

The same dye in the same solvent will always move the same distance in the same period of time. The individual components can therefore be characterized by the R_f value (R_f stands for "retention factor"):

Experiment 15.8 *Chromatographic separation of black felt pen ink*: Different black felt tip pens are used to make dots along a line near the bottom edge of a thin-layer chromatography (TLC) plate covered with silica gel for thin-layer chromatography. The bottom edge of this plate is then put into a TLC chamber (or a beaker) containing a few centimeters of water as solvent. A separation of the black ink into variously colored components (for example, violet, yellow, blue) can soon be observed. Instead of a TLC plate, a strip of filter or blotting paper can also be used.

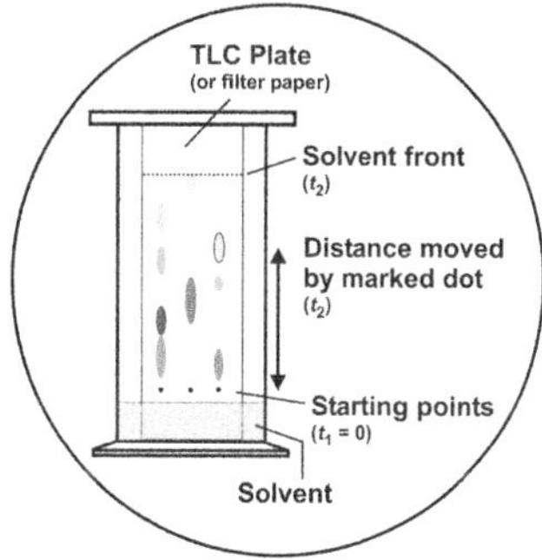

$$R_f = \frac{\text{Distance starting line} - \text{center of dot}}{\text{Distance starting line} - \text{solvent front}}.$$

Heterogeneous Catalysis Adsorption is the basis of heterogeneous catalysis, which makes it especially important to industrial production processes. This will be gone into more detail in Sect. 19.4.

Chapter 16
Basic Principles of Kinetics

The branch of matter dynamics called *chemical kinetics* will be the topic of the next four chapters. Chemical kinetics is concerned with the temporal course of chemical reactions, meaning one investigates how fast the reactants are consumed or the products are formed. The goal of such investigations is to provide the means for predicting the rate of processes and to find the influencing factors that promote a desired reaction or inhibit an undesired one. In this introductory chapter we will first get to know the fundamental quantities *conversion rate* and *rate density* as well as different methods for measuring them in slow and fast reactions. In the last part of the chapter, it will be shown how the dependence of the rate density on the concentrations of reactants (and products) can be summarized by mathematical expressions called *rate laws*. Subsequently, the relatively simple rate laws of different types of reactions taking place in only one single step will be discussed.

16.1 Introduction

Concept of Chemical Kinetics *Chemical kinetics* or simply *kinetics* is the area of chemistry that deals with the temporal course of transformations of substances and the intermediate steps involved in it, especially the

- Recording of the *temporal course* of chemical reactions,
- Determination of (differential) *rate laws* and the corresponding *integrated rate laws* under given conditions,
- Identification of *intermediate steps* (elucidation of the reaction mechanism),
- Investigation of *temperature dependency*,
- Investigation of *facilitating* and *inhibiting influences* (catalysis, inhibition).

The goal of investigations in this area is to provide the means for predicting the rate of processes and to find the influencing factors that promote a desired reaction or inhibit an undesired one.

© Springer International Publishing Switzerland 2016

G. Job, R. Rüffler, *Physical Chemistry from a Different Angle*,

DOI 10.1007/978-3-319-15666-8_16

Reaction Resistance We expect that the greater the drive $\mathcal{A}$ of a chemical transformation, meaning the greater the drop of potential from the reactants to the products, the faster the process will run. However, it would be a mistake to believe that the strength of the drive *alone* determines the speed, i.e., the rate at which the process runs. There are *inhibitions* in every kind of change of substances that must be overcome. They can be large or small, depending upon experimental conditions, and they influence the rate of the process as much as the drive does. In a system involving substances, the conversion in the direction of the drop in chemical potential is inhibited or even stopped by various resistances (Fig. 16.1). This is similar to an electrical circuit where the flow of charge in the direction of the drop in potential is hindered by high electrical resistance or totally stops when the circuit is interrupted. There are similar examples to be found in other fields such as mechanics. An example might be bicycling on sandy paths or stirring a thick soup. A simple idea describing the influences mentioned above would be that the rate of a transformation is directly proportional to the drive and inversely proportional to the resistance to be overcome:

$$\text{rate} = \frac{\text{drive}}{\text{resistance}}.$$

However, we do not encounter such simple circumstances very often. Ohm's law, i.e., current $I =$ voltage U/resistance R, is a well-known example of this idea, but even a lightbulb behaves differently in this case because the resistance is not constant but increases with temperature. It is not surprising, then, when in chemical processes such a simple law like Ohm's is only valid for drives $\mathcal{A} \ll RT$, meaning when equilibrium $\mathcal{A} = 0$ is almost attained.

Experiment 16.1 clarifies the concept of reaction resistance using a hydromechanical analogue for a reaction between dissolved substances A and B.

The reaction resistance can be changed analogously to the other examples (e.g., by catalysts).

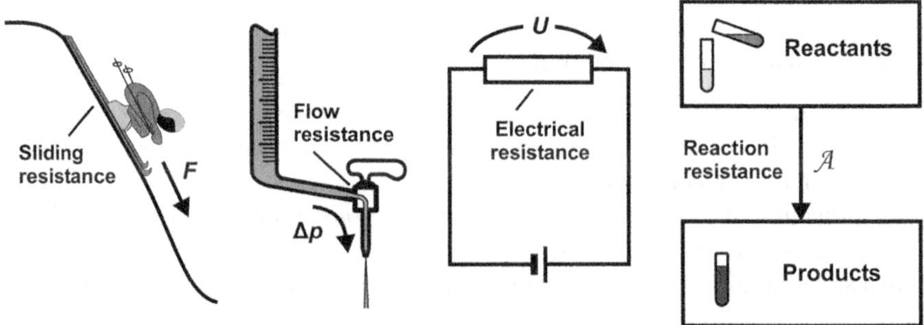

Fig. 16.1 Driving force and resistance to be overcome in different processes.

Experiment 16.1 *Hydromechanical analogue to reaction resistance*: Colored water is filled into one of the glass jars. (Their special form corresponds to the "exponential horn" in Sect. 6.7, i.e., it symbolizes the dependence of the chemical potential upon the amount of dissolved substance as in a potential diagram.) The initially closed stopcock is opened and the liquid distributes to both jars until equilibrium is established. The stopcock takes the role of reaction resistance inhibiting the reaction although it could proceed in principle.

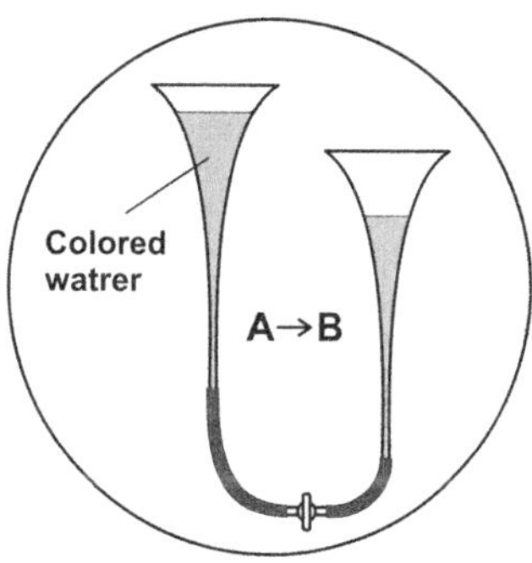

Duration of Conversion   The *duration of conversion*, meaning the time period $\mathcal{T}$ of substance transformations, is expressed by characteristic quantities such as half-life, lifetime, or response time. It spans many orders of magnitude in the range of $\mathcal{T} < 10^{-9}$s to $\mathcal{T} > 10^9$a. Compared to a usual observation period $\mathcal{T}_O$, a transformation is

- *Inhibited*, when $\mathcal{T} \gg \mathcal{T}_O$ (when $\mathcal{T}$ is much larger than $\mathcal{T}_O$)
- *Slow*, when $\mathcal{T} \approx \mathcal{T}_O$ (when $\mathcal{T}$ and $\mathcal{T}_O$ are of similar order of magnitude)
- *Fast*, when $\mathcal{T} \ll \mathcal{T}_O$ (when $\mathcal{T}$ is much smaller than $\mathcal{T}_O$)

The Experiments 16.2, 16.3, 16.4, and 16.5 illustrate the large interval of durations and rates of conversion in which chemical reactions occur.

The duration of conversion increases strongly in each experiment from Experiment 16.2 to Experiment 16.5 while the rate decreases. while the rate decreases. The role played by the chosen time frame can be seen in the transition "peat $\rightarrow$ coal": Compared to how long usual laboratory experiments take, it is inhibited. Compared to geological time frames, it is not.

Reaction Intermediates and Reaction Mechanisms   Closer examination has shown that chemical reactions are not as simple as the usual summary conversion formulas would lead us to believe, but mostly run in several steps or loops in a sequence of *reaction intermediates* quickly transforming into each other and reacting with each other. They often appear in low concentrations and remain inconspicuous and mostly undetected. These substances, appearing only in small or trace amounts, often form a "bottleneck" that limits the rate of a reaction. If it is possible to facilitate the formation of such intermediates, a reaction can be *promoted*. In contrast, it can be *hindered* by repressing production of certain intermediates. In order to approach this systematically, it is first necessary to know the steps

Experiment 16.2 *Precipitation of zinc carbonate*: If a solution containing CO_3^{2-} is added to a Zn^{2+} solution, white zinc carbonate precipitates immediately; the reaction proceeds very fast and is finished after a very short time (a period of maybe 0.3 s).

Experiment 16.3 *Reaction of iodate with iodide*: The reaction of IO_3^- and I^- in aqueous solution proceeds more slowly than the reaction of CO_3^{2-} with Zn^{2+}—the brown color of the iodine that forms makes this visible (the duration is about 30 s).

Experiment 16.4 *Rusting of iron*: Rusting of a piece of iron in humid air takes about 10 years.

Experiment 16.5 *Coal formation*: The formation of coal from peat takes millions of years.

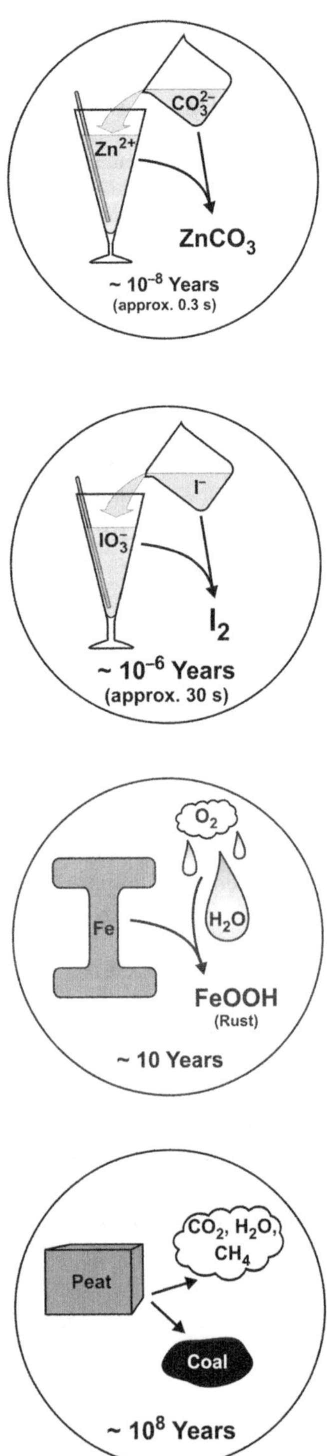

a reaction takes. In other words, we must know the *step-by-step sequence*, meaning the *reaction mechanism*. For this reason, elucidation of *reaction mechanisms* is an important part of chemical kinetics.

16.2 Conversion Rate of a Chemical Reaction

Conversion and Conversion Rate We have already learned the terms *extent of reaction* and *conversion* in Sect. 1.7. They are necessary to know for the following. Let us take another look at the example of rusting of a piece of iron in humid air. The conversion formula for this is:

$$4\ Fe + 3\ O_2 + 2\ H_2O \rightarrow 4\ FeOOH$$

or when we put all the substances on the right-hand side,

$$0 \rightarrow \underbrace{-4}_{\nu_{Fe}}\ Fe\ \underbrace{-3}_{\nu_{O_2}}\ O_2\ \underbrace{-2}_{\nu_{H_2O}}\ H_2O\ \underbrace{+4}_{\nu_{FeOOH}}\ FeOOH.$$

The coefficients ν_i represent the *conversion numbers*. The *extent of reaction* ξ can be calculated according to

$$\xi = \frac{\Delta n_i}{\nu_i} = \frac{n_i - n_{i,0}}{\nu_i} \qquad (16.1)$$

[compare Eq. (1.14)], where n_i represents the instantaneous amount of substance at time t, and $n_{i,0}$ represents the amount at initial time t_0. ξ as well as *conversion* $\Delta\xi$ are then functions of time. In our example, the value of $\xi(t)$ indicates what extent the rusting process has attained at the point in time t. Think about how old automobiles rust. When the weather is dry, the reaction comes almost to a standstill, due to a lack of water. The extent of reaction ξ is constant and the conversion $\Delta\xi$ during 1 day, for example, is just about zero. When the weather is wet or the car is in a damp garage, the reaction will proceed, but slowly. The extent of reaction will gradually increase, which can be seen in the growing formation of rust. If, after being out on winter streets, there is salt sticking to the car, the reaction will proceed especially fast. The daily conversion can reach considerable values, to the annoyance of the car's owner. This observation suggests that the *conversion rate* ω of a reaction should be described by the ratio "distance"/time span like it is done in mechanics, only that "distance" as used here means the change of extent of reaction (reaction coordinate) ξ by the continuation of the reaction during the short time span Δt (Fig. 16.2). (We will avoid using the obvious name *reaction rate* for ω because it can mean different things [more about this in Sect. 16.3)].

Mechanical: Chemical:

$$v := \frac{\Delta x}{\Delta t},\qquad\qquad \omega := \frac{\Delta\xi}{\Delta t}\quad \text{Unit: } \mathrm{mol\,s^{-1}}. \tag{16.2}$$

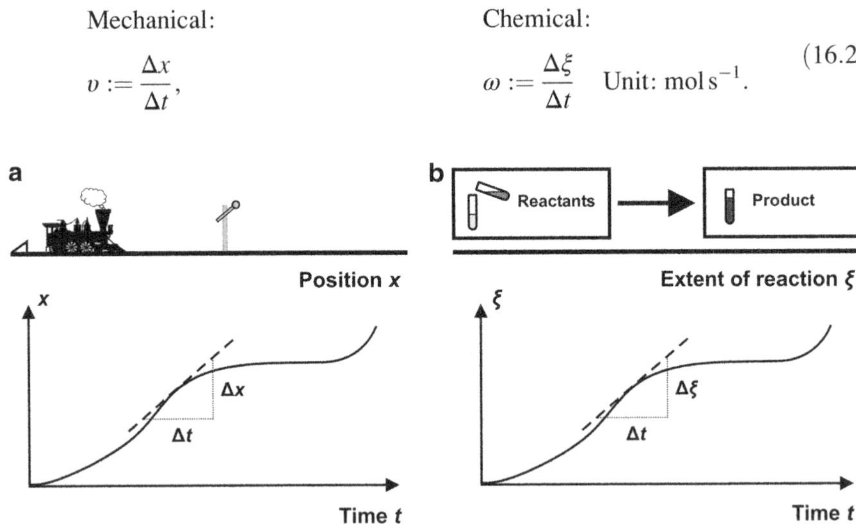

Fig. 16.2 "Distance" covered as a function of time using examples of (**a**) a locomotive and (**b**) a chemical reaction.

Experiment 16.6 *Oscillating reaction according to Briggs and Rauscher*: A solution of malonic acid, manganese sulfate, and starch as well as a solution of acidulated potassium iodate are filled into a beaker. A solution of hydrogen peroxide is then added. The color of the solution changes periodically from colorless to yellow brown, dark blue and then back to colorless, and so on and so forth

Instantaneous Rate   Just as the speed of a train constantly changes by accelerating and slowing down, the speed, the rate at which the reactants are used up and the products are formed, can also change during a reaction. Let us consider the example of a periodic reaction according to Thomas S. Briggs and Warren C. Rauscher, also known as the "oscillating iodine clock" (Experiment 16.6). The repeated color changes occur due to periodic concentration oscillations whereby the reaction does not oscillate forward and backward. Instead, a complex combination of slow and fast reactions in batch mode is taking place simultaneously.

In other words, we must consider an instantaneous value: the speed at the moment in question. In order to find this *instantaneous rate*, it is necessary to move to very small time spans, as is symbolically expressed by the differential quotient dξ/dt:

$$v := \frac{\mathrm{d}x}{\mathrm{d}t}, \qquad\qquad \omega := \frac{\mathrm{d}\xi}{\mathrm{d}t} \quad \text{(De Donder 1929).} \quad (16.3)$$

The instantaneous rate corresponds to the slope of the tangent to the $x(t)$ or $\xi(t)$ curve at this point. The steeper the slope, the higher the rate.

If the system in question is closed and there is only one reaction occurring inside it we can also write

$$\frac{\mathrm{d}n_i}{\mathrm{d}t} = v_i \underbrace{\frac{\mathrm{d}\xi}{\mathrm{d}t}}_{\omega} . \qquad (16.4)$$

This is so because of $n_i = n_{i,0} + v_i\xi$ [compare Eq. (1.16)]. Here, $n_{i,0}$ is the (time independent) initial amount of substance. This constant term vanishes when we take the derivative. After rearranging Eq. (16.4), the following results:

$$\omega = \frac{1}{v_i}\frac{\mathrm{d}n_i}{\mathrm{d}t}. \qquad (16.5)$$

16.3 Rate Density

Concept of Rate Density We know that chemical reactions do not take place at the same rate everywhere. For example, the flame of a candle will transform much more substance in the hot zones than in the cooler zones relative to a volume of, say, 1 mm^3 (Fig. 16.3). $\Delta\omega$ represents the contribution of a small volume ΔV of the reaction mixture to the total rate of conversion.

In order to characterize such local differences, we introduce a new quantity r that specifies conversion rate per volume (of the small volume considered) which we will call *density of conversion rate* or simply *rate density:*

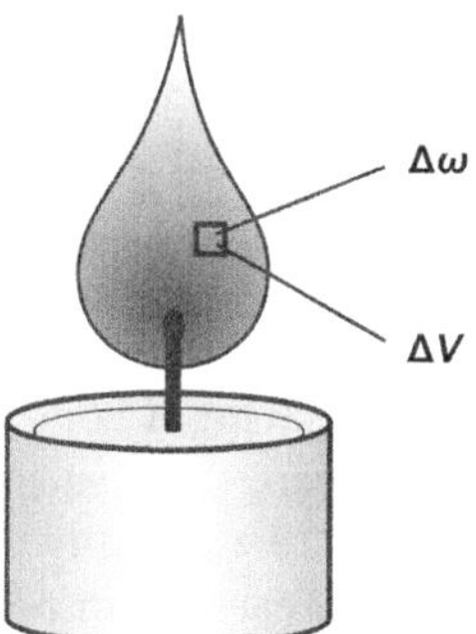

Fig. 16.3 Explanation of the concept of rate density using the example of a candle flame.

$$r := \frac{\Delta\omega}{\Delta V} \quad \text{or more precisely} \quad r := \frac{d\omega}{dV} \quad \text{Unit: } \mathrm{mol\, s^{-1}\, m^{-3}}. \tag{16.6}$$

The quotient $\Delta\omega/\Delta V$ expresses the average value of the conversion $\Delta\xi$ resulting in a short time span Δt in the small volume ΔV considered here. Again, it is necessary to use a very small section when trying to find the value of the rate density r at a given point. This is characterized by the differential notation.

For future considerations, it would be practical to divide chemical reactions into certain classes and to discuss them each separately (see Sect. 16.5). For the moment, though, we will deal with some basic concepts.

Homogeneous and Heterogeneous Reactions A reaction is called *homogeneous* when it takes place in a uniform mixture. An example of this from everyday life is a cup of sweetened tea with lemon where table sugar (sucrose) is gradually split into its two component parts (glucose and fructose). A reaction is called *heterogeneous* when the substances involved are distributed over regions having differing properties. Rusting is a heterogeneous reaction because it takes place in four different regions: the metallic iron, the layer of rust, the water in the fissures of the layer of rust, and the air above. The processes of precipitation or effervescence are both heterogeneous reactions as well.

When dealing with a homogeneous reaction where the reaction takes place uniformly everywhere, the size of the section is unimportant so the entire volume V of the reaction mixture can be used for calculating the rate density:

$$r = \frac{\omega}{V} \quad \text{for reactions in a homogeneous environment at constant volume.}$$

If we insert $\omega = \Delta\xi/\Delta t$ and $\Delta\xi = \Delta n_i/v_i$ as well as $c_i = n_i/V$, we obtain

$$r = \frac{1}{v_i} \cdot \frac{\Delta c_i}{\Delta t}$$

or in the limit of very small Δt

$$r = \frac{1}{v_i} \cdot \frac{dc_i}{dt}. \tag{16.7}$$

The rate density describes the change of concentration of a certain substance per unit of time (under the conditions mentioned above, which are a homogeneous system having a constant volume as well as only one reaction taking place and no exchange of substances with the environment). When the reaction runs forward, the quantity r is positive for all the substances participating—whether they are reactants or products. This is so because the concentration of the reactants decreases ($dc_i < 0$) and we divide by $v_i < 0$. r is negative when the reaction runs backward.

Let us take a look at one final example of this, ammonia synthesis, which is very important to industrial applications. We assume a closed system at constant volume:

$$N_2 + 3\,H_2 \rightarrow 2\,NH_3 \quad \text{or} \quad 0 \rightarrow -1\,N_2 - 3\,H_2 + 2\,NH_3.$$

We then obtain for rate density r

$$r = -\frac{dc_{N_2}}{dt} = -\frac{1}{3}\frac{dc_{H_2}}{dt} = +\frac{1}{2}\frac{dc_{NH_3}}{dt}.$$

The procedures for heterogeneous reactions are less uniform. Depending upon the application, the conversion rate can be related to various quantities, for example, for processes related to membranes

- To the membrane surface,
- To the amount of bound enzyme there,
- To the amount of pores in the membrane,
- To the amount of available carrier molecules, etc.

In closing, a few words about the concept of "reaction rate." Because homogeneous reactions at almost constant volume—possibly a reaction in a beaker or flask where the volume of solution remains almost unchanged in a reaction process—often get the most attention, the difference quotient $\Delta c_i/\Delta t$ or the differential quotient dc_i/dt for a product i or the quotient $-\Delta c_i/\Delta t$ or $-dc_i/dt$ for a chosen reactant i is defined as the reaction rate. These quantities are closely related to our rate density, differing only by the factor $1/|v_i|$, but due to this difference they are dependent upon the substances and therefore not useful for our purposes. This quantity does not come into question as a general measure for conversion rate because, for example, in reactions between pure substances such as $Fe + S \rightarrow FeS$ the concentrations of all substances remain constant even though the amounts vary, or in reactions between gases the concentrations can vary as a result of compression or expansion alone without any conversion taking place at all. For this reason, we will avoid the ambiguous term "reaction rate" and use the quantities introduced above.

16.4 Measuring Rate Density

Introduction When making a kinetic investigation of a reaction, its stoichiometry must first be determined and possible side reactions must be identified. Because the rate density is proportional to temporal change of concentration of the substances involved in the reaction, at least when the reaction is homogeneous and the volume is constant, the concentrations of reactants and products must be determined at various times during the course of the reaction. Because the rates of disappearance and formation of chemical species are related to each other, the measurement of the change of concentration of one of the reactants or products suffices. The temperature of a reaction mixture must be kept constant during the entire time of reaction due to the temperature dependence of chemical reactions (think of the slowing of biochemical processes of food in a freezer).

As we have seen, chemical reactions take place in time intervals that range from a fraction of a second to millions of years. The speed interval to be considered is just as great. For this reason, the experimental methods for determining rate density can vary greatly from case to case. The basic problem in analyzing reacting systems is that their composition is constantly changing and we do not have unlimited time to carry out an analysis. Depending upon the speed at which a reaction runs and the method used for analysis, there are different approaches for acquiring kinetic data.

Slow Reactions When investigating a *slow* reaction, a small sample of the reacting mixture can be taken to instantly determine the concentration of the reaction partners chemically by, for example, titration or gravimetry. This is called *real-time analysis*. As an example, we will investigate the decomposition of trichloroacetic acid by decarboxylation into chloroform according to

$$CCl_3-COOH|w \rightarrow CCl_3H|g + CO_2|g.$$

(see Experiment 16.7).

Physical analytic methods such as optical or electric measurement methods can be used as well. One must always make certain, though, that the total volume of a mixture does not change much. The sampling and analysis must also take place quickly in relation to how long the conversion takes. If this is not possible, the reaction continuing to take place in the sample must be stopped in some way, possibly by diluting or cooling.

Direct measurement of concentration by utilizing some physical characteristic of the reaction mixture is much more practical than taking and analyzing small samples. If, for example, one of the reaction partners is a gas, the total volume can change during the reaction taking place in a container under constant pressure (possibly air pressure). The progress of the reaction can be followed by measuring the change in volume over time. Let us consider the reaction of zinc with an acid (Experiment 16.8), where hydrogen is produced and zinc is dissolved:

Experiment 16.7 *Measuring rate density by titration*: A trichloroacetic acid solution is poured into slightly alkalized boiling water. The concentration of the acid left in the reaction mixture after a certain time interval can be determined by titrating a sample with a sodium hydroxide solution. When the titrated solution is transferred into test tubes, the reaction process can be followed by observing the fill level. As the acid concentration decreases, so does the amount of sodium hydroxide that needs to be added.

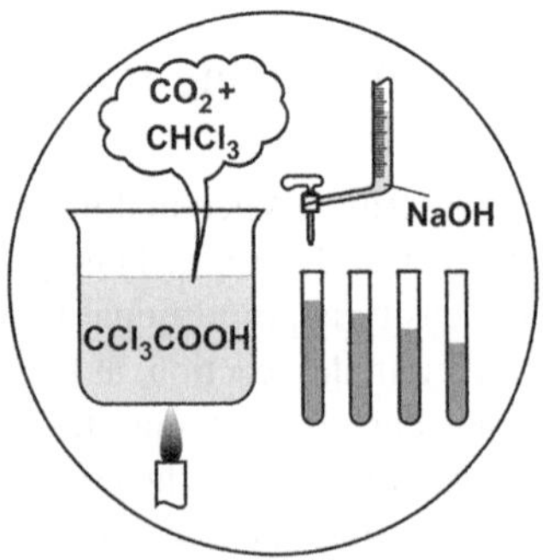

Experiment 16.8 *Volumetric determination of conversion*: Hydrochloric acid is trickled onto zinc granules. The hydrogen gas produced by this is caught pneumatically in a measuring cylinder or eudiometer. Water is used to trap the gas in the cylinder, and the graduation allows the change of gas volume to be measured.

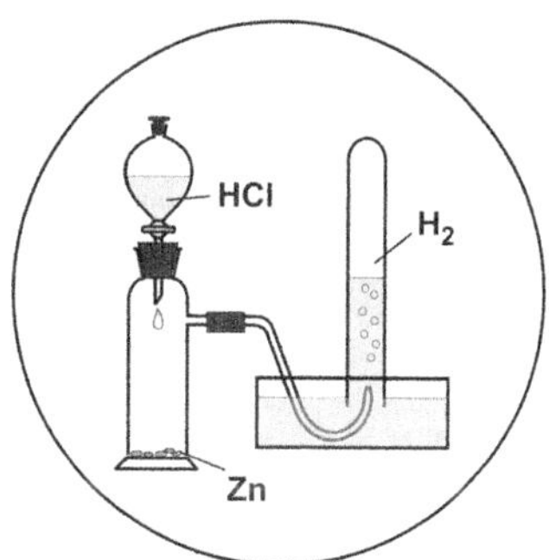

Experiment 16.9 *Conductometric determination of conversion*: A conductivity meter with a double platinum electrode is used for measuring conductivity. (We will go into conductivity and measuring it in more detail in Chap. 21.) Conductivity is temperature dependent so the use of a thermostat is recommended. To start the reaction, a known amount of tertiary butyl chloride is pipetted into the demineralized water in the measuring cell to start the reaction.

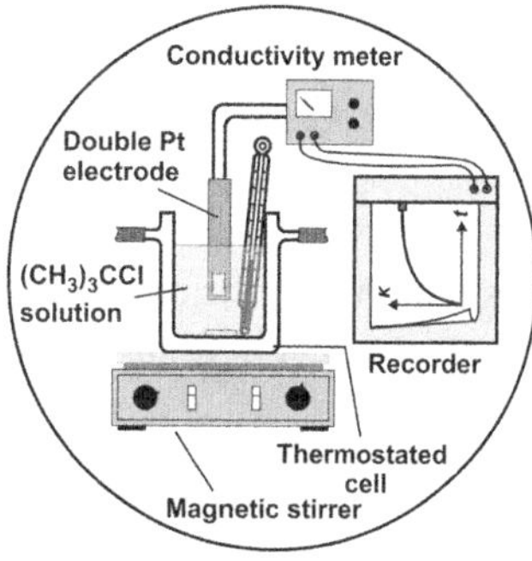

$$Zn|s + 2\,H^+|w \rightarrow Zn^{2+}|w + H_2|g.$$

The volume of the hydrogen gas can be measured easily.

Other useful characteristics for analysis are the pressure in a gas reaction at constant volume, the refraction index, or the electric conductivity. If a reaction changes the number or type of ions in a solution, its progress can be monitored utilizing conductivity. As a result of hydrolysis of tertiary butyl chloride for example, tertiary butanol is produced along with H^+ and Cl^- ions. The H^+ ions, in particular, strongly raise the solution's conductivity (Experiment 16.9):

$$(CH_3)_3C-Cl|l + H_2O|l \rightarrow (CH_3)C-OH|w + H^+|w + Cl^-|w.$$

A disadvantage of the methods mentioned above is their lack of specificity because, for one thing, all particles in the gas phase contribute to the change in volume. Therefore, *molecular specific* characteristics are better suited than characteristics related to the entire system. One method of measurement that is often applied in kinetics is *photometry*, the measurement of the absorption intensity of light in a given spectral range. The strength of absorption or transparency for light at a certain wavelength is a measure for the concentration of the reaction partner in

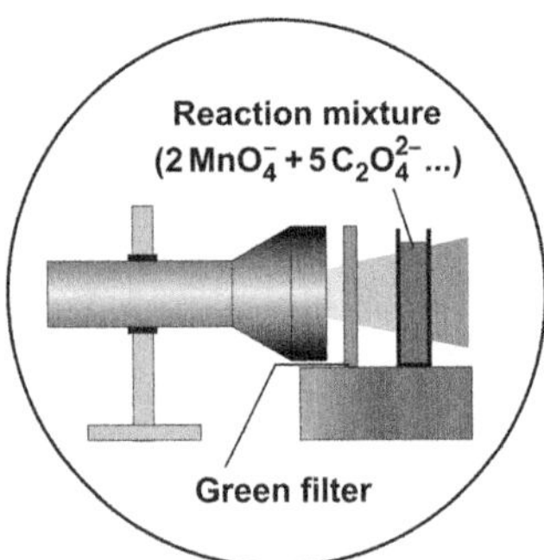

Experiment 16.10 *Photometric monitoring of the progress of a reaction*: A green filter is placed in front of a flashlight and the light is allowed to fall through a cuvette containing the reaction mixture. At first, the light can hardly be seen because it is almost totally absorbed by the intensely violet permanganate ions. Only as the reaction progresses, does the light eventually turn bright green. This is because the solution gradually loses its color. A spectrophotometer is necessary for quantitatively recording the changes of concentration.

question. When, for example, the reaction of a potassium permanganate solution with oxalic acid in a sulfuric acid solution is investigated (Experiment 16.10),

$$2\,MnO_4^-|w + 5\,C_2O_4^{2-}|w + 16\,H^+|w \rightarrow 2\,Mn^{2+}|w + 10\,CO_2|g + 8\,H_2O|l,$$

its progress can be followed by measurement of absorption in the visible range because the permanganate ion is colored.

Fast Reactions Chemists have lately become interested in especially fast reactions and great progress has been made in investigating them. We consider a reaction to be fast when the complete process takes less than about 1 s to occur. Special methods have been developed for these kinetic analyses. In the *flow method*, the reaction mixture does not remain in the reaction vessel over a longer time, as it does with the static methods discussed above, but flows through the reaction volume. The moment the reactants enter a mixing chamber, they are very quickly and thoroughly mixed. At that point the reaction starts; it progresses while the mixture is flowing through the outlet tube.

The distance covered by the reaction mixture in the tube is a measure of the time elapsed since the start of the reaction. If we observe the reaction at different positions along this tube using, for example, a moveable spectrophotometer, we see the mixture at different times of the reaction. The temporal coordinate of the reaction process is thereby mapped onto the spatial coordinate along the flow tube. When highly efficient mixing chambers are used, the flow tube technique is applicable for fast reactions with reaction times down to the millisecond range.

A disadvantage of the flow method is that relatively large volumina of the reaction mixture are necessary. Consumption of substances is especially great in very fast reactions because the flow rate needs to be very high so that the reaction process can be spread over a long enough length of the outlet tube. The "*stopped-flow method*" lets us avoid this disadvantage. Here, as well, the reactants are mixed very quickly and then let flow into a flow tube. However, this flow tube contains a piston that can abruptly stop the flow as soon as a given volume (mostly around 1 cm^3) is injected. The reaction then continues in the resting, well-mixed solution and can be followed from outside spectrometrically. The filling of the observation chamber corresponds to the sudden taking of a small initial sample of the reaction mixture, so the "stopped-flow" technique is much more economical than the flow method. It is especially useful in investigating biochemical reactions.

When reactions with durations of conversion shorter than 10^{-3} s are to be investigated, the *mixing methods* we have introduced no longer apply. *Relaxation methods* let us avoid the time-consuming mixing of reaction partners. Instead, they allow us to observe how a system in equilibrium reacts to an external perturbation of equilibrium. If, for example, parameters such as pressure or temperature are suddenly changed, a chemical reaction must take place in order to once again establish equilibrium. This time, however, pressure or temperature takes new values (pressure-jump or temperature-jump method). The reaction's return to equilibrium, called relaxation, can be followed spectroscopically.

16.5   Rate Laws of Single-Step Reactions

Basic Principles When investigated in detail, most chemical reactions can be broken down into different steps occurring either in sequence or in parallel. These kinds of reactions are called *multistep* or *composite* reactions. The smallest units involved in such an analysis are called *single-step* or *elementary* reactions. Such reactions take place in one single step. This means that all the particles appearing in a conversion formula react *simultaneously* with each other. We use the word *molecularity* to indicate the number of particles of reactants involved in a single-step reaction. When one, two, three, … particles interact, we speak of *mono-*, *di-*, *tri-*, … *molecular* reactions.

For the beginning, it will suffice to limit our investigation to only simple types of homogeneous reactions. We are most interested in what influence

- The concentrations (and types) of reaction partners B, B′, … and
- The temperature, as well as
- The presence and types of substances not appearing in the conversion formula (catalysts, inhibitors, solvents)

have upon the rate density r. Let us take a closer look at concentration dependency using the example of decoloration of a potassium permanganate solution by oxalic acid in a sulfur acid solution at various dilutions (Experiment 16.11). (The reaction itself was presented already in Experiment 16.10.)

The higher the dilution, meaning the lower the concentration of reactants, the more slowly the reaction proceeds. Obviously, the rate density depends directly upon the concentration.

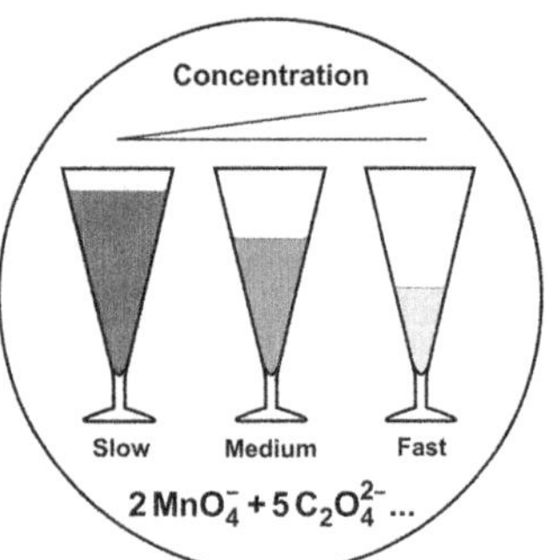

Experiment 16.11 *Concentration dependency of rate of conversion*: The same amount of acidulated oxalic acid is poured into each of three conical cups (goblets). Potassium permanganate solution along with a larger amount of water is added to the first goblet. The same amount of potassium permanganate solution is added to the second goblet but with less water. The third goblet receives only potassium permanganate solution. The three solutions decolorize from violet to wine red and then yellow brown until they are colorless. At the same time, some bubble formation due to the generation of carbon dioxide can be observed. The decolorizing appears just after a short time in the third goblet, while the reaction in the first goblet takes the most time.

The equation $r = f(c_B, \ldots, T)$ describing the relations above is called the *rate law* of the reaction. Most reactions take place in many steps, so the various dependencies can be rather complicated. However, under certain circumstances (certain ranges of concentration, moving away from equilibrium, etc.), there are simple relations that allow us to write the rate law in the form of a power function,

$$r = k(T) \cdot c_B^b \cdot c_{B'}^{b'} \ldots \,, \tag{16.8}$$

where the concentrations of reactants are usually found on the right-hand side (in rare cases, those of a product or another type of substance not found in the conversion formula are used). We call the exponents $b, b', \ldots$ the *order* of a reaction with respect to the individual reactants B, B', …. . Usually, the numbers 1 and 2 appear as exponents, rarely fractions like $\frac{1}{2}$ or $\frac{1}{3}$ and even more rarely negative numbers. If, for example, the exponent is $b = 1$, the reaction is said to be first order with respect to substance B. If $b = 2$ or $b = 0$, which can also occur, we speak of second- or zero-order reactions with respect to B. Moreover, it is usual to introduce an *overall order* of a reaction which is given by the sum $b + b' + \ldots$ of the exponents of all concentrations present. It is important to note that the exponents often do *not* agree with the conversion numbers. There are several relations, such as the mass action law where conversion numbers actually appear as exponents so that they might also be expected to appear in the rate law. However, this is not generally the case.

The proportionality factor, i.e., the *rate coefficient k*, generally depends strongly upon the conditions of the reaction, especially temperature, but also upon the types of reactants and the reaction medium. The usual name "*rate constant*" does not seem appropriate here because k is not a constant. The units of k are dependent upon the form of the rate law, as we will see.

Finally, we will take a brief look at the relation between the order of a reaction and the molecularity mentioned above. Reaction orders are experimentally determined quantities while molecularity (of a reaction step) is a theoretical quantity essential for the elucidation of a reaction mechanism. In single-step reactions, molecularity and reaction order (as well as conversion number sum) agree with each other because all the particles react *simultaneously* with each other according to their appearance in the conversion formula. Conversely, it is not necessarily possible to infer the molecularity of an arbitrary reaction from its order. This is because in complex reaction processes made up of several single-step reactions, simple rate laws might still be valid.

We will limit the following to homogeneous single-step reactions in a closed container at constant volume. Side reactions and exchange of substances with the environment are excluded (ξ', ξ'', $\ldots = \xi_{\text{other}} = \text{const}$).

First-Order Reactions   Consider a single-step reaction

$$B \rightarrow \text{products},$$

which comes about through random transformations of individual particles (meaning their decomposition or internal rearrangement) and is therefore *monomolecular*. Assuming that almost no backward (reverse) reaction takes place, it obeys a rate law in which the rate density r is directly proportional to the concentration c of the substance B at the time t:

$$r = k \cdot c_B. \tag{16.9}$$

This simply means that the more particles there are, the greater the number that transform per unit time. In this case, the rate coefficient k has the unit s^{-1} and therefore the characteristic of a decay frequency. Because the exponent of the concentration c_B equals 1, we have a reaction of first order in B. Moreover, the overall order of the reaction is equal to 1. Under the conditions that apply here ($V = \text{const}$, $\xi_{\text{other}} = \text{const}$), we also have (see Sect. 16.3):

$$r = -\frac{dc_B}{dt}. \tag{16.10}$$

In summary, we obtain the differential rate law

$$-\frac{dc_B}{dt} = kc_B. \tag{16.11}$$

The rate density at a given time corresponds to the (negative) slope of the experimentally determined $c_B(t)$ curve (Fig. 16.4). The rate density at the beginning of the reaction ($t = 0$; initial concentration $c_{B,0}$) is at its maximum value and approaches zero as substance B is used up.

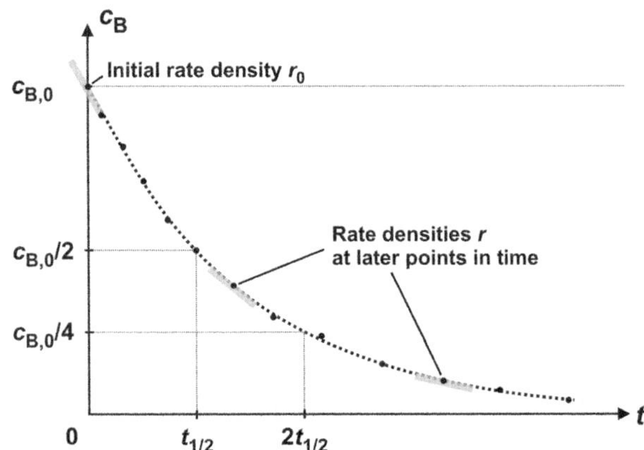

Fig. 16.4 Decrease in concentration of the reactant over time in a first-order reaction. The rate density at a given time can be determined from the slope of the tangent (*light gray*). $t_{1/2}$ illustrates the half-life (the time it takes to reduce the concentration of the reactant to half of its initial value).

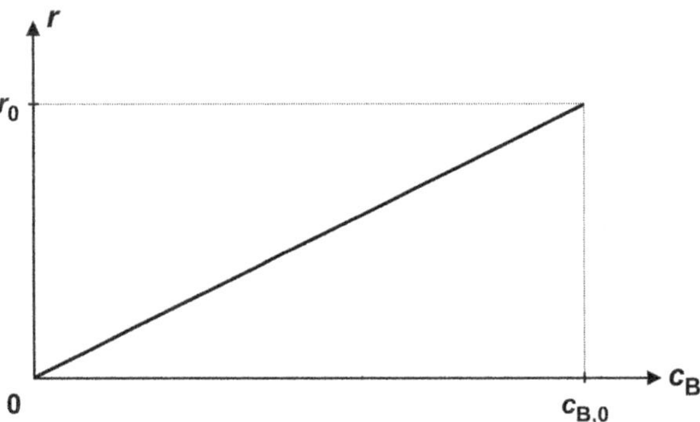

Fig. 16.5 Rate density as a function of the corresponding concentration of substance B.

If the rate densities r are plotted as functions of the concentration c_B (Fig. 16.5), we obtain straight lines as can be expected according to Eq. (16.9). The slope corresponds to the rate coefficient k.

Rate densities are, however, rarely determined directly because slopes can only be determined inexactly. It is, therefore, desirable to know the mathematical relation between the measureable quantities, i.e., concentration and time. Important parameters such as rate coefficient and half-life can be calculated using them. Moreover, if we have a relation for the concentration as a function of time, and if the initial concentration $c_{B,0}$ is given, we can predict the concentration of the substance for any point in time. This is of great importance for industrial processes.

The starting point of our considerations is Eq. (16.11) which, mathematically speaking, is a so-called differential equation. It is necessary to find a function $c_B(t)$ that satisfies this equation. The following relation shows a solution,

$$\ln\frac{c_{B,0}}{c_B} = kt, \tag{16.12}$$

that establishes the desired connection between concentration and time. An alternative form of Eq. (16.12) is obtained by applying the quotient rule of logarithms [see Eq. (A.1.2) in the Appendix]. However, it should be kept in mind that the argument of a logarithmic function must be dimensionless. This means that we have to divide by an arbitrary reference value $c^\dagger$ having the same dimension as c_B (e.g., the standard concentration $c^\ominus = 1\,\mathrm{kmol\,m^{-3}}$):

$$\ln(c_B/c^\dagger) - \ln(c_{B,0}/c^\dagger) = -kt.$$

In order to avoid unnecessarily complicated looking formulas, we will indicate division by a reference value with curly brackets from now on; we then obtain

$$\ln\{c_B\} = \ln\{c_{B,0}\} - kt. \tag{16.13}$$

An alternative way of expressing this is

$$c_B = c_{B,0} \cdot e^{-kt}. \tag{16.14}$$

The solution's correctness can be easily confirmed by taking the derivative:

$$\frac{dc_B}{dt} = -c_{B,0} \cdot k e^{-kt} = -k c_B.$$

This corresponds to the differential equation (16.11) from which we started.

Solving a differential equation—finding a formula for the desired function—simply means transforming it so that all the derivatives disappear. This requires using integral calculus. When we do this, we speak of the *integrated rate law*. Let us consider this method step by step. First, we "separate" the variables in the initial equation

$$-\frac{dc_B}{dt} = k c_B;$$

this means we rearrange the equation so that all terms with the independent variable appear on one side and all terms with the dependent variable on the other,

$$-\frac{1}{c_B}\, dc_B = k dt.$$

We then integrate on both sides of the equation between the limits $t = 0$ with corresponding initial concentration $c_{B,0}$, and an arbitrary later time t with corresponding concentration c_B:

$$-\int_{c_{B,0}}^{c_B} \frac{1}{c_B}\, dc_B = k \int_0^t dt.$$

Here, the following elementary indefinite integral that we have already applied in Sect. 5.5 serves us well:

$$\int \frac{1}{x}\, \mathrm{d}x = \ln x + \text{constant}.$$

This way we can directly obtain Eq. (16.12),

$$\ln \frac{c_{B,0}}{c_B} = kt,$$

which is the desired integrated rate law.

Equation (16.14) illustrates a characteristic of first-order reactions: the concentration of the reactant decreases *exponentially* with time (compare dashed curve in Fig. 16.4). The paradigm for this kind of process is radioactive decay, but all monomolecular elementary reactions such as the rearrangement of cyclopropane into propane in the gas phase are included in this. There are many further decomposition reactions to be found in "classical chemistry," such as decomposition of dinitrogen pentaoxide N_2O_5 in the gas phase according to

$$2\,N_2O_5|g \rightarrow 4\,NO_2|g + O_2|g$$

that follow a first-order rate law even if they proceed according to a complex mechanism, i.e., even if they are not monomolecular reactions.

With the help of relation (16.14) we can check whether we are actually dealing with a first-order reaction. However, the logarithmic relation (16.13) is more suitable in this case. If we do have a first-order reaction, we will obtain a straight line when $\ln\{c_B\}$ is plotted as a function of t (Fig. 16.6). Its slope gives us the rate coefficient k.

Another important quantity characterizing the rate of a reaction is the *half-life* $t_{1/2}$ (Fig. 16.4). It gives the time that elapses until the concentration of the reactant has been reduced by half of its initial value. Now, if $c_B = c_{B,0}/2$ and $t = t_{1/2}$ are inserted into Eq. (16.12), it follows that

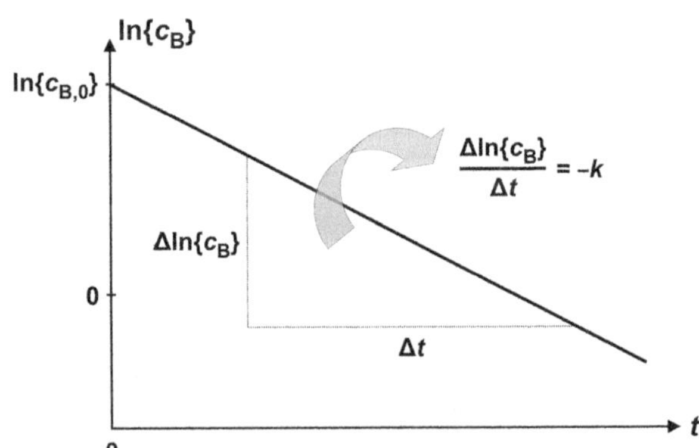

Fig. 16.6 Determining the rate coefficient k of a first-order reaction by drawing the logarithm of the concentration of the reactant as a function of time.

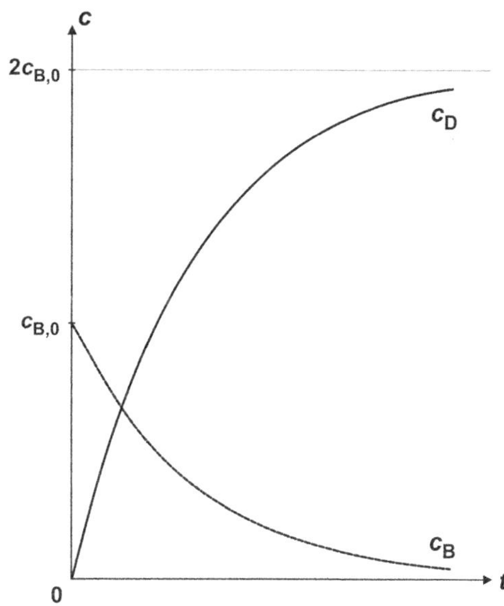

Fig. 16.7 Temporal change of concentration of reactant and product in a first-order reaction of type $B \to 2\,D$.

$$t_{1/2} = \frac{\ln 2}{k}. \tag{16.15}$$

The half-life for a first-order reaction is therefore independent of the initial concentration of the reactant.

There is an increase in the concentration of the reaction product that corresponds to the exponential decrease of the reactant concentration. It is equal to the product of the ratio of the conversion numbers and the decrease of concentration $(c_{B,0} - c_B)$. Let us consider the reaction

$$B \to 2\,D$$

(see Fig. 16.7). The concentration c_D of the reaction product D is:

$$c_D = 2(c_{B,0} - c_B).$$

When Eq. (16.14) is used instead of c_B, it follows that:

$$c_D = 2\left(c_{B,0} - c_{B,0}e^{-kt}\right) = 2c_{B,0}\left(1 - e^{-kt}\right). \tag{16.16}$$

When the reaction process has completely run its course, the concentration will be $c_D = 2c_{B,0}$ at time $t = \infty$.

Second-Order Reactions Single-step reactions

$$B + B' \to \text{products},$$

which result from random encounters of *two* particles, are called *bimolecular* reactions. They follow a rate law in which r is proportional to *both* concentrations c_B and $c_{B'}$:

$$r = k \cdot c_B \cdot c_{B'}. \tag{16.17}$$

This relation becomes clear when one considers that two particles of the types B and B$'$ will encounter each other more often, the more B there is and the more B$'$ is present. A rate law for the overall order of 2 is valid here. The rate coefficient k now has the unit $m^3\,mol^{-1}\,s^{-1}$.

Our analysis becomes simpler when the following conversion formula:

$$2\,B \rightarrow products$$

is taken as the basis. We then obtain

$$r = k \cdot c_B^2. \tag{16.18}$$

Taking the conversion number $v_B = -2$ into account, the following holds for the rate density:

$$r = -\frac{1}{2}\frac{dc_B}{dt}. \tag{16.19}$$

The differential equation to be considered is then

$$-\frac{dc_B}{dt} = 2k\,c_B^2. \tag{16.20}$$

The relation

$$\frac{1}{c_B} = \frac{1}{c_{B,0}} + 2kt \tag{16.21}$$

is a solution for this equation. Solving for c_B yields:

$$c_B = \frac{c_{B,0}}{1 + 2kt c_{B,0}}. \tag{16.22}$$

In order to find the integrated rate law, we proceed analogously to what we do with first-order reactions. Starting with the differential equation (16.20), we first separate the variables:

$$-\frac{dc_B}{c_B^2} = 2kdt.$$

In order to solve the equation, we will need the following standard integral (compare Sect. A.1.3 in the Appendix):

$$\int \frac{1}{x^2}\,dx = -\frac{1}{x} + \text{constant}.$$

The integration limits correspond to those of the first-order reaction, resulting in

$$-\int_{c_{B,0}}^{c_B} \frac{dc_B}{c_B^2} = 2k\int_0^t dt.$$

After integrating, we obtain

$$\frac{1}{c_B} - \frac{1}{c_{B,0}} = 2kt,$$

from which, after transformation, Eq. (16.21) results.

When c_B is plotted as a function of t (Fig. 16.8), we notice that for the same initial concentration $c_{B,0}$ and the same initial rate, the curve approaches zero much more slowly than it would in a first-order reaction.

In order to verify that we are really dealing with a second-order reaction, we plot $1/c_B$ as a function of t (Fig. 16.9). According to Eq. (16.21), this must result in a straight line from whose slope the rate coefficient can be determined.

We can obtain the half-life $t_{1/2}$ again by inserting the value $c_{B,0}/2$ for c_B:

$$t_{1/2} = \frac{1}{2kc_{B,0}}. \tag{16.23}$$

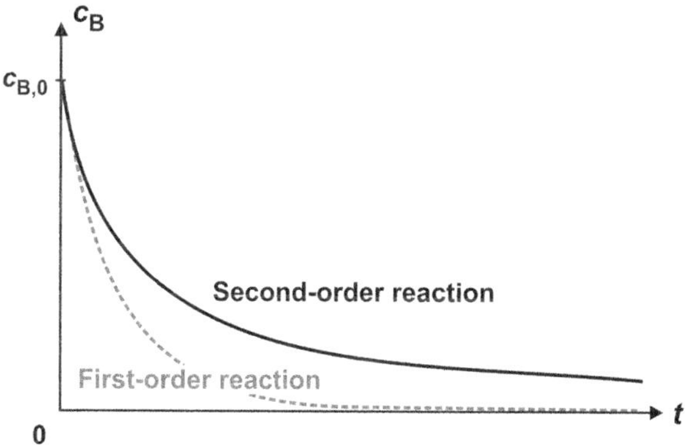

Fig. 16.8 Temporal progression of a second-order reaction (compared to a first-order reaction).

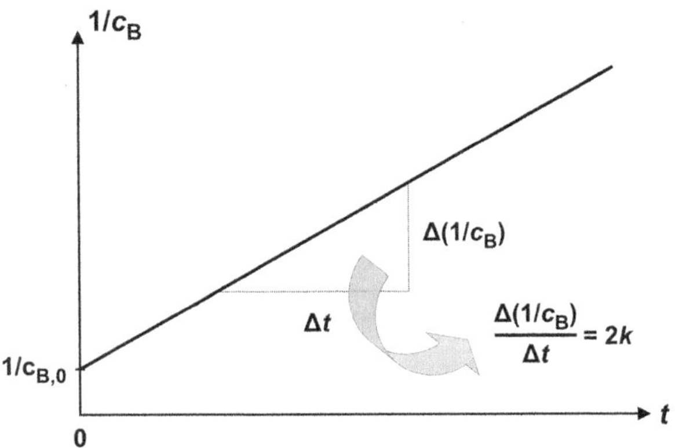

Fig. 16.9 Determining the rate coefficient k of a second-order reaction by representing $1/c_B$ as a function of t.

In contrast to a first-order reaction, the half-life depends upon the initial concentration of the reactant and is *not* characteristic for the reaction.

The general case

$$B + B' \rightarrow \text{products}$$

results in very similar relations when the initial concentrations $c_{B,0}$ and $c_{B',0}$ are identical, and both reaction partners participate equally in the reaction. The conversion number ν_B now equals -1, so the factor 2 drops out from all equations and we obtain

$$\frac{1}{c_B} = \frac{1}{c_{B,0}} + kt \tag{16.24}$$

for the integrated rate law. The half-life equals

$$t_{1/2} = \frac{1}{k\,c_{B,0}}. \tag{16.25}$$

If the initial concentrations of reactants B and B' are not identical, integration will require a partial fraction decomposition. This is why only the final result will be shown (for the sake of completeness):

$$\ln\frac{c_B \cdot c_{B',0}}{c_{B'} \cdot c_{B,0}} = (c_{B,0} - c_{B',0})kt. \tag{16.26}$$

Second-order reactions are relatively common. We could mention the reaction between hydrogen and iodine to form hydrogen iodide or the decomposition of nitrogen dioxide according to

$$2\,NO_2 \rightarrow 2\,NO + O_2$$

as examples of reactions in the gas phase. In addition, numerous reactions in solution such as the alkaline ester saponification

$$CH_3COOC_2H_5 + OH^- \rightarrow CH_3COO^- + C_2H_5OH$$

obey this rate law as well.

There are rate laws corresponding to *tri-* or *higher molecular reactions,* in which *three* or *more* particles must encounter each other. However, these types of reactions are so rare that we do not really need to go into them separately.

Zero-Order Reactions   Zero-order reactions have a rate that is independent of the concentration of the reactant(s), i.e., they are characterized by a constant rate density:

$$r = k. \tag{16.27}$$

The decrease of the reactant's concentration is then described by

$$-\frac{dc_B}{dt} = k. \tag{16.28}$$

The solution to this differential equation is:

$$c_B = c_{B,0} - kt. \tag{16.29}$$

Integration observing the known limits,

$$-\int_{c_{B,0}}^{c_B} dc_B = k \int_0^t dt,$$

yields the following integrated rate law:

$$c_B - c_{B,0} = -kt.$$

Such behavior is not possible in single-step reactions alone, but only when a process that acts as a kind of "bottleneck" precedes or follows the actual reaction, a process which runs at a constant rate or serves to keep the concentration constant. These kinds of processes can be:

- Adsorption or desorption processes such as those playing a role in heterogeneous catalysis,
- Diffusion processes,
- Radiation with constant light intensity in photochemical reactions,
- Dissolution processes.

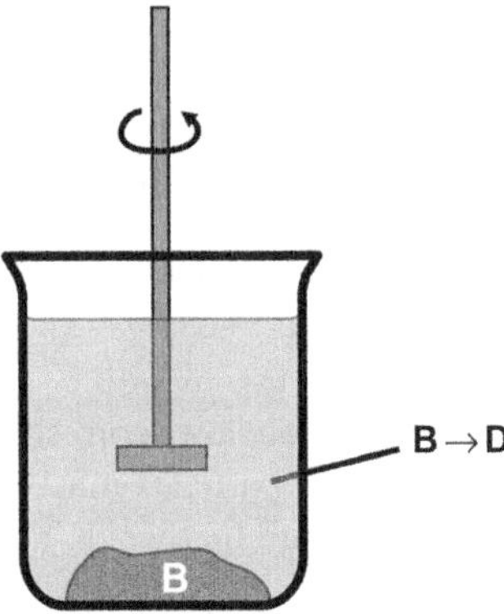

Fig. 16.10 Decomposition reaction of a reactant B in a saturated solution with excess solid solute as example of a (pseudo-) zero-order reaction.

Consider, for example, the decomposition of a reactant B in a saturated solution with excess solid solute (Fig. 16.10). The actual decomposition reaction is preceded by the solubility equilibrium

$$B|s \rightleftharpoons B|d$$

[where the dissolved form is characterized by the abbreviation |d (lat. dissolutus)]. This always keeps the concentration c_B constant:

$$r = k \cdot c_B (= \text{const.}) = k'. \tag{16.30}$$

This concentration can be combined with the actual rate coefficient k to produce a new rate coefficient k' so that the order is lowered. In this case, the term *pseudo-order* may be used.

A pseudo-order can occur when reactions take place in dilute solutions where the solvent (possibly water) simultaneously functions as the reaction partner. Because there is a great excess of it, the solvent concentration remains almost constant in comparison to the other substances and can, in turn, be included in the rate coefficient.

Chapter 17
Composite Reactions

Kinetic measurements show that the simple rate laws known from the last chapter are often not sufficient for a correct description of the temporal course of a reaction or the composition of a reaction mixture. Many reactions take place by mechanisms that involve several elementary steps. Three fundamental types of composite reactions are discussed in this chapter: opposing or equilibrium reactions, parallel reactions, and consecutive reactions. Composite reactions not only play a large role in industrial applications (e.g., heterogeneous catalysis) but are also very important in nature (e.g., enzyme reactions).

17.1   Introduction

Kinetic investigations have shown that the simple rate laws we have learned up until now are often not enough for describing a reaction process over time or the composition of a reaction mixture. This tells us that even reactions that can be described by simple overall conversion formulas often proceed according to more complicated mechanisms. These complex reactions can be divided into three basic types:

- Opposing or equilibrium reactions,
- Parallel reactions (also called competitive or side reactions),
- Consecutive reactions.

These basic types can also appear in combinations, such as consecutive reactions with preceding equilibrium.

Composite reactions not only play a large role in industrial applications (e.g., heterogeneous catalysis) but are also important in nature (e.g., enzyme reactions).

In the following derivations, we will limit ourselves to processes where all the substances are distributed homogeneously in a region whose volume is more or less constant.

© Springer International Publishing Switzerland 2016
G. Job, R. Rüffler, *Physical Chemistry from a Different Angle*,
DOI 10.1007/978-3-319-15666-8_17

17.2   Opposing Reactions

Until now we have assumed that in a reaction

$$\mathrm{B} + \mathrm{B}' + \ldots \xrightarrow{+1} \mathrm{D} + \mathrm{D}' + \ldots$$

the reactants B, B', … are always completely converted into the products D, D', …. However, homogeneous reactions are never complete, but run only until the chemical potentials of both the reactants and products are equal. Because $\mu(c)$ tends toward $-\infty$ for $c \to 0$, neither of the substances can totally disappear. This can be interpreted as follows: in a *backward* (also called *reverse*) *reaction* (–1) belonging to the *forward reaction* (+1) introduced above, the substances B, B', … can be produced from the substances D, D', … :

$$\mathrm{D} + \mathrm{D}' + \ldots \xrightarrow{-1} \mathrm{B} + \mathrm{B}' + \ldots .$$

The total rate density r which can be observed macroscopically is a result of the difference of rate densities of the forward and backward reactions:

$$r = r_{+1} - r_{-1}. \tag{17.1}$$

In elementary reactions, we expect the following rate laws for the rate densities of the forward and backward reactions:

$$r_{+1} = k_{+1} \cdot c_\mathrm{B} \cdot c_{\mathrm{B}'} \cdots \quad \text{and} \quad r_{-1} = k_{-1} \cdot c_\mathrm{D} \cdot c_{\mathrm{D}'} \cdots , \tag{17.2}$$

so that we obtain:

$$r = k_{+1} \cdot c_\mathrm{B} \cdot c_{\mathrm{B}'} \cdot \cdots - k_{-1} \cdot c_\mathrm{D} \cdot c_{\mathrm{D}'} \cdot \ldots . \tag{17.3}$$

k_{+1} and k_{-1} are the rate coefficients for the forward and backward reaction, respectively. At the beginning of the reaction, the rate density is much higher in the forward direction because there is little or no end product present. However, it falls along with decreasing concentration of the reactants. At the same time, the rate density in the backward direction increases along with the continuously increasing concentrations of the products. The total rate density continues to drop until it reaches a value of zero:

$$r = k_{+1} \cdot c_{\mathrm{B},\mathrm{eq.}} \cdot c_{\mathrm{B}',\mathrm{eq.}} \cdot \ldots - k_{-1} \cdot c_{\mathrm{D},\mathrm{eq.}} \cdot c_{\mathrm{D}',\mathrm{eq.}} \cdot \ldots = 0. \tag{17.4}$$

In this state of equilibrium, both forward and backward reactions occur at the same rate. In this case, we speak (incorrectly) of a *dynamic* (instead of the correct *kinetic*) *equilibrium* (compare Sect. 6.3, last paragraph). Outwardly, the total reaction has ceased and the equilibrium concentrations no longer change. Because of

$$k_{+1} \cdot c_{B,eq.} \cdot c_{B',eq.} \cdots = k_{-1} \cdot c_{D,eq.} \cdot c_{D',eq.} \cdots$$

we have

$$\frac{c_{D,eq.} \cdot c_{D',eq.} \cdots}{c_{B,eq.} \cdot c_{B',eq.} \cdots} = \frac{k_{+1}}{k_{-1}} = \overset{\circ}{K}_c \tag{17.5}$$

in equilibrium.

The ratio of the rate coefficients corresponds to the conventional equilibrium constant that we were introduced to in Sect. 6.4. Therefore, Eq. (17.5) is nothing more than the mass action law derived on the basis of another concept than what we used so far. Such simple kinetic derivations are only possible for elementary reactions or reactions that obey at least a rate law of the corresponding order. There is no such limitation, though, for the matter dynamical derivation [Eqs. (6.19) and (6.21)].

If the rate coefficient of the forward is much higher than that of the backward reaction ($k_{+1} \gg k_{-1}$) and therefore $\overset{\circ}{K} \gg 1$, equilibrium is strongly shifted toward the side of the products. Conversely, if $k_{+1} \ll k_{-1}$ and therefore $\overset{\circ}{K} \ll 1$, it will strongly favor the reactants.

The reaction of hydrogen with iodine to form hydrogen iodide,

$$H_2 + I_2 \rightleftharpoons 2\,HI,$$

which was investigated in detail by the physical chemist Max Bodenstein in 1894, can be described as an opposing reaction. Correspondingly, its equilibrium constant can be determined by measuring the formation and decomposition rates of the hydrogen iodide (although this is not a bimolecular reaction but a reaction mechanism of greater complexity).

When investigating the time dependencies of concentrations, we will limit ourselves to the simplest case, a reaction of the type

$$B \overset{+1}{\underset{-1}{\rightleftharpoons}} D$$

in which both the forward and backward reactions obey a first-order rate law. Some examples of such equilibrium reactions are isomerizations like the transition of α-D-glucose into β-D-glucose in an aqueous solution (Sect. 6.3). The following is valid for the forward reaction:

$$r_{+1} = k_{+1}c_B, \tag{17.6}$$

and correspondingly, for the backward reaction:

$$r_{-1} = k_{-1}c_D. \tag{17.7}$$

If we assume that at time $t = 0$, the initial concentration of the reactant B is $c_{B,0}$, and that there is no product, meaning $c_D = 0$, then stoichiometry tells us that for every moment during the reaction,

$$c_B + c_D = c_{B,0}. \tag{17.8}$$

For the decrease of concentration of B, we obtain

$$-\frac{dc_B}{dt} = r = r_{+1} - r_{-1} = k_{+1}c_B - k_{-1}c_D \tag{17.9}$$

or after inserting c_D according to Eq. (17.8)

$$-\frac{dc_B}{dt} = k_{+1}c_B - k_{-1}(c_{B,0} - c_B) = (k_{+1} + k_{-1})c_B - k_{-1}c_{B,0}. \tag{17.10}$$

Calculating the integrated rate law is rather complicated, so only the end result will be given here (it can be easily proven by taking the derivative):

$$c_B = \frac{k_{-1} + k_{+1} \cdot e^{-(k_{+1}+k_{-1})t}}{k_{+1} + k_{-1}}c_{B,0}. \tag{17.11}$$

The following is then valid for c_D:

$$c_D = c_{B,0} - c_B = c_{B,0} - \frac{k_{-1} + k_{+1} \cdot e^{-(k_{+1}+k_{-1})t}}{k_{+1} + k_{-1}}c_{B,0} \quad \text{or}$$

$$c_D = \frac{k_{+1}\left(1 - e^{-(k_{+1}+k_{-1})t}\right)}{k_{+1} + k_{-1}}c_{B,0}. \tag{17.12}$$

The behavior over time of the concentrations of both reaction partners is determined by both the rate coefficients of the forward and the backward reaction. In a graphic representation (Fig. 17.1), our expectations concerning the form of the functions are fulfilled. As time progresses, the concentrations increasingly approach their equilibrium values. We can determine what these are by finding the limit for $t \to \infty$, taking into account that $e^{-x} \to 0$ is valid for $x \to \infty$:

$$c_{B,eq.} = \frac{k_{-1}}{k_{+1} + k_{-1}}c_{B,0} \tag{17.13}$$

and

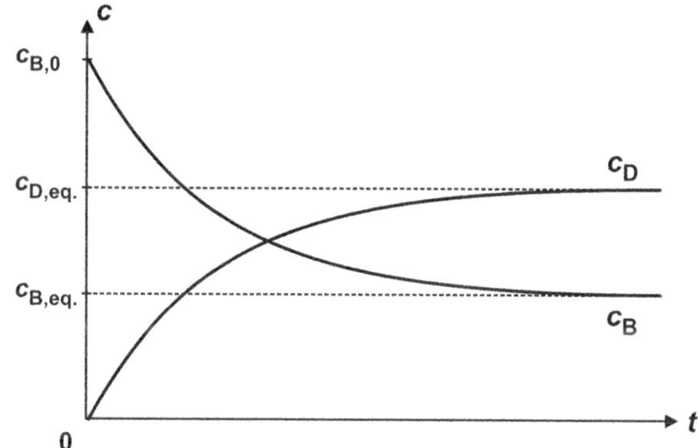

Fig. 17.1 Time dependence of substance concentrations in opposing reactions.

$$c_{D,eq.} = \frac{k_{+1}}{k_{+1} + k_{-1}} c_{B,0}. \tag{17.14}$$

The form of the functions becomes clearer when we rearrange Eq. (17.12) with the help of Eq. (17.14) and abbreviate the sum $(k_{+1} + k_{-1})$ to k. We then obtain:

$$c_D = c_{D,eq.} - \underbrace{c_{D,eq.} \cdot e^{-kt}}_{\Delta c_D(t)}.$$

The term $\Delta c_D(t)$ describes the difference between the instantaneous concentration c_D and the final value $c_{D,eq.}$: it decreases exponentially with a half-life of $t_{1/2} = \ln 2/k$.

For substance B, we correspondingly have:

$$c_B = c_{B,eq.} + \underbrace{\left(c_{B,0} - c_{B,eq.}\right) \cdot e^{-kt}}_{\Delta c_B(t)}.$$

The difference between the instantaneous concentration c_B and the final value $c_{B,eq.}$ also decreases exponentially and with the same half-life as in the case of D. The two curves are therefore mirror images of each other.

If the rate coefficient of the forward reaction is noticeably higher than that of the backward reaction, meaning $k_{+1} \gg k_{-1}$, then Eq. (17.10) transforms into the familiar Eq. (16.11) for first-order reactions. In this case, the equilibrium reaction can be described by a simple rate law. Moreover, at the beginning of the equilibrium reaction, when it is far away from equilibrium, the reaction can be treated as one-sided process because the rate of the backward reaction is very small due to the very low product concentration.

17.3 Parallel Reactions

Identical reactants can undergo simultaneous reactions leading to different products. These kinds of reactions are called *parallel*, *competitive*, or *side reactions*.

We will look at the simple case of two parallel monomolecular elementary reactions 1 and 2:

$$B \overset{1}{\underset{2}{\diagdown}} \begin{matrix} D \\ \\ D' \end{matrix} \quad .$$

The individual rate densities are

$$r_1 = k_1 c_B \quad \text{and} \quad r_2 = k_2 c_B, \tag{17.15}$$

and collectively, therefore:

$$r = r_1 + r_2 = k_1 c_B + k_2 c_B. \tag{17.16}$$

The decrease in concentration of B can be described by

$$-\frac{dc_B}{dt} = r = k_1 c_B + k_2 c_B = (k_1 + k_2) c_B = k c_B, \tag{17.17}$$

where the rate coefficients k_1 and k_2 can combine into one coefficient k. By integrating, we obtain analogously to Eq. (16.4),

$$c_B = c_{B,0} e^{-kt} = c_{B,0} e^{-(k_1 + k_2) t}. \tag{17.18}$$

If k is the same, the change of concentration of the reactants is independent of the number of products formed.

The formation of the product D satisfies:

$$\frac{dc_D}{dt} = r_1 = k_1 c_B. \tag{17.19}$$

Inserting Eq. (17.18) yields:

$$\frac{dc_D}{dt} = k_1 c_{B,0} e^{-kt}. \tag{17.20}$$

After separating the variables and then integrating while taking the initial condition $c_D = 0$ at $t = 0$ into account, we obtain:

$$\int\limits_{0}^{c_{\mathrm{D}}} c_{\mathrm{D}}\mathrm{d}c_{\mathrm{D}} = k_1 c_{\mathrm{B},0} \int\limits_{0}^{t} \mathrm{e}^{-kt}\mathrm{d}t.$$

While the calculation on the left-hand side of the equation does not pose much difficulty, integrating the nested function on the right needs a bit more skill. The result for the change of concentration over time of product D is:

$$c_{\mathrm{D}} = \frac{k_1}{k} c_{\mathrm{B},0}\left(1 - \mathrm{e}^{-kt}\right), \tag{17.21}$$

and correspondingly for product D′:

$$c_{\mathrm{D}'} = \frac{k_2}{k} c_{\mathrm{B},0}\left(1 - \mathrm{e}^{-kt}\right). \tag{17.22}$$

In the case of the integral

$$\int\limits_{0}^{t} \mathrm{e}^{-kt}\mathrm{d}t$$

it is a good idea to choose the function nested inside as the new variable (*substitution rule*) (Appendix A1.3):

$$g(t) = -kt = z.$$

By taking the derivative we obtain

$$g'(t) = \frac{\mathrm{d}z}{\mathrm{d}t} = -k.$$

The integration limits must be adapted to correspond to the substitution so that, in the end

$$\int\limits_{0}^{z} \mathrm{e}^{z} \cdot -\frac{1}{k}\mathrm{d}z = -\frac{1}{k}\int\limits_{0}^{z} \mathrm{e}^{z}\mathrm{d}z = -\frac{1}{k}(\mathrm{e}^{z} - 1)$$

is the result, whereby the standard integral

$$\int \mathrm{e}^{x}\mathrm{d}x = \mathrm{e}^{x} + \text{constant}$$

has served us well.

Figure 17.2 graphically illustrates the time dependency of the individual substance concentrations.

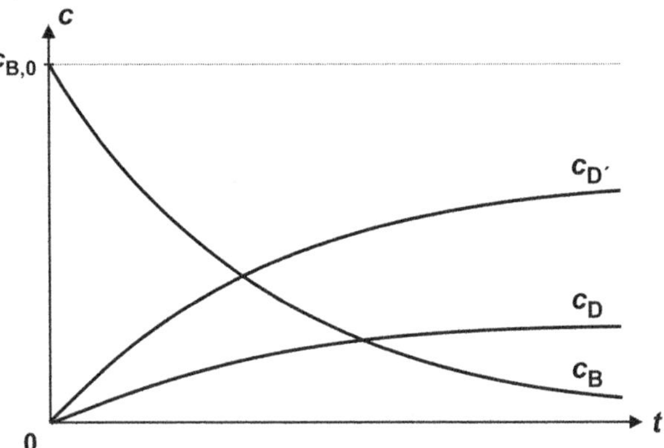

The various products compete proportionally to their rate coefficients for the reactant concentration:

$$c_D : c_{D'} = k_1 : k_2. \tag{17.23}$$

The ratio of products is therefore time independent. The fraction of a product is higher, the greater the corresponding rate coefficient is. The fastest of the parallel reactions determines the main product.

Parallel reactions can be differently influenced by various reaction conditions such as change of temperature, addition of catalysts, or the choice of appropriate solvents. This can result in differing product proportions.

Let us consider the example of chlorination of toluene:

The rate coefficient for the electrophilic substitution at the benzene nucleus (reaction 1) is strongly raised by polar solvents at low temperatures and by using catalysts (Lewis acids such as $FeCl_3$), where almost only o-chlorotoluene (or p- and m-chlorotoluene) is produced.

In contrast to this, reaction 2 is strongly accelerated by high temperatures and UV radiation, resulting in benzyl chloride as the main product.

17.4 Consecutive Reactions

Basic Principles Reaction products often do not result from single-step reactions of the starting substances but from consecutive elementary reactions via more or less stable intermediate substances (I).

Let us consider as the simplest case a series of monomolecular elementary reactions. We will ignore any backward reactions:

$$B \xrightarrow{1} I \xrightarrow{2} D.$$

The corresponding rate laws are:

$$r_1 = k_1 c_B \quad \text{and} \quad r_2 = k_2 c_I. \tag{17.24}$$

The following is valid for the decrease of reactant B:

$$-\frac{dc_B}{dt} = r_1 = k_1 c_B.$$

This equation is identical to Eq. (16.11). By integrating we obtain:

$$c_B = c_{B,0} e^{-k_1 t}. \tag{17.25}$$

As for every first-order reaction, the concentration of reactant B decreases exponentially in the course of time.

The intermediate substance I is formed by reaction 1 and then decomposed in the subsequent reaction 2, so that

$$\frac{dc_I}{dt} = r_1 - r_2 = k_1 c_B - k_2 c_I. \tag{17.26}$$

This equation is more difficult to solve, so only the result is shown:

$$c_I = \frac{k_1}{k_2 - k_1} c_{B,0} \left(e^{-k_1 t} - e^{-k_2 t} \right). \tag{17.27}$$

Formation of product D is then described by:

$$\frac{dc_D}{dt} = k_2 c_D. \tag{17.28}$$

The concentration of D can also be determined by mass balance:

$$c_D = c_{B,0} - c_B - c_I. \tag{17.29}$$

Then we obtain:

$$c_D = c_{B,0} - c_{B,0}e^{-k_1 t} - \frac{k_1}{k_2 - k_1}c_{B,0}\left(e^{-k_1 t} - e^{-k_2 t}\right) \tag{17.30}$$

or

$$c_D = c_{B,0}\left(1 - \frac{k_2 e^{-k_1 t} - k_1 e^{-k_2 t}}{k_2 - k_1}\right). \tag{17.31}$$

These rather complicated relations are illustrated in the following by plots for different proportions of rate coefficients k_1 and k_2 (Fig. 17.3):
We find

- The greater the rate coefficient k_1 is, the faster the concentration of reactant B will decay exponentially to zero.
- The concentration of intermediate substance I goes through a maximum that will be the lower, the greater the ratio k_2/k_1 is.

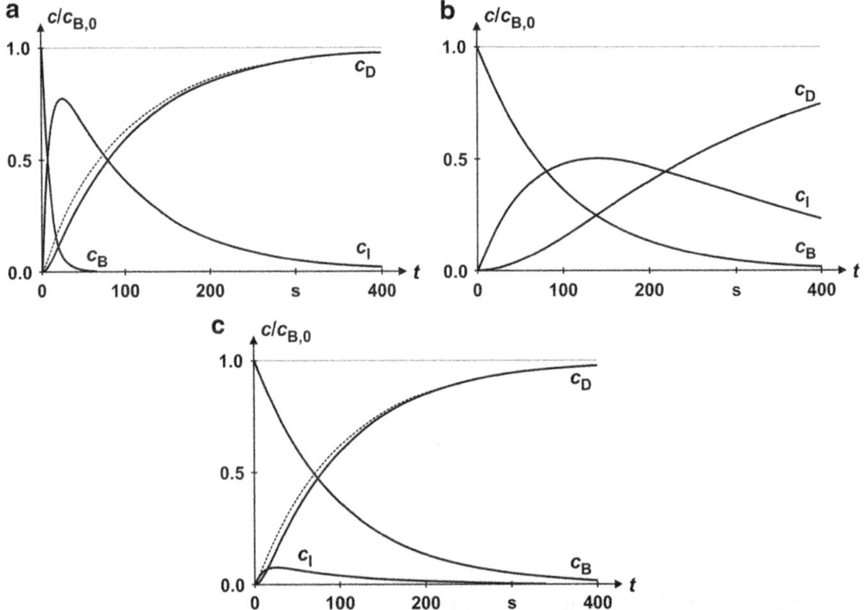

Fig. 17.3 Time dependence of substance concentrations in consecutive reactions for **(a)** $k_1 = 0.1\ \text{s}^{-1}$, $k_2 = 0.01\ \text{s}^{-1}$, **(b)** $k_1 = 0.01\ \text{s}^{-1}$, $k_2 = 0.005\ \text{s}^{-1}$, **(c)** $k_1 = 0.01\ \text{s}^{-1}$, $k_2 = 0.1\ \text{s}^{-1}$.

- The rate density r_D for product formation is proportional to the concentration of I [Eq. (17.28)]. At the beginning, it equals zero because there is no intermediate substance present yet. A noticeable formation of product D begins only after a certain start-up time (*induction period*). Induction periods are characteristic of consecutive reactions. The formation rate density of D grows with increasing concentration of I until the maximum value $c_{I,max}$ is attained, and then begins to fall. Correspondingly, the $c_D(t)$ curve is s-shaped where the inflection point appears at the point of time where c_I and, therefore, r_D have attained their maxima.

Rate-Limiting Step In the limit of $k_1 \gg k_2$ (Fig. 17.3a), the reactant B almost entirely converts to I, before any subsequent reaction of I to D occurs. The approximations $c_{I,0} \approx c_{B,0}$ and $k_1 c_B \approx 0$ can be used for the decomposition of the intermediate substance I. Equation (17.26) is simplified to:

$$-\frac{dc_I}{dt} = k_2 c_I.$$

(17.32)

Integrating yields:

$$c_I = c_{I,0} e^{-k_2 t}$$

or because of $c_{I,0} \approx c_{B,0}$,

$$c_I = c_{B,0} e^{-k_2 t}.$$

(17.33)

Because c_B is almost totally used up upon onset of the actual formation of the product ($c_B \approx 0$), the following is valid for concentration c_D:

$$c_D \approx c_{B,0} - c_I,$$

(17.34)

and therefore,

$$c_D = c_{B,0}\left(1 - e^{-k_2 t}\right)$$

(17.35)

(dashed line in Fig. 17.3a). Formation of the product D is then determined by the "slow" reaction 2 with its much lower rate coefficient k_2. The time-dependent concentration c_D now corresponds to that of a first-order reaction (compare Sect. 16.5).

Of course, we can come to the same conclusions by taking the condition $k_1 \gg k_2$, in the form of $k_2 - k_1 \approx -k_1$, and $e^{-k_1 t} \ll e^{-k_2 t}$, into account in Eqs. (17.27) and (17.31).

In general, the slowest step determines the total rate of a consecutive reaction. This can be compared to a convoy of automobiles where the slowest one determines how fast everyone travels. Kinetically speaking, when the rate coefficients differ from each other by at least one order of magnitude, we refer to a *rate-limiting step*.

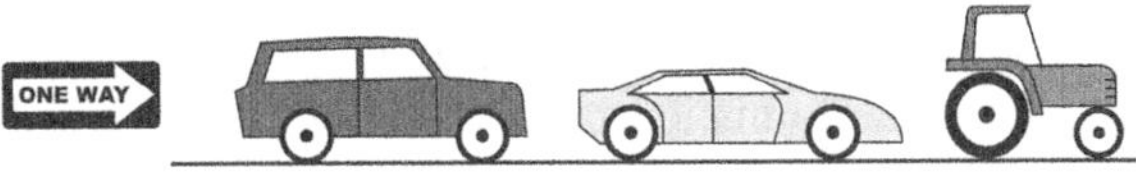

Steady-state Approximation In the reverse limit of $k_2 \gg k_1$, the intermediate substance I converts almost instantaneously into product D, which tells us that I has a low concentration and a relatively short life span. We call this a very *reactive* intermediate substance. After a short induction period, the low concentration of I does not change much in the course of time (see Fig. 17.3c) compared to the concentration changes of the other reaction participants. It can therefore be considered to be nearly constant (*quasi steady state*). Hence:

$$\frac{dc_I}{dt} \approx 0. \tag{17.36}$$

This method for deriving approximate solutions for kinetic equations, known as *steady-state approximation*, was first introduced by the German physicochemist Max Bodenstein. It is very helpful in simplifying complicated kinetic relations.

Applied to Eq. (17.26) the result is

$$\frac{dc_I}{dt} = k_1 c_B - k_2 c_I \approx 0. \tag{17.37}$$

Formation and decomposition rate densities of I are, as expected, approximately the same. Therefore, the result for formation of product D is

$$\frac{dc_D}{dt} = k_2 c_I = k_1 c_B \tag{17.38}$$

or after inserting the concentration c_B from Eq. (17.25)

$$\frac{dc_D}{dt} = k_1 c_{B,0} e^{-k_1 t}. \tag{17.39}$$

Integration once again proceeds according to the substitution method described in Sect. 17.3. By taking the initial condition $c_D = 0$ at $t = 0$ into account, we obtain the following relation:

$$c_D = c_{B,0}\left(1 - e^{-k_1 t}\right) \tag{17.40}$$

(dashed line in Fig. 17.3b). In this case, the slow first reaction having the rate coefficient k_1 determines the formation of the reaction product.

Here, as well, the results are identical to the approximations of the exact solutions of (17.27) and (17.31) for $k_2 \gg k_1$, but were obtained by much simpler mathematical methods.

There are numerous examples of consecutive reactions such as radioactive decay series, hydrolyses of dicarboxylic acid esters or tertiary alkyl halides, as well as nitrations of aromatics. Conversions of gases on catalyst surfaces are also examples of this.

Chain Reactions Chain reactions where reactive intermediate substances (*chain carriers* such as atoms, free radicals, or ions) are responsible for the permanent repetition of steps are consecutive reactions of a special type. We distinguish the following elementary steps in a chain reaction:

* *Chain initiation*: Formation of chain carriers,
* *Chain propagation*: Reaction of chain carriers with reactant molecules while forming new chain carriers,
* *Chain termination*: Recombination of chain carriers.

The resulting rate laws are often rather complicated and exhibit fractional reaction orders.

An example of a chain reaction is the formation of hydrogen chloride from chlorine and hydrogen gas which proceeds explosively while producing intense heat (chlorine–hydrogen reaction). Reactive chlorine atoms that we will call radicals due to their unpaired electron (indicated by a dot) are formed during dissociation of Cl_2 molecules when energy is added, possibly by a flash of light or by heating:

Chain initiation: $Cl_2 \rightarrow 2\ Cl^{\bullet}$.

They initiate the actual chain:

$$\text{\textit{Chain propagation:}} \quad \begin{aligned} Cl^{\bullet} + H_2 &\rightarrow HCl + H^{\bullet}, \\ H^{\bullet} + Cl_2 &\rightarrow HCl + Cl^{\bullet}. \end{aligned}$$

In order to achieve the termination of the chain, a collision partner X is necessary. This might be the wall of the reaction vessel or a nonreactive molecule to withdraw the released energy:

$$\text{\textit{Chain termination:}} \quad \begin{aligned} Cl^{\bullet} + Cl^{\bullet} + X &\rightarrow Cl_2 + X^{*}, \\ (H^{\bullet} + H^{\bullet} + X &\rightarrow H_2 + X^{*}), \\ (H^{\bullet} + Cl^{\bullet} + X &\rightarrow HCl + X^{*}). \end{aligned}$$

The elementary steps in the parentheses are unimportant. A large vessel surface obviously lowers the rate of the chain reaction, because chain termination is facilitated by it. The effect of tetraethyl lead that was used as an anti-knock agent in fuels is based on this principle. A porous layer of lead oxide forming on the walls of the pistons in the combustion engine (along with added chain terminating reagents) reduces "knocking," a premature ignition of the fuel–air mixture.

If a chain carrier attacks a molecule that was produced earlier in the reaction, another chain carrier will be formed, but the formation of product is slowed. In this case, we speak of an *inhibition reaction*. Here is an example of formation of hydrogen bromide from the elements that also proceeds as a chain reaction:

$$\text{Inhibition:}\quad \begin{aligned} &H^{\bullet} + HBr \rightarrow H_2 + Br^{\bullet}, \\ &(Br^{\bullet} + HBr \rightarrow Br_2 + H^{\bullet}). \end{aligned}$$

Polymerization of unsaturated organic compounds (*monomers*) is a special type of chain reaction. In a process of breaking of multiple bonds, new monomers continuously attach to the radical or ionic chain carriers. An example of this is the cationic polymerization of vinylchloride into polyvinylchloride (PVC):

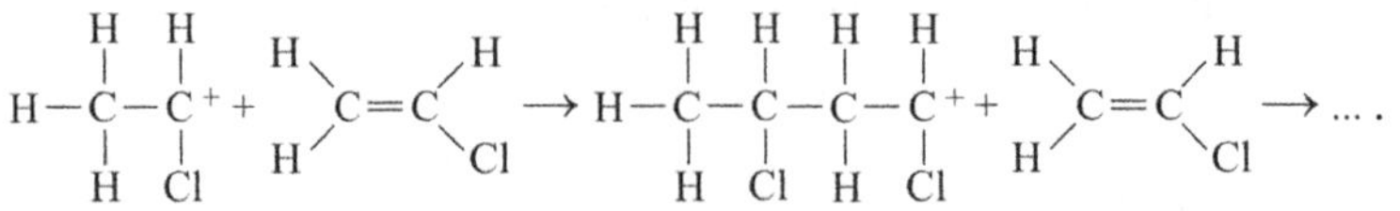

Chapter 18
Theory of Rate of Reaction

Everyday experience demonstrates that, most of the time, the rate of a chemical reaction will increase with a rise in temperature. Food, for example, will spoil outside on a hot summer day much faster than it would in a refrigerator. A simple but remarkably accurate relationship for the temperature dependence of reaction rates was empirically found by the Swedish chemist Svante Arrhenius in 1889. The interpretation of the parameters in the Arrhenius equation leads to the development of the idea that when reactants convert into products, they must go through an activated state that requires a characteristic energy. This was the basis of two of the most important theories of reaction rates, collision theory and transition state theory. Collision theory, which only suffices for simple gas phase reactions, essentially views reactants as if they were particles with a certain kinetic energy. Reactions can only occur if two molecules collide with a minimum energy necessary for rearranging the bonds. Matter dynamic considerations play no role here. In transition state theory, a more comprehensive theory that can, in principle, be applied to every possible type of reaction, the rate coefficient is expressed in terms of a difference in chemical potentials between the reactants and a kind of "transition substance" ("ensemble" of all activated complexes), a so-called "potential barrier." For a deeper understanding, the transition state can be interpreted on a molecular level with the help of potential energy surfaces and the "motion" of molecules through these surfaces.

18.1 Temperature Dependence of Reaction Rate

Everyday experience demonstrates that, most of the time, the rate of a chemical reaction will increase with a rise in temperature. Food will spoil outside on a hot summer day much faster than it would in a refrigerator. The decolorization of potassium permanganate by oxalic acid in a sulfuric acid solution (a reaction we already discussed in Chap. 16) is also strongly accelerated by heating (Experiment 18.1).

© Springer International Publishing Switzerland 2016

G. Job, R. Rüffler, *Physical Chemistry from a Different Angle*,

DOI 10.1007/978-3-319-15666-8_18

Experiment 18.1 *Temperature dependency of rate of reaction*: Potassium permanganate and acidulated oxalic acid solutions are brought to three different temperatures (in an ice bath at about 0 °C, at room temperature, and in a water bath of about 50 °C). Starting with the coldest oxalic acid solution, the potassium permanganate solution of the same temperature is added to each one. The hottest solution loses color by far the fastest.

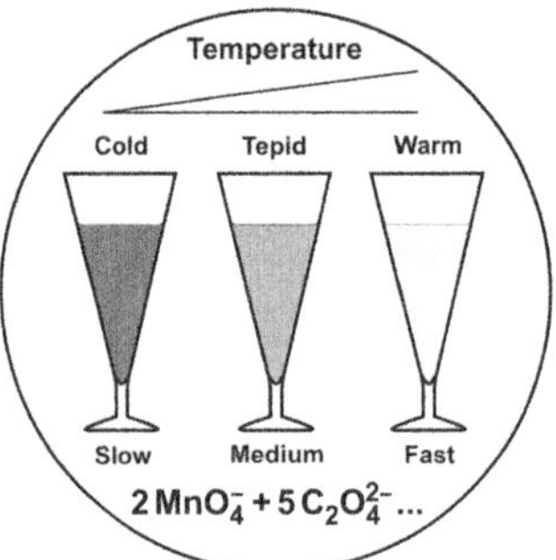

An old rule of thumb tells us that a temperature rise of 10 K doubles the rate of reaction. This rule is valid for slow reactions that take between 1 s and 1 a, at not too high temperatures. The actual factor lies between 1.5 and 4.

As we hinted at in Sect. 16.5, the influence of temperature upon the reaction rate is included in the rate coefficient k. Toward the end of the nineteenth century, the Swedish chemist Svante Arrhenius proposed after comparing the experimental kinetic data available at that time that in most reactions, the rate coefficient changes exponentially with the reciprocal of temperature:

$$k(T) = Ae^{-B/T}. \tag{18.1}$$

The parameters A and B that Arrhenius considered to be independent of temperature are characteristic of a reaction.

Arrhenius was also the first to interpret this result, especially the parameter B. In a chemical reaction, the arrangement of atoms in the starting substances must be transformed into that of the products. In this process, preexisting bonds are broken in order to form new ones. One can imagine that a certain minimum energy, the molar (Arrhenius) *activation energy* W_A of the given reaction, is necessary for this. By expanding the exponent by the gas constant R, Eq. (18.1) can be rewritten as follows:

$$k(T) = k_\infty e^{-W_A/RT} \quad \text{(Arrhenius equation)}. \tag{18.2}$$

(k_∞ corresponds to the parameter A and W_A/R corresponds to the parameter B [in all, the exponent is then dimensionless, as it should be]).

The *pre-exponential factor* k_∞ (also called the *frequency factor*) mathematically represents the limiting value of the rate coefficient for very high (and in practice impossible to realize) temperatures above 10^4 K ($T \to \infty$) (Fig. 18.1a).

In order to get an impression of the magnitude of activation energy in chemical reactions, let us return to the rule of thumb mentioned above. According to this rule, a temperature increase of 10 K from, for example, $T_1 = 298$ K to $T_2 = 308$ K, should result in a doubling of the rate coefficient. This means that

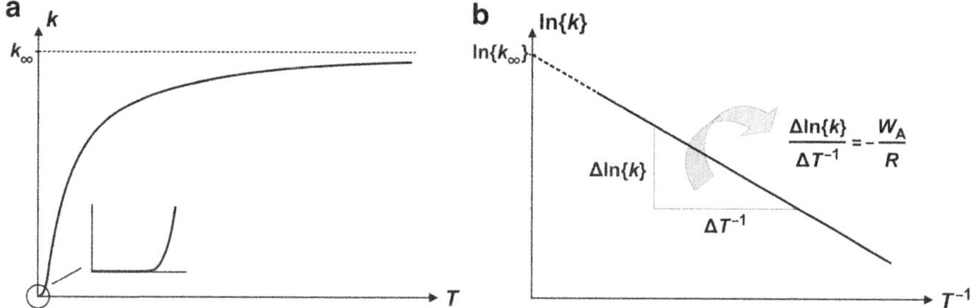

Fig. 18.1 (**a**) Temperature dependency of the rate coefficient k, (**b**) Determination of the activation energy W_A from the Arrhenius diagram.

$$2 \approx \frac{k_2}{k_1} = \frac{k_\infty e^{-W_A/RT_2}}{k_\infty e^{-W_A/RT_1}} = \exp \frac{W_A}{R}\left(\frac{1}{T_1} - \frac{1}{T_2}\right).$$

Taking the logarithm and solving for W_A yields:

$$W_A = \frac{\ln \dfrac{k_2}{k_1} \cdot R}{\dfrac{1}{T_1} - \dfrac{1}{T_2}} \approx \frac{\ln 2 \times 8.314 \ \mathrm{J\,mol^{-1}\,K^{-1}}}{\dfrac{1}{298\ \mathrm{K}} - \dfrac{1}{308\ \mathrm{K}}} \approx 53\ \mathrm{kJ\,mol^{-1}}.$$

In fact, the values for molar activation energies of many common reactions lie between 30 and 100 kJ mol^{-1}.

In order to find the molar activation energy for a certain reaction from experimental data, it is a good idea to first take the logarithm of Eq. (18.2):

$$\ln \frac{k}{k^\dagger} = \ln \frac{k_\infty}{k^\dagger} - \frac{W_A}{R} \cdot \frac{1}{T}.$$

$k^\dagger$ represents an arbitrarily chosen reference value with the same dimension as k or k_∞, which is introduced because the argument of a logarithm must be dimensionless. We will try to keep the equation from becoming unnecessarily complicated by using curly brackets to indicate the division by the reference value (compare Sect. 16.5):

$$\ln\{k\} = \ln\{k_\infty\} - \frac{W_A}{R} \cdot \frac{1}{T}. \tag{18.3}$$

If $\ln\{k\}$ is now plotted as a function of $1/T$ (Arrhenius diagram) (Fig. 18.1b), we obtain a straight line from whose slope $-W_A/R$, the molar activation energy results. The value of $\ln\{k_\infty\}$ and, therefore, k_∞ can be determined from the axis intercept after extrapolation to $1/T = 0$.

The steeper the straight line is, meaning the higher the activation energy of the reaction, the stronger its temperature dependency will be. When the temperatures of reactions with low activation energies (around 10 kJ mol^{-1}) are raised, they only accelerate a little. The rates of reactions with high activation energies (around 60 kJ mol^{-1}) increase strongly with rising temperatures.

If $\ln\{k\}$ drawn as a function of $1/T$ is not properly straight, the activation energy can be determined from the slope of the tangent for a section of the curve. W_A is then no longer constant but changes with temperature. In general, reactions with complex reaction mechanisms such as chain reactions, enzyme reactions, and heterogeneous catalytic reactions exhibit non-Arrhenius behavior. In the following, however, we will avoid such complications.

The Arrhenius equation is important for its development of the idea that when reactants convert into products, they must go through an activated state that requires a characteristic energy. This was the basis of two of the most important theories of reaction rates, collision theory and transition state theory.

18.2   Collision Theory

A deeper understanding of what the Arrhenius parameters mean can be developed from the *collision theory* of bimolecular gas phase reactions, which itself is based upon kinetic gas theory. The requirement for two particles like H_2 and I_2 or two HI particles to react with each other is that they encounter each other, i.e., that they collide. It has been found, however, that the frequency of collisions in an ideal gas (which is of the order of 10^{35} m^{-3} s^{-1} at standard conditions) by far surpasses the number of particles present so that any gas phase reaction should actually occur in fractions of a microsecond. However, this is not the case. Experimentally determined half-lives are much longer. For example, the reaction of H_2 and I_2 has a half-life of $t_{1/2} = 2 \times 10^{-2}$ s and that of 2 HI has a half-life of $t_{1/2} = 5 \times 10^{-3}$ s. Obviously not all collisions lead to reactions, but only those where the collision energy has exceeded a certain minimum value necessary for rearranging the bonds (Fig. 18.2).

Let us now consider a bimolecular gas phase reaction between particles of types A and B from this point of view. The more particles of one type there are, the more often collisions will occur between the different particles A and B. The *collision frequency* or *"collision density"* Z_{AB}, meaning the number of collisions between A and B (given in mol) per volume and time is directly proportional to the concentrations of both types of particles:

$$Z_{AB} \sim c_A \cdot c_B \quad \text{or} \quad Z_{AB} = \text{const.} \cdot c_A \cdot c_B. \tag{18.4}$$

The amount of energy available at collision for breaking the bonds of two particles A and B does not depend upon their speed v, but upon their speed relative to each other. The way in which they encounter each other (centrally or grazingly) and how

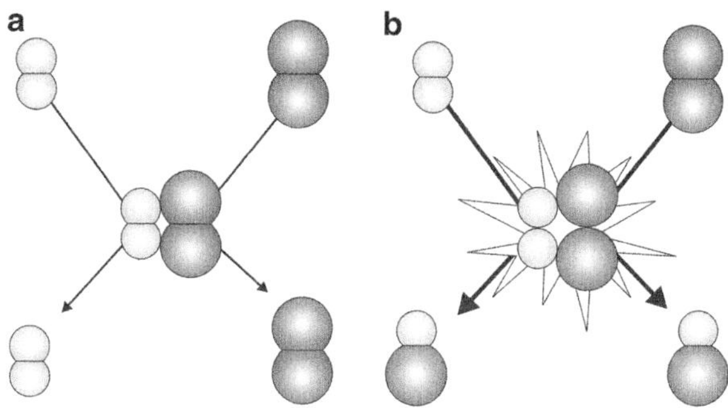

Fig. 18.2 (a) Absence of reaction due to too little collision energy, (b) Successful reaction in a collision with high enough energy (for the sake of simplicity, we will imagine that all colliding particles—despite their differences in size—are equally heavy and equally fast).

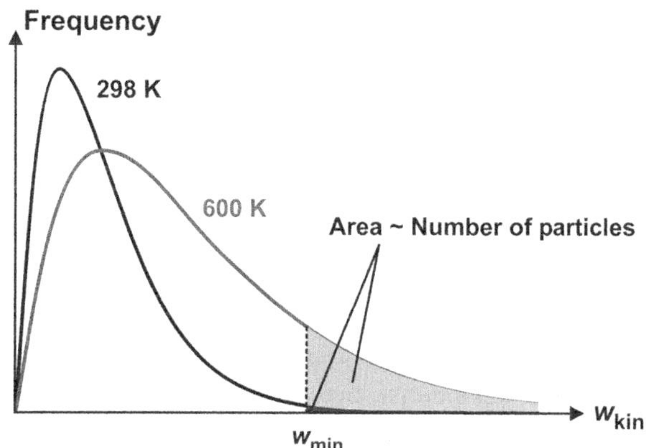

Fig. 18.3 Frequency of gas particles per energy interval dw as a function of kinetic energy w_{kin} at various temperatures. The factor $e^{-w/k_B T}$ with $w = w_{kin}$ stemming from Maxwell's distribution is responsible for the rapid fall of the distribution curve as energy increases.

they rotate or oscillate in the process also plays a role. It is plausible to note that as v increases, the other velocities also increase. We were introduced to Maxwell's distribution of speeds in Sect. 10.4. It indicates the frequency of gas particles per velocity interval dv as a function of velocity v. It is relatively easy to now convert the velocity distribution into a distribution of kinetic energy $w_{kin} = \frac{1}{2}mv^2$ (Fig. 18.3).

The shaded area underneath the corresponding curve indicates the number of gas particles having at least the kinetic energy w_{min}. As the temperature rises, the proportion of particles capable of reacting strongly increases. This is mainly due to the so-called "Boltzmann factor" $e^{-w/k_B T}$ in the energy distribution (compare Sect. 10.5). This factor remains when we integrate over the distribution between $w = w_{min}$ and $w = \infty$. Neglecting the modifying prefactors which enter into this, we obtain a surprising result. The fraction q of all particles having a minimum energy of w_{min} at temperature T equals

$$q = \frac{N(w \geq w_{min})}{N_{ges}} \approx e^{-w_{min}/k_B T} \quad \text{or} \tag{18.5}$$

$$q \approx e^{-W_{min}/RT}, \tag{18.6}$$

where the energy W_{min} refers to one mol of particles.

We obtain the rate density r by multiplying the collision density by the fraction of collisions having sufficient energy:

$$r = q \cdot Z_{AB} = e^{-W_{min}/RT} \cdot \text{const.} \cdot c_A \cdot c_B. \tag{18.7}$$

When this expression is compared to the second-order rate law (Eq. 16.17),

$$r = k \cdot c_A \cdot c_B,$$

the rate coefficient k turns out to be:

$$k = \text{const.} \cdot e^{-W_{min}/RT}. \tag{18.8}$$

This relation has exactly the same form as the Arrhenius equation (18.2). The Arrhenius parameters can then be interpreted as follows:

- The activation energy W_A corresponds to a minimum energy necessary for breaking existing bonds and forming new bonds when two gas particles collide.
- The pre-exponential factor k_∞ is the maximum possible rate coefficient that could be attained if all collisions were successful.

To illustrate this, we will take a look at what fraction of gas particles would even be capable of reacting at room temperature when we assume a typical activation energy of 50 kJ mol^{-1}:

$$q = \exp\left(-\frac{W_A}{RT}\right) = \exp\left(-\frac{50 \times 10^3 \ \text{J mol}^{-1}}{8.314 \ \text{J mol}^{-1} \text{K}^{-1} \times 298 \ \text{K}}\right) = 1.7 \times 10^{-9},$$

meaning that fewer than two collisions in a billion can lead to a reaction.

The magnitudes of pre-exponential factors k_∞, which can be calculated with the help of kinetic gas theory, generally agree with empirically determined values. However, experiments can also show values that are smaller by one or two orders of magnitude than the ones that are calculated. Obviously, a collision of two gas particles with sufficient energy alone, does not guarantee a successful conversion. The particles must be favorably oriented toward each other in order to make a bond between particular atoms possible (Fig. 18.4).

We introduce the so-called *steric factor* $p \leq 1$ in order to correct for this effect. Its numeric value indicates the fraction of collisions having favorable orientation. The more complicated the particles participating in the reaction are,

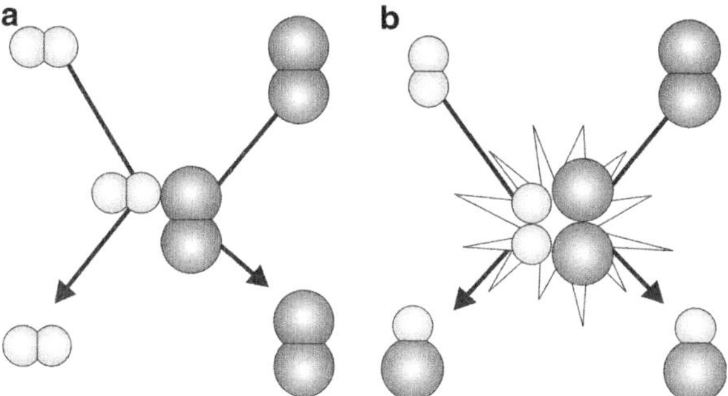

Fig. 18.4 Collision with (**a**) unfavorable orientation, (**b**) favorable orientation.

the higher the requirements for orientation will be and therefore the smaller p will be.

Let us summarize: There are essentially three things necessary for a chemical reaction:

- Collisions of gas particles A and B,
- Excess energy for rearrangement of bonds (activation),
- Favorable mutual orientation at collision (orientation).

18.3 Transition State Theory

Collision theory, which only suffices for simple gas reactions, essentially views reactants as if they were particles with a certain kinetic energy. Matter dynamic considerations play no role here. In the following, we will get to know a more comprehensive theory that can, in principle, be applied to every possible type of reaction.

We can imagine breaking up even single-step reactions into still smaller partial steps. A chemical reaction, as we imagine in atomic models, is a rearrangement of certain molecular components. Atoms making up a certain kind of molecule can be rearranged and combined into new kinds of molecules. For this to happen, existing bonds between atoms must be loosened or broken to form new ones. In a single-step process, all the participating atoms must be present at the same time. They form a so-called "*transition complex*" where the rearrangement takes place. This "complex" is a labile entity, a kind of *transition state* that has a well-defined composition and its own chemical potential, just like any substance. The configuration in the transition state is richer in energy than the particles at the beginning or end of the reaction. The atoms or molecules need to be in an "activated" (energy rich) state in order to form this configuration of higher energy. Therefore, the transition complex is also called *activated complex*.

It takes a certain amount of time to pass through the transition state. This can be regarded as a finite but extremely short lifetime. Despite their short lifetime, these complexes behave like a kind of particle and the ensemble of these labile "transition particles" behaves like a substance that is present in very small concentration in the reaction mixture. In order to emphasize this aspect, we will call the ensemble of such short-lived particles a *transition substance* and mark it by the symbol ‡. Formation of a transition substance can be expressed by the following formula:

$$\mathrm{A + BC} \rightleftharpoons \overbrace{\mathrm{A \cdots B \cdots C}}^{\ddagger} \rightarrow \mathrm{AB + C}.$$

$$\underbrace{\phantom{\mathrm{A + BC \rightleftharpoons A \cdots B \cdots C}}}_{\text{Activation}}$$

The first half-step of this transformation, which requires energy input, is called the *activation* or *activation reaction*. We will use the index $_\ddagger$ for the quantities belonging to this process as we do with all quantities related to the transition substance. In the second half-step, the transition substance decays into the products (this latter step is monomolecular).

The extremely short lifetime and maximum energy distinguish between this transition substance and the unstable intermediate substance of a consecutive reaction (compare Sect. 17.4). The latter has "normal" bonds and can be isolated and investigated, while the former cannot.

An optimally realistic description of transition substances based upon quantum mechanics form the core of the theory developed in the 1930s by Henry Eyring, Meredith Gwynne Evans, and Michael Polanyi.

A transformation of the starting substances into the final products must always proceed over a transition substance whose instantaneous amount $n_\ddagger$ as well as lifetime $\tau_\ddagger$ will determine the rate of the reaction:

$$\omega = \dot{\xi} = \frac{n_\ddagger}{\tau_\ddagger}. \tag{18.9}$$

In a homogeneous reaction, we obtain the rate density r from ω, as usual, by dividing the equation above by volume V (where $c_\ddagger = n_\ddagger/V$ is the concentration of the transition substance):

$$r = \frac{c_\ddagger}{\tau_\ddagger}. \tag{18.10}$$

According to considerations stemming most likely from Eyring in 1935, it can be assumed approximately that the amount of the short-lived transition substance present in a reaction mixture will reach a value that would form in equilibrium with the reactants. (This assumption is not strictly valid because there is no classical chemical equilibrium when the transition substance constantly decays into the products. We therefore also speak from the quasi-equilibrium assumption.)

However, if we proceed on the assumption that the transition substance exists in close to equilibrium concentration, this quantity can easily be calculated using the mass action law. For the reaction above we have:

$$\overset{\circ}{\mathcal{K}}_{\ddagger} = \frac{c_{\ddagger}/c^{\ominus}}{(c_A/c^{\ominus}) \cdot (c_{BC}/c^{\ominus})}. \tag{18.11}$$

If we solve the equation for the concentration $c_{\ddagger}$ of the transition substance, we obtain

$$c_{\ddagger} = \overset{\circ}{\mathcal{K}}_{\ddagger} \cdot c^{\ominus} \cdot \frac{c_A}{c^{\ominus}} \cdot \frac{c_{BC}}{c^{\ominus}}. \tag{18.12}$$

Eyring used quantum mechanics to derive a very simple expression for the lifetime of the transition state:

$$\tau_{\ddagger} = \frac{h}{k_B T}. \tag{18.13}$$

Here, h is Planck's constant with $h = 6.626 \times 10^{-34}$ J s and k_B is Boltzmann's constant with $k_B = 1.381 \times 10^{-23}$ J K^{-1}.

Equation (18.13) only deals with the decay of the transition substance into the products, because the reverse decay into the reactants is compensated for by constant formation.

The resulting order of magnitude of $\tau_{\ddagger}$ at room temperature is $\tau_{\ddagger}$ [$= 6.626 \times 10^{-34}$ J s/$(1.381 \times 10^{-23}$ J K$^{-1} \times 298$ K)] $\approx 10^{-13}$ s. The lifetime is, indeed, very short. As temperature rises, it gets even shorter. One of the reasons is that, on average, the transition state will be passed through more quickly because of greater particle velocity in a warmer environment. The best aspect of this equation is that all transition substances behave identically, *independent of their type*.

Because we lack the theoretical background to reason in detail about the two Eyring assumptions—namely those concerning concentration and lifetime of transition substances—we treat them as basic assumptions that can be justified by comparing their conclusions in retrospect with those of experience. But what are these conclusions?

By combining the equations for $c_{\ddagger}$ and $\tau_{\ddagger}$, we obtain the desired rate density r, which we will contrast with the corresponding rate law for a second order reaction:

$$r = \boxed{\frac{k_B T}{h} \cdot \overset{o}{\mathcal{K}}_{\ddagger} \cdot c^{\ominus} \cdot \frac{c_A}{c^{\ominus}} \cdot \frac{c_{BC}}{c^{\ominus}}} = k \cdot c_A \cdot c_{BC} . \qquad (18.14)$$

In the framed expression, the equilibrium number $\overset{o}{\mathcal{K}}_{\ddagger}$ is the only quantity that is dependent upon the *type* of reaction. As usual, this can be calculated from the relation

$$\overset{o}{\mathcal{K}}_{\ddagger} = \exp\left(\frac{\overset{o}{\mathcal{A}}_{\ddagger}}{RT}\right) = \exp\left(\frac{\overset{o}{\mu}_A + \overset{o}{\mu}_{BC} - \overset{o}{\mu}_{\ddagger}}{RT}\right) = \exp\left(-\frac{\Delta_{\ddagger}\overset{o}{\mu}}{RT}\right). \qquad (18.15)$$

The rate coefficient k then results in

$$k = \kappa_{\ddagger}\frac{k_B T}{h} \cdot \exp\left(\frac{\overset{o}{\mathcal{A}}_{\ddagger}}{RT}\right) = \kappa_{\ddagger}\frac{k_B T}{h} \cdot \exp\left(-\frac{\Delta_{\ddagger}\overset{o}{\mu}}{RT}\right) \qquad (18.16)$$

with the dimensional factor $\kappa_{\ddagger} = (c^{\ominus})^{-1}$. We will call the quantity $-\mathcal{A}_{\ddagger} = \Delta_{\ddagger}\mu = \mu_{\ddagger} - \mu_A - \mu_{BC}$ the *activation threshold* of the reaction and its special value $-\overset{o}{\mathcal{A}}_{\ddagger} = \Delta_{\ddagger}\overset{o}{\mu}$ its basic value. Note that because we assume equilibrium, we have $\mathcal{A}_{\ddagger} = 0$. However, the basic value $\overset{o}{\mathcal{A}}_{\ddagger}$ is not equal to zero.

This conclusion is rather remarkable. It tells us that the reaction resistance and therefore the individual differences in the rates of different reactions are solely dependent upon the height of the potential threshold $\Delta_{\ddagger}\overset{o}{\mu}$ between the starting substances and the transition substance. In order to clarify this statement, the potentials, i.e., the basic values $\overset{o}{\mu}$ and the actual values μ, will be represented graphically (Fig. 18.5).

Only in the case of basic potentials does an activation threshold $\Delta_{\ddagger}\overset{o}{\mu}$ appear in the form of a step ascending from the left toward the transition substance. The threshold is equal to zero in the case of actual potentials because of the assumed equilibrium:

$$A + BC \rightleftarrows \ddagger \qquad \Delta_{\ddagger}\mu = 0.$$

The potential threshold $\Delta_{\ddagger}\overset{o}{\mu}$ determines the conversion rate of the reaction proceeding from left to right. The higher the activation threshold is, the lower the rate coefficient, and the slower the reaction. The rate decreases very quickly, i.e., exponentially, with the height of the activation threshold.

When the chemical potential $\overset{o}{\mu}_{\ddagger}$ of the transition substance is at the level $\overset{o}{\mu}_A + \overset{o}{\mu}_{BC}$ of the reactants, meaning the activation threshold $\Delta_{\ddagger}\overset{o}{\mu}$ is zero, and all the substances are present in standard concentration, then our formula will yield

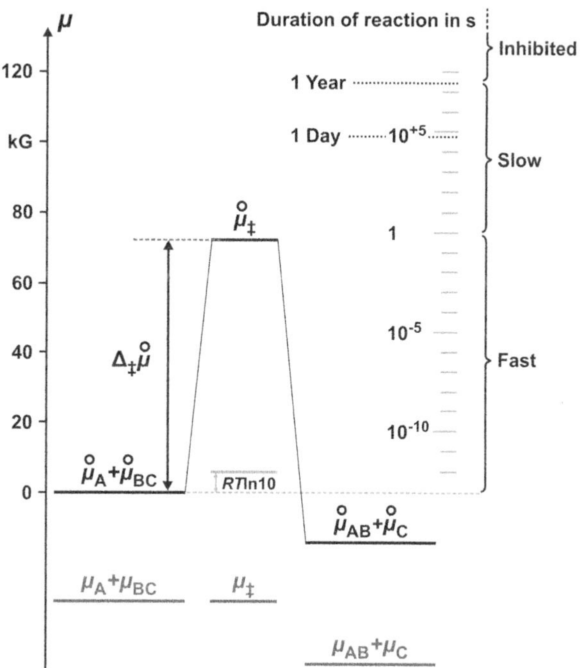

Fig. 18.5 Potential diagram for describing conversion rates. Basic values (*black bars*) and actual values (*gray bars*) are represented for starting substances and final products as well as for the transition substance ‡. (The level of the basic values of μ was arbitrarily chosen as the zero point of the potential scale).

$$r_0 \approx \frac{1}{10^{-13}\,\text{s}} \times 10^3 \ \text{mol\,m}^{-3} \times \exp(0) \times 1 \times 1 = 10^{16} \ \text{mol\,m}^{-3}\,\text{s}^{-1}$$

for the rate density r at normal laboratory temperatures. Since there are 10^3 mol of each substance in a cubic meter, they would be used up in 10^{-13} s if these conditions prevailed and the rate were constant.

The conversion rate will slow to one-tenth when $\Delta_{\ddagger}\overset{\circ}{\mu}$ grows by the decapotential $\mu_d = RT \ln 10 = 5.71$ kG [remember that $\exp(-RT \ln 10/RT) = 10^{-1}$] and it will take 10 times as long for the reactants to be used up. Repeatedly elevating the threshold by $RT \ln 10$ will lengthen the duration of reaction each time by a factor of 10. After 13 such steps, the duration of reaction will have reached about 1 s which is perceivable under normal laboratory conditions. The dividing line between *fast* and *slow* reactions can be drawn here. Past the twentieth step, the duration of reaction reaches 1 year, sorely taxing the stamina of even the most patient chemists. Such reactions should be considered to be *inhibited* because almost no conversion occurs within the period of observation.

In order to demonstrate the relation with Arrhenius's proposition $k = A\mathrm{e}^{-B/T}$, it is enough to apply the usual linear approximation $\mathcal{A}_{\ddagger} = \mathcal{A}_{\ddagger,0} + \alpha(T - T_0)$ for $\mathcal{A}_{\ddagger}(T)$:

$$\mathcal{K}_{\ddagger} = \exp\frac{\overset{\circ}{\mathcal{A}}_{\ddagger}}{RT} = \exp\frac{\overset{\circ}{\mathcal{A}}_{\ddagger,0} + \alpha\cdot(T-T_0)}{RT} = \exp\frac{\overbrace{(\overset{\circ}{\mathcal{A}}_{\ddagger,0} - \alpha\cdot T_0)/R}^{-B}}{T}\cdot\overbrace{\exp\frac{\alpha}{R}}^{A*} = A*\mathrm{e}^{-B/T}$$

and consequently,

$$k = \kappa_{\ddagger}\frac{k_{\mathrm{B}}T}{h}A*\mathrm{e}^{-B/T}.$$

$A*$ corresponds to Arrhenius's parameter A, except for the factor $\kappa_{\ddagger}\cdot k_{\mathrm{B}}T/h$. B is a constant as assumed by Arrhenius, but A is not. However, compared to the factor $\mathrm{e}^{-B/T}$, this temperature dependency is almost unnoticeable and can be ignored if the temperature range is not too large.

The temperature coefficient α, which we can write more elaborately as $\overset{\circ}{\alpha}_{\ddagger,0}$, agrees numerically with the activation entropy, so $\alpha = \overset{\circ}{\alpha}_{\ddagger,0} = \Delta_{\ddagger}\overset{\circ}{S}_0$. It is negative because the transition state $\ddagger$ is better ordered and lower in entropy than what it was formed from: separated, swarming, turbulent particles. When a certain orientation for the colliding particles is required, it is necessary for the transition state to be less arbitrary and have more order so that the activation entropy is more strongly negative. In collision theory, we used the steric factor to describe this characteristic.

18.4   Molecular Interpretation of the Transition State

Although the short characterization of the transition state in the last section would basically suffice for our future purposes, a more detailed description is often desired for a deeper understanding.

The rearrangement of atoms during a reaction does not take place all at once but extends over a certain time span. In the process, reactant particles change into product particles. We have seen an example of this in the reaction

$$\mathrm{A + BC \rightarrow AB + C}$$

where it is assumed that the centers of mass of all three atoms always lie in a straight line. During the reaction, when A approaches BC, the bond between B and C is loosened. (A very simple image for this bond might be a spring.) At the same time, a new bond between A and B starts to form. As the reaction progresses, it goes through the aforementioned transition state (activated complex) A⋯B⋯C which finally breaks apart forming molecule AB and atom C:

$$A + BC \rightleftarrows \underbrace{A \cdots B \cdots C}_{\ddagger} \rightarrow AB + C.$$

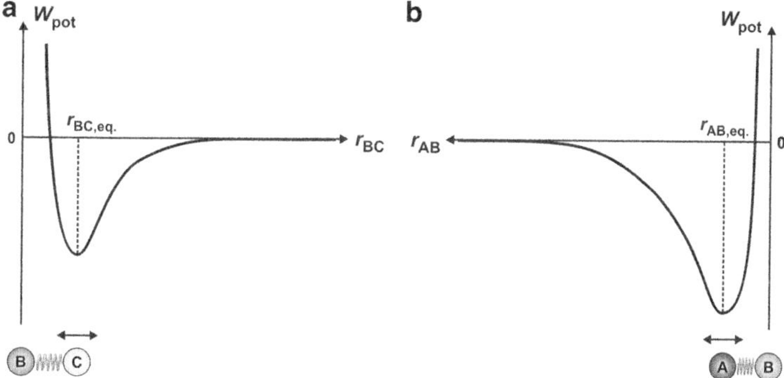

We can imagine both starting substances and final products to be extreme arrangements of the activated complex. In the initial state, atoms B and C are at bonding distance and atom A is far away. Figure 18.6a shows the potential energy W_{pot} as a function of distance r_{BC} between the nuclei in molecule BC (compare Sect. 11.1). It rises sharply when the bond is compressed relative to the equilibrium distance $r_{BC,eq.}$ ($r_{BC} < r_{BC,eq.}$). If the bond is elongated ($r_{BC} > r_{BC,eq.}$), W_{pot} increases as well due to subsiding of attractive forces. It asymptotically approaches a limiting value corresponding to the energy of the completely separated atoms B and C (dissociation energy). The potential energy for various distances between the nuclei of molecule BC can be calculated using quantum mechanics. An analogous diagram can also be created for molecule AB (Fig. 18.6b).

The minimum of potential energy at equilibrium distance $r_{BC,eq.}$ (the always present zero-point energy of the vibrational ground state will not be taken into consideration here) represents the initial state, meaning that atom A is at a great distance from molecule BC. If atom A approaches molecule BC, which will decay during the reaction, then for each instant of this rearrangement, the positions of the molecular components participating in the triatomic stretched "molecule" A⋯B⋯C can be characterized. In this way, we end up with a very large number of intermediate states for the reaction. Each possible intermediate state, including the initial and final states, has a certain potential energy assigned to it that is dependent upon the geometry of the given arrangement, meaning the distances r_{AB} and r_{BC} of the atoms. The energy can, in principle, be calculated quantum mechanically. If this energy is plotted in the z direction as a function of the distances of the nuclei (r_{AB} in the x and r_{BC} in the y direction), a three-dimensional representation will result

Fig. 18.6 Potential energy W_{pot} of (**a**) molecules BC and (**b**) molecules AB for large distances from the third partners A or C. $r_{BC,eq.}$ and $r_{AB,eq.}$ are the equilibrium distances.

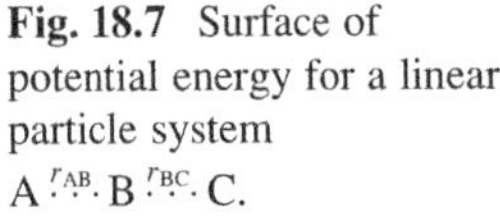

Fig. 18.7 Surface of potential energy for a linear particle system A $\overset{r_{AB}}{\cdots}$ B $\overset{r_{BC}}{\cdots}$ C.

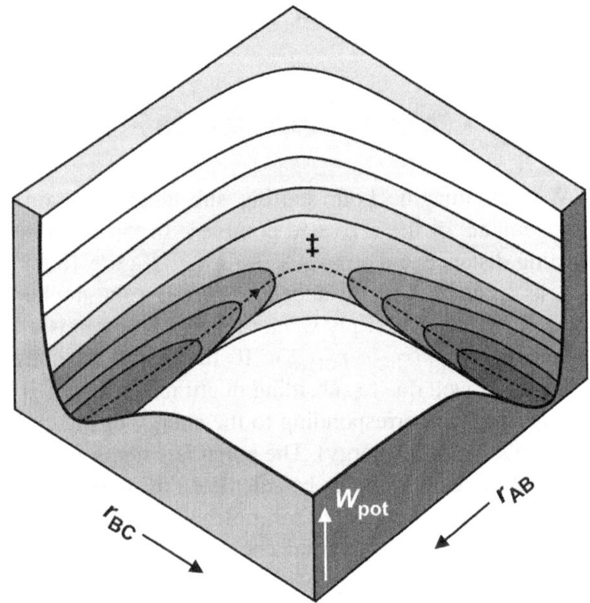

(potential energy surface; Fig. 18.7). The preceding diagrams 18.6a and b form the vertical side-walls. To make the energy surface clearer, the points of identical energies have been connected at given intervals by lines of constant energy (contour lines).

Initially, when A is still very far away, molecule BC is in a deep energy trough ("bottom of the valley" on the left). If atom A approaches molecule BC, the potential energy rises as a result of loosening of the BC bond ("movement up in the valley") until a maximum ("valley-ridge inflection point" or "saddle" ‡) is obtained that equals the energetically labile transition state A···B···C. As the distance between A and B continues to decrease, W_{pot} will fall again due to formation of the AB bond in the direction of the "bottom of the valley" on the right. At the same time, C moves away from the AB molecule that is forming. In the final state of the reaction, we have the molecule AB in a deep energy trough (equilibrium state $r_{AB,eq.}$) as well as the separated atom C.

A clearer image of this might be gained by imagining the three-dimensional potential energy surface (comparable to a mountainous landscape in which elevation corresponds to potential energy) to be projected upon the plane spanned by the r_{AB} and r_{BC} axes. This produces a two-dimensional contour diagram that is comparable to a topographic contour map (Fig. 18.8a).

Although the initial and the final states of the reaction are uniquely specified, the path by which the rearrangement converts the initial state into the final state is not. It is easy to imagine that the individual molecular components might take very different paths to achieve their stable final arrangements. There are any number of

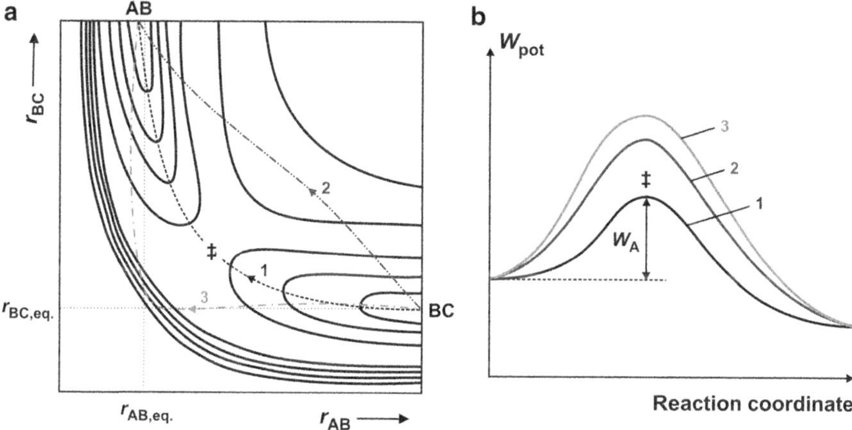

Fig. 18.8   (a) Contour map of the energy surface for the linear particle system A···B···C with three possible paths of reaction $A + BC \rightarrow AB + C$, (b) Corresponding energy profiles.

reaction paths the process of rearrangement may take. In general, there is at least one instantaneous arrangement upon every path whose energy exceeds the energy of the initial and final states. This state is usually called a *transition* or *activated state*. Hence, there is at least one activated state with *maximum energy* on every reaction path. Among all the reaction paths, one path stands out for which the maximum of energy is minimal. The corresponding atomic arrangement is the transition state in the strict sense (for short, *the* transition state).

An image that can make this clearer is a hiker crossing some mountains to reach a destination leaving from a starting point. Upon *each* path the hiker could take, he will find himself at a point of maximum elevation (potential energy) as he crosses the ridge of the mountains. Among these possible paths, there are those that go over a pass. The summit of the pass with the lowest elevation corresponds to the transition state (in the strict sense).

The potential energy contour map (Fig. 18.8a) shows three of the many possible paths leading from BC to AB. If we follow the change of potential energy along these paths with the help of an energy profile (Fig. 18.8b), we see that the path leading over the saddle point (the transition state in the strict sense) is the most economic because it requires the least energy. This special minimum energy path is called *reaction coordinate*. It is important to keep in mind that the transition state (in the strict sense) itself corresponds to an energy maximum along this coordinate, differentiating it from an intermediate product. The course of molecular energies of the reaction is reflected in the course taken by the chemical potentials (compare Fig. 18.5).

Chapter 19
Catalysis

Reactions can not only be accelerated by raising their temperatures, but also by addition of small amounts of a substance, a so-called catalyst, which is not consumed during the process. An everyday example of a catalyst is the exhaust gas catalytic converter in motor vehicles with gasoline engines, which eliminates combustion pollutants by accelerating subsequent reactions. But why do reactions proceed faster with a catalyst than without a catalyst? The catalyst lowers the reaction resistance by opening up more easily accessible bypasses with smaller activation thresholds. Enzymes, the vitally essential biological catalysts, and the kinetics of their reactions with structurally suitable substrates are discussed in detail. An enzyme can be compared to a lock into which only the proper key (substrate) can fit (key–lock principle). This is where the "key" for the exceedingly high substrate specificity of an enzyme lies. The chapter ends with the discussion of the technically important heterogeneous catalysis.

19.1 Introduction

We saw in the last chapter that reactions can be accelerated by raising their temperatures. Catalysis is an alternative method for raising the rate of a chemical reaction. The substance which is added in a small amount for this purpose and which is not consumed during the process is called a *catalyst*. The catalyst lowers the reaction resistance by opening up more easily accessible bypasses. An everyday example of a catalyst is the exhaust gas catalytic converter in motor vehicles with gasoline engines, which eliminates combustion pollutants by accelerating subsequent reactions.

There are various types of catalysis. If all the substances participating in a catalytic process form a uniform mixture, i.e., if they are all in the same phase

© Springer International Publishing Switzerland 2016
G. Job, R. Rüffler, *Physical Chemistry from a Different Angle*,
DOI 10.1007/978-3-319-15666-8_19

Experiment 19.1 *Decomposition of H_2O_2 by various catalysts*: When a solution of iron(III) chloride is added to the hydrogen peroxide solution, a noticeable development of oxygen can be observed after a while. When manganese dioxide is added instead, the reaction proceeds much faster than in the first case and also produces fog. (Therefore, the expression "genie in a bottle" is used for a variation of this experiment.) When the enzyme catalase is added, a vigorous reaction producing foam occurs.

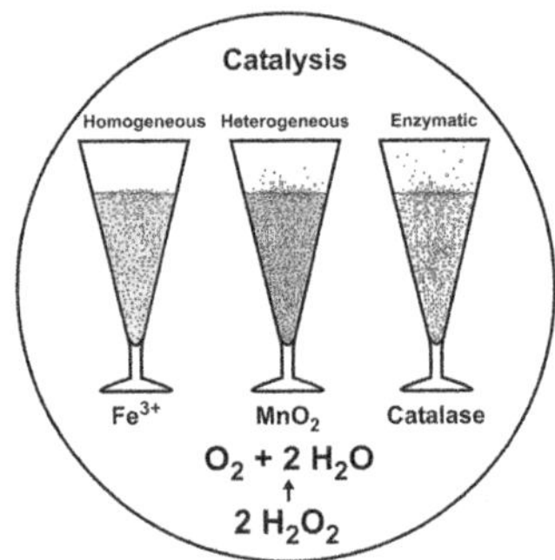

(gaseous or liquid), it is called a *homogeneous catalysis*. In *heterogeneous catalysis*, the catalyst and the reaction mixture are in different phases. Most heterogeneous catalysts are solid substances, while the substances in the reaction mixture are either gaseous or liquid. This type of catalysis plays an important role in industrial applications. *Enzymatic catalysis* lies somewhere in between the two. Enzymes are proteins, i.e., macromolecules with diameters of between 10 and 100 nm, that are present colloidally in a solution and are usually much larger than the substrate molecules. Therefore, this is also referred to as *microheterogeneous catalysis*.

As an example, let us investigate the different types of catalysis using decomposition of hydrogen peroxide into water and oxygen (Experiment 19.1):

$$2\ H_2O_2|w \rightarrow 2\ H_2O|l + O_2|g.$$

Without adding a catalyst, the decomposition rate at room temperature is immeasurably small. Fe^{3+} ions function as a homogeneous catalyst, whereas solid manganese dioxide (MnO_2) works as a heterogeneous catalyst. For enzymatic catalysis, we use the enzyme catalase.

Catalase is a biocatalyst that specializes in destroying H_2O_2 that is poisonous for cells. As we have seen in our experiment, it also possesses the highest efficiency.

We use the term *autocatalysis* if a catalyst is formed only during the reaction. An example of this is the reaction of permanganate with oxalic acid,

$$2\ MnO_4^-|w + 5\ C_2O_4^{2-}|w + 16\ H^+|w \rightarrow 2\ Mn^{2+}|w + 10\ CO_2|g + 8\ H_2O|l,$$

which we have already discussed in Chaps. 16 (Experiment 16.11) and 18 (Experiment 18.1) from different points of view. The Mn^{2+} ions that are produced represent the catalyst. The role these Mn^{2+} ions play can be easily demonstrated by adding them right at the beginning of the reaction (Experiment 19.2).

A catalytic reaction can also be inhibited or even stopped by adding a substance called an *inhibitor*. Such substances are used up in the process—in contrast to

Experiment 19.2 *Autocatalysis*: Decolorization of the solution begins slowly but accelerates as the reaction proceeds because more and more Mn^{2+} ions are produced. If the Mn^{2+} ions are added right at the beginning of the reaction, the decolorization begins immediately.

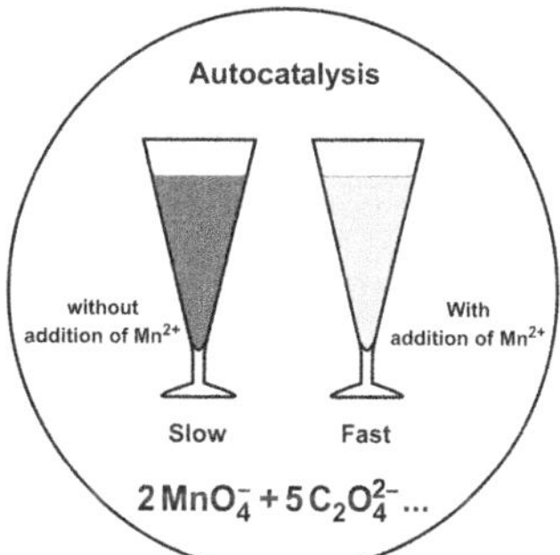

Experiment 19.3 *Inhibition of the enzyme catalase*: A piece of potato is laid in deionized water while another piece of nearly identical size is put into a mercury(II) chloride solution. Two Petri dishes are then filled with a solution of hydrogen peroxide and the two pieces of potato are put into the Petri dishes with tweezers. Immediately, bubbles start collecting on the surface of the untreated piece of potato, due to development of oxygen. The piece of potato treated with the $HgCl_2$ solution hardly produces any gas.

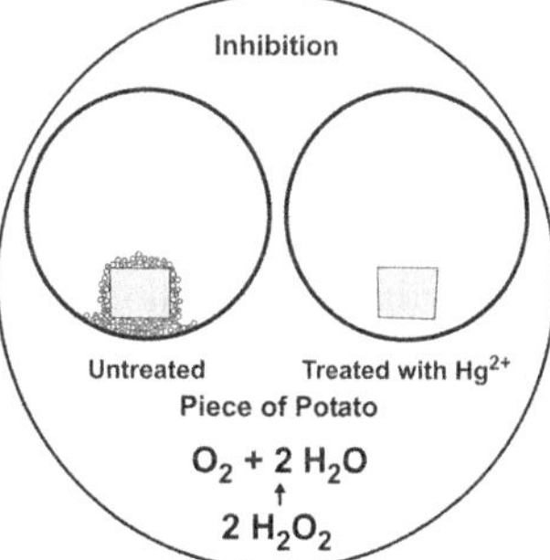

catalysts. When a catalyst is irreversibly deactivated, one refers to *catalyst poison* or *enzyme poison*. The transition between inhibition and poisoning is, however, smooth. Let us take another look at the strongly accelerated oxygen producing decomposition of hydrogen peroxide by the enzyme catalase (Experiment 19.3). Potatoes contain this enzyme. Its effect, which can be seen in the formation of gas bubbles, can be totally inhibited by pretreating the potato with a mercury chloride solution. The mercury ions alter the structure of the protein and destroy its enzyme function.

An everyday example of this is the poisoning of the previously mentioned exhaust catalyst in motor vehicles by the heavy metal lead (also compare Sect. 19.4). This is why lead-free fuels must always be used.

19.2   How a Catalyst Works

The Baltic German chemist and Nobel Prize laureate Wilhelm Ostwald found already in the 1880s that by bonding to a catalyst in catalytic reactions, intermediate substances are produced that subsequently decompose, thereby regenerating the catalyst. A simple chemical reaction such as

$$A + B \rightleftarrows P$$

can therefore be influenced by a catalyst C as follows:

$$C + A \overset{1}{\rightleftarrows} CA,$$

$$CA + B \overset{2}{\rightleftarrows} C + P.$$

Why then, do reactions proceed faster with a catalyst (that changes the reaction mechanism) than without a catalyst? The rate coefficient k of the uncatalyzed reaction that proceeds via the activated complex ‡,

$$A + B \rightleftarrows \ddagger \rightleftarrows P,$$

is determined only by the activation threshold $\Delta_\ddagger \overset{\circ}{\mu}$ for a given temperature (Sect. 18.3). The lower the activation threshold is, the faster the reaction will run. The higher rates of formation and decomposition of the intermediate substance CA— compared to those of the uncatalyzed reaction—can therefore only be explained by lower activation thresholds (Fig. 19.1):

$$C + A \rightleftarrows \ddagger' \rightleftarrows CA,$$

$$CA + B \rightleftarrows \ddagger'' \rightleftarrows C + P.$$

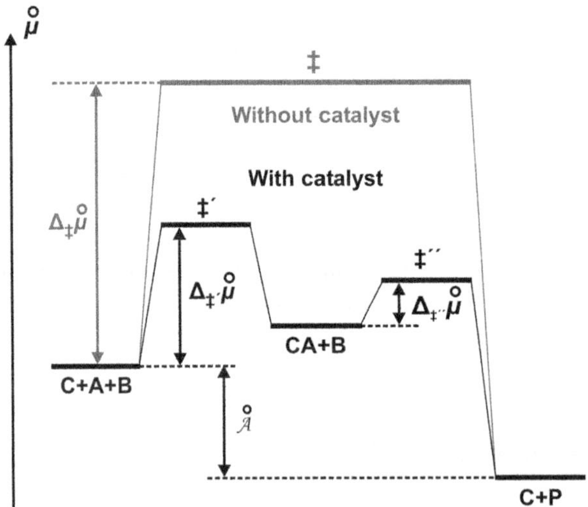

Fig. 19.1 Influence of catalyst C upon the activation thresholds.

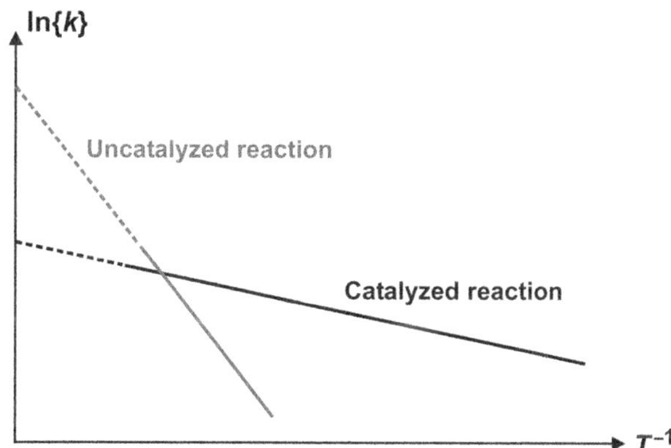

Fig. 19.2 Arrhenius diagram of the uncatalyzed and catalyzed reactions.

In the Arrhenius diagram, the lowering of the activation threshold of the catalyzed reaction is seen by the considerably smaller slope of the corresponding straight line (Fig. 19.2).

The decomposition of hydrogen peroxide, for instance, requires a molar activation energy of 76 kJ mol^{-1}, which is why it runs so slowly at room temperature. When the enzyme catalase is added, the threshold reduces to 6 kJ mol^{-1}, leading to extreme acceleration. However, the activation energy applies formally to the entire changed mechanism and cannot be attributed to an individual reaction step as before.

The temporal behavior of catalyzed reactions is determined by the rate of the elementary reactions 1 and 2. To simplify matters, we will assume that the intermediate substance CA is formed slowly and decays quickly ($k_2 \gg k_1$ or $\Delta_{\ddagger'}\overset{\circ}{\mu} > \Delta_{\ddagger''}\overset{\circ}{\mu}$ as in Fig. 19.1); we encountered a similar case in Sect. 17.4 dealing with consecutive reactions. The first step, the formation of CA, determines the overall rate:

$$r = -\frac{dc_A}{dt} = \frac{dc_P}{dt} = k_1 c_C c_A = k_C c_A. \tag{19.1}$$

Ideally, the concentration c_C of a catalyst will remain constant during the reaction, so there will be a linear relation between the rate density r and c_A. c_C can be combined with the rate coefficient k_1 to form a new coefficient k_C. This means that a reaction of the (pseudo) first order results (Fig. 19.3, catalyzed reaction).

We find a totally different temporal behavior in autocatalysis. Here, the catalyst is formed only during the reaction. Let us look at the following simple example:

$$A\,(+C) \rightarrow P + C\,(+C).$$

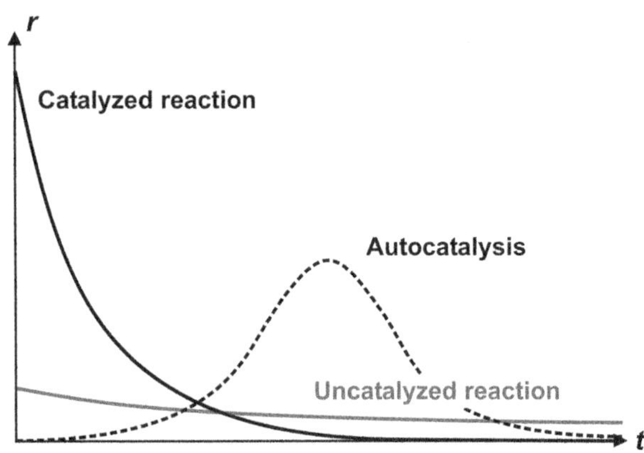

Fig. 19.3 Time dependency of rate density in an uncatalyzed, a catalyzed, and an autocatalyzed reaction.

The rate law is then:

$$r = -\frac{dc_A}{dt} = \frac{dc_P}{dt} = kc_C c_A. \tag{19.2}$$

At the beginning of the reaction, the concentration of catalyst and the rate density are both extremely small. As the formation of catalyst proceeds, the reaction rate also increases until consumption of A finally overcompensates for the increase of C. Consequently, the rate density passes through a maximum (Fig. 19.3, autocatalysis).

As we can see in Fig. 19.1, the presence of a catalyst in a reaction has an influence upon the activation threshold $\Delta_{\ddagger}\overset{\circ}{\mu}$, but not upon the drive $\overset{\circ}{\mathcal{A}}$. The drive is determined exclusively by the difference of the potentials of the reactants and products. Because the catalyst drops out of the overall conversion formula, it can play no role in determining this value. This also means, though, that a reaction which cannot run spontaneously due to a negative drive cannot be forced by using a catalyst because its drive does not change. Along with the drive, the equilibrium constant is the same for uncatalyzed and catalyzed reactions. Catalysts do not shift the position of chemical equilibrium, but make that it is achieved faster via an easier path for the reaction.

Catalysts not only accelerate a chemical reaction, but also help to channel a reaction to produce a desired product. This *selectivity* does not contradict the fact that the position of equilibrium itself cannot be influenced. It only means that under given circumstances, one of the many possible spontaneous parallel reactions will be considerably more accelerated than the others. For example, the process of hydrogenating carbon monoxide (Fischer–Tropsch synthesis) can produce methanol (catalysts: ZnO, Cr_2O_3) or unsaturated hydrocarbons (catalyst: Fe), depending upon the type of catalyst used and the reaction conditions. In contrast, we use the term *specificity* if a catalyst only affects certain substances. Very high selectivity and specificity can be found in reactions catalyzed by enzymes. These are very important reactions that will be gone into more detail in the next section.

19.3 Enzyme Kinetics

Enzymes Enzymes, the biocatalysts in living organisms, are almost exclusively proteins and belong to the group of macromolecules with diameters of between 10 and 100 nm. It is not the entire molecule that is catalytically effective, though. The actual location of a reaction is limited to a small area, the so-called *active site*. This can be made up of proteinogenic amino acids or nonprotein components (cofactors) such as heme or adenosine triphosphate. The spatial structure of the enzyme around the active site only allows adsorption of a structurally suitable *substrate* (i.e., the reactant in an enzyme-catalyzed reaction). An enzyme can be compared to a lock into which only the proper key (substrate) can fit (key–lock principle). This is where the "key" for the exceedingly high *substrate specificity* of an enzyme lies. The formation of the enzyme substrate complex alters the electron density distribution in the substrate and enhances its ability to continue reacting. A converted substrate molecule then leaves the active site, making space for the next unaltered one. This is how the enzyme urease catalyzes the hydrolysis of urea in which ammonia and carbon dioxide are produced:

$$(NH_2)_2CO|w + H_2O|l \rightarrow CO_2|w + 2\,NH_3|w.$$

An alkaline milieu forms due to the ammonia:

$$2\,NH_3|w + CO_2|w + 2\,H_2O|l \rightarrow 2\,NH_4^+|w + HCO_3^-|w + OH^-|w.$$

For this reason, the sudden change in color of the indicator phenolphthalein can serve to verify the hydrolysis (Experiment 19.4). Structurally related substances like thiourea, methylurea, or semicarbazide are, by contrast, not decomposed. This is an indication of the high substrate specificity of urease.

Michaelis–Menten Kinetics The cornerstone for describing simple enzyme-catalyzed reactions was laid in 1913 by the collaboration of the German biochemist and physician Leonor Michaelis and the Canadian physician-scientist Maud Leonora Menten. The proposed mechanism assumes that from enzyme E and

Experiment 19.4 *Catalytic decomposition of urea by the enzyme urease*: Some phenolphthalein solution is added to solutions of urea and methylurea which are then divided into three goblets (see illustration). The urea solution in the first glass serves as the reference. A suspension of urease is added to the second goblet containing urea solution as well as to the third containing methylurea solution. After a short while, the urea solution has a violet color due to formation of ammonia, while the methylurea solution remains unchanged.

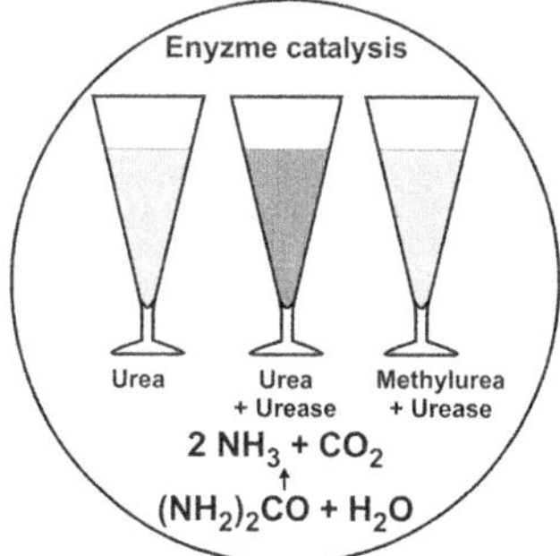

substrate S, an enzyme–substrate complex ES can be formed quickly and reversibly. Then, in a slow step, this complex decomposes irreversibly into a product P while the enzyme regenerates. The term irreversible means that the backward reaction of E and P into ES can be ignored (because $c_P \sim 0$ and/or $k_{-2} \ll k_2$):

$$E + S \underset{k_{-1}}{\overset{k_1}{\rightleftharpoons}} ES \overset{k_2}{\rightarrow} E + P.$$

The term *pre-equilibrium* can also be used in this case. While Michaelis and Menten assumed that the subsequent reaction creating the product runs so slowly as to be negligible, the extended approach of George Edward Briggs and John Burdon Sanderson Haldane (1925) also takes the rate coefficient of the consecutive reaction into account.

The following is valid for the rate density in product formation:

$$r = \frac{dc_P}{dt} = k_2 c_{ES}. \tag{19.3}$$

But how can we now obtain the required concentration of the intermediate substance c_{ES}? In order to do this, we will first set up the rate law for producing ES:

$$\frac{dc_{ES}}{dt} = k_1 c_E c_S - k_{-1} c_{ES} - k_2 c_{ES}. \tag{19.4}$$

Because the concentration of the unstable intermediate substance ES is very small compared to the concentration of the substrate that is to be converted, a quasi-steady state with $dc_{ES}/dt \approx 0$ can be assumed over long periods of time (compare Sect. 17.4). We then obtain

$$k_1 c_E c_S = k_{-1} c_{ES} + k_2 c_{ES},$$

meaning that the formation of the enzyme–substrate complex takes place in a second-order reaction having nearly the same rate as its decomposition into either the reactants or the product following first-order kinetics. After rearranging, we obtain the following for the desired concentration c_{ES}:

$$c_{ES} = \frac{k_1}{k_{-1} + k_2} c_E c_S. \tag{19.5}$$

The rate coefficients can now be combined into the Michaelis constant K_M (in the Briggs–Haldane version) which is a quantity characteristic of a given enzyme and given substrate,

$$K_M \equiv \frac{k_{-1} + k_2}{k_1}, \tag{19.6}$$

and we obtain

$$c_{\mathrm{ES}} = \frac{c_{\mathrm{E}}c_{\mathrm{S}}}{K_{\mathrm{M}}}. \tag{19.7}$$

The concentration c_{E} of the *free* enzyme in Eq. (19.7) can be expressed by the difference of the initial concentration $c_{\mathrm{E},0}$ and the concentration c_{ES} of the intermediate substance. Moreover, the concentration of free substrate corresponds very closely to its total concentration because only very small amounts of enzyme are used. We then obtain for the (steady-state) concentration of c_{ES}

$$c_{\mathrm{ES}} = \frac{\left(c_{\mathrm{E},0} - c_{\mathrm{ES}}\right) \cdot c_{\mathrm{S}}}{K_{\mathrm{M}}} \tag{19.8}$$

or, after solving for c_{ES},

$$c_{\mathrm{ES}} = \frac{c_{\mathrm{E},0}c_{\mathrm{S}}}{K_{\mathrm{M}} + c_{\mathrm{S}}}. \tag{19.9}$$

After inserting the expression for c_{ES} into Eq. (19.3), we obtain the following for the rate density of enzymatic catalysis:

$$r = \frac{\mathrm{d}c_{\mathrm{P}}}{\mathrm{d}t} = k_2 \frac{c_{\mathrm{E},0}c_{\mathrm{S}}}{K_{\mathrm{M}} + c_{\mathrm{S}}}. \tag{19.10}$$

This relation is called the *Michaelis–Menten equation*.

Initial Rate Density In order to minimize the bothersome effects of a backward reaction of E and P into ES, of an inhibition of the enzyme by products, or of a gradual inactivation of the enzyme, etc., we will consider the *initial rate density* r_0, because product concentration c_{P} plays no role at the beginning of the reaction:

$$r_0 = \left(\frac{\mathrm{d}c_{\mathrm{P}}}{\mathrm{d}t}\right)_{t=0} = k_2 \frac{c_{\mathrm{E},0}c_{\mathrm{S},0}}{K_{\mathrm{M}} + c_{\mathrm{S},0}}. \tag{19.11}$$

This approach is based upon the work of Michaelis and Menten. In order to determine the initial rate density r_0 as a function of initial concentration $c_{\mathrm{S},0}$ of substrate (at identical constant enzyme concentration), we measure the initial increase of product concentration c_{p} or a quantity directly proportional to c_{p} such as the absorption of VIS or UV radiation or the conductance. r_0 corresponds to the slope of the $c_{\mathrm{P}}(t)$ curve at the beginning of the reaction, meaning at $t = 0$ (Fig. 19.4).

When the initial rate density is now represented in function of the corresponding substrate concentration $c_{\mathrm{S},0}$, we obtain a characteristic curve (Fig. 19.5) which we will look into more detail in the following.

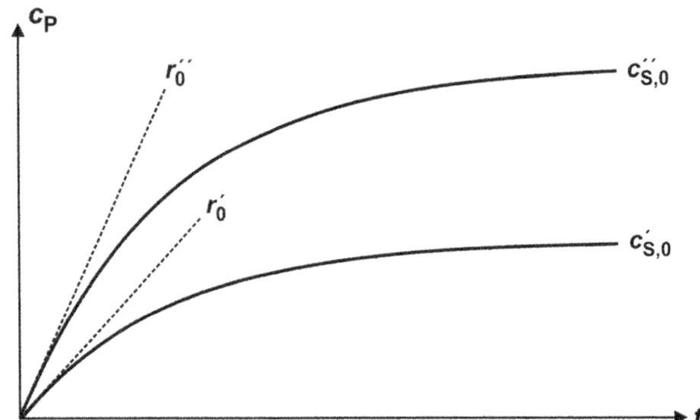

Fig. 19.4 Determining the initial rate density for differing initial concentrations $c_{S,0}'$ and $c_{S,0}''$ of substrate (at identical constant enzyme concentration).

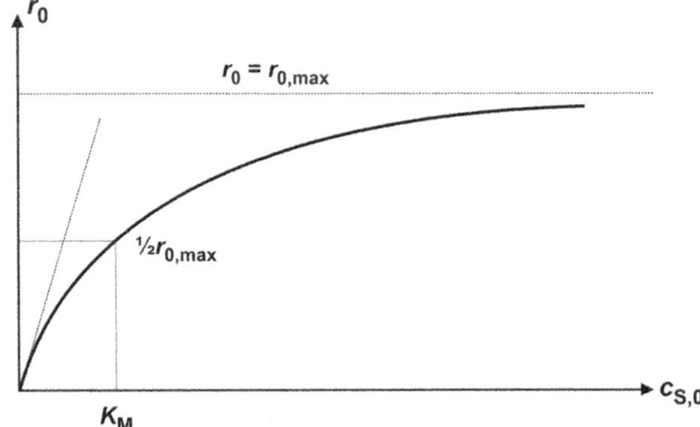

Fig. 19.5 Dependency of initial rate density r_0 of an enzyme-catalyzed reaction upon the initial substrate concentration $c_{S,0}$ according to the Michaelis–Menten mechanism.

If the substrate concentration is small ($c_{S,0} \ll K_M$), the rate density at constant enzyme concentration $c_{E,0}$ will be proportional to the substrate concentration and the reaction will follow first-order kinetics:

$$r_0 = \left[\frac{k_2 c_{E,0}}{K_M}\right] c_{S,0}. \tag{19.12}$$

Consequently, the plot of the initial rate density as a function of the initial substrate concentration shows a line through the origin for very small concentrations.

As the concentration of substrate increases, the reaction order begins to take fractional values. Finally, for high substrate concentration ($c_{S,0} \gg K_M$), the reaction order is almost zero (i.e., r_0 is no longer dependent upon $c_{S,0}$). The rate density approaches a maximum value:

$$r_0 = k_2 c_{E,0} = r_{0,\max}. \tag{19.13}$$

Almost the entire amount of enzyme has been transformed into the ES complex and the active sites of the enzyme are now "saturated" with substrate. In this case, the coefficient k_2 corresponds to the maximum number of substrate molecules that can be transformed into the product by one enzyme molecule (or more exactly, by an active site) per time unit. This number is called *turnover number* or *molecular activity* or *catalytic constant*. Typical values lie between 1 and 10^5 s^{-1}. The name turnover number is unfortunate, though, since k_2 represents a frequency (with unit s^{-1}) and not a number (unit 1).

If we take Eq. (19.13) into account, Eq. (19.10) yields a modified version of the Michaelis–Menten equation that is valid for arbitrary substrate concentration:

$$r_0 = \frac{r_{0,\max} c_{S,0}}{K_M + c_{S,0}}. \tag{19.14}$$

An advantage of this version is that it can be applied to cases in which the molar mass of the enzyme (and therefore its concentration) is unknown.

The Michaelis constant K_M corresponds to the substrate concentration where the enzyme works at only half the maximum possible rate, i.e., when half the active sites are occupied. K_M can also be interpreted as follows: If the rate coefficient of product formation (k_2) is much lower than k_{-1}, which is often the case, Eq. (19.6) simplifies to $K_M = k_{-1}/k_1$. In this case, K_M represents the equilibrium constant of dissociation of the enzyme–substrate complex. It is therefore a measure of the substrate affinity of the enzyme where low values indicate a high affinity. Typical K_M values lie between 10^{-1} and 10^{-7} mol L^{-1}.

There is a remarkable similarity between the curve in Fig. 19.4 and the Langmuir adsorption isotherm (Fig. 15.9). This similarity is not a purely formal coincidence, but has a real physicochemical basis. In each case, we are dealing with the bonding of a substance (substrate, adsorptive) on a certain number of sites (active sites, adsorption sites) defined by experiment.

Characteristic Quantities In principle, the quantities K_M and $r_{0,\max}$ that characterize an enzyme can be determined by directly fitting the Michaelis–Menten equation to the measured data using computer supported methods of nonlinear regression. We can simplify the analysis in a manner suggested by Hans Lineweaver and Dean Burke in 1934 by linearizing the relation. In order to do this, we must find the reciprocal of the Michaelis–Menten equation. After transforming, we have:

$$\frac{1}{r_0} = \frac{K_M}{r_{0,\max}} \cdot \frac{1}{c_{S,0}} + \frac{1}{r_{0,\max}}. \tag{19.15}$$

If we now plot $1/r_0$ versus $1/c_{S,0}$ (Fig. 19.6), we obtain a straight line from whose extrapolated intersection points with the ordinate and abscissa we can determine the values of $r_{0,\max}$ and K_M. The slope $K_M/r_{0,\max}$ can also be used to determine K_M.

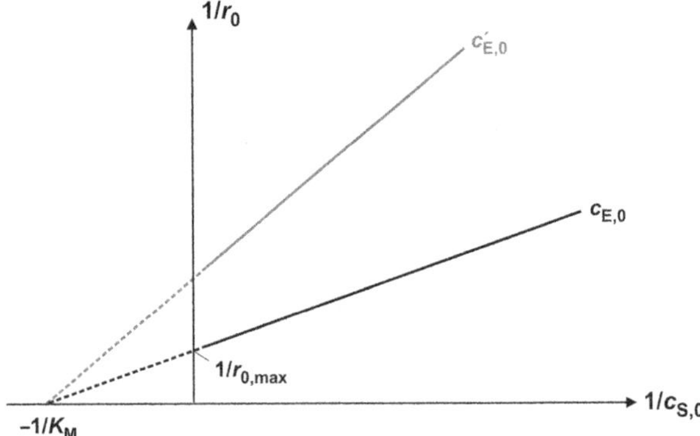

Fig. 19.6 Lineweaver–Burk diagram for two different enzyme concentrations $c_{E,0}$.

Table 19.1 Characteristic quantities of the enzymes urease and catalase (from: Voet D, Voet JG, Pratt CW (2002) Lehrbuch der Biochemie, 2nd edn. Wiley VCH, Weinheim).

Enzyme	Substrate	K_M (mol L^{-1})	k_2 (s^{-1})	k_2/K_M (mol L^{-1} s^{-1})
Urease	$(NH_2)CO$	2.5×10^{-2}	1.0×10^4	4.0×10^5
Catalase	H_2O_2	2.5×10^{-2}	1.0×10^7	4.0×10^8

The linear relation is only valid for a fixed total concentration $c_{E,0}$ of enzyme. If we change the enzyme concentration, the data points will lie along a second straight line with a changed slope which intersects the abscissa at the same point as the first line.

We can use the Michaelis–Menten constant K_M and the rate coefficient k_2 that we determine from $r_{0,\text{max}}$ to compare the catalytic effectiveness of various enzymes, or rather, the conversion of various substrates by the same enzyme. If we limit ourselves to low substrate concentrations ($c_{S,0} \ll K_M$) that are often encountered under physiological conditions especially with enzymes of high turnover number, we have according to Eq. (19.14):

$$r_0 = \frac{k_2}{K_M} c_{E,0} c_{S,0}.$$

The quotient k_2/K_M represents the "apparent" rate coefficient of a second-order reaction whose rate density is determined by the frequency of (effective) collisions between enzyme and substrate molecules. For the same concentrations of enzyme and substrate, respectively, the *catalytic efficiency* can be described by the quotient k_2/K_M. With the help of the values of k_2/K_M, we can investigate which among several substrates the enzyme prefers; k_2/K_M is therefore a measure of the substrate specificity of an enzyme where high values indicate high specificity. Typical values lie between 10^6 and 10^9 mol L^{-1} s^{-1}. Table 19.1 is a compilation of characteristic quantities of the enzymes urease and catalase.

There is an upper limit for the value of k_2/K_M which is contingent upon the rate of diffusion-controlled encounters of enzyme and substrate molecules, i.e., the rate of product formation is no longer limited by the reaction rate but by the diffusion rate. In an aqueous solution, this limiting value lies between 10^8 and $10^9\,\text{mol}\,\text{L}^{-1}\,\text{s}^{-1}$ (compare Sect. 20.2). Enzymes such as catalase that exhibit a value of k_2/K_M of this order of magnitude are considered to be (almost) *catalytically perfect* because (almost) every contact between enzyme and substrate leads to a reaction.

After considering enzymatic (or microheterogeneous) catalysis, we will now take a look at the heterogeneous catalysis that is so important to industrial applications.

19.4 Heterogeneous Catalysis

As we mentioned in the introduction, in heterogeneous catalysis, the catalyst and the converted substances are in different phases. Catalysts in the solid state are by far the most commonly used. They are also called *contacts*—especially in industrial technology. In this case, the reaction takes place on the surface of the catalyst. An example of this is the oxidation of acetone vapor by atmospheric oxygen on a copper wire spiral acting as the catalyst. In the process, acetaldehyde is formed (Experiment 19.5).

Because the accelerating effect of the solid catalyst comes from the surface atoms, it is desirable to have the largest surface possible, meaning a high degree of dispersion of the substance in question. In most cases, very small particles of catalytically active material such as platinum or rhodium are applied for stability to highly porous carrier materials with specific surfaces of several hundred square meters per gram. Aluminum oxide, silicon dioxide, activated carbon, as well as zeolites (crystalline aluminosilicates with numerous submicroscopic pores and canals) are all suitable for this purpose. Examples of such *supported catalysts* are

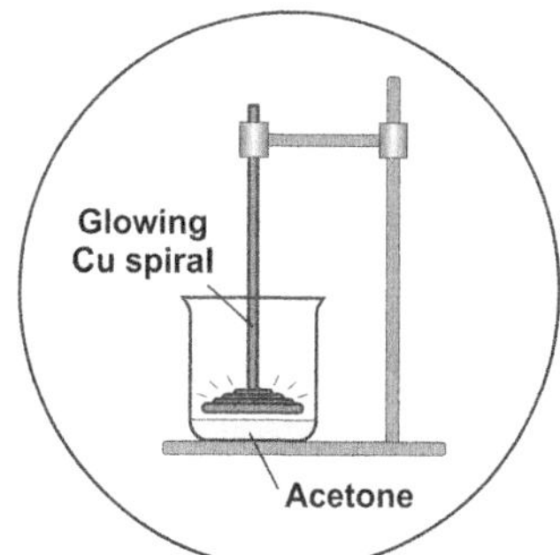

Experiment 19.5 *Catalysis of acetone oxidation by copper*: A copper spiral is heated by a laboratory gas burner until all coils glow red. It is then positioned closely over the surface of acetone in a beaker. The processes taking place upon the surface can be seen by the periodic glowing of parts of the spiral. There is also the typical stinging smell of aldehyde.

Experiment 19.6 *Catalyzed gas ignition*:
Hydrogen gas is passed over a small heap of
platinized activated carbon in a Petri dish or over
a pad of platinized quartz wool that is being held
with tweezers. After a short time, the catalyst
begins to glow, and the gas ignites with a gentle
bang and burns with an almost colorless flame.

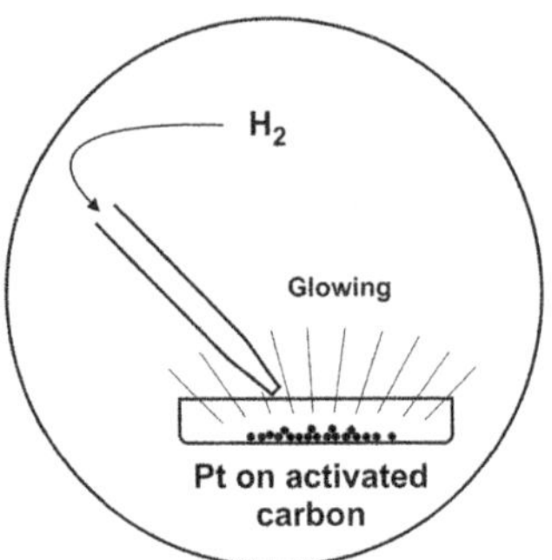

activated carbon or quartz wool covered with highly dispersed platinum. In a
current of hydrogen, they begin to glow until the gas finally ignites (Experiment
19.6). The lighter developed by Johann Wolfgang Döbereiner in 1823 is based upon
this principle.

The mechanism of heterogeneous catalysis is made up of a complex series of
individual processes:

- Diffusion of reactants toward the catalyst,
- Adsorption on the catalyst surface (The bonds in the reactant molecules can be
  weakened or even broken, which facilitates the subsequent reaction.),
- Surface reaction,
- Desorption of products from the catalyst surface,
- Diffusion of products away from the catalyst.

A quantitative description of heterogeneous catalysis is very difficult because
the concentrations of the reaction partners in the adsorption layer cannot be directly
determined in most cases. In a very complicated way, they are related to the
measurable concentrations in the liquid or gas phase through the adsorption or
desorption equilibrium. This is why the mechanisms of many heterogeneously
catalyzed reactions have not yet been clarified in detail. However, common bimolecular gas reactions according to

$$A|g + B|g \rightarrow P|g$$

can often be divided into two principally different types:

- The *Langmuir–Hinshelwood mechanism* assumes that on a surface, adsorbed
  neighboring fragments or atoms of reaction partners A and B react with each
  other to form the product P (Fig. 19.7).

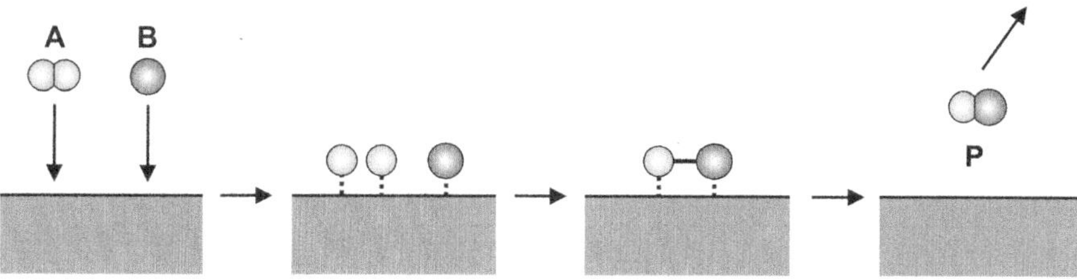

Fig. 19.7 Reaction of two chemisorbed components A and B to form product P (Langmuir–Hinshelwood mechanism).

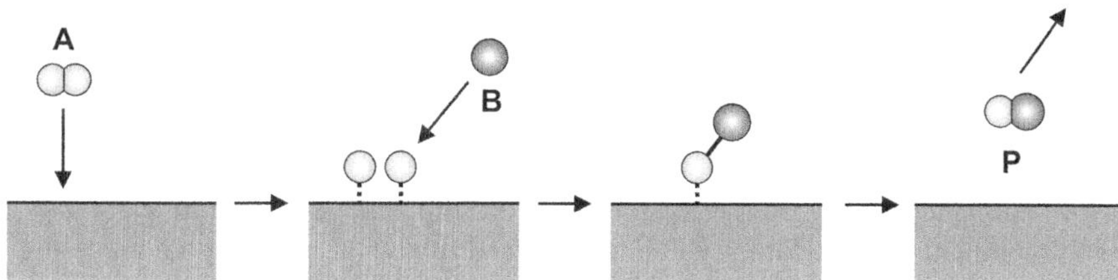

Fig. 19.8 Reaction of an adsorbed component A with the free component B (Eley–Rideal mechanism).

If we make the simplifying assumption that the surface reaction between the two components represents the limiting step, the rate density for forming product P then depends upon the number of adsorbed reactant molecules A and B and, therefore, the corresponding degrees of coverage,

$$r_P = \frac{\mathrm{d}p_P}{\mathrm{d}t} = k \cdot \Theta_A \cdot \Theta_B,$$

where p_P represents the partial pressure of the product.

In the simplest case of adsorption without dissociation, degrees of coverage can be determined with the help of Langmuir isotherms.

- According to the *Eley–Rideal mechanism*, only component A is chemisorbed. There is then a subsequent reaction with the free gas component B forming the initially adsorbed product P (Fig. 19.8).

Here, as well, we will only consider the case where the surface reaction determines the rate of reaction. Only the degree of coverage relating to component A plays a role in the corresponding kinetic considerations, while component B enters through its partial pressure:

$$r_P = \frac{\mathrm{d}p_P}{\mathrm{d}t} = k \cdot \Theta_A \cdot p_B.$$

The most well-known type of heterogeneous catalyst is the *exhaust-gas catalytic converter* found in vehicles with gasoline engines. It eliminates combustion

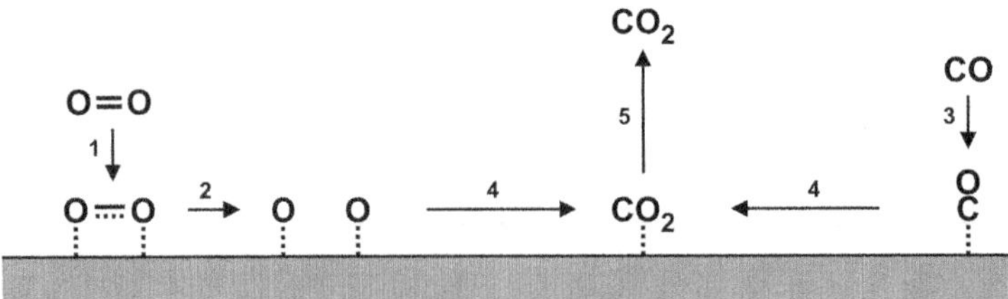

Fig. 19.9 Oxidation of carbon monoxide on a platinum catalyst according to the Langmuir–Hinshelwood mechanism (strongly simplified).

pollutants by the use of catalyzed secondary reactions. These pollutants are primarily carbon monoxide CO, nitrogen oxides NO_x, and unburned hydrocarbons C_xH_y. The catalyst makes use of different reaction paths to convert them to the nonpoisonous substances carbon dioxide, water, and nitrogen. The common name three-way catalyst refers to the three most important groups of pollutants mentioned above. The catalytically active substance is an alloy of platinum group metals, mostly platinum and rhodium, which is dispersed finely over a fine-pored metallic or, more commonly, ceramic honeycomb body acting as the carrier. However, in order for the catalytic reactions to run as desired, the supply of oxygen must be optimized. A so-called Lambda sensor is used for this. It measures the concentration of O_2 in the exhaust gas and controls the optimal composition of the fuel–air mixture going into the engine. In closing, we will look more closely at the chemism of CO oxidation (Fig. 19.9). On platinum catalysts, this takes place according to the Langmuir–Hinshelwood mechanism.

Chapter 20
Transport Phenomena

Diffusion can play an important role for the kinetics of chemical reactions in solutions. We use that as an opportunity to discuss this process of molecular motion more closely. The migration velocity is determined by a gradient of chemical potential and therefore eventually by a concentration gradient. This leads us to the quantitative description with the help of Fick's law of diffusion. But not only matter can be transported from one place to another but also some other properties such as entropy or momentum. *Entropy conduction* is determined by the migration of entropy down a temperature gradient and *viscosity* by a migration of linear momentum down a velocity gradient. In order to carve out the commonalities and differences of the transport phenomena discussed, they are summed up in the last section and compared with the transport of electric charge, because the latter is the best known of these phenomena.

20.1 Diffusion-Controlled Reactions

In order for bimolecular or trimolecular reactions to even occur, the reacting particles must collide with each other. As we have learned, not every collision leads to a reaction. The particles involved must bring the necessary energy for forming the generally energy-rich transition complex, and not all of them do. The microscopic velocity of the particles is basically the same in all states of aggregation at the same temperature. However, compared to gases, their mobility in liquids, and especially in solids, is very low. In the extreme case of crystals, their motion is limited to a fast oscillation around a position of rest. Only occasionally does a switching to an interstitial site or a change of position occur. As the particles transition from gas to solid, i.e., with increasing condensation, the frequency of collisions and the average residence time of one particle in the neighborhood of another rise. Gas particles colliding with each other is a very fleeting occurrence, after which the partners immediately separate. In liquids, though, the particles

© Springer International Publishing Switzerland 2016
G. Job, R. Rüffler, *Physical Chemistry from a Different Angle*,
DOI 10.1007/978-3-319-15666-8_20

colliding with each other are crowded by their neighbors and often remain together for a good while.

If the activation threshold $\Delta_{\ddagger}\overset{\circ}{\mu}$ vanishes or is very low, just about every encounter will lead to a reaction. Hence, it is not the height of this potential threshold but the frequency of collisions that then determines the conversion rate. In this case, the concentration of the transition complex can remain far below its equilibrium value because continued supply is stalled, while decomposition continues taking place. Reactions of this kind are said to be *diffusion-controlled* (or *diffusion-limited*) because their collision frequency is dependent upon the diffusion rate (diffusion velocity) of the partners involved. Bimolecular reactions in water and similarly viscous liquids are of this type if the activation threshold sinks under the third or fourth rung of our "potential ladder," meaning that $\Delta_{\ddagger}\overset{\circ}{\mu} < 20$ kG (see Fig. 18.2). Because diffusion in solid substances proceeds incomparably slowly, almost all the bimolecular reactions in such an environment are diffusion-controlled.

We will take this opportunity to discuss the transport of substances by diffusion and the transport phenomena associated with it. Entropy transport is often closely coupled with transport of substances so we will include it as well.

20.2 Rate of Spreading of Substances

Mobility All dissolved substances migrate under the influence of external forces. Gravitational influences will cause them to sink; they will follow the centrifugal force in a centrifuge or diffuse due to "chemical forces." (In Sect. 12.2, we were introduced to the term *diffusion* as spreading of substances caused by a potential gradient and therefore essentially by differences of concentration.) There is a unified approach that can describe all these processes. Measurements show that the migration velocity of a substance B, or the particles it is made up of, does not depend upon types of forces (be they mechanical, chemical, or electrical) but only upon their magnitude. If the forces are not too great, we can assume the migration velocity v_B to be proportional to the effective force F_B:

$$v_B = \omega_B \cdot \frac{F_B}{n_B}. \tag{20.1}$$

n_B is the amount of substance B. The fact that v_B is proportional to the quotient F_B/n_B and not to F_B makes sense because if the amount of migrating substance is doubled, the force upon a given portion of the substance will be halved. The migration velocity then behaves no differently than the velocity of a train of tugged barges with the number n of barges (Fig. 20.1). We call the factor ω_B the mechanical *mobility* of substance B. The quantity has the unit $(\mathrm{m\,s^{-1}})/(\mathrm{N\,mol^{-1}}) = (\mathrm{m\,s^{-1}})/(\mathrm{kg\,m\,s^{-2}\,mol^{-1}}) = \mathrm{s\,mol\,kg^{-1}}$.

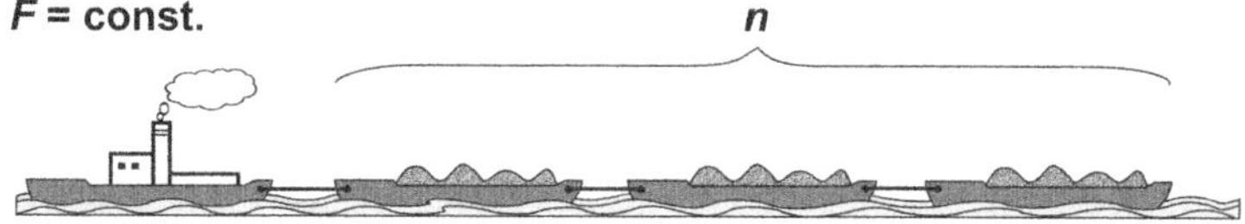

Fig. 20.1 Velocity v of a train of barges with n tugged barges to demonstrate the concept of migration velocity v of a substance if the amount n to be moved is increased while force F is kept constant.

Experiment 20.1 *Unevenly colored crystals as an example of an extremely small diffusion velocity*: Even after millions of years, amethysts, a violet variety of quartz, can exhibit uneven violet coloring due to diffusing Fe^{3+} ions. In spite of its great age, an even distribution has not been achieved. The color itself results from Fe^{4+} ions which are formed by irradiation of Fe^{3+} ions (e.g., with naturally occurring isotopes).

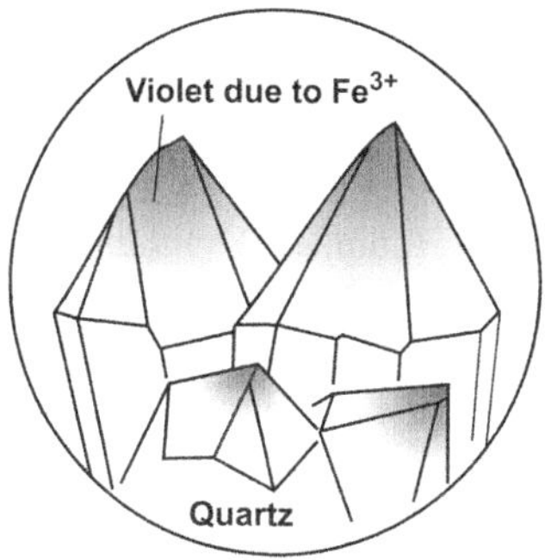

Experiment 20.2 *Diffusion of permanganate in silica gel*: Sodium silicate is acidulated, whereby clear silica gel is produced. One half of the gel is colored by permanganate and poured into a large test tube. The colorless half is then also poured into the test tube. After a week, the MnO_4^- ions have already covered a distance of a few centimeters. The process can be followed easily using the violet color as indicator.

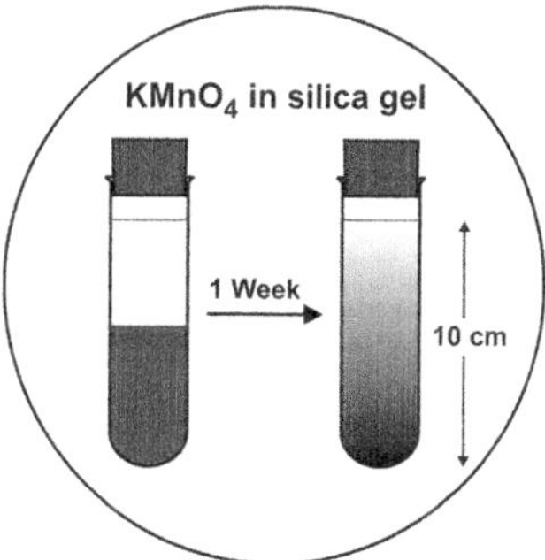

Mobility of particles in crystals [like for example quartz (SiO_2)] at room temperature is extremely low (Experiment 20.1).

The electrons in metals are an exception. In good conductors, they are around ten times as mobile as gas molecules in air.

But let us now have a look at the spreading of particles in liquids or gels. In silica gel, which is mostly made up of water, violet-colored MnO_4^- ions require about one week to move a distance of a few centimeters (Experiment 20.2). We use gel instead of a liquid to avoid convective disturbances in diffusion.

We find similar values of diffusion velocities and relatively unified conditions in liquids. Table 20.1 shows mobilities for various types of particles in water. The values are valid for very small concentrations, meaning practically pure water,

Table 20.1 Mobility of various kinds of particles in water. The values are valid for 298 K and for very small concentrations.

Substance	Ar	CO_2	D_2O	Na^+	Cl^-
ω_0 (10^{-12} s mol kg^{-1})	0.91	0.73	1.00	0.538	0.820

ω values in other environments are presented for comparison (Table 20.2)

Table 20.2 Mobility of various kinds of particles in differing environments at a temperature of 298 K.

Substance	e^- (in silver)	Cu (in silver)	Na^+ (in table salt)	CO_2 (in air)
ω_0 (s mol kg^{-1})	6×10^{-8}	10^{-41}	10^{-35}	6×10^{-9}

which is indicated by the index $_0$. (We use the slashed zero for clarity.) ω values in other environments are presented below for comparison (Table 20.2).

Despite the low viscosity of water, the mobility of particles in aqueous solutions is surprisingly low. The weight (force of gravity) in the Earth's gravitational field, for example, is much too small to evoke any noticeable effect. Only the enormous forces in ultracentrifuges can cause an observable spreading of substances. If all other influences could be ignored, carbon dioxide in water in the Earth's gravitational field would sink at the extremely small velocity of about 1 mm per century. Because $F_B = m_B g = n_B M_B g$, we have:

$$v_B = \omega_B M_B g$$
$$= 0.73 \times 10^{-12} \, \text{s mol kg}^{-1} \times 0.044 \, \text{kg mol}^{-1} \times 9.81 \, \text{m s}^{-2}$$
$$= 0.32 \times 10^{-12} \, \text{m s}^{-1} \approx 10 \, \mu\text{m a}^{-1}.$$

In contrast, particles in gases have a high mobility. We showed this in Experiment 12.3 (spreading of Br_2 in air).

Diffusion In Chap. 12, we were introduced to the driving force behind the spreading of substances that we call diffusion. This is the diffusion force F_B, which is dependent upon the gradient of the chemical potential:

$$F_B = -n_B \cdot \frac{d\mu_B}{dx}.$$

If substance B is mobile, it will migrate in the direction of the potential gradient at a velocity which results from combining Eqs. (12.1) and (20.1):

$$v_B = -\omega_B \cdot \frac{d\mu_B}{dx}. \tag{20.2}$$

A flow of matter is generated in the direction of a drop in chemical potential. v_B is a (virtual) drift velocity, which results from the much faster, mostly erratic and random, actual (Brownian) molecular motion. In order to achieve a handier description of this process, we will introduce the following established quantities.

The *throughput* (of matter) is the amount of substance Δn_B passing through a given cross-sectional area A in the time span Δt (Fig. 20.2). This includes all the particles that are not farther away than $\Delta x = v_B \cdot \Delta t$ from area A, i.e., all the particles inside the (dashed) cuboid with the volume $A \cdot \Delta x = A \cdot v_B \cdot \Delta t$. We find the amount of substance when we multiply this volume by the substance concentration c_B:

$$\textit{Throughput: } \Delta n_B = c_B \cdot A \cdot v_B \cdot \Delta t. \tag{20.3}$$

We can derive other useful quantities from the throughput such as *matter flux* (or *current of amount of substance*) J_B as well as *flux density* (or *current density*) (of matter) j_B, which denotes the flux per area through which the flow passes:

$$\textit{Matter flux: } J_B = \frac{\Delta n_B}{\Delta t} = c_B \cdot A \cdot v_B, \tag{20.4}$$

$$\textit{Flux density} \text{ (of matter): } j_B = \frac{J_B}{A} \text{ or } j_B = c_B \cdot v_B. \tag{20.5}$$

The relation with the chemical potential gained from Eq. (20.2) can be used for v_B, where $c_B \cdot \omega_B$ appears as a kind of conductivity σ_B for substance B:

$$j_B = -c_B \cdot \omega_B \frac{d\mu_B}{dx} \text{ with "matter conductivity" } \sigma_B = c_B \cdot \omega_B. \tag{20.6}$$

Law of Diffusion The flow of matter in a homogeneous environment at low concentration c_B represents an important special case. Here, the concentration, and indirectly the position dependency of the chemical potential, can be expressed by the mass action equation:

$$\mu_B(x) = \mu_{B,0} + RT\ln\frac{c_B(x)}{c_{B,0}}. \tag{20.7}$$

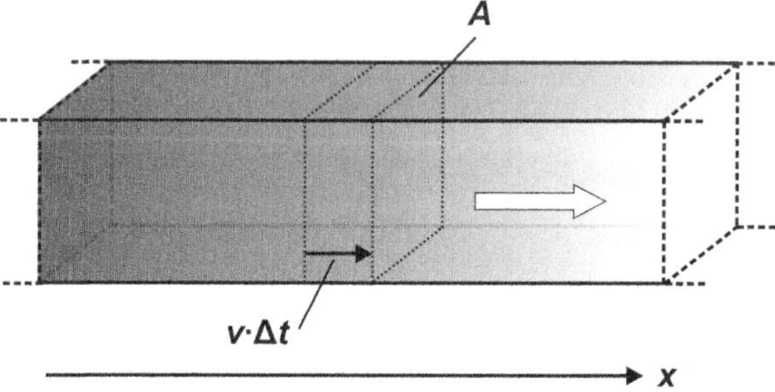

Fig. 20.2 Schema of the relationship of migration velocity v, concentration c, and matter flux J or flux density j.

In order to obtain an expression for the potential gradient $d\mu_B(x)/dx$, we must take the derivative with respect to x at constant T. Taking the chain rule into account, we obtain (compare Sect. A.1.2 in the Appendix):

$$\frac{d\mu_B(x)}{dx} = \frac{RT}{c_B(x)} \cdot \frac{dc_B(x)}{dx}. \tag{20.8}$$

A few words about the calculation. The term $\mu_{B,0}$, which is constant, drops out when taking the derivative. The constant factor RT remains. The derivative of the logarithmic function $y = \ln x$ yields the reciprocal of its argument, meaning $1/x$, whereby a constant factor there (in this case, the factor $1/c_{B,0}$) can be omitted. The reason for this omission is simply that $\ln(ax) = \ln a + \ln x$, and in taking the derivative, the constant expression $\ln a$ disappears. The intermediate result is $RT/c_B(x)$. According to the chain rule, we must still multiply this result by the derivative of the "inner" function $c_B(x)$. This means multiplying by $dc_B(x)/dx$, as in Eq. (20.8) above.

With the help of this equation, we obtain the following expression for the drift velocity v_B and the flux density j_B:

$$v_B = -\omega_B \cdot \frac{d\mu_B}{dx} = -\omega_B \cdot \frac{RT}{c_B} \frac{dc_B}{dx} \quad \text{or} \quad j_B = c_B \cdot v_B = -\omega_B RT \frac{dc_B}{dx}. \tag{20.9}$$

The flux density is proportional to the concentration gradient. The expression ωRT is usually abbreviated to D and called the *diffusion coefficient* (unit $\mathrm{m^2\ s^{-1}}$). This relation was first set up in this form by the German physiologist Adolf Eugen Fick in the year 1855:

$$j_B = -D_B \cdot \frac{dc_B}{dx} \quad \text{Fick's (first) law of diffusion.} \tag{20.10}$$

The relation between ω_B (or a corresponding quantity) and D_B was only discovered in 1908, almost simultaneously by Albert Einstein and Marian von Smoluchowski:

$$D_B = \omega_B RT \quad \text{Einstein-Smoluchowski equation.} \tag{20.11}$$

Table 20.3 shows some values for diffusion coefficients in water.

Table 20.3 Diffusion coefficients of some substances in water at 298 K (catalase at 293 K) in the limit of vanishing concentration. The value for water can be determined experimentally by using H_2O molecules marked with isotopes (such as $H_2{}^{17}O$).

Substance	D_0 ($10^{-9}\ \mathrm{m^2\ s^{-1}}$)
Water	2.26
Hydrogen	5.11
Carbon dioxide	1.91
Acetate ion	1.29
Glucose	0.67
Catalase	0.041

While D_B in liquids at room temperature and standard pressure is of the order of 10^{-9} m^2 s^{-1}, the value for gases lies much higher, at around 10^{-4} m^2 s^{-1}. It is, however, many orders of magnitude lower for solid substances.

The diffusion coefficient depends upon the temperature, not so much because of the factor T appearing in Eq. (20.11), but because ω_B itself is not constant. In liquids and especially in solids, moving particles have considerable attractive forces to overcome—meaning, there will be an activation threshold. If we assume a temperature dependency with a molar activation energy W_A corresponding to the Arrhenius equation, we obtain

$$D_B \sim e^{-W_A/RT}. \tag{20.12}$$

This strong temperature dependency, which is caused by the exponential expression that is "hidden" behind ω_B in the equation above, masks all the other influences such as the linear dependency of T. For this reason, a diffusion coefficient of a magnitude of 10^{-24} m^2 s^{-1} in a solid substance at low temperatures can climb to 10^{-8} m^2 s^{-1} at temperatures of 1,500 K.

Duration of Concentration Equalization   In order to get an impression of how fast a uniform distribution of substances at room temperature in a limited space can happen, we will consider a container in which an artificial linear concentration gradient along its largest diameter l has been created for a substance B contained in it (Fig. 20.3). Take $c_{B,0}$ as the concentration of B at uniform distribution in the container. The concentration gradient is then $dc_B/dx = 2c_{B,0}/l$. At the first moment

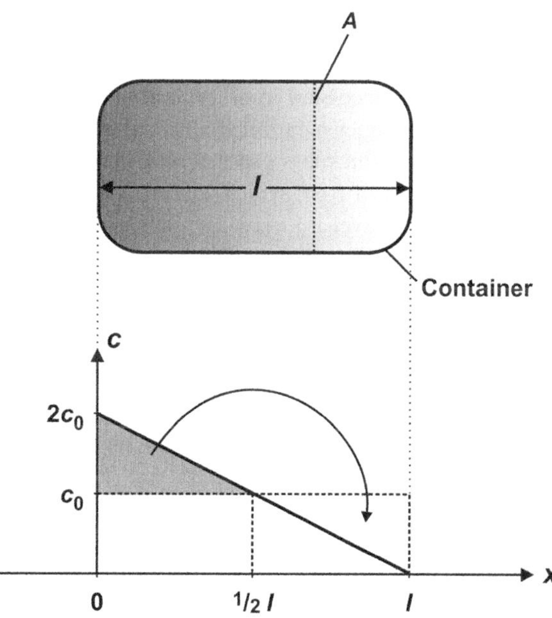

Fig. 20.3 Concentration equalization in a container in which a linear concentration gradient is initially assumed. To achieve equalization, it is enough to transfer the amount of substance $n = \frac{1}{2} \cdot c_0 \cdot (\frac{1}{2}l) \cdot A$ indicated by the *gray triangle* in the lower partial image from left to right.

t_0, the flux density j_B of B is the same everywhere. The diffusion law tells us that it is $j_B = D_B \cdot 2c_{B,0}/l$. We assume this value to remain unchanged for a cross section exactly through the middle of the container for a time span Δt. The amount of substance flowing through this cross section in this time span is then $\Delta n_B = A \cdot \Delta t \cdot j_B$. The amount of B flowing from left to right (in order to establish a uniform concentration) is indicated by the dark triangle in the figure. It is exactly $\frac{1}{4}$ of the amount of B in the container, $A \cdot l \cdot c_{B,0}$, that is represented by the large triangle. If Δn_B should equal this amount, then

$$\Delta n_B = A \cdot \Delta t \cdot j_B = A \cdot \Delta t \cdot D_B \frac{2c_{B,0}}{l} = \frac{1}{4} A \cdot l \cdot c_{B,0} \quad \text{and} \quad \Delta t = \frac{l^2}{8D}. \quad (20.13)$$

Δt roughly represents the time needed to approximately equalize the concentration in the container. We see that the time decreases by l^2, becoming very short for small diffusion lengths. Hence, the differences in concentration equalize very quickly. This is why for small dimensions, diffusion is a very efficient method of distribution. We see it in the exchange of substances in living cells and in the equalization of residual differences of concentration after two different solutions have been stirred together.

Here is an example: In liquids, low-molecular substances have a diffusion coefficient of the order of 10^{-9} m^2 s^{-1}. In a red blood cell with a diameter of about 10^{-5} m, the time necessary for diffusion of these substances is approximately

$$\Delta t = \frac{l^2}{8D} = \frac{\left(10^{-5}\,\mathrm{m}\right)^2}{8 \times 10^{-9}\,\mathrm{m^2\,s^{-1}}} \approx 10^{-2}\,\mathrm{s}.$$

In normal cells having a size of about 10^{-4} m, the time needed is $\Delta t \approx 1$ s, while in a 1 L glass beaker (diameter 10^{-1} m), 10^6 s ≈ 2 weeks are generally required. For this reason, stirring is necessary in order to homogenize solutions in large containers.

While the container mentioned above should be closed on all sides, let us now imagine a layer of thickness l so that a substance B can escape on either side. This may be a plastic foil from which a plasticizer gradually diffuses. Starting on the surface, the entire layer is gradually affected and a sinusoidal concentration profile quickly emerges (Fig. 20.4). This profile then flattens slowly until it finally

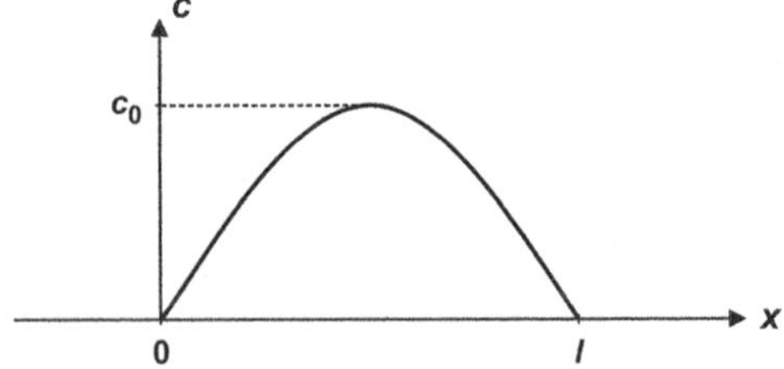

Fig. 20.4 Sinusoidal concentration profile on a layer open on both sides.

disappears altogether. At any position x, the concentration decreases exponentially with time t:

$$c_B = c_{B,0} \sin \frac{\pi x}{l} \cdot e^{-t/\tau_1} \quad \text{with} \quad \tau_1 = \frac{l^2}{\pi^2 D_B}. \tag{20.14}$$

The decay time τ_1 agrees with our estimate (20.13), but instead of the 8 in the denominator, we now have $\pi^2 \approx 9.9$. Any concentration profile can be represented by a superposition of such sinusoidal profiles that change independently of each other in the course of time. Profiles having a "wavelength" $\lambda_n = \lambda_1/n$ that is shorter by $1/n$ $(n = 1, 2, 3, \ldots)$ than that of the "fundamental wave" $\lambda_1 = 2l$ have a decay time $\tau_n = \tau_1/n^2$ that is shorter by the factor $1/n^2$ as we can see in Eq. (20.14). The corresponding theory was developed by the French mathematician and physicist Jean Baptiste Fourier at the beginning of the nineteenth century using heat conduction that obeys similar laws.

Diffusion Control   Let us now return to the subject of Sect. 20.1. We wish to make a rough estimate of the rate density r for a reaction $A + B \rightarrow \ldots$, where the collision of two particles A and B is the step that determines the rate. For the sake of simplicity, we imagine the particles to be spheres where the type A is mobile and the type B is not. As soon as an A particle approaches a B particle (or more exactly, the centers of the particles) to a distance of a_0, a reaction is assumed to occur where A disappears. In the process, an A-poor environment is produced around B, which we will imagine as spherically symmetrical in shape and into which the A diffuses from the surroundings. If c_A^∞ is the concentration of A in the environment, meaning far away from B, the following equation should be valid for c_A in the depletion zone at a distance a from B:

$$c_A = c_A^\infty \cdot \left(1 - \frac{a_0}{a}\right) \quad \text{and therefore} \quad \frac{dc_A}{da} = c_A^\infty \cdot \frac{a_0}{a^2}.$$

This approach seems reasonable because for large a, c_A tends toward c_A^∞ and at a distance, $a = a_0$, we have $c_A = 0$, because A disappears. On the other hand, the matter flux J_A out of the environment in the B direction must be independent of a for $a \geq a_0$ because A is neither formed nor destroyed under way. This applies to the approach above because according to Fick's law $j_A = -D_A \cdot (dc_A/da)$, we find for the flux J_A through the surface $4\pi a^2$ of a sphere with a radius a around the center of B:

$$J_A = 4\pi a^2 \cdot |j_A| = 4\pi a^2 \cdot D_A \cdot c_A^\infty \frac{a_0}{a^2} = 4\pi D_A a_0 c_A^\infty = \text{const}.$$

J_A multiplied by the number $N_B = n_B/\tau = c_B^\infty V/\tau$ of all the B particles in a volume V yields the conversion rate ω. When we finally divide by V, the result is the desired rate density r:

$$r = k \cdot c_A^\infty \cdot c_B^\infty \quad \text{with} \quad k = 4\pi D_A \frac{a_0}{\tau}.$$

If B is also mobile, r will be somewhat greater, because $D_A + D_B$ will appear in the formula on the right instead of D_A. This detail is unimportant to our estimate, so we can omit it. According to Table 20.3, D is of the order of 10^{-9} $\mathrm{m^2\ s^{-1}}$ for low-molecular substances in an aqueous solution, and a_0 corresponds to the particle diameters and is approximately 10^{-9} m. These values yield

$$k = 4\pi \times 10^{-9}\,\mathrm{m^2\,s^{-1}} \frac{10^{-9}\,\mathrm{m}}{1.66 \times 10^{-24}\,\mathrm{mol}} \approx 10^7\,\mathrm{m^3\,s^{-1}\,mol^{-1}}.$$

Compared to this, the k value for a bimolecular reaction (Sect. 18.3) with a vanishing activation threshold, $\Delta_\ddagger \overset{\circ}{\mu} = 0$, is 1,000 times greater:

$$k = \frac{k_B T}{hc^\ominus} = \frac{1.38 \times 10^{-23}\,\mathrm{J\,K^{-1}} \times 300\,\mathrm{K}}{6.63 \times 10^{-34}\,\mathrm{J\,s} \times 10^3\,\mathrm{mol\,m^{-3}}} \approx 10^{10}\,\mathrm{m^3\,s^{-1}\,mol^{-1}}.$$

Only at an activation threshold $\Delta_\ddagger \overset{\circ}{\mu}$ considerably above $3\mu_d$, does this step—rather than diffusion—determine the conversion rate.

Due to their larger particle diameters and decreased diffusion coefficients, high-molecular substances such as enzymes lead to considerably lower k values of the order of 10^5 to 10^6 $\mathrm{m^3\ s^{-1}\ mol^{-1}}$.

20.3   Fluidity

Viscosity Gases and liquids can flow more or less easily. Shear forces cause the particles to rub and slide against each other. The more viscous the liquid or gas is under equal driving forces.

- The lower the flow velocity and therefore for example the outflow velocity associated with it (Experiment 20.3),
- The slower the migration of dissolved substances in it (diffusion, sedimentation, ion migration).

Flow of substances is always hindered by inner friction. The driving force needs to overcome this resistive force. Let us imagine two parallel plates with surfaces of A at a set distance of d from each other. There is a gas or liquid between these plates acting as a kind of lubricant (Fig. 20.5). The upper plate is now passed over the stationary lower plate in the x direction at a constant speed v_0. If we were to look at

Experiment 20.3 *Outflow of various glycerin–water mixtures*: Three glycerin–water mixtures with different compositions are filled into separating funnels. The stopcocks are then quickly opened, one after the other. The lower the fraction of viscous glycerin, the faster the mixture flows out.

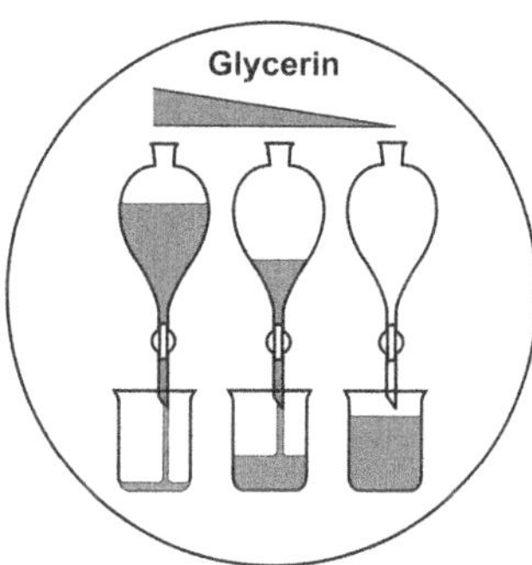

Fig. 20.5 Linear velocity profile of a liquid or, in this case, a gas between one moveable and one stationary plate. A change of particles between the layers is related to a momentum transport.

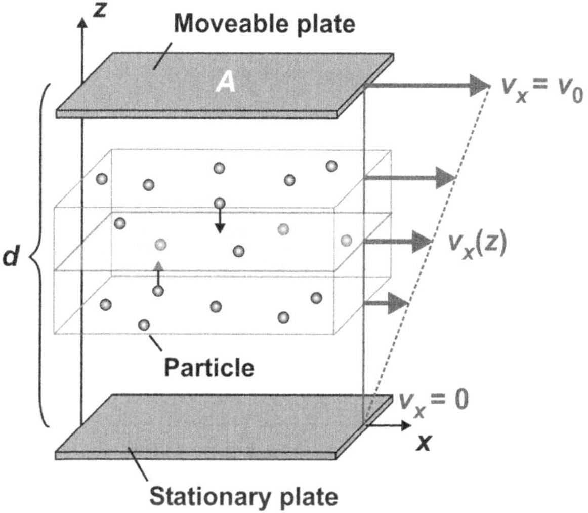

the liquid at microscopic scale, we would see that due to adhesion forces, its particles adhere to the plates in a thin layer. The particles are motionless on the stationary lower plate, while the particles on the upper plate are carried along at the speed v_0. The liquid lying in between can be imagined as divided into thin, flat layers where the speed will be higher, the closer it is to the upper plate. A linear velocity profile $v_x(z)$ emerges that is dependent upon the vertical distance z from the lower plate. A stack of glass plates held together by honey can serve as a model (Experiment 20.4).

The particles do not adhere to their original layer but can leave it by taking their p_x momentum (in the x direction) so that a transport of momentum takes place perpendicular to the direction of motion. When particles move from a lower to a higher layer, they will have a braking effect because they have a lower momentum in the x direction than the other particles. Conversely, an acceleration will occur due to particles moving down from a higher layer. These interactions are noticeable as

Experiment 20.4 *Honey experiment*: About ten glass plates (microscope slides, for example) are smeared with honey and then stacked one upon the other. The slow slide of the topmost plate leads to a movement by all of them (except the lowest one which is held in place).

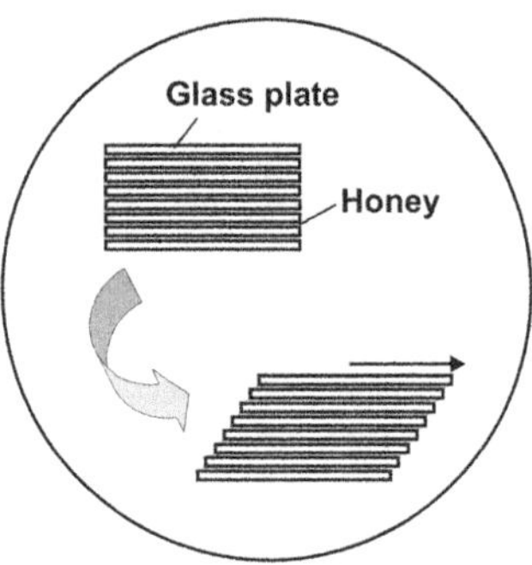

Table 20.4 Dynamic and kinematic viscosity, η and v ($= \eta/\rho$), of various substances at 293 K.

Substance	η (10^{-3} Pa s)	v (10^{-6} m^2 s)
Water	1.002	1.004
Mercury	1.526	0.115
Diethyl ether	0.243	0.340
Benzene	0.604	0.736
Ethylene glycol	19.9	17.9
Glycerin	1,412	1,120
Honey	~10^4	~10^4
Tar	>10^5	>10^5
Hydrogen	0.89×10^{-2}	106
Oxygen	2.03×10^{-2}	15.3

so-called *inner friction* (as opposed to outer friction between two solid bodies such as static friction and rolling friction where it is mostly only surfaces that are affected). The friction force F_f to be overcome is not only dependent upon the surface A, but also upon the transfer of momentum between the layers and therefore the velocity gradient $\Delta v_x/\Delta z$,

$$F_f = -\eta \cdot A \cdot \frac{\Delta v_x}{\Delta z} \quad \text{Newton's law of friction (global form),} \quad (20.15)$$

where, in our case, we simply have $\Delta v_x/\Delta z = v_0/d$. The proportionality factor η is called the (dynamic) *viscosity*. It is a property of a substance and its SI unit is N s m^{-2} = Pa s. Table 20.4 contains viscosities for some substances at 293 K. In addition, the ratio $v = \eta/\rho$, the so-called *kinematic viscosity*, has been included. We will return to this further down.

A force F can be considered a momentum current (a flux of momentum) J_p (Sect. 2.7). Consider one-dimensional motion, say pushing a motor vehicle (Fig. 2.12). If p indicates the momentum of the object being moved and F is the only force acting upon it (i.e., if F is the only path over which the object exchanges momentum with its environment), we have:

$$F = J_p = \frac{\mathrm{d}p}{\mathrm{d}t} \quad \left(\text{for several paths we have} : \sum_i F_i = \sum_i J_{p,i} = \frac{\mathrm{d}p}{\mathrm{d}t} \right).$$

If we imagine momentum flowing via a straight rod having a uniform cross section of A into an object being pushed in the x direction, the *momentum current density* everywhere in the rod will equal $j_p = J_p/A$. The transport of momentum here occurs at the speed of sound, without loss and without energy being dissipated and entropy being generated.

The momentum transport we are interested in here differs in several ways from the one just discussed. It is not one-dimensional, so for vector quantities such as momentum and velocity we need to distinguish the components in the x, y, and z directions. In Fig. 20.4, these are the directions left to right, front to back, and bottom to top. Our case is comparably simple because only the x components p_x and v_x of momentum and velocity do not equal zero. Unlike before, the momentum is not conveyed in the x direction but perpendicularly to it, following the velocity gradient $\mathrm{d}v_x/\mathrm{d}z$, from top to bottom. The fact that we are dealing with the z-component of the flux of p_x may be expressed by notations such as $J_{px,z}$ or $(J_p)_{x,z}$. Since this is the only relevant component here, we can forgo the indices and simply write J_p, which is identical to the force appearing in Newton's law of friction (20.15). If we introduce this quantity or, rather, the related momentum flux density $j_p = J_p/A$ into the friction law, and replace the difference quotient $\Delta v_x/\Delta z$ by the differential quotient $\mathrm{d}v_x/\mathrm{d}z$, it turns into

$$j_p = -\eta \frac{\mathrm{d}v_x}{\mathrm{d}z} \quad \text{Newton's law of friction (local form).} \tag{20.16}$$

The velocity v_x appears here in the role of a potential belonging to the substance-like quantity momentum p_x. We describe it as "kinetic" or "kinematic" to distinguish it from other potentials such as the chemical potential μ or the "thermal potential" T.

Duration of Velocity Equalization If the upper plate is not constantly pushed forward giving it a constant supply of momentum p_x, the fluid flow between the plates will soon come to a stop. This also happens when the lower plate is not held so that momentum cannot flow off there. In this situation, the momentum behaves like a diffusing substance in a closed container. For the sake of simplicity, we will imagine the plates to have no mass so that they can adapt without inertia to the velocity of the adjacent layer of liquid. All that remains is the redistribution of momentum p_x in the liquid where the excess in the upper half is to be moved to the lower half. If ρ is the density of the liquid, mass (first term), average velocity (second term), and momentum (third term) are:

$$\text{above}: \quad \tfrac{1}{4}\,\rho \cdot A \cdot d, \quad \tfrac{3}{4}v_0, \quad \tfrac{3}{8}\,\rho \cdot A \cdot d \cdot v_0,$$
$$\text{below}: \quad \tfrac{1}{4}\,\rho \cdot A \cdot d, \quad \tfrac{1}{4}v_0, \quad \tfrac{1}{8}\,\rho \cdot A \cdot d \cdot v_0.$$

In order to attain a uniform distribution, it is enough to relocate $\Delta p_x = \frac{1}{8}\rho \cdot A \cdot d \cdot v_0$ from above to below. Having an initial velocity gradient of v_0/d, the amount $\Delta p_x = |j_p| \cdot A \cdot \Delta t = \eta \cdot (v_0/d) \cdot A \cdot \Delta t$ can be conveyed downward in the time span Δt. If we equate both Δp_x values and solve for Δt, we obtain a relation corresponding to Eq. (20.13) where v is the *kinematic* viscosity mentioned above:

$$\Delta t = \frac{d^2}{8v} \quad \text{with} \quad v = \frac{\eta}{\rho}. \tag{20.17}$$

v has the same SI unit as the diffusion coefficient D, namely $\text{m}^2\,\text{s}^{-1}$. The statements above about concentration profiles and their decomposition into sinusoidal contributions and their decay times are correspondingly valid for velocity profiles $v_x(z)$.

Particles in a Viscous Medium In many examples, particles such as molecules or macromolecules or ions move in a medium having a viscosity of η. However, let us first consider the motion of a macroscopic sphere with radius r at a velocity of v in a liquid or gas. This motion is caused by a force such as the force of gravity. A frictional force acts against it and this force will be greater the more viscous the medium is and the larger the sphere. It also rises as velocity increases. From the field of hydrodynamics, we obtain

$$F_f = 6\pi \cdot \eta \cdot r \cdot v \quad \text{Stokes' law.} \tag{20.18}$$

This equation can also be used approximately for microscopic particles like the molecules and ions mentioned above. We expect that the diffusion coefficient D_B of a substance B will be the smaller, the more viscous the medium is in which it migrates. Let us consider a simple example of a rigid, spherical particle with the radius r. The diffusion force [expressed by Eq. (20.1)] is counteracted by the Stokes frictional force [Eq. (20.18)]:

$$F_B = \tau \cdot \frac{v_B}{\omega_B} = 6\pi \cdot \eta \cdot r_B \cdot v_B = F_f,$$

where $n_B = \tau$ because we are only considering a single B particle, meaning an amount of substance B exactly corresponding to the elementary amount of substance τ. The result for mobility ω_B is then

$$\omega_B = \frac{\tau}{6\pi \cdot \eta \cdot r_B} \tag{20.19}$$

and because of $D_B = \omega_B RT$ (Einstein–Smoluchowski equation), the diffusion coefficient equals

$$D_B = \frac{k_B T}{6\pi \cdot \eta \cdot r_B}, \tag{20.20}$$

where $k_B = \tau R$ is the Boltzmann constant. Because of this relation, we expect a temperature dependency for the viscosity of liquids which is the reciprocal of that of the diffusion coefficient [Eq. (20.12)]. If we take into account only the most important contribution, the exponential expression, we obtain:

$$\eta \sim \eta_\infty \cdot e^{+W_A/RT}, \quad \text{where, for example,} \, W_A(H_2O) \approx 16\,\text{kJ}\,\text{mol}^{-1}; \quad (20.21)$$

therefore, we should expect a temperature dependency for the viscosity of liquids which is the reciprocal of that of the diffusion coefficient [Eq. (20.12)].

Viscosity should strongly reduce as temperature rises. Boiling water has four times lower viscosity than water at room temperature, allowing it to run correspondingly faster through a filter. This is good for laboratory work as well as making coffee.

20.4 Entropy Conduction

Entropy Conductivity Most of our experiences of entropy are gained in everyday life, even if unconsciously. The coffee in a thermos remains hot because it is difficult for entropy to penetrate the vacuum jacket, while coffee in a cup cools down fairly quickly because entropy leaves along with the steam rising from it. Gases and foam materials whose volumes are up to 97 % gas-filled pockets strongly hinder the flow of entropy, while metal exhibits a conductivity of about 1,000 times that much. We will quantify this phenomenon in the following (Fig. 20.6).

We will consider the simplest case, a *homogeneous, isotropic* body with a constant cross section A. Isotropic means that—in contrast to a block of wood, for example—all the directions are equal. If the temperature drops from left to right, entropy moves in this direction. Because most substance properties more or less depend upon temperature, and this is also true for the quantity that interests us here, we will only consider a small temperature difference $\Delta T \ll T$ between the left and right sides of the body. The entropy flux (or current) J_S or the entropy flux

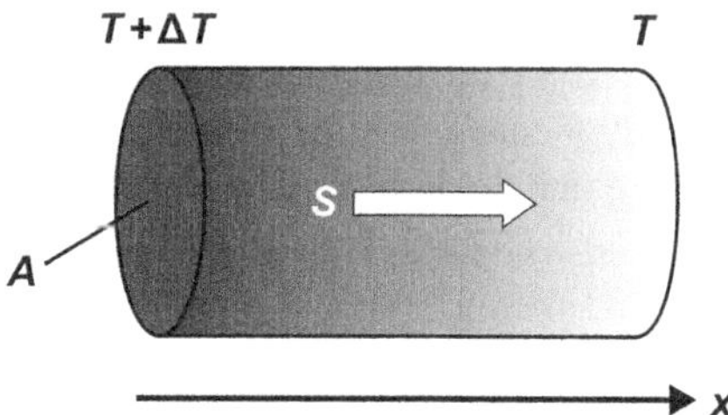

Fig. 20.6 Entropy conduction.

Table 20.5 Entropy conductivity σ_S and thermal conductivity λ at room temperature, when not otherwise indicated.

Substance	σ_S (Ct K^{-1} s^{-1} m^{-1})	λ (J K^{-1} s^{-1} m^{-1})
Diamond	8	2,300
Copper	1.3	400
Brass (40 % Zn)	0.2	113
Stainless steel	0.05	15
Glass	0.003	0.8
Water (0 °C)	0.00205	0.56
Water (100 °C)	0.00182	0.68
Foam materials	0.00015	0.04
Air	0.000087	0.026

Experiment 20.5 *Entropy conduction in solid substances*: A cross made of a copper rod, a brass rod, and a steel rod with wax spheres attached to their ends is heated in the middle. The wax spheres fall off one after another according to the different entropy conductivities: first copper, then brass, and finally, steel.

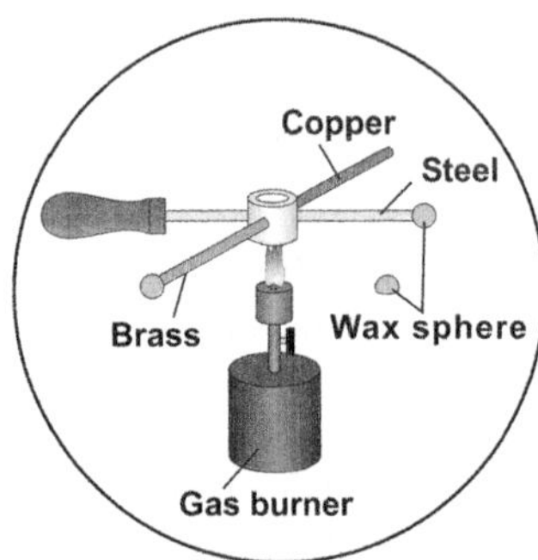

(or current) density $j_S = J_S/A$ are proportional to the temperature gradient $\Delta T/\Delta x$ or more precisely dT/dx:

$$J_S = \sigma_S \cdot A \cdot \frac{\Delta T}{\Delta x} \quad \text{or} \quad j_S = -\sigma_S \cdot \frac{dT}{dx} \quad \text{Fourier's law.} \tag{20.22}$$

These are two simple versions, a global one and a local one, of a law mentioned at the end of Sect. 20.2 for which Fourier introduced solution methods in 1822 which have proven applicable to many areas of physics and mathematics. σ_S is the entropy conductivity. Instead of σ_S, in tables we usually find the entropy conductivity multiplied by T, the "heat conductivity" (or thermal conductivity) $\lambda = T \cdot \sigma_S$. Table 20.5 contains some values for these two quantities.

The example of water shows us that neither λ nor σ_S can be considered constant. In the case of monoatomic or biatomic gases such as air, σ_S changes in proportion to $T^{-1/2}$ and therefore $\lambda \sim T^{+1/2}$. This dependency is mostly neglected because any mistakes involving it are unimportant compared to other less controllable influences.

Let us take a look at an experiment showing different entropy conductivities in various solid substances (Experiment 20.5).

Duration of Temperature Equalization   When making a boiled egg for breakfast we want to know how long it needs to boil to reach the soft or hard consistency that we desire. This question is similar to the one about how long it takes for an equalization of concentration to be achieved in a closed container (Fig. 20.3). We will refer to the same image, but will now plot temperature T instead of concentration c along the ordinate. In place of the equalized concentration c_0 obtained in the end, we have the corresponding temperature T_0. We assume a linear temperature gradient where, on the far left, we have a temperature of $T_0 + \Delta T$, and on the far right, $T_0 - \Delta T$, with $\Delta T \ll T$. The gray triangle on the left corresponds to an excess of entropy of $\Delta S = \frac{1}{2}\, \mathcal{C} \cdot \frac{1}{2}\Delta T$ in the left part of the body and for which there is a corresponding lack of entropy on the right. $\mathcal{C}$ is the entropy capacity $\mathcal{C} = m \cdot \boldsymbol{c}$ of the entire body, $m = \rho \cdot A \cdot l$ its mass, ρ is the density, and $\boldsymbol{c} = \mathcal{C}/m$ the *specific* entropy capacity. While $\mathcal{C}$, A, l, and m are characteristics of the body, ρ and $\boldsymbol{c}$ indicate properties of the substance it is made up of. In summary:

$$\Delta S = \frac{1}{2}\, \mathcal{C} \cdot \frac{1}{2}\, \Delta T = \frac{1}{4}\boldsymbol{c}\,\rho A l \cdot \Delta T.$$

If we imagine a constant temperature gradient of $2\Delta T/l$, then an amount of entropy equal to

$$\Delta S = J_S \Delta t = \sigma_S \cdot A \cdot \frac{2\Delta T}{l}\, \Delta t$$

will flow from left to right during the short time span Δt (see Fourier's law [Eq. (20.22)]. If we equate both ΔS values and solve for Δt, we obtain an approximate value for the duration of temperature equalization:

$$\Delta t = \frac{l^2}{8a} \quad \text{with} \quad a = \frac{\sigma_S}{\rho \boldsymbol{c}} = \frac{\lambda}{\rho c_s}. \tag{20.23}$$

The coefficient a in the denominator is called the "temperature conductivity," although this is somewhat misleading because it is not the temperature which is conducted. In the formula on the right, a is not only expressed by the entropic quantities σ_S and $\boldsymbol{c}$, but also by their energetic opposites λ and c_s, thermal conductivity and specific heat capacity. (We use the symbol c_s instead of the usual c to avoid any confusion with the symbol c for concentration.)

Coupling of Matter and Entropy Currents   What happens when a substance is simultaneously subjected to a gradient of its potential μ and the temperature T? Let us imagine a pond where the water below is cold, T_1, and heats up in the Sun becoming warm above with $T_2 > T_1$. Then $\mu_2 = \mu_1 + \alpha \cdot (T_2 - T_1) < \mu_1$ would be the case because $\alpha = -S_m$ is negative. Consequently, there is a drive for water to rise upward. The entropy in the water strives downward, and in the process, both influences cancel each other. The amount of energy $W_n + W_S$ expended for shifting an amount of water n upward is:

$$W_n + W_S = (\mu_2 - \mu_1)n + (T_2 - T_1)S = -S_m(T_2 - T_1)n + (T_2 - T_1)S_m \cdot n = 0.$$

This is not always the case. When a dissolved substance diffuses, the molar entropy carried along with it, the so-called transfer entropy S_m^*, is greater or smaller than S_m. S_m^* is about 2.7 Ct mol^{-1} greater in a cane sugar solution with a concentration of 1 kmol L^{-1}. The influence of entropy gains the upper hand here so the sugar is forced to the cooler side. At a temperature difference of 10 K, and an enrichment of about 1.1 %, the two effects just compensate for each other. Let us calculate the energy expenditure for the transport of a small amount of sugar n from the cooler side, T_1, to the warmer side, $T_2 > T_1$:

$$W_n + W_S = (\mu_2 - \mu_1) \cdot n + (T_2 - T_1) \cdot S_m^* \cdot n,$$
$$= \left[((\mu_1 - S_m(T_2 - T_1)) - (\mu_1 + RT_1 \ln 1.011) + (T_2 - T_1)S_m^* \right] \cdot n,$$
$$= \left[(S_m^* - S_m)(T_2 - T_1) - RT_1 \ln 1.011 \right] \cdot n,$$
$$= \left[2.7\,\text{Ct mol}^{-1} \times 10\,\text{K} - 8.3\,\text{Ct mol}^{-1} \times 300\,\text{K} \times \ln 1.011 \right] \cdot n = [27 - 27] \cdot n = 0.$$

Differences in the chemical potential caused by a temperature gradient are called *thermodiffusion*, whereas the inverse effect where temperature differences are caused by a gradient of the chemical potential is called the *diffusion-thermo effect*. Such couplings between currents are common. The most well known is the *thermo-electric effect* which is caused by a coupling of entropy and charge currents.

20.5   Comparative Overview

Transport Equations In order to carve out the commonalities and differences of the phenomena we have discussed, we will sum them up here. In doing so, we will add another phenomenon: transport of electric charge (Fig. 20.7), because it is the best known of these. Its concepts can serve as an example or help in orientation. In all four cases, diffusion, viscous flow, entropy conduction, and conduction of electricity, there is a substance-like quantity (substance B, momentum p_x, entropy S, and charge Q) being conveyed along the gradients of the corresponding potentials (chemical μ_B, "kinetic" v_x, "thermal" T, and electric φ). In order to look at all this uniformly, we can imagine a small cuboid shaped section of a larger area with the base area A and the height l. The height should be so small that despite an assumed potential difference, the cuboid can be considered homogeneous.

A drop in potential of $\Delta\mu_B$, Δv_x, ΔT, $\Delta\varphi$ from the upper surface to the lower surface drives a current of J_B, J_p, J_S, J_Q downward through the cuboid as long as there is a corresponding conductivity σ_B, σ_p, σ_S, σ_Q to allow for this. The current is proportional to the cross section A and the potential difference and inversely

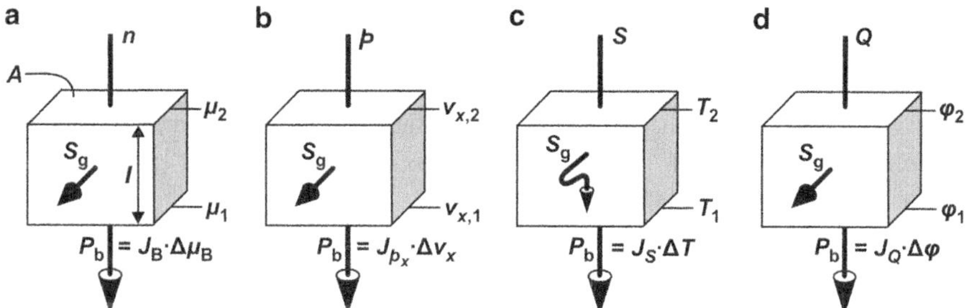

Fig. 20.7 Comparison of different transport processes: (**a**) substance, (**b**) momentum, (**c**) entropy, (**d**) charge. P_b is the dissipation rate, $\dot{S}_g = P_b/T$ the generation rate of entropy, if T is the instantaneous average temperature of the body. In the case of (**c**), the generated entropy leaves along the same path as the one entering from above (indicated by the *arrow bending backward*).

Table 20.6 Comparison of related formulas for different transport processes.

	Diffusion (Fick)	Viscous flow (Newton)	Entropy conduction (Fourier)	Electric conduction (Ohm)
Global form	$J_B = -\sigma_B \cdot A \frac{\Delta\mu_B}{l}$	$J_{px} = -\sigma_p \cdot A \frac{\Delta v_x}{l}$	$J_S = -\sigma_S \cdot A \frac{\Delta T}{l}$	$I = -\sigma \cdot A \frac{\Delta\varphi}{l}$
Local form	$j_B = -\sigma_B \cdot \frac{d\mu_B}{dz}$	$j_{px} = -\sigma_p \cdot \frac{dv_x}{dz}$	$j_S = -\sigma_S \cdot \frac{dT}{dz}$	$j = -\sigma \cdot \frac{d\varphi}{dz}$
Special form (example)	$j_B = -D_B \cdot \frac{dc_B}{dz}$	$F_f = -\eta \cdot A \cdot \frac{dv_x}{dz}$	$J_Q = -\lambda \frac{dT}{dz}$	$I = \frac{U}{R_Q}$
Conductivity	$\sigma_B = c_B D_B / RT$	$\sigma_p = \eta$	$\sigma_S = \lambda / RT$	$\sigma = 1/\rho_Q$
"Diffusivity"	D_B	$v = \eta/\rho$	$a = \lambda/(c_s\rho)$	–

The formulas in the first line are analogous to each other as are the ones in the second line. In the third line, they are adapted to special applications and contain quantities that only correspond to each other to a certain extent. In the fourth line the "conductivities" and in the fifth line the corresponding "equalization coefficients" ("diffusivities") are expressed by the preferred quantities of their respective domains. (The index Q in the fourth row corresponds to "heat" and in the fifth row, however, to "charge")

proportional to the length l through which the current flows whereby the conductivity σ_B, σ_p, σ_S, σ_Q appears as a proportionality factor. The formulas have been compiled in Table 20.6, where in the case of charge, the usual symbols I, j, and σ have been used instead of J_Q, j_Q, and σ_Q. The names of the persons for whom these laws have been named are at the top of the table. In the line below that are the analogous "local" laws in which the current J is replaced by the corresponding current density $j = J/A$ and the quotient "potential difference/length" is formulated as the derivative of the potential in question with respect to the spatial coordinate z.

The most well-known version of *Ohm's law* is $I = U/R_Q$, where I generally means "current," $U = -\Delta\varphi$ "voltage," $R_Q = \rho_Q \cdot l/A$ "(electric) resistance," and $\rho_Q = 1/\sigma$ "(electric) resistivity." The "conductance" $G_Q = 1/R_Q$ is often used instead of R_Q, which gives Ohm's law the following form: $I = G_Q \cdot U$. For clarity's sake, we have included the index $_Q$ because we are already using the symbols without indexes for other purposes. There is a comparable diversity for the formulas in the other columns, so it is difficult to find their common pattern. To give an example, the inverse $\varphi = \eta^{-1}$ of the viscosity η is in use as well. It is called *fluidity* and corresponds to the resistivity ρ_Q.

"Diffusivities" Despite their very different names, the "diffusion coefficient" D_B, the "kinematic viscosity" v, and the "temperature conductivity" a are analogous quantities. They all have the SI unit $m^2\,s^{-1}$ and quantify properties of substances with respect to different transport processes that determine the durations of equalization processes. They are called *diffusivities* (or *"equalization coefficients"*).

They can be written as the quotient "conductivity/capacity density," as would be expected. The more conductive the medium, the more quickly the potential differences equalize. The larger the amount to be transported, i.e., the higher the capacities at equal potential differences, the longer it will take. Because $\sigma_B = c_B \omega_B$ describes the "matter conductivity" [Eq. (20.6)] and $b_B = c_B/RT$ the "matter capacity density" (Sect. 6.7), the following is valid:

$$D_B = \omega_B RT = \frac{\omega_B c_B}{c_B/RT} = \frac{\sigma_B}{b_B} = \frac{\text{"matter conductivity"}}{\text{"matter capacity density"}}.$$

In the case of momentum, the mass $m = \rho \cdot V$ plays the role of "capacity" and because $p_x = m v_x$, taking the derivative with respect to the proper "potential" v_x simply results in $\mathrm{d}p_x/\mathrm{d}v_x = m$. The corresponding "capacity density" is therefore the mass density $m/V = \rho$. Viscosity η corresponds to momentum conductivity σ_p, so the following holds:

$$v_B = \frac{\eta}{\rho} = \frac{\text{"momentum conductivity"}}{\text{"momentum capacity density"}}.$$

In Eq. (20.23) for the temperature conductivity (thermal diffusivity) $a = \sigma_S/(\rho \cdot c), c$ indicates the specific or entropy capacity per mass and $\rho \cdot c$ is the corresponding volumic quantity, the corresponding "capacity density," so that:

$$a = \frac{\sigma_S}{\rho c} = \frac{\text{"entropy conductivity"}}{\text{"entropy capacity density"}}.$$

Accompanying Energy Currents Each of the currents J_B, J_p, J_S, J_Q is accompanied by an energy current $J_W = \mu_B \cdot J_B$, $J_W = v_x \cdot J_p$, $J_W = T \cdot J_S$, $J_W = \varphi \cdot J_Q$. More energy flows in than flows out because the potentials at the inflow are higher than at the outflow. The excess energy could be used for other purposes if there were an apparatus for doing so. This does not happen here. Instead, as the energy is dissipated while entropy is generated, the energy is lost or becomes "useless." The *dissipation power* P_b (the part of the energy current that is dissipated or "burnt") is given in Fig. 20.7 for the cases illustrated there. If T is the temperature, the generation rate of entropy resulting from it is:

$$\dot{S}_{g} = \frac{P_{b}}{T}.$$

If this entropy is not drawn off, the body will heat up continuously. However, if the entropy is allowed to flow away at a constant temperature T, energy, namely P_{b}, will also flow off along the same path. The energy current then bifurcates as part of it flows out with the entropy S_{g}.

This looks a little different in case c of Fig. 20.7, because the entropy S_{g} generated there flows away along the same path as the one flowing in above. While the entropy continually grows as it moves through the temperature gradient, the energy remains the same. For this reason, there is a preferred description of this case in which the energy itself is considered the flowing substance. The corresponding current density $j_{W} = T \cdot j_{S}$ is obtained by multiplying Eq. (20.22) for entropy by the temperature T, where we also replace $T \cdot \sigma_{S}$, as is usual in this case, by λ:

$$J_{W} = \lambda \cdot A \cdot \frac{\Delta T}{\Delta x} \quad \text{or} \quad j_{W} = -\lambda \cdot \frac{dT}{dx} \quad \text{``heat conduction equation''.} \quad (20.24)$$

Chapter 21
Electrolyte Solutions

A discussion of the chemical drive of solvation and hydration processes, respectively, leads to the introduction of the basic concept of *electrolytic dissociation*, the disintegration of a substance in solution into mobile ions. Subsequently, we learn about the migration of these ions along an electric potential gradient as a special case of spreading of substances in space. The ionic mobilities provide a link to conductance and the related quantities conductivity as well as molar and ionic conductivity. For determining the conductivity of ions experimentally, the introduction of the term transport number which indicates the different contribution of ions to the electric current in electrolytes is very useful. In the last section, the technique for measuring conductivities is presented as well as its application in analytical chemistry where conductometric titration is a routine method.

21.1 Electrolytic Dissociation

Electrolytes are substances in solid, melted, or dissolved states that either fully or partly disintegrate into mobile ions. In the first chapter, we were introduced to the concept of ions as "substances" according to

$$[\mathrm{Cl}]^- = \mathrm{Cl}_1 \mathrm{e}_1 \quad \text{or} \quad [\mathrm{Na}]^+ = \mathrm{Na}_1 \mathrm{e}_{-1}.$$

Actual electrolytes already contain mobile ions such as molten salts or salt solutions (molten table salt or a solution of table salt, for example), but in some cases also solids (solid electrolytes in fuel cells). The substances referred to are already composed of ions in the solid state. Almost all salts are like this; one example is

$$\mathrm{NaCl}|s \equiv [\mathrm{Na}^+\mathrm{Cl}^-]|s \rightarrow \mathrm{Na}^+|w + \mathrm{Cl}^-|w.$$

In this case we speak of *true electrolytes*.

© Springer International Publishing Switzerland 2016
G. Job, R. Rüffler, *Physical Chemistry from a Different Angle*,
DOI 10.1007/978-3-319-15666-8_21

Table 21.1 Lowering of the chemical potential during transition of a substance from a gaseous state into an aqueous solution ($B|g \rightarrow B|w$). The values are valid for 298 K and equal concentrations in the gas and solution.

Substance	H^+	H_3O^+	OH^-	Cl^-	Na^+	Mg^{2+}	Al^{3+}
$\mu_g - \mu_w$ (kG)	1,090	418	460	333	411	1,893	4,621
Substance	Hg	Ar	H_2	CO_2	HCl	NH_3	D_2O
$\mu_g - \mu_w$ (kG)	3	-8	10	0	9	18	27

Substances that only form mobile ions when they are dissolved are generally considered electrolytes, in a wider sense. Among the so-called *potential electrolytes* are acids and organic bases such as

$$HCl|g + H_2O|l \rightarrow H_3O^+|w + Cl^-|w.$$

These types of decomposition processes are called *electrolytic dissociation.* As in the previous examples, this kind of dissociation often appears in aqueous solutions but hardly plays a role in gaseous states or in other solvents. The reason for this is that ions in water—in contrast to other environments—have an unusually low chemical potential. Let us consider the lowering of potential that results when ions are transferred from a gaseous state into an aqueous solution (Table 21.1, top). Just how dramatic this effect is can be appreciated when it is compared to a corresponding process using neutral particles (Table 21.1, bottom).

The difference $\mu_g - \mu_w$ is simply the drive of the process of dissolution

$$B|g \rightarrow B|w.$$

If the sequence of numbers at the top referring to ions is carefully compared to the one below for neutral particles, we find that the changes of potential differ by several orders of magnitude. The values for ions reach or even exceed the drive for forming covalent bonds. If atomic hydrogen unites with atomic chlorine, oxygen, nitrogen, carbon, etc.,

$$H + R \rightarrow H-R,$$

the following would be valid:

$$\mathcal{A} = \mu_H + \mu_R - \mu_{H-R} = 300\ldots400 \text{ kG}.$$

Generally, the drive of the following reaction:

$$A + B + C + \ldots \rightarrow ABC\ldots$$

is considered to be the measure of the *bonding strength* between two or more substances, or in other words, it is the measure of the *affinity* between the substances:

$$\mathcal{A}_{\mathrm{ABC}...} := \mu_\mathrm{A} + \mu_\mathrm{B} + \mu_\mathrm{C} + \ldots - \mu_{\mathrm{ABC}...}.$$

We should remember that $\mathcal{A}_{\mathrm{ABC}...}$ is not constant, but depends upon the milieu the participating substances are in. The values applying in the limit state $T, p \to 0$ are usually chosen for theoretical considerations and are called *bonding energies*.

The concept of bonding strength is not only used for the substances but also for their atomic or molecular building blocks. We speak of the bonding strength of atoms in a molecular or crystal structure (or the bonding energy when T and p vanish).

In this sense, the values in our table describe the bonding strength of ions onto water under the conditions described there. This becomes clearer when we rewrite the process of dissolving in terms of the general conversion formula above:

$$\mathrm{B}^z|\mathrm{g} + \nu\,\mathrm{H_2O}|\mathrm{l} \to \left[\mathrm{B(H_2O)}_\nu\right]^z|\mathrm{w} \equiv \mathrm{B}^z|\mathrm{w} + \nu\,\mathrm{H_2O}|\mathrm{l}.$$

Ions cannot carry just any amount of charge, but only a single *electric elementary charge* ($e_0 = 1.602 \times 10^{-19}$ C) or integer multiples of it. This is expressed by the *charge number* (or *valence*) z. The number of water molecules in the complex $[\mathrm{B(H_2O)}_\nu]^z$ is high and not sharply defined, so the bonded $\mathrm{H_2O}$ molecules are usually formally separated and included in the water of the solution. The leftover ion $\mathrm{B}^z|\mathrm{w}$ is assigned a value of the chemical potential so that the sum of the potentials does not change:

$$\mu(\mathrm{B}^z|\mathrm{w}) \equiv \mu\big(\left[\mathrm{B(H_2O)}_\nu\right]|\mathrm{w}\big) - \nu \cdot \mu(\mathrm{H_2O}|\mathrm{l}).$$

Formally, the water in the conversion formula can be canceled so that the bonding strength is simply the difference of potential values of the ions in the gas and dissolved state, as seen in the table. Although $\mathrm{H^+}|\mathrm{w}$ and $\mathrm{H_3O^+}|\mathrm{w}$ both describe the same kind of ion, the values of the corresponding chemical potentials are not equal. The relation results from

$$\mu(\mathrm{H_3O^+}|\mathrm{w}) = \mu(\mathrm{H^+}|\mathrm{w}) + \mu(\mathrm{H_2O}|\mathrm{l}).$$

The way ions bond to water can be most easily described as an ordered agglomeration of water dipoles on a central ion. As a whole, a water molecule is neutral, but due to its bent structure, the positive and negative charge within it is unevenly distributed, causing it to have a positive as well as a negative pole which represent an electric *dipole*. The negative pole of this bent molecule is found on the side of the oxygen atom, and the positive pole on the side of the hydrogen atoms (Fig. 21.1).

Such molecular dipoles are attracted by ions exactly like a magnetic dipole—say a magnetic needle—is attracted by the pole of a magnet. This is how an ion surrounds itself with numerous water molecules that are sometimes strongly and sometimes loosely bonded. This is not much different from a magnetic pole that is put into a box of needles. This process, which creates a more or less ordered and not

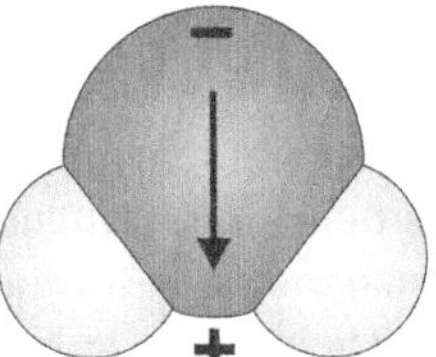

Fig. 21.1 Dipole characteristics of a bent water molecule (valence angle between the two hydrogen atoms bonded to oxygen: 104°).

Fig. 21.2 Hydration of ions while dissolving a NaCl crystal in water.

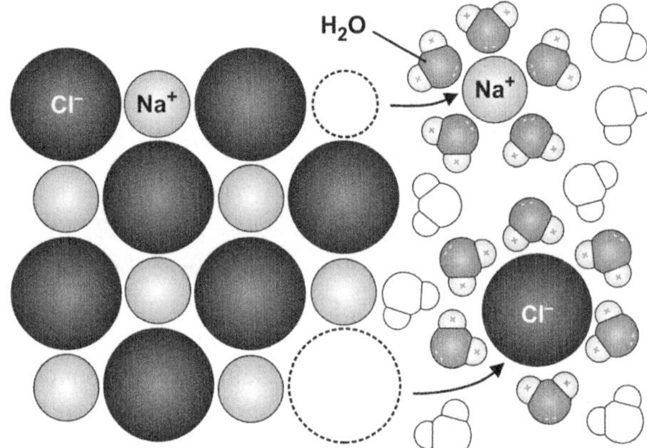

very sharply bounded shell of solvent particles around an ion, is generally called *solvation* or, if the solvent is water as it mostly is, *hydration* (Fig. 21.2).

This process releases a rather large amount of energy which roughly grows with the square of the charge number, but it is also dependent upon other characteristics of the ion such as the ionic radius. This quadratic relation is reflected in the bonding strength to the water. Compare for example the values for Na^+, Mg^{2+}, and Al^{3+} in Table 21.1.

Hydration also explains the fact that table salt and hydrogen chloride form freely moving ions in water but not in air. Dissolving in water takes place by dissociation into ions because the bonding of the ions in the assemblies formed by them, whether lattice or molecule, is weaker than their bonding to the water. The ions are virtually torn out of their old bonds by the formation of many new bonds to water molecules.

In the case of electrolytic dissociation, which can proceed in several steps and in many ways, different particles can be produced, for example:

$$NaHCO_3 \rightarrow Na^+ + HCO_3^-$$
$$HCO_3^- \rightarrow H^+ + CO_3^{2-}$$
$$HCO_3^- \rightarrow \qquad\quad OH^- + CO_2$$

| Charge number z: | $+1$ | -1 | $+1$ | -2 | -1 | 0 |

Each type of particle basically behaves like an independent substance with its own concentration and its own chemical potential. However, dissociation will produce equal numbers of positive and negative charges, so the solution remains neutral overall. The so-called *charge neutrality rule* (*electroneutrality rule*) is valid here:

$$z_A c_A + z_B c_B + \cdots = 0 \quad \text{short} \quad \sum_i z_i c_i = 0. \tag{21.1}$$

We can violate this rule by adding ions of a uniform sign from outside. An addition of ions of this type is only possible in extremely small amounts because the solution becomes charged in the process and the strong electric field it produces does not allow more to come in (compare Chap. 22). For this reason, although the electroneutrality rule is not strictly obeyed, it is to a high degree in all electrolyte solutions. We are therefore not completely free in choosing ion concentrations because field forces enforce electroneutrality of the solution.

21.2 Electric Potential

In the previous chapter we dealt with rates of spreading of substances and their causes. Migration of ions is influenced by additional forces caused by the electric charge of the particles. The force F acting upon a charge Q in an electric field E is given by

$$F = Q \cdot E. \tag{21.2}$$

Similar to the "chemical" forces in diffusion, we can imagine electrostatic forces being produced by a drop in a potential, in this case, the *electric potential* φ. We can construct a quantity $\varphi(x)$ for any field $E(x)$ so that

$$E(x) = -\frac{d\varphi(x)}{dx}. \tag{21.3}$$

If the field E has a constant value of E_0, the potential $\varphi(x)$ will be a linear function of position x, $\varphi(x) = \varphi(0) - E_0 \cdot x$, whose graph is a "ramp" with a constant gradient (Fig. 21.3). In general, it is also possible to give a three-dimensional field distribution $\vec{E}(x, y, z)$ a potential $\varphi(x, y, z)$ whose gradient at position x, y, z is exactly equal to the field $\vec{E}(x, y, z)$ there. In this case, we must write the field E as a vector because its complete determination requires magnitude and direction or the values of its three components E_x, E_y, E_z. However, the one-dimensional description is enough for our purposes here.

In a (chemically uniform) electric conductor such as a piece of copper or a salt solution, initial differences of potential are immediately compensated for by charge transfer if the conductor does not have a potential gradient forced upon it by an outside source—for instance, in a current carrying copper wire or a working

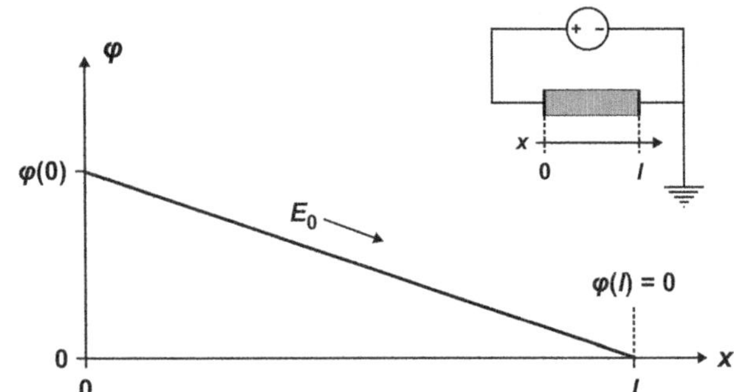

Fig. 21.3 Form of the electric potential φ in a uniform current carrying conductor of the length l. The end on the right is grounded so that $\varphi(l) = 0$, while $\varphi(0)$ is positive. A uniform potential gradient from left to right is formed, meaning an electric field with constant field strength E_0.

electrolytic cell. The interior electric field disappears, $E(\text{inside}) = 0$. It is then possible to assign a uniform and location independent potential $\varphi(C)$ to a (chemically uniform) conductor C that does not carry a current. This simplifies the description considerably. In the field of electrical engineering, the ground is considered to be a conductor with approximately uniform potential.

The zero point of the potential φ can be arbitrarily chosen by assigning a random point, possibly $x = 0$, the value 0, $\varphi(0) = 0$. The potential of the ground terminal in electrical devices or circuits is usually set at zero, $\varphi(\text{ground}) = 0$.

The values of electric potentials are given in Volts (V) (like voltages). Therefore, the electric potential of the positive pole of a typical three-cell lantern is $\varphi(\text{positive pole}) = +4.5$ V if the negative pole is *grounded*, $\varphi(\text{negative pole}) = 0$. The difference between the electric potentials $\Delta\varphi = \varphi_2 - \varphi_1$ at two different positions 1 and 2, is called the *voltage U*. The sign of voltage is defined as follows. Let us utilize an index of the type $_{1\to2}$ where the numbers refer to the initial and end points. $U_{1\to2}$ counts as positive if $U_{1\to2}$ attempts to drive a *positive* charge from position 1 to position 2, meaning when the potential φ_1 lies higher than φ_2. Therefore, we have

$$U_{1\to2} = \varphi_1 - \varphi_2 = -\Delta\varphi. \tag{21.4}$$

We need the minus sign here if we wish to hold to the general convention that for an arbitrary quantity $\mathcal{G}$, $\Delta\mathcal{G}$ symbolizes the difference $\mathcal{G}_{\text{final}} - \mathcal{G}_{\text{initial}}$. Thus, $U_{1\to2}$ represents the drive for the transport of charge from position 1 to position 2, very similarly to how $\mathcal{A}_{1\to2}$ represents the drive for a substance flowing from 1 to 2:

$$\mathcal{A}_{1\to2} = \mu_1 - \mu_2 = -\Delta\mu. \tag{21.5}$$

Because of this analogy, $\mathcal{A}$ is sometimes referred to as *chemical tension* (just like voltage refers to the notion of electric tension). In order to avoid always having to write the index $_{1\to2}$, we will tacitly agree to take the direction of the coordinate axes as the reference direction. For two points lying approximately on a line parallel to the x-axis going from left to right, we will say that $U = \varphi(\text{lefthand position}) - \varphi(\text{righthand position})$; the same holds true for the other coordinate directions.

The electric potential allows us the advantage of using similar descriptions for both chemical and electrical phenomena. The equation for the force upon a small sample of amount n ("test" amount) in a nonhomogeneous environment, meaning with position dependent chemical potential $\mu(x)$, $F = -n \cdot (d\mu/dx)$ [Eq. (12.1)] can be taken as the analogue of an equation for the force on a "test" charge Q in an electric field. Such a field is an environment having a position dependent electric potential $\varphi(x)$. We only need to insert Eq. (21.3) into Eq. (21.2):

$$F = -Q \cdot \frac{d\varphi}{dx}. \tag{21.6}$$

21.3 Ion Migration

The force upon a small sample of amount n of ions is made up of at least two components,

$$F = -n \cdot \frac{d\mu}{dx} - Q \cdot \frac{d\varphi}{dx} + \ldots \quad \text{(e.g., additional forces caused by counter-movement of ions of opposite charge).} \tag{21.7}$$

The electric potential gradient is produced by two *electrodes* immersed in an electrolyte solution. These electrodes are connected to a direct-current voltage source and they establish the electronic conducting contact to the differently conducting phase (electrolyte solution). They are very often made up of metal sheets or wires. Depending upon the sign of the ionic charge, the resulting electric force will lead to a migration in the direction of the electric field or opposite to it. The positively charged *cations* migrate toward the minus pole. This is the electrode with the lower electric potential which functions here as a *cathode*. Cathodes are the electrodes where electrons are transferred to the reacting substances, i.e., where a process of reduction takes place. (We will deal in more detail with redox reactions in Sect. 22.4). On the other hand, negatively charged *anions* migrate toward the positive pole (the electrode with the higher electric potential). The *anode* absorbs electrons from the reacting substances and an oxidation process occurs there. Each ion carries a charge of ze_0 causing a transport of charge, an *electric current*. This electric current caused by ions is easy to demonstrate (Experiment 21.1).

Similar to the case of diffusion, we can now, in principle, calculate the migration velocity v of the ions using the equation $v = \omega \cdot F/n$ [Eq. (20.1)] and ignoring any additional forces:

$$v = \omega \cdot \frac{F}{n} = -\omega \cdot \frac{d\mu}{dx} - \omega \cdot \frac{Q}{n} \cdot \frac{d\varphi}{dx}. \tag{21.8}$$

The charge of an ion ensemble (all the ions of one type) results from its amount of substance n and the charge number z of each individual ion:

Experiment 21.1 *Charge transport through saline solutions*: Two cables are connected to each other over a voltage source and a light bulb. The ends of the cables are immersed in a Petri dish filled with distilled water. The bulb remains dark. When table salt is sprinkled into the water, the light bulb begins to glow.

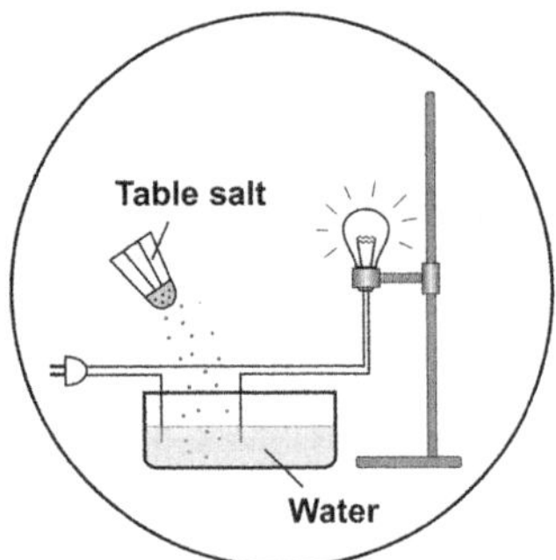

$$Q = z\mathcal{F}n. \tag{21.9}$$

The *Faraday constant* $\mathcal{F} = e_0/\tau$ gives the amount of charge of 1 mol of monovalent cations and is equal to $96{,}485\ \mathrm{C\ mol^{-1}}$. The relation in Eq. (21.9) was discovered in 1831 by the eminent English experimental physicist and chemist Michael Faraday, and formulated in the so-called *Faraday laws*.

Inserting into the equation above yields

$$\upsilon = -\omega \cdot \frac{\mathrm{d}\mu}{\mathrm{d}x} - \omega z\mathcal{F} \cdot \frac{\mathrm{d}\varphi}{\mathrm{d}x}. \tag{21.10}$$

The product $\omega z\mathcal{F}$ is referred to as *electric mobility* and is abbreviated with the symbol u. The unit for electric mobility $\mathrm{m^2\ V^{-1}\ s^{-1}}$ results from the unit $\mathrm{m\ s^{-1}}$ of the migration velocity and $\mathrm{V\ m^{-1}}$ of the potential gradient.

If there is no gradient of the chemical potential μ, the result for the speed of the ions is:

$$\upsilon = -u \cdot \frac{\mathrm{d}\varphi}{\mathrm{d}x} = u \cdot E. \tag{21.11}$$

Numerical values for the electric mobility of selected ions are compiled in Table 21.2. The index $_0$ indicates strongly diluted solutions, meaning solutions with vanishingly small concentrations in which no interaction between the ions takes place.

The migration velocity of ions appears relatively small at first glance. For example, in the case of a voltage drop of 1 V and a 1 cm thick layer of solution and therefore a field of $100\ \mathrm{V\ m^{-1}}$, it will be between 10^{-5} and $10^{-6}\ \mathrm{m\ s^{-1}}$. Even so, in 1 s, the ion will cover a distance of almost 10,000 times its diameter.

In closing this section, we will look more closely at the motion of one individual ion with the charge ze_0. The ion is initially accelerated by the electric force $F = Q \cdot E$, i.e., by

$$F = ze_0 E. \tag{21.12}$$

Table 21.2 Electric mobility of ions at 298 K in water (strongly diluted) (from: Lide D R (ed) (2008) CRC Handbook of Chemistry and Physics, 89th edn. CRC Press, Boca Raton).

Ion	u_0 (10^{-8} m^2 V^{-1} s^{-1})	Ion	u_0 (10^{-8} m^2 V^{-1} s^{-1})
H^+	36.2	OH^-	-20.5
Li^+	4.0	F^-	-5.7
Na^+	5.2	Cl^-	-7.9
K^+	7.6	Br^-	-8.1
Rb^+	8.1	I^-	-8.0
Cs^+	8.0	NO_3^-	-7.4
Ag^+	6.4	CH_3COO^-	-4.2
NH_4^+	7.6	MnO_4^-	-6.4
$N(CH_3)_4^+$	4.7	HCO_3^-	-4.6
$N(C_2H_5)_4^+$	3.4	CO_3^{2-}	-7.2
$N(C_3H_7)_4^+$	2.4	SO_4^{2-}	-8.3
Mg^{2+}	5.5		
Ca^{2+}	6.2		
Ba^{2+}	6.6		
Cu^{2+}	5.6		

The faster it moves, though, the greater the frictional force due to viscosity η that acts in the opposite direction. If we now assume that the ions behave almost like rigid spheres with radius r, we can make use of the Stokes law ($F_R = 6\pi\eta r v$) of hydrodynamics we were introduced to in the last chapter. After a short response time, the opposing forces affecting the ion become equal,

$$ze_0 E = 6\pi\eta r v,$$

and the ion will move at a constant velocity of migration:

$$v = \frac{ze_0}{6\pi\eta r} E. \tag{21.13}$$

A comparison of this expression with Eq. (21.11) shows that

$$u = \frac{ze_0}{6\pi\eta r}. \tag{21.14}$$

The electric mobility is therefore inversely proportional to the viscosity of the medium. Because η is strongly dependent upon temperature (compare Sect. 20.3), the temperature must be taken into account when determining u. According to Eq. (21.14), the smaller the ion radius is, the greater the electric mobility should be. Let us now take a look at the series $Li^+ \rightarrow Rb^+$ and $NH_4^+ \rightarrow N(C_3H_7)_4^+$ in Table 21.2. We clearly see that these theoretical predictions hold for the large tetraalkylammonium ions but not for small alkali ions. Making use of the ion radii determined by the lattice dimensions of solid salts, we see that the K^+ ion is almost twice the size of the Li^+ ion, but it exhibits a much greater mobility. Until now, though, we have only considered "naked" ions. We have not taken into account the fact that ions have a solvation shell or, in an aqueous solution, a hydration shell which they must "drag" along. The ions will "puff up" and the smaller they are, the more they puff up. More precisely: Small ions form a much stronger electric field $\left(E \sim 1/r^2\right)$ and accumulate more water dipoles, which leads to increasing radii of the *hydrated ions* in the series $Li^+ \rightarrow Rb^+$. If we take the radius of the hydrated ions into account in Eq. (21.14), the ratio of mobilities agrees with the theoretical expectations. In the case of the large tetraalkylammonium ions, hydration does not play a decisive role.

What surprises us, though, is the unusually high mobility of the very small protons. This is due to the special structure of water which exhibits a relatively high order even in its liquid state due to formation of "*hydrogen bridge bonds.*" Naked protons do not persist in water because of the strong electric fields emanating from them, so they immediately accumulate on the negative side of the water dipoles and form H_3O^+ ions. Consequently, the positive charge is not concentrated on the original proton, but is symmetrically distributed over all three protons. Correspondingly, a separation of a proton on the opposite side of the H_3O^+ ion becomes possible so that the positive charge appears to have "moved" across the diameter of the ion without a true ion migration in the sense discussed so far, having occurred. The separated proton can reattach to another water molecule, etc. Ultimately, an efficient "proton transport" along a chain of water molecules takes place through rearrangement of bonds (Grotthus mechanism). Applying an electric field will cause the previously arbitrary "migration" of positive charge to be directed toward the negative electrode (Fig. 21.4).

The high mobility of the proton is connected to the special transport mechanism and the structure of the water molecule. Protons in other types of solvents exhibit a mobility similar to other ions. However, the special transport mechanism can be transferred to migration of hydroxide ions in water because they are also part of the

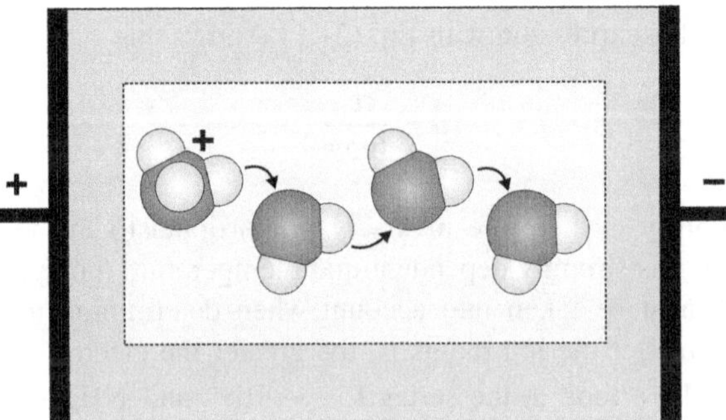

Fig. 21.4 Simplified representation of charge transport by protons in water.

Experiment 21.2 *Ion migration*: A U-tube contains agar–agar gel mixed with a universal indicator and table salt. Hydrochloric acid is poured into the left leg and sodium hydroxide is poured into the leg on the right. Afterward, an electrode is dipped into each leg and then attached to a direct-current voltage source (the acidic side with the positive terminal and the alkaline side with the negative terminal). Two discoloration zones emerge and slowly spread. Because of the different ion mobilities, the discoloration zone caused by migration of OH^- ions is only half as large as the zone produced by migration of H_3O^+ ions.

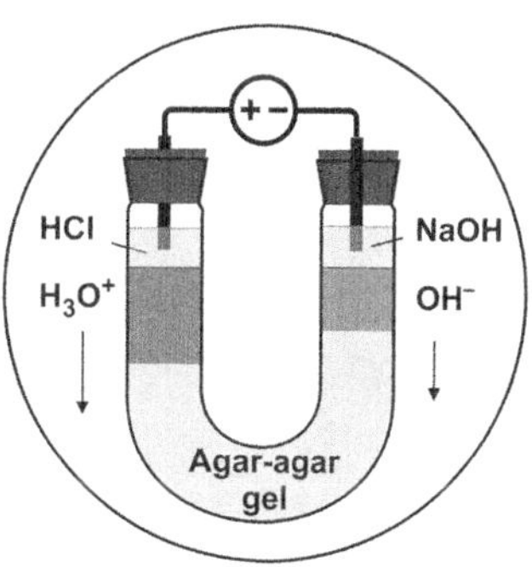

solvent, i.e., components of a water molecule. This is why these ions also show relatively high mobility. The migration of protons and hydroxide ions in an electric field produced by a direct-current voltage source can be made visible by discoloring the column of agar–agar gel containing an indicator (Experiment 21.2).

21.4 Conductivity of Electrolyte Solutions

Now we will consider a special case that is particularly suited to visualization and calculation, the processes in a cuboid electrolytic trough with a cross section A and a length l and with two electrodes attached to its end walls (Fig. 21.5). There is an electrolyte solution with a concentration c in the trough. A direct-current voltage U is established across the electrodes.

Initially, there is no gradient of the chemical potential μ available, but there is a uniform gradient of the electric potential φ,

$$\frac{d\mu}{dx} = 0, \qquad \frac{d\varphi}{dx} = \frac{\Delta\varphi}{\Delta x} = \frac{\varphi_2 - \varphi_1}{x_2 - x_1} = \frac{-U}{l}\,. \tag{21.15}$$

The mobile ions begin to migrate perpendicularly to the electrodes in the constant electric field causing, as stated above, a charge flow or *electric current*. Equation (21.11) gives us the migration velocity under these conditions:

$$v = u \cdot \frac{U}{l}. \tag{21.16}$$

In order to calculate the charge ΔQ moving through the cross-sectional area A of the cuboid, we will limit ourselves to just one type of mobile ion. We will imagine the corresponding counter-ion to be immobile, i.e., it contributes almost nothing to charge transport. The charge transported in relation to the amount Δn of ions transferred is obviously

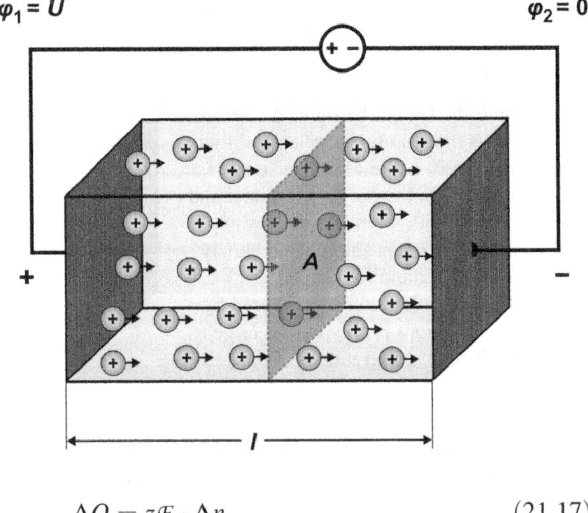

Fig. 21.5 Electrolytic trough filled with the solution of an electrolyte in which only the positive ions are considered sufficiently mobile.

$$\Delta Q = z \mathcal{F} \cdot \Delta n. \tag{21.17}$$

If we divide this equation by the time span Δt (which we must imagine as very short so that no noticeable shifts of concentration occur), we obtain the electric current I,

$$I = \frac{\Delta Q}{\Delta t} = z \mathcal{F} \frac{\Delta n}{\Delta t} = z \mathcal{F} \cdot J. \tag{21.18}$$

We were introduced to Eq. (20.4) for describing a matter flux (or current of amount of substance) J in the last chapter,

$$J = c \cdot A \cdot v.$$

If we take Eq. (21.16) above for the migration velocity v into account, we obtain:

$$J = c \cdot A \cdot u \cdot \frac{U}{l}. \tag{21.19}$$

The electric current is therefore equal to

$$I = z \mathcal{F} \cdot c \cdot A \cdot u \cdot \frac{U}{l}. \tag{21.20}$$

Usually there are several types of mobile ions present. As long as they remain undisturbed in their migration, which is only the case in strongly diluted solutions, the contributions of the individual types of ions, whether cations or anions, add up to the electric current I. Instead of Eq. (21.20), we then have the following relation:

$$I = U \cdot \frac{A}{l} \cdot \sum_i z_i \mathcal{F} \cdot c_i \cdot u_i. \tag{21.21}$$

An electrolyte solution of a given composition will have a current I proportional to voltage U. This reminds us of the well-known Ohm's law for homogeneous conductors:

$$U = R \cdot I \quad \text{or} \quad I = G \cdot U. \tag{21.22}$$

R is the Ohmic resistance and G is the *conductance* with

$$G = \frac{1}{R}. \tag{21.23}$$

Conductance is usually given in the unit Siemens ($S = \Omega^{-1}$). The validity of Ohm's law for the case of electrolyte solutions can be experimentally proven (Experiment 21.3 showing electrolysis of a copper sulfate solution with copper electrodes).

If Eq. (21.22) is compared to Eq. (21.21), we see that the conductance of an electrolyte solution depends upon the types of ions (via z_i, u_i), their concentrations (c_i), as well as the dimension of the ionic conductor (cross-sectional area A and distance between electrodes l):

$$G = \frac{A}{l} \sum_i z_i \mathcal{F} \cdot c_i \cdot u_i. \tag{21.24}$$

In analogy to resistivity ρ, the *conductivity* σ (unit $S\ m^{-1}$) is a quantity characteristic of an electrolyte solution of given composition. It is independent of the geometry of the cell:

$$R = \rho \cdot \frac{l}{A} \quad \text{or} \quad G = \sigma \cdot \frac{A}{l}. \tag{21.25}$$

If we compare this last expression to Eq. (21.24), we obtain an important equation that is valid for any electrolyte solution such as for example also seawater or gastric juice:

$$\sigma = \sum_i z_i \mathcal{F} \cdot c_i \cdot u_i \quad \text{"Four-factor formula."} \tag{21.26}$$

Experiment 21.3 *Ohm's law for electrolytes*: Two Cu electrodes are immersed in a $CuSO_4$ solution and are connected to a direct-current voltage source. When the voltage is raised in steps (up to approx. 2 V), the current rises proportionally.

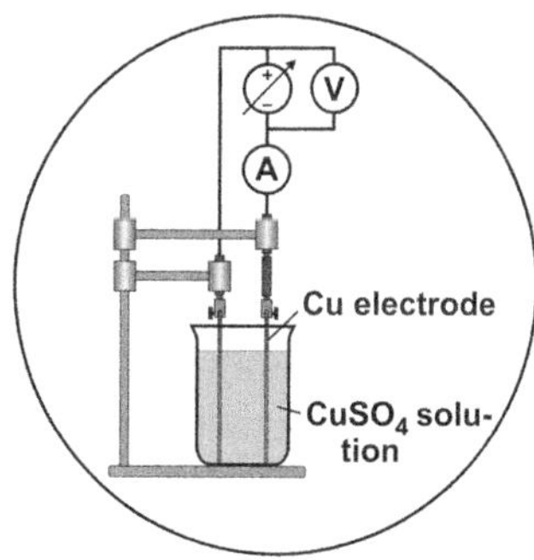

In the following we will limit ourselves to solutions of just one electrolyte having the concentration c. We will assume that it completely dissociates which means we will have v_i ions of charge number z_i ($c_i = v_i c$) per formula unit. The following is then valid:

$$\sigma = \sum_i z_i \mathcal{F} \cdot v_i c \cdot u_i = c \sum_i v_i z_i u_i \mathcal{F}. \qquad (21.27)$$

The conductivity σ generally increases with concentration c of the dissolved electrolytes. However, mobility u is markedly dependent upon c as well so the two influences overlap. In order to separately investigate the second one, we consider the *molar conductivity* Λ,

$$\Lambda = \frac{\sigma}{c} = \sum_i v_i z_i u_i \mathcal{F}. \qquad (21.28)$$

The fraction of molar conductivity attributed to the ion of type i is called the (molar) *ionic conductivity* Λ_i:

$$\Lambda_i = z_i u_i \mathcal{F}. \qquad (21.29)$$

This allows us to simplify Eq. (21.28) to

$$\Lambda = \sum_i v_i \Lambda_i. \qquad (21.30)$$

According to the equations above, the molar conductivity Λ should be independent of the concentration. However, this is only true in the limit of infinite dilution. In order to ensure the validity of the equation, we replace Λ by Λ^0, which is the molar conductivity at infinitely low concentrations (*molar conductivity at infinite dilution* or *limiting molar conductivity*):

$$\Lambda^0 = \sum_i v_i \Lambda_i^0. \qquad (21.31)$$

Only in strongly diluted solutions where there are no noticeable interactions between the ions do the individual ions move in the electric field independently of the type of counter-ions. This *law of independent migration of ions* was found by the German physicist Friedrich Kohlrausch in the nineteenth century.

The cations and anions migrating in opposite directions are hindered in the same way one is hindered in a moving crowd of people on a sidewalk. The denser the crowd is, the stronger the effect will be. The hydration shell the ion drags with it plays a role as well as its tendency to surround itself with a cloud of oppositely charged ions. We will return to the latter effect in the next section.

In conclusion, we will look at the simple case of an electrolyte made up of only two ions $C_{v+}^{z+}A_{v-}^{z-}$ which decomposes into v_+ cations C of charge number z_+ and v_- anions A of charge number z_-. The resulting molar conductivity is then:

$$\Lambda = (v_+z_+u_+ + v_-z_-u_-)\mathcal{F}. \tag{21.32}$$

The fractions of molar conductivity Λ_+ and Λ_- belonging to the cations and anions are:

$$\Lambda_+ = z_+u_+\mathcal{F}; \quad \Lambda_- = z_-u_-\mathcal{F}. \tag{21.33}$$

For Eq. (21.30) we can write:

$$\Lambda = v_+\Lambda_+ + v_-\Lambda_-. \tag{21.34}$$

The corresponding limiting molar conductivity results in:

$$\Lambda^0 = v_+\Lambda_+^0 + v_-\Lambda_-^0. \tag{21.35}$$

Finally, a reminder that there is no unified usage of the term molar conductivity in the literature. For example, we often find the value of ionic conductivity for the fraction $1/|z_i|$ of an ion of charge number z_i in tables, for example, $\Lambda\left(\tfrac{1}{2}Ba^{2+}\right)$, $\Lambda\left(\tfrac{1}{3}La^{3+}\right)$, etc. This is done in order to obtain expressions independent of charge number. To avoid errors we should always keep in mind what types of particles the listed values are for and the calculations should be adapted accordingly.

21.5 Concentration Dependence of Conductivity

In more strongly concentrated solutions, the electrostatic attractive forces of the ions cause the ions to prefer to be surrounded by counter-ions (Fig. 21.6a). Applying a field results in the counter-ions migrating in the opposite direction (Fig. 21.6b) so that as concentration c grows, there is—simply speaking—an increasing obstruction of ion migration. The Debye–Hückel theory describes the different types of mutual obstruction of the ions in more detail.

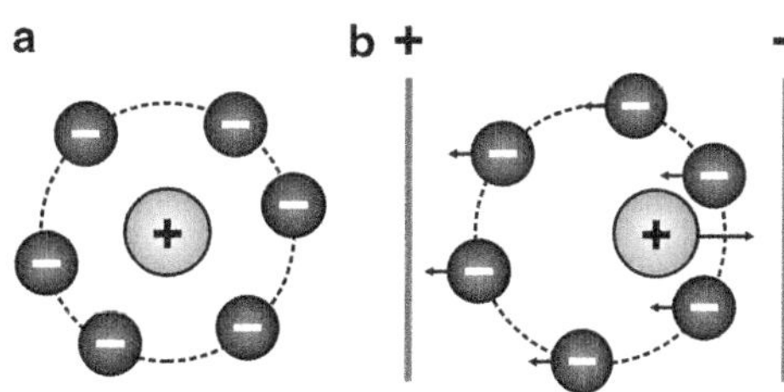

Fig. 21.6 Positive ion with counter-ions (**a**) without external electric field, (**b**) in an external electric field; the positive ion migrates to the right, the counter-ions to the left.

Therefore, the electric mobility and the corresponding molar conductivity Λ are lower than they would be at infinite dilution. The concentration dependency of Λ can be demonstrated using an experimental setup analogous to Experiment 21.3 (Experiment 21.4).

Addition of water increases the vertical distance between the charge carriers migrating in different horizontal levels so that they do hinder each other less, but it does not change their numbers, the forces of electric field driving them or the distances covered by them.

Two different cases can be discerned in the concentration dependency of Λ (Fig. 21.7):

- *Strong electrolytes.* These electrolytes completely dissociate independent of their concentration (potassium chloride, KC1, for example); they exhibit only a slight drop of molar conductivity with increased concentration.

Experiment 21.4 *Concentration dependency of molar conductivity*: The copper sulfate solution is successively diluted at constant voltage. The corresponding current I is measured at each step. This is proportional to Λ and independent of the volume of solution. (Doubling the amount of water halves the concentration, but doubles the cross-sectional area the current flows through so that the influences compensate for each other.) We observe an increase of conductivity with decreasing concentration.

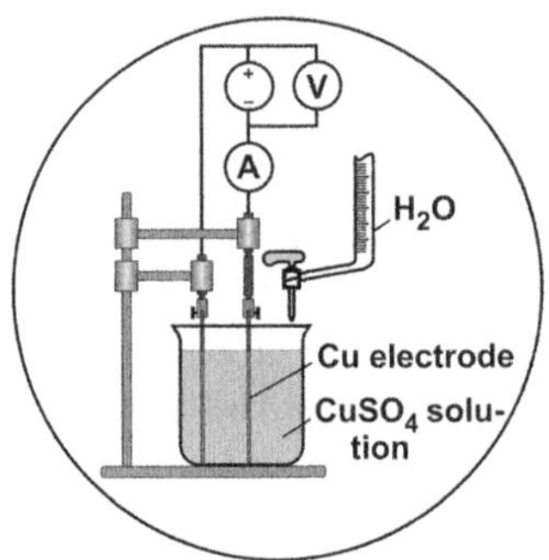

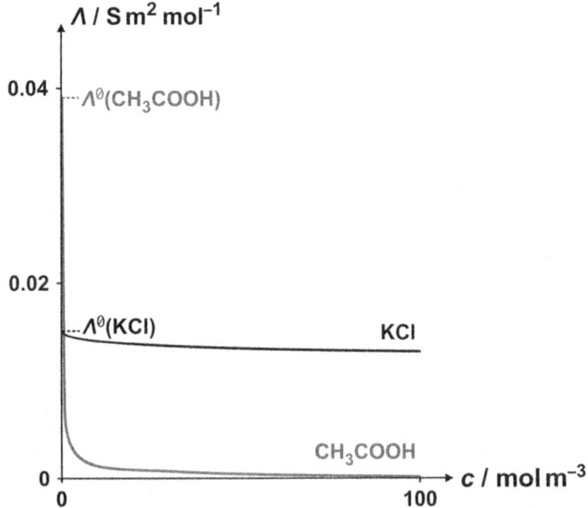

Fig. 21.7 Concentration dependency of molar conductivity of strong and weak electrolytes using KCl (*black*) and CH₃COOH (*gray*) as examples (at 298 K).

- *Weak electrolytes.* In the case of weak electrolytes (acetic acid, CH_3COOH, for example), the extent to which the electrolyte dissociates into ions strongly depends upon the total concentration c. The fraction

$$\alpha = \frac{c_{\text{diss}}}{c} \tag{21.36}$$

of all molecules that are dissociated is referred to as the *degree of dissociation* α. The degree of dissociation drastically reduces as concentration rises. Undissociated molecules contribute nothing to conductivity. Only the ions ($c_i = v_i \alpha c$) do contribute, so the concentration dependency of the molar conductivity

$$\Lambda = \sum_i v_i \alpha \Lambda_i \tag{21.37}$$

is mostly determined by the degree of dissociation. When the total concentration c is increased, Λ will decrease to relatively low values already at low concentrations.

We will now consider strong electrolytes which can easily be investigated experimentally. Around 1900, the German physicist Friedrich Kohlrausch empirically found the following relation between molar conductivity Λ, concentration c, and limiting molar conductivity Λ^0:

$$\Lambda = \Lambda^0 - b\sqrt{c}. \tag{21.38}$$

This relation is called *Kohlrausch's square root law*. It can be theoretically supported with the help of the Debye–Hückel theory. The limiting molar conductivity Λ^0 is impossible to measure directly because at infinite dilution, the solution does not conduct electricity. However, if Λ is plotted as a function of $\sqrt{c}$ at concentrations that are not too high, we obtain a linear relation (Fig. 21.8) and Λ^0 can be determined from extrapolation to the intercept of the straight line.

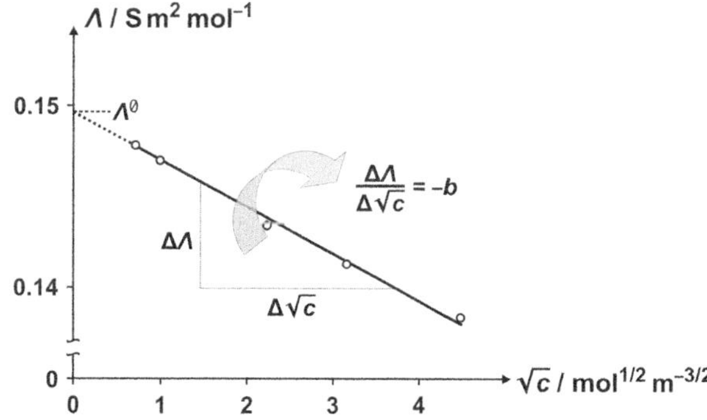

Fig. 21.8 Molar conductivity of aqueous KCl solutions at 298 K as a function of $\sqrt{c}$ (experimental data from Hamann CH, Vielstich W (1998) Elektrochemie. Wiley-VCH, Weinheim).

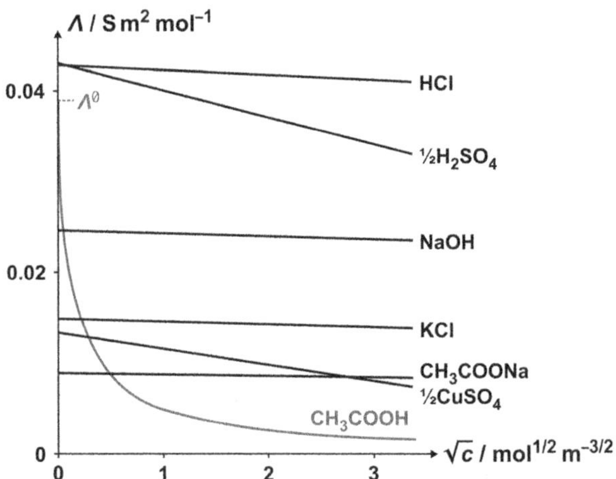

Fig. 21.9 Molar conductivity of some electrolyte solutions at 298 K as functions of $\sqrt{c}$.

As the charge numbers of the ions increase, the constant b assumes higher values—evidenced by the increasing gradients of the corresponding straight lines (Fig. 21.9, H_2SO_4 and $CuSO_4$ [$\Lambda(\frac{1}{2}H_2SO_4)$ and $\Lambda(\frac{1}{2}CuSO_4)$ are shown for easier comparison)], because polyvalent electrolytes exhibit stronger interaction between the ions.

Figure 21.9 also shows the behavior of the molar conductivity of acetic acid as an example of weak electrolytes (which are principally organic acids and bases). The strong decrease of conductivity, which resembles a hyperbola, is caused by the dissociation equilibrium between the ions and the uncharged molecules. We can use the following formula for the case of acetic acid:

$$CH_3COOH|w \rightleftarrows H^+|w + CH_3COO^-|w$$

or abbreviated to

$$HAc|w \rightleftarrows H^+|w + Ac^-|w.$$

Equilibrium can be described by the conventional equilibrium constant $\overset{\circ}{K}_c$ (compare Sect. 6.5):

$$\overset{\circ}{K}_c = \frac{c(H^+) \cdot c(Ac^-)}{c(HAc)}. \tag{21.39}$$

If c is the total concentration of the acetic acid and c_{diss} is the equilibrium concentration of the ions, the following is valid if degree of dissociation α is included:

$$c(H^+) = c(Ac^-) = \alpha c \quad \text{and} \quad c(HAc) = (1-\alpha)c. \tag{21.40}$$

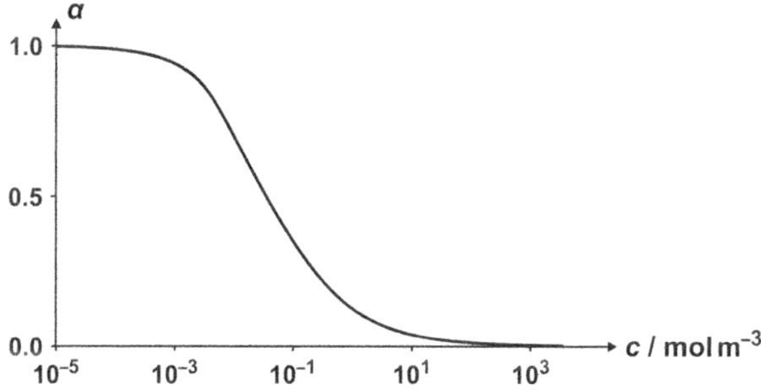

Fig. 21.10 Degree of dissociation α as a function of the total concentration of acetic acid.

Inserting into Eq. (21.39) yields *Ostwald's dilution law*:

$$\overset{\circ}{K}_c = \frac{\alpha^2}{1-\alpha} \cdot c. \tag{21.41}$$

When we calculate the degree of dissociation α for various total concentrations c of acetic acid [$\overset{\circ}{K}_c$ (acetic acid) $= 1.74 \times 10^{-5}$ kmol m^{-3}], we obtain the graph of Fig. 21.10.

A weak electrolyte only fully dissociates into ions ($\alpha \approx 1$) at infinite dilution. As concentration c grows, α drastically decreases. If we assume that the strong decrease of molar conductivity is caused mostly by reduction of the degree of dissociation, then both quantities should be proportional to each other. If we further assume that at infinite dilution, i.e., for a degree of dissociation α equal to 1, the limiting molar conductivity Λ^0 will be attained, we have:

$$\frac{\Lambda}{\Lambda^0} = \alpha. \tag{21.42}$$

By inserting this expression into Eq. (21.41), we find for low degrees of dissociation ($\alpha \ll 1$, meaning $(1-\alpha) \approx 1$):

$$\overset{\circ}{K}_c = \left(\frac{\Lambda}{\Lambda^0}\right)^2 \cdot c \quad \text{or} \quad \Lambda = \Lambda^0 \sqrt{\overset{\circ}{K}_c} \cdot \frac{1}{\sqrt{c}}. \tag{21.43}$$

If we plot Λ as a function of $\sqrt{c}$, this corresponds to the equation of a simple hyperbola ($y = a/x$), as we would expect from the graph (Fig. 21.9).

Conversely, Eq. (21.42) can be used to determine the degree of dissociation α of a weak electrolyte at a given concentration c by measuring the molar conductivity. Moreover, with the help of Eq. (21.41), the equilibrium constant of the substance becomes accessible. However, for these calculations we need the limiting molar conductivity Λ^0. This quantity is very difficult to find experimentally because the steep rise of the Λ at low concentrations makes an extrapolation to infinite dilution very uncertain. The law of independent migration of ions [Eq. (21.35)] offers a way out. In the case of infinite dilution, the limiting molar conductivity of acetic acid is the sum of the contributions of cation and anion:

$$\Lambda^0(\text{HAc}) = \Lambda^0(\text{H}^+) + \Lambda^0(\text{Ac}^-). \tag{21.44}$$

By clever combination

$$\Lambda^0(\text{HAc}) = \Lambda^0(\text{H}^+) + \Lambda^0(\text{Ac}^-) + \Lambda^0(\text{Na}^+) - \Lambda^0(\text{Na}^+) + \Lambda^0(\text{Cl}^-)$$
$$- \Lambda^0(\text{Cl}^-), \tag{21.45}$$

the corresponding value can be determined from the limiting molar conductivities for the strong electrolytes HCl, NaCl, and sodium acetate (NaAc), which are easily accessible experimentally by extrapolation using Kohlrausch's square root law:

$$\Lambda^0(\text{HAc}) = \Lambda^0(\text{HCl}) + \Lambda^0(\text{NaAc}) - \Lambda^0(\text{NaCl}). \tag{21.46}$$

21.6   Transport Numbers

So far, we have delved into the mobility and limiting molar conductivity of ions. We have shown the usefulness of these values that are characteristic of individual types of ions [compare Eq. (21.44)]. But how do we obtain this experimental data? Measuring conductivity is obviously not sufficient because electrolyte solutions always contain cations and anions.

One possibility is to directly measure an ion's migration velocity v. Two electrolyte solutions of differing densities are necessary for this. Each should have one type of ion in common and one with a different color, and be carefully layered on top of each other so that the interface between them is as sharp as possible (Fig. 21.11). The colored indicator ion in the solution below should also exhibit lower mobility. A suitable combination for this would be a colorless KNO_3

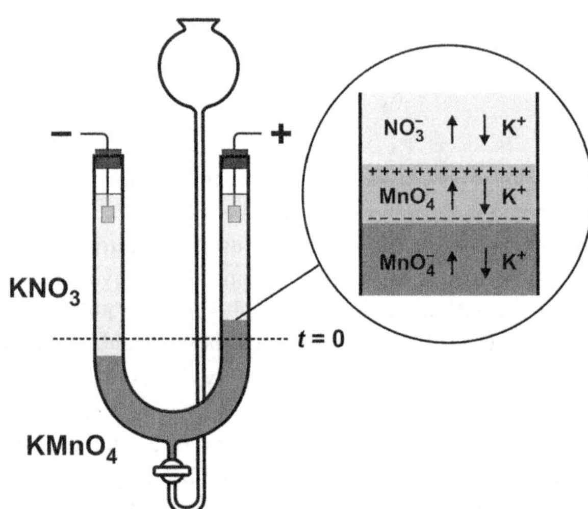

Fig. 21.11 Experiment for directly measuring migration velocity. We consider only the interface in the right limb because it remains sharp, while the interface in the left limb becomes blurred.

solution and a violet-colored $KMnO_4$ solution. When a direct-current voltage is applied to this combination, the ions will begin to migrate and the interface will shift ("method of migrating interface"). In the right-hand limb of the U-tube (anode limb) the less mobile permanganate ions can never overtake the nitrate ions. However, they also cannot lag behind the nitrate ions because the solution must remain electrically neutral. A slightly diluted zone is formed with a somewhat enhanced field, assuring that the permanganate ions can keep up with the migrating nitrate ions.

As a result, the interface migrates with the velocity of the nitrate ions. This is easy to determine from the distance covered s for a given time t:

$$v = \frac{s}{t}. \tag{21.47}$$

When the applied voltage and the distance between the electrodes are known, Eq. (21.16) can be transformed to find the electric mobility,

$$u = v \cdot \frac{l}{U}, \tag{21.48}$$

or the ion conductivity,

$$\Lambda = z \cdot u \cdot \mathcal{F}. \tag{21.49}$$

The method described above is not universally applicable because most ions are colorless.

We will therefore introduce another experimentally accessible quantity which is dependent upon ion mobility or ion conductivity. This is the so-called *transport number* t. Depending upon their mobility and charge, the individual ions will contribute to a larger or smaller extent to the total current I. In a solution where only one electrolyte is dissolved, the resulting transport number for the cations is

$$t_+ = \frac{I_+}{I}, \tag{21.50}$$

where I_+ is the current caused by the migration of cations. The transport number for anions is analogous to this:

$$t_- = \frac{I_-}{I}. \tag{21.51}$$

Because the total current is the sum of the currents transported by the cations and anions, it immediately follows that

$$t_+ + t_- = 1. \tag{21.52}$$

Therefore, the transport number of an ion, by definition, depends upon the type of counter-ion involved.

By inserting Eq. (21.21) into (21.50), and taking the electroneutrality rule into account [Eq. (21.1)] with $c_+z_+ = -c_-z_-$, we obtain

$$t_+ = \frac{I_+}{I_+ + I_-} = \frac{c_+z_+\mathcal{F}\cdot u_+}{c_+z_+\mathcal{F}\cdot u_+ + c_-z_-\mathcal{F}\cdot u_-} = \frac{c_+z_+\mathcal{F}\cdot u_+}{c_+z_+\mathcal{F}(u_+ - u_-)}$$
$$= \frac{u_+}{u_+ - u_-}. \tag{21.53}$$

(Remember that the electric mobilities u_- of anions are negative.) Equation (21.33) and the relation $c_i = v_i c$ can be used to find a relationship between the transport number of an ion and the ion conductivity:

$$t_+ = \frac{c \cdot v_+z_+\mathcal{F}\cdot u_+}{c \cdot v_+z_+\mathcal{F}\cdot u_+ + c \cdot v_-z_-\mathcal{F}\cdot u_-} = \frac{v_+\Lambda_+}{v_+\Lambda_+ + v_-\Lambda_-} = \frac{v_+\Lambda_+}{\Lambda}. \tag{21.54}$$

The following is correspondingly valid for vanishingly small concentrations:

$$t_+^0 = \frac{v_+\Lambda_+^0}{\Lambda^0}. \tag{21.55}$$

Of course, there are completely analogous relations for anions. If the transport numbers t_+^0 or t_-^0 can be determined experimentally, the limiting molar conductivity of the electrolytes—accessible via Kohlrausch's square root law for strong electrolytes—can be used to calculate the individual ion mobilities, which is the goal of this procedure.

How can we measure the transport numbers? Formally, we first divide the electrolysis trough in Fig. 21.5 into two separate compartments, a cathode compartment (CC) and an anode compartment (AC). When a current I is allowed to flow through the cell for a time span t, a total charge of $Q = I \cdot t$ will be transported. The cations take over the part Q_+ of the transport, and the anions transport the part Q_-. Therefore, Q_+/z_+e_0 cations migrate out of the anode compartment into the cathode compartment in the time t. In the same time span, during electrolysis, Q/z_+e_0 cations are discharged at the cathode and "disappear" from the cathode compartment. The total change of the number of cations in the cathode compartment (N_{CC}^+) is then:

$$\Delta N_{CC}^+ = \frac{Q_+}{z_+e_0} - \frac{Q}{z_+e_0} = -\frac{Q_-}{z_+e_0}. \tag{21.56}$$

The amount of substance of cations decreases correspondingly in the cathode compartment:

$$\Delta n_{CC}^{+} = \frac{\Delta N_{CC}^{+}}{\tau} = -\frac{Q_-}{z_+ \mathcal{F}}.$$ (21.57)

It is easy to quantitatively determine this change of amount of substance by titrating before and after the passage of the current. This makes the transported amount of charge Q_- and, ultimately, the transfer number t_- accessible. The change of amount of substance in the cathode compartment gives us the transport number of the anions!

A totally analogous balance can be made for the anions in the anode compartment:

$$\Delta N_{AC}^{-} = -\frac{Q_-}{z_- e_0} + \frac{Q}{z_- e_0} = \frac{Q_+}{z_- e_0}$$ (21.58)

and therefore:

$$\Delta n_{AC}^{-} = \frac{\Delta N_{AC}^{-}}{\tau} = \frac{Q_+}{z_- \mathcal{F}}.$$ (21.59)

The amount of substance of anions in the anode compartment also decreases (because z_- is negative). This yields the transport number of the cations.

We have not yet dealt with the behavior of the counter-ions in the two compartments. Not only cations migrate into the cathode compartment in time t, but also

$$\Delta N_{CC}^{-} = \frac{Q_-}{z_- e_0}$$ (21.60)

anions from the cathode compartment migrate into the anode compartment, which corresponds to a change of amount of substance

$$\Delta n_{CC}^{-} = \frac{\Delta N_{CC}^{-}}{\tau} = \frac{Q_-}{z_- \mathcal{F}}.$$ (21.61)

We can formulate the change of numbers of cations in the anode compartment analogously:

$$\Delta N_{AC}^{+} = -\frac{Q_+}{z_+ e_0}$$ (21.62)

or

$$\Delta n_{AC}^{+} = \frac{\Delta N_{AC}^{+}}{\tau} = -\frac{Q_+}{z_+ \mathcal{F}}.$$ (21.63)

Comparing Eqs. (21.57) and (21.61) as well as (21.59) and (21.63) shows that

$$\Delta n_{CC}^{+} \cdot z_{+} + \Delta n_{CC}^{-} \cdot z_{-} = 0 \quad \text{and} \quad \Delta n_{AC}^{-} \cdot z_{-} + \Delta n_{AC}^{+} \cdot z_{+} = 0. \qquad (21.64)$$

The electroneutrality rule is therefore satisfied in both the cathode and anode compartments.

We will now consider the determination of the transport number using the example of a 1:1 electrolyte (this is an electrolyte where the two ions both have a charge number of 1). We choose hydrochloric acid and pour it into the divided trough already mentioned (Fig. 21.12 top). We also assume that the cations exhibit four times the mobility of the anions which is approximately the case with H^+ and Cl^- ions. A total of 5 mol of charge should be transported by the current. This means that 5 mol of cations are discharged at the cathode and 5 mol of anions at the anode. At the same time, 4 mol of cations are transported from the anode compartment into the cathode compartment, while only 1 mol of anions are transported in the opposite direction (Fig. 21.12, center). In all, 5 mol of electrolytes "disappear" in the form of gases H_2 and Cl_2, from the electrolysis trough. In the process, a deficit of just 1 mol results in the cathode compartment while the deficit in the anode compartment is 4 mol (Fig. 21.12, bottom).

The ratio of changes of amount of substance Δn_{CC} and Δn_{AC} of 1:4 (or the corresponding ratio of changes of concentration) equals the ratio of mobilities of anions and cations and the ratio of the transport numbers:

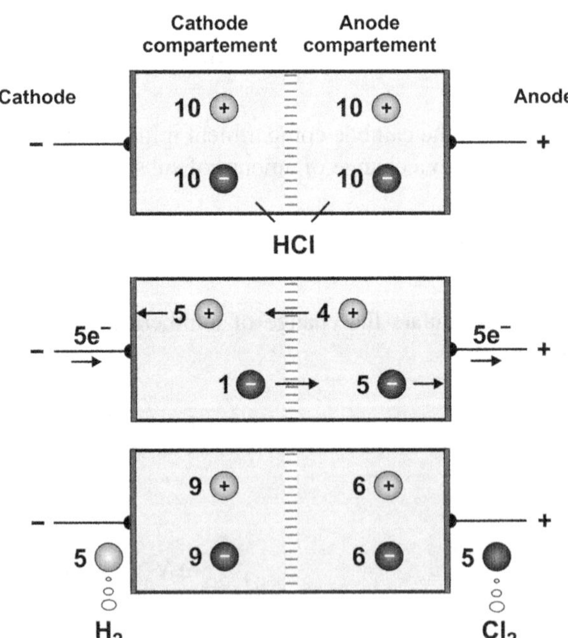

Fig. 21.12 Schema to illustrate the transport numbers: Change of amount of substance in the cathode and anode compartments during electrolysis.

$$\frac{\Delta n_{CC}}{\Delta n_{AC}} = \frac{\Delta c_{CC}}{\Delta c_{AC}} = \frac{-u_-}{u_+} = \frac{t_-}{t_+}. \tag{21.65}$$

Using Eq. (21.52), we find:

$$\frac{\Delta c_{CC}}{\Delta c_{AC}} = \frac{t_-}{1 - t_-} \tag{21.66}$$

or

$$t_- = \frac{\Delta c_{CC}}{\Delta c_{CC} + \Delta c_{AC}}. \tag{21.67}$$

Analogously,

$$t_+ = \frac{\Delta c_{AC}}{\Delta c_{CC} + \Delta c_{AC}}. \tag{21.68}$$

Figure 21.13 shows the experimental setup of a special electrolytic cell introduced by the German physicist Johann Wilhelm Hittorf. One platinum sheet serves as the cathode and one serves as the anode. After a given duration of electrolysis, the stopcock is closed and the solutions from both the anode compartment and cathode compartment are drained off and then titrated. Subsequently, the transport numbers can be determined from the reduction of concentration in both electrode compartments (so-called Hittorf method).

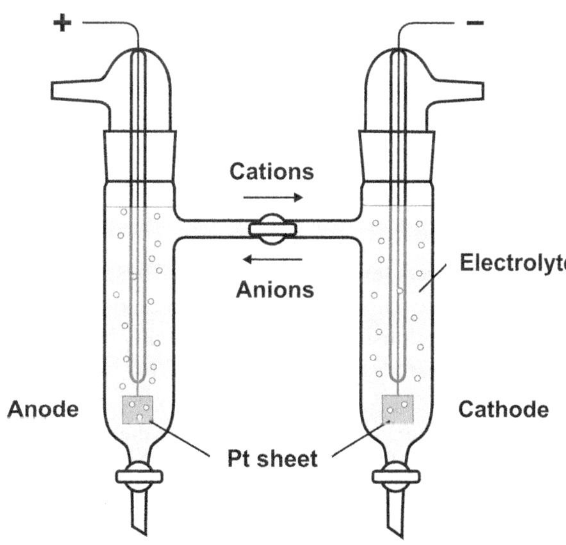

Fig. 21.13 Experimental setup of a Hittorf cell.

There is still one last problem to discuss: Although we are actually interested in the transport numbers t_+^0 or t_-^0 at infinite dilution in order to calculate the characteristic molar conductivities for individual types of ions, nature only allows us to experiment using a real solution having finite dilution. As concentration increases, the ionic conductivity as well as the molar conductivity of the electrolytes continuously decrease. As a result, the concentration dependency largely cancels out when we calculate the ratio. At concentrations that are not too high (below 10 mol m^{-3}), we have approximately $t_+ = t_+^0$ or $t_- = t_-^0$.

21.7 Conductivity Measurement and Its Applications

As stated in Sect. 21.4, the conductivity of an electrolyte solution is closely related to the conductance and, therefore, the electric resistance measured between the electrodes of a conductometric cell (cross-sectional area A, distance l) when a voltage of U is applied:

$$\sigma = G \cdot \frac{l}{A} = \frac{1}{R} \cdot \frac{l}{A}. \tag{21.69}$$

Platinated platinum or graphite is generally used as the inert electrode material. There is, however, a metrological problem that must be avoided in electrolysis, meaning the decomposition of the electrolyte and the polarization of the electrodes related to it. For this reason, we use a high-frequency alternating voltage because only then is the assumed ohmic behavior [Eq. (21.22)] of the electrolyte resistance guaranteed. Typically, a Wheatstone bridge circuit (Fig. 21.14) can be employed for measuring the resistance of the electrolyte cell. A resistor having a variable resistance R_v is adjusted so that the ammeter shows no current. A current equal to zero means:

$$\frac{R}{R_v} = \frac{R_1}{R_2} \quad \text{or} \quad R = \frac{R_1 \cdot R_v}{R_2}.$$

R_1 and R_2 are predetermined resistances. These days, Wheatstone's bridge circuit is only used for precision measurements, if at all. Modern instruments for measuring conductivity work using automatic balancing by a complex electronic circuitry and display the desired resistance directly (or, when appropriately calibrated, the conductivity).

The surface area of the electrodes and their arrangement in the conductometric cell influence the electric resistance via the quotient l/A, the so-called *cell constant*. However, these geometric quantities are often difficult to investigate, especially in the case of platinated electrodes. Therefore, the cell constant is determined by using a calibrating solution of known σ value (usually a solution of potassium chloride). These days, commercial equipment for measuring conductivity (conductometer) is

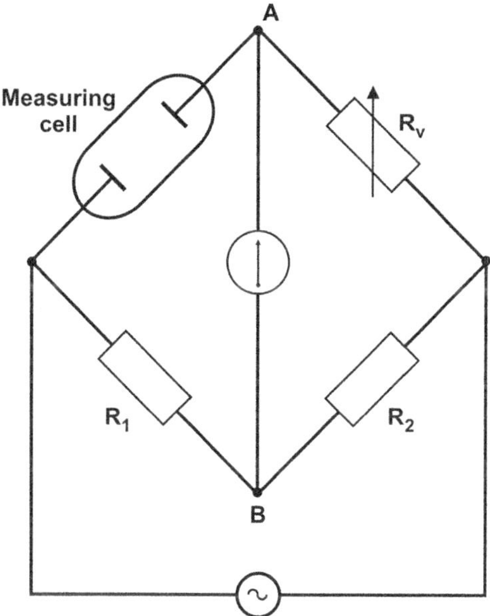

Fig. 21.14 Experimental setup (Wheatstone's bridge circuit) for determining the electric resistance of an electrolyte solution.

delivered with the cell constant already defined and tested so that it can immediately show conductivity. The accuracy of the device can be checked and calibrated, if need be, by using a calibration solution.

It is important that the temperature has to be carefully adjusted each time conductivity is measured, because viscosity will, for example, noticeably decrease in aqueous solutions as temperature increases (compare Sect. 20.3). Correspondingly, conductivity will increase by ca. 2 % with a temperature change of 1 °C [due to relations (20.21), (21.14), and (21.26)].

Several matter dynamical constants can be determined by measuring conductivities, such as molar conductivity at infinite dilution or dissociation constants of weak electrolytes (compare Sect. 21.5). Conductivity measurements are also useful for kinetic investigations (see Experiment 16.9).

When determining concentrations in analytic chemistry, measurement of conductivity is the preferred method for indexing the end point of a titration (*conductometric titration*). A change of conductivity during titration can be caused by a change of number of ions or their type. An example of this is the neutralizing of a strong acid (hydrochloric acid, HCl, for example) with a strong base (possibly sodium hydroxide, NaOH). When the conductivity (or a quantity proportional to it such as the conductance) is represented as a function of the amount of added base (Fig. 21.15), the reaction

$$(H^+|w + Cl^-|w) + (Na^+|w + OH^-|w) \rightarrow Na^+|w + Cl^-|w + H_2O|l$$

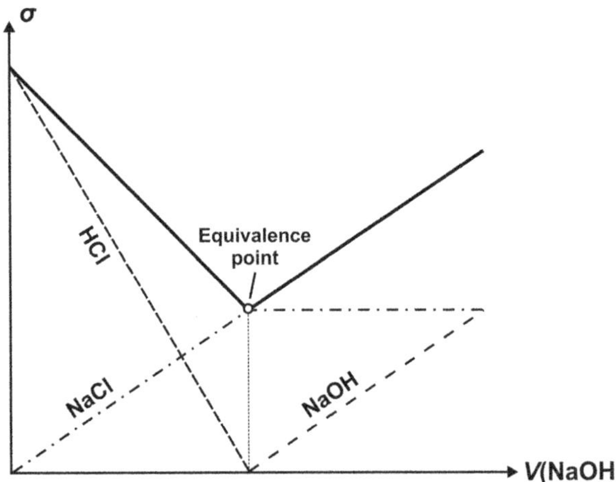

Fig. 21.15 Conductometric titration of a strong acid (hydrochloric acid, for example) with a strong base (possibly sodium hydroxide, NaOH) (The contributions of the various electrolytes to the conductivity are represented by *dashed lines.*).

initially causes a lowering of the measured value because highly mobile H^+ ions are replaced by much less mobile Na^+ ions (compare Sect. 21.3). However, the total concentration of ions does not change up to the equivalence point! Only after exceeding the equivalence point, does the conductivity begin to increase again because the total ionic concentration continuously rises by further addition of Na^+ and OH^- ions.

Precipitation titrations can also be indexed conductometrically.

Chapter 22
Electrode Reactions and Galvani Potential Differences

Initially, the terms Galvani potential for the electric potential in the bulk of a phase, electrochemical potential $\tilde{\mu}$, and electron potential μ_e are introduced to characterize processes in which charge-carrying species are involved. The electrochemical potentials can be used to determine the Galvani potential difference between two phases in equilibrium, as an especially simple example between two different metals. The formation of an electric double layer at the interface of both metals as well as the corresponding Galvani potential difference, the so-called *contact voltage*, will be presented. More important for practical use like that in galvanic cells is, however, the Galvani potential difference between a metal electrode and an electrolyte solution. The electrochemical potentials and their possible composition dependence are used to describe the underlying charge transfer reaction and to derive Nernst's equation. This type of charge transfer reaction can be regarded from a formal point of view as a special kind of a so-called *redox reaction*. Redox reactions in which electrons are transferred from one species to another are together with the proton transfer typical of acid–base reactions central to chemistry and its applications. Subsequently, different types of half-cells such as redox electrodes, gas electrodes, as well as film electrodes and the corresponding Galvani potential differences are discussed. The Galvani potential differences across liquid–liquid interfaces and membranes will be the topic of the last section. Such membrane voltages described by Donnan's equation play an important role in biological membranes, for example, for information transfer in nerve cells.

© Springer International Publishing Switzerland 2016
G. Job, R. Rüffler, *Physical Chemistry from a Different Angle*,
DOI 10.1007/978-3-319-15666-8_22

22.1   Galvani Potential Difference and Electrochemical Potential

In the first chapter, we were introduced to the concept of *phase* for a homogeneous region. If such a phase is electrically conductive due to its mobile electrons or ions, and is isolated from the environment, all differences of the electric potential will quickly equalize. The entire interior of the phase will then have a uniform value of the potential—except for a microscopically thin surface layer. We then simply speak of the inner electric potential (or *Galvani potential*) of the phase. However, the electric potential will generally *not* equalize between two chemically different phases that border on each other, such as two metals or a metal and a solution. In fact, a very definite electric potential difference meaning an electric voltage will usually form. These kinds of voltages appearing at phase boundaries are referred to as *Galvani potential differences* (or *Galvani voltages*). (In literature, sometimes the term Galvani potential is used for this kind of voltage instead of Galvani potential difference. But this could lead to misunderstandings and should therefore be avoided.) The rules of signs for voltages discussed in Sect. 21.2 are also valid here,

$$U_{1 \to 2} = \varphi_1 - \varphi_2 = -\Delta\varphi.$$

φ_1 and φ_2 are the inner electric potentials (Galvani potentials) of both phases.

The interiors of such phases are electrically neutral because the electroneutrality rule is valid for any arbitrary part of the phase [compare Eq. (21.1)]:

$$Q = \sum_i z_i \mathcal{F} n_i = 0. \tag{22.1}$$

Excess positive or negative charge carriers accumulate in the boundary layer. However, compared to the amount of substance in the interior, their quantity is so small (in the order of 10^{-10} mol per cm^2 of interface) that they are of no account in the balance (22.1). Although the charge carriers in the boundary layer determine the electric potential, they will not change the composition or chemical processes in the interior of the phase. We will go into this again later on.

If a small amount Δn_i of a substance i having a charge of $\Delta Q = z_i \mathcal{F} \Delta n_i$ is added to a phase with a Galvani potential of $\varphi = 0$, or alternatively to a second chemically homogeneous phase with a potential of $\varphi \neq 0$, the expended energy will differ by $\varphi \cdot \Delta Q$. This difference is due to the energy

$$W_t = \varphi \cdot \Delta Q \tag{22.2}$$

that has to be expended when transferring the charge ΔQ against the electric field between two locations of unequal electric potential (in this case $0 \to \varphi$) or is released (depending upon the signs of z_i and φ).

What are the consequences of this for the chemical potential of substance i? Let us think back to the hypothetical measurement method of Sect. 4.8. If we assign the value μ_i to the chemical potential in the phase with the Galvani potential $\varphi = 0$, its value $\widetilde{\mu}_i$ in the phase having $\varphi \neq 0$ must be different by the energy change $\varphi \cdot \Delta Q$ per transferred amount of substance Δn_i, i.e., by $\varphi \cdot \Delta Q/\Delta n_i = \varphi \cdot z_i \mathcal{F}$:

$$\widetilde{\mu}_i = \underbrace{\mu_i}_{\text{chemical}} + \underbrace{z_i \mathcal{F} \varphi}_{\text{electrical term}} . \tag{22.3}$$

$\widetilde{\mu}_i$ is called the *electrochemical potential* to distinguish it from μ_i. Because the location where the quantity φ has the value 0 can be chosen at will, the division of $\widetilde{\mu}_i$ into a chemical and an electric term in Eq. (22.3) is more or less arbitrary. If we assign different names to $\widetilde{\mu}_i$ and μ_i in the following, this is for practical reasons only and is of no principal importance. When it is unnecessary to distinguish between the two, we will simply speak of the "potential of a substance." These expressions were also used by Josiah Willard Gibbs. Equation (22.3) succinctly describes the simple φ dependency of this potential. In the case of charged substances $(z_i \neq 0)$, the electric expression is proportional to φ, while it drops out for uncharged substances.

What effect does the electric term $z_i \mathcal{F} \varphi$ have upon the chemical behavior of a substance? We were dealing with charged substances—different types of ions—before without having to consider their charge. Why wasn't this necessary? The answer is simple. As long as there is no charge being moved between regions of differing potentials in a process, the contributions by the electric terms cancel each other out in the calculations of drives.

Let us consider the homogeneous reaction of iron (II) ions with permanganate ions in an acidic solution as well as a heterogeneous reaction of dissolving calcium sulfate in water. For clarity, we will omit the chemical terms and insert only the electric ones below the substances into the conversion formula:

$$5\,Fe^{2+}|w + MnO_4^-|w + 8\,H^+|w \rightarrow 5\,Fe^{3+}|w + Mn^{2+}|w + 4\,H_2O|l$$

$$\underbrace{\underbrace{5\cdot 2\mathcal{F}\varphi}\quad \underbrace{(-\mathcal{F}\varphi)}\quad \underbrace{8\mathcal{F}\varphi}}_{17\mathcal{F}\varphi}\qquad \underbrace{\underbrace{5\cdot 3\mathcal{F}\varphi}\quad \underbrace{2\mathcal{F}\varphi}\quad \underbrace{4\cdot 0}}_{17\mathcal{F}\varphi}$$

and

$$CaSO_4|s \rightarrow Ca^{2+}|w + SO_4^{2-}|w$$

$$\underbrace{0 \qquad \underbrace{2\mathcal{F}\varphi}\quad \underbrace{(-2\mathcal{F}\varphi)}}_{0} .$$

In the first case, charge switches from one substance to the other—as is the case in all homogeneous reactions where ions participate—but remains at the same level of potential φ. In the second case, Ca^{2+} and SO_4^{2-} migrate from the solid substance into the solution, but the charges carried along compensate for each other so that the total charge transferred from the solid into the liquid phase becomes zero. The drive for the dissolution process and the solubility of the salt are not influenced by any differences of Galvani potential φ between the two phases.

However, things are different if charge is, in fact, transferred between regions of different electric potentials. In this case, the electrochemical potential $\tilde{\mu}_i$ should be considered instead of the chemical potential μ_i of the participating charge-carrying substances. Reactions correspondingly run in the direction of decreasing electrochemical potential (just as reactions between electrically neutral substances always occur in the direction of a drop in chemical potential). We will deal with these kinds of reactions in the following.

22.2   Electron Potential in Metals and Contact Potential Difference

An especially simple example of Galvani potential differences is the so-called *contact potential difference* (or *contact voltage*) between two metals. The electrons e^- in a metal can be assigned a chemical potential, similar to ions in a solution. In this case it is the *electron potential* μ_e. (We will only attach an index indicating charge such as $^+$, $^{2-}$, etc., to formulas and names of substances when it is necessary or useful for clarity. Expressions like e^- and e will be treated as equal.) μ_e will be different depending upon the metal in question.

The electron potential in alkaline and alkaline-earth metals is comparatively high, meaning the tendency to emit electrons is high. When this kind of metal is heated, the electrons evaporate out of it easily and can be removed by an electrode that is positively charged against the metal. The electrons can then be collimated and accelerated using auxiliary electrodes and can be used for an array of purposes. The electrons in these metals are referred to as weakly bound. We can express the bonding strength of the electrons to a metal in the same way we express the bonding strength of ions to water as a solvent, i.e., as the difference of electron potential in the metal and the gaseous state: $\Delta\mu_e = \mu_e(\text{metal}) - \mu_e(\text{gas})$. Table 22.1 shows corresponding values for some typical metals as well as graphite.

The bonding strength is of the same magnitude as the bonding strength of monovalent ions to water (compare Table 21.1). The electrons in platinum are especially strongly bound. It is very difficult for them to move from the metal into the gaseous area, but they can move freely in the interior of the metal.

When two different metals touch each other, the one that binds electrons more strongly will extract electrons from the other one. In other words, the electrons will

Table 22.1 Examples of differences of electron potentials in metal and gaseous state (at 298 K and 100 kPa).

Substance	Na	Zn	Cu	Fe	Ag	Pt	C|graphite
$\Delta\mu_e$ (kG)	-212	-404	-424	-439	-446	-509	-412

follow the chemical potential from the metal with higher electron potential into the one with lower potential. In the process, one metal will be positively charged and the other negatively. Copper, for example, binds electrons somewhat more strongly than zinc ($\Delta\mu_e = -424\,\text{kG}$ as opposed to -404 kG according to Table 22.1). When copper and zinc are brought into contact, electrons flow from the zinc into the copper and the copper charges negatively and the zinc positively.

The electric field that forms between the separated charges causes the depletion and enrichment zones in each metal to be limited to a thin boundary layer. The two boundary layers together are called an *electric double layer* where the distance between the opposite charges is of an order of magnitude of 10^{-10} m. The electric field emanating from the boundary layer of one of the phases is shielded by the opposite charge of the other layer so that no field exists beyond the double layer. As a result of charging, a potential difference (electric voltage) forms between the metals. In this case, the Galvani potential difference corresponds to the drop of potential from the *interior* (meaning outside of the boundary layer) of phase I to the *interior* of phase II. The voltage $U_{\text{I}\rightarrow\text{II}}$ works here as the electric drive for transporting the negative (!) charge carriers. Let us return to our example of contact between copper and zinc: the electric potential of the copper decreases and the electric potential of the zinc increases. An electric drive is formed that opposes the chemical drive.

This transport process continues until equilibrium is established between the chemical drive as a result of a drop in chemical potential and the opposing electric drive due to a drop in electric potential. These two opposing tendencies are summed up in the electrochemical potential mentioned above. *Electrochemical equilibrium* then generally means that the difference of electrochemical potentials of charge carriers of type i in two phases I and II, $\Delta\tilde{\mu}_i$, will equal zero:

$$\Delta\tilde{\mu}_i = 0 \quad \text{electrochemical equilibrium.} \tag{22.4}$$

This is equivalent to the electrochemical potential in both phases I and II being equal:

$$\tilde{\mu}_i(\text{I}) = \tilde{\mu}_i(\text{II}). \tag{22.5}$$

When Eq. (22.3) is inserted into the condition for equilibrium, it follows that

$$\mu_i(\text{I}) + z_i \mathcal{F}\varphi(\text{I}) = \mu_i(\text{II}) + z_i \mathcal{F}\varphi(\text{II}) \quad \text{or} \quad \mu_i(\text{II}) - \mu_i(\text{I}) = -z_i \mathcal{F}[\varphi(\text{II}) - \varphi(\text{I})].$$

Finally we obtain:

$$\Delta\varphi = \varphi(\text{II}) - \varphi(\text{I}) = -\frac{\mu_i(\text{II}) - \mu_i(\text{I})}{z_i \mathcal{F}} = -\frac{\Delta\mu_i}{z_i \mathcal{F}}. \tag{22.6}$$

Because $\mathcal{A}_{i,\text{I}\rightarrow\text{II}} = -\Delta\mu_i$ and $\Delta\varphi = -U_{\text{I}\rightarrow\text{II}}$, the Galvani voltage $U_{\text{I}\rightarrow\text{II}}$ in equilibrium between the metals results in

$$U_{\text{I}\rightarrow\text{II}} = -\frac{\mathcal{A}_{i,\text{I}\rightarrow\text{II}}}{z_i \mathcal{F}}, \qquad \text{or abbreviated,} \qquad U = -\frac{\mathcal{A}_i}{z_i \mathcal{F}}. \qquad (22.7)$$

We will again emphasize the fact that the electrochemical equilibrium is in no way identical to simultaneously existing chemical and electric equilibria. Rather, the effects of the chemical "tension" $\mathcal{A}$ and the electric tension (or voltage) U are in balance. Neither $\mathcal{A}$ nor U need to vanish in order for the condition for equilibrium $\Delta\tilde{\mu}_i = 0$ to be satisfied, as we can see in Eq. (22.7).

In the example of two metals in contact, the exchanged charge carriers are electrons. The electron potential μ_e for the metal in question should be inserted for the chemical potential, and the charge number z_i should receive a value of $z_e = -1$. We then obtain

$$\Delta\varphi = -\frac{\mu_e(\text{II}) - \mu_e(\text{I})}{z_e \mathcal{F}} = \frac{\Delta\mu_e}{\mathcal{F}} \qquad (22.8)$$

or for the contact voltage

$$U_{\text{I}\rightarrow\text{II}} = -\frac{\mathcal{A}_{e,\text{I}\rightarrow\text{II}}}{z_e \mathcal{F}}, \qquad \text{abbreviated} \qquad U = \frac{\mathcal{A}_e}{\mathcal{F}}. \qquad (22.9)$$

Figure 22.1 graphically illustrates this relation.

The Galvani potential difference formed between copper and zinc is not very high—it is about 0.2 V—but it is high enough to be easily measured in principle. However, this voltage is unnoticeable when the piece of copper and the piece of zinc are attached to the two cables of a voltmeter. The entire circuit made up of copper, zinc, cables, and voltmeter behaves as if the contact voltage between the two metals did not exist. Why is this? First we must keep in mind that a contact voltage does not form at only one location but at every location of contact between different metals. To simplify this, let us imagine all the cables and the wires inside

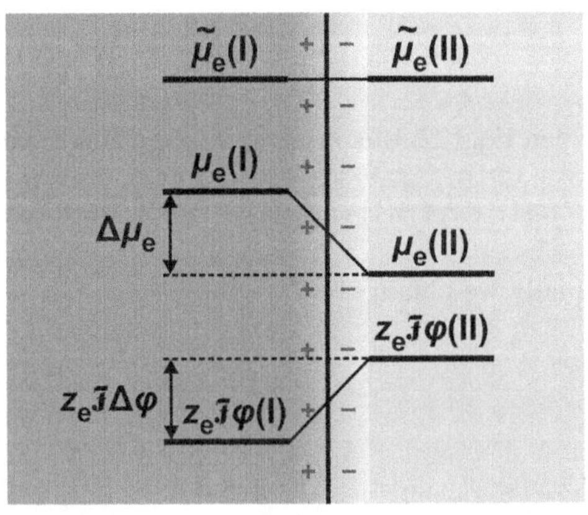

Fig. 22.1 Illustration of contact voltage and electrochemical equilibrium for two metals I and II in contact.

our voltmeter to be made of copper. Along with the contact between the pieces of copper and zinc, a second such contact will appear where the cable is attached to the piece of zinc. A Galvani potential difference is formed here as well and its magnitude will be the same as that of the first contact. The electric potential rises by 0.2 V at the transition from copper to zinc but falls again by the same amount when we go from zinc on the other side to the copper cables. The voltmeter does not indicate these steps of potential. Even when any number of electron conductors are joined together forming a circuit, the Galvani voltages will cancel each other. Voltage sources cannot be built this way. This changes when ion conductors are added to the electron conductors in the circuit. When this happens, the Galvani potential difference no longer needs to cancel each other and a voltage source will form. This voltage source is called a galvanic cell. We will go more deeply into galvanic cells in Chap. 23.

22.3 Galvani Potential Difference Between Metal and Solution

It will change nothing if, instead of electrons, we consider ions J as the charge-carrying particles exchanged between phases, for example, between a solution and a solid such as an ion-exchange resin or a metal. A metal (abbreviation: Me) can be imagined as basically made up of freely moving negative electrons and positive metal ions arranged on lattice sites. The chemical potential of the metal ions together with those of the electrons yields the potential of the metal as a whole:

$$\mu(\mathrm{Me}) = \mu(\mathrm{J}|\mathrm{m}) + z_{\mathrm{J}}\mu(\mathrm{e}|\mathrm{m}). \tag{22.10}$$

z_{J} is the charge number of the metal ions. If two of these potentials are known, the third can be calculated from them. Let us look at the example of copper:

$$\mu(\mathrm{Cu}) = \mu\!\left(\mathrm{Cu}^{2+}|\mathrm{m}\right) + 2\mu(\mathrm{e}|\mathrm{m}).$$

The abbreviation $|\mathrm{m}$ indicates a metallic phase, generally the pure metal phase which is pure copper in this case, but it can also be an alloy. In order to give the formulas more clarity, we will write them as follows:

$$\mu(\mathrm{Me}) = \mu_{\mathrm{J}}(\mathrm{Me}) + z_{\mathrm{J}}\mu_{\mathrm{e}}(\mathrm{Me}).$$

A copper sheet (phase II), dipped into a copper-salt solution (e.g.. a copper sulfate solution) (phase I), can exchange Cu^{2+} ions with the solution [abbreviated: d (for dissolved); compare Sect. 1.6]:

$$Cu^{2+}|d \rightarrow Cu^{2+}|s.$$

Because the chemical potential of the Cu^{2+} ions in the metal lies generally much lower than that in a solution, ions migrate out of the solution and into the metal. In the process, charge passes through the interface. This is called a *charge transfer reaction*. Water is, by far, the most common solvent, so we will assume aqueous solutions (abbreviation: |w) in the following. We will generally use the term *electrodes* for metals (in this case, copper) or other types of electron conductors (graphite, among others), which have the purpose of transferring electric charge between the usual conductors in circuits and other mediums—a solution, for example (compare Sect. 21.3).

The deposition of the metal ions, in our example copper ions, charges the metal positively relative to the electrolyte solution. This positive charge is limited to a thin boundary layer at the surface of the metal. It attracts anions toward the neighborhood of the electrode which leads to an excess of negative charge in the boundary layer of the solution. An electric potential difference will then form between the two phases. The positive charges remaining on the surface of the metallic phase form an electric double layer together with the anions which are enriched near the phase boundary because of electrostatic forces. This double layer is composed of a *"rigid" double layer* (*Helmholtz double layer*) and a *"diffuse" double layer* (*Gouy–Chapman double layer*) (Fig. 22.2).

The strong field near the metallic phase and the ensuing strong interaction forces cause the solvated anions to arrange themselves fairly rigidly like "pearls on a

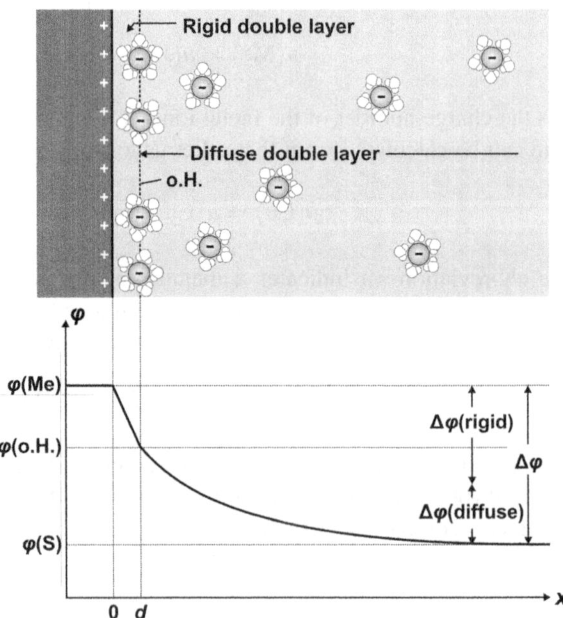

Fig. 22.2 Electric double layer at the phase boundary between a metal (Me) and an electrolyte solution (S) as well as the corresponding potential (o.H.: outer Helmholtz plane, d: thickness of the rigid double layer).

string" along the surface of the electrode—as far as their solvation shells allow; at least this is the picture we form. The ions in the solution form the outer Helmholtz layer and the charges on the surface of the electrode the inner layer. The distance between the layers is determined by the plane going through the cores of the solvated ions—the outer Helmholtz plane. The electric potential varies linearly between the two charge layers.

As the distance to the phase boundary grows, thermal motion increasingly disturbs the ordering of the ions. In this diffuse double layer, which extends rather far into the solution, the solvated ions are present in higher concentrations than in the interior of the electrolyte solution (abbreviation: S). The concentration as well as the electric potential decay almost exponentially to the values inside the phase $[\varphi(S)$ in the case of the potential].

Even the smallest amount of ions is enough to charge the metal sheet so strongly that the chemical forces are no longer sufficient to continue forcing more cations into the metal against the electric field. The resulting voltage can be calculated analogously to the contact voltage between metals. We will need to insert $\Delta\varphi = \varphi(\text{metal}) - \varphi(\text{solution})$ and $\Delta\mu = \mu(\text{ion in metal}) - \mu(\text{ion in solution})$ as well as the ionic charge number into Eqs. (22.6) and (22.7). The equation for ion type J with the charge number z_J is then:

$$\varphi(\text{Me}) - \varphi(\text{S}) = -\frac{\mu_J(\text{Me}) - \mu_J(\text{S})}{z_J \mathcal{F}}, \quad \text{or more simply} \quad \Delta\varphi = -\frac{\Delta\mu_J}{z_J \mathcal{F}} \quad (22.11)$$

or, if we replace the potential differences (electric and chemical) by the corresponding drives,

$$U_{\text{S}\to\text{Me}} = -\frac{\mathcal{A}_J(\text{J}|\text{d} \to \text{J}|\text{m})}{z_J \mathcal{F}}, \quad \text{abbreviated} \quad U = -\frac{\mathcal{A}_J}{z_J \mathcal{F}}. \quad (22.12)$$

$\mathcal{A}_J$ indicates the chemical drive of the process of metal ion J in solution (which is almost always aqueous) $\to$ metal ion J in metal.

We can write the following for our example of a copper sheet dipped into an aqueous copper-salt solution:

$$\Delta\varphi = -\frac{\mu_{\text{Cu}^{2+}}(\text{Me}) - \mu_{\text{Cu}^{2+}}(\text{S})}{2\mathcal{F}}.$$

We obtain charge transfer in the opposite direction if the chemical potential of the ions in the solution predominates. This is, for example, generally the case with a zinc rod immersed in a zinc-salt solution. The Zn^{2+} ions migrate out of the metal into the solution and, as a result, the metal charges negatively relative to the solution.

The chemical potential of the metal ions in solution phase S depends upon their concentration. We will refer to the mass action equation (Sect. 6.2) to calculate this concentration dependency:

$$\mu_{\mathrm{J}}(\mathrm{S}) = \overset{\circ}{\mu}_{\mathrm{J}}(\mathrm{S}) + RT\ln\frac{c_{\mathrm{J}}(\mathrm{S})}{c^{\ominus}} = \overset{\circ}{\mu}_{\mathrm{J}}(\mathrm{S}) + RT\ln c_{\mathrm{r,J}}(\mathrm{S}). \tag{22.13}$$

In this case, $\overset{\circ}{\mu}_{\mathrm{J}}(\mathrm{S})$ is the basic value of the chemical potential of the metal ions in the solution at the standard concentration $c^{\ominus} = 1\,\mathrm{kmol\,m^{-3}}(= 1\,\mathrm{mol\,L^{-1}})$ and arbitrary temperature T and pressure p.

The chemical potential of the metal ions in pure metal is naturally independent of concentration. According to Eq. (22.10), at corresponding temperature T and corresponding pressure p, it is:

$$\mu_{\mathrm{J}}(\mathrm{Me}) = \mu(\mathrm{Me}) - z_{\mathrm{J}}\mu_{\mathrm{e}}(\mathrm{Me}). \tag{22.14}$$

Inserting into Eq. (22.11) yields:

$$\Delta\varphi = -\frac{\left[\mu(\mathrm{Me}) - z_{\mathrm{J}}\mu_{\mathrm{e}}(\mathrm{Me})\right] - \left[\overset{\circ}{\mu}_{\mathrm{J}}(\mathrm{S}) + RT\ln c_{\mathrm{r,J}}(\mathrm{S})\right]}{z_{\mathrm{J}}F}$$

and finally,

$$\Delta\varphi = \frac{\overset{\circ}{\mu}_{\mathrm{J}}(\mathrm{S}) - \mu(\mathrm{Me}) + z_{\mathrm{J}}\mu_{\mathrm{e}}(\mathrm{Me})}{z_{\mathrm{J}}F} + \frac{RT}{z_{\mathrm{J}}F}\ln c_{\mathrm{r,J}}(\mathrm{S}). \tag{22.15}$$

Here, we are mainly interested in the concentration dependency of $\Delta\varphi = \varphi(\mathrm{Me}) - \varphi(\mathrm{S})$. The first term in Eq. (22.15) still depends upon temperature T and pressure p, but not upon the concentration of metal ions in the solution. For this reason, we abbreviate it just like the corresponding chemical quantities to $\Delta\overset{\circ}{\varphi} = -\overset{\circ}{U}$:

$$\Delta\varphi = \Delta\overset{\circ}{\varphi} + \frac{RT}{z_{\mathrm{J}}F}\ln\frac{c_{\mathrm{J}}(\mathrm{S})}{c^{\ominus}} \quad \text{or} \quad U = \overset{\circ}{U} - \frac{RT}{z_{\mathrm{J}}F}\ln\frac{c_{\mathrm{J}}(\mathrm{S})}{c^{\ominus}}. \tag{22.16}$$

Again, we call $\Delta\overset{\circ}{\varphi}$ and $\overset{\circ}{U}$ the basic values. According to Eq. (22.16), $\Delta\varphi$ increases with the concentration of metal ions: The more concentrated the solution, the higher the potential $\varphi(\mathrm{Me})$ compared to $\varphi(\mathrm{S})$. This equation, which describes the concentration dependency of $\Delta\varphi$ or U, is also called *Nernst's equation*.

If the electrode is not made up of pure metal, but is a homogeneous alloy (possibly an amalgam), the concentration dependency in the metallic phase must also be taken into account by the corresponding mass action equation. We then obtain the following relation instead of Eq. (22.16):

$$\Delta\varphi = \Delta\overset{\circ}{\varphi} + \frac{RT}{z_J \mathcal{F}} \ln \frac{c_J(S)}{c_J(Me)}. \qquad (22.17)$$

The following formulation of the exchange reaction,

$$Cu|s \rightleftarrows Cu^{2+}|w + 2\,e^-,$$

illustrates that the case of a metal in contact with a solution of its ions can be regarded from a formal point of view as a special kind of a so-called redox reaction. Redox reactions will be the subject of the next section.

22.4 Redox Reactions

One essential characteristic of *redox reactions* is that the participating substances exchange electrons e^- and therefore electric charge as well. Because of this charge transfer, charged particles, meaning ions, always appear as reaction partners in this kind of transformation. We will treat ensembles of a type of ion as we would treat ensembles of neutral particles—namely as substances, even when it is not possible to isolate these substances into pure form. This means that we consider hydrogen carbonate (HCO_3^-) and calcium(II) (Ca^{2+}) as substances (quasi "charged substances") like the "neutral" substances carbonic acid (H_2CO_3) and calcium(II) carbonate ($CaCO_3$). Redox reactions are transfer reactions like the acid–base reactions in Chap. 7; only instead of an exchange of protons, we have an exchange of electrons. Conceptually, they can then be treated in the same way.

When a substance B emits electrons, it leaves behind a new substance D. B is referred to as the *reducing agent* (or reductant) (abbreviation: Rd) and D as the *oxidizing agent* (or oxidant) (abbreviation: Ox). We use the abbreviations Rd and Ox for any type of particle whether it is positive, neutral, or negative, freely mobile in a solution or in a gas, or is only a component in a crystal. The simplest form of electron donation can be expressed as follows:

$$\underbrace{Rd \rightarrow Ox}_{\text{redox pair}} + e. \qquad (a)$$

The substances Rd and Ox together form a so-called *redox pair* or *redox system* Rd/Ox (equivalent to an acid–base pair Ad/Bs). Ox is the oxidizing agent *belonging* (*corresponding, conjugated*) to Rd, and Rd is the reducing agent *belonging* to Ox. If the process runs from left to right, we say that the reducing agent is being oxidized, and if it runs in the opposite direction, we say that the oxidizing agent is being

reduced. It is often advantageous to speak of the oxidation or the reduction of a
redox pair as a unit. An *oxidation* is equivalent to electron release out of the redox
system and a *reduction* is equal to electron acceptance. An example of this is the
oxidation of Fe^{2+} ions to Fe^{3+} ions in an aqueous solution:

$$Fe^{2+}|w \rightarrow Fe^{3+}|w + e^-.$$

Free electrons are extremely reactive so, under normal laboratory conditions, they
cannot accumulate anywhere in noticeable amounts. As they form, or rather in
avoiding the free state, they are immediately used up. Process (a) never appears
alone, but always paired with a second one of the same kind (a'):

$$Rd^* \rightarrow Ox^* + e. \quad (a')$$

When the first process (a) runs forward, the second one (a') is driven backward, and
vice versa. This reaction pair, made up of one oxidation and one reduction process,
is called a *redox reaction*, and the processes (a) and (a') the respective *half-reactions*.

Let us return to the simple basic process (a) and generalize it. On the one hand,
several electrons will often be exchanged simultaneously and, on the other, several
substances can appear in place of the simple substances Rd and Ox. We will
indicate the conversion number of the electrons with v_e and allow that Rd and Ox
can mean combinations of substances. The generalized process is then:

$$\overbrace{v_{Rd'}Rd' + v_{Rd''}Rd'' + \ldots}^{Rd} \rightarrow \overbrace{v_{Ox'}Ox' + v_{Ox''}Ox'' + \ldots}^{Ox} + v_e e.$$

The following is an example of a half-reaction in which composite reducing and
oxidizing agents participate:

$$Mn^{2+}|w + 12\ H_2O|l \rightarrow MnO_4^-|w + 8\ H_3O^+|w + 5\ e^-.$$

A redox pair behaves like an *electron reservoir*. When the reservoir is completely
filled, the pair is in its totally reduced form, Rd. If it is completely empty, only the
oxidized form Ox will be present. The different redox pairs and electron reservoirs,
respectively, have a greater or a smaller tendency to donate electrons. We can
describe this by a chemical potential we will call the *electron potential* of the redox
pair Rd/Ox:

$$\mu_e(Rd/Ox) := \frac{1}{v_e}[\mu_{Rd} - \mu_{Ox}]. \tag{22.18}$$

This equation is very similar to the definition of the proton potential μ_p [Eq. (7.1)]. This one as well results from the equilibrium condition for a transformation, in this case for the reaction

$$\mathrm{Rd} \rightleftarrows \mathrm{Ox} + \nu_e e$$

according to

$$\mu_{\mathrm{Rd}} = \mu_{\mathrm{Ox}} + \nu_e \mu_e.$$

The electron potential, like the proton potential, describes the strength of a tendency to transfer (in this case) electrons to other substances. If the chemical potential μ_e of the electrons in the surroundings is lower than $\mu_e(\mathrm{Rd/Ox})$, i.e., if $\mu_e < \mu_e(\mathrm{Rd/Ox})$, electrons will be released to the environment and the electron reservoir will be discharged. However, if μ_e is higher on the outside, $\mu_e > \mu_e(\mathrm{Rd/Ox})$, the redox pair will accept electrons, and the reservoir will begin to fill up and be charged. The limiting value $\mu_e(\mathrm{Rd/Ox})$ indicates the chemical potential of the electrons up to where donating electrons is just possible. In other words, the electron potential measures the maximum "electron pressure" the redox pair is able to produce. Acceptance of electrons is equivalent to a reduction, so $\mu_e(\mathrm{Rd/Ox})$ is also a measure of how strongly reducing a pair Rd/Ox is, meaning its *reductive capacity*.

Because the electron potential of a redox pair is characteristic of it, it is necessary to state which pair μ_e refers to. If Rd or Ox stands for a composite agent, we stipulate:

$$\mu_{\mathrm{Rd}} = \nu_{\mathrm{Rd}'}\mu_{\mathrm{Rd}'} + \nu_{\mathrm{Rd}''}\mu_{\mathrm{Rd}''} + \ldots \quad \text{and} \quad \mu_{\mathrm{Ox}} = \nu_{\mathrm{Ox}'}\mu_{\mathrm{Ox}'} + \nu_{\mathrm{Ox}''}\mu_{\mathrm{Ox}''} + \ldots .$$

Table 22.2 shows some examples of values where water appears as both reducing and oxidizing agent.

Table 22.2 Standard values of electron potentials of some redox systems (298 K, 100 kPa, 1 kmol m^{-3} in aqueous solution).

Reducing agent/Oxidizing agent	$\mu_e^{\ominus}$ (kG)
K$\|$s/K$^+\|$w	+283
½ H$_2\|$g + OH$^-\|$w/H$_2$O$\|$l	+80
Fe$\|$s/Fe$^{2+}\|$w	+39
½ H$_2\|$g + H$_2$O$\|$l/H$_3$O$^+\|$w	0
Sn$^{2+}\|$w/Sn$^{4+}\|$w	−14
2 OH$^-\|$w/½O$_2\|$g + H$_2$O$\|$l	−39
I$^-\|$w/½I$_2\|$s	−52
Fe$^{2+}\|$w/Fe$^{3+}\|$w	−74
3 H$_2$O$\|$l/½ O$_2\|$g + 2 H$_3$O$^+\|$w	−119
Mn$^{2+}\|$w + 12 H$_2$O$\|$l/MnO$_4^-\|$w + 8 H$_3$O$^+\|$w	−146
HF$\|$g + H$_2$O$\|$l/½ F$_2\|$g + H$_3$O$^+\|$w	−275

Experiment 22.1 *Reduction of Fe^{3+} by Sn^{2+} ions*: An iron (III) nitrate solution is combined with a tin (II) chloride solution. The progress of the reaction can be easily followed if a few drops of thiocyanate solution are added to the iron (III) salt solution at the beginning. The initially strong red color caused by the iron (III) thiocyanate complex disappears a few minutes after Sn^{2+} is added.

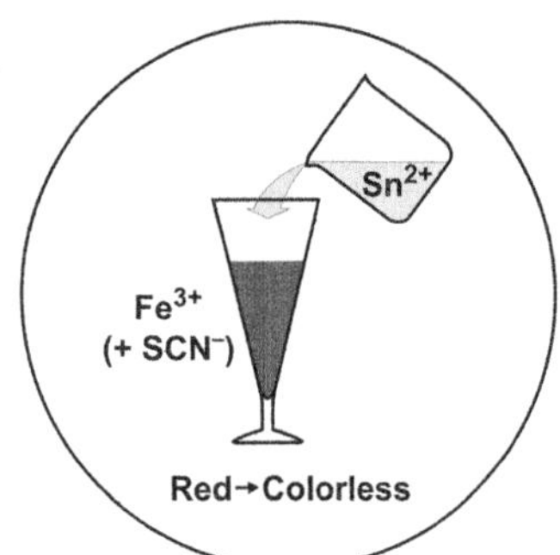

If an exchange of electrons between two redox pairs is allowed to take place, the "stronger" pair, meaning the one with the higher potential, will give off electrons to the "weaker" pair, which will then be reduced. If, for example, an Sn^{2+} solution is added to an Fe^{3+} solution, the Fe^{3+} will be reduced to Fe^{2+} and the Sn^{2+} will be oxidized to Sn^{4+}, because according to the levels of the electron potentials $(\mu_e^\ominus(Sn^{2+}/Sn^{4+}) = -14\,kG; \mu_e^\ominus(Fe^{2+}/Fe^{3+}) = -74\,kG)$, the redox pair Sn^{2+}/Sn^{4+} is more strongly reducing than the redox pair Fe^{2+}/Fe^{3+} (Experiment 22.1).

Electron transfer is often strongly inhibited and does not run as promptly as proton exchange. For this reason, it is often possible to manipulate redox pairs in aqueous solutions that, depending upon the levels of their electron potentials, should reduce the water or its components to H_2 or oxidize it to O_2. For example, in an acidic solution, the $\mu_e^\ominus$ value of the Mn^{2+}/MnO_4^- system is so low at $-146\,kG$ that it can draw electrons out of water (or more exactly, the H_2O/O_2 pair) having a $\mu_e^\ominus$ of $-119\,kG$. Even so, permanganate solutions can be kept for months.

22.5   Galvani Potential Difference of Half-Cells

Redox Electrodes When a chemically indifferent metal like platinum is immersed in a solution of a redox pair (meaning a homogeneous redox system), it will exchange almost no metal ions with the solution but will accept (or release) electrons. For the example of an aqueous solution of a redox pair in combination with a Pt electrode, we can formulate the process as follows:

$$Rd|w \rightleftarrows Ox|w + \nu_e e|Pt.$$

In the process, an electrochemical equilibrium will be established whereby the metal sheet or rod is charged and an electric double layer is formed. A well-defined

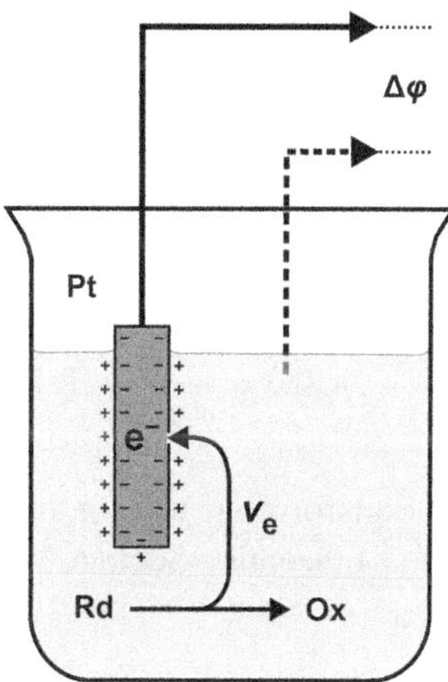

Fig. 22.3 Formation of an electric potential difference in the case of a homogeneous redox system (phase I) in contact with a noble metal electrode (phase II).

Galvani voltage $U_{\mathrm{I}\to\mathrm{II}}$ then forms between the solution (phase I) and the inert metal (phase II), exactly as it would form between two metals (Fig. 22.3). We can calculate the Galvani potential difference or the Galvani voltage, respectively, according to Eq. (22.8) or Eq. (22.9), by inserting the value of the electron potential in the metal, here the commonly used platinum (Pt), and of the redox pair Rd/Ox in the solution (S), $\Delta\mu_{\mathrm{e}} = \mu_{\mathrm{e}}(\mathrm{Pt}) - \mu_{\mathrm{e}}(\mathrm{S})$:

$$\Delta\varphi = \varphi(\mathrm{II}) - \varphi(\mathrm{I}) = -\frac{\mu_{\mathrm{e}}(\mathrm{Pt}) - \mu_{\mathrm{e}}(\mathrm{S})}{z_{\mathrm{e}}\mathcal{F}} = \frac{\Delta\mu_{\mathrm{e}}}{\mathcal{F}} \tag{22.19}$$

or

$$U_{\mathrm{I}\to\mathrm{II}} = -\frac{\mathcal{A}_{\mathrm{e},\mathrm{I}\to\mathrm{II}}}{z_{\mathrm{e}}\mathcal{F}}, \qquad \text{abbreivated to} \qquad U = \frac{\mathcal{A}_{\mathrm{e}}}{\mathcal{F}}, \tag{22.20}$$

where the charge number of the electrons $z_{\mathrm{e}} = -1$ was inserted on the right in each case.

If we allow for the concentration dependency of the chemical potential in relation (22.19), we obtain another version of Nernst's equation. Thus, according to Eq. (22.18), the following is valid for the electron potential of the redox pair in the solution:

$$\mu_{\mathrm{e}}(\mathrm{S}) = \mu_{\mathrm{e}}(\mathrm{Rd/Ox}) = \frac{1}{\nu_{\mathrm{e}}}[\mu(\mathrm{Rd}) - \mu(\mathrm{Ox})],$$

and therefore,

$$\mu_e(S) = \frac{1}{\nu_e}\left[\left(\overset{\circ}{\mu}(Rd) + RT\ln c_r(Rd)\right) - \left(\overset{\circ}{\mu}(Ox) + RT\ln c_r(Ox)\right)\right]$$

or

$$\mu_e(S) = \frac{1}{\nu_e}\left[\left(\overset{\circ}{\mu}(Rd) + RT\ln\frac{c(Rd)}{c^{\ominus}}\right) - \left(\overset{\circ}{\mu}(Ox) + RT\ln\frac{c(Ox)}{c^{\ominus}}\right)\right].$$

By transforming we obtain [remembering the rules for logarithms: $\ln a - \ln b = \ln(a/b)$ (see also Sect. A.1.1 in the Appendix)]:

$$\mu_e(S) = \left[\frac{1}{\nu_e}\overset{\circ}{\mu}(Rd) - \overset{\circ}{\mu}(Ox)\right] + RT\ln\frac{c(Rd)}{c(Ox)}. \tag{22.21}$$

Inserting into Eq. (22.19) yields:

$$\Delta\varphi = \frac{\mu_e(Pt) - \dfrac{1}{\nu_e}\left[\left[\overset{\circ}{\mu}(Rd) - \overset{\circ}{\mu}(Ox)\right] + RT\ln\dfrac{c(Rd)}{c(Ox)}\right]}{\mathcal{F}}$$

or

$$\Delta\varphi = \frac{\nu_e\mu_e(Pt) + \overset{\circ}{\mu}(Ox) - \overset{\circ}{\mu}(Rd)}{\nu_e\mathcal{F}} + \frac{RT}{\nu_e\mathcal{F}}\ln\frac{c(Ox)}{c(Rd)}. \tag{22.22}$$

If we again abbreviate the first term in Eq. (22.22) to $\Delta\overset{\circ}{\varphi}$, we obtain:

$$\Delta\varphi = \Delta\overset{\circ}{\varphi} + \frac{RT}{\nu_e\mathcal{F}}\ln\frac{c(Ox)}{c(Rd)}. \tag{22.23}$$

This version of Nernst's equation describes the dependency of the electric potential difference $\Delta\varphi = \varphi(Pt) - \varphi(S)$ upon the concentrations of the oxidized and reduced forms of the redox pair in question. The higher the concentration $c(Ox)$ of the oxidizing agent and the lower the concentration $c(Rd)$ of the corresponding reducing agent, the higher $\varphi(Pt)$ will be compared to $\varphi(S)$. We expect this because when the concentration $c(Ox)$ is high, the electron reservoir representing the redox pairs will be almost completely empty. Hence, there is the strong tendency to extract electrons from the noble metal which then becomes positively charged relative to the solution. However, if the concentration $c(Ox)$ is low compared to $c(Rd)$, the electron reservoir will be almost full and will have a strong tendency to donate electrons to the noble metal which will then become negatively charged. The electron gas of the metal acts as a kind of reservoir with an extremely small capacity that can accept or release electrons.

If we apply Eq. (22.23) to our first example of the simple redox pair of bivalent and trivalent iron, we then have:

$$\Delta\varphi = \Delta\overset{\circ}{\varphi}\left(Fe^{2+}/Fe^{3+}\right) + \frac{RT}{\mathcal{F}}\ln\frac{c(Fe^{3+})}{c(Fe^{2+})}.$$

If we have a composite redox system $Rd \rightarrow Ox + \nu_e e$, in which Rd stands for the combination of substances $\nu_{Rd'}Rd' + \nu_{Rd''}Rd'' + \ldots$ and Ox stands for the combination $\nu_{Ox'}Ox' + \nu_{Ox''}Ox'' + \ldots$, we can derive the generalized form of Nernst's equation analogously and then obtain:

$$\Delta\varphi = \frac{\nu_e\mu_e(Pt) + \nu_{Ox'}\overset{\circ}{\mu}(Ox') + \nu_{Ox''}\overset{\circ}{\mu}(Ox'') + \ldots - \left(\nu_{Rd'}\overset{\circ}{\mu}(Rd') + \nu_{Rd''}\overset{\circ}{\mu}(Rd'') + \ldots\right)}{\nu_e\mathcal{F}}$$

$$+ \frac{RT}{\nu_e\mathcal{F}}\left[\nu_{Ox'}\ln c_r(Ox') + \nu_{Ox''}\ln c_r(Ox'') + \ldots - \left(\nu_{Rd'}\ln c_r(Rd') + \nu_{Rd''}\ln c_r(Rd'') + \ldots\right)\right]$$

or

$$\Delta\varphi = \Delta\overset{\circ}{\varphi} + \frac{RT}{\nu_e\mathcal{F}}\ln\frac{c_r(Ox')^{\nu_{Ox'}} \cdot c_r(Ox'')^{\nu_{Ox''}} \cdot \ldots}{c_r(Rd')^{\nu_{Rd'}} \cdot c_r(Rd'')^{\nu_{Rd''}} \cdot \ldots}$$

or in abbreviated form

$$\Delta\varphi = \Delta\overset{\circ}{\varphi} + \frac{RT}{\nu_e\mathcal{F}}\ln\frac{\displaystyle\prod_{i=1}^{k}c_r(Ox_i)^{\nu_{Ox_i}}}{\displaystyle\prod_{j=1}^{l}c_r(Rd_j)^{\nu_{Rd_j}}}. \tag{22.24}$$

The product sign ($\prod$) is defined similarly to the summation sign ($\sum$), but in this case, a multiplication of factors follows.

This looks rather complicated at first, but the method quickly becomes clearer when we use our second example of the oxidation of Mn^{2+} ions:

$$\Delta\varphi = \Delta\overset{\circ}{\varphi}\left(Mn^{2+}/Mn_4^-\right) + \frac{RT}{5\mathcal{F}}\ln\frac{c_r(Mn^{2+}) \cdot c_r(H_3O^+)^8}{c_r(MnO_4^-)}.$$

Because of its high concentration, we will again use the potential of the pure solvent $\overset{\circ}{\mu}(H_2O)$ for water (compare Sect. 6.3) and factor it into the concentration-independent term $\Delta\overset{\circ}{\varphi}$.

The piece of metal (in this case, platinum) conveying electric charge between the usual conductors in a circuit and another medium (in this case, a solution) is called as already mentioned an *electrode*. We will call the combination of an electrode and a redox pair a *galvanic half-cell* (or an electrode in a wider sense).

In contrast to the homogeneous redox systems discussed above, there are numerous systems where the participating substances are not all together in one solution but are distributed over different phases. An example of this is when one of the partners is present in the form of a gas or a solid substance. If we combine such a redox pair with a metal like platinum, we not only have a phase boundary metal/solution, but several other similar interfaces, say, between platinum and gas or between platinum and solid substance. Now the question arises of which of these interfaces the Galvani potential difference forms across that corresponds to the electron transition between redox pair and platinum. This can only be the phase where the positive excess charge remains after the electrons separate from the redox pair. Exactly this phase will be charged in opposition to the metal. If the redox pair is made up of substances in two or three phases, we want to assign the exchangeable electrons to one certain phase. We choose the one where the positive charge remains after donating electrons. We will call this phase the *corresponding electrolyte*.

Gas Electrodes Let us look at an example of a gas electrode, a so-called *hydrogen electrode*: A platinum sheet is immersed in a solution containing hydrogen ions and is bathed in hydrogen gas (Fig. 22.4). In the case of the redox pair $H_2|g$ and $2H^+|w$, the electrons are not considered to come from the gas phase, which remains neutral when releasing electrons, but to come out of the solution because the positive H^+ ions accumulate there. The donated electrons are missing in this phase.

The conversion formula is:

$$H_2|g \rightarrow 2\,H^+|w + 2\,e^-|Pt.$$

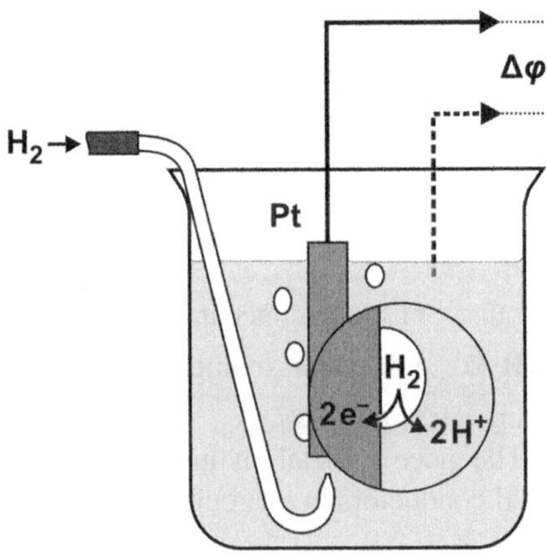

Fig. 22.4 A very simple hydrogen electrode as an example of a heterogeneous redox system. The enlarged section shows the electrode processes more clearly.

The electric potential difference in equilibrium can be calculated analogously to how homogeneous redox systems are calculated, but the equation

$$\mu(H_2) = \overset{\circ}{\mu}(H_2) + RT\ln\frac{p(H_2)}{p^{\ominus}} = \overset{\circ}{\mu}(H_2) + RT\ln p_r(H_2)$$

is applied to the chemical potential of the hydrogen gas because the concentration of a dissolved gas is proportional to its pressure in the gas phase (compare Sect. 6.6). The electron potential of the redox pair H_2/H^+ then results in:

$$\mu_e(H_2/H^+) = \frac{1}{2}[\mu(H_2) - 2\mu(H^+)].$$

Using the hydrogen electrode as an example of a gas electrode, we finally arrive at

$$\Delta\varphi = \frac{2\mu_e(Pt) + 2\overset{\circ}{\mu}(H^+) - \overset{\circ}{\mu}(H_2)}{2\mathcal{F}} + \frac{RT}{2\mathcal{F}}\ln\frac{c_r(H^+)^2}{p_r(H_2)} \qquad (22.25)$$

or abbreviated

$$\Delta\varphi = \Delta\overset{\circ}{\varphi}(H_2/H^+) + \frac{RT}{\mathcal{F}}\ln\frac{c_r(H^+)}{\sqrt{p_r(H_2)}}. \qquad (22.26)$$

Metal–Metal Ion Electrodes Until now, we have only discussed half-cells with so-called inert electrodes that can exchange electrons but not ions with other substances. The noble metal platinum is the preferred material for inert electrodes. In contrast to this, a metal can itself be part of a (heterogeneous) redox pair. In such a case, ions are exchanged between the solution and the metal. (We discussed this in detail in Sect. 22.3.) But formally, we can also imagine that the metal which is part of the redox pair is conductively connected to a piece of platinum and immersed in the corresponding electrolyte. Then we can deal with the metal–metal ion electrodes in the same way as with the redox electrodes discussed above, meaning we imagine that the conversed electrons are finally taken from the platinum ($Me|s \rightleftarrows Me^{z+}|w + \nu_e|Pt$). This approach has the advantage that the contact potential differences which appear by connecting the electrodes with a measuring device are made equal, meaning they cancel each other and can therefore be neglected. As an example, we will consider a silver–silver ion electrode (Fig. 22.5), which is composed of a piece of silver connected to a piece of platinum and immersed in a solution of Ag^+ ions, possibly a silver nitrate solution. Ag and the Ag^+ ions in the solution then form the redox pair.

Also in this case, the Galvani potential difference in equilibrium can be calculated analogously to how homogeneous redox systems are calculated, and finally we obtain the equation

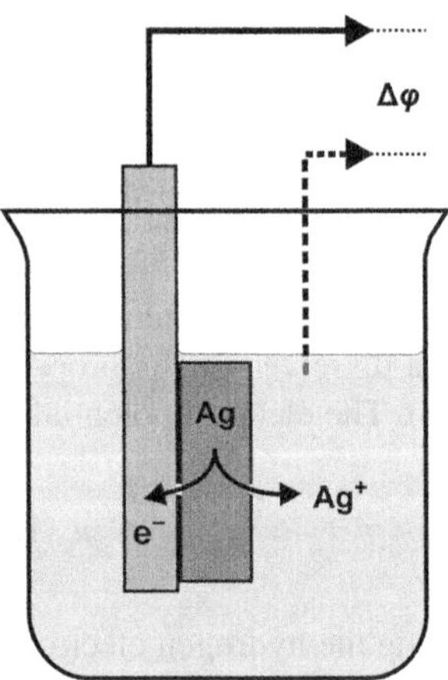

Fig. 22.5 Silver–silver ion electrode as an example of a metal–metal ion electrode.

$$\Delta\varphi = \Delta\overset{\circ}{\varphi} + \frac{RT}{v_e \mathcal{F}} \ln\frac{c_r(\mathrm{Me}^{z+})}{c_r(\mathrm{Me})}. \tag{22.27}$$

In the case of a pure solid metal phase, the mass action term drops out (compare Sect. 6.6) and therefore we have

$$\Delta\varphi = \Delta\overset{\circ}{\varphi} + \frac{RT}{v_e \mathcal{F}} \ln c_r(\mathrm{Me}^{z+}). \tag{22.28}$$

Both equations show, as expected, great similarities with Eq. (22.16) and Eq. (22.17). According to the relation $\mathrm{Me}|s \rightleftarrows \mathrm{Me}^{z+}|w + v_e|\mathrm{Pt}$, we have always $z_J = v_e$.

If we apply Eq. (22.28) to our example, the silver–silver ion electrode, this results in:

$$\Delta\varphi = \Delta\overset{\circ}{\varphi}\,(\mathrm{Ag/Ag}^+) + \frac{RT}{\mathcal{F}} \ln c_r(\mathrm{Ag}^+).$$

As we see, the Galvani potential difference depends solely upon the amount of Ag^+ ions in the solution. This is a fact that can also be used for analytic purposes (see Sect. 23.4).

Film Electrodes There are special cases where an ion electrode can react to ions other than its corresponding ones. We can have electrodes that react to Cl^- ions when silver is coated with a thin layer of low-soluble silver chloride (Fig. 22.6). This type of electrode is called a *film electrode*. Although the silver chloride is not a

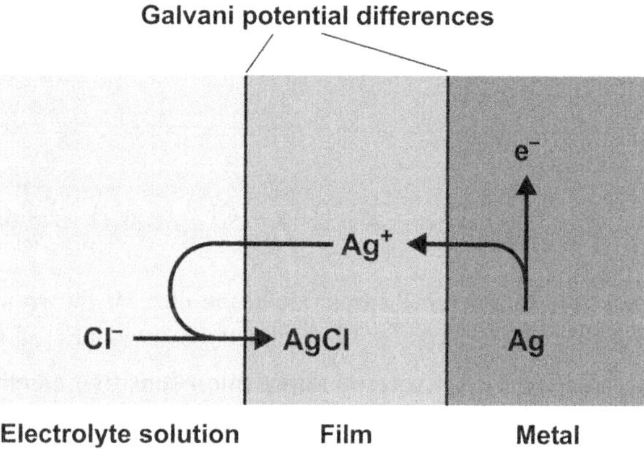

Fig. 22.6 Cross section through the surface of a electrode, in this case, a silver–silver chloride electrode.

metal, it is about as conducting as water. It is a solid electrolyte in which the silver ions have a certain mobility. The process of conduction can be illustrated by imagining the Ag^+ ions to be forcing their way through the spaces between the much larger spherical Cl^- ions. This is possible because the ions are not rigid spheres but elastic entities. They are never at rest. At room temperature, they oscillate and collide with each other at speeds comparable to gas molecules, meaning several hundred m s^{-1}. The Ag^+ ions move particularly easily along the grain boundaries where differently oriented crystalline parts meet.

Let us imagine a piece of silver with an AgCl film to be immersed in a Cl^- solution. Two phase boundaries will form, one between the metal and the film and the other between the film and the solution. The interface metal/film is permeable for Ag^+ but not for Cl^- ions or electrons. It is impermeable to Cl^- ions because they cannot be inserted into the metal lattice and electrons cannot pass through because silver chloride does not conduct electrons. Therefore, a Galvani potential difference will form there that is determined only by the chemical potentials of the Ag^+ ions in both phases. These are fixed values, so the Galvani potential difference will also have a fixed value. The interface film/solution is permeable to both Cl^- and Ag^+ so that Ag^+ and Cl^- ions compete to set the Galvani potential difference. However, free Ag^+ ions in a Cl^- solution can only be present in extremely low concentrations, so they are hopelessly outnumbered by the Cl^- ions. This is why only the Cl^- ions determine the potential difference at this interface.

The electrode reaction can be described by

$$Ag|s + Cl^-|w \rightarrow AgCl|s + e^-.$$

In this case, we must consider the composite redox pair $(Ag + Cl^-)/AgCl$. We obtain the corresponding Galvani potential difference $\Delta\varphi$ of the silver–silver chloride electrode by inserting the information about the redox pair into Eq. (22.24):

$$\Delta\varphi = \Delta\overset{\circ}{\varphi}\left((Ag + Cl^-)/AgCl\right) + \frac{RT}{\mathcal{F}}\ln\frac{1}{c_r(Cl^-)}$$

or

$$\Delta\varphi = \Delta\overset{\circ}{\varphi}\left((Ag + Cl^-)/AgCl\right) - \frac{RT}{\mathcal{F}}\ln\frac{c(Cl^-)}{c^\ominus}. \qquad (22.29)$$

The mass action term drops out in the case of the solid substances Ag and AgCl (compare Sect. 6.6), but the related basic values of the chemical potential are included in $\Delta\overset{\circ}{\varphi}$. Corresponding anion-sensitive electrodes can be produced for $Br^-, I^-, S^{2-}, SCN^-, \ldots$ by using silver with films of AgBr, AgI, Ag_2S, AgSCN, $\ldots$.

22.6   Galvani Potential Difference Across Liquid–Liquid Interfaces

A Galvani potential difference can form not only across the interface electrode/ electrolyte solution but also across the interface between two electrolyte solutions. There is usually a fine-pored wall (made of sintered glass or ceramic), a so-called *diaphragm*, to stabilize the phase boundary so that the solutions do not mix too quickly.

A Galvani potential difference is caused by the different chemical potentials of the various types of ions in the two neighboring phases. The potential gradient causes the ions to diffuse through the phase boundary. Because they have different mobility, they also migrate at different rates so that the charge separates and there is a jump of electric potential across the interface. This Galvani potential difference— called *diffusion (Galvani) voltage* (or *liquid junction voltage*) U_{diff}—is generally very difficult to calculate. However, there is a rather simple equation for the special case where only two ions can be exchanged. This special case occurs, for example, when two solutions of different concentrations $c(I)$ and $c(II)$ of the same *binary* electrolyte are allowed to border each other. Different concentrations of sodium chloride can be used as an example for this.

Both types of ions migrate out of the more concentrated solution into the more diluted one as they are driven by the gradients of their chemical potentials. In the process, the more mobile ion (let it be the negative ion as would be the case for Na^+ and Cl^-) moves a bit more quickly, so that the more dilute solution is charged negatively due to the retardation of the positive ions. Different electric potentials then develop in the two solutions. As a result, an electric potential gradient exists in the interface so that the ions there are not only subject to chemical forces but to electric ones as well. These forces cause the ions hurrying ahead to slow down and the slower ones behind to accelerate, so that both types of ions finally move through the phase boundary at the same rate. The potential difference $\Delta\varphi_{\text{diff.}}$ in the steady state for a 1–1 electrolyte turns out to be

$$\Delta\varphi_{\text{diff.}} = \varphi(\text{II}) - \varphi(\text{I}) = -\frac{(t_+ - t_-)}{\mathcal{F}} RT\ln\frac{c(\text{II})}{c(\text{I})} = -U_{\text{diff.}}, \qquad (22.30)$$

where t_+ and t_- are the transport numbers of the cations or anions, respectively (compare Sect. 21.6).

The equal migration velocities of the positive and negative ions serve as the point of departure for our calculation. We will use the expression [Eq. (21.8)] we were introduced to in our discussion of ion migration, for our approach to velocity. We indicate the quantities having to do with positive ions with the index +, and we use the index − for the corresponding quantities of negative ions. We use the letter x for the position coordinate that is perpendicular to the boundary layer of our electrolytes. We speak of a boundary layer rather than an interface because such layers have finite thickness that constantly widens by diffusion. We will consider a small volume element in this boundary layer, containing equal amounts n_+ and n_- of positive and negative ions. Depending on the gradients of chemical and electric potentials present, chemical and electric forces simultaneously act upon these ions

$$v_+ = \omega_+ \cdot \frac{F_+}{n_+} = -\omega_+ \cdot \frac{\mathrm{d}\mu_+}{\mathrm{d}x} - \omega_+ z_+ \mathcal{F} \cdot \frac{\mathrm{d}\varphi}{\mathrm{d}x},$$

$$v_- = \omega_- \cdot \frac{F_-}{n_-} = -\omega_- \cdot \frac{\mathrm{d}\mu_-}{\mathrm{d}x} - \omega_- z_- \mathcal{F} \cdot \frac{\mathrm{d}\varphi}{\mathrm{d}x}.$$

By applying the mass action equation, we obtain $\mathrm{d}\mu/\mathrm{d}x = RT/c \cdot \mathrm{d}c/\mathrm{d}x$ [Eq. (20.8)]. Since $c_+(x) = c_-(x) = c(x)$ must always hold because of charge neutrality, we have $\mathrm{d}\mu_+/\mathrm{d}x = \mathrm{d}\mu_-/\mathrm{d}x = \mathrm{d}\mu_\pm/\mathrm{d}x$. If we take both $z_+ = +1$ and $z_- = -1$ and $v_+ = v_-$ into account, we can write:

$$v_+ = -\omega_+ \cdot \frac{\mathrm{d}\mu_\pm}{\mathrm{d}x} - \omega_+ \mathcal{F} \cdot \frac{\mathrm{d}\varphi}{\mathrm{d}x} = v_- = -\omega_- \cdot \frac{\mathrm{d}\mu_\pm}{\mathrm{d}x} + \omega_- \mathcal{F} \cdot \frac{\mathrm{d}\varphi}{\mathrm{d}x}.$$

By putting all the expressions with $\mathrm{d}\mu_\pm/\mathrm{d}x$ on the right side of the equation, and the others on the left, we obtain

$$-(\omega_+ + \omega_-) \cdot \mathcal{F} \cdot \frac{\mathrm{d}\varphi}{\mathrm{d}x} = (\omega_+ - \omega_-) \cdot \frac{\mathrm{d}\mu_\pm}{\mathrm{d}x} \quad \text{and therefore}$$

$$\frac{\mathrm{d}\varphi}{\mathrm{d}x} = -\frac{\omega_+ - \omega_-}{\omega_+ + \omega_-} \cdot \frac{1}{\mathcal{F}} \cdot \frac{\mathrm{d}\mu_\pm}{\mathrm{d}x}.$$

If we consider the quotient formed from the mobilities as independent of concentration (which is only possible to a certain extent because the ω values are influenced by the composition of the solution), the differentials can be replaced by finite differences. Δx drops out of the equation and we obtain

$$\Delta\varphi_{\text{diff.}} = -\frac{\omega_+ - \omega_-}{\omega_+ + \omega_-} \cdot \frac{\Delta\mu_\pm}{\mathcal{F}}.$$

With the help of the mass action equations, and assuming $c(\mathrm{I}), c(\mathrm{II}) \ll c^{\ominus}$, the result is

$$\Delta\mu = \mu(\mathrm{II}) - \mu(\mathrm{I}) = \overset{\circ}{\mu}(\mathrm{II}) + RT\ln\frac{c(\mathrm{II})}{c^{\ominus}} - \overset{\circ}{\mu}(\mathrm{I}) - RT\ln\frac{c(\mathrm{I})}{c^{\ominus}} = RT\ln\frac{c(\mathrm{II})}{c(\mathrm{I})}.$$

Because the solution is uniform, the basic values of the potentials cancel each other. Keeping in mind that $\omega_+/(\omega_+ + \omega_-)$ equals $u_+/(u_+ - u_-)$ (because $u = \omega z \mathcal{F}$) and therefore equals t_+, and that $\omega_-/(\omega_+ + \omega_-)$ equals t_-, we obtain Eq. (22.30).

The more strongly the mobilities of the two ions differ, the higher the diffusion voltage will be. Conversely, it disappears when the anions and cations possess equal mobility. This is a fact that can be utilized for practical applications, as we will see in the next chapter (Sect. 23.1).

22.7 Galvani Potential Difference Across Membranes

Let us now modify the setup in the last section. We separate the two electrolyte solutions of different concentrations, $c(\mathrm{I})$ and $c(\mathrm{II})$, by an ion-selective *membrane*. This membrane allows only one type of ion J to permeate it. We will again use the example of two NaCI solutions separated by a membrane only permeable for Na^+ cations.

The difference of chemical potentials on either side of the membrane causes a natural tendency toward equalization of concentration for all the ions. However, because the membrane is only permeable for Na^+ ions, only these can migrate from the side having higher concentration [say, $c_\mathrm{J}(\mathrm{II})$] to the side of lower concentration [$c_\mathrm{J}(\mathrm{I})$] (Fig. 22.7), resulting in an excess of positive charge on the side of the diluted

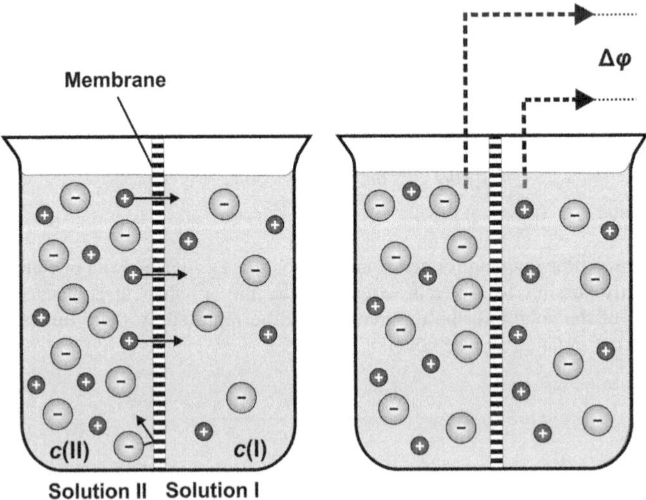

Fig. 22.7 Formation of membrane voltage.

solution as well as an excess of negative charge on the side of the more highly concentrated solution. An electric potential difference $\Delta\varphi$ then develops between the two regions. As the potential difference develops, it generates an electric field with a strong backward driving force. This causes the flow of ions to cease even before weighable amounts of Na^+ have been transported. In contrast to diffusion voltage forming because of constant substance transport, a *membrane voltage* is an effect related to equilibrium. Analogous to Eq. (22.11), we obtain the following for the potential difference $\Delta\varphi$:

$$\Delta\varphi = -\frac{\Delta\mu_J}{z_J \mathcal{F}}. \tag{22.31}$$

Applying the mass action equations under the same conditions as in the last section, meaning dilute solutions ($c_J(I), c_J(II) \ll c^{\ominus}$) and the same solvent on both sides of the membrane $\left[\overset{\circ}{\mu}_J(I) = \overset{\circ}{\mu}_J(II)\right]$, results in:

$$\Delta\mu_J = \mu_J(II) - \mu_J(I) = \overset{\circ}{\mu}_J(II) + RT\ln\frac{c_J(II)}{c^{\ominus}} - \overset{\circ}{\mu}_J(I) - RT\ln\frac{c_J(I)}{c^{\ominus}} = RT\ln\frac{c_J(II)}{c_J(I)}.$$

The difference of electric potential between the two solutions, the so-called membrane voltage, equals in equilibrium

$$\Delta\varphi = \varphi(II) - \varphi(I) = -\frac{RT}{z_J \mathcal{F}}\ln\frac{c_J(II)}{c_J(I)} = -U \tag{22.32}$$

Extremely thin layers of glass containing Na_2O or Li_2O are very suitable membranes for Na^+ or Li^+ ions. The ions are just slightly mobile in the amorphous SiO_2 structure, just enough for us to measure a membrane voltage with high resistance voltmeters. A thin LaF_3 layer is suitable as a membrane for F^- ions and specifically pretreated ZrO_2, as a membrane for O^{2-} ions at higher temperature. Membrane voltages play an important role in biological membranes, for example, for information transfer in nerve cells.

If the concentration $c_J(I)$ of the ion on one side is kept constant, the membrane voltage will be determined by $c_J(II)$ alone. It can therefore be understood as a measure of the ion concentration c_J. The most well-known application of this type is the measurement of proton potentials, i.e., pH values by *glass electrodes.*

When a piece of soda-rich glass (solidified SiO_2–CaO–Na_2O melt) is immersed in water, extremely thin "swollen" layers will gradually begin to form at the surface where the cations (Na^+) bonded to the SiO_2 structure are largely replaced by (hydrated) hydrogen ions (H_3O^+). This type of glass acts as a kind of membrane permeable for hydrogen ions or hydroxide ions. The mechanism for ion transfer is indicated in Fig. 22.8. For this reason, a membrane voltage should develop across the glass membrane, which, according to Eq. (22.31), is determined by the difference of proton potential on both sides of the membrane, $\Delta\mu_p = \mu_p(II) - \mu_p(I)$.

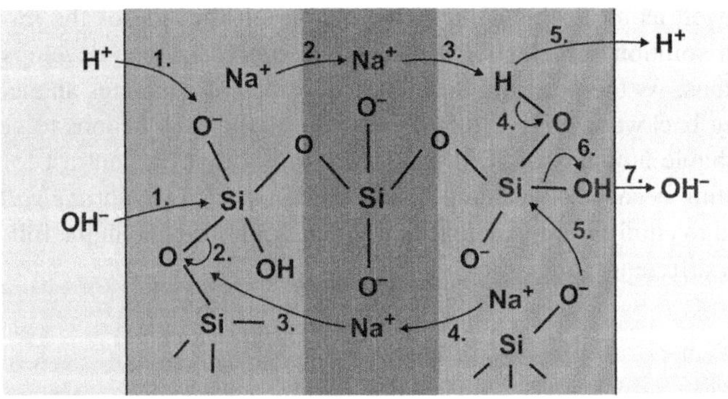

Fig. 22.8 Cross section through a membrane of a glass electrode (*above*, "apparent" transport of H⁺ ions using several steps, *below* that of OH⁻ ions).

When a buffer solution is put on one side that keeps the $\mu_p(I)$ value constant, only the proton potential $\mu_p(II) = -\mu_d \cdot \mathrm{pH}$ [Eq. (7.10)] will determine the measurable voltage. The decapotential μ_d is an abbreviation for $RT\ln 10$. Under these conditions, the membrane voltage is also a measure of the pH value of the solution on the other side of the membrane. We then have an arrangement that is suitable for measuring pH values and proton potentials, respectively.

If a membrane is permeable for several types of ions, Eq. (22.32) must be satisfied individually for each type of ion J, K, ... :

$$\Delta\varphi = -\frac{RT}{z_J \mathcal{F}}\ln\frac{c_J(II)}{c_J(I)} = -\frac{RT}{z_K \mathcal{F}}\ln\frac{c_K(II)}{c_K(I)} = \dots . \qquad (22.33)$$

According to the rules for logarithms, we can add the factors $1/z$ as exponents to the argument of the logarithmic function: $a \cdot \log b = \log b^a$ (compare Sect. A.1.1 in the Appendix):

$$-\frac{RT}{\mathcal{F}}\ln\left(\frac{c_J(II)}{c_J(I)}\right)^{\frac{1}{z_J}} = -\frac{RT}{\mathcal{F}}\ln\left(\frac{c_K(II)}{c_K(I)}\right)^{\frac{1}{z_K}} = \dots .$$

Multiplying by $-\mathcal{F}/RT$ and taking the exponential function leads to

$$\left(\frac{c_J(II)}{c_J(I)}\right)^{\frac{1}{z_J}} = \left(\frac{c_K(II)}{c_K(I)}\right)^{\frac{1}{z_K}} = \dots . \qquad (22.34)$$

This equation, called the *Donnan equation*, is named after the British chemist Frederick George Donnan who published his important paper about the theory of membrane equilibrium in 1911.

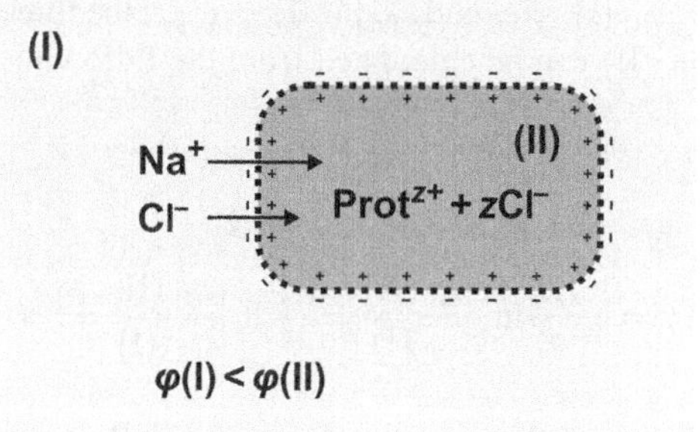

Fig. 22.9 Protein solution—the protein here is assumed to be a cation having charge number z and Cl$^-$ as the counter-ion—contained in a membrane permeable only to small ions and surrounded by a NaCl solution.

There is a better known special form of Eq. (22.34) for $z_+ = 1$ and $z_- = -1$, meaning two oppositely charged monovalent ions:

$$\frac{c_+(\text{II})}{c_+(\text{I})} = \left(\frac{c_-(\text{II})}{c_-(\text{I})}\right)^{-1} = \frac{c_-(\text{I})}{c_-(\text{II})}$$

or

$$c_+(\text{I}) \cdot c_-(\text{I}) = c_+(\text{II}) \cdot c_-(\text{II}). \tag{22.35}$$

We arrive at the surprisingly simple conclusion that the product of the concentrations of permeable cations and anions in both electrolyte solutions must be the same in equilibrium.

If the membrane is permeable to all the types of ions present, the concentrations will equalize and the electric potential difference will disappear. However, if the membrane is impermeable to a type of ion B and c_B on both sides is unequal, there will be an electric voltage across the membrane even at equilibrium.

Here is an example illustrating the case we just discussed. There are numerous higher-molecular substances in a cell. They carry charged groups such as proteins and nucleic acids that the cell membrane is practically impermeable to. We can imagine a solution of such a substance Prot^{z+} having a positive charge number z and the corresponding number of small counter-ions Cl$^-$ contained in a cell membrane that allows small ions to permeate it. The system floats in a table salt solution (Fig. 22.9). Assume the Cl$^-$ concentrations inside (I) and outside (II) to be the same initially, $c_{\text{Cl}^-}(\text{I}) = c_{\text{Cl}^-}(\text{II}) = c_0$. A steep $\mu(\text{Na}^+)$ gradient then exists from outside inward because there should not be any Na$^+$ inside. Na$^+$ begins to flow in, charging the interior (II) of the cell positively. This, in turn, elicits an inflow of the Cl$^-$ ions attracted to the positive interior. This happens *against* the gradient of concentration c_{Cl^-}! A noticeable amount of table salt ends up in the cell as a result of this. The drive for the inflow of Na$^+$ continuously decreases as the content of Na$^+$ continues to increase. Likewise, the inflow of Cl$^-$ becomes increasingly difficult as the amount of Cl$^-$ grows until the whole process finally stops. Because of $c_{\text{Na}^+}(\text{I})$

$= c_{Cl^-}(I) \approx c_0$ and $c_{Prot^{z+}}(II) = c_0/z$, the three unknowns $\Delta\varphi$ as well as $c_{Na^+}(II)$ and $c_{Cl^-}(II)$ can be calculated from the following equations:

$$c_{Na^+}(I) \cdot c_{Cl^-}(I) = c_{Na^+}(II) \cdot c_{Cl^-}(II) \qquad \text{Donnan equation,}$$

$$z\,c_{Prot^{z+}}(II) + c_{Na^+}(II) = c_{Cl^-}(II) \qquad \text{Electroneutrality rule,}$$

$$\Delta\varphi = -\frac{RT}{\mathcal{F}}\ln\frac{c_{Na^+}(II)}{c_{Na^+}(I)} = \frac{RT}{\mathcal{F}}\ln\frac{c_{Cl^-}(II)}{c_{Cl^-}(I)} = -U \qquad \text{Membrane voltage (Donnan volt.).}$$

However, these types of equilibria almost never occur in living cells. Pumping processes and other activities cause finite flows of ions.

Chapter 23
Redox Potentials and Galvanic Cells

In the last chapter, we learned a lot about Galvani potential differences across different individual interfaces and the usefulness of these potential differences, but we did not get to know how they can be measured. The problem is that it is impossible to measure the Galvani potential difference across a single interface in a half-cell directly, because an electrolyte's contact to the conductors of an electric measuring device requires a second electrode and this produces a new interface with an additional Galvani potential difference. The way out of the dilemma is the use of always the same reference half-cell, the standard hydrogen electrode, so that the measured voltage of the galvanic cell is only determined by the measuring half-cell. In this way, we obtain the so-called *redox potential* of a half-cell which represents, just like the electron potential, a measure of the strengths of reducing or oxidizing agents. The redox potentials under standard conditions are often compiled according to their values in a table, the *electrochemical series*. The combination of two arbitrary half-cells results in a galvanic cell. The reversible cell voltage of such a cell, meaning the cell voltage in equilibrium, can be described by Nernst's equation and used to predict the chemical drive, the equilibrium constant, and other thermodynamic properties of chemical reactions. Subsequently, some technically important galvanic cells will be discussed, which yield usable energy due to the spontaneous chemical reactions running inside them. In closing, the technique of potentiometry and the corresponding potentiometric titration is presented. This electroanalytical method uses the concentration dependence of the reversible cell voltage for quantitative analysis of ions.

23.1 Measuring Redox Potentials

Standard Hydrogen Electrode It is impossible to measure the Galvani potential difference across a single individual interface directly, because an electrolyte's contact to the conductors of an electric measuring device requires a second

© Springer International Publishing Switzerland 2016
G. Job, R. Rüffler, *Physical Chemistry from a Different Angle*,
DOI 10.1007/978-3-319-15666-8_23

electrode and this produces a new interface with an additional Galvani potential difference. The usual measuring device is a voltmeter showing the sum of the two (or more) Galvani potential differences forming across the different interfaces. The total arrangement, a combination of two half-cells, is a so-called *galvanic cell*. If all the potential differences except one are kept constant, the total voltage will be influenced only by changes to this single Galvani potential difference. Measurement techniques often make use of this principle. It is possible for the Galvani potential difference to remain constant at one electrode by keeping the conditions there constant. In other words, we always use the same *reference half-cell*. In general, when reporting measured values, a *standard hydrogen electrode* (abbreviated to SHE) functions as this reference half-cell. The redox pair involved is made up of hydrogen gas at standard pressure 100 kPa, and a solution of hydrogen ions with a pH value of 0.

In Chap. 22 (Sect. 22.4), we were introduced to a simple version of such a half-cell. In the case of the reference half-cell, it is made up of a platinum sheet electrode that is immersed in an acid solution and bathed in hydrogen gas. The electrode has undergone a special platinization process that gives it a rough, strongly enlarged surface (up to 500 times). This large surface facilitates the exchange of charge between metal and solution. It should be remembered that $pH = 0$ means that it is not the concentration which is prescribed, but the proton potential μ_p. Depending upon the acid used and other substance present, the $c(H^+)$ value can be a little greater or smaller than 1 kmol m^{-3}, but μ_p must correspond exactly to the basic value!

Salt Bridges The different *measuring half-cells* must be conductively connected to the reference half-cell. This electric contact between the two electrolyte solutions could be accomplished with a diaphragm. In this case, however, a noticeable Galvani potential difference (up to a few 10 mV) may form across the interface between the two solutions (diffusion voltage), disturbing the measurement (compare Sect. 22.6). We use a trick in order to reduce the diffusion voltage close to a value of zero. The electrolyte solutions are not brought directly in contact but separated by a third electrolyte solution with high concentration. This third electrolyte should have only two types of ions with equal but opposite charge, and mobilities that are as similar as possible. Two examples of such electrolytes are KCl und NH$_4$NO$_3$. The reason for this is that only a comparably small diffusion voltage will form across the interface of such a highly concentrated electrolyte and a less concentrated ion solution. If the ions are equally mobile, one type will not move faster than the other and no charging and no electric potential difference will occur. When there is a great excess of both ions, the other ions will be unable to act against these superior numbers. The diffusion voltage forming will not be exactly 0, but certainly much smaller than it would be without the intermediate electrolyte. This type of almost zero voltage connection between two electrolytes is called a *salt bridge*. The residual voltage is approximately 1 mV.

Figure 23.1 shows a practical example. The two legs of the H-shaped container are filled with saturated KCl solution. There are diaphragms at the bottom of each

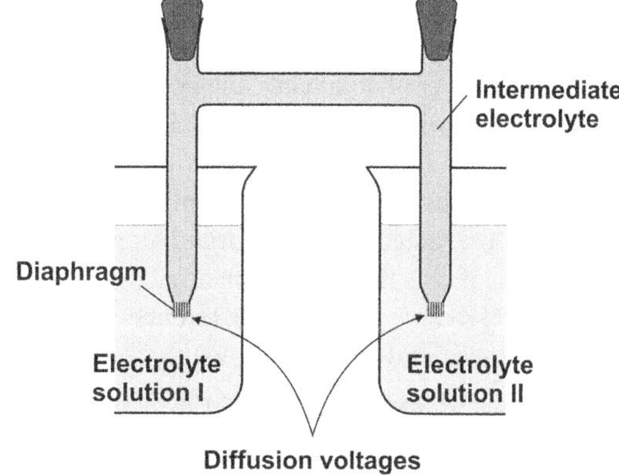

Fig. 23.1 Example of a salt bridge for producing an (almost) voltage-free connection between two electrolytes.

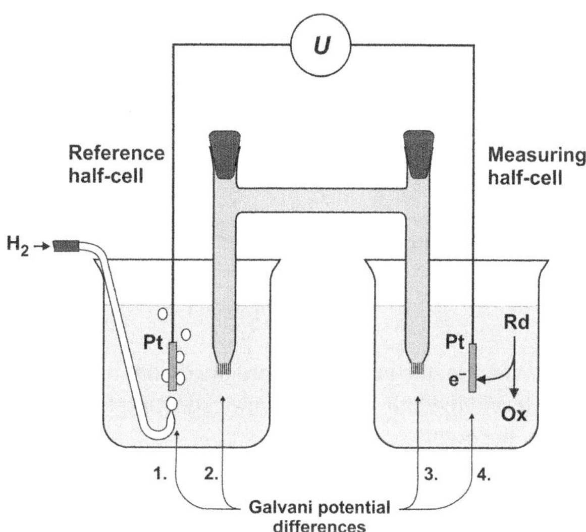

Fig. 23.2 Galvanic cell made up of a reference half-cell (standard hydrogen electrode) and a measuring half-cell containing the redox pair Rd/Ox dissolved in the corresponding electrolyte.

leg (a wad of cotton wool, a frit, a capillary, etc.). Diffusion voltages form across the diaphragms at the junctions to the different electrolyte solutions. As a rule, these voltages can be ignored, because they are small and tend to cancel each other as they have opposite signs relative to the positive direction of the current (which goes from left to right).

This arrangement of a fixed reference half-cell and salt bridge (Fig. 23.2) is frequently used in chemical measurements. It is important to perform the measurements with a current as weak as possible (ideally with no current) so that the electrochemical equilibria remain undisturbed. This can be accomplished by applying an equal counter-voltage from an external voltage source in combination with an adjustable resistor. It is much simpler and more precise, though, to use a voltmeter with high internal resistance.

Redox Potentials The total voltage U of the cell is essentially determined by the Galvani potential difference of the electrode in the measuring half-cell. U is actually the sum of four Galvani potential differences where the contributions of the two in the middle can generally be ignored. Because the first Galvani potential difference (for $T = \text{const.}$) remains constant for all the measurements, the fourth potential difference is almost the only influence upon U which is then a measure of how high the electrode is charged compared to the electrolyte. If the copper conductors of the measurement device are connected to the platinum electrode, two new Galvani potential difference will appear (so-called contact voltages between the metals), which exactly compensate for each other. The voltage between the copper conductors—shown on the measurement device—will agree with the voltage U. Therefore, U serves as the measure of the fourth Galvani potential difference which cannot be measured directly. If we keep in mind the cell notation convention from left to right when adding the four Galvani potential differences (i.e., electric potential differences), we have:

$$-U = \Delta\varphi(\text{left half-cell}) + \underbrace{\Delta\varphi(\text{left salt br.}) + \Delta\varphi(\text{right0 salt br.})}_{\approx 0}$$

$$+ \Delta\varphi(\text{right half-cell}).$$

If we use the abbreviations "l" and "r" for left and right, we obtain:

$$-U = \left[\varphi(\text{S})_\text{l} - \varphi(\text{Pt})_\text{l}\right] + \left[\varphi(\text{Pt})_\text{r} - \varphi(\text{S})_\text{r}\right].$$

By convention, the half-cell where the oxidation takes place has to be written on the left side and the half-cell where the reduction takes place on the right side ("reduction on the **right**").

We form the difference $\Delta\varphi$ as we did before, meaning that the electric potential of the solution is subtracted from that of the metallic phase, and we obtain:

$$-U = -\left[\varphi(\text{Pt})_\text{l} - \varphi(\text{S})_\text{l}\right] + \left[\varphi(\text{Pt})_\text{r} - \varphi(\text{S})_\text{r}\right] = \Delta\varphi(\text{meas.}) - \Delta\varphi(\text{ref.}).$$

This measure, which has been adapted for practical purposes, is frequently used in chemistry under the name of *redox potential E* of a redox pair Rd/Ox or as *electrode potential* or *half-cell potential*. (In the following, however, we will avoid use of the term electrode potential, because it could be confused with electron potential.):

$$E = -U = \Delta\varphi(\text{meas.}) - \Delta\varphi(\text{ref.}). \qquad (23.1)$$

Except for one summand, the redox potential describes the potential difference that develops between an indifferent electrode and its corresponding electrolyte in a measuring half-cell. If we use a standard hydrogen electrode as the reference electrode, we obtain

$$E = \Delta\varphi(\text{meas.}) - \Delta\overset{\circ}{\varphi}\,(\text{H}_2/\text{H}^+). \tag{23.2}$$

Due to the prevailing conditions for the hydrogen electrode, only the basic value needs to be considered. If we choose the Galvani potential difference of the standard hydrogen electrode at standard temperature $T^{\ominus} = 298\,\text{K}$ as the zero point, i.e., $\Delta\varphi^{\ominus}(\text{H}_2/\text{H}^+) = 0$, the definition $E = \Delta\varphi(\text{meas.})$ is valid. This means that the redox potential agrees with the Galvani voltage $U_{\text{Me}\rightarrow\text{S}}$ across the measuring electrode.

The redox potential measured in this way is independent of the material the indifferent electrode is made up of. If we totally replace the platinum with a different metal, possibly nickel or palladium, there will be no change to the voltage shown by the measuring device, although the Galvani voltage between electrode and electrolyte does not remain the same. The reason for this becomes clear when we replace the redox potential E by the chemical potentials of the participating substances, i.e., their electron potentials, by referring back to Eq. (22.19):

$$E = \Delta\varphi(\text{meas.}) - \Delta\varphi(\text{re f.}) = \frac{\mu_e(\text{Pt}) - \mu_e(\text{Rd}/\text{Ox})}{\mathcal{F}} - \frac{\mu_e(\text{Pt}) - \mu_e(\text{H}_2/\text{H}^+)}{\mathcal{F}}.$$

The chemical potential of the electrons in the metal drops out so that E is independent of whatever the electrode is made of:

$$E = -\frac{\mu_e(\text{Rd}/\text{Ox}) - \mu_e(\text{H}_2/\text{H}^+)}{\mathcal{F}}. \tag{23.3}$$

If we imagine the transferred electrons to be taken from the platinum, the metal ion electrodes conductively connected to the platinum of the reference electrode can be formally considered as redox electrodes. The redox potential of a copper ion electrode would then be:

$$\begin{aligned}
E &= \frac{\mu_e(\text{Pt}) - \mu_e(\text{Cu}/\text{Cu}^{2+})}{\mathcal{F}} - \frac{\mu_e(\text{Pt}) - \mu_e(\text{H}_2/\text{H}^+)}{\mathcal{F}} \\
&= -\frac{\mu_e(\text{Cu}/\text{Cu}^{2+}) - \mu_e(\text{H}_2/\text{H}^+)}{\mathcal{F}}.
\end{aligned}$$

The redox potential indicates up to a factor of $-\mathcal{F}$, how high the level of the electron potential $\mu_e(\text{Rd}/\text{Ox})$ is, compared to the level $\mu_e(\text{H}_2/\text{H}^+)$ of a fixed reference redox pair (where the minus sign before $-\mathcal{F}$ comes from the electrons' charge number -1). So seen, E only appears to represent an electric quantity. Actually, E describes a chemical quantity, a chemical potential difference in the same way that the difference in levels of mercury in a mercury manometer does not represent a geometric quantity but a dynamic quantity: the difference in pressure.

We were introduced to the redox pair

$$H_2|g \rightarrow 2\,H^+|w + 2e^-|Pt$$

as a general system of reference for the chemical potential of electrons under standard conditions (compare Sect. 4.4). If the temperature is allowed to vary, the chemical potentials of the substances will correspond to their basic values. In order to find the electron potential of this reference redox pair that is only dependent upon temperature, we can use the defining equation (22.18) for $\mu_e(\text{Rd}/\text{Ox})$:

$$\mu_e(H_2/H^+) = \overset{\circ}{\mu}_e(H_2/H^+) = \tfrac{1}{2}\left[\overset{\circ}{\mu}(H_2;T) - 2\overset{\circ}{\mu}(H^+;T)\right]. \qquad (23.4)$$

In particular, for $T^{\ominus} = 298\,\mathrm{K}$ we have $\mu_e^{\ominus}(H_2/H^+) = 0$. At standard temperature, Eq. (23.3) therefore simplifies to

$$E = -\frac{\mu_e(\text{Rd}/\text{Ox})}{\mathcal{F}}. \qquad (23.5)$$

Nernst's Equation In closing, we will talk about the concentration dependency of the redox potential. We obtain the following relation for a simple redox pair $\text{Rd} \rightarrow \text{Ox} + \nu_e e$ by inserting the $\mu_e(\text{Rd}/\text{Ox})$ defining equation into Eq. (23.3):

$$E = -\frac{\dfrac{1}{\nu_e}[\mu(\text{Rd}) - \mu(\text{Ox})] - \overset{\circ}{\mu}_e(H_2/H^+)}{\mathcal{F}}.$$

Allowing for the mass action equation $\mu_B = \overset{\circ}{\mu}_B + RT\ln(c_B/c^{\ominus})$ for $\mu(\text{Rd})$ and $\mu(\text{Ox})$ yields

$$E = -\frac{\left[\overset{\circ}{\mu}(\text{Rd}) - \overset{\circ}{\mu}(\text{Ox})\right] - \overset{\circ}{\mu}_e(H_2/H^+)}{\nu_e\,\mathcal{F}} + \frac{RT}{\nu_e\,\mathcal{F}} \cdot \ln\frac{c(\text{Ox})}{c(\text{Rd})}. \qquad (23.6)$$

By abbreviating the first term (the *basic value* of the redox potential) to $\overset{\circ}{E}$, we obtain the corresponding Nernst equation:

$$E = \overset{\circ}{E} + \frac{RT}{\nu_e\,\mathcal{F}} \cdot \ln\frac{c(\text{Ox})}{c(\text{Rd})}. \qquad (23.7)$$

As we would expect from relation (23.2), the redox potential shows the same concentration dependency as the Galvani potential difference [Eq. (22.23)]. The various versions of Nernst's equation generally exhibit great similarity to the proton level equation [Eq. (7.12)]. This equation describes the dependency of proton potential μ_p upon the ratio of acid and base concentrations, i.e., upon $c(\text{Ad})/c(\text{Bs})$. The similarity results from the direct relation of the Galvani potential difference of a half-cell or its redox potential with the electron potential μ_e [Eqs. (22.19) or (23.3)], which is formally closely related to the proton potential.

The frequently used "*decadic*" version of Eq. (23.7) can be obtained by going from the natural logarithm to the logarithm to base 10 in the second term (the so-called *mass action term* of the redox potential):

$$E = \overset{\circ}{E} + \frac{E_N(T)}{v_e} \cdot \lg \frac{c(\text{Ox})}{c(\text{Rd})}. \tag{23.8}$$

The following is valid for the factor

$$E_N(T) := \frac{RT\ln 10}{\mathcal{F}} \tag{23.9}$$

at the standard temperature $T^{\ominus} = 298\,\text{K}$:

$$E_N^{\ominus} = E_N(T^{\ominus}) = \frac{8.314\,\text{J K}^{-1}\,\text{mol}^{-1} \times 298\,\text{K} \times 2.303}{96,485\,\text{C mol}^{-1}} = 0.059\,\text{V}.$$

In the case of the composite redox pair $\text{Rd} \to \text{Ox} + v_e\text{e}$, where Rd stands for the combination of substances $\text{Rd}' + \text{Rd}'' + \ldots$ and Ox stands for the combination $\text{Ox}' + \text{Ox}'' + \ldots$, we must correspondingly alter the Nernst equation [compare derivation of Eq. (22.24)]:

$$E = \overset{\circ}{E}(\text{Rd}/\text{Ox}) + \frac{E_N(T)}{v_e} \cdot \lg \frac{c_r(\text{Ox}')^{v_{\text{Ox}'}} \cdot c_r(\text{Ox}'')^{v_{\text{Ox}''}} \cdot \ldots}{c_r(\text{Rd}')^{v_{\text{Rd}'}} \cdot c_r(\text{Rd}'')^{v_{\text{Rd}''}} \cdot \ldots}.$$

or abbreviated

$$E = \overset{\circ}{E}(\text{Rd}/\text{Ox}) + \frac{E_N(T)}{v_e} \cdot \lg \frac{\prod_{i=1}^{k} c_r(\text{Ox}_i)^{v_{\text{Ox}_i}}}{\prod_{j=1}^{l} c_r(\text{Rd}_j)^{v_{\text{Rd}_j}}}.$$

Using the redox potential of the pair made up of bivalent and trivalent iron as a simple example we obtain with $v_e = 1$:

$$E = \overset{\circ}{E}\left(\text{Fe}^{2+}/\text{Fe}^{3+}\right) + E_N \cdot \lg \frac{c(\text{Fe}^{3+})}{c(\text{Fe}^{2+})}$$

or, at the standard temperature $T^{\ominus} = 298\,\text{K}$,

$$E = E^{\ominus}\left(\text{Fe}^{2+}/\text{Fe}^{3+}\right) + 0.059\,\text{V} \cdot \lg \frac{c(\text{Fe}^{3+})}{c(\text{Fe}^{2+})}.$$

Increasing $c(\text{Fe}^{3+})/c(\text{Fe}^{2+})$ tenfold leads to a 59 mV increase of E. The concentration dependency of the redox potential $E(\text{Fe}^{2+}/\text{Fe}^{3+})$ is shown in Fig. 23.3.

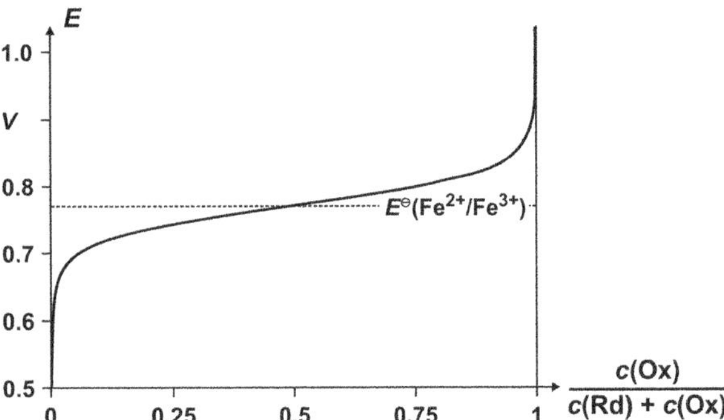

Fig. 23.3 Concentration dependency of the redox potential E for the pair Fe^{2+}/Fe^{3+} at 298 K. [The abscissa shows the fraction in oxidized form, meaning the ratio $c(Ox)/(c(Rd)+c(Ox))$].

If the concentrations of the oxidized and reduced forms of the redox pair are equal $[c(Ox) = c(Rd)$ or $c(Ox)/(c(Rd) + c(Ox)) = 0.5]$, then according to Eq. (23.7), the redox potential will correspond to the standard value $E^{\ominus}$ (because $\ln 1 = 0$). However, if the concentration of the oxidizing agent is greater than that of the corresponding reducing agent, the redox potential will be shifted to higher values, and if it is smaller, there will be a shift to lower values. This is expected because when we have a high concentration $c(Ox)$, there will be a strong tendency to "snatch" electrons from the measuring electrode, thereby giving it a positive potential. On the other hand, if $c(Rd)$ is high, the tendency will be to emit electrons to the measuring electrode, resulting in a lowering of its potential [$\ln(c(Ox)/c(Rd))$ will be negative for $c(Ox) < c(Rd)$].

In contrast, the redox potential of the hydrogen electrode results in

$$E = \overset{\circ}{E}(H_2/H^+) + \frac{E_N}{2} \cdot \lg \frac{(c_{H^+}/c^{\ominus})^2}{p_{H_2}/p^{\ominus}}$$

with $v_e = 2$. Because of $E^{\ominus}(H_2/H^+) = 0$, the equation simplifies to

$$E = \frac{0.059\,V}{2} \cdot \lg \frac{(c_{H^+}/c^{\ominus})^2}{p_{H_2}/p^{\ominus}}.$$

Electrochemical Series Some redox potentials under standard conditions ($T^{\ominus} = 298\,K$, $p^{\ominus} = 100\,kPa$, $c^{\ominus} = 1$ $kmol\,m^{-3}$ in aqueous solutions) have been compiled in the following table (Table 23.1). This kind of sequence of standard redox potentials is also called the *electrochemical series*. The half-cells in question have been characterized by the corresponding "Stockholm convention" of 1953. In this convention, a phase boundary is denoted by a single vertical line.

Equation (23.6) tells us that although the standard value of a redox potential $E^{\ominus}$ depends upon the type of reference electrode being used, it is otherwise a characteristic quantity for the redox pair because of its direct dependency upon the electron potential. Just like the electron potential, it represents a measure of

Table 23.1 Redox potentials at standard conditions (298 K, 100 kPa, 1 kmol m^{-3} in an aqueous solution), based upon a standard hydrogen electrode (from: Lide D R (ed) (2008) CRC Handbook of Chemistry and Physics, 89th edn. CRC Press, Boca Raton).

Half cell	Electrode reaction	$E^{\ominus}$ (V)
$Li\|Li^+$	$Li^+ + e^- \rightleftarrows Li$	-3.0401
$Cs\|Cs^+$	$Cs^+ + e^- \rightleftarrows Cs$	-3.026
$Rb\|Rb^+$	$Rb^+ + e^- \rightleftarrows Rb$	-2.98
$K\|K^+$	$K^+ + e^- \rightleftarrows K$	-2.931
$Ca\|Ca^{2+}$	$Ca^{2+} + 2\,e^- \rightleftarrows Ca$	-2.868
$Na\|Na^+$	$Na^+ + e^- \rightleftarrows Na$	-2.71
$Mg\|Mg^{2+}$	$Mg^{2+} + 2\,e^- \rightleftarrows Mg$	-2.372
$Al\|Al^{3+}$	$Al^{3+} + 3\,e^- \rightleftarrows Al$	-1.662
$Zn\|Zn^{2+}$	$Zn^{2+} + 2\,e^- \rightleftarrows Zn$	-0.7618
$Fe\|Fe^{2+}$	$Fe^{2+} + 2\,e^- \rightleftarrows Fe$	-0.447
$Cd\|Cd^{2+}$	$Cd^{2+} + 2\,e^- \rightleftarrows Cd$	-0.4030
$Ni\|Ni^{2+}$	$Ni^{2+} + 2\,e^- \rightleftarrows Ni$	-0.257
$Pb\|Pb^{2+}$	$Pb^{2+} + 2\,e^- \rightleftarrows Pb$	-0.1262
$Cu\|Cu^{2+}$	$Cu^{2+} + 2\,e^- \rightleftarrows Cu$	$+0.3419$
$Ag\|Ag^+$	$Ag^+ + e^- \rightleftarrows Ag$	$+0.7996$
$2\,Hg\|Hg_2^{2+}$	$Hg_2^{2+} + 2\,e^- \rightleftarrows 2\,Hg$	$+0.7973$
$Au\|Au^+$	$Au^+ + e^- \rightleftarrows Au$	$+1.692$
$Pt\|H_2\|OH^-$	$2\,H_2O + 2\,e^- \rightleftarrows H_2 + 2\,OH^-$	-0.8277
$Pt\|Cr^{2+}, Cr^{3+}$	$Cr^{3+} + e^- \rightleftarrows Cr^{2+}$	-0.407
$Pt\|H_2\|H^+$	$2H^+ + 2\,e^- \rightleftarrows H_2$	0.00000
$Pt\|Sn^{2+}, Sn^{4+}$	$Sn^{4+} + 2\,e^- \rightleftarrows Sn^{2+}$	$+0.151$
$Pt\|Cu^+, Cu^{2+}$	$Cu^{2+} + e^- \rightleftarrows Cu^+$	$+0.153$
$Pt\|\left[Fe(CN)_6\right]^{4-}, \left[Fe(CN)_6\right]^{3-}$	$\left[Fe(CN)_6\right]^{3-} + e^- \rightleftarrows \left[Fe(CN)_6\right]^{4-}$	$+0.358$
$Pt\|O_2\|OH^-$	$O_2 + H_2O + 2\,e^- \rightleftarrows H_2 + 2OH^-$	$+0.401$
$Pt\|I_2\|I^-$	$I_2 + 2\,e^- \rightleftarrows 2I^-$	$+0.5355$
$Pt\|Fe^{2+}, Fe^{3+}$	$Fe^{3+} + e^- \rightleftarrows Fe^{2+}$	$+0.771$
$Pt\|O_2\|H^+$	$\frac{1}{2}O_2 + 2H^+ + 2\,e^- \rightleftarrows H_2O$	$+1.229$
$Pt\|Cl_2\|Cl^-$	$Cl_2 + 2\,e^- \rightleftarrows 2Cl^-$	$+1.35827$
$Pt\|F_2\|F^-$	$F_2 + 2\,e^- \rightleftarrows 2F^-$	$+2.866$
$Pb\|PbSO_4\|SO_4^{2-}$	$PbSO_4 + 2\,e^- \rightleftarrows Pb + SO_4^{2-}$	-0.3588
$Ag\|AgI\|I^-$	$AgI + e^- \rightleftarrows Ag + I^-$	-0.15224
$Ag\|AgCl\|Cl^-$	$AgCl + e^- \rightleftarrows Ag + Cl^-$	$+0.22233$
$Hg\|Hg_2Cl_2\|Cl^-$	$Hg_2Cl_2 + 2\,e^- \rightleftarrows 2\,Hg + 2Cl^-$	$+0.26808$

the strengths of the reducing or the oxidizing agents. A strongly negative value of the redox potential means that the corresponding redox pair is highly reductive and will have a strong tendency to "press" electrons onto the measuring electrode, producing a high "electron pressure." Correspondingly, a more strongly reductive redox pair with a lower redox potential can force electrons onto a "weaker" pair in contact with it. A look at the redox potentials tells us that the redox pair Sn^{2+}/Sn^{4+} is more strongly reductive than the redox pair Fe^{2+}/Fe^{3+} $\left[E^{\ominus}(Sn^{2+}/Sn^{4+}) = +0.151\,V;\ E^{\ominus}(Fe^{2+}/Fe^{3+}) = +0.771\,V\right]$ (compare Sect. 22.4).

Reference Electrodes The advantage of the standard hydrogen electrode used as the reference for a given redox potential is that its equilibrium Galvani potential difference adjusts quickly and reproducibly. It is, however, rather complicated to deal with because, among other things, a bottle containing oxygen-free (and highly explosive) hydrogen gas is necessary.

A film electrode made up of silver–silver chloride also has a well-reproducible Galvani potential difference, as we saw in Sect. 22.5. Because it is much easier to handle than a standard hydrogen electrode, it is the preferred reference electrode. Its structure is rather simple: Silver chloride is precipitated directly onto a silver wire as a thin layer by electrolysis—the wire is immersed in a highly concentrated chloride solution—mostly potassium chloride (saturated or 3 kmol m^{-3}). The concentration of Cl^{-} ions determines the Galvani potential difference. In practice and for simplicity (Fig. 23.4), the electrolyte solutions in the reference half-cell and the (integrated) salt bridge are made equal. The diaphragm between these two parts can therefore be left out and only one diaphragm at the measuring cell is necessary. The potassium and chloride ions are about equally mobile and highly concentrated, so no noticeable diffusion voltage builds up at this point of contact.

Because of the constant potential differences at the charge exchanging interfaces, the *silver–silver chloride reference electrode* forms a kind of "pluggable connector" between the metallic conductors and the outer solution. This makes it possible to easily determine the variable potential differences in the circuit. This

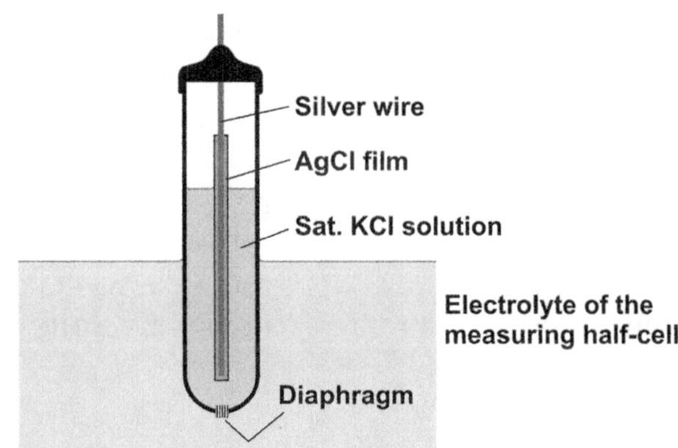

Fig. 23.4 Practical construction of a silver–silver chloride reference electrode.

allows us to determine the Galvani potential difference of a measuring half-cell relative to this new reference point and then convert to the standard hydrogen electrode (SHE). However, the potential differences of the silver–silver chloride electrode relative to the SHE must be known. At 298 K and using saturated KCl solution as the electrolyte, it is +0.1976 V.

In principle, a *calomel electrode* is very similar. Here, mercury (Hg) and low soluble mercury(I)chloride (calomel, Hg_2Cl_2) replace Ag and AgCl. The liquidity of mercury necessitates a somewhat different arrangement, but the process is the same.

23.2   Cell Voltage

Until now, we have been dealing mostly with reactions of the type $Rd \rightarrow Ox + \nu_e e$. Since free electrons cannot be obtained under the usual laboratory conditions, these kinds of processes never happen alone, but always occur in pairs—as we have mentioned already. Therefore, a complete *redox reaction* is composed of two *half-reactions*,

$$Rd \rightarrow Ox + \nu_e e \quad \text{and} \quad Rd^* \rightarrow Ox^* + \nu_e e,$$

where the first runs backward and the second runs forward or vice versa. In order for coupling to be possible, the two half-reactions must be formulated so that the conversion number ν_e of the electrons is the same in both of them. If necessary, this can be accomplished by multiplying the conversion formulas by the appropriate numerical factors. The resulting total reaction would then be:

$$Rd + Ox^* \rightarrow Ox + Rd^*.$$

The half-reactions of the reaction $2\,Fe^{3+}|w + Sn^{2+}|w \rightarrow 2\,Fe^{2+}|w + Sn^{4+}|w$ (compare Experiment 22.1) must be correspondingly formulated as follows:

$$Sn^{2+}|w \rightarrow Sn^{4+}|w + 2\,e^- \quad \text{and} \quad 2\,Fe^{2+}|w \leftarrow 2\,Fe^{3+}|w + 2\,e^-.$$

This example shows how a reaction can run under conditions where the electrons are directly exchanged between the substances. A further example is the reaction of zinc cuttings with a copper sulfate solution (Experiment 23.1),

$$Zn|s + Cu^{2+}|w \rightarrow Zn^{2+}|w + Cu|s,$$

where the following half-reactions,

$$Zn|s \rightarrow Zn^{2+}|w + 2\,e^- \quad \text{and} \quad Cu|s \leftarrow Cu^{2+}|w + 2\,e^-,$$

Experiment 23.1 *Reduction of Cu^{2+} ions by zinc*: When zinc cuttings are poured into a solution containing copper ions, they precipitate densely and immediately turn black. The precipitate then slowly turns coppery brown while the solution's color successively changes from blue to green to brown and, finally, colorless. Simultaneously, the temperature rises considerably.

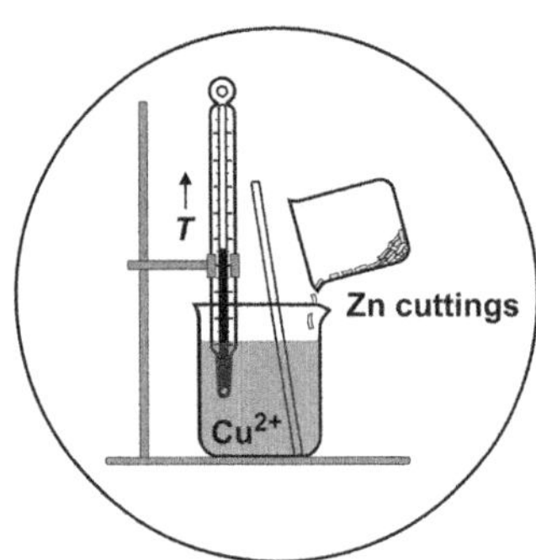

must be considered [respective redox potentials: $E^{\ominus}\left(Zn/Zn^{2+}\right) = -0.7618\,V$, $E^{\ominus}(Cu/Cu^{2+}) = +0.3419\,V$].

According to Eq. (8.18), the energy W_{ξ} necessary for a small conversion $\Delta\xi$ of the total reaction is

$$W_{\xi} = -\mathcal{A}\cdot\Delta\xi = \left[\mu(Cu) + \mu(Zn^{2+}) - \mu(Zn) - \mu(Cu^{2+})\right]\cdot\Delta\xi, \qquad (23.10)$$

in which energy $W_{\xi} < 0$ is released during the spontaneous process ($\mathcal{A} > 0$). This released energy is dissipated and entropy is generated, which is expressed by a warming of the reaction mixture.

The two half-reactions can also be spatially separated from each other by dividing them into the two half-cells of a galvanic cell where they are connected to each other by an exterior circuit. For example, the so-called *Daniell cell* (Fig. 23.5) is composed of a Zn and a Cu electrode that are immersed in corresponding Zn^{2+} or Cu^{2+} solutions whereby these electrolyte solutions are in contact with each other through a diaphragm. To avoid diffusion voltages, a salt bridge can be used instead.

The gradient of the chemical potential continues to drive the reaction

$$Zn|s + Cu^{2+}|w \rightarrow Zn^{2+}|w + Cu|s.$$

However, the reactants can no longer reach each other so easily because they are separated by a "wall" (the electrolyte solutions) that ions can permeate but electrons cannot. The only possibility is for the ions and electrons to go "separate ways." While the ions can migrate into the electrolyte solution, the electrons must be diverted through the external circuit. Zinc ions at the zinc electrode go into the solution, meaning that oxidation takes place, so we know we are dealing with an anode (Fig. 23.6). The accumulation of electrons caused by the electrons left behind gives this electrode a negative charge. At the copper electrode, on the other hand,

Fig. 23.5 Schematic of a Daniell cell.

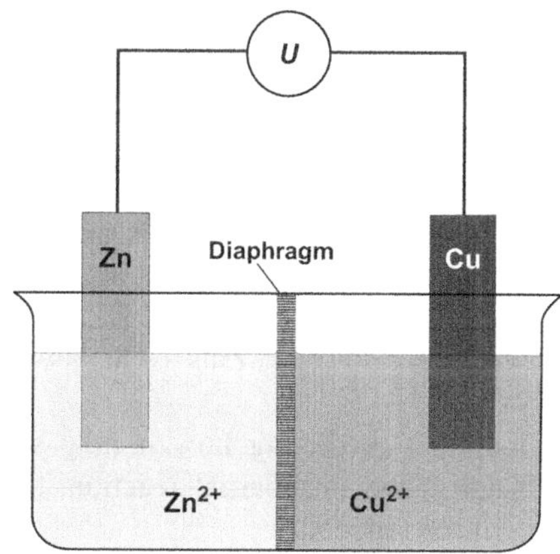

Fig. 23.6 Electrode processes in a Daniell cell and nomenclature.

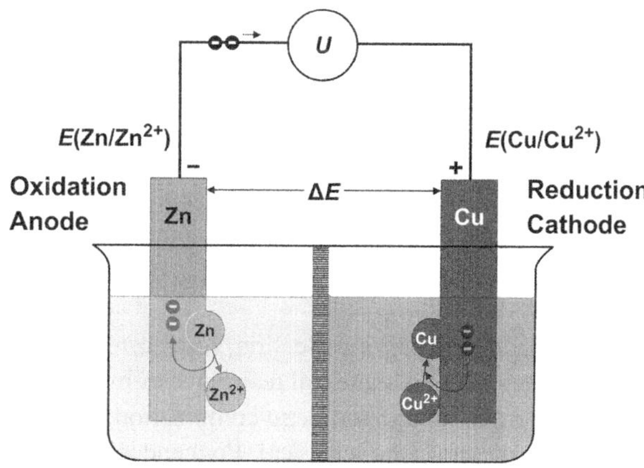

copper ions are deposited in the form of neutral copper. This means that the ions are reduced (cathode). The "electron suction" caused by the consumption of electrons gives this electrode a positive charge and an electric voltage then develops between the two electrodes. However, immediately after even extremely small amounts of ions have transferred, electrochemical equilibrium occurs at the electrodes (compare Sect. 22.3).

Using the example of a Daniell cell, we will consider the conventions used in representing galvanic cells. In a cell diagram, as previously mentioned, a vertical line symbolizes a phase boundary. A dashed line represents a diaphragm through which both electrolyte solutions in the half-cell are in contact with each other. If the diffusion voltage is minimized, possibly by the use of a salt bridge, a dashed double line will stand for this. Therefore, depending upon the setup, the abbreviation for a Daniell cell is

$$\text{Zn}|\text{Zn}^{2+} \,\vdots\, \text{Cu}^{2+}\,|\text{Cu} \quad \text{or} \quad \text{Zn}|\text{Zn}^{2+} \,\vdots\,\vdots\, \text{Cu}^{2+}\,|\text{Cu}.$$

If the direction of a reaction is arbitrarily chosen when writing the conversion formula of a cell reaction, for example,

$$\text{Zn}|\text{s} + \text{Cu}^{2+}|\text{w} \rightarrow \text{Zn}^{2+}|\text{w} + \text{Cu}|\text{s},$$

without first predicting how it may spontaneously run, the redox pair will be oxidized in the first partial reaction, and in the other partial reaction, the other pair will be reduced:

Oxidation(anode): $\text{Zn}|\text{s} \rightarrow \text{Zn}^{2+}|\text{w} + 2\,\text{e}^-$ (left, in the cell diagram),
Reduction(cathode): $\text{Cu}^{2+}|\text{w} + 2\,\text{e}^- \rightarrow \text{Cu}|\text{s}$ (right, in the cell diagram).

If the reaction actually does run spontaneously in the chosen direction, positive charge will be transported through the cell from left to right (and because of this, from right to left through the external part of the electric circuit). The electrode on the right is the positive terminal. When we measure the voltage of this cell for zero current flowing in order not to disturb the equilibrium, it corresponds to the difference of the cathode and anode potentials:

$$-U = \Delta E = E(\text{cathode}) - E(\text{anode}) = E(\text{Cu/Cu}^{2+}) - E(\text{Zn/Zn}^{2+}). \quad (23.11)$$

As mentioned in the previous section, zero current flowing can be realized by using of a voltmeter with high internal resistance or by applying an equal counter-voltage from an external voltage source in combination with an adjustable resistor. Even if the second method, the classical Poggendorff compensation method, is more complicated it is interesting from a theoretical point of view because it illustrates the thermodynamic relevance of ΔE. A very slight, in principle infinitesimally small, decrease of the opposing voltage will allow the reaction to proceed in the spontaneous direction; an infinitesimal increase of the opposing voltage, however, will force the reaction backward. Only if the opposing voltage exactly corresponds to the potential difference generated by the cell, the cell properties are determined under reversible conditions and, therefore, in electrochemical equilibrium.

ΔE is often called the *electromotive force* (*EMF*). But we will avoid using this term because it is potentially misleading: An electric potential difference is not a force! Instead, we will use the term "reversible cell voltage" ("zero-current cell voltage", "open circuit voltage"). We will consider ΔE to be positive in the case discussed here. At standard conditions, the resulting value for the Daniell cell is

$$\Delta E^{\ominus} = E^{\ominus}(\text{Cu}/\text{Cu}^{2+}) - E^{\ominus}(\text{Zn}/\text{Zn}^{2+}) = +0.3402\,\text{V} - (-0.7628\,\text{V})$$
$$= +1.103\,\text{V}.$$

However, if the reaction does not run spontaneously in the chosen direction, as would happen in the case of

$$\text{Zn}^{2+}|\text{w} + \text{Cu}|\text{s} \rightarrow \text{Zn}|\text{s} + \text{Cu}^{2+}|\text{w},$$

the cell voltage defined by convention changes its sign and ΔE becomes negative.

In summary: In a spontaneous reaction, the electrode with the higher redox potential is the cathode and the one with lower redox potential is the anode. In other words, the redox pair with the higher redox potential will be reduced. The redox pair with the lower redox potential, however, will be oxidized.

We know (compare Sect. 4.7) that there is a close relation between the reversible cell voltage ΔE and the chemical drive $\mathcal{A}$ of the underlying total reaction. We will use our example to see how this works. We will rearrange Eq. (23.10):

$$-\mathcal{A} \cdot \Delta \xi = \left[\left(\mu(\text{Cu}) - \mu(\text{Cu}^{2+}) \right) - \left(\mu(\text{Zn}) - \mu(\text{Zn}^{2+}) \right) \right] \cdot \Delta \xi. \tag{23.12}$$

According to the defining equation for the redox potential, the following holds for both redox pairs:

$$\mu(\text{Cu}) - \mu(\text{Cu}^{2+}) = -2 \cdot \mathcal{F} \cdot E\left(\text{Cu}/\text{Cu}^{2+}\right)$$

or

$$\mu(\text{Zn}) - \mu(\text{Zn}^{2+}) = -2 \cdot \mathcal{F} \cdot E\left(\text{Zn}/\text{Zn}^{2+}\right).$$

By inserting into Eq. (23.12) and dividing both sides by $\Delta \xi$, we obtain

$$-\mathcal{A} = -2 \cdot \mathcal{F} \cdot E(\text{Cu}/\text{Cu}^{2+}) + 2 \cdot \mathcal{F} \cdot E(\text{Zn}/\text{Zn}^{2+}) = -2 \cdot \mathcal{F} \cdot \Delta E \tag{23.13}$$

and finally

$$\Delta E = -U = \frac{\mathcal{A}}{2\mathcal{F}}. \tag{23.14}$$

We can obtain the same result more simply if we equate the energy released during the reaction process with the energy emitted electrically $W_{\xi} = -\mathcal{A} \cdot \Delta \xi = U \cdot v_e \mathcal{F} \Delta \xi$:

$$\Delta E = -U = \frac{\mathcal{A}}{v_e \mathcal{F}}, \tag{23.15}$$

where v_e is the number of electrons exchanged in a formula conversion.

Measurement of cell voltages for zero current flowing through the cell can therefore be used to determine the drive of a reaction. In practice, it is usually the standard values of these quantities that are determined and tabulated.

The reversible cell voltage is concentration dependent, just as the redox potential is. We will consider the general cell reaction

$$Rd + Ox^* \rightarrow Rd^* + Ox.$$

If we now apply Nernst's equation for redox potentials,

$$E(Ox/Rd) = \overset{\circ}{E}(Ox/Rd) + \frac{RT}{v_e \mathcal{F}} \cdot \ln \frac{c_r(Ox)}{c_r(Rd)}$$

or

$$E(Ox^*/Rd^*) = \overset{\circ}{E}(Ox^*/Rd^*) + \frac{RT}{v_e \mathcal{F}} \cdot \ln \frac{c_r(Ox^*)}{c_r(Rd^*)},$$

we obtain the following relation:

$$\Delta E = E(Ox^*/Rd^*) - E(Ox/Rd)$$

$$= \overset{\circ}{E}(Ox^*/Rd^*) + \frac{RT}{v_e \mathcal{F}} \cdot \ln \frac{c(Ox^*)}{c(Rd^*)} - \overset{\circ}{E}(Ox/Rd) - \frac{RT}{v_e \mathcal{F}} \cdot \ln \frac{c(Ox)}{c(Rd)}$$

and then Nernst's equation for the total reaction

$$\Delta E = \Delta \overset{\circ}{E} + \frac{RT}{v_e \mathcal{F}} \cdot \ln \frac{c(Ox^*) \cdot c(Rd)}{c(Rd^*) \cdot c(Ox)} \tag{23.16}$$

with $\Delta \overset{\circ}{E} = \overset{\circ}{E}(Ox^*/Rd^*) - \overset{\circ}{E}(Ox/Rd)$ as the basic value for the reversible cell voltage ΔE. The concentration dependency of ΔE in the case of a Daniell cell, for example, yields

$$\Delta E = \Delta \overset{\circ}{E} + \frac{RT}{2\mathcal{F}} \cdot \ln \frac{c_r(Cu^{2+})}{c_r(Zn^{2+})}.$$

If we replace the voltmeter in the experiment above (Fig. 23.5) with a load having a finite resistance R, possibly a small motor (Experiment 23.2), an electric current I will flow through the cell and the external circuit.

The Daniell cell discussed here is just one of many possible designs of galvanic cells (we will see more in Sect. 23.3). What they all have in common is that a chemical reaction (divided into two partial reactions) is used to drive an electron current. These cells make it possible to electrically utilize the energy released by

Experiment 23.2 *Daniell
cell*: A small motor with a
black and white disk
(to show motion) can be
driven using a Daniell cell.

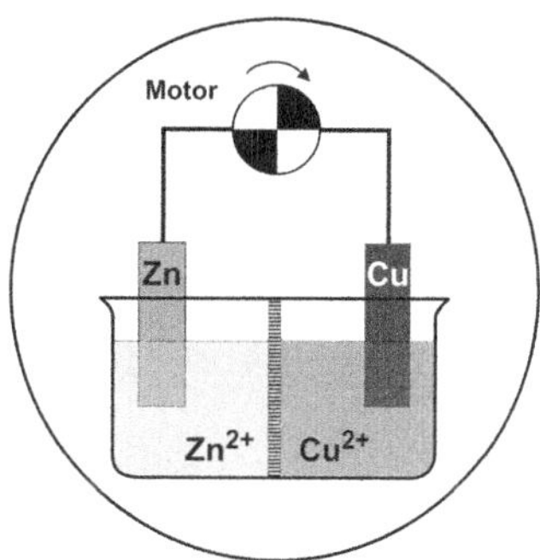

chemical reactions. In the ideal case, there will be no disturbing side reactions
occurring in the cell, so that chemical changes are only possible by the one reaction
and only simultaneously with an exchange of electrons through the electrodes. In
such cells, electron flow and chemical reaction are tightly coupled.

However, the electric voltage falls when the current flows through a load
because the electrochemical equilibrium is disturbed. Electric and chemical potential differences are no longer in balance: More zinc ions at the zinc electrode go into
the solution and copper ions discharge at the copper electrode.

Galvanic cells yield usable energy due to the spontaneous chemical reactions
running inside them, but reversing the cell reaction must be forced by adding
energy. This process, which is connected to chemical changes in the electrolyte,
is called *electrolysis*.

23.3   Technically Important Galvanic Cells

In closing we will discuss some technically important galvanic cells. We will
differentiate between primary cells, secondary cells, and fuel cells.

Primary Cells The processes running at the electrodes of primary cells (also
called *primary batteries*) are irreversible, meaning that these batteries are not
intended to be recharged. A well-known example of such a battery is the *zinc–
manganese dioxide cell*, also known as the *zinc–carbon dry battery*, which is a
further development of the galvanic cell patented by Georges Leclanché in 1866. It
is made up of a zinc can acting as the anode and a cathode composed of a carbon rod
surrounded by a mixture of manganese dioxide and carbon black (Fig. 23.7). The
carbon black is added to increase the low electric conductivity of the manganese
dioxide. The electrolyte is a paste of a 20 % solution of ammonium chloride
thickened by starch or sawdust. This is why such galvanic elements are called dry
batteries. The following simplified processes take place at the electrodes and in the
electrolyte:

Fig. 23.7 Schematic cross section of a zinc–carbon dry battery.

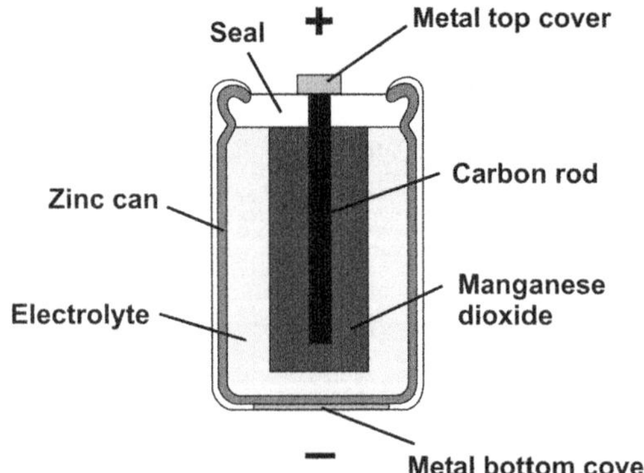

$\text{Anode}(-): \text{Zn}|\text{s} \rightarrow \text{Zn}^{2+}|\text{w} + 2\,\text{e}^-$

$\text{Cathode}(+): 2\,\text{MnO}_2|\text{s} + 2\,\text{H}_2\text{O}|\text{l} + 2\,\text{e}^- \rightarrow 2\,\text{MnOOH}|\text{s} + 2\,\text{OH}^-|\text{w}$

$\text{Electrolyte}: \text{Zn}^{2+}|\text{w} + 2\,\text{NH}_4^+|\text{w} + 2\,\text{Cl}^-|\text{w} + 2\,\text{OH}^-|\text{w} \rightarrow \text{Zn(NH}_3)_2\text{Cl}_2|\text{s} + 2\,\text{H}_2\text{O}|\text{l}$

$\text{Total reaction}: \text{Zn}|\text{s} + 2\,\text{MnO}_2|\text{s} + 2\,\text{NH}_4\text{Cl}|\text{w} \rightarrow 2\,\text{MnOOH}|\text{s} + \text{Zn(NH}_3)_2\text{Cl}_2|\text{s}$

When a current flows, the manganese dioxide is reduced to MnOOH ($\text{Mn}^{4+} \rightarrow \text{Mn}^{3+}$), and the primarily formed zinc ions react with the electrolyte to form the hardly soluble complex $\text{Zn(NH}_3)_2\text{Cl}_2$.

At standard conditions, the redox potential of the zinc electrode equals -0.76 V. The redox potential of the manganese dioxide electrode is about $+1.1$ V and the nominal voltage at normal usage is around 1.5–1.6 V.

A fundamental improvement upon the zinc–carbon battery is the *alkaline-manganese dioxide battery* (short *alkaline battery*), which has a higher current-carrying capability and capacity and a longer shelf life. In this type of battery, potassium hydroxide solution is the electrolyte and the anode is zinc powder mixed with the electrolyte and a thickening agent to form a paste. This paste is put into a hollow cylinder made up of a compressed mixture of manganese dioxide and graphite. The cylinder lies against the inner wall of a steel can, forming the cathode. The electrode arrangement is opposite to that of the zinc–carbon battery.

The *zinc–mercury oxide button cell* (Fig. 23.8) uses a pellet of mercury oxide with a little graphite added to it for better conductivity as cathode. The anode of this battery is zinc powder (pressed or amalgamated). The electrolyte, a concentrated ZnO saturated potassium hydroxide solution, is on a cellulose felt. The following shows the simplified processes at the electrodes:

$\text{Anode}(-): \text{Zn}|\text{s} + 2\,\text{OH}^-|\text{w} \rightarrow \text{Zn(OH)}_2|\text{s} + 2\,\text{e}^-$

$\text{Cathode}(+): \text{HgO} + \text{H}_2\text{O} + 2\,\text{e}^- \rightarrow \text{Hg} + 2\,\text{OH}^-$

$\text{Total reaction}: \text{Zn}|\text{s} + \text{HgO} + \text{H}_2\text{O} \rightarrow \text{Zn(OH)}_2|\text{s} + \text{Hg}$

These button cells have a very long shelf life (up to 10 years) and their nominal voltage during discharge remains practically constant at 1.35 V. They have been normally used in small devices with low power demand such as watches, pocket

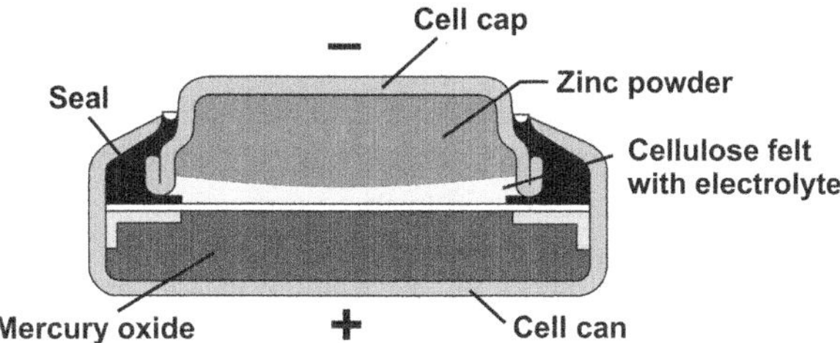

Fig. 23.8 Cross section of a zinc–mercury oxide button cell.

calculators, hearing aids, and cardiac pacemakers. However, due to the toxicity of the mercury, they are now replaced by the similar zinc–silver oxide cells or zinc–air cells.

Secondary Cells In contrast to what is the case in primary cells, it is mostly reversible processes that run at the electrodes of secondary cells (also called *secondary batteries* or *accumulators*). These cells are rechargeable.

The most common of this type of cell is the *lead-acid battery* that was developed in the nineteenth century. In its charged state, it is, in principle, made up of a lead anode and a lead oxide cathode. The electrolyte is sulfuric acid (25–30 %) saturated with lead sulfate. A simplified representation of the battery's charging and discharging processes shows the following reactions:

$$\text{Negative electrode: } Pb|s + SO_4^{2-}|w \underset{\text{charge}}{\overset{\text{discharge}}{\rightleftharpoons}} PbSO_4|s + 2\,e^-$$

$$\text{Positive electrode: } PbO_2|s + SO_4^{2-}|w + 4\,H_3O^+|w + 2\,e^- \underset{\text{charge}}{\overset{\text{discharge}}{\rightleftharpoons}} PbSO_4|s + 6\,H_2O|l$$

$$\text{Total reaction: } Pb|s + PbO_2|s + 2\,H_2SO_4|w \underset{\text{charge}}{\overset{\text{discharge}}{\rightleftharpoons}} 2\,PbSO_4|s + 2\,H_2O|l$$

When the battery is discharging, metallic lead oxidizes into lead sulfate and lead oxide is reduced to lead sulfate. Therefore, lead sulfate forms at both electrodes. At the same time, sulfuric acid is consumed and water is produced so that the density of sulfuric acid falls as discharging progresses. The state of the battery's charge can then be determined by this change of density.

Under standard conditions, the redox potential of the half-cell $Pb|PbSO_4|SO_4^{2-}$ is -0.36 V, and that of the half-cell $PbO_2|PbSO_4|SO_4^{2-}$ is $+1.69$ V. The nominal voltage of the cell is about 2 V. Depending upon the charge state and charging and discharging current, it can fluctuate between 1.75 and 2.4 V.

In order to recharge the battery, i.e., to "force" the reaction in the opposite direction, energy input is necessary. In the process, lead or lead oxide will reform at the lead sulfate electrodes.

When constructing a lead-acid battery it is important to remember that both electrode types need the largest surfaces possible so that the electrochemical reaction can occur at the highest possible rate. This can be attained, for example, by coating lead grids alloyed with antimony or calcium with a paste of Pb, PbO, and $PbSO_4$. The plates then receive opposite charges in acid, forming lead sponge on one side and porous lead oxide on the other (formation process). In order to avoid short circuiting through physical contact, the electrodes are separated by microporous plastic. Rechargeable lead-acid batteries can be produced in a closed construction (with valve), which is advantageous for transport and maintenance. In this case, the electrolyte is not liquid but fixed in a gel or a fibrous mat.

The most common use for rechargeable lead batteries in everyday life is for starting combustion engines in motor vehicles. Six cells in series deliver a terminal voltage of about 12 V. The constant running of the vehicle's generator keeps recharging the battery.

The negative electrodes in *nickel–cadmium batteries* are made up of finely distributed cadmium. The positive electrodes are composed of Ni(III) oxide hydroxide (with graphite or Ni powder added to enhance the conductivity). The electrolyte is usually a 20 % potassium hydroxide solution.

As the battery charges and discharges, the following simplified reactions take place at the electrodes:

$$\text{Negative electrode: } Cd|s + 2\,OH^-|w \underset{\text{charge}}{\overset{\text{discharge}}{\rightleftharpoons}} Cd(OH)_2|s + 2\,e^-$$

$$\text{Positive electrode: } 2\,NiOOH|s + 2\,H_2O|l + 2\,e^- \underset{\text{charge}}{\overset{\text{discharge}}{\rightleftharpoons}} 2\,Ni(OH)_2|s + 2\,OH^-|w$$

In all, during discharging, metallic cadmium is transformed into the bivalent state (solid cadmium(II) hydroxide) by oxidation and trivalent nickel is reduced into bivalent nickel (solid nickel(II) hydroxide):

$$\text{Total reaction: } Cd|s + 2\,NiOOH|s + 2\,H_2O|l \underset{\text{charge}}{\overset{\text{discharge}}{\rightleftharpoons}} Cd(OH)_2|s + 2\,Ni(OH)_2|s$$

The nominal voltage of a nickel–cadmium battery is 1.2 V.

Gas-proof constructions are often designed like commercial batteries (including button cells), so they can replace the primary cells in portable electric and microelectric devices. Meanwhile, usage of nickel–cadmium batteries is strongly limited by law because of the dangers related to toxic cadmium.

A more environmentally friendly replacement for the Ni–Cd battery is the *nickel–metal-hydride battery* (NiMH), in which cadmium is replaced by a metal alloy that is able to store hydrogen reversibly. In the charged state, we have an anode made of metal hydride which is produced during the charging process by storing atomic hydrogen in a crystal lattice of the alloy (e.g., $La_{0.8}Nd_{0.2}Ni_{2.5}Co_{2.4}Si_{0.1}$). As the cell discharges, the stored hydrogen oxidizes on the electrode's surface:

$$\text{Negative electrode}: MH|s + OH^-|w \underset{\text{charge}}{\overset{\text{discharge}}{\rightleftharpoons}} M|s + H_2O|l + e^-.$$

The cathode and electrolyte are identical to those in the NiCd battery. The total reaction is then:

$$\text{Total reaction}: MH|s + NiOOH|s \underset{\text{charge}}{\overset{\text{discharge}}{\rightleftharpoons}} M|s + Ni(OH)_2|s.$$

Again, the cell voltage is about 1.2 V, so the voltage of the NiMH cell is compatible with that of the NiCd cell. Along with its environmental advantages, it also has greater capacity and durability. However, its load capacity is lower.

Special accumulators for hybrid (electric) vehicles have been developed on the basis of nickel–metal hydride cells, i.e., cars powered by electric engines in combination with other energy converters (mostly combustion engines).

Fuel Cells In contrast to the galvanic cells we have discussed up to now, fuel cells have a constant feed of the substances consumed at the electrodes. This means that, theoretically, a current can be made to flow indefinitely. The term "fuel cell" actually means that substances can be converted there that are normally considered fuels and are otherwise burned for heating or for gaining energy. One well-known example of this is the so-called *hydrogen–oxygen fuel cell* (Fig. 23.9), which is used in manned space flight. This kind of cell uses hydrogen as the fuel and oxygen as the oxidizing agent. The electrolyte is a concentrated aqueous solution of KOH at elevated temperature and pressure [hence the alternative name alkaline fuel cell (AFC)]. The gases are conducted through porous plate electrodes (possibly made up of sinter nickel or pressed active carbon powder), the so-called gas diffusion

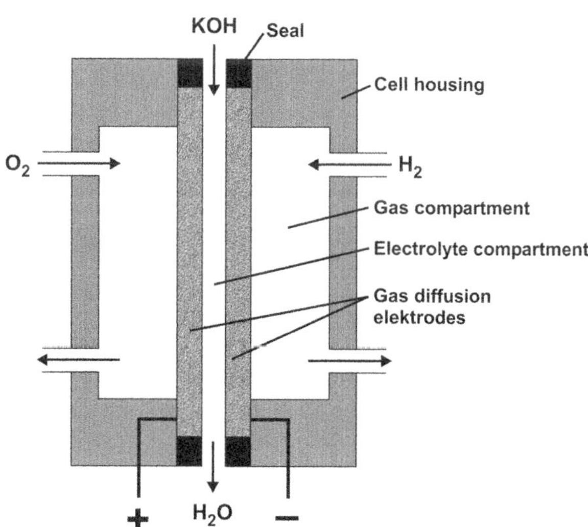

Fig. 23.9 Schematic of a hydrogen–oxygen fuel cell.

electrodes. In order to promote the splitting of hydrogen and oxygen molecules which precedes the actual reaction, the plates are coated with a small amount of a catalytic substance such as platinum or palladium.

The following simplified reactions then take place at the electrodes:

$$\text{Anode}(-)\colon 2\,H_2|g + 4\,OH^-|w \rightarrow 4\,H_2O|l + 4\,e^-$$
$$\text{Cathode}(+)\colon O_2|g + 2\,H_2O|l + 4\,e^- \rightarrow 4\,OH^-|w$$
$$\text{Total reaction}\colon 2\,H_2|g + O_2|g \rightarrow 2\,H_2O|l$$

The fuel oxidizes at the fuel cell's anode and oxygen reduces at the cathode. In total, the hydrogen undergoes a process of "cold combustion" where it turns into water. The cell voltage (at 200 °C and 4 MPa) is about 1.2 V.

Carbon monoxide as well as low-molecular organic compounds such as methane (CH_4) or methanol (CH_3OH) can also serve as fuels, and along with oxygen, air can also be an oxidizing agent. In place of KOH, solid electrolytes such as polymer membranes permeable only for protons [polymer electrolyte membrane fuel cells (PEMFC)] or oxygen-ion conducting oxide ceramics (yttria-stabilized zirconia) [solid oxide fuel cells (SOFC)] can also be used. One more possibility is melted salt as the electrolyte.

23.4 Cell Voltage Measurement and Its Application

We mentioned in Sect. 23.2 that galvanic cells can make the energy released by a chemical reaction usable, but they can also be utilized as a "measuring instrument" for the differences of redox potentials and therefore the electron potentials of various redox pairs. Moreover, because the electron potential itself is determined by the chemical potentials of the substances making up the redox pair, it is also possible to find the μ values as well as the drive $\mathcal{A}$ of the underlying total reaction with the help of galvanic cells. Reversible cell voltages measured with zero current can be used to determine these quantities and derived ones such as equilibrium constants.

Chemical potentials also depend upon the concentrations of substances, and in many cases, these dependencies are well known. Therefore, it is possible to utilize the measured voltages to find ion concentrations, especially solubilities and pH values (see Sect. 22.7). This method is called *potentiometry* and is widely applied in analytic chemistry.

Let us consider the simple example of determining the concentration of Ag^+ ions in an aqueous solution. We use a piece of silver as the electrode sensitive to Ag^+ ions, which we combine with the solution into a half-cell. Ag and the Ag^+ ions together form a redox pair to which we can assign the following electron potential:

$$\mu_{\mathrm{e}}(\mathrm{Ag}/\mathrm{Ag}^{+}) = \mu_{\mathrm{Ag}} - \mu_{\mathrm{Ag}^{+}} = \mu_{\mathrm{Ag}} - \overset{\circ}{\mu}_{\mathrm{Ag}^{+}} - RT\ln\frac{c_{\mathrm{Ag}^{+}}}{c^{\ominus}}.$$

Because, at a given temperature, μ_{Ag} and $\overset{\circ}{\mu}_{\mathrm{Ag}^{+}}$ have constant values, the electron potential and the corresponding Galvani potential difference developing between the metal and the solution depend only upon the concentration of Ag^{+} ions in the solution (see also Sect. 22.5). If we now measure the Galvani potential difference relative to the standard hydrogen electrode to find the redox potential, we can apply Nernst's equation to determine the concentration $c_{\mathrm{Ag}^{+}}$ in the solution.

Measuring chemical potentials or determining concentrations using galvanic cells looks very promising at first. However, at many electrodes, potential differences develop only very gradually or can be disturbed by secondary reactions. These may consist of the formation of cover layers or may be the result of other redox pairs participating in the exchange of charge. Moreover, considerable deviations from the mass action equation—and therefore Nernst's equation—appear for ion concentrations above $10\ \mathrm{mol\ m^{-3}}$ (see also Sect. 6.2), so that the concentrations determined from measured cell voltage are often inexact. For this reason, potentiometry is combined with titration (*potentiometric titration*) because here the precision of the absolute value is unimportant. For example, the concentration of the ion that determines the potential may be changed by *precipitation titration*. We shall take a closer look at this by considering the determination of the silver content of a solution by titration with a KCl standard solution:

$$\mathrm{Ag}^{+}|\mathrm{w} + \mathrm{Cl}^{-}|\mathrm{w} \to \mathrm{AgCl}|\mathrm{s}.$$

In this case, we are following the redox potential of the $\mathrm{Ag}/\mathrm{Ag}^{+}$ electrode $(E = E^{\ominus} + (RT^{\ominus}/\mathcal{F}) \cdot \ln c_{\mathrm{Ag}^{+}} = 0.7996 + 0.059 \cdot \lg c_{\mathrm{Ag}^{+}}\,\mathrm{V}\ \mathrm{SHE}$ at $T^{\ominus} = 298\,\mathrm{K})$ as a function of the addition of titrator (Fig. 23.10). When KCl standard solution is added drop by

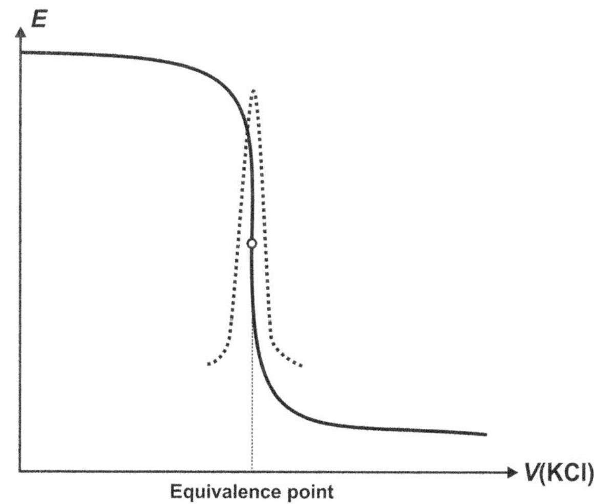

Fig. 23.10 Change of redox potential at an $\mathrm{Ag}/\mathrm{Ag}^{+}$ electrode during potentiometric titration of a silver-containing solution with a KCl standard solution (*solid line*) and first derivative of the titration curve (*dotted line*).

drop to the Ag^+ solution, the AgCl immediately precipitates and the concentration of Ag^+ ions in the recipient steadily decreases. This means that with each decrease of one decade of Ag^+ ion concentration, there is a lowering of potential of 59 mV. At the equivalence point, there is an abrupt fall of Ag^+ concentration and the corresponding potential, because at this point, one drop of KCl is enough to pretty much remove all the remaining Ag^+ ions. As a consequence, the concentration of Ag^+ ions is determined only by the solubility equilibrium of the silver chloride $[\overset{\circ}{\mathcal{K}}_{sd}(AgCl) = c_r(Ag^+) \cdot c_r(Cl^-) = 1.78 \times 10^{-10}$; compare also Sect. 6.6] and is therefore very low. According to the solubility product, a further addition of KCl solution continues to lower the Ag^+ concentration but, in parallel with the steady addition of KCl, the Ag^+ concentration drops now also steadily.

The equivalence point is determined by the inflection point of the titration curve. The first derivative of the titration curve is often used for a more exact determination of this point because a maximum is easier to localize than an inflection point. In practice, the quotient $\Delta E/\Delta V$ is plotted as a function of the given volume of titrator V. (ΔE corresponds to the difference of two consecutive measured values.)

The end point of *acid–base titrations*, *complexometric titrations*, or *redox titrations* can also be potentiometrically indexed by the use of suitable electrodes, for example, the glass electrode discussed in Sect. 22.7 in the case of acid–base reactions. The advantage of this method is that colored or cloudy solutions can also be titrated and it is simple to automate because an easily measured electric quantity participates.

Chapter 24
Thermodynamic Functions

In addition to the terms discussed so far a number of other quantities and functions are used in thermodynamics without which textbooks that follow the conventional concept cannot manage. Because knowing these additional terms is essential for understanding traditional textbooks and the corresponding data collections, we will deal with the most important of them in this chapter and establish the relations to the concept chosen in this book. The major subsidiary terms are the four energetic quantities inner energy U, enthalpy H, Helmholtz energy A, and Gibbs energy G. The same quantity can serve different purposes depending on the variables chosen. The function $U(S, V, \ldots)$ characterizes the system under consideration. It is almost never explicitly stated (the abstractness of the variable S can be considered the underlying cause for this), but its differential plays a central role for all derivations. The functions $U(T, V, \ldots)$ and $H(T, p, \ldots)$ serve to describe the heat exchanged between system and surroundings under different experimental conditions (the first at constant volume, the second at constant pressure). The functions $A(T, V, \ldots)$ and $G(T, p, \ldots)$ play a similar role. Both are used to calculate the energy released during the considered process. This enables us to predict whether or not the process may run spontaneously. In the last section, we will discuss quantities such as activity, fugacity, etc. These quantities are used for describing deviations from what is considered ideal behavior of dissolved substances and gases.

24.1 Introduction

The subject of thermodynamics is a prime example of an axiomatic science whose basic assumptions are gained from everyday experiences. These basic assumptions first lead to *fundamental laws* from which a large number of other laws and relations are derived. The modest number of assumptions on the one hand and the abundance of derived results on the other is a widely admired characteristic of this science. Thermal effects are a part of almost every process dealt with in everyday life,

© Springer International Publishing Switzerland 2016
G. Job, R. Rüffler, *Physical Chemistry from a Different Angle*,
DOI 10.1007/978-3-319-15666-8_24

including technological ones. These effects are often unnoticeable or undesirable, so that we are prone to overlook them. However, they do often rule the processes, making it necessary to deal with them. The ubiquity of thermal effects indicates the special role of heat in nature. It is therefore of great importance to know what this role is.

Thermodynamics, as it has developed in the last 150 years, is criticized for its lack of tangibility. Not only beginners complain of this, but sometimes even professionals do as well. This lack of tangibility makes it difficult to evaluate results qualitatively for their relevancy, algebraic signs, or orders of magnitude. Although many relations can be formally derived, one does not "understand" them as one would understand them in the field of mechanics, for example. Intuition is hindering and even misleading. Results must simply be accepted and the arguments that are allowed must simply be memorized. With growing routine, one will gradually forget one's concerns.

The attempt has been made to compensate for this lack of tangibility by using an expanded formalism. The most noticeable differences between this expanded formalism and the presentation in this book are additionally introduced quantities such as enthalpy and free energy (in various forms), without which textbooks that follow the conventional concept cannot manage. Because knowing these new quantities is essential for understanding traditional textbooks and the corresponding data collections, we will deal with the most important of them in the following.

24.2 Heat Functions

Preliminary Remarks Thermodynamics is considered conceptually very difficult—even if this is not true for the mathematics involved. For example, Arnold Münster wrote 1969 in his textbook about chemical thermodynamics: "In contrast [to the mathematical formalism], the conceptualization in thermodynamics is especially abstract and this abstract conceptualization is the core of the difficulty of this scientific area." This is due to the awkwardness of attribution of the heat quantities. Intuition and language from everyday experience give us preconceived structures. Because of their fuzziness, we are free to some degree in attributing everyday and physical concepts. However, ill-judged arbitrariness leads to difficulties.

The everyday expression "heat" has many meanings. In the field of thermodynamics, there are at least three quantities to which that name suits:

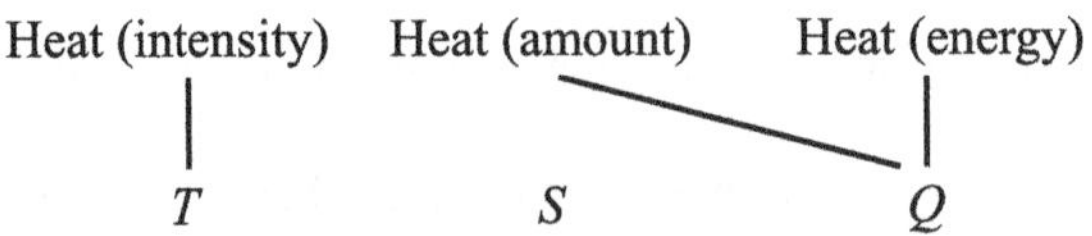

The name "heat (energy)" (or "thermal energy") for Q is admittedly too general but for the moment it should suffice to grasp the underlying idea. The correlation of the terms shown vertically on top of each other should have been easy and natural. However, in 1850, a different decision was made indicated by the lines. The results of this decision can be summed up as follows:

- The quantity S cannot be interpreted macroscopically. Therefore one tends to avoid this quantity and to replace it with energetic terms.
- The quantity Q is automatically connected to characteristics that Q doesn't actually possess, leading to misconceptions. We will come back to this below. Q is mathematically inconvenient as a process quantity (see Sect. 1.6) so it also seems advantageous to avoid Q and to rewrite it using other quantities.
- In order to mitigate the conceptual difficulties and to bridge gaps in understanding, a number of new quantities have been introduced.

The question of how, in history, it could come to such a decision and why subsequent changes are difficult, if not impossible, would be a chapter in itself. Although this is an interesting and important question, we must exclude it here.

Let us now take a look at the most important of the additionally introduced quantities. In the conventional terminology of thermodynamics, *heat* (symbol Q) stands for the energy exchanged thermally between system and surroundings and *work* (symbol W) for the one exchanged mechanically. These terms will now and then appear in the following whereas we avoided them so far. In these and similar cases we spoke neutrally about expended (or released) energy. Therefore, a few words of explanation:

Work W and Heat Q As yet, we mentioned work W only shortly in Sect. 2.1. In more detail, we presented *mechanical work* by means of the relation "work = force times distance" as access route to an indirect introduction of the term energy. In mechanics, work generally describes a quantity which is defined as product of a displacement Δx (e.g., along the x-axis) with a force F_x, causing the displacement:

$$W_{\to x} = F_x \cdot \Delta x.$$

The same applies to displacements in arbitrary directions. Already in mechanics, the term is generalized. One considers an increase of volume ΔV, an increase of surface ΔA, etc., also as "displacement" and the pressure p in hydraulics (Sect. 2.5), the surface tension σ (Sect. 15.2), etc., as corresponding "force" causing the change:

$$W_{\to V} = p \cdot \Delta V,$$
$$W_{\to A} = \sigma \cdot \Delta A, \text{etc.}$$

The volume V, the surface A, ... are regarded as "generalized coordinates" and correspondingly p, σ, ... as "generalized forces."

In thermodynamics, one encounters the term "pressure–volume work" which relates to the compressibility of an elastic body (e.g., also of a gas). The more an object is pressed from all sides, meaning the more work has to be done, the more strongly volume V decreases. Therefore, a minus sign appears in the expression:

$$W_{\to V} = -p \cdot \Delta V. \tag{24.1}$$

As yet, also *heat Q* was only marginally mentioned such as in Sects. 3.1, 3.11, and 8.7. According to the traditional view, it characterizes the energy transferred thermally between a system and its surroundings because of differences in temperature. As a consequence of this type of energy exchange, the entropy of the system changes.

Process Quantities The phrase "work has to be done" in the case of something being displaced against a counteracting force or inhibition already expresses that work describes a certain aspect of a process. It causes therefore no difficulties to accept that work represents a so-called *process quantity* or corresponds to a process.

This is completely different in the case of the term heat. In everyday language but also in many fields of science and technology, one tends to the suggestion that the heat that is added to the body or generated in it is contained in the body (and therefore represents a state function). As long as energy exchange on other paths than the thermal one is negligibly small this simple suggestion ("The heat that goes in can only come out again as heat") actually suffices to explain qualitatively and quantitatively a lot of effects in context with heat in everyday life or in other fields such as in building industry. The suggestion fails, however, if the bodies in consideration can exchange energy on others than thermal paths. In this case, it makes no sense to speak of "heat content." In most of the textbooks on thermodynamics, the term *"heat"* as quantity Q indicates, as mentioned above, something different, namely a *mode of energy transfer*. It depends on the conditions under which the transfer process acting on energy takes place. According to this view, heat is like work a process quantity.

In order to avoid misunderstandings, we will use the symbol "δ" instead of the simple "d" for the cases of small changes of the process quantities *exchanged* heat Q_e and *expended* work W_e (see Sect. 1.6).

Internal Energy U Kinetic and potential energies W_{kin} and W_{pot} contribute to the total energy W_{total} of a body. It possesses these energies when it moves or when it is raised or lowered as a whole. These energies are mostly insignificant in chemistry. They are additive terms that depend upon the (linear) velocity v (such as W_{kin}) or elevation h (such as W_{pot}) and can be easily split off so that, as a rule, instead of W_{total}, only the residual, the so-called *internal energy U*, is considered. The internal energy of a body that can only exchange energy as heat Q_e (see Sect. 8.7) and work W_e (in this case "pressure–volume work" by increasing or decreasing its volume V [see Eq. (24.1)] with the surroundings, can change as follows:

$$dU = \delta Q_e + \delta W_e = TdS - pdV \quad \left(\text{where } dS_g = 0\right). \tag{24.2}$$

The energy absorbed by a body in different ways (mechanically, thermally, chemically, electrically, etc.) is *not* stored in these different forms. Rather, it forms a common energy supply (Fig. 24.1).

Fig. 24.1 Pond as model. The idea that a body would contain the added heat Q as such is unsupportable in the same way that it is impossible to see in a pond how much of the water came from rain, dew, or groundwater.

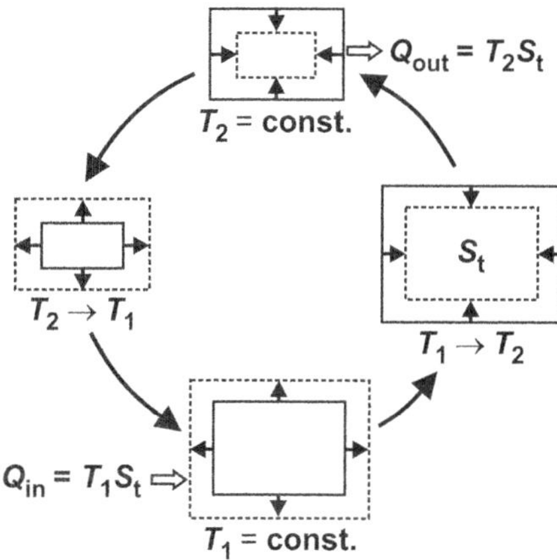

Fig. 24.2 Transfer of entropy S_t in a reversible cycle from a cold to a warm reservoir. Changes of volume are indicated by *arrows* (initial state: *contour line solid*, final state: *contour line dashed*). More heat Q flows off with the entropy S_t than in $Q_{out} > Q_{in}$, even though the body completely reverts to its initial state after every cycle and does not cool down at all. This means that energy is emitted as heat, which was not present in that form before but is generated. The question remains: what phase of the process does this happen in and how?

In the case of a body that can be expanded and heated, there is no quantity describing what might be called the amount of heat contained in it or simply its *"heat content"* (Fig. 24.2). One says that heat is a *process variable* and not a *state variable*. If we wish to describe the state of a body by using its temperature T and its volume V or its pressure p, for instance, in mathematical terms this means that the functions $Q(T, V)$ or $Q(T, p)$ or derivatives thereof do not exist. The consequences of this are aggravating, as we will soon see.

When there is only a single path available for an exchange of energy with the surroundings, such as in the case of "heat reservoirs" often used in thought experiments, energy can only be exchanged over this path. The heat Q that goes in can only come out again as heat. The situation is similar for a body having a pressure p that is either constant or is only dependent upon its volume V, $p = f(V)$ (Fig. 24.3). Whether or not it expands in the process is unimportant. However, if it does expand, a part of the energy which flowed in as heat is diverted to the outside over the mechanical path, but it flows back unchanged when energy is retrieved over the thermal path.

Such examples erroneously lead us to think of heat as being an entity contained in bodies (see above). Expressions such as "heat capacity" or "conduction of heat" support these images. They are holdovers of the time before heat was deemed a special mode of energy transfer.

Including Irreversible Processes Equation (24.1) changes when friction plays a role as it does in Fig. 24.4. Keep in mind that the *exchanged* entropy S_e, the *generated* entropy S_g, and the external pressure p_e are not state variables of the gas, but entropy S and internal pressure p are:

$$\underbrace{dU = \delta Q_e + \delta W_e}_{T dS - p dV} = T \delta S_e - p_e dV \; .$$

$$(24.3)$$

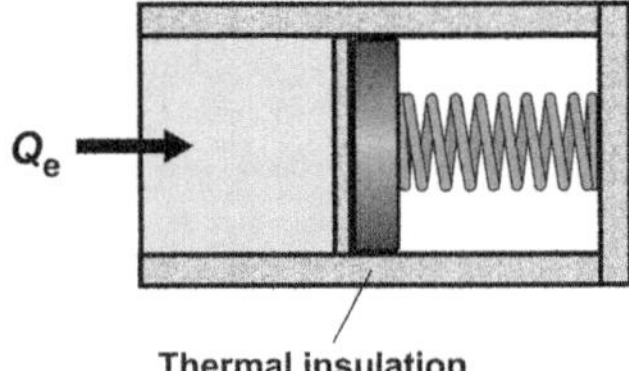

Fig. 24.3 Cylinder that uses a spring-loaded piston to press upon a gas contained in it.

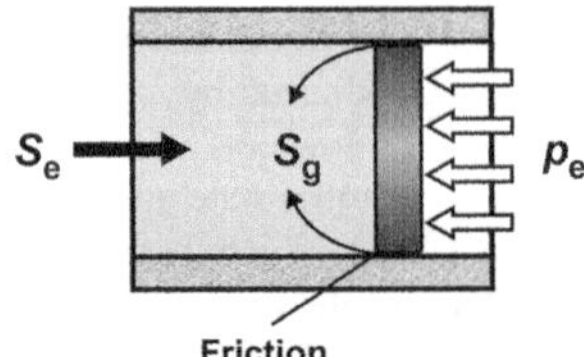

Fig. 24.4 Gas in a cylinder whose piston cannot move without friction. The work $dW_e = -p_e dV$ performed upon the piston by the external pressure p_e ($>$internal pressure p) when the volume changes, $dV < 0$, is greater due to friction, while the amount of heat $\delta Q_e = T \delta S_e$ needed for same change of state is correspondingly smaller.

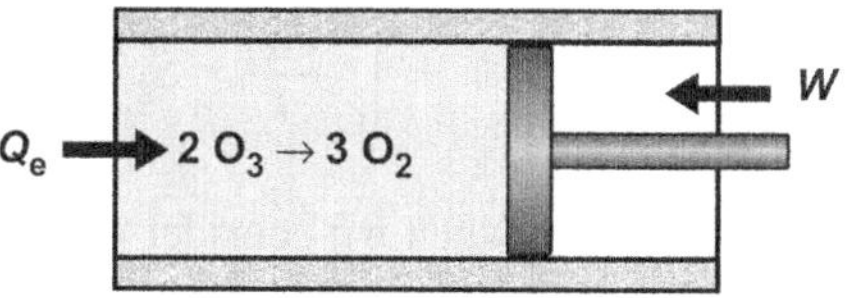

Fig. 24.5 Cylinder (with frictionless piston). A chemical reaction takes place inside the cylinder. Extent and drive of the reaction are described by ξ and $\mathcal{A}$, respectively.

Consider now not friction but a chemical reaction continuously running against inhibitions causing generation of entropy (Fig. 24.5). This is common in chemistry. The following is valid for the change of the internal energy:

$$\mathrm{d}U = \underbrace{\delta Q_e + \delta W_e}_{T\mathrm{d}S - p\mathrm{d}V - \mathcal{A}\mathrm{d}\xi} = T\delta S_e - p\,\mathrm{d}V \; . \tag{24.4}$$

The last two formulas, Eqs. (24.2) and (24.3), specify an increase of energy $\mathrm{d}U$ which is caused by a small change of state. The expression under the curly brackets shows which state parameters are responsible and to what extent, while the expression on the right side describes how this increased demand is met. If we insert $\mathrm{d}S = \delta S_e + \delta S_g$ (see Sect. 8.7), meaning $\delta S_e = \mathrm{d}S - \delta S_g$, and there and then solve for $T\delta S_g$, we obtain formal expressions that we can easily visualize:

$$\underbrace{T\delta S_g}_{\delta Q_g} = \underbrace{-(p_e - p)\cdot \mathrm{d}V}_{\delta W_b}\,, \quad \text{and} \quad \underbrace{T\delta S_g}_{\delta Q_g} = \underbrace{\mathcal{A}\mathrm{d}\xi}_{\delta W_b}\; . \tag{24.5}$$

Just as frictional work $-(p_e - p)\cdot \mathrm{d}V$ is used to generate entropy, the equivalent is true for the work against reaction inhibitions.

Generated Heat Heat is generated by friction. Around the middle of the nineteenth century, this important insight led to upheaval in the field of thermodynamics. The effects of the energy contribution $T\delta S_g$ upon the bodies involved are the same as the effects of heat added to them. It seems logical, therefore, to call this kind of contribution *heat* or, more exactly, *generated heat* δQ_g. Hence, it appears justified to call the expression $T\delta S_g$ on the left of Eq. (24.5)—as is customary—the (generated) *frictional heat*, and, correspondingly, to call the one on the right the *generated reaction heat*.

The contribution δQ_g, which is always positive and only disappears in the limiting case, stands in opposition to the quantity $\mathrm{d}Q$ as the *exchanged heat*. For clarity, the latter is symbolized by δQ_e already used in this chapter. δQ_e can be either positive or negative, because during heat exchange heat can be added but also removed. The sum $\delta Q_e + \delta Q_g = \delta Q_{total}$ would represent the *total amount* of heat collected. There appears to be no problem with adding up infinitesimally small contributions going over some path from a state I to a state II or to calculate the corresponding entropies after dividing the contributions by T:

$$S_e = \int_I^{II} \frac{\delta Q_e}{T}, \quad S_g = \int_I^{II} \frac{\delta Q_g}{T}, \quad \Delta S = \int_I^{II} \frac{\delta Q_{total}}{T}. \tag{24.6}$$

A conflict arises when we consider entropy generation in conduction of heat. When we have heat Q_t coming out of a hot reservoir (2) into a cold one (1) with a temperature of $T_2 > T_1$, the entropy generated by this is a result of the sums of the entropy changes $\Delta S_1 = Q_t/T_1$ and $\Delta S_2 = -Q_t/T_2$ of the two reservoirs: $S_g = Q_t/T_1 - Q_t/T_2 \cdot S_g$ appears in the cold reservoir, so $Q_g = T_1 S_g$ should be the generated heat. This result seems ridiculous because heat conductance is considered a process where heat is conserved and remains constant.

However, this conclusion is not as absurd as it may appear at first glance. It can be justified if Q_g is considered compensation for the reduction of "free energy." If this energy is used in a "heat engine," it does so at the cost of the transferred heat Q, causing less of it to arrive at the cold reservoir. If this energy remains unused, it is "burnt" (dissipated) and the generated heat Q_g compensates for the expected decrease so that the same amount of heat arrives at the cold reservoir as is emitted from the hotter reservoir.

Two Prototypical Examples In chemistry, we are primarily interested in systems in which at least one chemical transformation takes place. The case in Fig. 24.5 gives us a concrete example where, instead of the slow decay of ozone, we can imagine some other kind of gas reaction. We will choose two special cases. The first one is the reaction of formation of water where the process can be controlled by turning a catalyst on or off. The second case is the dimerization of nitrogen dioxide (see Experiment 9.3) in which the participating gases, the brown nitrogen dioxide and its colorless dimer, are permanently in a strongly temperature dependent equilibrium:

$$2\,H_2|g + O_2|g \rightarrow 2\,H_2O|g, \quad \text{inhibited},$$
$$2\,NO_2|g \rightleftarrows N_2O_4|g, \quad \text{uninhibited}.$$

At first, our second example appears to be a rarely seen exception and hardly worth mentioning. It is, surprisingly, extremely common. An ensemble of molecules in a certain state of excitation, association, or conformation, can be considered an independent substance. Experiment 9.3 offers an example of such a case. Conversely, a mixture of chemicals where equilibrium between the individual components is established very quickly can be treated as one substance and calculations can be performed accordingly.

In order to characterize the state of the system in our simple prototypical examples, we need three parameters. Along with entropy S and volume V, we can use the extent of reaction ξ. The main equation for systems of this type (see Sect. 9.1), formulated with the help of the internal energy U, is:

$$dU = T\mathrm{d}S - p\mathrm{d}V - \mathcal{A}\mathrm{d}\xi. \tag{24.7}$$

In order to understand the conventional approach, we must first think of this equation as being unknown to us. It serves only as a background that allows us to consider the following development from another viewpoint.

Applying the First and Second Laws (of Thermodynamics) The equation $dU = \delta Q_e + \delta W_e$ generally serves as a first step toward the calculus of thermodynamics to be created. It is considered an application of the First Law where the new state variable U can be constructed with the help of the two measurable process quantities Q_e and W_e. At first, we will limit ourselves to simple, *closed* systems, meaning systems without any exchange of substance with the surroundings and in which temperature and pressure are the same everywhere. Except as heat, energy can only be transferred in or out, without friction, by changes to the volume: $\delta W_e = -p\mathrm{d}V$ and therefore

$$dU = \delta Q_e - p\mathrm{d}V. \tag{24.8}$$

In the second step, the process quantity δQ_e is replaced by $T\mathrm{d}S$ as an application of the Second Law. S is considered abstract but, like U, is actually a quantity that is measurable or obtainable from measured data. Because only processes for which $\delta Q_e \leq T\mathrm{d}S$ can take place, Eq. (24.8) converts to:

$$dU \leq T\mathrm{d}S - p\,\mathrm{d}V \quad \text{for spontaneous processes.} \tag{24.9}$$

This relation concisely summarizes the two Laws. Note that it only contains state variables. This is an important and helpful characteristic to keep in mind. Starting from this equation, a specific formalism with its own new quantities and terms is developed. These new quantities and terms are not necessary for understanding physical chemistry, but are crucial for understanding the pertinent literature.

Let us first consider changes of state that do not generate entropy. In this case, instead of the inequality above, the following equation is valid:

$$dU = T\mathrm{d}S - p\,\mathrm{d}V \quad \text{for reversible processes.} \tag{24.10}$$

We know that the state of our model systems is determined by three parameters such as S, V, and ξ. Therefore, the formula above appears incomplete. It can be easily completed as follows:

$$dU = T\mathrm{d}S - p\,\mathrm{d}V + ?\mathrm{d}\xi. \tag{24.11}$$

At this stage of the thermodynamic calculus formulated according to the traditional concept, the variable represented by the question mark is still unknown. It is important to close this gap. This is mathematically simple because except for the quantity we are seeking, all the others can be measured. The missing quantity equals the derivative of internal energy with respect to the extent of reaction at constant S and V:

$$? = \left(\frac{\partial U}{\partial \xi}\right)_{S,V}.$$

Nonetheless, one commonly sees a problem whose solution requires a specially created formalism. Why is this?

The abstractness of the variable S can be considered the underlying cause for this. When it appears as a function of quantities like the more or less familiar $T, p, V, \xi, \ldots$, it is already almost incomprehensible. The situation becomes even more difficult when S appears as a parameter upon which other variables depend in as yet unknown ways. The manipulation necessary to overcome this obstacle will be described in Sect. 24.3.

Equation (24.10) is a special case of Eq. (24.11) in which the expression $?\mathrm{d}\xi$ is left out. Under which conditions is this allowed? In anticipation of the subsequent derivation we equate $?\mathrm{d}\xi$ with $-\mathcal{A}\mathrm{d}\xi$. The answer to this will be different in each of our examples. The summand $-\mathcal{A}\mathrm{d}\xi$ disappears, because in the first example, ξ is constant so that $\mathrm{d}\xi = 0$. In the second example, equilibrium remains unaltered through all changes, so $\mathcal{A} = 0$ is always the case.

"Heat Content" It has already been mentioned at the beginning that there is no quantity of this type. There are replacements for this quantity in the traditional concept, though. For instance, internal energy U at constant volume (*isochoric* processes) can play this role. This results from the equation $\mathrm{d}U = \delta Q_\mathrm{e} - p\mathrm{d}V$ if we set $\mathrm{d}V = 0$. This is succinctly expressed in the formulas:

$$(\mathrm{d}U)_V = (\delta Q_\mathrm{e})_V \quad \text{or} \quad (\Delta U)_V = \Delta_V U = Q_{\mathrm{e},V}. \tag{24.12}$$

This relation can be used to define various other *isochoric* heat quantities such as integral and differential, molar and specific heats of reaction and the corresponding heat capacities. The most well known of these quantities is the "(global or integral) heat capacity at constant volume" or *isochoric heat capacity*, which we got to know briefly in Sect. 9.1:

$$C_V = \left(\frac{\partial U}{\partial T}\right)_V \quad \text{or more precisely,} \quad C_V = \left(\frac{\partial U}{\partial T}\right)_{V,\mathrm{rev}}. \tag{24.13}$$

The expression on the right specifically expresses what the one on the left implies. This is the fact that entropy may not be generated because this would reduce the amount of heat being supplied, falsifying the result. In the case of water formation, reversibility requires $\xi = \mathrm{const.}$, and in the case of NO_2 dimerization, $\mathcal{A} = 0$:

$$\text{Case 1: } C_V = \left(\frac{\partial U}{\partial T}\right)_{V,\xi}, \quad \text{Case 2: } C_V = \left(\frac{\partial U}{\partial T}\right)_{V,\mathcal{A}=0}.$$

From another common point of view "heat capacity" is only a name for the expression on the left in line (24.13), regardless of whether or not it describes in reality a temperature-related quantity. We will return to the subject in a somewhat different context in the subsection "*Heat capacities*" below.

Enthalpy The quantity U can only appear in the role of "heat content" when volume V is constant. The most important case in practice, however, is the transfer of heat Q when pressure p is kept constant, instead of V (*isobaric* processes). In everyday life, but also in science and technology, many processes take place under conditions where the atmosphere ensures an approximately constant pressure (e.g., reactions in open flasks in the laboratory). A state quantity conceived for exactly this purpose is *enthalpy H*. Translated from the Greek it means "in-heat" or, more extensively, "heat content." It is defined as being derived from internal energy:

$$H := U + pV \qquad \text{with the differential} \qquad \underbrace{\mathrm{d}H = \delta Q_\mathrm{e} + V\mathrm{d}p}_{T\mathrm{d}S + V\mathrm{d}p - \mathcal{A}\mathrm{d}\xi}. \tag{24.14}$$

This formula is equivalent to Eq. (24.4)

> The expression above, on the right, is the formal result of the defining equation when we first use the sum rule for the formation of differentials from two (and more) functions (see Sect. A.1.2 in the Appendix),
>
> $$\mathrm{d}H = \mathrm{d}(U + pV) = \mathrm{d}U + \mathrm{d}(pV),$$
>
> subsequently the corresponding product rule,
>
> $$\mathrm{d}H = \mathrm{d}U + V\mathrm{d}p + p\mathrm{d}V,$$
>
> and then insert $\mathrm{d}U = \delta Q_\mathrm{e} - p\mathrm{d}V$,
>
> $$\mathrm{d}H = \delta Q_\mathrm{e} - p\mathrm{d}V + V\mathrm{d}p + p\mathrm{d}V = \delta Q_\mathrm{e} + V\mathrm{d}p.$$
>
> The expression below the parentheses results when the main equation $\mathrm{d}U = T\mathrm{d}S - p\mathrm{d}V - \mathcal{A}\mathrm{d}\xi$ [Eq. (24.7)] is used instead:
>
> $$\mathrm{d}H = T\mathrm{d}S - p\mathrm{d}V - \mathcal{A}\mathrm{d}\xi + V\mathrm{d}p + p\mathrm{d}V = T\mathrm{d}S + V\mathrm{d}p - \mathcal{A}\mathrm{d}\xi.$$

The expression above the parentheses in Eq. (24.14) describes what one notices of the action in the system in the surroundings. The expression below describes what is actually happening in the system itself. As before, we have everything we need mathematically to calculate the missing quantity (represented again by the question mark and later identified as $\mathcal{A}$) from the measured data:

$$? = \left(\frac{\partial H}{\partial \xi}\right)_{S,p}.$$

Here the problem is the same as with internal energy: a quantity we do not understand (entropy) as an independent variable. For this reason, the purpose of the variable H is perceived differently, namely in its suitability for calculating *isobaric* heat effects. Equation (24.14) simplifies at constant pressure ($\mathrm{d}p = 0$):

$$(\mathrm{d}H)_p = (\delta Q_\mathrm{e})_p \quad \text{or} \quad (\Delta H)_p = \Delta_p H = Q_{\mathrm{e},p}. \tag{24.15}$$

As we have seen in the case above of internal energy [Eq. (24.12)], this relation can be useful for defining various *isobaric* heat quantities such as integral and differential, molar and specific heats of reaction, transition, solution, mixing, etc. These are all produced similarly at constant p and T and, depending upon the process in question, each one can have various symbols and names. We will be content with only two examples, one integral quantity and one differential quantity:

$$(\Delta H)_{T,p} \equiv \Delta_{T,p} H \quad \text{general isothermal-isobaric change of enthalpy,}$$
$$\left(\frac{\partial H}{\partial \xi}\right)_{T,p} \equiv \Delta_\mathrm{R} H \quad \text{(differential molar) enthalpy of reaction.} \tag{24.16}$$

The word "heat" in names of quantities is almost always left out. One reason for this is that reaction enthalpies $\Delta_\mathrm{R} H$ can be defined for both spontaneous as well as forced processes; however, they appear as heat only in spontaneous processes. In the process $2\,H_2 + O_2 \rightarrow 2\,H_2O$, it would make sense to speak of the "heat of formation of water," while in the opposite case of $2\,H_2O \rightarrow 2\,H_2 + O_2$, it would make rarely sense to speak of "heat of decomposition of water."

We will take a closer look at the molar enthalpy of reaction of a spontaneously running process based upon Eq. (24.14). We will keep an eye on the effects on the surroundings (upper line) as well as what is happening inside the system (lower line). In both cases, we will make use of the possibility of writing a derivative as a differential quotient and to convert it using the rules of fractions:

$$\Delta_\mathrm{R} H = \left(\frac{\partial H}{\partial \xi}\right)_{T,p} = \left(\frac{\mathrm{d}H}{\mathrm{d}\xi}\right)_{T,p} = \left(\frac{\delta Q_\mathrm{e} - \cancel{V\mathrm{d}p}}{\mathrm{d}\xi}\right)_{T,p} = \left(\frac{\delta Q_\mathrm{e}}{\mathrm{d}\xi}\right)_{T,p}$$
$$= \left(\frac{T\mathrm{d}S - \cancel{V\mathrm{d}p} - \mathcal{A}\mathrm{d}\xi}{\mathrm{d}\xi}\right)_{T,p} = T\left(\frac{\mathrm{d}S}{\mathrm{d}\xi}\right)_{T,p} - \mathcal{A}\left(\frac{\cancel{\mathrm{d}\xi}}{\cancel{\mathrm{d}\xi}}\right)_{T,p} = T\cdot\Delta_\mathrm{R}S - \mathcal{A}.$$

($V\mathrm{d}p$ disappears because $p = \text{const.}$, therefore $\mathrm{d}p = 0$, and the $\mathrm{d}\xi$ cancels out). The result in the upper line tells us that $\Delta_\mathrm{R} H$ describes an effect noticeable in the surroundings as exchanged heat. The lower line states that the effect in the system is made up of two contributions, "latent heat" and released energy; the latter can be arbitrarily made use of. In particular, it can be dissipated. There is more about this in Sects. 8.6 and 8.7. We will omit it here because in the traditional structure of thermodynamics, it can only be discussed later on (see Sect. 24.4).

Heat Capacities We will call the amount of entropy necessary for heating a body by 1 K the entropy capacity $\mathcal{C}$, while heat capacity C will describe the necessary heat (Q_e) for the same process. We also assume that no entropy (S_g) and therefore no heat (Q_g) is generated in the interior. The amount of entropy or heat the body can absorb depends upon whether it can expand or not—whether or not pressure p or volume V is constant. There may be additional conditions which need to be met. To be more mathematically correct, we might write:

$$\mathcal{C}_V = \left(\frac{dS}{dT}\right)_V = \left(\frac{\partial S}{\partial T}\right)_V, \quad C_V = \left(\frac{\delta Q_e}{dT}\right)_V = \left(\frac{\partial Q_e}{\partial T}\right)_V.$$

The second to last expression in parentheses can be understood as a quotient of the *differential form* $\delta Q_e = TdS$ and the differential dT subject to the side condition $V = \text{const.}$ or $dV = 0$. The last expression, however, requires that the function whose derivative is to be taken, i.e., $Q(T, V)$, actually exists, which is not the case, as we have seen above. However, if we insert $\delta Q_e = TdS$ into the second to last differential quotient, we obtain:

$$C_V = \underbrace{\left(\frac{\delta Q_e}{dT}\right)_V = \left(\frac{TdS}{dT}\right)_V = T\left(\frac{dS}{dT}\right)_V}_{\text{not mandatory}} = T\mathcal{C}_V.$$

Equation $C_V = T\mathcal{C}_V$ appears so self-evident as to make the intermediate steps above superfluous. We also see that heat and entropy capacities only differ from each other by the factor T not only at constant volume but at constant pressure p as well, or if another quantity X is kept constant:

$$C_p = T\mathcal{C}_p, \quad C_X = T\mathcal{C}_X, \text{etc.}$$

Entropy is usually considered to be an abstract and especially difficult quantity. The attempt has therefore been made to perform calculations and derivations using other quantities, especially energetic quantities and to express the chemical data using those quantities. This works very well in the case of C_V because internal energy U at constant volume can play the role of "heat content" [cf. Eq. (24.12)]. If we write all the intermediate steps as above, we have:

$$C_V = \underbrace{\left(\frac{\delta Q_e}{dT}\right)_V = \frac{(\delta Q_e)_V}{dT} = \frac{(dU)_V}{dT} = \left(\frac{dU}{dT}\right)_V}_{\text{not mandatory}} = \left(\frac{\partial U}{\partial T}\right)_V,$$

in short,

$$C_V = \left(\frac{\partial U}{\partial T}\right)_V \qquad \text{``(integral) heat capacity at constant volume.''}$$

When pressure remains constant, enthalpy H plays the role of "heat content" [cf. Eq. (24.15)], and we can forgo writing the intermediate steps:

$$C_p = \left(\frac{\partial H}{\partial T}\right)_p \qquad \text{``(integral) heat capacity at constant pressure.''}$$

In fact, heat capacities are commonly not defined in terms of exchanged heat (Q_e), but are directly used as derivatives of internal energy U and enthalpy H. The disadvantage here is that, for every side condition (constant volume, constant pressure, constant X, etc.), a different quantity is necessary for the role of "heat content."

The fact that we have not addressed all the different types of heat capacities became evident at the end of the subsection on *"Heat content"* where a certain difficulty became apparent in our two prototypical example systems. Along with the integral quantities dealt with above, we need various specific (related to the mass) and molar (related to the amount of substance) quantities derived from them. We can omit them here because their definitions and applications follow known patterns.

24.3   Free Energy

Basic Idea   Already in the nineteenth century it was assumed that the energy W_f released in a chemical transformation as well as the heat Q_g generated by it were a measure of the "driving force" of such a process. Energy was considered to be *free* when, for given conditions, it could be used for some other purpose, especially generation of entropy. W_f increases proportionally to the conversion $\Delta\xi$, so W_f itself is not the correct measure, but W_f, relative to the conversion, is: $W_f/\Delta\xi$ or, more exactly, $\delta W_f/d\xi$.

If the available energy W_f can be calculated for the process, then the "driving forces" can be derived from it and we can predict whether or not the process may run spontaneously. However, W_f cannot be determined only from a change to the total energy. The amount of it that can be released depends upon the particular circumstances. Depending upon the general conditions, different types of positive and negative contributions must be considered (see Fig. 24.6 and Experiment 24.1).

In Experiment 24.1 the free energy W_f of a raised body, which is lowered in water (case 1) and in air (case 2), is used to lift a second object. In the second case the lifting height is considerably greater. This energy W_f can be put to any number of other uses, especially for generation of entropy.

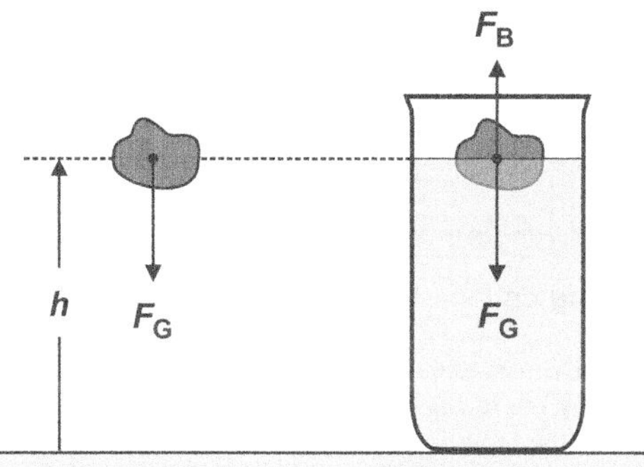

Fig. 24.6 A body sinking in air (*left*) and in water (*right*). The free and available energy W_f on the left is the total energy $W_f = F_G \cdot h$ originally expended to raise it in the air. Only a part of this, $W_f = (F_G - F_B) \cdot h$, is available on the right because energy is needed for pushing the body downward against the effect of buoyancy (F_G force of gravity, F_B buoyancy, h height).

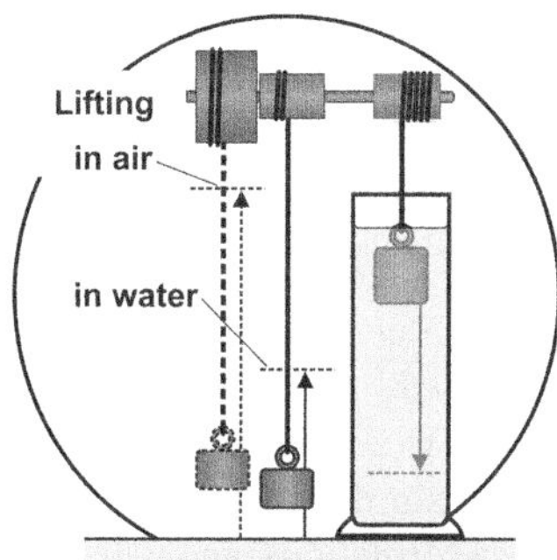

Experiment 24.1 Using the free energy W_f of a raised body (on the *right*) to lift a second object (*middle*) using ropes and hoisting drums. The *dashed line* shows the change of position of the same object if the first object is lowered in air. W_f is greater here so that the attainable lifting height increases correspondingly.

***U* as Free Energy** Let us again consider the concrete system of water formation in our prototypical example system, where the process is controlled by a catalyst. Volume V and entropy S should be kept constant. In the case of volume, this can be accomplished by blocking the piston. It is more difficult to do for entropy because although the entropy leaves the system, all other exchanges of entropy should also be prevented. This can be done by not dissipating the released energy W_f in the system, but to first remove it electrically and to only produce the heat $Q_g - W_f$ outside of the system (Fig. 24.7).

In systems of this type, the internal energy U at constant S and V appears as stored free energy, which we can simply express in one or the other of the following ways:

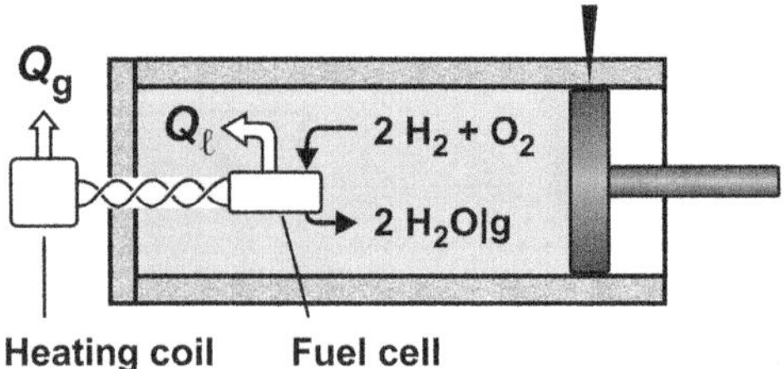

Fig. 24.7 A cylinder thermally insulated on all sides with built-in fuel cell. The fuel cell serves to remove the energy released by the reaction of H$_2$ and O$_2$ to H$_2$O|g out of the cylinder, while the latent heat Q_ℓ that develops in the process of water formation cannot leave the cylinder. The piston remains stationary and therefore the volume constant.

$$W_f = (\Delta U)_{S,V} = \Delta_{S,V} U \quad \text{or} \quad \delta W_f = (dU)_{S,V}.$$

Above, we have attempted to demonstrate such an implementation of an isentropic-isochoric process. This is not the point, though. All the necessary quantities are measurable so that the missing measure of the drive could be calculated:

$$\left(\frac{\delta W_f}{d\xi}\right)_{S,V} = \left(\frac{dU}{d\xi}\right)_{S,V} = \left(\frac{\cancel{TdS} - \cancel{pdV} - \mathcal{A}d\xi}{d\xi}\right)_{S,V} = \left(\frac{-\mathcal{A}\cancel{d\xi}}{\cancel{d\xi}}\right)_{S,V} = -\mathcal{A}. \quad (24.17)$$

(Because S and V are constant, TdS and pdV disappear, so that $d\xi$ cancels.) In order to get a feel for the traditional way of thinking about this, we must ignore the expression on the right. The goal is to develop a "deeper understanding" from the expression on the left (and ones similar to it, which we will go into later) of the actual causes of chemical transformations and what parameters can be used to influence them.

In this context, it is remarkable that Josiah Willard Gibbs chose this method of using energy at constant entropy to derive numerous results for the behavior of homogeneous and heterogeneous chemical systems. Except for the path for outflow of entropy, all other energy paths should be blocked. The actual trick of Gibbs' method is to *remove* entropy generation (S_g) from the system and the energy W_b that is used and dissipated along with it. Neither S_g nor W_b feed back upon the system; it is as if they did not exist. Under these conditions, if there is an entropy generating process, the internal energy U decreases. Equilibrium is reached if U has a minimum. The quantity U plays a role analogous to potential energy in mechanics. This applies to stability or lability of equilibria as well.

There is another notable point. The temperature in the system need not be temporally nor spatially constant. The transfer of a quantity of entropy S_t from a hot to a cold subarea by use of an auxiliary body that repeatedly undergoes a reversible cyclic process (see Fig. 24.2) delivers useful work to be stored in the system because all the possible paths for energy outflow have been blocked. Energy U remains constant in the process. Transferring the same amount of entropy S_t by

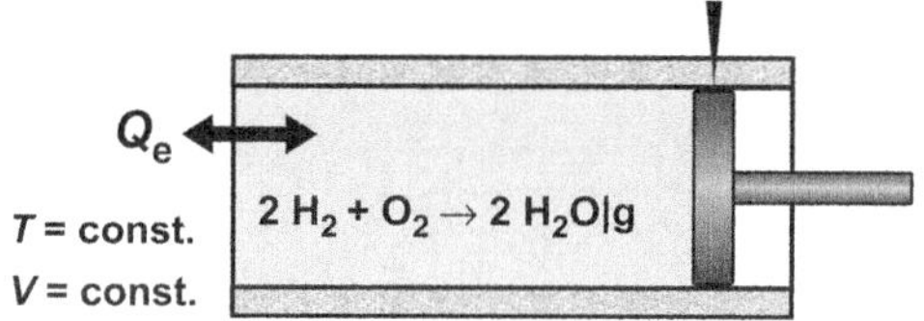

Fig. 24.8 Cylinder exchanging heat with a reservoir at constant temperature T. The piston is fixed so that V remains constant.

heat conduction, however, causes a decrease of U, while the same amount of entropy S_t as before is transferred from the hotter subarea to the colder subarea. U decreases, because under the conditions chosen by Gibbs also energy "flows out" together with the removal of entropy S_g from the location where it was generated.

Helmholtz Energy Let us again imagine a system at constant volume, but now at fixed temperature T, instead of fixed entropy S (Fig. 24.8). This may again be our example of the formation of water in a cylinder with a blocked piston, but in this case, heat exchange takes place with an exterior reservoir. In contrast to our previous case, the outflow of released and dissipated energy $W_f = Q_g$ during a given conversion $\Delta\xi$ requires no special measures because it runs by itself. The generated entropy $S_g = W_f/T$ does not enter into the entropy balance ΔS of our system because whatever is generated inside flows out. Not only $-W_f$ appears in the energy balance ΔU, but the (negative) term $T\Delta S$ does as well, which is caused by the change of chemical composition of our system:

$$\Delta U = T\Delta S - W_f \quad \text{or} \quad \text{more detailed,} \quad U_2 - U_1 = TS_2 - TS_1 - W_f.$$

The equation, solved for W_f, results first in

$$W_f = -(U_2 - TS_2) + (U_1 - TS_1) \quad \text{and finally in} \quad W_f = -\Delta\underbrace{(U - TS)}_{A}.$$

Here, T plays the role of a constant parameter.

The additional quantity $A := U - TS$ introduced in the traditional concept of thermodynamics actually appears as a reservoir of free energy W_f under the given conditions (constant temperature and volume, no exchange of substances with the surroundings). When the supply of A decreases, energy $W_f = -\Delta A$ becomes available for any purpose—depending upon the equipment being used. In general, without proper equipment, W_f will be dissipated. This is no different than when energy is used from other sources (sun, wind, water, coal). However, the contribution $T\Delta S$ cannot just be used freely. It is *"earmarked"* for a specific purpose—in this case, for shifting entropy between the system and the surroundings.

In the past, the quantity A was called *"free energy."* The fact that, under different conditions, there are other quantities that play the same role (for instance, U in closed systems at constant S and V) makes the name too general. It is therefore recommended calling the quantity A the *Helmholtz free energy* or just *Helmholtz energy*.

The fact that A can appear as "free energy" and under which circumstances this happens can be formally expressed similarly to how we did this in the case of the quantity U:

$$W_f = (\Delta A)_{T,V} = \Delta_{T,V} A \quad \text{or} \quad \delta W_f = (dA)_{T,V}.$$

The definition of the quantity $A := U - TS$ and the main equation $dU = TdS - pdV - \mathcal{A}d\xi$ of our example system lead to the following expression for the differential dA [by using the product rule for the term $d(TS)$]:

$$dA = d(U - TS) = dU - d(TS) = (TdS - pdV - \mathcal{A}d\xi) - (SdT + TdS)$$

or

$$dA = -S\,dT - pdV \underbrace{- \mathcal{A}\,d\xi}_{(dA)_{T,V} = \delta W_f} . \tag{24.18}$$

The idea behind dealing with "free energy" W_f was the possibility of gleaning a measure for the "driving force" of chemical transformation; actually, not W_f itself, but δW_f relative to the conversion $d\xi$ was the sought-after measure:

$$\left(\frac{\delta W_f}{d\xi}\right)_{T,V} = \left(\frac{dA}{d\xi}\right)_{T,V} = \left(\frac{-SdT - pdV - \mathcal{A}d\xi}{d\xi}\right)_{T,V} = \left(\frac{-\mathcal{A}d\xi}{d\xi}\right)_{T,V} = -\mathcal{A}. \tag{24.19}$$

(The terms $-SdT$ and $-pdV$ vanish because according to the chosen conditions, temperature T as well as volume V should be constant, meaning $dT = 0$ and $dV = 0$. Subsequently, $d\xi$ cancels.)

We see that the quantity intended to be the "driving force"—in this case of the reaction of formation of water—can be written as a derivative of the state function $A(T, V, \xi)$. We already arrived at an equivalent result in Eq. (24.17). The essential difference here in contrast to before is that entropy S does *not* appear as an independent variable.

Gibbs Energy In practical cases, it happens much more often that not volume V but pressure p is kept constant along with temperature T. It is easy to include such cases (Fig. 24.9). Another pathway over which energy can be exchanged is the mechanical one through a moving piston. This does not change anything about the balance of entropy, but it does affect the balance of energy which now looks like this:

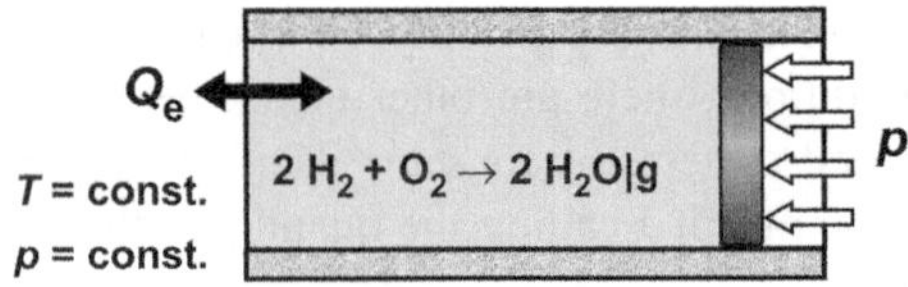

Fig. 24.9 Cylinder exchanging heat with a reservoir at constant temperature T. The piston is freely moveable at constant external pressure p.

$$\Delta U = T\Delta S - p\Delta V - W_f \quad \text{or} \quad U_2 - U_1 = TS_2 - TS_1 - pV_2 + pV_1 - W_f.$$

The equation, solved for W_f, results first in

$$W_f = -(U_2 - TS_2 + pV_2) + (U_1 - TS_1 + pV_1)$$

$$\text{and finally in} \quad W_f = -\Delta(\underbrace{\overbrace{U + pV}^{H} - TS}_{G}).$$

Here, T as well as p appear in the role of a constant parameter.

As before the quantity A, the quantity $G := U + pV - TS = H - TS$, a further quantity introduced in the conventional concept of thermodynamics, acts as a reservoir of free energy W_f, but here under the changed conditions p, $T = $ const. (instead of V, $T = $ const.). If G decreases under these conditions, energy $W_f = -\Delta G$ becomes available for many uses, especially for dissipation.

In the past, the quantity $G = H - TS$ was called "*free enthalpy*," analogous to $A = U - TS$, which was called "free energy." Today *Gibbs free energy* or just *Gibbs energy* is the recommended expression (IUPAC). How and when G can appear in the role of "free energy" can be formally expressed very similarly to the cases of U and A:

$$W_f = (\Delta G)_{T,V} = \Delta_{T,V}G \quad \text{or} \quad \delta W_f = (dG)_{T,V}.$$

We obtain the following expression for the differential dG from the definition $G := U + pV - TS$ and the main equation $dU = TdS - pdV - \mathcal{A}d\xi$:

$$dG = d(U + pV - TS) = dU + d(pV) - d(TS)$$
$$= (TdS - pdV - \mathcal{A}d\xi) + (Vdp + pdV) - (SdT + TdS)$$

and therefore

$$dG = -SdT + Vdp \underbrace{- \mathcal{A}d\xi}_{(dG)_{T,p} = \delta W_f.} \, , \tag{24.20}$$

from which we can conclude that the quantity we are seeking as the "driving force" of a reaction can be expressed in numerous ways as the derivative of a state function, in the present case, as the derivative of the function $G(T, p, \xi)$ (same procedure as in the case of the derivative of the state function $A(T, V, \xi)$):

$$\left(\frac{\delta W_{\mathrm{f}}}{\mathrm{d}\xi}\right)_{T,p} = \left(\frac{\mathrm{d}G}{\mathrm{d}\xi}\right)_{T,p} = \left(\frac{-S\mathrm{d}T + V\mathrm{d}p - A\mathrm{d}\xi}{\mathrm{d}\xi}\right)_{T,p} = \left(\frac{-A\mathrm{d}\xi}{\mathrm{d}\xi}\right)$$

$$= -A. \tag{24.21}$$

As we have also seen with enthalpy H, there are numerous other quantities that can be derived using G as the basis. Here are just two examples, an integral and a differential quantity:

$$(\Delta G)_{T,p} \equiv \Delta_{T,p}G \quad \text{general isothermal-isobaric change of Gibbs energy,}$$

$$\left(\frac{\partial G}{\partial \xi}\right)_{T,p} \equiv \Delta_{\mathrm{R}}G \quad \text{(differential molar) Gibbs energy of reaction.}$$

$$\tag{24.22}$$

Spontaneous Process The most commonly observed transformations in chemistry are, as already mentioned, those at constant temperature and constant pressure, which is why the mathematical tools are generally oriented toward these conditions. The heat function that we deal with the most is enthalpy $H(T, p, \xi, \ldots)$ and the quantity most often playing the role of free energy is the Gibbs energy $G(T, p, \xi, \ldots)$. Until now, we have only allowed a single parameter ξ, but it is also possible to observe two or more transformations or other types of change at the same time. Spontaneous changes in closed systems at constant T and p are only possible in the direction in which the free energy decreases, in this case, the Gibbs energy:

$$
\begin{array}{lll}
\mathrm{d}G < 0 & \text{spontaneously possible in closed} & \left\{\begin{array}{l}\text{isothermal-isobaric,}\\ \text{isothermal-isochoric,}\\ \text{isentropic-isochoric}\end{array}\right\} \text{systems.}\\
\mathrm{d}A < 0 & & \\
\mathrm{d}U < 0 & &
\end{array}
$$

We have included two further possibilities. These are different state functions depending upon the side conditions that must be met. For chemical transformations of all types, meaning the processes chemists are most interested in, all these conditions for spontaneous processes can be simply summed up as one, we already know very well and used very often (see e.g. Sect. 4.6): $A > 0$ [cf. Eqs. (24.17), (24.19), and (24.21)].

If we have $(\mathrm{d}G)_{T,p} = 0$ for small changes of state, there will be no preferred direction and the system will be in equilibrium. Correspondingly, under other side conditions, $(\mathrm{d}A)_{T,V} = 0$, $(\mathrm{d}U)_{S,V} = 0$, etc., are valid. For chemical transformations this means simply $A = 0$.

Coupling The thermodynamic functions $G(T, p, \xi)$, $A(T, V, \xi)$, $U(S, V, \xi)$, …, allow for another type of application if we keep in mind that the mixed second derivatives are independent of the order in which the derivatives are taken

(Schwarz' theorem). Here, it is taken for granted that these derivations exist and are continuous.

In the case of a state function $Z = f(x, y)$ Schwarz' theorem states that

$$\left(\frac{\partial}{\partial y}\left(\frac{\partial Z}{\partial x}\right)_y\right)_x = \left(\frac{\partial}{\partial x}\left(\frac{\partial Z}{\partial y}\right)_x\right)_y \quad \text{or alternatively formulated}$$

$$\left(\frac{\partial^2 Z}{\partial x \partial y}\right) = \left(\frac{\partial^2 Z}{\partial y \partial x}\right).$$

This can be used to derive several important relations between different coefficients. At first, Eq. (24.20) yields:

$$S = -\left(\frac{\partial G}{\partial T}\right)_{p,\xi}, \quad V = +\left(\frac{\partial G}{\partial p}\right)_{T,\xi}, \quad \mathcal{A} = -\left(\frac{\partial G}{\partial \xi}\right)_{T,p}. \tag{24.23}$$

If one takes derivates of the quantities S, V, $\mathcal{A}$, represented themselves as derivatives, one can use Schwarz' theorem:

$$\left(\frac{\partial S}{\partial p}\right)_{T,\xi} = -\left(\frac{\partial^2 G}{\partial p \partial T}\right)_\xi = -\left(\frac{\partial^2 G}{\partial T \partial p}\right)_\xi = -\left(\frac{\partial V}{\partial T}\right)_{p,\xi},$$

$$\left(\frac{\partial S}{\partial \xi}\right)_{T,p} = -\left(\frac{\partial^2 G}{\partial \xi \partial T}\right)_p = -\left(\frac{\partial^2 G}{\partial T \partial \xi}\right)_p = +\left(\frac{\partial \mathcal{A}}{\partial T}\right)_{p,\xi},$$

$$\left(\frac{\partial V}{\partial \xi}\right)_{T,p} = +\left(\frac{\partial^2 G}{\partial \xi \partial p}\right)_T = +\left(\frac{\partial^2 G}{\partial p \partial \xi}\right)_T = -\left(\frac{\partial \mathcal{A}}{\partial p}\right)_{T,\xi}.$$

For better understanding, let us take a look at the first line. The parameter ξ should be always constant meaning it is enough if we focus on the function $G = f(T, p)$ and their differential $dG = -S dT + V dp$. To begin with, the derivative of this function is firstly taken with respect to T at constant p and secondly with respect to p at constant T:

$$\left(\frac{\partial G}{\partial T}\right)_p = -S \quad \text{and} \quad \left(\frac{\partial G}{\partial p}\right)_T = V.$$

Subsequently, the derivative of the expression on the left is taken with respect to p at constant T and that on the right with respect to T at constant p:

$$\left(\frac{\partial^2 G}{\partial T \partial p}\right) = \left(\frac{\partial}{\partial p}\left(\frac{\partial G}{\partial T}\right)_p\right)_T = -\left(\frac{\partial S}{\partial p}\right)_T, \quad \left(\frac{\partial^2 G}{\partial p \partial T}\right) = \left(\frac{\partial}{\partial T}\left(\frac{\partial G}{\partial p}\right)_T\right)_p = \left(\frac{\partial V}{\partial T}\right)_p.$$

Because the expressions on the left in both equations are equal according to Schwarz' theorem, this also holds true for the expressions on the right.

This same pattern can be used for finding numerous other relations. However, we do not need this method because the flip rule leads directly to the same result without taking the detour over a second derivative of an appropriately chosen thermodynamic function. The relation in the first line, for example, is already well known from Sect. 9.2 [Eq. (9.7)].

24.4   Partial Molar Quantities

Molar Enthalpy When dealing with enthalpy, it is common to use the same procedure as the one applied for quantifying a substance's volume demand and associating the volume demand of a mixture to the individual components (see Sect. 8.2). Enthalpy at fixed T and p for pure substances increases proportionally to the amount of substance n, so the characteristic molar quantity is the enthalpy relative to n:

$$H_m = \frac{H}{n} \quad molar \ enthalpy. \tag{24.24}$$

For a substance in a mixture of other substances, this is defined correspondingly [cf. Eq. (8.2)]:

$$H_m = \left(\frac{\partial H}{\partial n}\right)_{T,\,p,n',n'',\dots} \quad (partial) \ molar \ enthalpy \ of \ a \ substance. \tag{24.25}$$

The enthalpy of the total mixture equals the sum of the contributions by the individual components A, B, C, ... We were introduced to this for volume and entropy before [cf. Eqs. (8.3) and (8.12)]:

$$H = n_A H_A + n_B H_B + n_C H_C + \dots . \tag{24.26}$$

If we are interested in the (differential) molar enthalpy of reaction $\Delta_R H(\xi)$ of a transformation which takes place in the system under consideration, we again start from the general conversion formula for an arbitrary reaction between pure or dissolved substances (like in Chap. 8):

$$|v_B|B + |v_{B'}|B' + \dots \rightarrow v_D D + v_{D'}D' + \dots .$$

For better understanding of the following approach it is recommendable to read the short Sect. 8.3 again. Analogously to the molar volume of reaction $\Delta_R V(\xi)$ and the molar entropy of reaction $\Delta_R S(\xi)$ discussed in the mentioned Section, we obtain for the molar enthalpy of reaction $\Delta_R H(\xi)$ in the case of small conversions $\Delta\xi$ (when p, T, ξ', ξ'', ... are kept constant):

$$\Delta_R H = \frac{\Delta H}{\Delta \xi} = v_B H_B + v_{B'} H_{B'} + \ldots + v_D H_D + v_{D'} H_{D'} + \ldots = \sum_i v_i H_i. \quad (24.27)$$

Applied to our prototypical example reaction $2\,H_2|g + O_2|g \rightarrow 2\,H_2O|g$, the equation becomes:

$$\Delta_R H = -2H(H_2|g) - H(O_2|g) + 2H(H_2O|g).$$

Because the conversion numbers of the reactants are negative and those of the products positive, the expression can be read as a difference:

$$\Delta_R H = \underbrace{2H(H_2O|g)}_{products} - \underbrace{(2H(H_2|g) + H(O_2|g))}_{reactants}.$$

This means we follow the familiar schema when calculating $\Delta_R H$: "The sum of the molar characteristic quantities of the products minus the sum of the molar characteristic quantities of the reactants."

In the limit, we require the $\Delta \xi$ to be infinitesimally small in Eq. (24.27). This is again expressed formally by using the symbol ∂ instead of the difference Δ. If we now introduce all the quantities that are to be kept constant as indices of the differential quotient, the equation takes the following form:

$$\Delta_R H = \left(\frac{\partial H}{\partial \xi}\right)_{p,T,\xi',\xi'',\ldots} = \sum_i v_i H_i. \quad (24.28)$$

This is the (differential) molar enthalpy of reaction $\Delta_R H$ mentioned above [see Eq. (24.16)].

Molar Gibbs Energy   Other extensive thermodynamic quantities are dealt with in the same way. In the case of pure substances, they are considered a function of T, p, n, and for a substance in a mixture with other substances, as a function of T, p, n, n', n'', $\ldots$. The Gibbs energy G is especially interesting in this context because in the conventional thermodynamic calculations, it is very closely connected with the chemical potential. In the case of a pure substance at fixed T and p, G is proportional to the amount of substance n. Therefore, G itself does not serve as the substance-specific characteristic, but the quotient G/n:

$$G_m = \frac{G}{n} \quad molar\ Gibbs\ energy. \quad (24.29)$$

We proceed accordingly for a substance in a mixture with other substances:

$$G_\mathrm{m} = \left(\frac{\partial G}{\partial n}\right)_{T,\, p,\, n',\, n'',\, \ldots} \qquad \textit{(partial) molar Gibbs energy of a substance.}$$

$$(24.30)$$

The values for the mixture as a whole are added up from the contributions by the individual components A, B, C, ... just as they are for volume, entropy, enthalpy:

$$G = n_\mathrm{A} G_\mathrm{A} + n_\mathrm{B} G_\mathrm{B} + n_\mathrm{C} G_\mathrm{C} + \ldots . \tag{24.31}$$

The (differential molar) *Gibbs energy of reaction* $\Delta_\mathrm{R} G$ of a transformation can be expressed analogously to how we dealt with the enthalpy of reaction $\Delta_\mathrm{R} H$. In the case of small conversions $\Delta \xi$, when p, T, ξ', ξ'', ... are again kept constant, we obtain:

$$\Delta_\mathrm{R} G = \frac{\Delta G}{\Delta \xi} = v_\mathrm{B} G_\mathrm{B} + v_{\mathrm{B}'} G_{\mathrm{B}'} + \ldots + v_\mathrm{D} G_\mathrm{D} + v_{\mathrm{D}'} G_{\mathrm{D}'} + \ldots = \sum_i v_i G_i \tag{24.32}$$

or more precisely for vanishingly small conversions $\mathrm{d}\xi$:

$$\Delta_\mathrm{R} G = \left(\frac{\partial G}{\partial \xi}\right)_{p,T,\xi',\xi'',\ldots} = \sum_i v_i G_i. \tag{24.33}$$

This is the (differential molar) Gibbs energy of reaction already mentioned [cf. Eq. (24.22)].

Chemical Potential We obtain the differential $\mathrm{d}G$ from the definition for Gibbs energy $G := U + pV - TS$ and the main equation for a mixture $\mathrm{d}W = \mathrm{d}U = T\mathrm{d}S - p\mathrm{d}V + \mu_\mathrm{A}\mathrm{d}n_\mathrm{A} + \mu_\mathrm{B}\mathrm{d}n_\mathrm{B} + \ldots$ [see Eq. (9.2)]:

$$\mathrm{d}G = -S\mathrm{d}T + V\mathrm{d}p + \mu_\mathrm{A}\mathrm{d}n_\mathrm{A} + \mu_\mathrm{B}\mathrm{d}n_\mathrm{B} + \ldots . \tag{24.34}$$

Because we consider the material system in question to be at rest and weightless, we can equate the total energy W to the internal energy U.

Let us have a closer look at the derivation of Eq. (24.34). For the differential $\mathrm{d}G$ we obtain from the definition of G:

$$\mathrm{d}G = \mathrm{d}U + \mathrm{d}(pV) - \mathrm{d}(TS) = \mathrm{d}U + V\mathrm{d}p + p\mathrm{d}V - S\mathrm{d}T - T\mathrm{d}S.$$

Substitution of the differential $\mathrm{d}U$ according to the main equation results in

$$\mathrm{d}G = \left(\cancel{T\mathrm{d}S} - \cancel{p\mathrm{d}V} + \mu_\mathrm{A}\mathrm{d}n_\mathrm{A} + \mu_\mathrm{B}\mathrm{d}n_\mathrm{B} + \ldots\right) + \left(V\mathrm{d}p + \cancel{p\mathrm{d}V} - S\mathrm{d}T - \cancel{T\mathrm{d}S}\right)$$

and therefore

$$dG = -SdT + Vdp + \mu_A dn_A + \mu_B dn_B + \ldots.$$

For a substance B as a component of a mixture of other substances A, C, …, Eq. (24.34) formally results in the following:

$$G_B = \left(\frac{\partial G}{\partial n_B}\right)_{T,\,p,\,n_A,\ldots} = \left(\frac{dG}{dn_B}\right)_{T,\,p,\,n_A,\ldots}$$

$$= \left(\frac{-\cancel{SdT} + \cancel{Vdp} + \cancel{\mu_A dn_A} + \mu_B dn_B + \ldots}{\cancel{dn_B}}\right)_{T,\,p,\,n_A,\ldots} = \mu_B.$$

"Chemical potential" and "partial molar Gibbs energy" of a substance are identical! This makes it possible to construct simple translation rules between conventional formalisms and the one used by us:

$$G_B = \mu_B, \qquad\qquad H_B = \mu_B + TS_B,$$

$$\Delta_R G = -\mathcal{A}, \qquad\qquad \Delta_R H = -\mathcal{A} + T\Delta_R S.$$

If we insert $G_B = \mu_B$ into Eq. (24.33), we obtain:

$$\Delta_R G = v_B \mu_B + v_{B'} \mu_{B'} + \ldots + v_D \mu_D + v_{D'} \mu_{D'} + \ldots = \sum_i v_i \mu_i.$$

This is nothing else than $-\mathcal{A}$ (see Sect. 8.6). The expression for H_B is obtained by using the definition for $G := H - TS$,

$$H_B = G_B + TS_B = \mu_B + TS_B,$$

and that for $\Delta_R H$ by inserting the equation above into Eq. (24.28):

$$\Delta_R H = v_B(\mu_B + TS_B) + v_{B'}(\mu_{B'} + TS_{B'}) + \ldots + v_D(\mu_D + TS_D) + v_{D'}(\mu_{D'} + TS_{D'}) + \ldots$$

and therefore

$$\Delta_R H = (v_B \mu_B + v_{B'} \mu_{B'} + \ldots + v_D \mu_D + v_{D'} \mu_{D'})$$
$$+ T(v_B S_B + v_{B'} S_{B'} + \ldots + v_D S_D + v_{D'} S_{D'} + \ldots)$$

The expression in the first set of parentheses corresponds again to $-\mathcal{A}$ and that in the second set of parentheses to $\Delta_R S$ [according to Eq. (8.13)].

In books of tables, usually the standard values of molar Gibbs free energies of formation $\Delta_f G$ of substances are listed. $\Delta_f G$ is the change of Gibbs free energy that accompanies the formation of 1 mol of the substance in question, pure or dissolved, from its elements under standard conditions. Because $\Delta_f G$ is nothing else than a special case of $\Delta_R G$, it corresponds to the negative "drive of formation" $(-\mathcal{A})$ or the positive "drive of decomposition" $\mathcal{A}$, respectively. In Sect. 4.6, however,

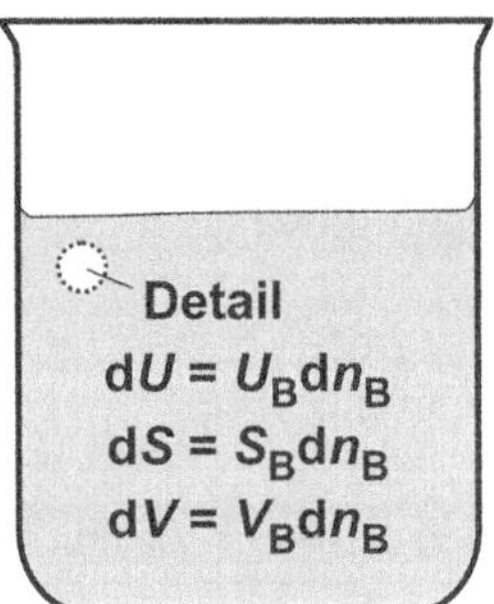

Fig. 24.10 Chemical potential μ_B visualized as the energy released when a small amount of dn_B disappears, shown here as a section of a larger area of a pure substance B. As B disappears in the section, the energy dU in it is released. The volume dV of the part shrinks down to a point while the entropy dS in it is moved to the surrounding matter. The shrinking down to a point causes a contribution to the released energy of $+p \cdot dV$ and the displacement of entropy causes one of $-T \cdot dS$.

we have learned that the chemical drive to decompose corresponds to the chemical potential of the substance. Therefore, the tabulated $\Delta_f G$ values are nothing else than the standard values of the chemical potential we have used in this book!

The "chemical potential μ" and "drive (affinity) $\mathcal{A}$" are not considered independent concepts in the conventional thermodynamic formalism. Therefore, one does not directly define the temperature and pressure coefficients α and β, as well as α and β; rather, they are always expressed in terms of different quantities:

$$\alpha_B = -S_B, \qquad \beta_B = V_B \qquad \text{[see Eqs.(9.11) and (9.16)]},$$

$$\alpha = \Delta_R S, \qquad \beta = -\Delta_R V \quad \text{[see Eqs.(9.13) and (9.18)]}.$$

If one is interested, for example, in the temperature coefficient of the chemical potential of a substance, it is only necessary to find the value of the corresponding molar entropy in an appropriate table book and to change the sign.

If the equation $H_B = \mu_B + TS_B$ is solved for μ_B, and $H_B = U_B + pV_B$ is taken into consideration, we obtain a relation that can be interpreted descriptively (Fig. 24.10):

$$\mu_B = H_B - TS_B = U_B + pV_B - TS_B.$$

24.5   Activities

Basic Idea The quantities used for describing deviations from what is considered ideal behavior of gases and dissolved substances are another characteristic feature of the traditional formalism. We prefer to take the discrepancies into account by additional terms in the chemical potential because these quantities can be seamlessly inserted into the thermodynamic apparatus. By contrast, in the traditional approach, it is customary to introduce the necessary corrections as correcting factors to the measures of composition (concentration, etc.) and to use these modified quantities instead of the actual ones.

The basic idea is the same one as for mass action. The higher the concentration c_B of a dissolved substance B, the stronger its influence upon the formation of some product will be. The simplest case is a dissolved substance D. When compared to its content, the effect is simply proportional to the concentration c_B. We imagine this is valid as long as c_B remains small and the B atoms are therefore far enough apart from each other. At higher concentrations, the atoms begin to influence each other. This can either strengthen or weaken their influence upon the formation of product, just as if the concentration of B had increased or decreased. This apparent increase or decrease is described by a factor γ_B, the so-called *activity coefficient*, which multiplies c_B. $\gamma_B c_B$ is basically the "chemically effective" or "chemically active" concentration of B that can be greater or smaller than the actual c_B. The concentration c_B itself does not appear as the argument of a logarithmic function in the mass action equation, but rather the relative concentration $c_{r,B} = c_B/c$ or the *active relative concentration* $\gamma_B c_B/c$, which is usually but inaccurately called the *activity* of B:

$$\mu_B = \overset{\circ}{\mu}_B + RT\ln \underbrace{\frac{\gamma_B c_B}{c^\pi}}_{a_B} \quad \text{with } a_B \text{ as activity of B (on the } c \text{ scale)}. \quad (24.35)$$

Activities and activity coefficients are commonly used in order to rewrite some frequently used equations into more pleasing form. In this manner, some relations can be written in especially short form. There is a certain difficulty, though, in the fact that these quantities are introduced and applied in many variations. To introduce the subject, we will choose the most general form that is less often used but can be most easily understood.

Chemical Activity This quantity (symbol λ_B), which is assigned to a substance B, is formed by a simple scale transformation from the chemical potential μ_B:

$$\mu_B = RT\ln\lambda_B \quad \text{or} \quad \lambda_B = \exp\left(\frac{\mu_B}{RT}\right).$$

We call the quantity λ_B the *chemical activity*. This is based upon the name "chemical potential" for the quantity μ_B, from which it stems. The recommended name "absolute activity" is unfitting because, depending upon the choice of zero point for the scale of μ, other "relative" λ_B values can result that differ by fixed factors.

λ_B results from μ_B by transforming into an exponential scale. Conversely, the potential can be regained when the activities are transferred into a corresponding logarithmic scale. It becomes easy to understand why (except for some special cases) every statement that can be formulated with chemical potential can also be expressed by chemical activities and vice versa. Qualitatively seen, activities are measures of a substance's tendency to transform, just as potentials are.

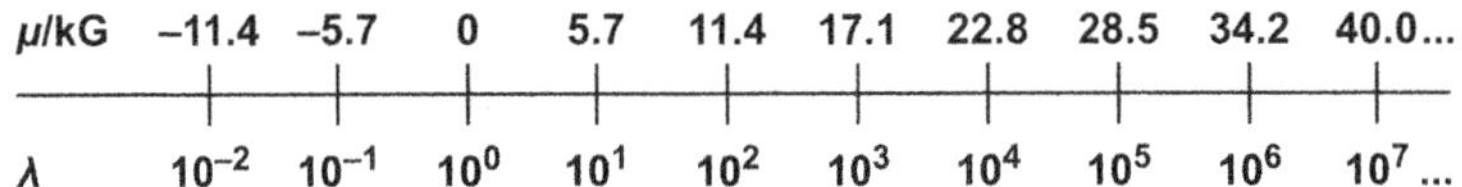

Fig. 24.11 Relationship between μ and λ scales at 298 K.

Such coexistence of various scales is not uncommon in science and technology. If a quantity changes by many orders of magnitude, it is common to start using logarithmic scales in order to represent the entire range of values more easily. For example, the concentration c_{H^+} of hydrogen ions changes by 14 orders of magnitude from that in a strongly acidic solution to that in a strongly basic one; and the acoustic power of an acoustic source changes by 13 orders of magnitude from the auditory threshold to the pain threshold. In the first case, instead of the c_{H^+} value, the preferred one is the pH value, which was originally introduced as a logarithmic measure for hydrogen ion concentration $pH = -\lg(c_{H^+}/c)$ (see Sect. 7.3). In the second case, instead of the *acoustic power* P, it is the *acoustic power level* (or *sound power level*) $\lg(P/P_0)$ with the reference value $P_0 = 10^{-12}$ W that is used. Figure 24.11 illustrates how the μ and λ scales relate to each other.

Residual Activities In Chap. 13, we were introduced to the first steps of a kind of series expansion for the content dependence of chemical potential. The description can be refined by repeated splitting into a *basic value* describing a main effect and a *residual value* summing up the side effects:

$$\text{value} = \underbrace{\text{basic value} + \text{residual value}} \qquad\qquad (\text{step}\,1)$$
$$\text{value*} = \text{basic value*} + \underbrace{\text{residual value*}} \qquad\qquad (\text{step}\,2)$$
$$\text{value**} = \text{basic value**} + \underbrace{\text{residual value**}} \quad (\text{step}\,3)$$
$$\text{value***} = \dots \qquad\qquad (\text{ "})$$

As a result, the quantity μ is split into a sum $\mu = \mu^\circ + \mu^* + \mu^{**} + \mu^{***} + \dots$ which, depending upon how accurate it must be, can have more or fewer terms. The sum transforms into a product if the potentials are transformed into activities:

$$\lambda = \lambda^\circ \cdot \lambda^* \cdot \lambda^{**} \cdot \lambda^{***} \cdot \dots .$$

At that point (Chap. 13), we only used a two-step approach [see Eq. (13.2)], so that $\mu(x)$ appeared to split into three terms: basic value $\overset{\bullet}{\mu}$ + basic value* $\overset{\times}{\mu}$ + residual value* $\overset{+}{\mu}$. Three factors then correspondingly appear in the activity scale:

$$
\begin{array}{ccc}
\overbrace{\text{basic term}} \;\; \overbrace{\text{residual term}} & \qquad & \overbrace{\text{"basic activity"}} \;\; \overbrace{\text{"(residual) activity"}} \\[4pt]
\mu(x) \;=\; \overset{\bullet}{\mu} \;+\; \overset{*}{\mu}(x) & & \lambda(x) \;=\; \overset{\bullet}{\lambda} \;\times\; \overset{*}{\lambda}(x) \\[4pt]
\underbrace{\overset{\times}{\mu}(x) + \overset{+}{\mu}(x)} & & \underbrace{x} \;\times\; \underbrace{\overset{+}{\lambda}(x)} \\[4pt]
\underbrace{\text{mass action term}} \;\; \underbrace{\text{extra term}} & & \underbrace{\text{measure of composition}} \;\; \underbrace{\text{activity coeff.}}
\end{array}
$$

For clarity's sake, we will only use "basic term" and "residual term" for expressions

Fig. 24.12 Potential μ and activity λ for cane sugar in a glass of Turkish tea ($\vartheta = 50\,°C$, $c = 1{,}000\ \text{mol m}^{-3}$), divided into basic values $\overset{\circ}{\mu}$ and $\overset{\circ}{\lambda}$, mass action contributions $\overset{\times}{\mu}$ and $\overset{\times}{\lambda}$, as well as extra values $\overset{+}{\mu}$ and $\overset{+}{\lambda}$:

$$\mu = \overset{\circ}{\mu}_{\mathrm c} + \overset{\times}{\mu}_{\mathrm c} + \overset{+}{\mu}_{\mathrm c} \;=\; (-1{,}575.59 + 0.00 \; + 0.65)\text{kG},$$
$$\lambda = \overset{\circ}{\lambda}_{\mathrm c} \cdot \overset{\times}{\lambda}_{\mathrm c} \cdot \overset{+}{\lambda}_{\mathrm c} = 2.03 \times 10^{-245} \times 1.00 \times \; 1.27.$$

At the taste threshold at approximately 5 mol m^{-3}, the values are:

$$\mu = \overset{\circ}{\mu}_{\mathrm c} + \overset{\times}{\mu}_{\mathrm c} + \overset{+}{\mu}_{\mathrm c} = (-1{,}575.59 - 14.23 \; + 0.003)\ \text{kG},$$
$$\lambda = \overset{\circ}{\lambda}_{\mathrm c} \cdot \overset{\times}{\lambda}_{\mathrm c} \cdot \overset{+}{\lambda}_{\mathrm c} = 2.03 \times 10^{-245} \times 5 \times 10^{-3} \times 1.001.$$

of the first step. Otherwise we will use appropriate alternative names such as "mass action term" for the basic term of step 2 and "extra term" for the corresponding residual term. The mass action term $\overset{\times}{\mu}\,(x)$ is given by the relation $\overset{\times}{\mu} = RT\ln x$. Correspondingly, the $\overset{\times}{\lambda}\,(x)$ value results in $\overset{\times}{\lambda} = \exp[(RT\ln x)/RT] = \exp[\ln x] = x$.

Going from potentials to exponentially growing activities leads to unwieldy values (see Fig. 24.12) which are not suitable for numerical calculations and tabulating of chemical data. This is why the basic values of λ are rarely used but usually only the residual values $\overset{*}{\lambda}$. The latter are commonly called "*activity*" and an independent symbol is introduced $a\ (\equiv \overset{*}{\lambda} = \overset{\times}{\lambda}\,(x) \cdot \overset{+}{\lambda}\,(x) = \overset{+}{\lambda}\,(x) \cdot x$ with $\overset{+}{\lambda}\,(x)$ in the role of an activity coefficient).

When dealing with mixtures and solutions, there are different approaches to separating basic and residual values (see Sect. 1.5). We tend to assume that the participating substances in a mixture can be dealt with uniformly; in particular, a substance can appear in a pure state in the same liquid or solid α-, β-, γ- ... phase. In a solution, however, we contrast the *solvent* as the main component and the *solutes* B, C, ... as the others. The sweetened tea in Fig. 24.12 is an example of such a solution. Considered as a mixture, this would mean that the sugar should be treated as a liquid component in the entire range of 0 to 100 %. For a solvent and all components of mixtures, we always use $\overset{\circ}{\mu}_\bullet \left(\equiv \overset{\bullet}{\mu}\right)$ or $\overset{\circ}{\lambda}_\bullet$ in the pure state as the basic value for potential μ or activity λ. For dissolved substances, in contrast, we choose $\overset{\circ}{\mu}$ or $\overset{\circ}{\lambda}$ which is extrapolated to the standard value $c, x(= 1), b, \ldots$ starting from very low concentrations along an imagined ideal curve. In Sects. 6.2 and 13.2, we took a closer look at the method where the concentration c or the molar fraction x served as a measure of composition. We deal similarly with other measures of composition, at least with those that change proportionally to each other for small values. The basic values of the chemical potential for a substance B dissolved in a solvent A, $\overset{\circ}{\mu}_{\bullet B}$, $\overset{\circ}{\mu}_{c,B|A}$, $\overset{\circ}{\mu}_{x,B|A}$, $\overset{\circ}{\mu}_{b,B|A}$, $\ldots$ and the corresponding λ values $\overset{\circ}{\lambda}_{\bullet B}$, $\overset{\circ}{\lambda}_{c,B|A}$, $\overset{\circ}{\lambda}_{x,B|A}$, $\overset{\circ}{\lambda}_{b,B|A}$, $\ldots$ are all different. We will not discuss here how to convert one into another.

Activity Coefficients The (residual) activities a_B are themselves decomposed into a product of the particular concentration (or mole fraction) and its activity coefficient, as follows:

$$a_{\bullet B} = x_B \gamma_{\bullet B}, \quad a_{c,B} = c_{r,B} \gamma_{c,B}, \quad a_{x,B} = x_B \gamma_{x,B}, \quad a_{b,B} = b_{r,B} \gamma_{b,B}, \ldots .$$

Again, $c_{r,B}$ describes the relative concentration c_B/c, while $b_{r,B}$ is the relative molality b_B/b. Depending upon the chosen basic value, the resulting activities will vary and can be distinguished by indices. If it is clear which alternative is meant, extra identifiers can be omitted.

We use the abbreviation "Suc" for cane sugar (sucrose) in our example in Fig. 24.12. The following holds for the activity coefficient at the standard concentration of 1,000 mol m^{-3}: $\gamma_c(\text{Suc}) \equiv \lambda_c^+(\text{Suc}) = 1.27$. This value indicates graphically that the sugar in the tea glass behaves as if its concentration were 27 % higher than it actually is. Correspondingly, $\gamma_c(\text{Suc}) = 1.001$ at the taste threshold, in other words at a concentration of only 5 mol m^{-3}, means that the deviations from ideal behavior are immeasurably small for such dilutions. This is true for neutral substances, while charged (ionic) ones still display noticeable deviations at concentrations below 10 mol m^{-3}.

In the limit of "infinite" dilution, when the content of B in A (but also the content of all other substances C, D, $\ldots$ in A, if they exist) tends to 0, $\gamma_{c,\emptyset} = \gamma_{x,\emptyset} = \gamma_{b,\emptyset} = \ldots = 1$ is valid for the activity coefficients. $\gamma_{\bullet\emptyset} = 1$ is correspondingly valid for the solvent. We have chosen the "slashed zero" $\emptyset$ as the index, as we did in Sect. 13.3 in order to characterize this state.

Viewing the (residual) activities a as modified measures of composition is quite graphic: the corresponding activity coefficient can be simply understood as a fitting correction factor. This approach can be applied independently of which of the usual measures of composition are being used, whether it is concentration c or c_r, mole fraction x, molality b or b_r, etc. It is also simpler than using the extra potential $\overset{+}{\mu}$, especially when we try to understand the basic quantities themselves as "partial molar Gibbs energies." However, understanding becomes more difficult when trying to capture and calculate the influence of parameters like pressure, temperature, contents of components in mixtures, etc.

Drives Activities are usually introduced to describe deviations of the functions $\mu(x)$ or $\mu(c)$, etc., from values considered ideal at the same temperature and pressure, $\mu = \overset{\circ}{\mu}_{\bullet} + RT\ln x$ or $\mu = \overset{\circ}{\mu}_{c} + RT\ln c_{r}$ as well as $\mu = \overset{\circ}{\mu}_{x} + RT\ln x$. As mentioned right at the start, to correct for this, the actual measure of composition is replaced by the "active" one $a_{\bullet}, a_{c}, a_{x}, a_{b}, \ldots$ which results from the mutual interaction of the substances. A mixed approach is employed here in which we apply the basic values in the scale of the potential and the residual values in the activity scale. This is a compromise that lets us avoid the clumsy and extremely large or extremely small basic values of λ, while allowing for the graphic appeal of the λ residuals.

Let us now consider the familiar example of cane sugar decomposing into glucose and fructose: $\mathrm{Suc|w} + \mathrm{H_2O|l} \to \mathrm{Glc|w} + \mathrm{Fru|w}$ (see Sect. 6.3). The process runs slowly in our tea glass if the tea is slightly acidified (maybe with some lemon juice). We can write the drive $\mathcal{A}$ for this process using the approach discussed in Sect. 6.3: the relative concentrations c_{r} of the dissolved substances are replaced by the activities a_{c}. This time, though, we must remember that the potential of solvent A (in this case, water) can be markedly different due to possible higher concentrations of dissolved B, C, $\ldots$. We can achieve this formally by including the residual value $RT \ln a_{\bullet A}$ for the solvent as well:

$$\mathcal{A} = \overset{\circ}{\mathcal{A}} + \overset{*}{\mathcal{A}}$$

$$\underbrace{\overset{\circ}{\mu}(\mathrm{Suc|w}) + \overset{\circ}{\mu}(\mathrm{H2O|l}) - \overset{\circ}{\mu}\left(\mathrm{Glc|w}\right) - \overset{\circ}{\mu}\left(\mathrm{Fru|w}\right)} \qquad \underbrace{RT\ln\frac{a_{c}(\mathrm{Suc|w}) \cdot a_{\bullet}(\mathrm{H_2O|l})}{a_{c}(\mathrm{Glc|w}) \cdot a_{c}(\mathrm{Fru|w})}}$$

Notice that we write the activity $a_{\bullet}$ and not a_{c} for the solvent. This is important because the indexes $_{\bullet}$ and $_{c}$, etc., are almost always omitted since it is generally clear from the context which one of these is meant. As before, the contributions of $\overset{*}{\mu}_{\bullet}$ or $a_{\bullet}$ are omitted for substances being converted in their pure states. Although they can be written for pure substances, $\overset{*}{\mu}_{\bullet} = 0$ and $a_{\bullet} = 1, \overset{*}{\mu}_{\bullet}$ as a summand or $a_{\bullet}$ as a factor does not affect the result. A simple example of this is dissolving cane sugar in water: $\mathrm{Suc|s} \to \mathrm{Suc|w}$:

$$\mathcal{A} = \overset{\circ}{\mathcal{A}} + \overset{*}{\mathcal{A}}$$

$$\underbrace{\overset{\circ}{\mu}_{\bullet}\left(\mathrm{Suc|s}\right) - \overset{\circ}{\mu}_{c}\left(\mathrm{Suc|w}\right)} \quad \underbrace{RT\ln\frac{a_{\bullet}(\mathrm{Suc|s})}{a_{c}(\mathrm{Sac|w})} = RT\ln\frac{1}{a_{c}(\mathrm{Suc|w})}}$$

The activity of the pure solid substance sucrose is, as mentioned, $a_{\bullet}(\mathrm{Suc|s}) = 1$.

The above can easily be transferred to other chemical transformations. For example:

$$\mathrm{B} + \mathrm{B}' + \ldots \to \mathrm{D} + \mathrm{D}' + \ldots \quad \text{or} \quad 0 \to v_{\mathrm{B}}\mathrm{B} + v_{\mathrm{B}'}\mathrm{B}' + \ldots + v_{\mathrm{D}}\mathrm{D} + v_{\mathrm{D}'}\mathrm{D}' + \ldots .$$

The conversion numbers $v_{\mathrm{B}}, v_{\mathrm{B}'}, \ldots, v_{\mathrm{D}}, v_{\mathrm{D}'}, \ldots$ are always negative for reactants and positive for products. In our example on the left they are only -1 or $+1$, while on the right, they can be random or even fractions. In our example and especially further below, we should note that the more general way of writing on the right, despite its very different appearance, leads to the simpler one on the left if the conversion number $+1$ for the reactants and -1 for the products are used and the quantities suitably transposed.

The negative drive $-\mathcal{A}$ can be written as a sum of the potentials μ of the participating substances, weighted with the conversion numbers [cf. Eq. (4.3)]:

$$-\mathcal{A} = -\mu_B - \mu_{B'} - \ldots + \mu_D + \mu_{D'} + \ldots$$

or in general

$$-\mathcal{A} = \nu_B \mu_B + \nu_{B'} \mu_{B'} + \ldots + \nu_D \mu_D + \nu_{D'} \mu_{D'} + \ldots \, .$$

If we decompose the potentials into basic and residual terms, $\mu = \overset{\circ}{\mu} + RT\ln a$, we obtain:

$$-\mathcal{A} = -\overset{\circ}{\mathcal{A}} - \overset{*}{\mathcal{A}}$$

$$= \left(-\overset{\circ}{\mu}_B - \overset{\circ}{\mu}_{B'} - \ldots + \overset{\circ}{\mu}_D + \overset{\circ}{\mu}_{D'} + \ldots\right) + RT\ln\frac{a_D a_{D'}\ldots}{a_B a_{B'}\ldots}. \qquad (24.36)$$

or in general

$$-\mathcal{A} = -\overset{\circ}{\mathcal{A}} - \overset{*}{\mathcal{A}}$$

$$= \left(\nu_B \overset{\circ}{\mu}_B + \ldots + \nu_D \overset{\circ}{\mu}_D + \ldots\right) + RT\ln\left(a_B{}^{\nu_B} \cdot \ldots \cdot a_D{}^{\nu_D} \cdot \ldots\right). \qquad (24.37)$$

"Reactivities" Just as we can convert the chemical potentials μ into chemical activities λ, we can transform sums of potentials $\mu_B + \mu_C + \mu_D + \ldots$ into products of activities $\lambda_B \cdot \lambda_C \cdot \lambda_D \cdot \ldots$:

$$\exp\frac{\mu_B + \mu_C + \mu_D + \ldots}{RT} = \exp\frac{\mu_B}{RT} \cdot \exp\frac{\mu_C}{RT} \cdot \exp\frac{\mu_D}{RT} \ldots = \lambda_B \cdot \lambda_C \cdot \lambda_D \cdot \ldots \, .$$

Multiples of potentials $\nu\,\mu$ can be similarly converted into powers of activities λ^ν:

$$\exp\frac{\nu\mu}{RT} = \left(\exp\frac{\mu}{RT}\right)^\nu = \lambda^\nu.$$

Drives $\mathcal{A}$ can also be similarly converted into activities, either in their entireties or decomposed into basic and residual values, $\mathcal{A} = \overset{\circ}{\mathcal{A}} + \overset{*}{\mathcal{A}}$, or into basic, mass action, and extra terms $\mathcal{A} = \overset{\circ}{\mathcal{A}} + \overset{\times}{\mathcal{A}} + \overset{+}{\mathcal{A}}$,

$$\underbrace{\exp\frac{\mathcal{A}}{RT}}_{\mathcal{K}} = \underbrace{\exp\frac{\overset{\circ}{\mathcal{A}}}{RT}}_{\overset{\circ}{\mathcal{K}}} \cdot \underbrace{\exp\frac{\overset{*}{\mathcal{A}}}{RT}}_{\overset{*}{\mathcal{K}}} = \underbrace{\exp\frac{\overset{\circ}{\mathcal{A}}}{RT}}_{\overset{\circ}{\mathcal{K}}} \cdot \underbrace{\exp\frac{\overset{\times}{\mathcal{A}}}{RT}}_{\overset{\times}{\mathcal{K}}} \cdot \underbrace{\exp\frac{\overset{+}{\mathcal{A}}}{RT}}_{\overset{+}{\mathcal{K}}} \cdot \tag{24.38}$$

Basically, the quantity $\mathcal{K}$ is a measure of the "drive" or "strength" of a reaction as much as the quantity $\mathcal{A}$, from which it stems. The only difference is that the scales are different and certain conditions are formulated differently. While $\mathcal{A} > 0$ means that the process runs forward, $\mathcal{A} < 0$ denotes a backward tendency, and $\mathcal{A} = 0$ indicates equilibrium, the corresponding conditions in the new, exponential scale are $\mathcal{K} > 1, \mathcal{K} < 1$, and $\mathcal{K} = 1$, respectively. In order to emphasize its relation to activity, $\mathcal{K}$ can be called "reactivity," but the name is actually unnecessary.

Mass Action Law The quantity $\mathcal{K}$ is itself extremely unusual, but the factors $\overset{\circ}{\mathcal{K}} \cdot \overset{*}{\mathcal{K}}$ or $\overset{\circ}{\mathcal{K}} \cdot \overset{\times}{\mathcal{K}} \cdot \overset{+}{\mathcal{K}}$, into which it can be decomposed, are not. When discussing the mass action law in Sect. 6.4, we encountered the quantity $\overset{\circ}{\mathcal{K}}$ as a "(numerical) equilibrium constant" or "equilibrium number" [Eq. (6.18)]. We are also familiar with $\overset{\times}{\mathcal{K}}$, not as itself but as the reciprocal value $\overset{\times}{\mathcal{K}}{}^{-1}$ (but we did not use this symbol as yet). Generally, $\overset{\times}{\mathcal{K}}{}^{-1}$ appears in the mass action law as a quotient where the numerator corresponds to the products and the denominator corresponds to the reactants.

This can be shown as follows. When the drive disappears, $\mathcal{A} = 0$, equilibrium is established, so that $\mathcal{K} = \overset{*}{\mathcal{K}} \cdot \overset{\circ}{\mathcal{K}} = 1$ or $\overset{\circ}{\mathcal{K}} = \overset{*}{\mathcal{K}}{}^{-1}$ (or $\overset{\circ}{\mathcal{K}} = \overset{\times}{\mathcal{K}}{}^{-1}\overset{+}{\mathcal{K}}{}^{-1}$) is valid. If we insert $\overset{*}{\mathcal{A}}$ from Eq. (24.36) or Eq. (24.37) into the expression for $\overset{*}{\mathcal{K}}$ in Eq. (24.38), we obtain the conditions for equilibrium in the following form:

$$\overset{\circ}{\mathcal{K}} = \left(\frac{a_{\mathrm{D}} \cdot a_{\mathrm{D}'} \cdot \ldots}{a_{\mathrm{B}} \cdot a_{\mathrm{B}'} \cdot \ldots}\right)_{\mathrm{eq.}} \quad \text{or} \quad \overset{\circ}{\mathcal{K}} = \left(a_{\mathrm{B}}{}^{\nu_{\mathrm{B}}} \cdot \ldots \cdot a_{\mathrm{D}}{}^{\nu_{\mathrm{D}}} \cdot \ldots\right)_{\mathrm{eq.}}$$

If all the substances B, B', …, D, D', … are dissolved components in a dilute solution, the activities $a = \gamma_c\, c_{\mathrm{r}}$ can be replaced by the relative concentrations c_{r}, because we have in the case of strong dilution $\gamma_c = 1$. The conditions above then give way to the equations familiar to us from Sect. 6.4:

$$\overset{\circ}{\mathcal{K}} = \left(\frac{c_{\mathrm{r}}(\mathrm{D}) \cdot c_{\mathrm{r}}(\mathrm{D}') \cdot \ldots}{c_{\mathrm{r}}(\mathrm{B}) \cdot c_{\mathrm{r}}(\mathrm{B}') \cdot \ldots}\right)_{\mathrm{eq.}} \quad \text{or} \quad \overset{\circ}{\mathcal{K}} = \left(c_{\mathrm{r}}(\mathrm{B})^{\nu_{\mathrm{B}}} \cdot \ldots \cdot c_{\mathrm{r}}(\mathrm{D})^{\nu_{\mathrm{D}}} \cdot \ldots\right)_{\mathrm{eq.}}$$

The expression in parentheses on the right side of the equation is just $\overset{\times}{\mathcal{K}}{}^{-1}$. In this case, $\overset{+}{\mathcal{K}} = 1$ is valid for the extra factor $\overset{+}{\mathcal{K}}$ in which the activity coefficients are summed up, because all activity coefficients, as mentioned, should be equal to 1. In the general case, however, we have:

$$\overset{+}{\mathcal{K}}{}^{-1} = \frac{\gamma_c(\mathrm{D}) \cdot \gamma_c(\mathrm{D}') \cdot \ldots}{\gamma_c(\mathrm{B}) \cdot \gamma_c(\mathrm{B}') \cdot \ldots} \quad \text{or} \quad \overset{+}{\mathcal{K}}{}^{-1} = \gamma_{c,\mathrm{B}}{}^{\nu_\mathrm{B}} \cdot \ldots \cdot \gamma_{c,\mathrm{D}}{}^{\nu_\mathrm{D}} \cdot \ldots \ .$$

The Special Case of Gases For a gas B in a mixture of gases, the partial pressure $p_\mathrm{B} = x_\mathrm{B} \cdot p$ or the relative partial pressure $p_{\mathrm{r},\mathrm{B}} = x_\mathrm{B} \cdot p/p$ are preferred measures of composition. Neither p_B nor $p_{\mathrm{r},\mathrm{B}}$ are themselves "chemically active" but a changed value $a_{p,\mathrm{B}} = \gamma_{p,\mathrm{B}} \cdot p_{\mathrm{r},\mathrm{B}}$ (according to general belief, this is the result of the interaction of the particles). Similar to the activities discussed above, $a_{c,\mathrm{B}|\mathrm{A}}$, $a_{x,\mathrm{B}|\mathrm{A}}, a_{b,\mathrm{B}|\mathrm{A}}, \ldots$ and $\gamma_{c,\mathrm{B}|\mathrm{A}}, \gamma_{x,\mathrm{B}|\mathrm{A}}, \gamma_{b,\mathrm{B}|\mathrm{A}}, \ldots$, the same scale transformation also produces $a_{p,\mathrm{B}}$ and $\gamma_{p,\mathrm{B}}$ from the corresponding potentials: $a_{p,\mathrm{B}} = \exp(\overset{*}{\mu}_{p,\mathrm{B}}/RT)$ and $\gamma_{p,\mathrm{B}} = \exp(\overset{+}{\mu}_{p,\mathrm{B}}/RT)$.

Still, there is an important difference here. A vacuum plays the role of solvent A as a medium in which the substances are dispersed. While A continues to exist even at finite pressures when no other substances are distributed within it, this is not true for a vacuum. Starting from a state of very low total pressure p_0, we extrapolate from $p_\mathrm{B} = x_\mathrm{B}\,p_0$ to $p_\mathrm{B} = x_\mathrm{B}\,p$ at constant temperature T and obtain a value which is used as basic value $\overset{\circ}{\mu}_{p,\mathrm{B}}$ of the chemical potential μ_B. Extrapolation is done along the ideal logarithmic curve $\mu = \mu_0 + RT\ln(p/p_0)$. Differently from the basic values $\overset{\circ}{\mu}_{c,\mathrm{B}|\mathrm{A}}, \overset{\circ}{\mu}_{x,\mathrm{B}|\mathrm{A}}, \overset{\circ}{\mu}_{b,\mathrm{B}|\mathrm{A}}, \ldots$ of the potential μ_B of a substance B in a solid or liquid mixed phase, $\overset{\circ}{\mu}_{p,\mathrm{B}}$ is not dependent upon pressure. It is also not dependent upon a solvent, because there is none.

In 1901, Gilbert Lewis suggested the name *"fugacity"*—meaning volatility—for the modified pressure $\gamma_{p,\mathrm{B}} \cdot p_\mathrm{B}$ (not $\gamma_{p,\mathrm{B}} \cdot p_{\mathrm{r},\mathrm{B}}$!) and gave it its own symbol f_B. Correspondingly, the quantity $\gamma_{p,\mathrm{B}}$ is also known as the *fugacity coefficient* (symbol ϕ_B). Lewis described this quantity as a "tendency to transition" or "tendency to escape." He was describing the tendency of a substance in one phase to go into another one, and especially as a gas to volatilize. The concept of *activity* also came from Lewis, who introduced it in 1908 as modified concentration. This made it possible to treat substances that are not noticeably volatile but are easily dissolved in water (urea, glycerin, cane sugar, etc.) according to the same paradigm used for fugacious substances.

Appendix

A.1. Foundations of Mathematics

A.1.1. Linear, Logarithmic, and Exponential Functions

In physical chemistry, *functions* generally describe the relations between various quantities. The simplest case would be how a quantity y depends upon another quantity x: $y = f(x)$. Here, x represents the *independent variable* that is pre-determined and varied in an experiment. y is the *dependent variable* (dependent upon x) whose changes are measured. A good image of the function can be gained by plotting a selection of pairs (x, y) resulting from $y = f(x)$ as points in an (x, y) coordinate system.

We can say that y is *linearly* dependent upon x when an increase of x by a fixed amount a (the starting value of x can be arbitrarily chosen) causes y to increase by a fixed amount b (Fig. A.1).

In order to make it easier to compare with other dependencies, we will sum it up briefly in one line:

$$y = f(x) \quad \text{is } linear \text{ if:} \quad x \to x + a \Rightarrow y \to y + b.$$

However, y is *logarithmically* dependent upon x if an increase of x ($x > 0$, but otherwise arbitrary) by a fixed factor α causes an increase of y by a fixed amount b (Fig. A.2); in short:

$$y = f(x) \quad \text{is } logarithmic \text{ if:} \quad x \to x \cdot \alpha \Rightarrow y \to y + b.$$

We can supplement these statements with another useful one. We say that y is exponentially dependent upon x when an increase of x by a fixed amount a causes an increase of y by a fixed factor β (Fig. A.3); in short:

© Springer International Publishing Switzerland 2016
G. Job, R. Rüffler, *Physical Chemistry from a Different Angle*,
DOI 10.1007/978-3-319-15666-8

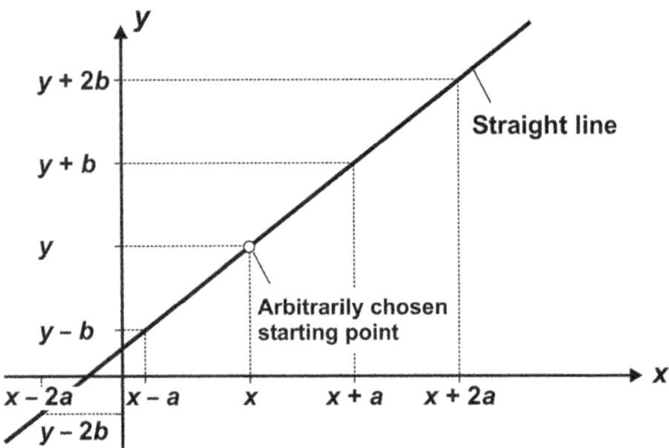

Fig. A.1 Linear relation between y and x.

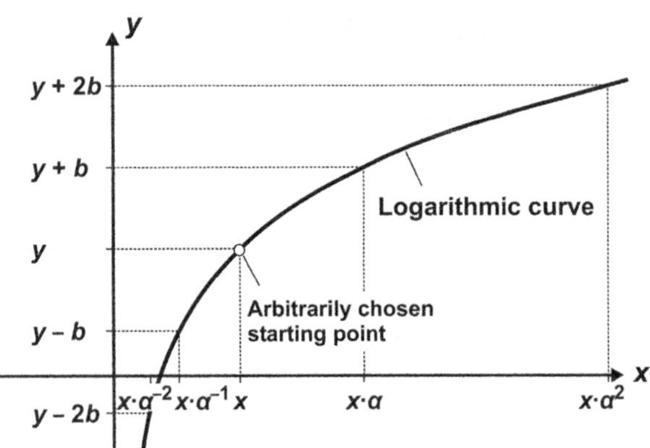

Fig. A.2 Logarithmic relation between y and x.

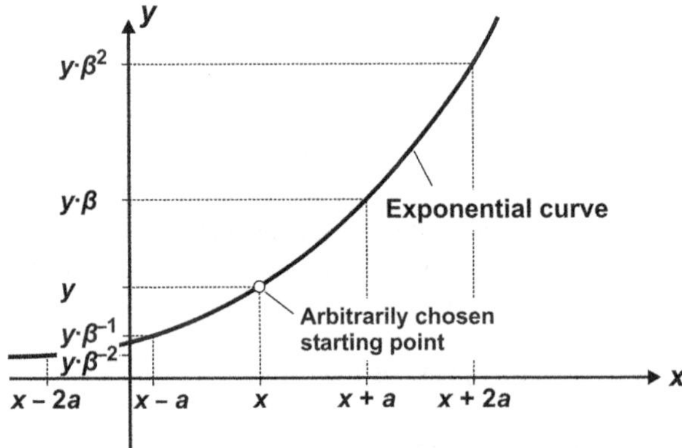

Fig. A.3 Exponential relation between y and x.

$$y = f(x) \quad \text{is exponential if:} \quad x \rightarrow x + a \Rightarrow y \rightarrow y \cdot \beta.$$

While we can usually distinguish between a linear "curve" (straight line) and a curve just by glancing at them, classifying curves into categories is much more difficult. The easy to check characteristics mentioned above help in determining what type of curve we are dealing with. For example, we can quickly check whether or not a given curve is logarithmic by choosing the most simple $\alpha = 2$ and, using an arbitrarily chosen starting point, repeatedly increasing or decreasing the abscissa by the same factor (as in Fig. A.2). Conversely, this characteristic can also serve to quickly sketch the curve.

Unfortunately, the mathematical expressions for the linear, logarithmic, and exponential relation do not allow us to see the similarity that is noticeable in the expressions mentioned above. We write

$$\begin{aligned}
y &= mx + n, &&\text{if } y \text{ depends linearly upon } x, \\
y &= \log_m \frac{x}{n}, &&\text{if } y \text{ depends logarithmically upon } x, \\
y &= m^x \cdot n, &&\text{if } y \text{ depends exponentially upon } x.
\end{aligned}$$

In the case of straight lines, m represents the *slope* and n is the *y-intercept*.

By contrast, m represents the *base* in logarithmic and exponential relations. The irrational *Euler's number* $e = 2.7182\ldots$ is often used as the base. When this is done, we speak of *natural logarithms*, abbreviated to ln, or (natural) *exponential functions*, which we call e-functions because of their relation to the number e. We then obtain:

$$\begin{aligned}
y &= \ln x \quad \text{or} \\
y &= e^x \equiv \exp x.
\end{aligned}$$

In closing, we will repeat a few rules for logarithmic expressions. When the base m is fixed (omitted for simplicity's sake), the following is valid:

$$\log(x \cdot y) = \log x + \log y, \tag{A.1}$$

$$\log(x/y) = \log x - \log y, \tag{A.2}$$

$$\log(x^a) = a \cdot \log x. \tag{A.3}$$

When the base changes, the general rule (with bases b and c) is

$$\log_b x = \log_b c \cdot \log_c x \tag{A.4}$$

which results in the special case for bases e (Euler's number) and 10

$$\log_e x = \log_e 10 \cdot \log_{10} x$$

or using different notation

$$\ln x = \ln 10 \cdot \lg x, \tag{A.5}$$

where the abbreviation lg stands for the *decadic logarithm* or *common logarithm* (logarithm to base 10).

A.1.2. Dealing with Differentials

Most of the functions we deal with in physical chemistry are, casually expressed, "user friendly." Their graphs are almost always smooth curves without jumps, bends, or gaps so that a few points are enough to indicate a complete picture (compare the graphs in Sect. A.1.1). The interesting points on such graphs are often the zero points, maxima, minima, and inflection points as well as intersection points with other curves. Elementary mathematics gives us the tools to calculate the coordinates of such points, at least in simpler cases.

One important intermediate step to getting there is calculating the slope m of a graph at a given point x (Fig. A.4a). The "user-friendly" (differentiable) functions which we almost always deal with are represented by graphs that, if we look at them with a magnifying glass, are almost straight in a small neighborhood around a point x. Enlarging this section even more would show no discernible curvature at all (Fig. A.4b). In physical chemistry, we generally choose a representation where the curvature is still discernible, but so weak that the observer can easily imagine what the results would be if it disappeared altogether.

This trick proves to be extremely useful. We know how to calculate the slope of a straight line: $m = \Delta y / \Delta x$, where $\Delta y = y_2 - y_1$ indicates the increase in "height" if one proceeds a distance $\Delta x = x_2 - x_1$ from location x_1 to x_2. In order to express that, for a curve, the increase in the x direction must be very small or even infinitesimally

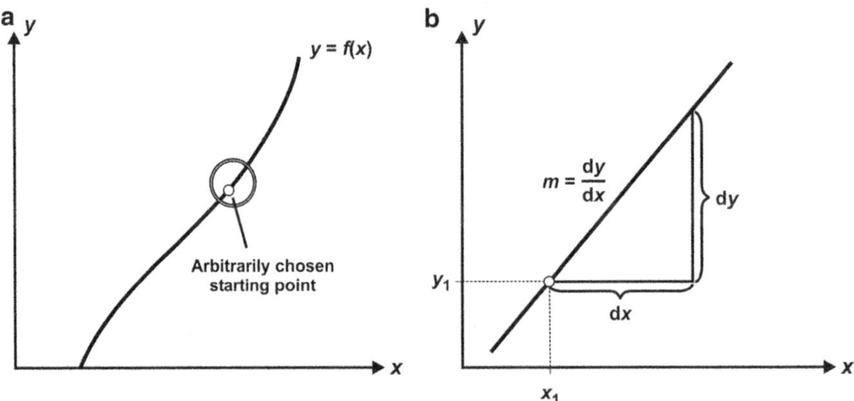

Fig. A.4 (a) Graphic representation of an arbitrary functional relation $y = f(x)$ in the (x, y) coordinate system, (b) Strongly magnified section around the point (x_1, y_1).

small, the symbol Δ for the difference is replaced by the symbol d for the differential: $m = \mathrm{d}y/\mathrm{d}x$. This was Gottfried Wilhelm Leibniz's original idea when he chose this suggestive syntax. We do not need to be bothered by the fact that differentials in mathematics today are introduced as limits. It is enough for us to utilize the procedures described in order to gain quick access to the necessary sophisticated mathematical "tools."

A specific slope m corresponds to each value of x, so the slope itself represents a function of x which is naturally different from $f(x)$. In order to distinguish the two, as well as to show where it comes from, it is indicated by an apostrophe $'$ (Langrange's notation). We then write $y' = f'(x)$, by replacing m by $y' (= \mathrm{d}y/\mathrm{d}x)$. $f'(x)$ is the *derivative* of the function $f(x)$. There are numerous rules in mathematics for finding the *derived function* $f'(x)$ corresponding to a given $f(x)$ as well as methods for finding the *antiderivative* $f(x)$ for a given function $f'(x)$ (compare Sect. A.1.3).

Let us remember the most important derivatives and rules for calculating them. When taking a derivative, the degree of a power function is always lowered by one,

$$y = x^n \quad \Rightarrow \quad y' = nx^{n-1}. \tag{A.6}$$

The exponential function, on the other hand, agrees with its derivative,

$$y = e^x \quad \Rightarrow \quad y' = e^x, \tag{A.7}$$

and in the case of the natural logarithm, we obtain

$$y = \ln x \quad \Rightarrow \quad y' = \frac{1}{x} \quad (x > 0). \tag{A.8}$$

A *constant factor* k is preserved when taking the derivative:

$$y = k \cdot f(x) \quad \Rightarrow \quad y' = k \cdot f'(x). \tag{A.9}$$

Sums (as well as differences) of two or more functions can be derived term by term:

$$y = f(x) \pm g(x) \quad \Rightarrow \quad y' = f'(x) \pm g'(x). \tag{A.10}$$

The so-called *product rule* is applied to the product of two functions,

$$y = f(x) \cdot g(x) \quad \Rightarrow \quad y' = f'(x) \cdot g(x) + f(x) \cdot g'(x), \tag{A.11}$$

and the *quotient rule* is applied to quotients,

$$y = \frac{f(x)}{g(x)} \quad \Rightarrow \quad y' = \frac{f'(x) \cdot g(x) - f(x) \cdot g'(x)}{(g(x))^2}. \tag{A.12}$$

The *chain rule* describes how derivatives of composite (or nested) functions can be taken. The simplest chain is composed of an *outer function* $f(z)$ with an *inner function* $z = g(x)$, which can be written as $y = f(g(x))$. The following is valid:

$$y = f(g(x)) \quad \Rightarrow \quad y' = f'(z) \cdot g'(x). \tag{A.13}$$

More simply: The derivative of the total function is the product of the derivative of the outer function and the derivative of the inner function.

Instead of using Langrange's notation, the above rules are often formulated in thermodynamics by the use of differentials. We then have for two functions f and g:

$$d(f \pm g) = df \pm dg,$$
$$d(f \cdot g) = g\,df + f\,dg,$$
$$d\left(\frac{f}{g}\right) = \frac{g\,df - f\,dg}{g^2},$$

and finally for a function $f = f(g)$, where $g = g(x)$:

$$\frac{df}{dx} = \frac{df}{dg} \cdot \frac{dg}{dx}.$$

In this (Leibniz) form, the structure of the chain rule is emphasized more clearly.

In the field of physical chemistry, the quantity y we have been considering is generally dependent upon *several* quantities $y = f(x_1, x_2 \ldots)$, but in the simplest case, just two: $y = f(u, v)$. The graph of such a function, which is represented in a triaxial (u, v, y) coordinate system, is no longer a curve but a surface (Fig. A.5a). If we are dealing with "user-friendly" functions, and we almost always are, the surface in question will be smooth although bent but without holes, folds, or jumps. In an arbitrarily chosen point P, this kind of surface can rise sharply in the direction parallel to the u-axis. At the same time, it might only rise slightly, run horizontally, or even fall in the direction parallel to the v-axis. If we look at it through a magnifying glass, a small enough section around point P appears plane, with no visible curvature (Fig. A.5b). If the gradients $m_{\to u}$ and $m_{\to v}$ in the direction of the u- and v-axes are known (the arrow in the index indicates an increase in a certain direction), it is then possible to calculate the increase of Δy when proceeding in the u direction by Δu as well as in the v direction by Δv, at least when the increase is so small that the surface in the area can be considered flat. This is illustrated in Fig. A.5b. In order to indicate the increase, we have again replaced the differences with differentials:

$$dy = m_{\to u} \cdot du + m_{\to v} \cdot dv.$$

If we are interested in the increase in the u direction, we need to insert $dv = 0$ into the expression above and obtain $dy = m_{\to u} \cdot du$ or, when transformed, $m_{\to u} = dy/du$. In order to express that the result is valid only when v is kept constant, we write:

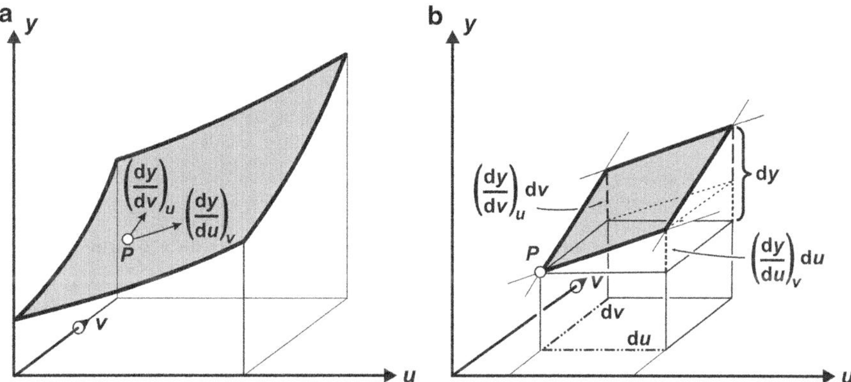

Fig. A.5 (**a**) Graphic representation of the functional relation $y = f(u, v)$ as a surface in the (u, v, y) coordinate system, (**b**) Magnified section around point P.

$$m_{\to u} = \left(\frac{\mathrm{d}y}{\mathrm{d}u}\right)_{v} \quad \text{and correspondingly} \quad m_{\to v} = \left(\frac{\mathrm{d}y}{\mathrm{d}v}\right)_{u}.$$

(The expression in parentheses is to be read as "dy with respect to du at constant v").

The use of the rounded parentheses has become standard in physical chemistry. They can also be used for the differential dy where we then write $(\mathrm{d}y)_v = m_{\to u} \cdot (\mathrm{d}u)_v$ to save us the trouble of writing $\mathrm{d}v = 0$. In our example, both u and v appear as independent variables, so $(\mathrm{d}u)_v \equiv \mathrm{d}u$. In this case, the two expressions with and without parentheses are the same.

If the function $y = f(u, v)$ is known, the two slopes $m_{\to u}$ and $m_{\to v}$ can be calculated by referring to the usual rules for determining derivatives. In the case of $m_{\to u}$, only u is considered variable, while v is treated as a constant parameter. We are used to this with functions of one variable when fixed parameters appear in the formulas.

What appears here to be a slope of a graph is, as a rule, in physical chemistry a quantity that quantifies an observable characteristic and for which there is already a symbol. Such signs might be $\alpha \equiv m_{\to u}$ and $\beta \equiv m_{\to v}$. The quantities α and β are themselves functions of u and v. We could therefore write $\alpha = f'_u(u, v)$ and $\beta = f'_v(u, v)$ to show that they derive from the function $y = f(u, v)$.

$f'_u(u, v)$ and $f'_v(u, v)$ are called *partial derivatives* of the function $f(u, v)$. There are numerous similar, less clear-cut ways of expressing this, including ones that relate to calculations of slopes as quotients of two differentials. They are all called *partial differential quotients*:

$$\alpha = \frac{\partial f(u,v)}{\partial u} \;\; = \frac{\partial y(u,v)}{\partial u} \;\; = \left(\frac{\partial y}{\partial u}\right)_v \;\; = \frac{(\mathrm{d}y)_v}{\mathrm{d}u},$$

$$\beta = \frac{\partial f(u,v)}{\partial v} \;\; = \frac{\partial y(u,v)}{\partial v} \;\; = \left(\frac{\partial y}{\partial v}\right)_u \;\; = \frac{(\mathrm{d}y)_u}{\mathrm{d}v}.$$

Mathematics   Physics   Chemistry

Writing the rounded ∂ instead of the straight d should remind us that in the numerator of the expressions above, we mean only the increase $(\mathrm{d}y)_v$, i.e., when u changes by $\mathrm{d}u$, while all other variables (only v in this case) remain unchanged. This is correspondingly valid for the expression below. In mathematics, the first notation is the preferred one. The second one is preferred in physics, and in (physical) chemistry, it is the third one. In the latter notation, the rounded ∂ can be replaced by the straight d, without anything changing. This fourth form is especially suited as an intermediate step in converting various differential quotients one into another. In such calculations, one starts with the *complete* differential $\mathrm{d}y$, only later taking the appropriate side conditions into account:

$$\mathrm{d}y = \left(\frac{\mathrm{d}y}{\mathrm{d}u}\right)_v \mathrm{d}u + \left(\frac{\mathrm{d}y}{\mathrm{d}v}\right)_u \mathrm{d}v. \tag{A.14}$$

We will close our discussion of this here. We will have the opportunity of dealing with examples of this method often enough. Some important rules for converting differential quotients have been compiled in Sect. 9.4.

A.1.3. Antiderivatives and Integration

Knowing how to deal with differentials plays an essential role in physical chemistry because many quantities there are related to each other through expressions containing them. It is therefore important to not deal only with taking derivatives, but with the inverse operation as well. This means finding—for a given function $f(x)$—the *antiderivative* $F(x)$ whose derivative results in $f(x)$,

$$\frac{\mathrm{d}F(x)}{\mathrm{d}x} = f(x).$$

We obtain first examples of antiderivatives by simply reversing the arrows in the expressions (A.6) to (A.8). $F(x) = \ln x$ is then an antiderivative of $f(x) = 1/x$. However, it is only possible to determine an antiderivative up to a constant additive term C because it drops out when taking the derivative. Along with $F(x)$, all functions $F(x) + C$ are an antiderivative of $f(x)$.

For reasons we will go into below, an antiderivative is also called an *indefinite integral*. Determining it by reversing the derivative is known as *indefinite integration*. The summand C is the so-called *constant of integration*.

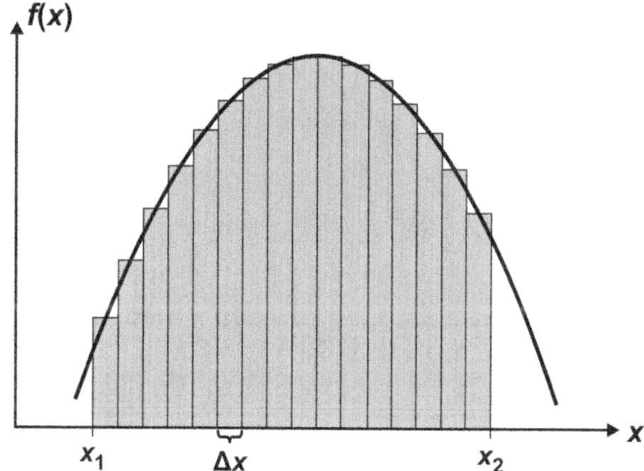

Fig. A.6 Approximation of the definite integral of $f(x)$ from x_1 to x_2 as the area under the curve.

Let us consider a problem that, at first glance, does not appear to have anything to do with finding an antiderivative, namely determining the "area under a curve" of an arbitrary function $y = f(x)$. This is the area that is bordered by the curve between two limits x_1 and x_2 and the x-axis (Fig. A.6). We obtain approximations of surface area A by dividing the surface into strips having width Δx, each one limited by a horizontal line at its function value and then adding the areas $f(x)\Delta x$ of these strips:

$$A \approx \sum f(x)\Delta x.$$

We use the Greek letter $\sum$ as the symbol for calculating a sum.

If very small interval widths dx are used, we obtain the exact value for A:

$$A = \int_{x_1}^{x_2} f(x)dx.$$

Here, we are speaking of a *definite integral* of the function between the limits x_1 and x_2 (in short: "Integral of $f(x)$ with respect to x from x_1 to x_2"). The elongated S used as the symbol indicates the underlying summation and is based upon the work of the German mathematician and philosopher Gottfried Wilhelm von Leibniz.

Let us return to a "user-friendly" function, i.e., the differentiable function $y = F(x)$ (Fig. A.7a). We will connect the function values at intervals Δx by straight lines and obtain an approximation (gray dotted line) of the function. The gradient m of such a connecting line then results in

$$m = \frac{\Delta y}{\Delta x},$$

where Δy indicates the function's increase of "height" in the chosen interval.

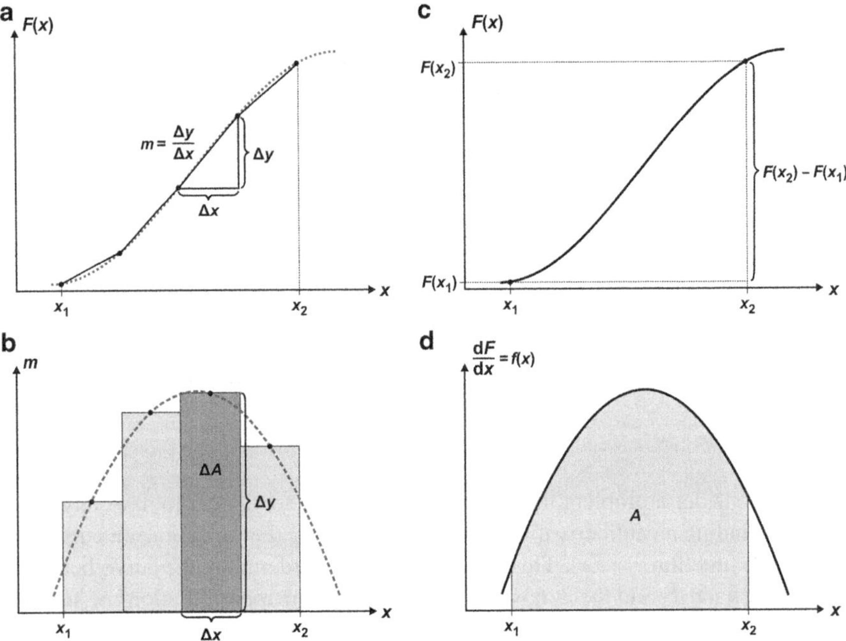

Fig. A.7 Relation between the increase of the antiderivative $F(x)$ (*above*) and the area (integral) under the curve $dF/dx = f(x)$ (*below*); (**a, b**) Approximate, (**c, d**) After the limiting process.

As we see in Fig. A.7b, a "column" is erected above each interval with the width Δx. Its height corresponds to the appropriate gradient m in graph A.7a. We then obtain the following for the area of the rectangle:

$$\Delta A = m \cdot \Delta x,$$

this means that it corresponds to the distance Δy (the terms "surface area" and "distance" are meant figuratively because the variables, and the functions, are normally linked to units). If all the intervals between x_1 and x_2 are combined, the entire area under the *histogram* (compare Sect. A.1.4 for the term) equals the increase $F(x_2) - F(x_1)$ of the function $F(x)$.

If we finally allow the interval width to decrease, the broken line will change into the correct curve $F(x)$. Its gradient will correspond to the derivative $F'(x) = dF/dx$ (Fig. A.7c). The surface area of the histogram merges with the surface under the curve $dF/dx = f(x)$ and can be described by the definite integral (Fig. A.7d):

$$\int_{x_1}^{x_2} f(x)dx.$$

The surface continues to correspond to the increase of the function $F(x)$. This means that the definite integral of $f(x)$ with respect to x between the limits x_1 and x_2 corresponds to the difference of function values $F(x)$ at these integration limits:

$$\int_{x_1}^{x_2} f(x)dx = F(x_2) - F(x_1),$$

where $f(x) = dF(x)/dx$. At the end, the definite integral is a real number value. $F(x)$ is nothing more than the antiderivative of the function $f(x)$, which has been integrated. The relation between the terms antiderivative and (definite) integral now becomes clear. The indefinite integral, however, in which no limits of integration are specified (therefore the name) is a function, more precisely, an infinite number of functions which differ only by a constant.

We will briefly compile the antiderivatives of some elementary functions in the following.

When we integrate a power function we obtain

$$y = x^n \quad \Rightarrow \quad \int ydx = \frac{1}{n+1}x^{n+1} + C \quad (n \neq 1), \tag{A.15}$$

so that, for example,

$$\int \frac{1}{x^2}dx = -\frac{1}{x} + C.$$

In the rule above, the exponent $n = 1$ is not allowed because the function $f(x) = 1/x$ is not the derivative of the function $F(x) = x^0$, but is the derivative of the function $F(x) = \ln x$ as we have seen in Sect. A.1.2. More precisely,

$$y = \frac{1}{x} \quad \Rightarrow \quad \int ydx = \ln|x| + C, \tag{A.16}$$

since this rule is also valid for negative x values. The antiderivative for $x > 0$ is $F(x) = \ln x$, and for $x < 0$, it is $F(x) = \ln(-x)$.

The following holds for the exponential function:

$$y = e^x \quad \Rightarrow \quad \int ydx = e^x + C. \tag{A.17}$$

There are general rules for integration, just as there are general rules for differentiation. We will discuss the most important ones in closing.

A *constant factor k* is conserved during integration:

$$y = k \cdot f(x) \quad \Rightarrow \quad \int y\,dx = k \int f(x)\,dx. \qquad (A.18)$$

In cases of *sums* (or differences) of two or more functions, each term is integrated individually:

$$y = f(x) \pm g(x) \quad \Rightarrow \quad \int y\,dx = \int f(x)\,dx \pm \int g(x)\,dx. \qquad (A.19)$$

When integrating nested functions, it is often a good idea to define the inner function as a new variable (*substitution rule*). We wish to determine the indefinite integral of the function

$$y = f(g(x)) \quad \text{so we insert} \quad g(x) = z$$

as a variable. The inverse function belonging to $z = g(x)$ is then $x = g^{-1}(z) = \varphi(z)$. The differential dx must also be replaced by a differential of the new variable, meaning dz. The necessary relation is achieved by taking the derivative:

$$\varphi'(z) = \frac{dx}{dz} \quad \text{and therefore} \quad dx = \varphi'(z)dz \,.$$

We then obtain:

$$\int y\,dx = \int f(z) \cdot \varphi'(z)\,dz. \qquad (A.20)$$

The substitution rule is the reverse of the chain rule of differential calculus. The method looks more complicated than it actually is. Let us consider an example: We are looking for an antiderivative of

$$\int (3x + 4)^2 dx.$$

We now set $z = 3x + 4$ and by differentiation obtain either $dx/dz = 1/3$ or $dx = dz/3$. According to the substitution rule, the result is:

$$\int (3x + 4)^2 dx = \int \tfrac{1}{3}z^2 dz = \tfrac{1}{9}\, z^3 = \tfrac{1}{9}(3x + 4)^3.$$

These general rules are also applicable when calculating definite integrals, but in the case of substitution, the integration limits must be adapted accordingly.

A.1.4. Short Detour into Statistics and Probability Calculation

Statistics is a body of mathematical methods used for analyzing large amounts of data. The goal of statistics is to condense the data in a way that allows us to arrive at statements regarding the basic principles and structures of the data.

Important parameters are *average values* which can be calculated from a series of attribute values x_i such as measured values of a (random) sample. The most commonly used average value is the *arithmetic mean* $\bar{x}_{\mathrm{arithm.}}$, which is defined as the sum of all values divided by the number N of the values:

$$\bar{x}_{\mathrm{arithm.}} = \frac{x_1 + x_2 + \ldots + x_N}{N}.$$

This can be expressed with the help of the sigma sign:

$$\bar{x}_{\mathrm{arithm.}} = \frac{1}{N}\sum_{i=1}^{N} x_i. \tag{A.21}$$

Let us take a closer look at the measured values of the random sample. There is a certain scatter caused by statistical errors, but most lie around the mean value with only a few exhibiting greater deviations. In order to gain a better idea of this distribution of measured values, we divide the data field between x_{min} and x_{max} into uniformly wide bins (*classes*) and assign each value to its corresponding class. This method allows us to find the *frequency distribution* of the attribute values described by the *absolute frequency* N_i of the results in class i, or their relative frequency p_i,

$$p_i = \frac{N_i}{N}.$$

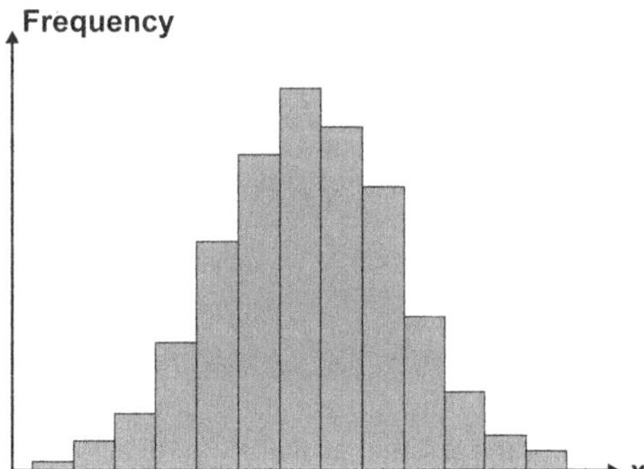

Fig. A.8 Histogram of a discrete frequency distribution.

The form of distribution is often illustrated by a *histogram* (Fig. A.8) where a "column" is erected over each class, whose area is proportional to the frequency in question.

We will now turn to the question of what kind of frequency distribution theoretically results from random (statistical) fluctuations. To do this, we need some basic concepts from the theory of *probability*, which deal with randomly distributed quantities and events. Each result of a measurement can be understood as an event. In order to quantify a random result or event E—meaning an event whose occurrence under given circumstances is uncertain, i.e., neither certain nor impossible—it is assigned a certain value, i.e., its *probability* p(E). This indicates the relative frequency of an event occurring if there are enough trials (in the limit, an infinite number, $n \to \infty$, law of large numbers). p(E) lies between 0 (impossible) and 1 (100 %, meaning certain) including the limits, $0 \le p(E) \le 1$. The sum of the probabilities of all possible events must equal 1.

If measured quantities are *continuously variable*, which is often the case, the concept of frequency must be adapted accordingly. To do this, we can imagine the range of variable x being divided into small intervals $[x, x+\Delta x]$. The relative frequency then results in

$$p(x) = \frac{1}{N} \frac{\Delta N(x)}{\Delta x},$$

where $\Delta N(x)$ is the number of results in the corresponding interval. If we assume a quantity having a dimension, $p(x)$ would have a dimension of $1/x$. In this case, it is better to speak of *density* of the relative frequency. Let us now make the interval Δx very small ($\Delta x \to 0$ in the limit), which we express by using the differential quotient instead of the difference quotient:

$$p(x) = \frac{1}{N} \frac{dN(x)}{dx}.$$

The now *continuous* distribution is represented by the probability density function $p(x)$.

The probability density function of the most well-known and most utilized distribution, the so-called *normal distribution*, is given by

$$p(x) = \frac{1}{\sqrt{2\pi} \cdot \sigma} \exp\left(-\frac{(x-\mu)^2}{2\sigma^2} \right), \tag{A.22}$$

where μ is the expected value (expectation) and σ the standard deviation. This is based upon work by the German mathematician and physical scientist Carl Friedrich Gauss in the nineteenth century and is also called the Gaussian distribution.

The graph of the probability density function is a *bell-shaped curve* that is symmetrical to the value of μ and whose form is determined by the parameter σ (Fig. A.9). The maximum of the curve is found at μ and has a height given by the

Fig. A.9 Normal distribution: probability density for a continuously variable quantity.

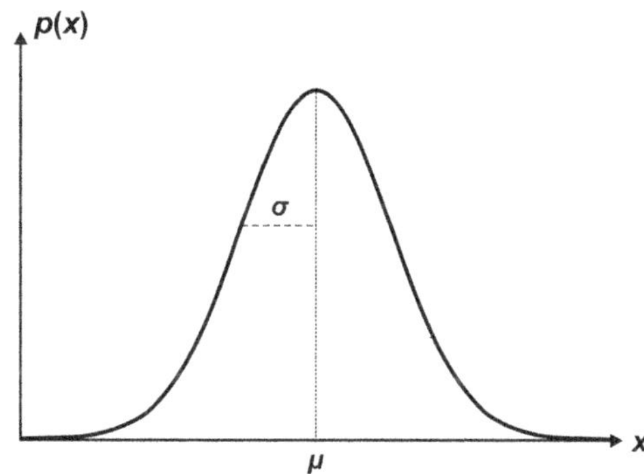

prefactor. The two inflection points of the curve are found at locations $\mu - \sigma$ and $\mu + \sigma$. Because σ is in the denominator of the prefactor, the curve becomes wider and flatter as σ increases. The area under the curve is constant and equals 1 because it is the sum of the relative frequencies of all possible events and therefore corresponds to the probability of a definite event.

Under certain conditions that are approximately fulfilled in practice, we can *expect* that the distribution of continuously variable measured values of x around a central value μ can be described by a normal distribution as a result of numerous independent random disturbances. This is where the expression "expected value" for the center of the distribution comes from. $x - \mu$ represents the deviation of the measured value from the expected value whose spread is determined by the parameter σ (called the "standard deviation"). A low standard deviation, for example, indicates that the measured values tend to be very close to the expected value; a high standard deviation, however, indicates that the measured values are widely spread.

The normal distribution is valid if there are many (theoretically infinitely many) measurements. In practice, there is always a finite number N of measured values. They represent a random selection of the (infinite) number of possible values of the entire population. In statistics, we refer to this as a "random sample." The arithmetic mean of a random sample taken from a normally distributed population (discussed at the beginning of the section) is a suitable estimate of the expected value.

A.2. Tables

A.2.1. Table of Chemical Potentials

The following Table contains values of the chemical potentials μ and corresponding temperature coefficients α for more than 400 inorganic and organic substances, compiled from the sources listed in the bibliography.

The scale is defined by:

- $\mu = 0$ for the elements in their most stable modification (except phosphorus) in the standard state, omitting nuclear entropy.
- $\mu = 0$ for $H^+|w$ in the standard state, entropy of $H^+|w$ set at zero.

The data are valid

- At standard conditions ($T = 298.15$ K, $p = 100$ kPa),
- For a dissolved substance with the standard value of concentration $(1{,}000$ mol m$^{-3})$,
- For a gaseous or dissolved substance in the ideal state without interaction of the spatially distributed molecules,
- For all substances for the elements in their natural isotopic compositions.

Some tips: The element symbols in the content formulas are arranged in the following sequence (with decreasing rank):

- Electropositive elements (metals, noble gases),
- Electronegative elements (nonmetals except noble gases, O, H),
- Oxygen,
- Hydrogen.

Water can be found under OH_2, for example, and sulfuric acid under SO_4H_2.

References

1. Landolt-Börnstein, New Series (1999–2001) Thermodynamical Properties of Inorganic Materials, Part 1 to 4, Vol IV/19. Springer, Berlin
2. Wagman DW, Evans WH, Parker VB, Schumm RH, Halow I, Bailey SM, Churney KL, Nuttal RL (1982) The NBS Tables of Chemical Thermodynamical properties. J Phys Chem Ref Data 11 (Suppl 2)
3. Bahrin I (1995) Thermochemical Data of Pure Substances. VCH, Weinheim
4. Alberty RA (1998) Calculation of Standard Transformed Gibbs Energies and Standard Transformed Enthalpies of Biochemical Reactants. Arch Biochem Biophys 353:116–130 and corresponding references
5. Chase MW, Davies CA, Downey JR, Frurip DJ, McDonald RA, Syverud AN (1985) JANAF Thermochemical Tables, 3rd edn. J Phys Chem Ref Data 14 (Suppl 1)
6. Landolt-Börnstein (1961) Kalorische Zustandsgrößen, 6th edn, Vol II/4. Springer, Heidelberg
7. Calculated from various data (acid constants ...)

Substance	Phase	μ (kG)	α (G K^{-1})	Refs.
Ag	g	246.01	−173.00	[1]
Ag	s	0.00	−42.55	[1]
Ag$^+$	w	77.11	−72.68	[2]
AgBr	s	−96.97	−107.11	[1]
AgCl	s	−109.82	−96.23	[1]
AgI	s	−66.35	−115.48	[1]
AgNO$_2$	s	19.13	−128.20	[2]
AgNO$_3$	s	−33.41	−140.92	[2]
Ag$_2$O	s	−11.25	−121.00	[1]
Ag$_2$S	s	−40.46	−142.89	[1]
Ag$_2$SO$_4$	s	−617.95	−200.41	[1]
Al	g	289.38	−164.55	[1]
Al	l	7.20	−39.55	[5]
Al	s	0.00	−28.30	[1]
Al^{3+}	w	−485.00	+321.70	[2]
AlCl$_3$	s	−630.01	−109.29	[1]
Al$_2$O$_3$	s, α, corundum	−1,582.26	−50.94	[1]
Ar	g	0.00	−154.84	[1]
As	g	260.46	−174.21	[3]
As	s, α, gray	0.00	−35.69	[1]
Au	g	328.84	−180.51	[1]
Au	s	0.00	−47.49	[1]
Au$_2$O$_3$	s	77.86	−130.33	[3]
B	g	521.01	−153.44	[1]
B	s	0.00	−5.90	[1]
Ba	g	146.94	−170.25	[5]
Ba	s	0.00	−62.50	[1]
Ba^{2+}	w	−560.77	−9.60	[2]
BaCO$_3$	s	−1,135.33	−112.10	[1]
BaCl$_2$	s	−806.94	−123.70	[1]
BaI$_2$	s	−602.00	−165.20	[1]
BaO	s	−520.25	−72.00	[1]
BaS	s	−456.00	−78.20	[2]
BaSO$_4$	s	−1,347.86	−132.10	[1]
Be	g	286.20	−136.27	[1]
Be	s	0.00	−9.50	[1]
Bi	g	169.90	−187.01	[1]
Bi	s	0.00	−56.74	[1]
Br	g	82.38	−175.02	[1]
Br$^-$	w	−104.00	−82.40	[2]
BrH	g	−53.40	−198.70	[1]
Br$_2$	g	3.11	−245.47	[1]
Br$_2$	l	0.00	−152.21	[1]

(continued)

Substance	Phase	μ (kG)	α (G K^{-1})	Refs.
C	g	671.26	−158.10	[1]
C	s, diamond	2.90	−2.36	[1]
C	s, graphite	0.00	−5.74	[1]
CCl_2O	g, carbonyl dichloride (phosgene)	−204.6	−283.53	[2]
CCl_4	g, tetrachloromethane	−58.15	−310.23	[3]
CCl_4	l, tetrachloromethane	−62.54	−216.19	[3]
CF_4	g, tetrafluoromethane	−888.52	−261.45	[1]
$CHCl_3$	l, trichloromethane (chloroform)	−73.66	−201.70	[2]
CH_2Cl_2	l, dichloromethane	−67.26	−177.80	[2]
CH_2O	g, methanal (formaldehyde)	−109.87	−218.77	[3]
CH_2O	g, methanoic acid (formic acid)	−350.97	−248.85	[3]
CH_2O	l, methanoic acid (formic acid)	−361.37	−128.95	[3]
CH_3	g, methyl	148.63	−194.00	[1]
CH_3Cl	g, monochloromethane	−58.34	−234.39	[1]
CH_4	g, methane	−50.53	−186.37	[1]
CH_4N_2O	s, diaminomethanal (urea)	−197.33	−104.60	[2]
CH_4O	g, methanol	−162.30	−239.87	[1]
CH_4O	l, methanol	−166.25	−126.70	[1]
CN^-	w	172.40	−94.10	[2]
CNH	g, hydrogen cyanide	124.70	−201.78	[2]
CNH	l, hydrogen cyanide	124.97	−112.84	[2]
CO	g	−137.17	−197.67	[1]
CO_2	g	−394.37	−213.78	[1]
CO_2	w	−385.98	−117.60	[2]
CO_3^{2-}	w	−527.81	+56.90	[2]
CO_3H^-	w	−586.77	−91.20	[2]
CO_3H_2	w	−623.08	−187.40	[2]
CS_2	l	65.13	−151.36	[1]
C_2H_2	g, ethyne (acetylene)	209.88	−200.93	[1]
$C_2H_2O_2$	s, ethanedioic acid (oxalic acid)	−697.97	−120.10	[6]
$C_2H_3O_2^-$	w, acetate anion	−369.31	−86.60	[2]
C_2H_4	g, ethene (ethylene)	68.36	−219.32	[1]
C_2H_4O	g, ethanal (acetaldehyde)	−133.24	−264.33	[3]
C_2H_4O	l, ethanal (acetaldehyde)	−128.12	−160.20	[2]
$C_2H_4O_2$	l, ethanoic acid (acetic acid)	−389.23	−159.83	[1]
$C_2H_4O_2$	w, ethanoic acid (acetic acid)	−396.46	−178.70	[2]
C_2H_5Cl	g, monochloroethane	−60.39	−276.00	[2]
$C_2H_5O_2N$	s, 2-aminoethanoic acid (glycine)	−368.44	−103.51	[2]
C_2H_6	g, ethane	−32.01	−229.16	[1]
C_2H_6O	g, ethoxyethane(dimethylether)	−112.59	−266.38	[2]
C_2H_6O	g, ethanol	−167.87	−281.62	[1]
C_2H_6O	l, ethanol	−174.63	−160.71	[1]
$C_2H_6O_2$	l, ethane-1.2-diol (ethylene glycol)	−323.23	−166.94	[1]

(continued)

Substance	Phase	μ (kG)	α (G K^{-1})	Refs.
C_3H_4	g, propyne	194.49	-248.22	[3]
$C_3H_5O_3{}^-$	w, lactate anion	-516.72	-146.44	[4]
C_3H_6	g, propene	62.82	-267.05	[3]
C_3H_6	g, cyclopropane	104.28	-238.01	[3]
C_3H_6O	l, 2-propanone (acetone)	-155.26	-200.41	[3]
$C_3H_6O_3$	w, 2-hydroxypropanoic acid (lactic acid)	-538.77	-221.75	[4]
C_3H_8	g, propane	-23.37	-270.02	[3]
C_4H_8	g, 1-butene	203.11	-290.90	[3]
$C_4H_8O_2$	l, ethyl ethanoate (ethyl acetate)	-323.19	-259.00	[6]
C_4H_{10}	g, butane	-16.99	-310.23	[3]
C_5H_{10}	l, cyclopentane	-36.49	-204.10	[6]
C_5H_{12}	g, pentane	-8.18	-349.06	[3]
C_5H_{12}	l, pentane	-9.21	-262.70	[6]
C_6H_5Cl	l, chlorobenzene	-93.65	-194.10	[6]
C_6H_6	g, benzene	129.79	-269.31	[1]
C_6H_6	l, benzene	125.05	-171.54	[1]
C_6H_6O	s, benzenol (phenol)	-50.22	-144.01	[3]
C_6H_6N	l, benzeneamine (aniline)	147.58	-192.00	[6]
C_6H_{12}	l, cyclohexane	26.89	-204.35	[3]
$C_6H_{12}O_6$	s, β-fructose (fruit sugar)	-905.65	-212.74	[4]
$C_6H_{12}O_6$	w, fructose (fruit sugar)	-915.51	-279.65	[4]
$C_6H_{12}O_6$	s, α-D-glucose (grape sugar)	-910.56	-212.13	[4]
$C_6H_{12}O_6$	s, β-D-glucose (grape sugar)	-908.89	-228.03	[4]
$C_6H_{12}O_6$	w, α-D-glucose (grape sugar)	-914.54	-264.01	[4]
$C_6H_{12}O_6$	w, β-D-glucose (grape sugar)	-915.79	-264.01	[4]
$C_6H_{12}O_6$	w, α, β-D-glucose (grape sugar)	-916.97	-269.45	[4]
C_6H_{14}	l, hexane	-4.04	-296.02	[3]
$C_7H_6O_2$	s, benzenecarboxylic acid (benzoic acid)	-245.20	-167.60	[6]
C_7H_8	g, methylbenzene (toluene)	122.19	-320.77	[3]
C_7H_8	l, methylbenzene (toluene)	113.96	-220.96	[3]
C_8H_{18}	l, octane	6.71	-361.21	[3]
$C_{12}H_{22}O_{11}$	s, sucrose (cane sugar)	$-1,557.60$	-392.40	[4]
$C_{12}H_{22}O_{111}$	w, sucrose (cane sugar)	$-1,564.70$	-435.40	[4]
Ca	g	144.02	-154.89	[1]
Ca	s, α	0.00	-41.59	[1]
Ca^{2+}	w	-553.58	$+53.10$	[2]
$CaBr_2$	s	-664.78	-130.00	[1]
$CaCO_3$	s, aragonite	$-1,127.85$	-88.70	[2]
$CaCO_3$	s, calcite	$-1,128.79$	-92.70	[2]
CaC_2	s	-64.55	-70.29	[1]
$CaCl_2$	s	-748.79	-108.37	[1]
CaF_2	s	$-1,175.55$	-68.45	[1]
CaO	s	-603.30	-38.10	[1]

(continued)

Substance	Phase	μ (kG)	α (G K^{-1})	Refs.
Ca(OH)$_2$	s	−898.24	−83.40	[1]
CaSO$_4$	s	−1,325.14	−106.69	[1]
Cd	g	77.23	−167.75	[1]
Cd	s	0.00	−51.80	[1]
Cd^{2+}	w	−77.61	73.20	[2]
CdCO$_3$	s	−670.53	−92.47	[1]
CdO	s	−229.72	−54.81	[1]
Cl	g	105.31	−165.19	[1]
Cl$^-$	g	−240.17	−153.36	[5]
Cl$^-$	w	−131.23	−56.50	[2]
ClH	g	−95.30	−186.90	[1]
ClH	w	−97.00		[7]
ClO$_2$	g	122.83	−256.88	[1]
ClO$_4{}^-$	w	−8.52	−182.00	[2]
ClO$_4$H	w	48.56		[7]
Cl$_2$	g	0.00	−223.08	[1]
Co	g	379.47	−179.52	[1]
Co	s, α, hexagonal	0.00	−30.04	[1]
Cr	g	352.20	−174.31	[1]
Cr	s	0.00	−23.54	[1]
CrO$_4{}^{2-}$	w	−727.75	−50.21	[2]
Cr$_2$O$_3$	s	−1,058.99	−81.10	[1]
Cr$_2$O$_7{}^{2-}$	w	−1,301.10	−261.90	[2]
Cs	g	49.56	−175.60	[1]
Cs	s	0.00	−85.23	[1]
Cs$^+$	w	−292.02	−133.05	[2]
Cu	g	298.31	−166.29	[1]
Cu	s	0.00	−33.15	[1]
Cu$^+$	w	49.98	−40.60	[2]
Cu^{2+}	w	65.49	99.60	[2]
CuCl$_2$	s	−173.73	−108.07	[1]
CuO	s	−128.08	−42.74	[1]
CuS	s	−53.47	−66.48	[3]
CuSO$_4$	s	−660.78	−109.25	[1]
CuSO$_4$·H$_2$O	s	−914.76	−145.10	[1]
CuSO$_4$·5H$_2$O	s	−1,876.83	−301.25	[1]
Cu$_2$O	s	−147.84	−92.55	[1]
D	g, deuterium	206.55	−123.35	[1]
DH	g	−1.46	−143.80	[1]
DOH	g	−233.09	−199.51	[1]
D$_2$	g	0.00	−144.96	[1]
F	g	62.28	−158.75	[1]
F$^-$	g	−262.00	−145.58	[5]

(continued)

Substance	Phase	μ (kG)	α (G K^{-1})	Refs.
F^-	w	-278.79	$+13.80$	[2]
FH	g	-275.40	-173.78	[1]
F_2	g	0.00	-202.79	[1]
Fe	g	368.32	-180.49	[1]
Fe	l	5.34	-35.55	[5]
Fe	s, α, cubic	0.00	-27.28	[1]
Fe^{2+}	w	-78.90	137.70	[2]
Fe^{3+}	w	-4.70	315.90	[2]
$Fe(OH)_2$	s	-496.98	-88.00	[3]
$Fe(OH)_3$	s	-708.98	-105.00	[1]
FeS	s	-101.97	-60.32	[3]
$FeSO_4$	s	-824.89	-120.96	[1]
FeS_2	s, pyrite	-166.90	-52.93	[2]
Fe_2O_3	s, hematite	-741.04	-87.40	[1]
Fe_3O_4	s, magnetite	$-1{,}017.48$	-145.27	[1]
Ga	g	233.74	-169.04	[1]
Ga	s	0.00	-40.73	[1]
Ge	g	333.68	-167.90	[3]
Ge	s	0.00	-31.09	[1]
H	g	203.28	-114.72	[1]
H^+	w	0.00	0.00	[2]
H_2	g	0.00	-130.68	[1]
He	g	0.00	-126.15	[1]
Hg	g	32.46	-174.97	[1]
Hg	l	0.00	-75.90	[1]
Hg^{2+}	w	164.40	32.20	[2]
$HgCl_2$	s	-183.44	-144.49	[1]
HgI_2	s, red	-101.70	-180.00	[2]
HgI_2	s, yellow	-101.15	-186.29	[7]
HgO	s, red	-58.54	-70.29	[2]
HgO	s, yellow	-58.41	-71.10	[2]
HgS	s, black	-47.70	-88.30	[2]
HgS	s, red	-50.60	-82.40	[2]
Hg_2^{2+}	w	153.52	-84.50	[2]
Hg_2Cl_2	s	-209.33	-192.54	[1]
Hf	g	579.62	-186.90	[1]
Hf	s	0.00	-43.56	[1]
I	g	70.17	-180.78	[1]
I^-	w	-51.57	-111.30	[2]
IH	g	1.70	-206.59	[1]
I_2	g	19.32	-260.68	[1]
I_2	l	3.32	-150.36	[5]
I_2	s	0.00	-116.14	[1]

(continued)

Substance	Phase	μ (kG)	α (G K^{-1})	Refs.
I$_2$	w	16.40	−137.20	[2]
In	g	206.08	−173.78	[1]
In	s	0.00	−57.65	[1]
Ir	g	622.87	−193.58	[1]
Ir	s	0.00	−35.51	[1]
K	g	60.48	−160.34	[1]
K	s	0.00	−64.68	[1]
K$^+$	w	−283.27	−102.50	[2]
KBr	s	−380.07	−95.92	[1]
KCl	s	−408.76	−82.56	[1]
KF	s	−538.93	−66.55	[1]
KI	s	−324.32	−106.05	[1]
KOH	s	−379.46	−81.25	[1]
K$_2$O	s	−321.17	−96.00	[1]
K$_2$SO$_4$	s	−1,319.59	−175.54	[1]
Kr	g	0.00	−164.09	[1]
La	g	392.59	−182.38	[1]
La	s	0.00	−56.90	[1]
Li	g	126.66	−138.77	[2]
Li	s	0.00	−29.12	[1]
Li$^+$	w	−293.31	−13.40	[2]
LiH	g	116.47	−170.91	[1]
LiH	s	−68.63	−20.60	[1]
Mg	g	115.98	−148.65	[1]
Mg	s	0.00	−32.67	[1]
Mg^{2+}	w	−454.80	+138.10	[2]
MgCO$_3$	s	−1,012.21	−65.09	[1]
MgCl$_2$	s	−594.77	−89.62	[1]
MgO	s	−569.31	−26.95	[1]
MgS	s	−343.70	−50.33	[1]
MgSO$_4$	s	−1,174.48	−91.60	[1]
Mn	g	238.50	−173.70	[2]
Mn	s	0.00	−32.22	[1]
Mn^{2+}	w	−228.10	73.60	[2]
MnO$_2$	s	−465.08	−53.05	[1]
MnO$_4{}^-$	w	−447.20	−191.20	[2]
Mo	g	611.88	−181.95	[1]
Mo	s	0.00	−28.56	[1]
N	g	455.55	−153.30	[1]
NH$_3$	g	−16.45	−192.45	[2]
NH$_3$	l	−10.16	−103.90	[7]
NH3	w	−26.59	−111.30	[2]
NH$_4{}^+$	w	−79.31	−113.40	[2]

(continued)

Substance	Phase	μ (kG)	α (G K^{-1})	Refs.
NH$_4$Cl	s	−203.09	−94.86	[1]
NO	g	87.59	−210.74	[1]
NOCl	g	67.11	−261.58	[1]
NO$_2$	g	52.31	−240.17	[1]
NO$_3$$^-$	w	−108.74	−146.40	[2]
NO$_3$H	g	−73.69	−266.88	[1]
NO$_3$H	l	−80.71	−155.60	[2]
N$_2$	g	0.00	−191.61	[1]
N$_2$H$_4$	l	149.34	−121.21	[2]
N$_2$H$_4$O$_3$	s, ammonium nitrate	−183.76	−150.81	[1]
N$_2$O	g	104.20	−219.85	[2]
N$_2$O$_4$	g	97.89	−304.29	[2]
N$_2$O$_4$	l	97.54	−209.20	[2]
N$_2$O$_5$	g	115.10	−355.70	[2]
N$_2$O$_5$	s	113.90	−178.20	[2]
N$_3$H	g	328.10	−238.97	[2]
N$_3$H	l	327.30	−140.60	[2]
Na	g	76.96	−153.72	[1]
Na	s	0.00	−51.30	[1]
Na$^+$	w	−261.91	−59.00	[2]
NaBr	s	−349.09	−86.93	[1]
NaCl	s	−384.07	−72.12	[1]
NaI	s	−286.41	−98.56	[1]
NaOH	s	−379.65	−64.43	[1]
NaSO$_4$H	s	−992.80	−113.00	[2]
Na$_2$O	s	−379.18	−75.04	[1]
Na$_2$SO$_4$	s	−1,270.02	−149.58	[1]
Nb	g	678.42	−186.27	[1]
Nb	s	0.00	−36.27	[1]
Ne	g	0.00	−146.33	[1]
Ni	g	384.50	−182.19	[2]
Ni	s	0.00	−29.80	[1]
Ni^{2+}	w	−45.60	128.90	[2]
NiCl$_2$	s	−258.65	−98.10	[1]
NiO	s	−211.59	−38.07	[1]
NiSO$_4$	s	−759.70	−92.00	[2]
O	g	231.74	−161.06	[1]
OD$_2$	g	−234.54	−198.34	[3]
OD$_2$	l	−243.40	−75.94	[3]
OD$_2$	s	−242.89	−50.59	[7]
OH$^-$	w	−157.24	10.75	[2]
OH$_2$	g	−228.58	−188.83	[1]

(continued)

Substance	Phase	μ (kG)	α (G K^{-1})	Refs.
OH$_2$	l	-237.14	-69.95	[1]
OH$_2$	s	-236.55	-44.81	[7]
OH$_3{}^+$	w	-237.14	-69.95	[1]
O$_2$	g	0.00	-205.15	[1]
O$_2$	w	16.40	-110.9	[2]
O$_2$H$_2$	g	-105.45	-233.00	[1]
O$_2$H$_2$	l	-120.42	-109.62	[1]
O$_2$H$_2$	w	-134.03	-143.90	[2]
O$_3$	g	163.29	-239.01	[1]
Os	g	740.31	-192.58	[1]
Os	s	0.00	-32.64	[1]
P	g	280.09	-163.20	[1]
P	s, red	-12.02	-22.85	[1]
P	s, white	0.00	-41.09	[1]
PCl$_3$	g	-269.61	-311.68	[5]
PCl$_3$	l	-274.04	-218.49	[1]
PCl$_5$	g	-305.00	-364.58	[2]
PH$_3$	g	13.55	-210.31	[3]
PO$_4{}^{3-}$	w	$-1,018.70$	222.00	[2]
PO$_4$H^{2-}	w	$-1,089.15$	33.50	[2]
PO$_4$H^{2-}	w	$-1,130.28$	-90.40	[2]
PO$_4$H$_3$	l	$-1,123.60$	-150.78	[5]
PO$_4$H$_3$	w	$-1,018.70$	222.00	[2]
P$_4$O$_{10}$	g	$-2,671.28$	-402.09	[1]
P$_4$O$_{10}$	s	$-2,724.15$	-231.00	[1]
Pb	g	162.23	-175.37	[1]
Pb	l	2.22	-71.71	[5]
Pb	s	0.00	-64.80	[1]
Pb^{2+}	w	-24.43	-10.50	[2]
PbCO$_3$	s	-625.41	-130.96	[1]
PbI$_2$	s	-173.57	-174.84	[1]
PbI$_2$	s	-173.57	-174.84	[1]
PbO	s, yellow	-188.68	-68.70	[1]
PbO	s, red	-188.92	-67.84	[1]
PbO$_2$	s	-215.39	-71.80	[1]
PbS	s	-97.77	-91.20	[1]
PbSO$_4$	s	-816.20	-148.49	[1]
Pd	g	338.028	-167.06	[3]
Pd	s	0.00	-37.82	[1]
Pt	g	520.05	-192.41	[1]
Pt	s	0.00	-41.63	[1]

(continued)

Substance	Phase	μ (kG)	α (G K^{-1})	Refs.
Re	g	729.42	−118.93	[1]
Re	s	0.00	−36.48	[1]
Rh	g	509.01	−185.83	[1]
Rh	s	0.00	−31.56	[1]
Ru	g	604.92	−186.51	[1]
Ru	s	0.00	−28.61	[1]
S	g	236.70	−167.83	[1]
S	s, α, rhombic	0.00	−32.07	[1]
S	s, β, monoclinic	0.07	−33.03	[5]
S^{2-}	w	85.80	+14.60	[2]
SF_6	g	−1,115.42	−291.67	[1]
SH^-	w	12.08	−62.80	[2]
SH_2	g	−33.44	−205.80	[1]
SH_2	w	−27.83	−121.00	[2]
SO_2	g	−300.12	−248.21	[1]
SO_3	g	−371.01	−256.77	[1]
SO_3^{2-}	w	−486.50	29.00	[2]
SO_3H^-	w	−527.78	−139.70	[2]
SO_3H_2	w	−537.81	−232.20	[2]
SO_4^{2-}	w	−744.53	−20.10	[2]
$SO_4{}^{H-}$	w	−755.91	−131.80	[2]
SO_4H_2	l	−689.92	−156.90	[1]
SO_4H_2	w	−738.79		[7]
S_2Cl_2	g	−28.66	−327.22	[1]
Sb	g	227.00	−180.27	[1]
Sb	s	0.00	−45.52	[1]
Sc	g	335.92	−174.79	[1]
Sc	s	0.00	−34.64	[1]
Se	g	197.41	−176.73	[1]
Se	s	0.00	−42.00	[1]
Si	g	405.53	−168.00	[1]
Si	s	0.00	−18.81	[1]
$SiCl_4$	g	−622.39	−331.45	[1]
SiO_2	s, α, cristobalite	−855.43	−42.68	[2]
SiO_2	s, α, quartz	−856.29	−41.46	[1]
Sn	g	266.22	−168.49	[1]
Sn	s, β, white	0.00	−51.18	[1]
Sn^{2+}	w	−27.2	17.0	[2]
Sn^{4+}	w	+1.9		[7]
SnO	s	−251.91	−57.17	[1]
SnO_2	s	−515.82	−49.01	[1]
Sr	g	128.02	−164.64	[1]
Sr	s	0.00	−55.69	[1]

(continued)

Substance	Phase	μ (kG)	α (G K^{-1})	Refs.
Te	g	169.65	−182.71	[1]
Te	s	0.00	−49.22	[1]
Ti	g	429.12	−180.30	[1]
Ti	s	0.00	−30.72	[1]
TiCl$_4$	l	−737.20	−252.34	[2]
TiO$_4$	s, rutile	−888.77	−50.62	[1]
Tl	g	146.22	−181.00	[1]
Tl	s	0.00	−64.30	[1]
U	g	490.40	−199.79	[1]
U	s	0.00	−50.20	[1]
V	g	472.19	−182.30	[1]
V	s	0.00	−30.89	[1]
W	g	809.11	−174.00	[1]
W	s	0.00	−32.62	[1]
Xe	g	0.00	−169.58	[1]
Zn	g	94.81	−161.00	[1]
Zn	s	0.00	−41.63	[1]
Zn^{2+}	w	−147.06	112.1	[2]
ZnCO$_3$	s	−731.45	−82.43	[1]
ZnCl$_2$	s	−370.32	−111.50	[1]
ZnI$_2$	s	−209.26	−161.50	[3]
ZnO	s	−320.37	−43.16	[1]
ZnS	s, zinc blende	−198.52	−58.66	[1]
ZnSO$_4$	s	−871.45	−110.50	[3]
Zr	g	556.91	−181.34	[1]
Zr	s	0.00	−39.18	[1]

Index

© Springer International Publishing Switzerland 2016
G. Job, R. Rüffler, *Physical Chemistry from a Different Angle,*
DOI 10.1007/978-3-319-15666-8